ASYMPTOTICS,
NONPARAMETRICS,
AND TIME SERIES

STATISTICS: Textbooks and Monographs

A Series Edited by

D. B. Owen, Founding Editor, 1972–1991

W. R. Schucany, Coordinating Editor
Department of Statistics
Southern Methodist University
Dallas, Texas

W. J. Kennedy, Associate Editor
for Statistical Computing
Iowa State University

A. M. Kshirsagar, Associate Editor
for Multivariate Analysis and for
Experimental Design
University of Michigan

E. G. Schilling, Associate Editor
for Statistical Quality Control
Rochester Institute of Technology

1. The Generalized Jackknife Statistic, *H. L. Gray and W. R. Schucany*
2. Multivariate Analysis, *Anant M. Kshirsagar*
3. Statistics and Society, *Walter T. Federer*
4. Multivariate Analysis: A Selected and Abstracted Bibliography, 1957–1972, *Kocherlakota Subrahmaniam and Kathleen Subrahmaniam*
5. Design of Experiments: A Realistic Approach, *Virgil L. Anderson and Robert A. McLean*
6. Statistical and Mathematical Aspects of Pollution Problems, *John W. Pratt*
7. Introduction to Probability and Statistics (in two parts), Part I: Probability; Part II: Statistics, *Narayan C. Giri*
8. Statistical Theory of the Analysis of Experimental Designs, *J. Ogawa*
9. Statistical Techniques in Simulation (in two parts), *Jack P. C. Kleijnen*
10. Data Quality Control and Editing, *Joseph I. Naus*
11. Cost of Living Index Numbers: Practice, Precision, and Theory, *Kali S. Banerjee*
12. Weighing Designs: For Chemistry, Medicine, Economics, Operations Research, Statistics, *Kali S. Banerjee*
13. The Search for Oil: Some Statistical Methods and Techniques, *edited by D. B. Owen*
14. Sample Size Choice: Charts for Experiments with Linear Models, *Robert E. Odeh and Martin Fox*
15. Statistical Methods for Engineers and Scientists, *Robert M. Bethea, Benjamin S. Duran, and Thomas L. Boullion*
16. Statistical Quality Control Methods, *Irving W. Burr*
17. On the History of Statistics and Probability, *edited by D. B. Owen*
18. Econometrics, *Peter Schmidt*
19. Sufficient Statistics: Selected Contributions, *Vasant S. Huzurbazar (edited by Anant M. Kshirsagar)*
20. Handbook of Statistical Distributions, *Jagdish K. Patel, C. H. Kapadia, and D. B. Owen*

Additional Volumes in Preparation

Multivariate Analysis, Design of Experiments, and Survey Sampling, *edited by Subir Ghosh*

Statistical Process Monitoring and Control, *S. Park and G. Vining*

ASYMPTOTICS, NONPARAMETRICS, AND TIME SERIES

edited by

SUBIR GHOSH

University of California, Riverside
Riverside, California

A Tribute to Madan Lal Puri

CRC Press
Taylor & Francis Group
Boca Raton London New York

CRC Press is an imprint of the
Taylor & Francis Group, an **informa** business

CRC Press
Taylor & Francis Group
6000 Broken Sound Parkway NW, Suite 300
Boca Raton, FL 33487-2742

First issued in paperback 2019

© 1999 by Taylor & Francis Group, LLC
CRC Press is an imprint of Taylor & Francis Group, an Informa business

No claim to original U.S. Government works

ISBN-13: 978-0-8247-0051-5 (hbk)
ISBN-13: 978-0-367-39992-4 (pbk)

Visit the Taylor & Francis Web site at
http://www.taylorandfrancis.com

and the CRC Press Web site at
http://www.crcpress.com

Madan Lal Puri

Preface

Asymptotic and nonparametric methods are widely used in statistics; time series is an important area of statistics. Many researchers have contributed and others are working to develop these subjects. As a result, we observe a profusion of research. This reference book is a collection of articles describing some of the recent developments and surveying some important topics. The book is a tribute to Professor Madan Lal Puri, who has contributed vigorously to asymptotic and nonparametric methods and their application in time series. This collection of articles is a present to Professor Puri for his 70th birthday to celebrate his contributions, leadership, and dedication to our profession. This is a collection not just by his friends but by the world leaders in their special research areas. The topics covered are broader than the title describes. Parametric, semiparametric, frequentist, bayesian, bootstrap, adaptive, univariate and multivariate methods, Markov chain models, and many others are also discussed in this book. All the articles have been refereed and are in general expository. The book should be of value to students, instructors, and researchers at colleges and universities, as well as in businesses, industries, and government organizations.

The following individuals were truly outstanding for their cooperation and help in reviewing the articles: Shun-ichi Amari, Gopal K. Basak, Jan Beran, Johanne F. Böhme, Dennis D. Boos, Jack Cuzick, Rainer Dahlhaus, Clive W. J. Granger, Cindy Greenwood, Wouter Den Hann, Nancy Heckman, Lajos Horvath, Irene Hueter, Harry Hurd, Wesley O. Johnson, Jerry H. Klotz, Eric D. Kolaczyck, Masao Kondo Johannes Ledolter, Ernst Linder, Olive Linton, Richard Lockhart, Bani K. Mallick, Marianthi Markatou, Michael A. Martin, Jean Meloche, Serena Ng, Dimitris N. Politis, Gregory C. Reinsel, Moshe Shaked, David Stoffer, Arnold J. Stromberg, Winfried Stute, Robert L. Taylor, Ram C. Tiwari, Stephen G. Walker, Edward C. Waymire, Granville T. Wilson, Wayne A. Woodward, Daming Xu, and G. Alastair Young. I am grateful to all our distinguished reviewers.

My deep appreciation and heartfelt thanks go to our renowned contributors, who I hope forgive me for not telling them in advance about some details regarding this book. But then, a surprise for Professor Madan Lal Puri and our contributors will uplift our spirits and stimulate us to contribute more to our society.

My sincere thanks go to Russell Dekker, Maria Allegra, and others at Marcel Dekker, Inc. I would like to thank my wife, Susnata, and our daughter, Malancha, for their support and understanding of my efforts in completing this project.

Subir Ghosh

Contents

Contributors

J. Antoch Department of Statistics, Charles University, Prague, Czech Republic

Josu Arteche Department of Econometrics, University of the Basque Country, Bilbao, Spain

Gutti Jogesh Babu Department of Statistics, Pennsylvania State University, University Park, Pennsylvania

Ishwar V. Basawa Department of Statistics, The University of Georgia, Athens, Georgia

R. J. Bhansali Department of Mathematical Sciences, University of Liverpool, Liverpool, England

Rabi Bhattacharya Department of Mathematics, Indiana University, Bloomington, Indiana

Robert R. Blandford Center for Monitoring Research, Arlington, Virginia

David R. Brillinger   Department of Statistics, University of California, Berkeley, California

Kamal C. Chanda   Department of Mathematics and Statistics, Texas Tech University, Lubbock, Texas

Richard A. Davis   Department of Statistics, Colorado State University, Fort Collins, Colorado

William T. M. Dunsmuir   School of Mathematics, University of New South Wales, Sydney, Australia

Alan E. Gelfand   Department of Statistics, University of Connecticut, Storrs, Connecticut

S. Ghosal   Faculty of Mathematics and Computer Science, Vrije Universiteit, Amsterdam, The Netherlands

J. K. Ghosh   Indian Statistical Institute, Calcutta, India

Marc Hallin   Institute of Statistics and Operations Research, European Center for Advanced Research in Economics, and Department of Mathematics, Université Libre de Bruxelles, Brussels, Beligum

T. P. Hettmansperger   Department of Statistics, Pennsylvania State University, University Park, Pennsylvania

M. Hušková   Department of Statistics, Charles University, Prague, Czech Republic

Jürg Hüsler   Department of Mathematical Statistics, University of Bern, Bern, Switzerland

Sung-Eun Kim   Department of Civil and Environmental Engineering, University of California, Davis, California

S. N. Lahiri   Department of Statistics, Iowa State University, Ames, Iowa

Bruce G. Lindsay   Department of Statistics, Pennsylvania State University, University Park, Pennsylvania

Robert B. Lund   Department of Statistics, The University of Georgia, Athens, Georgia

Mukul Majumdar   Department of Economics, Cornell University, Ithaca, New York

Ömer Öztürk   Department of Statistics, The Ohio State University, Marion, Ohio

B. L. S. Prakasa Rao   Indian Statistical Institute, New Delhi, India

R. V. Ramamoorthi   Department of Statistics and Probability, Michigan State University, East Lansing, Michigan

T. Subba Rao   Department of Mathematics, University of Manchester Institute of Science and Technology, Manchester, England

Peter M. Robinson   Department of Economics, London School of Economics and Political Science, London, England

George G. Roussas   Division of Statistics, University of California, Davis, California

Frits H. Ruymgaart   Department of Mathematical Statistics, Texas Tech University, Lubbock, Texas

Anton Schick   Department of Mathematical Sciences, Binghamton University, Binghamton, New York

Robert H. Shumway   Division of Statistics, University of California, Davis, California

Masanobu Taniguchi   Department of Mathematical Science, Faculty of Engineering Science, Osaka University, Toyonaka, Japan

Dag Tjøstheim   Department of Mathematics, University of Bergen, Bergen, Norway

Arnoud C. M. van Rooij   Department of Mathematics, University of Nijmegen, Nijmegen, The Netherlands

Ying Wang   Department of Statistics, Colorado State University, Fort Collins, Colorado

Richard Waterman   Department of Statistics, University of Pennsylvania, Philadelphia, Pennsylvania

Wolfgang Wefelmeyer   Department of Mathematics, University of GH Siegen, Siegen, Germany

Bas J. M. Werker   Institute of Statistics, Université Libre de Bruxelles Brussels, Belgium

W. K. Wong*   Department of Mathematics, University of Manchester Institute of Science and Technology, Manchester, England

*Current affiliation: University of Stirling, Scotland

Madan Lal Puri: Life and Contributions of a Mathematical Statistician

Madan Lal Puri was born in Sialkot (then in India, now in Pakistan) on February 20, 1929. In 1947, when India gained her independence and Pakistan was created, his family migrated to Delhi as refugees. He received a B.A. degree in 1948 and an M.A. degree in 1950, both in mathematics, from Panjab University in India. From January 1951 to August 1957, he served as a Lecturer in Mathematics in different colleges of Panjab University.

In September 1957, he came to the United States as an instructor and a graduate student in mathematics at the University of Colorado in Boulder. In September 1958, he moved to the University of California at Berkeley as a research assistant in the Department of Statistics and then received his Ph.D. in statistics in 1962. His dissertation was in the area of nonparametric inference under the guidance of Professor Erich L. Lehmann. As a world center of statistics in the 1950s and 1960s University of California at Berkeley hosted a number of world renowned experts in probability and statis-

The publications of Madan Lal Puri are given in the Appendix.

tics: Cox, Cramer, Doob, Feller, Hoeffding, Hotelling, Kiefer, Robbins, Kendall, Rao, and Wolfowitz, to name only a few. The list of faculty members during this golden period includes, among others, the prominent statisticians Blackwell, LeCam, Lehman, Loeve, Neyman, and Scheffe. Madan received his statistics education in this exciting environment.

In 1962, Dr. Madan L. Puri joined the Courant Institute of Mathematical Sciences at New York University as an Assistant Professor and became an Associate Professor in 1965. He joined Indiana University at Bloomington in 1968 as a full Professor of Mathematics and remains there to this day.

Professor Puri is one of the most versatile and prolific researchers in mathematical statistics. His research areas include nonparametric statistics, order statistics, limit theory under mixing, time series, splines, tests of normality, generalized inverses of matrices and related topics, stochastic processes, stotistics of directional data, random sets, fuzzy sets and fuzzy measures, among others. His fundamental contributions in developing rank-based methods and precise evaluation of the standard procedures, asymptotic expansions of distributions of rank statistics, as well as large deviation results concerning them, span various areas, such as analysis of variance, analysis of covariance, multivariate analysis, and time series. His in-depth analysis resulted in many pioneering research contributions in prominent journals which have substantial impact on current research.

Professor Puri has done joint work with many researchers from different countries. His numerous joint contributions with Professor P. K. Sen in the 1960s and 1970s on rank-based procedures and their asymptotic properties are greatly valued in our profession. Researchers in many other disciplines use these statistical procedures in their everyday work. This joint collaboration resulted in two advanced books, *Nonparametric Methods in Multivariate Analysis of Variance* and *Nonparametric Methods in General Linear Models*, which are still the leading books on these topics. In the 1980s and 1990s, his pioneering contributions with Professor M. Hallin resulted in new rank-based methods for time series analysis. His many joint contributions on convergence rates with Professors R. N. Bhattacharya, M. Harel, S. Ralescue, M. Seoh, and T. J. Wu; on generalized inverses with Professor S. K. Mitra; on fuzzy sets and measures with Professor D. A. Ralescu; on invariance principles for stochastic processes with Professors M. Denker, G. Haiman, and H. Harel; and on rank-based methods in the analysis of designed experiments with Professor E. Brunner are just a few of his many such accomplishments.

Until the early 1980s the theory and practice of rank-based inference was essentially limited to analysis with independent observations. This limitation on independent or, at least, exchangeable observations was more or less inherent in rank-based inference. The papers published between 1985 and

1994 by Hallin and Puri, as well as Hallin, Ingenbleek, and Puri, present detailed rank-based methods for the analysis of the popular autoregressive-moving average (ARMA) and other models. This much-needed development was made possible through the use of a new type of rank statistics, called the *serial linear rank statistics*, introduced by Hallin and Puri.

Professor Puri was the Alexander von Humboldt Guest Professor at the University of Göttingen in West Germany during 1974–75 and Guest Professor at many other universities in Germany, with German National Science Foundation grants. He was a Distinguished Visitor at the London School of Economics and Political Science, and Visiting Professor at the University of Auckland in New Zealand, the Universities of Bern and Basel in Switzerland, the University of New South Wales in Australia, the University of Goteborg and Chalmers University of Technology in Sweden, and Université des Sciences et Technologies de Lille France. He was invited to lecture at the Japanese Society for the Promotion of Science in 1971. He has been an invited speaker as well as plenary speaker at many international conferences all over the world.

Professor Puri has received numerous honors and awards. He is an elected member of the International Statistical Institute, and a Fellow of the Institute of Mathematical Statistics, the American Statistical Association, and the Royal Statistical Society. In 1975, he was honored with the D.Sc. degree from Panjab University in India. He twice received the Senior U.S. Scientist Award from the Alexander von Humboldt Foundation, in 1974 and 1983. In 1974, he was honored by the government of the Federal Republic of Germany, "In recognition of past achievements in research and teaching." In 1984, he received the best paper award from the Seventh European Meeting on Cybernetics and Systems Research, Vienna, Austria. In 1991, he received the Rothrock Faculty Teaching Award in recognition of outstanding teaching in the Department of Mathematics of Indiana University. He was ranked as the fourth most prolific author in 1997 and the ninth most in 1993 in top statistical journals in the world.

Professor Puri served on organizing committees of many international conferences in addition to those of the Institute of Mathematical Statistics and the American Statistical Association. He also served as Editor-in-Chief of the *Journal of Statistical Planning and Inference* during 1984–1988.

Professor Puri has directed 16 Ph.D. dissertations. Most of his former Ph.D. students are in research and teaching positions at good universities. A few of them hold responsible positions in industry.

Professor Puri is truly an international academician and a peripatetic scholar who works with missionary zeal. Many scientists from different countries visit him regularly and do research with him while staying at his home. He is a caring colleague with warmest affection, an international host, a persuasive communicator, a dedicated as well as an outstanding teacher, and a versatile statistician whose work continues to inspire the scientific community.

With great pleasure, pride, and admiration, we dedicate this book in honor of Professor Madan Lal Puri on his 70th birthday. The age of seventy is a time of liberation, a time to realize that there is more to do, more to see, and more reasons to be around for the people who really appreciate you. Madan, our best wishes to you on this happy occasion and in years to come.

Subir Ghosh
George G. Roussa

1

Some Examples of Empirical Fourier Analysis in Scientific Problems

DAVID R. BRILLINGER   University of California, Berkeley, California

> *"One can FT anything—often meaningfully."*
>
> J. W. Tukey

1. INTRODUCTION

As a concept and as a tool, the Fourier transform is pervasive in applied mathematics, computing, mathematics, probability and statistics as well as in substantive sciences such as chemistry, geophysics and physics. This chapter presents a review of such applications and then four personal analyses of scientific data based on Fourier transforms. Specific points made include: Fourier analysis is conceptually simple, its concepts often have direct physical interpretations, useful statistical properties are available, and there are various interesting connections between the mathematical and physical concepts.

By Fourier analysis is meant the study of spaces and functions, making substantial use of sine and cosine functions. The subject has a long and glorious history. In particular, fundamental work has been carried out by both mathematicians and applied scientists. Fourier analysis remains of interest to mathematicians because generalizations seem inexhaustible and because there are continual surprises. Classic works by mathematicians include: Wiener (1933), Bochner (1959, 1960) and Zygmund (1968). These particular authors are concerned with functions on the line or on a general Euclidian space. Works on extensions to general groups include: Loomis (1953), Rudin (1962), Hewitt and Ross (1963), Katznelson (1976). More recent books are Terras (1988) and Körner (1989), the former particularly addressing the nonabelian case, the latter presenting a variety of historical examples and essays on specific topics.

In contrast, the Fourier transform is of interest to statisticians because it proves inordinately useful in the analysis of data and eases the development of various theoretical results. Noteworthy contributions to statistics have been made by Slutsky (1934), Cramér (1942), Good (1958), Yaglom (1961), Tukey (1963), Hannan (1965, 1966), Priestley (1965), Bloomfield (1976), Diaconnis (1988, 1989). Slutsky developed some of the statistical properties of the Fourier transform of a stretch of time series values. Cramér set down a Fourier representation (see Sec. 4.1) for stationary processes. The representation admitted many extensions and made transparent the effect of a variety of operations on processes. Good and Tukey indicated how the transform could be computed recursively and hence quickly in certain circumstances. Yaglom extended the domain of application to processes defined on compact and locally compact groups. Hannan considered problems for other groups than Yaglom and presented material relevant to practical applications. Priestley provided a frequency domain representation to describe nonstationary processes. Bloomfield made complicated results available to a broad audience. Diaconnis considered symmetric and permutation groups and the analysis of ordered data.

Particular uses of the empirical Fourier transform include: the development of estimates of finite dimensional parameters appearing in time series models (Whittle (1952), Dzhaparidze (1986), Feuerverger (1990)), the assessment of goodness of fit of models (Feigin and Heathcote (1976)), and the deconvolution of random measurements (Fan (1992)). Fourier analysis has a special place amongst the tools of statistics for the concepts often have their own physical existence.

There are special computational, mathematical and statistical properties and surprises associated with the Fourier transform. These include: the central limit theorems for the stationary case with approximate independence at particular frequencies, the existence of fast Fourier transforms, (Good (1958), Tukey (1963), Cooley and Tukey (1965), Rockmore (1990)) the need for convergence factors, the ideas of aliasing.

Section 2 concerns some particular physical situations. Section 3 contains pertinent background from analysis. Section 4 contains stochastic background. Section 5 presents analyses of four data sets from the natural sciences and the author's experience. The examples highlight approximation, shrinkage estimation, the method of stationary phase, central limit theorems and uncertainty estimation. The first example, concerning crystalographic data, involves the empirical representation of a basic function on the plane by an expansion in sines and cosines. This makes sense because of periodicities inherent in the crystal structure. The example also involves shrinkage of the coefficients of the expansion in order to obtain improved estimates. The second analysis is of a record of an earthquake that took place in Siberia as recorded at Uppsala, Sweden. The oscillatory character of the data may be understood heuristically via the method of stationary phase, to be described below. A model of the transmission medium is constructed and model assessment carried out by a sliding or dynamic Fourier analysis. This last may be viewed as a form of wavelet analysis. The third analysis, concerned with nuclear magnetic resonance (NMR) spectroscopy, employs Fourier analysis to obtain physical insight into the behavior of an input-output system, and then makes use of cross-spectral analysis to estimate the transfer function of the system. The periodogram of the residuals is employed to assess the fit. The final example involves both wavelet and Fourier analysis. It is concerned with the question of whether a microtubule moves steadily or via jumps. The Fourier analysis is employed in this case to obtain uncertainty estimates in the presence of stationary noise. Section 6 contains conclusions and indicates open problems.

2. SOME PHYSICAL EXAMPLES OF FOURIER ANALYSIS

Cycles, periods, and resonances have long been noted in scientific discussions of astronomy, vibrations, oceanography, sound, light and crystalography amongst other fields. In technology oscillations occur often for example in telephone, radio, TV and laser engineering. Natural operations occur commonly that correspond with linear and time invariant systems as defined in Section 3 below. These are the eigenoperations of Fourier analysis.

Fourier analysis is sometimes tied specifically to the physics of a problem. For example Bazin et al. (1986) physically demonstrate the operations/concepts of translation, linearity, similarity, convolution and Parseval's theorem for the Fourier transform via diffraction experiments with laser light. The Fourier transform here is formed via a lens. See Goodman (1968) Shankar et al. (1982), Glaeser (1985) for a discussion of the optics involved.

An important example arises in radio astronomy. Suppose there is an array of receivers. Suppose there is a small incoherent source, at great distance, producing a plane travelling wave. If $Y(x, y, t)$ denotes the radio field measurement made at time t on a telescope located at position (x, y), then

$$E\{Y(x+u, y+v, t)\overline{Y(x,y,t)}\} = \int\int f(\alpha,\beta)e^{i(u\alpha+v\beta)}d\alpha\, d\beta \qquad (2.1)$$

where (α, β) are the coordinates of the source of interest in the sky and $f(\alpha, \beta)$ is its brightness distribution as a function of (α, β). In other words, the Fourier transform is the quantity observed. The result Eq. (2.1) is known as the van Cittert–Zernike Theorem, see Born and Wolf (1964).

Linear time invariant systems abound in nature. They have the property of carrying cosinusoids into cosinusoids. Nowadays in science there is much concern with nonlinear operations and phenomena. Impressively, the classic trigonometric identity

$$[\cos \lambda t]^2 = \tfrac{1}{2}\cos 2\lambda t + \tfrac{1}{2} \qquad (2.2)$$

is "demonstrated" in Yariv (1975) via a color plate showing red laser light becoming blue on passing through a crystal. The crystal involved squares the signal as in Eq. (2.2). A wavelength of 6940 Å (red) becomes one of 3970 Å (blue). Bloembergen (1982), Moloney and Newell (1989) discuss such nonlinear aspects of light. The appearance of harmonics such as in Eq. (2.2) leads to a consideration of higher-order spectra.

The Fourier transform is continually employed in the solution of equations of motion associated with physical phenomena and mathematicians have focussed on consequent cycles and harmonics. For example, Hirsch (1984) has remarked that "Dynamicists have always been fascinated (not to say obsessed) by periodicity." In that connection Ruelle (1989) makes effective use of the Fourier transform in the study of dynamic systems, specifically addressing aspects of chaos, periods and scaling.

The Fourier transform leads to entities with direct physical interpretations. One can point to a variety of success stories of the application of Fourier analysis. Michaelson (1891a, b) measured visibility curves, essentially the modulus of a Fourier transform, and after an inversion thereby inferred that the red hydrogen line was a doublet. This inference of splitting ultimately led to important developments in quantum mechanics. Tidal components caused by the sun, moon and planets have been isolated by Fourier analysis, see Cartwright (1982), Båth (1974), Bracewell (1989). Katz and Miledi (1971) inferred the mechanism of acetylcholine release via a Fourier analysis. Bolt et al. (1982) saw a fault rupturing in an earthquake by a frequency-wavenumber spectral analysis. Finally it may be noted that R. R. Ernst received the 1991 Nobel Prize in Chemistry for developing the technique of Fourier transform spectroscopy, see Amato (1991). A discussion of a variety of other physical examples may be found in Lanczos (1966), Båth (1974), Bracewell (1989).

3. SOME ANALYTIC BACKGROUND

3.1 The Fourier Case

Consider a square integrable function $g(x), 0 \leq x < 2\pi$. In this simple case Fourier analysis is built upon the values

$$c_k = \frac{1}{2\pi} \int_0^{2\pi} e^{-ikx} g(x)\, dx \tag{3.1}$$

$k = 0, \pm 1, \pm 2, \ldots$, and Fourier synthesis on expansions

$$g(x) \approx \sum_{k=-\infty}^{\infty} c_k e^{ixk} \tag{3.2}$$

The functions $\exp\{ikx\}$, $k = 0, \pm 1, \pm 2, \ldots$ here are orthogonal on $[0, 2\pi)$ and this connects Eqns. (3.1) and (3.2).

One important use of Fourier methods is the approximation of functions. If the values $c_k, k = 0, \pm 1, \ldots \pm K$ of Eq. (3.1) are available, a naive approximation to $g(x)$ is provided by

$$\sum_{k=-K}^{K} c_k e^{ixk} \tag{3.3}$$

However early researchers found that the approximation of Eq. (3.3) was often improved by inserting multipliers, w_k^K, such as $1 - |k|/K$, into the expansion and employing

$$g^K(x) = \sum_{k=-K}^{K} w_k^K c_k e^{ixk}. \tag{3.4}$$

instead of Eq. (3.3). Defining the kernel

$$W^K(x) = \sum_{k=-K}^{K} w_k^K e^{ixk}$$

Eq. (3.4) can be written

$$\int_0^{2\pi} W^K(y - x) g(x)\, dx \tag{3.5}$$

and one sees that Eq. (3.4) is a weighted average of the desired $g(\cdot)$. The effect of the multipliers, in some cases, is to improve the approximation by damping down the more rapidly oscillating terms in the expansion. This idea of damping down will recur below in the consideration of shrinking to improve estimates. The expression of Eq. (3.5) may be used to study directly the effect of the kernel function on the approximation. Timan (1963), Butzer and Nessel (1971) are books specifically concerned with approximations based on Fourier expressions.

In work with data values Y_t observed at $t = 0, \ldots, T - 1$ one might replace Eq. (3.1) with

$$\frac{1}{T} \sum_{t=0}^{T-1} \exp\left\{\frac{-i2\pi k t}{T}\right\} Y_t$$

having written $g(2\pi t/T) = Y_t$. As referred to earlier there are fast algorithms to evaluate this.

A second important use of Fourier analysis is in the study of time invariant systems. A simple linear time invariant system is described by

$$Y_t = \sum_{t=-\infty}^{\infty} c_{t-s} X_s.$$

i.e., a convolution. The response of this system to the input $X_t = \exp\{i\lambda t\}$ is

$$Y_t = C(\lambda) X_t \tag{3.6}$$

with $C(\lambda)$ the Fourier transform

$$C(\lambda) = \sum_{s=-\infty}^{\infty} e^{-i\lambda s} c_s$$

for $0 \leq \lambda < 2\pi$. This function is referred to as the transfer function of the system. Cosinusoids, $\exp\{i\lambda t\}$, are seen to be carried into cosinusoids. A variety of physical systems have this property to a good approximation.

Nonlinear time invariant systems may sometimes be approximated by Volterra expansions of the form

$$Y_t = \sum_{s=-\infty}^{\infty} c_{t-s} X_s + \sum_{s=-\infty}^{\infty} \sum_{s'=-\infty}^{\infty} d_{t-s,t-s'} X_s X_{s'} + \cdots$$

The input $X_t = \exp\{i\lambda t\}$ here leads to the output

$$C(\lambda) e^{i\lambda t} + D(\lambda, \lambda) e^{i2\lambda t} + \cdots$$

where $C(\lambda)$ is given above and

$$D(\lambda, \mu) = \sum_{s} \sum_{s'} e^{-i\lambda s - i\mu s'} d_{s,s'}$$

In such a nonlinear system one sees harmonics of the frequencies in the input appearing in the output. The laser example of Sec. 2 involved a system that was quadratic.

Fourier analysis is useful in work with constant coefficient differential equations. These show the occurrence of oscillations and are often effective models of physical systems. Consider for example the linear system

$$\frac{d\mathbf{S}(t)}{dt} = \mathbf{A}\mathbf{S}(t) + \mathbf{B}X(t)$$

with $\mathbf{S}(\cdot)$ vector-valued and $X(\cdot)$ scalar. Supposing

$$\mathbf{S}(t) = \int e^{it\lambda}\mathbf{s}(\lambda)\,d\lambda$$

and

$$X(t) = \int e^{it\lambda}x(\lambda)\,d\lambda$$

by Fourier analysis one has the solution directly as

$$\mathbf{S}(\lambda) = (i\lambda\mathbf{I} - \mathbf{A})^{-1}\mathbf{B}x(\lambda)$$

Supposing $x(\lambda)$ constant and the latent values, μ_j, of $\mathbf{A}$ to be distinct this may be written

$$\mathbf{S}(t) = \sum_j \mathbf{a}_j e^{i\mu_j t}$$

for some vectors $\mathbf{a}_j$. One sees the occurrence of oscillations at frequencies Re μ_j. One reference concerning such differential equations is Hochstadt (1964).

Turning to a further technique of Fourier analysis, that will be basic in one of the examples below, suppose that one is considering, for large x, an integral of the form

$$\int e^{ik(\lambda)x}R(\lambda)\,d\lambda$$

The method of stationary phase approximates this by

$$e^{\operatorname{sgn} k''(\lambda_0)i\pi/4}\sqrt{2\pi/(x|k''(\lambda^0|)}R(\lambda_0)e^{ik(\lambda_0)x}.$$

where λ_0 satisfies $k'(\lambda_0) = 0$. References include Barndorff-Nielsen and Cox (1989) and Aki and Richards (1980). The idea is that unless the $k(\lambda)$ is near 0 the rapidly oscillating multipliers $\cos k(\lambda), \sin k(\lambda)$ will give the integral value 0.

3.2 The Wavelet Case

Wavelet analysis is enjoying a surge of contemporary investigation and is a competitor of Fourier analysis. It may be viewed as Fourier analysis with the sine and cosine functions replaced by other families of (orthogonal) functions. There are many similarities between Fourier and wavelet analysis. Consider the expansion in Eq. (3.2) with the coefficients in Eq. (3.1). The expansion is based on the fact that the sine and cosine functions provide a basis for $L_2[0, 2\pi)$. In wavelet analysis other systems of functions are used, see e.g., Strichartz (1993), Benedetto and Frazier (1994). Wavelets are of practical importance because they can sometimes provide more parsimonious descriptions than Fourier ones.

Wavelets often focus on local versus global behavior and in particular can pick up transient behavior. Basic is a (mother) wavelet $\psi(\cdot)$ nonzero only on say the unit interval $[0, 1)$. Given a square-integrable function $g(x)$, one considers an expansion

$$g(x) = \sum_{j=-\infty}^{\infty} \sum_{k=-\infty}^{\infty} \beta_{jk} \psi_{jk}(x) \tag{3.7}$$

with

$$\psi_{jk}(x) = 2^{j/2} \psi(2^j x - k)$$

and

$$\beta_{jk} = \int \psi_{jk}(x) g(x)\, dx \tag{3.8}$$

The family $\{\psi_{jk}(\cdot)\}$ is taken to be orthonormal and complete, see e.g., Daubechies (1992), Walter (1992, 1994), Strichartz (1993), Benedetto and Frazier (1994).

The expansion in Eq. (3.7) represents $g(\cdot)$ in terms of functions with support individually on dyadic intervals $[k/2^j, (k+1)/2^j]$ for j, k integers. It suggests an approximation

$$g^{JK}(x) = \sum_{|j| \leq J} \sum_{|k| \leq K} \beta_{jk} \psi_{jk}(x) \tag{3.9}$$

to $g(x)$. This may be written as

$$g^{JK}(x) = \int W^{JK}(x, y) g(y)\, dy \tag{3.10}$$

the kernel being

$$W^{JK}(x, y) = \sum_{j,k} \psi_{jk}(x) \psi_{jk}(y) \tag{3.11}$$

This kernel will tend to a delta function in various circumstances, see Walter (1992). Equation (3.10) can be used to study the degree of approximation directly as could Eq. (3.5) in the Fourier case. Equations (3.10) and (3.11) are wavelet analogs of Eqns. (3.4) and (3.5).

In the case of a discontinuous function, as will occur in Example 5.4, a particular wavelet analysis is especially suitable, namely Haar wavelet analysis. This analysis is based on the function

$$\psi(x) = 1 \quad \text{for } 0 \leq x < \tfrac{1}{2}$$
$$\qquad -1 \quad \text{for } \tfrac{1}{2} \leq x < 1$$
$$\qquad 0 \quad \text{otherwise}$$

In the Haar case the kernel is

$$W_n(x,y) = 2^n \phi(2^n y - [2^n x]) \tag{3.12}$$

with $[\cdot]$ here referring to integral part and $g_n(x)$ of Eq. (3.10) a local mean, $g_n(x) = |I|^{-1} \int_I g(y)\, dy$, x being in the particular interval $I = [m/2^n, (m+1)/2^n)$, see Fine (1949), Walter (1992).

There are empirical versions of Eq. (3.8) for use when discrete time data $Y_t, t = 0, \ldots, T-1$ are available. One computes for example

$$\hat{\beta}_{jk} = \frac{1}{T} \sum_{t=0}^{T-1} \psi_{jk}(t/T) Y_t \tag{3.15}$$

Just as there are fast Fourier transforms, there are fast wavelet transforms, Strang (1993). Also one can write $p2^j$ for 2^j above, with no real change in concept, but improved approximations in practice. The dynamic spectrum analysis of Example 5.2 is one type of wavelet analysis with $j = j_0$ and $\psi(x) = \exp\{-i2\pi x\}$.

Insertion of multipliers, as in Eq. (3.4) for Fourier approximation, is fundamental. This will be discussed later.

4. STOCHASTICS AND STATISTICS

In this section the quantities being transformed will be random.

4.1 Stationary Processes

Fourier analysis is basic to dealing with stationary random processes. A process, Y_t, is said to be second-order stationary if $\text{cov}\{Y_{t+u}, Y_t\}$ exists for $t, u = 0, \pm 1, \pm 2, \ldots$ and does not depend on t. In practice this often appears a reasonable working assumption. In the case of

$Y_t, t = 0, \pm 1, \pm 2, \ldots$ a second-order stationary process, following Cramér (1942), one has the Fourier representation

$$Y_t = \int_{-\pi}^{\pi} e^{it\lambda} \, dZ(\lambda) \tag{4.1}$$

with $Z(\cdot)$ a random function such that

$$\mathrm{cov}\{dZ(\lambda), dZ(\mu)\} = \delta(\lambda - \mu) f(\lambda) \, d\lambda \, d\mu$$

$-\pi < \lambda, \mu \leq \pi, f(\cdot)$ being the power spectrum of Y and $\delta(\cdot)$ the Dirac delta function. The Cramér representation has the advantage of taking one directly to the Fourier domain and thereby making some operations on the process clearer. The series Y_t may be vector-valued. Then the cross-spectral density matrix, $\mathbf{f}(\cdot)$, is given by

$$\mathrm{cov}\{d\mathbf{Z}(\lambda), d\mathbf{Z}(\mu)\} = \delta(\lambda - \mu)\mathbf{f}(\lambda) \, d\lambda \, d\mu$$

Cross-spectrum analysis is useful for system analysis, i.e., estimating for example the transfer function of a linear time invariant system.

Higher-order spectra may be defined directly via $Z(\cdot)$, e.g., the bispectrum $f(\lambda, \mu)$ at frequency λ, μ is given by

$$\mathrm{cum}\{dZ(\lambda), dZ(\mu), dZ(\nu)\} = \eta(\lambda + \mu + \nu)f(\lambda, \mu)d\lambda \, d\mu \, d\nu$$

where $\eta(\lambda)$ is the 2π periodic extension of the Dirac delta function.

Empirical Fourier analysis, e.g., of residuals of a fit, provides a diagnostic using in particular the result that if the process is white noise, the power spectrum is constant in frequency, λ.

Blackman and Tukey (1959), Båth (1974), Brillinger (1975) and Bloomfield (1976) are books focussing on the empirical Fourier analysis of time series

4.2 Central Limit Theorems

In classic forms the central limit theorem is concerned with the distributions of sums of independent random variables

$$S_T = Y_0 + Y_1 + \cdots + Y_{T-1}$$

and their approximate normality with variance $T\sigma^2$ for large T. It is usual to assume that the Y's are identically distributed.

At some point engineers began promulgating a folk theorem to the effect that narrow-band noise is approximately Gaussian, [see Leonov and Shiryaev (1960), Picinbono (1960), Rosenblatt (1961)]. One fashion to formulate this remark is as a statement that

$$S_T(\lambda) = Y_0 + e^{-i\lambda}Y_1 + \cdots + e^{-i\lambda(T-1)}Y_{T-1} \tag{4.2}$$

$0 \leq \lambda < 2\pi$, is approximately (complex) normal for each λ. Under stationarity and mixing assumptions for the series Y_t, the variance of Eq. (4.2) is approximately

$$2\pi T f(\lambda) \tag{4.3}$$

with $f(\lambda)$ the power spectrum of Eq. (4.1) at frequency λ. Surprisingly, the values of $S_T(\lambda)$ at distinct frequencies of the form $\lambda = 2\pi j/T$, are approximately independent. Problems involving stationary mixing processes may thus be converted into ones involving (approximately) independent normal random variables. Empirical Fourier transforms such as Eq. (4.2) have many uses and several are indicated in this paper. A fundamental use is to estimate a power spectrum by smoothing the squared-modulus.

Early work on the asymptotic properties of finite Fourier transforms includes that of Slutsky (1934), Leonov and Shiryaev (1960), Rosenblatt (1961), Good (1963), Hannan (1969), Brillinger (1969), Hannan and Thomson (1971), Hannan (1972).

There has been some consideration of the cases of long range dependence and stable distributions. References include: Rosenblatt (1981), Freedman and Lane (1981), Fox and Taqqu (1986), Yajima (1989), Shao and Nikias (1993). The case of random generalized functions, which includes for example point processes and random measures, is considered in Brillinger (1982).

In the case of wavelets and a model

$$Y_t = g(t/T) + \varepsilon_t \tag{4.4}$$

with ε_t stationary noise having power spectrum $f(\lambda)$, under regularity conditions, the statistic $\hat{\beta}_{jk}$ of Eq. (3.13) may be shown to be asymptotically normal with mean β_{jk} and variance

$$\frac{2\pi}{T} f(0)$$

see Brillinger (1996). The variance is the same as that of Eq. (4.4). Further when the functions $\psi_{jk}(\cdot)$ and $\psi_{j'k'}(\cdot)$ are orthogonal, the coefficients $\hat{\beta}_{jk}, \hat{\beta}_{j'k'}$ are approximately independent for distinct (j,k) and (j',k'). This last results suggests that an estimate of $f(0)$ may be obtained by averaging the values $T|\hat{\beta}_{jk}|^2/T$ for which $\beta_{jk} = 0$.

4.3 Shrinking

Among surprises in working with Fourier transforms is the importance of convergence factors. These are the w_k^K of Eq. (3.4). In Eq. (3.4) they shrink the coefficients of the $\exp\{ixk\}$ towards 0 as k increases. Such multipliers are also important in the stochastic case, see: Tukey (1959), Brillinger (1975), Bloomfield (1976), Dahlhaus (1984, 1989).

A related concept is shrinking. In a regression context Tukey (1979) distinguishes three types of shrinking. Crudely: "first shrinkage" corresponds to pretesting and selection of regressor variables, "second shrinkage" corresponds to a type of Wiener filtering and "third shrinkage" corresponds to borrowing strength from other coefficients to improve the collection of coefficients. In this last case the multipliers are not meant for attenuating high frequencies, rather they are meant for attenuating uncertain terms. A common characteristic is that the estimates become biased; however, biased estimates have long been dominant in time series analysis.

Second shrinkage plays an important role in two of the examples that follow. A particular second shrinkage estimate, introduced in Tukey (1979), may be motivated as follows. Consider a classic simple regression model

$$y = \beta x + \varepsilon$$

with b an estimate of β and s an estimate of its standard error. Seek a multiplier m such that mbx is an improved estimate of βx. The mean-squared error of the new estimate is

$$x^2 E\{(\beta - mb)^2\}$$

which may be estimated by

$$x^2\{(1-m)^2[b^2 - s^2] + m^2 s^2\}$$

This is minimized by the choice $m = 1 - s^2/b^2$. One would prefer to take m to be the positive part

$$(1 - s^2/b^2)_+ \tag{4.5}$$

This multiplier has the reasonable property of being 0 for b less than its standard error.

In Sec. 3.1 convergence factors, w_k^K, were inserted into trigonometric expressions to obtain improved approximation. In Example 5.1 such multipliers based on the reliability of estimated coefficients $\hat{c}_k$ will be inserted to obtain an improved estimate. To estimate $\hat{g}(x)$ of Eq. (4.4) one considers, for example,

$$\hat{g}(x) = \sum_k w(\hat{c}_k/s_k)\hat{c}_k e^{ixk} \tag{4.6}$$

where s_k^2 is an estimate of the variance of $\hat{c}_k$ and $w(u)$ is a function that is near 1 for large u and near 0 for small u. Examples of functions $w(\cdot)$ are given in Fig. 1.

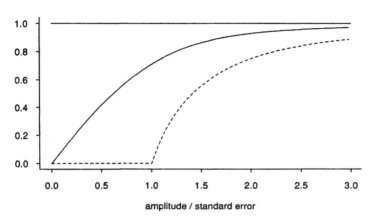

Figure 1. Graph of the multipliers Eqns (5.7) and Eq. (5.8), as a functions of the amplitude of the estimate divided by its estimated standard error.

In work to obtain improved wavelet-based estimates, Donoho and Johnstone (1990), Hall and Patil (1995) create shrinkage estimates involving multipliers, there referred to as "thresholders". The estimates take the form

$$\sum_{|j| \leq J} \sum_{|k| \leq K} w(|\hat{\beta}_{jk}|/s_{jk})\hat{\beta}_{jk}\psi_{jk}(x) \qquad (4.7)$$

where s_{jk} is an estimate of var $\hat{\beta}_{jk}$ and $0 \leq w(\cdot) \leq 1$.

There are many classical references to selection of variables and pretesting, i.e., first shrinkage. References to second shrinkage include: Whittle (1962), Thompson (1968), King (1972), Ott and Kronmall (1976), Tukey (1979), Zidek (1983), Donoho and Johnstone (1990), Stoffer (1991), Hall and Patil (1993), Donoho et al. (1995). References to third shrinkage include: Stein (1955), Efron and Morris (1977), Copas (1983), Saleh (1992).

5. EXAMPLES

In this Section four biological and physical examples are presented.

5.1 Electron Microscopy

Electron microscopy is a tool for studying the placements of atoms within molecules. It is mainly carried out with crystalline (periodic) material. One problem is to obtain improved images and that is the concern of the present example. Glaeser (1985), Henderson et al. (1986), Hovmöller (1990) are references describing the basics of electron microscopy.

In the planar case, the principal theoretical concept is the projected (Coulomb) density distribution

$$V(x,y) = \sum_{h,k} F_{h,k} e^{2\pi i(hx+ky)/\Delta} \tag{5.1}$$

$h, k = 0, \pm 1, \pm 2, \ldots$ with (x, y) planar coordinates and with Δ the period of the crystal. The function $V(\cdot)$ is real-valued and has various symmetries. The h, k in Eq. (5.1) are referred to as the Miller indices, while the $F_{h,k}$ are referred to as structure factors. One wishes to estimate $V(x, y)$ over $0 \leq x, y < \Delta$.

The datum is an image, $Y(x, y)$, with $0 \leq x < X, 0 \leq y < Y$. The image may be written as

$$Y(x,y) = V(x,y) + \text{noise} \tag{5.2}$$

The empirical Fourier transform is

$$\hat{F}_{h,k} = \int_0^Y \int_0^X Y(x,y) e^{-2\pi i(hx+ky)/\Delta} dx \ dy \tag{5.3}$$

which may be written

$$\int_0^\Delta \int_0^\Delta \sum_{m,n} Y(x+m\Delta, y+n\Delta) e^{-2\pi i(hx+ky)/\Delta} dx \ dy \tag{5.4}$$

The synthesis corresponding to the analysis Eq. (5.3) is

$$\sum_{h,k} \hat{F}_{h,k} e^{2\pi i(hx+ky)/\Delta} \tag{5.5}$$

$0 \leq x < \Delta, \ 0 \leq y < \Delta$.

There has been concern to form an improved image. In this connection Blow and Crick (1959), Hayward and Stroud (1981) introduced "multipliers", $w(\cdot)$, into expressions like Eq. (5.5), forming

$$\hat{V}(x,y) = \sum_{h,k} w(|\hat{F}_{h,k}|\hat{\sigma}_{h,k})\hat{F}_{h,k} e^{2\pi i(hx+ky)/\Delta} \tag{5.6}$$

where the $\hat{\sigma}_{h,k}$ are estimates of the standard errors of the $\hat{F}_{h,k}$. This is a second shrinkage estimate. Consideration of the mean-squared error, as in Eq. (4.5), leads to the multiplier

$$w(|\hat{F}|/\hat{\sigma}) = \left(1 - \frac{\hat{\sigma}^2}{|\hat{F}|^2}\right)_+ \tag{5.7}$$

which by analogy with Wiener filtering will be called the Wiener multiplier. By Bayesian arguments Blow and Crick (1959) and Hayward and Stroud (1981) were lead to the multiplier

$$w(\gamma) = \frac{\sqrt{\pi}}{2}\gamma\left[I_0\left(\frac{\gamma^2}{2}\right) + I_1\left(\frac{\gamma^2}{2}\right)\right]e^{-\gamma^2/2} \qquad (5.8)$$

with $\gamma = |\hat{F}|/\hat{\sigma}$, and I_0, I_1 modified Bessel functions, see Brillinger et al. (1989, 1990). It and Eq. (5.7) are graphed in Fig. 1. These multipliers approach 1 as the uncertainty approaches 0.

Estimates employing Eqns (5.7) and (5.8) are illustrated in Fig. 2 for images of the protein bacteriorhodopsin. This substance occurs naturally

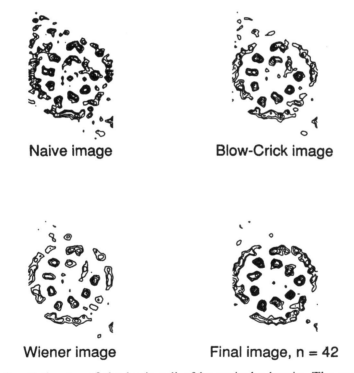

Naive image   Blow-Crick image

Wiener image   Final image, n = 42

Figure 2. Estimates of the basic cell of bacteriorhodopsin. The upper left panel is the naive estimate as shown in Eq. (5.5). The upper right panel is the estimate Eq. (5.6) with the multiplier, Eq. (5.8). The bottom left panel is the estimate Eq. (5.6) with the multiplier, Eq. (5.7). The last panel is Eq. (5.6), with Eq. (5.8), obtained by combining 42 individual images.

as a two-dimensional crystalline array within the cell membrane of *Halobacterium halobrium.* Together with accompanying lipid molecules, it is known as "purple membrane". This crystal is based on a hexagonal lattice. In Fig. 2 only the positive contours are shown. (Negative density features signify the absence of atoms and thus have no direct usefulness when the density map is interpreted.) The first panel of Fig. 2 shows the elementary estimate of Eq. (5.5). The top right shows Eq. (5.6) with $w(\cdot)$ of Eq. (5.7). The third, lower left, shows Eq. (5.6) with $w(\cdot)$ of (5.8). The final panel provides an estimate based on combining 42 individual images. This last image may be viewed as what the earlier estimates based on a single image ascribe to be.

Through the inclusion of the multipliers, the peaks have become more substantial and better separated. Also, the estimates show better approximations to a three-fold symmetry. Details of the data collection and further details of the analysis may be found in Brillinger et al. (1989, 1990).

The Fourier transform is useful in this example firstly because of the lattice periodicities and secondly for the central limit theorem result suggesting specific estimates of the s_{hk} of Eq. (5.6) namely for s_{hk}^2 one takes the average of the squared moduli of Fourier coefficients thought to be signal free.

There are extensions to the 3D case, see Henderson et al. (1990), Wenk et al. (1992).

5.2 Seismic Surface Waves

Various sound waves are transmitted through the Earth following a seismic disturbance, in particular surface (or Rayleigh) waves. These are vibrations whose energy is trapped and propagated just under the surface. The waves have sinusoidal form and are prominent in the later part of a seismogram. For example see Fig. 3 for an event that was recorded in Uppsala, Sweden. These waves have the interesting aspect of having been discovered mathematically. For basic details see Aki and Richards (1980) and Bullen and Bolt (1985).

Consider modelling that part of a seismogram where the Rayleigh waves occur. Let $Y(x,t)$ denote the vibrations recorded at distance x from the earthquake source, as a function of time t. With a layered crust model the theoretical seismogram is a solution of a system of differential equations with associated boundary conditions and may be represented as

$$\int e^{-i(\lambda t - k(\lambda)x)} R(\lambda) \, d\lambda$$

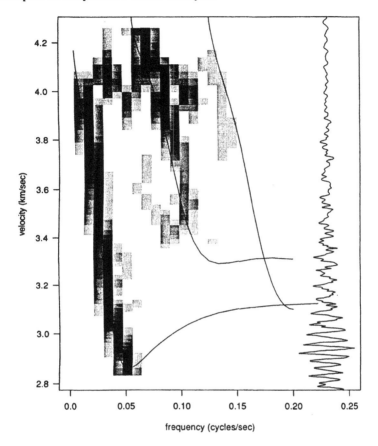

Figure 3. The Siberia–Upsalla dynamic spectrum as a function of frequency and velocity as computed from Eq. (5.11). The vertical trace is the seismogram as a function of velocity.

Here, when $x = 0$

$$\int e^{-i\lambda t} R(\lambda) \, d\lambda$$

represents the vibrations at the earthquake source. The solution in Eq. (5.9) comes from substituting a particular solution $\exp\{-i(\lambda t - kx)\}$ into the differential equations and matching boundary conditions, see Aki and Richards (1980). One writes $k(\lambda) = \lambda/c(\lambda)$ with $c(\lambda)$ the (phase) velocity with which the wave of frequency λ travels. The functions $k(\cdot)$ and $c(\cdot)$ depend on the transmission medium.

In the case that x is large one can use the method-of-stationary phase, described in Section 3.1, to see the sinusoidal form of the waves. Specifically for large x, Eq. (5.9) is approximately

$$R(\lambda_t) \exp\{-i(\lambda_t t - k(\lambda_t)x)\} \tag{5.10}$$

with λ_t the solution of

$$\frac{d}{d\lambda}\{\lambda t - k(\lambda)x\} = 0$$

that is $k'(\lambda_t) = t/x = 1/U(\lambda_t)$. Here $U(\lambda)$ is the group velocity, the velocity with which the energy travels, at frequency λ. The phenomenon of waves with different frequencies travelling with different velocities, as occurs here, is called dispersion.

Given an earth model, θ, that is a collection of layer depth, velocity and density parameters, one can compute the group velocity $U(\lambda|\theta)$, see Bolt and Butcher (1960), Aki and Richards (1980). For frequency λ and parameter θ there may be several possible dispersion curves $U_n(\lambda|\theta), n = 0, 1, 2, \ldots$ called modes. Dynamic Fourier analysis provides a way to see these modes, and is presented in Fig. 3. The concern of the example of this section is to estimate θ.

The event studied originated in Siberia, 20 April 1989, and the trace was recorded at Uppsala, Sweden. Figure 3 provides a grey scale display of energy as a function of velocity and frequency. It is computed as

$$\left| \sum_{s=-S}^{S} h(s/S)\, Y(t-s)e^{-is\lambda} \right|^2 \tag{5.11}$$

with $t = x_0/v$, v velocity, x_0 distance to source and $h(\cdot)$ a convergence factor. One sees waves of about 0.07 cycles/second arriving first. Figure 3 also shows the dispersion curves $U_n(\lambda|\hat{\theta})$ for one fitted earth model. Some further details are given in Brillinger (1993).

The velocity-frequency curves, $U_n(\lambda|\theta)$, may be inverted to frequency-time curves $\lambda = \lambda_t(t|\theta)$. To estimate θ one can then consider choosing θ, α to minimize

$$\sum_t \left| Y(t) - \int e^{-i(\lambda t - k(\lambda|\theta)x_0)} R(\lambda|\alpha)\, d\lambda \right|^2$$

where α is some parametrization of the source motion. One approach is to approximating the integral in Eq. (5.9), is to take $R(\cdot)$ piecewise constant, linear in α. Figure 4 provides the results of such an analysis. Graphed are

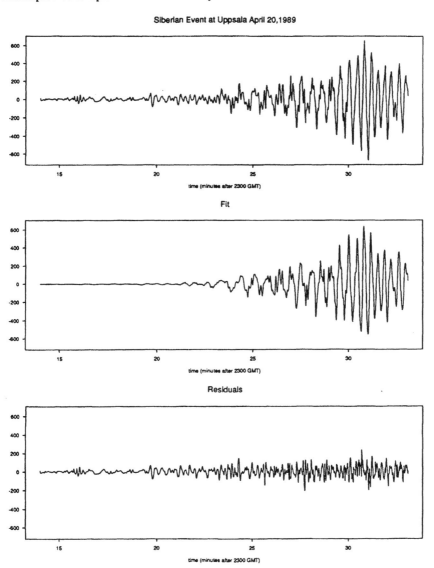

Figure 4. The top trace is the seismogram as a function of time. The middle is the fit based on Eq. (5.9). The bottom is the difference of these two.

the series, the fit and the residuals. The standard errors might be computed as in Richards (1961), focussing on the nonlinear parameters θ and acting as if the noise series was white. An improved estimation procedure is needed, for the residual series of Fig. 4 suggests the presence of signal-generated noise.

Even though this particular situation is clearly nonstationary, Fourier analysis has been basic to addressing it. This is a consequence of the presence of dispersion. The example is also of additional interest since one has a Fourier transform of two variables whose support lies on several curves, see Fig. 3. This type of plot allows inference of the presence of higher modes and assessment of the fit as well.

5.3 NMR Spectroscopy

Nuclear magnetic resonance is a quantum mechanical phenomenon employed to study the structure of various molecules. In an experiment, one creates a fluctuating magnetic field, $X(t)$, encompasing a substance and then observes an induced voltage, $Y(t)$. Hennel and Klinowski (1993) is one general reference.

If $S(t)$ is a vector describing the state of the system at time t, then the fluctuations are described by the Bloch equations

$$\frac{d\mathbf{S}(t)}{dt} = \mathbf{a} + \mathbf{A}\mathbf{S}(t) + \mathbf{B}\mathbf{S}(t)X(t) \tag{5.12}$$

and the measurements by

$$Y(t) = \mathbf{c}^T\mathbf{S}(t) + \text{noise} \tag{5.13}$$

with $\mathbf{c}$ depending on the geometry of the experiment. The principal parameters are frequencies of oscillation and decay rates. The parameters of interest sit in the matrices $\mathbf{A}$ and $\mathbf{B}$, see Brillinger and Kaiser (1992). The entries of $\mathbf{A}$ and $\mathbf{B}$ have physical interpretations, e.g., the diagonal entries of $\mathbf{A}$ represent occupancy probabilities.

Equations (5.12) is interesting for being bilinear. It can be solved, symbolically, by successive substitutions, obtaining

$$\mathbf{S}(t) = \mathbf{C} + \int^t e^{A(t-s)}\mathbf{C}X(s)\,ds + \int^t \int^s e^{A(t-s)}\mathbf{B}e^{A(s-r)}\mathbf{C}X(r)X(s)\,dr\,ds + \cdots$$

with $\mathbf{C} = -\mathbf{A}^{-1}\mathbf{a}$. If $\mathbf{A}$ is written $Ue^\Lambda U^{-1}$ with Λ diagonal, then the pulse response, $\mathbf{S}(t)$, is seen to be a sum of complex exponentials and various of their powers and products. The real parts of the entries of Λ will lead to the decay of these components while the imaginary parts represent resonance frequencies.

The problem is to estimate the parameters of Eq. (5.12) and thereby, to characterize the substance. Some of the parameters may be estimated by cross-spectral analysis and others by likelihood analysis.

Brillinger and Kaiser (1992) present results from a study of 2,3-dibromothiophene. The matrices A and B are 4×4 with complex-valued entries. The parameters include a coupling constant, J and frequencies ω_A and ω_B. In the experiment the input employed was a sequence of pulses

$$X(t) = \sum_j M_j \delta(t - j\Delta)$$

with $\Delta = 1/150\,\text{s}$, t in seconds and M_j the m-sequence given by $M_j = M_{j-1} M_{j-4} M_{j-8} M_{j-12}$ starting at $M_j = -1$ for $j = 1, \ldots, 12$.

Figure 5 presents corresponding stretches of input and output together with the results of a cross-spectral analysis. Specifically the first-order transfer function estimate

$$\hat{A}(\lambda) = \frac{\operatorname*{smooth}_{\mu \approx \lambda} \{[\sum_t Y(t)e^{-i\mu t}]\overline{[\sum_t X(t)e^{-i\mu t}]}\}}{\operatorname*{smooth}_{\mu \approx \lambda} \{|\sum_t X(t)e^{-i\mu t}|^2\}} = \hat{f}_{YX}(\lambda).\hat{f}_{XX}(\lambda)^{-1}$$

is given in Fig. 5. Theoretically its peaks are located at the frequencies

$$(\omega_A + \omega_B)/2 \pm J \pm \sqrt{J^2 + (\omega_A - \omega_B)^2/2}$$

and the widths of the peaks relate to a damping constant T_2.

In a more detailed analysis the parameters of the model, including initial state values, were estimated by least squares seeking

$$\min_\theta \sum_t |Y(t) - \mathbf{c}^T \mathbf{S}(t|\theta)|^2 \tag{5.14}$$

θ referring to the unknown parameters. In the computations the state vector, $\mathbf{S}(t|\theta)$ was evaluated recursively. Figure 6 shows the amplitude of the Fourier transform of the data and of the corresponding fit. (It is usual to graph an unsmoothed estimate in the NMR literature in order to obtain high resolution of peaks.) There is an intriguing small peak just above 60 Hz which recurs when the time series is broken down into contiguous segments. NMR researchers refer to such a phenomenon as a "birdie", but had no explanation for its source in the present case. Further details may be found in Brillinger and Kaiser (1992).

There are extensions of the cross-spectral approach to the 2, 3, 4, ... and higher dimensional cases, see Blümich (1985).

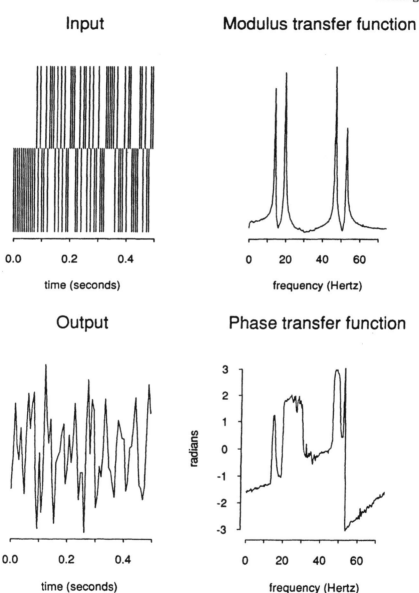

Figure 5. Results of a nuclear magnetic resonance study of 2,3-dibro-mothiophene. The top left is a segment of the input and below is the corresponding output. The right column provides the estimated amplitude and phase of the (linear) transfer function.

log10 |FT| fit

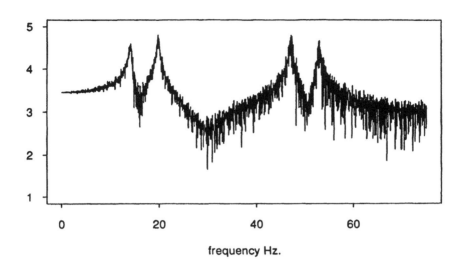

log10 |FT| data

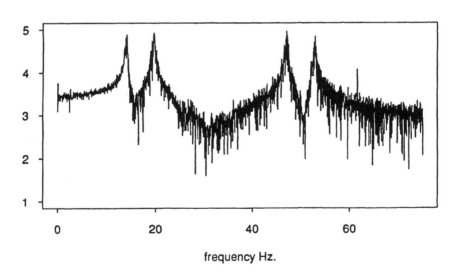

Figure 6. The modulus of the Fourier transform of the output and of the corresponding fit derived from Eq. (5.14).

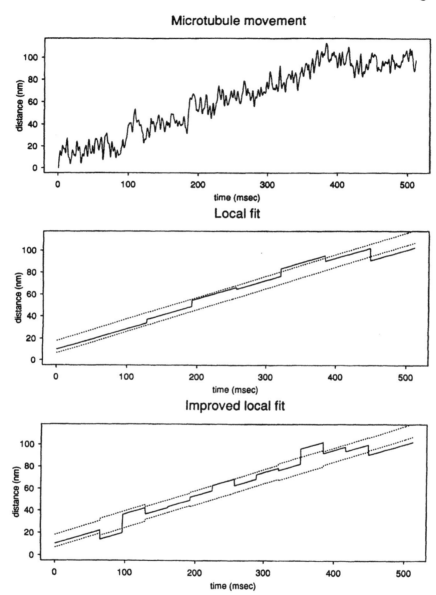

Figure 7. The top trace is the estimated movement of a microtubule as a function of time. The middle provides the fit with no shrinkage. The bottom panel provides a shrunken fit. The dashed lines provide approximate ±2 standard error limits.

In this example the Fourier transform is useful for examining resonance, for assessing goodness of fit and for understanding the nonlinearity involved.

5.4 Microtubule Movement

As an illustration of wavelet analysis, consider the problem of searching for jumps in records of microtubule movement. Microtubules are linear polymers basic to cell motility. A concern is whether movement is smooth, or rather via a series of jumps, see Malik et al. (1994).

If $Y(t)$ denotes the distance a microtubule has travelled at time t, the model considered is

$$Y(t) = \alpha t + g(t/T) + \text{noise} \tag{5.15}$$

$t = 0, \ldots, T - 1$ with α a parameter related to diffusion motion and $g(\cdot)$ a step function. The model in Eq. (5.15) will be approximated by

$$Y(t) = \alpha t + \sum_k \gamma_{nk} \phi_{nk}(t/T) + \text{noise} \tag{5.16}$$

for some n. Because of the presence of the term αt in Eq. (5.16) the analysis in the present case is not so immediate, but still all that one needs are local means. The least squares estimates are obtained by regression of Y on the $\phi_{nk}(t/T)$ and on t made orthogonal to the ϕ_{nk}. Further details on the fitting are given in the Appendix to this chapter.

In the experiments of concern samples were taken from the bovine brain. Specifics may be found in Malik et al. (1994). The top panel of Fig. 7 provides a data trace. Next is an estimate $\hat{g}_n(t/T)$ with $w(u)$, of Eq. (4.7), identically 1. The final panel an improved estimate based on the multiplier $w(u) = (1 - 1/u^2)_+$. The value of $n = 3$ was chosen having in mind a search for isolated jumps for this particular data set. Also indicated are approximate ± 2 standard error limits around the fitted straight line. There is little evidence for the presence of isolated jumps. The construction of the standard error estimate is described in the Appendix to this chapter.

The Fourier transform was used here to develop uncertainty estimates, following on an assumption that the noise was stationary.

6. SOME OPEN PROBLEMS

This Section briefly lists a number of topics, motivated by the examples of the paper, that appear fruitful for more development.

Foremost among the topics calling out for further research is the theoretical and practical development of shrinkage estimates. The ideas are basic.

The effects are substantial, see Fig. 2 for example. One wonders about "optimal" choice of the multipliers/shrinkage factors. Perhaps optimal rates of convergence may be determined and then it be checked which multipliers lead to those. This paper has focused on second shrinkage. Berger and Wolpert (1983) develop third shrinkage estimates in random function cases. Lillestol (1977) studies time series in one case.

In both the surface wave and nuclear magnetic resonance examples, examination of residuals suggests the presence of signal-generated noise. Better estimates are needed. Either because the ones used are inefficient or because the signal-generated noise is basic. In the latter case an appropriate likelihood function needs to be developed. Ihaka (1993) does this for one case in seismology. If the noise is indeed nonstationary and autocorrelated, then a novel form of uncertainty estimation technique will be needed. In the case of the "improved" wavelet estimate, the uncertainty was estimated as if the shrinkage factors were constant, see Appendix to this chapter. Perhaps a useful bootstrap procedure could be developed, based on an assumption of stationary innovations being present.

Quite a different type of problem is the following: develop the aliasing structure for higher-order spectra in the case of a process observed on a lattice. This will be particularly complicated in the case of lattices in R^p with $p > 1$. Another problem in the case of image estimates, is how to visualize the associated uncertainty.

The Fourier transforms studied have all been scalar-valued. There are central limit theorems for processes taking on values in a group. It would be of interest to obtain corresponding results for group-valued Fourier transforms, e.g., in the p-adic case.

7. DISCUSSION AND SUMMARY

The principal interest of the examples of the paper has been in problem formulation and in addressing particular scientific questions. In each of the examples, an empirical Fourier transform has played a central role. With its broad collection of understood properties this transform has assisted the analyses greatly. The usefulness of second shrinkage, analogous to the use of convergence factors in Fourier approximation, is also noteworthy.

The particular groups of the examples have been abelian. General group theoretic ideas and empirical Fourier analysis have been discussed for other groups. For the case of the symmetric group see Diaconnis (1988, 1989) and Kim and Chapman (1993). For the locally compact abelian case see Brillinger (1982). For p-adics see Brillinger (1992). The use of p-adics in signal processing is discussed in Gorgui-Naguib (1990). For other cases see Hannan (1969). Key distinctions that arise are abelian versus nonabelian groups, compact versus locally compact groups, and whether t is in a group or Y is in a group.

There are other transforms that are useful in practice. These include: the Laplace, Hilbert, Stieltjes, Mellin, with some work having been done for abstract groups, see Loomis (1953).

The case of lacunary trigonometric series is somewhat like the case of point processes. Here the Fourier transform has a different form, e.g., for point process data $\{\tau_1 < \tau_2 < \cdots < \tau_N\}$ it is given by

$$\sum_{j=1}^{N} \exp\{-i\lambda\tau_j\}$$

$-\infty < \lambda < \infty$. Such a transform is used in Rosenberg et al. (1989) for example.

Unemphasized, but important, topics include: the Poisson summation formula useful in understanding aliasing and the sampling theorem (Hannan (1965)), abstract fast algorithms (Rockmore (1990)), spherical functions (Terras (1988)), uncertainty principles (Smith (1990)).

In conclusion we quote J. B. Fourier (1822), *Théorie Analytique de la Chaleur*: "L'' étude approfondie de la nature est la source la plus féconde des découvertes mathématiques." There is so much evidence in favor of this remark today.

Acknowledgments

The author thanks his collaborators in the examples of Sec. 5: Ken Downing, Bob Glaeser, Guy Perkins, Bruce Bolt, Reinhold Kaiser, Fady Malik and Ron Vail respectively. This paper is based on talks presented at the 1993 opening ceremonies of the Fields Institute for Research in Mathematical Sciences at the University of Waterloo, Ontario, Canada and at the 1993 Winter Meeting of the Canadian Mathematical Society at Ottawa. The work was partially supported by the NSF Grants DMS-9208683, DMS-9300002, DMS-9625774 and the Office of Naval Research Grant N00014-94-1-0042. Comments of Bob Kass and the referee helped improve the layout and style of the paper.

REFERENCES

Aki, K. and P. G. Richards, *Quantitative Seismology I & II*. Freeman, San Francisco, 1980.

Amato, I., Nobel prizes 1991 *Science 254*: 518–519, 1991.

Barndorff-Nielsen, O. E. and D. R. Cox, *Asymptotic Techniques for Use in Statistics*, Chapman and Hall, London, 1989.

Båth, M., *Spectral Analysis in Geophysics*. Elsevier, Amsterdam, 1994.

Bazin, M. J., P. H. Lucie, and S. Moss de Olivierra, Experimental demonstrations of the mathematical properties of Fourier transforms using diffraction phenomena, *Eur. J. Phys. 7*: 183–188 (1986).

Benedetto, J. J. and M. W. Frazier, (eds), *Wavelets*, CRC Press, Boca Raton, 1994.

Berger, J. and R. Wolpert, Estimating the mean function of a Gaussian process and the Stein effect, *J. Mult. Anal. 13*: 401–424, (1983).

Blackman, R. B. and J. W. Tukey, *The Measurement of Power Spectra*. Dover, New York, 1959.

Bloembergen, N., Nonlinear optics and spectroscopy, *Science 215*: 1057–1064 (1982).

Bloomfield, P., *Fourier Analysis of Time Series: An Introduction*, Wiley, New York, 1976.

Blow, D. M. and F. H. C. Crick, The treatment of errors in the isomorphous replacement method, *Acta Cryst. 12*: 794–802 (1959).

Blümich, B., Stochastic nmr spectroscopy, *Bull. Magnet. Resonance 7*: 5–26 (1985).

Bochner, S., *Lectures on Fourier Integrals*, Princeton Press, Princeton, 1959.

Bochner, S., *Harmonic Analysis and the Theory of Probability*, Univ. Calif. Press, Berkeley, 1960.

Bolt, B. A. and J. Butcher, Rayleigh wave dispersion for a single layer on an elastic half space, *Australian J. Physics 13*: 498–504 (1960).

Bolt, B. A., Y. B. Tsai, K. Yeh and M. K. Hsu, Earthquake strong motions recorded at a large near-source array of digital seismographs, *Earthquake Eng. Structural Dynam. 10*: 561–573 (1982).

Born, M. and Wolf, E. (1964). Principles of Optics. Second Edition. Macmillan, New York.

Bracewell, R. N., The Fourier transform, *Scientific Ameri.*, June: 86–95 (1989).

Brillinger, D. R., A search for a relationship between monthly sunspot numbers and certain climatic series, *Bull. Inter. Statist. Inst. 43*: 293–306 (1969).

Brillinger, D. R., *Time Series: Data Analysis and Theory*, Holt, New York, 1975.

Brillinger, D. R., Asymptotic normality of finite Fourier transforms of stationary generalized processes, *J. Mult. Analysis 12*: 64–71 (1982).

Brillinger, D. R., Some asymptotics of finite Fourier transforms of a stationary p-adic process, *J. Comb. Inf. Sys. Sci. 16*: 155–169 (1991).

Brillinger, D. R., An application of statistics to seismology: dispersion and modes. In *Developments in Time Series Analysis* (ed. T. Subba Rao), Chapman and Hall, London, 1993, pp. 331–340.

Brillinger, D. R., Some uses of cumulants in wavelet analysis, *Nonpar. Statist. 6*: 93–114 (1996).

Brillinger, D. R., K. H. Downing and R. M. Glaeser, Some statistical aspects of low-dose electron imaging of crystals, *J. Stat. Planning Inf. 25*: 235–259 (1990).

Brillinger, D. R., K. H. Downing, R. M. Glaeser and G. Perkins, Combining noisy images of small crystalline domains in high resolution electron microscopy, *J. App. Stat. 16*: 165–175 (1989).

Brillinger, D. R. and R. Kaiser, Fourier and likelihood analysis in NMR spectroscopy. In *New Directions in Time Series I* (eds. D. Brillinger, P. Caines, J. Geweke, E. Parzen, M. Rosenblatt and M. Taqqu), Springer, New York, 1992, pp. 41–64.

Bullen, K. E. and B. A. Bolt, *An Introduction to the Theory of Seismology*, Cambridge Univ. Press, Cambridge, 1985.

Butzer, P. L. and R. J. Nessel, *Fourier Analysis and Approximation*, Academic, New York, 1971.

Cartwright, D. E., Tidal analysis—a retrospect in *Time Series Methods in Hydrosciences* (eds A. H. El-Shaarawi and S. R. Esterby), Elsevier, Amsterdam, 1982, pp. 170–188.

Cooley, J. W. and J. W. Tukey, An algorithm for the machine calculation of complex Fourier series, *Math. Comp. 19*: 297–301 (1965).

Copas, J. B., Regression, prediction and shrinkage, *J. Roy. Statist. Soc.* B 45: 311–335 (1983).

Cramér, H., On harmonic analysis in certain function spaces, *Arkiv Math. Astr. Fysik. 28*: 1–7 (1942).

Dahlhaus, R., Parameter estimation of stationary processes with spectra containing strong peaks in *Robust and Nonlinear Time Series Analysis* (eds J. Franke, W. Haerdle and D. Martin), Springer, New York, 1984, pp. 50–67.

Dahlhaus, R., Efficient parameter estimation for self-similar processes, *Ann. Statist. 17*: 1749–1766 (1989).

Daubechies, I., *Ten Lectures on Wavelets*, SIAM, Philadelphia, 1992.

Diaconnis, P., *Group Representations in Probability and Statistics*, Institute of Mathematical Statistics, Hayward, 1988.

Diaconnis, P., A generalization of spectral analysis with applications to ranked data, *Ann. Statist. 17*: 949–979 (1989).

Donoho, D. L. and I. M. Johnstone, *Wavelets and optimal nonlinear function estimates*, Tech. Report 281, Statistics Dept. Univ. California, Berkeley, 1990.

Donoho, D. L., I. M. Johnstone, G. Kerkyacharian and D. Picard. Wavelet shrinkage: asymptopia? *J. Roy. Statist. Soc.* B *57*: 301-369 (1995).

Dzhaparidze, K., *Parameter Estimation and Hypothesis Testing in Spectral Analysis of Stationary Time Series*, Springer, New York, 1986.

Efron, B. and C. Morris, Stein's paradox in statistics, *Scientific Amer. 236*: 119–127 (1977).

Fan, J., Deconvolution with supersmooth distributions, *Canadian J. Statist. 20*: 155–169 (1992).

Feuerverger, A., Efficiency in time series, *Canadian J. Statist. 18*: 155–162. (1990).

Feigin, P. D. and C. R. Heathcote, The empirical characteristic function and the Cramér-von Mises statistic, *Sankhya A 38*: 309–325 (1976).

Fine, N. J., On Walsh functions, *Trans. Amer. Math. Soc. 65*: 372–414 (1949).

Fox, R. and M. S. Taqqu, Large-sample properties of parameter estimates for strongly dependent stationary time series, *Ann. Statist. 14*: 517–532 (1986).

Freedman, D. and D. Lane, The empirical distribution of the Fourier coefficients of a sequence of independent, identically distributed long-tailed random variables, *Z. Wahrschein. Ver. Geb. 55*: 123–132 (1981).

Glaeser, R. M., Electron crystallography of biological macromolecules, *Ann. Rev. Phys. Chem. 36*: 243–275 (1985)

Good, I. J., The interaction algorithm and practical Fourier series, *J. Roy. Statist. Soc. B 20*: 361–372 (1958).

Good, I. J., Weighted covariance for detecting the direction of a Gaussian source in *Time Series Analysis* (ed. M. Rosenblatt), Wiley, New York, 1963, pp. 447–470.

Goodman, J. W., *Introduction to Fourier Optics*, McGraw-Hill, San Francisco, 1968.

Gorgui-Naguib, R. N., p-adic transforms in digital signal processing in *Mathematics in Signal Processing II* (ed. J. G. McWirter), Clarendon, Oxford, 1990, pp. 43–53.

Hall, P. and P. Patil, On wavelet methods for estimating smooth functions, *Bernoulli 1* (1995), pp. 41–58.

Halvorson, C., A. Hays, B. Kraabel, R. Wu, F. Wudl and A. J. Heeger, A 160-femtosecond optical image processor based on a conjugated polymer, *Science 265*: 1215–1216 (1994).

Hannan, E. J., *Aliasing*. Tech. Report 25, Statistics Dept., Johns Hopkins University, 1965.

Hannan, E. J., *Group Representations and Applied Probability*, Methuen, London, 1966.

Hannan, E. J., Fourier methods and random processes, *Bull. Internat. Statist. Inst. 42(1)*: 475–496 (1969).

Hannan, E. J., Spectra changing over narrow bands in *Statistical Models and Turbulence* (eds M. Rosenblatt and C. Van Atta), Springer, New York, 1972, pp. 460–469.

Hannan, E. J. and P. J. Thomson, Spectral inference over narrow bands, *J. Appl. Prob. 8*: 157–169 (1971).

Hayward, S. B. and R. M. Stroud, Projected purple membrane determined to 3.7 Å resolution by low temperature electron microscopy, *J. Molec. Biol. 151*: 491–517 (1981).

Henderson, R., J. M. Baldwin, K. H. Downing, J. Lepault, and F. Zemlin, Structure of purple membrane from *Halobacterium halobium*, *Ultramicroscopy 19*: 147–178 (1986).

Henderson, R., J. M. Baldwin, T. A. Ceska, F. Zemlin, E. Beckmann and K. H. Downing, Model for the structure of bacteriorhodopsin based on high-resolution electron cryo-microscopy, *J. Mol. Biol. 213*: 899–929 (1990).

Hennel, J. W. and J. Klinowski, *Fundamentals of Nuclear Magnetic Resonance*, Wiley, New York, 1993.

Hewitt, E. and K. A. Ross, *Abstract Harmonic Analysis I&II*, Academic, New York, 1963.

Higgins, J. R., Five short stories about the cardinal series, *Bull. Amer. Math. Soc. 12*: 45–89 (1985).

Hirsch, M. W., The dynamical systems approach to differential equations, *Bull. Amer. Math. Soc. 11*: 1–64 (1984).

Hochstadt, H., *Differential Equations, A Modern Approach*, Holt-Rinehart, New York, 1964.

Hovmoller, S., Structure analysis by crystallographic image processing—Hommage à Jean Baptiste Joseph Fourier (1768–1830), *Microsc. Microanal. Microstruct. 1*: 423–431 (1990).

Ihaka, R., Statistical aspects of earthquake source parameter estimation in the presence of signal generated noise, *Commun. Statist. A 22*: 1425–1440 (1993).

Katz, B. and R. Miledi, Further observations on acetylcholine noise, *Nature 232*: 124–126 (1971).

Katznelson, Y., *An Introduction to Harmonic Analysis*, Dover, New York, 1976.

Kim, P. and G. R. Chapman, Group action on a lattice and an application to time series, *J. Stat. Planning Inf. 34*: 183–195 (1993).

King, N., An alternative for the linear regression equation when the predictor variable is uncontrolled and the sample size is small, *J. Amer. Statist. Assoc. 67*: 217–219 (1972).

Korner, T. W., *Fourier Analysis*. Cambridge Univ. Press, Cambridge, 1989.

Lanczos, C., *Discourse on Fourier Series*, Hafner, New York, 1966.

Leonov, V. P. and A. N. Shiryaev, Some problems in the spectral theory of higher moments, *Theory Prob. Appl. 5*: 460–464 (1960).

Lillestol, J., Improved estimates for multivariate complex normal regression with application to analysis of linear time-invariant relationships. *J. Mult. Anal. 7*: 512–524 (1977).

Loomis, L., *An Introduction to Abstract Harmonic Analysis*, Van Nostrand, New York, 1953.

Malik, F., D. R. Brillinger, and R. D. Vale, High resolution tracking of microtubule motility driven by a single kinesin motor, *Proc. Natl. Acad. Sci.* USA 91, pp. 4584–4588 (1994).

Michaelson, A. A., On the application of interference methods to spectroscopic methods—I. *Phil. Mag. 33*: 338–346 (1891a).

Michaelson, A. A., On the application of interference methods to spectroscopic methods—II. *Phil. Mag. 34*: 280–299 (1891b).

Moloney, J. V. and A. C. Newell, *Nonlinear optics*. Tech. Report 574. IMA, University of Minnesota, Minneapolis, 1989.

Ott, J. and R. A. Kronmall, Some classification procedures for multivariate binary data using orthogonal functions, *J. Amer. Statist. Assoc. 71*: 391–399 (1976).

Picinbono, M. B., Tendance vers le caractère gaussien par filtrage sélectif, *Comptes Rendus Acad. Sci.*, 1174–1176 (1960).

Priestley, M. B., Evolutionary spectra and non-stationary processes, *J. Roy. Statist. Soc. B 27*: 204–237 (1965).

Richards, F. S. G., A method of maximum-likelihood estimation, *J. Roy. Statist. Soc. B 23*: 469–475 (1961).

Rockmore, D., Fast Fourier analysis for abelian group extensions, *Adv. Appl. Math. 11*: 164–204 (1990).

Rosenberg, J. R., A. M. Amjad, P. Breeze, D. R. Brillinger and D. M. Haliday, The Fourier approach to the identification of functional coupling between neuronal spike trains, *Prog. Biophys. Molec. Biol. 53*: 1–31 (1989).

Rosenblatt, M., Some comments on narrow band-pass filters, *Quart. Appl. Math. 18*: 387–394 (1961).

Rosenblatt, M., Probability limit theorems and some questions in fluid mechanics in *Statistical Models and Turbulence* (eds M. Rosenblatt and C. Van Atta), Springer, New York, 1972, pp. 27–40.

Rosenblatt, M., Limit theorems for Fourier transforms of functionals of Gaussian sequences, *Z. Wahrsch. Verw. Gebiete. 55*: 123–132 (1981).

Rudin, W., *Fourier Analysis on Groups*, Wiley, New York, 1962.

Ruelle, D., *Chaotic Evolution and Strange Attractors*, Cambridge Univ. Press, Cambridge, 1989.

Saleh, A. K., *Contributions to Preliminary Test and Shrinkage Estimation*, Department of Mathematics and Statistics, Carleton University, Canada, 1992.

Shao, M. and C. L. Nikias, Signal processing with fractional lower order moments: stable processes and their applications, *Proc. IEEE 81*: 986–1010 (1993).

Shankar, P. M., S. N. Gupta, and H. M. Gupta, Applications of coherent optics and holography in biomedical engineering, *IEEE Trans. Biomed. Eng. BME-29*: 8–15 (1982).

Slutsky, E., Alcuni applicazioni di coefficienti di Fourier al analizo di sequenze eventuali coherenti stazionari, *Giorn. d. Instituto Italiano degli Atuari 5*: 435–482 (1934).

Smith, K. T., The uncertainty principle on groups, *SIAM J. Appl. Math. 50*: 876–882 (1990).

Stein, C., Inadmissibility of the usual estimator for the mean of a multivariate normal distribution in *Proc. Third Berk. Symp. Math. Statist. Prob. Vol. 1*, Univ. Calif. Press, Berkeley, 1955, pp. 197–206.

Stoffer, D. S., Walsh-Fourier analysis and its statistical applications, *J. Amer. Statist. Assoc. 86*: 481–482 (1991).

Strang, G., Wavelet transforms versus Fourier transforms, *Bull. Amer. Math. Soc. 28*: 288–305 (1993).

Strichartz, R. S., How to make wavelets, *Amer. Math. Monthly 100*: 539–556 (1990).

Tarter, M. E. and M. D. Lock, *Model-free Curve Estimation*, Chapman and Hall, New York, 1993.

Terras, A., *Harmonic Analysis on Symmetric Spaces and Applications I & II*, Springer, New York, 1988.

Thompson, J. R., Some shrinkage techniques for estimating the mean, *J. Amer. Statist. Assoc. 63*: 113–122 (1968).

Timan, A. F., *Theory of Approximation of Functions of a Real Variable,* Pergamon, Oxford, 1963.

Tukey, J. W., An introduction to the frequency analysis of time series in *The Collected Works of John W. Tukey I* (1984). (ed. D. R. Brillinger), Wadsworth, Pacific Grove, 1963, pp. 503–650.

Tukey, J. W., Equalization and pulse shaping techniques applied to determination of initial sense of Rayleigh waves in *The Collected Works of John W. Tukey I* (1984). (ed. D.R. Brillinger), Wadsworth, Pacific Grove, 1959, pp. 309–358.

Tukey, J. W., *Introduction to the dilemmas and difficulties of regression.* Unpublished, 1979.

Walter, G. G., Approximation of the delta function by wavelets, *J. Approx. Theory 71*: 329–343 (1992).

Walter, G. G. (1994). Wavelets and Other Orthogonal Systems with Applications. CRC Press, Boca Raton.

Wenk H. R., K. H. Downing, M. S. Hu and M. A. Okeefe, 3D structure determination from electron-microscope images—electron crystallography of staurolite, *Acta Crystallographica A1, 48*: 700–716 (1992).

Whittle, P., Estimation and information in time series analysis, *Skand. Aktuar. 35*: 48–60 (1952).

Whittle, P., Discussion of C. M. Stein "Confidence sets for the mean of a multivariate normal distribution", *J. Roy. Statist. Soc. B 24*: 294 (1962).

Wiener, N., *The Fourier Integral and Certain of Its Applications*, Dover, New York, 1933.

Yaglom, A. M., Second-order homogeneous random fields in *Proc. Fourth Berkeley Symp. Math. Statist. Prob. 2*, Univ. Calif. Press, Berkeley, 1961, pp. 593–622.

Yajima, Y., A central limit theorem of Fourier transforms of strongly dependent stationary processes, *J. Time Series Analysis 10*: 375–384 (1989).

Yariv, A., ·*Quantum Electronics*, Second Edition, Wiley, New York, 1975.

Zidek, J., Discussion of Copas (1983), Regression, prediction and shrinkage, pp. 347–48. *J. Roy. Statist. Soc. B 45*: 311–335 (1983).

Zygmund, A., *Trigonometric Series*, Cambridge Univ. Press, Cambridge, 1968.

APPENDIX

The estimate presented in the middle panel of Fig. 7 is ordinary least squares. (In many time series situations such estimates are asymptotically efficient.)

The model shown in Eq. (5.16) is linear in α and the γ_{nk}. It may be written

$$y = X\gamma + Z\alpha + \varepsilon$$

taking $Z = [t - \bar{t}]$ and $X = [X_{tk}]$, with $X_{tk} = 1$ for $k/2'' \leq t/T < (k + 1)/T$ and 0 otherwise. It is seen to have the form of an analysis of covariance model. The least squares estimates may be written

$$\hat{\alpha} = (Z'PZ)^{-1}Z'Py \tag{A.1}$$

$$\hat{\gamma} = (X'X)^{-1}X'(y - Z\hat{\alpha}) \tag{A.2}$$

with $P = I - X(X'X)^{-1}X'$.

2

Modeling and Inference for Periodically Correlated Time Series

ROBERT B. LUND and **ISHWAR V. BASAWA** The University of Georgia, Athens, Georgia

1. INTRODUCTION

This chapter overviews general modeling and analysis problems with periodically correlated (PC) time series. The definition of a PC time series is first presented and some properties of these series are discussed. A frequency domain test to detect periodicities in the autocovariance structure of an observed series is then presented. Next, periodic autoregressive moringaverage (PARMA) models are introduced as a useful class of PC time series models; comparisons to seasonal Box–Jenkins models are made. The problem of parsimoniously fitting a PARMA model to a PC series is then addressed. Moment, maximum likelihood, and estimating equation techniques of estimation are considered; limit distributions of the parameter estimates are discussed.

Due to the cyclic nature of solar radiation, tides, economic activities, meteorological processes, etc., many observed time series exhibit a periodic statistical structure. Accordingly, time series models with periodic properties have received much attention in the literature. Applications of such models include studies in hydrology: Lawrance and Kottegoda (1977), Vecchia (1985a and 1985b), Anderson and Vecchia (1993), and McLeod (1993); meteorology: Hannan (1955), Monin (1963), Jones and Brelsford (1967), Bloomfield et al. (1994) and Lund et al. (1995); economics: Parzen and Pagano (1979); and electrical engineering: Gardner and Franks (1975).

In this chapter, we overview general analysis and modeling of periodically correlated (PC) time series. The rest of this article proceeds as follows. In Sec. 2, PC time series are defined and general properties of these series are discussed. Sec. 3 derives a frequency domain test to detect periodicities in the autocovariance structure of an observed series. The test indicates whether a stationary or PC model should be employed; however, the test does not suggest what type of PC model to use should one be deemed necessary. The question of how to model a PC series is subsequently examined in Sec. 4 where periodic autoregressive moving-average (PARMA) models are introduced. The next three Sections study parameter estimation in PARMA models. Sections 5, 6, and 7 consider moment, maximum likelihood, and estimating equation approaches, respectively, to this problem.

2. DEFINITION AND PROPERTIES OF PC TIME SERIES

The definition of a PC time series broadens the notion of stationarity to allow for periodicities in the mean and covariance structure of the series. Formally, a series $\{X_n\}$ with finite second moments is said to be PC with period T if

$$E[X_{n+T}] = E[X_n] \quad \text{and} \quad \text{Cov}[X_{m+T}, X_{n+T}] = \text{Cov}[X_m, X_n] \quad (2.1)$$

for all integer m and n. To avoid ambiguity, T is taken as the smallest positive integer satisfying Eq. (2.1). Equation (2.1) allows each "season" of the series to have a different mean and autocovariance function, but stipulates that mean and autocovariance properties must cycle every T time units. When $T = 1$, a PC series is equivalent to a stationary series.

The term periodic correlation was coined by Gladyšhev (1961) to describe correlation properties of the frequency increment process associated with PC series. In the time domain, other terminologies such as cyclostationary, Gardner and Franks (1975) and periodically stationary, Monin (1963), have been employed.

Suppose that $\{X_n\}$ is a PC series with period T. The seasonal mean and autocovariance function of $\{X_n\}$ will be denoted by

$$\mu_\nu = E[X_{nT+\nu}] \quad \text{and} \quad \gamma_\nu(h) = \text{Cov}[X_{nT+\nu}, X_{nT+\nu-h}], \quad (2.2)$$

where ν is a seasonal suffix satisfying $1 \leq \nu \leq T$. From Eq. (2.1), one sees that μ_ν and $\gamma_\nu(h)$ depend on the season ν and lag h, but not on the "year" n. In contrast to the covariance function of a stationary series, $\gamma_\nu(h)$ is not symmetric in h; however, $\gamma_\nu(-h) = \gamma_{\nu+h}(h)$ for $h \geq 0$ follows from Eq. (2.1). Here, $\nu + h$ is interpreted periodically with period T. The seasonal autocorrelation function of $\{X_n\}$ will be denoted by

$$\rho_\nu(h) = \text{Corr}[X_{nT+\nu}, X_{nT+\nu-h}] = \frac{\gamma_\nu(h)}{\sqrt{\gamma_\nu(0)\gamma_{\nu-h}(0)}}, \tag{2.3}$$

where $\gamma_j(0)$ is interpreted periodically in j with period T.

There is an intuitive equivalence between multivariate stationarity and periodic correlation. First, suppose that $\{X_n\}$ is PC with period T and consider the T variate series $\{\mathbf{Y}_n\}$ obtained by setting $Y_{n\nu} = X_{(n-1)T+\nu}$ for $1 \le \nu \le T$ (Y_{ij} denotes the jth component of $\mathbf{Y}_i$). It follows from Eq. (2.1) that $\{\mathbf{Y}_n\}$ is T-variate stationary. Conversely, for a T-variate stationary series $\{\mathbf{Y}_n\}$, define for each n the "year" m and "season" ν via $m = \lfloor (n-1)/T \rfloor$ and $\nu = n - mT$; here $\lfloor x \rfloor$ denotes the greatest integer less than or equal to x. Then a PC univariate series $\{X_n\}$ with period T is obtained by setting $X_n = Y_{m\nu}$.

Whereas the equivalence between periodic correlation and multivariate stationarity is a useful mathematical tool, analysis of PC series through multivariate time series techniques is seldom practical due to the large dimensionality frequently encountered in applications. For example, in a monthly series, $T = 12$ and a first order periodic autoregressive model with general monthly means has 36 parameters (see Sec. 4).

Perhaps the most simplistic subclass of PC models are those obtained by periodically modulating a mean zero stationary series $\{Z_n\}$. Suppose that $\{P_n\}$ and $\{\mu_n\}$ are nonrandom periodic sequences with period T and set $X_n = \mu_n + P_n Z_n$. Straightforward computations show that

$$E[X_n] = \mu_n \quad \text{and} \quad \text{Cov}[X_n, X_m] = P_n P_m \gamma_Z(|m - n|), \tag{2.4}$$

where $\gamma_Z(h) = \text{Cov}(Z_n, Z_{n+h})$. Since Eq. (2.1) follows directly from Eq. (2.4), $\{X_n\}$ is PC with period T. Note also that $\gamma_\nu(h) = P_\nu P_{\nu-h} \gamma_Z(h)$. While a periodically modulated PC series has an autocovariance function that depends on the season ν, the autocorrelation function is free of ν when $P_\nu > 0$ for all ν:

$$\rho_\nu(h) = \frac{\gamma_Z(h)}{\gamma_Z(0)}. \tag{2.5}$$

It is not true that every PC series can be expressed as a periodically modulated stationary series. To see this, interweave two independent mean zero stationary series $\{X_n^{(1)}\}$ and $\{X_n^{(2)}\}$ by setting $X_{2n-1} = X_n^{(1)}$ and $X_{2n} = X_n^{(2)}$. It is easy to see that the interwoven series $\{X_n\}$ is PC with period $T = 2$. Suppose further that $\{X_n^{(1)}\}$ is a first order non-degenerate auto-regression and that $\{X_n^{(2)}\}$ is a first order moving average, with both super-scripted series having a unit variance. If the representation $X_n = P_n Z_n$ were possible, one would obtain $1 = P_n^2 \text{Var}(Z_n)$ for $n = 1, 2$ by equating variances. Hence, P_1^2, P_2^2 and $\text{Var}[Z_n]$ are all nonzero and $P_1^2 = P_2^2$. Now observe that $\text{Cov}[X_6, X_2] = \text{Cov}[X_3^{(2)}, X_1^{(2)}] = P_2^2 \gamma_Z(4) = 0$ since the autocovariance of a first order moving average is zero when the lag exceeds 1. A contradiction is achieved by noting that $\text{Cov}[X_5, X_1] = \text{Cov}[X_3^{(1)}, X_1^{(1)}] = P_1^2 \gamma_Z(4) = P_2^2 \gamma_Z(4)$ which cannot be zero since the first order autoregression is non-degenerate.

Another large subclass of PC models can be generated by applying a periodic causal linear filter to a white noise sequence. Suppose that $\{Z_n\}$ is mean zero white noise with a unit variance. Define, in a mean square sense,

$$X_{nT+\nu} = \sum_{k=0}^{\infty} \psi_k(\nu) Z_{nT+\nu-k}. \tag{2.6}$$

In Eq. (2.6), a seasonal notation has been employed that will be used throughout: $X_{nT+\nu}$ refers to the series during season ν, $1 \leq \nu \leq T$, of "year" n. We shall continue to use the notations $\{X_n\}$, $\{Z_n\}$, etc. in preference to $\{X_{nT+\nu}\}$, $\{Z_{nT+\nu}\}$, etc. whenever emphasis on seasonality is not paramount. The weights $\psi_k(\nu)$ vary with ν in general, but are assumed absolutely summable for every season:

$$\sum_{k=0}^{\infty} |\psi_k(\nu)| < \infty \tag{2.7}$$

for $1 \leq \nu \leq T$. Noncausal expansions in Eq. (2.6) are of course possible. If the variance of $\{Z_n\}$ is not unity, then the $\psi_k(\nu)$s and the white noise $\{Z_n\}$ can be rescaled to produce new seasonal weights and unit variance white noise. Anderson and Meerschaert (1997) study properties of Eq. (2.6) when $\{Z_n\}$ has heavy tails.

From Eq. (2.6), one can verify that $\{X_n\}$ is a mean zero PC series with period T and covariance structure

$$\gamma_\nu(h) = \sum_{k=0}^{\infty} \psi_{k+h}(\nu) \psi_k(\nu - h). \tag{2.8}$$

In Eq. (2.8), $\psi_k(j)$ is interpreted periodically in j with period T.

While the main portion of this article focuses on the time domain, the spectral theory of PC series is by now well understood, see Gladyšhev (1961), Hurd (1989). In the next Section, we will use spectral properties of PC series to devise a test to detect periodic correlation.

An extension of the PC model class worth mentioning here is the almost periodically correlated (APC) model class, see Hurd and Leskow (1992). APC models are encountered when, for example, two independent PC series with incommensurate periods are added. APC models will not be considered further here.

3. DETECTION OF PERIODIC CORRELATION

This Section discusses how to detect periodic correlation in an observed series. Frequently, it is easy to visually detect periodicities in mean levels of a series by plotting the data. Figure 1 plots 40 years of monthly averaged temperatures from Juneau, Alaska for the years 1944–1983 inclusive. One easily sees the seasonal cycle in temperatures: high during summer months and low during winter months. Figure 2 shows the same series after a period $T = 12$ sample mean was removed from the series. In Fig. 2, it is more difficult to visually assess whether a PC model for this mean adjusted series is needed (a keen eye will detect a periodic variance in Fig. 1: wintertime troughs are more variable than their summertime counterparts) or if a stationary model is sufficient. Hence, it is desirable to develop a test to detect autocovariance periodicities.

Vecchia and Ballerini (1991) and Anderson and Vecchia (1993) consider the problem of detecting periodic correlation in an observed series with time domain methodology. Their methods are based on asymptotic properties of the sample periodic autocorrelation function. Hurd and Gerr (1991) and

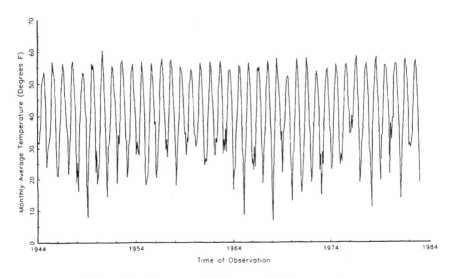

Figure 1. Juneau, Alaska monthly temperatures.

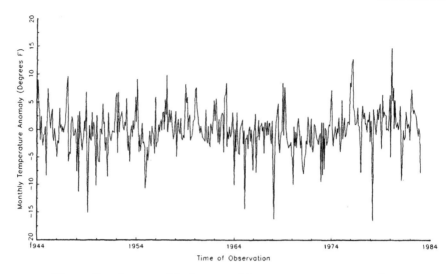

Figure 2. Juneau, Alaska monthly temperatures anomalies.

Leskow (1997) attack the same problem from the frequency domain. Their methods exploit the spectral support set properties, discussed further below, of PC series. To reach a conclusion about autocovariance periodicities in the series using any of the above methods, one must consider many sample autocorrelations or spectral estimates jointly. This, of course, is a difficult task statistically. Modifications of Hurd's and Gerr's spectral based methods were considered in Bloomfield et al. (1994) and Lund et al. (1995). These methods yield a graphical check for periodic correlation that is visually easy to interpret. We now elaborate on this graphical "test".

Suppose that $\{X_n\}$ is a mean zero PC series with period T. Then $\{X_n\}$ is harmonizable in the sense that it has the spectral representation

$$X_n = \int_{[0.2\pi)} e^{in\lambda} dZ(\lambda), \qquad (3.1)$$

where $i = (-1)^{1/2}$ and $\{Z(\lambda), 0 \le \lambda < 2\pi\}$ is a mean zero complex valued random process called the frequency process associated with $\{X_n\}$, Gladyšhev (1961). If $\{X_n\}$ is stationary ($T = 1$), then $\{Z(\lambda)\}$ has orthogonal increments in the sense that

$$E[dZ(\lambda_1) \overline{dZ(\lambda_2)}] = 0, \qquad (3.2)$$

unless $\lambda_1 = \lambda_2$. In Eq. (3.2) $\lambda_1, \lambda_2 \in [0, 2\pi)$ and the overline denotes complex conjugation which is needed to make $E[dZ(\lambda_1) \overline{dZ(\lambda_2)}]$ interpretable as a covariance when $\{Z(\lambda)\}$ is complex valued. If $\{X_n\}$ is PC with period T, then $\{Z(\lambda)\}$ has periodically correlated increments:

$$E[dZ(\lambda_1)\,\overline{dZ(\lambda_2)}] = 0, \tag{3.3}$$

unless $\lambda_1 = \lambda_2 + 2\pi k/T$ for some $k \in \{0, \pm1, \pm2, \ldots, \pm(T-1)\}$ (see Gladyšhev (1961)). Hence, the spectral support set of a PC series lies on periodic diagonal lines in the bifrequency square $[0, 2\pi) \times [0, 2\pi)$.

It follows that a test for periodic correlation can be devised by checking the increments of $\{Z(\lambda)\}$ for periodic correlations. Suppose that $\{X_n\}$ is observed for $1 \leq n \leq N$. To estimate the increments of $\{Z(\lambda)\}$, we will use the discrete Fourier transform of $\{X_n\}$ defined at the Fourier frequency $\lambda_j = 2\pi j/N$ by

$$I_j = \frac{1}{\sqrt{2\pi N}} \sum_{n=0}^{N-1} X_{n+1} e^{-in\lambda_j}. \tag{3.4}$$

The increments of $\{Z(\lambda)\}$ are estimated at the Fourier frequencies via

$$\widehat{dZ(\lambda_j)} = I_j. \tag{3.5}$$

To assess whether or not $E[dZ(\lambda_j)\,\overline{dZ(\lambda_{j-h})}] = 0$, one can compute a squared sample correlation between the two pairs of M-dimensional frequency increment estimates $\{I_j, \ldots, I_{j+M}\}$ and $\{I_{j+h}, \ldots, I_{j+h+M-1}\}$:

$$\gamma_h^2(j) = \frac{|\sum_{m=0}^{M-1} I_{j+m}\overline{I_{j+h+m}}|^2}{\sum_{m=0}^{M-1} |I_{j+m}|^2 \sum_{m=0}^{M-1} |I_{j+h+m}|^2}. \tag{3.6}$$

We call $\gamma_h^2(j)$ the squared coherence of $\{X_n\}$ at frequency λ_j and lag h; the absolute values in Eq. (3.6) denote squared length. Of course, one could employ frequency estimates centered around λ_j and λ_{j+h} rather than the "one-sided" versions used, but Eq. (3.6) will prove adequate for our purposes. Selection of the bandwidth parameter M is seldom crucial; however, as with any smoothing problem, one does not want to choose M excessively large or small. General guidelines for selecting M are not yet available.

There are redundancies in Eq. (3.6) as N^2 squared coherences can be made from a series of length N: one squared coherence for each h and $j \in \{0, 1, 2, \ldots, (N-1)\}$. To summarize the squared coherences, the frequency is averaged out via

$$\overline{\gamma}_h^2 = \frac{1}{N} \sum_{j=0}^{N-1} \gamma_h^2(j). \tag{3.7}$$

We call $\overline{\gamma}_h^2$ the averaged squared coherence at lag h. Whereas other methods could be used to summarize the squared coherences, averaging out the frequency is natural. For as noted above, if $\{X_n\}$ is PC with period T, the spectral support set of $\{Z(\lambda)\}$ is contained on the parallel diagonal lines $\lambda_1 = \lambda_2 + 2\pi k/T, k \in \{0, \pm 1, \ldots, \pm(T-1)\}$ in the bifrequency square $[0, 2\pi) \times [0, 2\pi)$. Hence, a large average squared coherence $\overline{\gamma}_h^2$ suggests that the spectral line $\lambda_1 = \lambda_2 + 2\pi h/N$ is an important component of the spectral support set of $\{Z(\lambda)\}$.

If $\{X_n\}$ is stationary, there should be no correlations in the increments of $\{Z(\lambda)\}$ and $\overline{\gamma}_h^2$ should be statistically small for all h except when $h = 0$ ($\overline{\gamma}_0^2 = 1$). Hence, if $\{X_n\}$ is PC with period T, one should see large $\overline{\gamma}_h^2$s only when h is an integer multiple of N/T—the observed number of years of data. Thus, a simple graphical check for periodic correlation merely examines a plot of h versus $\overline{\gamma}_h^2$ for periodic large values.

From the DFT periodicity and symmetry properties

$$I_{j+N} = I_j \qquad \text{and} \qquad I_{N-j} = \overline{I}_j, \tag{3.8}$$

one can derive some similar properties in the average squared coherences. We list these properties below; throughout, N is assumed to be an even integer for simplicity.

LEMMA 3.1

(i) $0 \leq \overline{\gamma}_h^2 \leq 1$ and $\overline{\gamma}_0^2 = 1$.

(ii) $\overline{\gamma}_{h+N}^2 = \overline{\gamma}_h^2$.

(iii) $\overline{\gamma}_{N/2-h}^2 = \overline{\gamma}_{N/2+h}^2$.

Proof. Property (i) follows from $0 \leq \gamma_h^2(j) \leq 1$ for all j which is the Cauchy-Schwarz inequality. Property (ii) follows from $\gamma_{h+N}^2(j) = \gamma_h^2(j)$ for all j which follows from Eqns. (3.8) and (3.6). For Property (iii), one uses Eq. (3.8) to get $\gamma_h^2(j) = \gamma_{-h}^2(j+h)$ for all j which gives

$$\overline{\gamma}_{N/2+h}^2 = N^{-1} \sum_{j=0}^{N-1} \gamma_{-N/2-h}^2(j + N/2 + h). \tag{3.9}$$

Using $\gamma_{h+N}^2(j) = \gamma_h^2(j)$ again yields

$$\overline{\gamma}_{N/2+h}^2 = N^{-1} \sum_{j=0}^{N-1} \gamma_{N/2-h}^2(j + N/2 + h) \tag{3.10}$$

when used in Eq. (3.9). Finally, Eq. (3.8) gives $\gamma_h^2(j+N) = \gamma_h^2(j)$ which proves Property (iii) when the index of summation is changed in Eq. (3.10).

One implication of Lemma 3.1 is that it is only necessary to consider $\overline{\gamma}_h^2$ for $1 \leq h \leq N/2$.

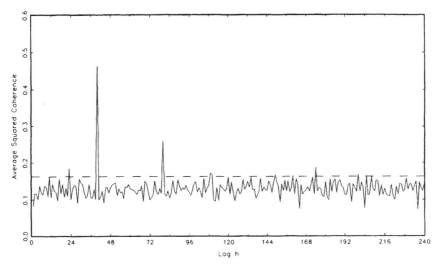

Figure 3.   Juneau average squared coherences.

Figure 3 plots average squared coherences for the 40 years ($N = 480$) of mean removed Juneau, Alaska data (the Fig. 2 series) with $M = 8$. The plot reveals two large average squared coherences at lags $h = 40$ and $h = 80$. Hence, one concludes that the Juneau series, in addition to its periodic mean, has autocovariance periodicities with period $T = 12$.

The dashed line in Fig. 3 is a 99% pointwise upper confidence band for the average squared coherences assuming that the series is Gaussian white noise. Specifically, there is a 0.01 probability that $\overline{\gamma}_h^2$ exceeds this confidence level for any fixed h. When the series is Gaussian white noise, $\overline{\gamma}_h^2$ is asymptotically normal as $N \to \infty$ with $E[\overline{\gamma}_h^2] = M^{-1}$ and $\mathrm{Var}[\overline{\gamma}_h^2] = N^{-1}\kappa_M$. Values of κ_M are tabulated in Lund et al. (1995). Lund et al. (1995) also shows that these confidence limits are reasonably robust for stationary Gaussian series other than white noise. Other values of M were studied for the Juneau series—none change our conclusion that this series has periodic autocovariances. It is worth noting that M is only needed to make a periodic/stationary model form conclusion; in particular, M is not used in model development.

The above average squared coherence test was studied via simulation in Chu (1996). There, a variety of periodically modulated and interwoven PC series were examined. In all simulations, Chu found values of M that detected the periodic correlation in the simulated series; in almost all cases, the choice of M was not crucial.

4. PARMA MODELS

Once a PC model is judged appropriate for a series, the frequent next question to address is how to fit such a model to the series. The stationary autoregressive moving-average (ARMA) model class can be enlarged to a PC class by allowing model parameters to vary periodically. Formally, $\{X_n\}$ is said to be a periodic autoregressive moving-average (PARMA) series with period T and mean μ_ν during season ν if it is a solution to the periodic linear difference equation

$$(X_{nT+\nu} - \mu_\nu) - \sum_{k=1}^{p(\nu)} \phi_k(\nu)(X_{nT+\nu-k} - \mu_{\nu-k}) = \sum_{k=0}^{q(\nu)} \theta_k(\nu)Z_{nT+\nu-k}, \qquad (4.1)$$

where $\{Z_n\}$ is white noise with a unit variance. During season ν, $p(\nu)$ and $q(\nu)$ are the AR and MA model orders respectively and $\phi_1(\nu),\ldots,\phi_{p(\nu)}(\nu)$ and $\theta_0(\nu),\ldots,\theta_{q(\nu)}(\nu)$ are the AR and MA model coefficients. In general, model orders and coefficients vary with ν. The seasonal means μ_ν are interpreted periodically in ν with period T if necessary. It is sufficient to consider unit variance white noise in the PARMA model; if $\{Z_n\}$ has a non-unit variance, then one can rescale the white noise and the $\theta_k(\nu)$s to produce unit variance white noise. As in the ARMA case, it is sufficient to consider the mean zero case where $\mu_\nu = 0$ for $1 \leq \nu \leq T$.

The first question we address is whether or not solutions to Eq. (4.1) are PC. Sufficient conditions guaranteeing the existence of a unique causal PC solution to Eq. (4.1) can be obtained by transforming Eq. (4.1) into a T-variate ARMA difference equation and applying multivariate ARMA causality results, see Vecchia (1985a) and (1985b). To do this, the multivariate form of Eq. (4.1) will be needed. Straightforward but tedious computations show that Eq. (4.1) can be written as

$$\Phi_0 \mathbf{Y}_n - \sum_{k=1}^{p*} \Phi_k \mathbf{Y}_{n-k} = \Theta_0 \varepsilon_n + \sum_{k=1}^{q*} \Theta_k \varepsilon_{n-k}, \qquad (4.2)$$

where

$$\mathbf{Y}_n = (X_{nT+1}, X_{nT+2}, \ldots, X_{nT+T})'$$

and

$$\varepsilon_n = (Z_{nT+1}, Z_{nT+2}, \ldots, Z_{nT+T})' \qquad (4.3)$$

are the "yearly" series and white noise processes respectively (′ denotes transpose). The model orders $p*$ and $q*$ in Eq. (4.2) are computed from the seasonal model orders via

$$
p^* = \max_{1 \leq \nu \leq T} \left\lceil \frac{p(\nu) - (\nu - 1)}{T} \right\rceil, \qquad q^* = \max_{1 \leq \nu \leq T} \left\lceil \frac{q(\nu) - (\nu - 1)}{T} \right\rceil. \tag{4.4}
$$

Here, $\lceil x \rceil$ denotes the smallest integer greater than or equal to x. The $T \times T$ matrix coefficients in Eq. (4.2) have (i, j)th entries

$$
(\Phi_0)_{ij} = \begin{cases} 1 & i = j \\ 0 & i < j, \\ \phi_{i-j}(i) & i > j \end{cases} \tag{4.5}
$$

$$
(\Theta_0)_{ij} = \begin{cases} 0 & i < j \\ \theta_{i-j}(i) & i \geq j \end{cases}, \tag{4.6}
$$

and

$$
(\Phi_m)_{ij} = \phi_{mT+i-j}(i), 1 \leq m \leq p^*
$$
$$
(\Theta_m)_{ij} = \theta_{mT+i-j}(i), \ 1 \leq m \leq q^*. \tag{4.7}
$$

The conventions $\phi_k(\nu) = 0$ for $k > p(\nu)$ and $\theta_k(\nu) = 0$ for $k > q(\nu)$ are used in Eq. (4.7).

THEOREM 4.1   A unique causal PC solution with period T exists to Eq. (4.1) whenever

$$
\det\left(\Phi_0 - \sum_{k=1}^{p^*} \Phi_k z^k \right) \neq 0 \tag{4.8}
$$

for all complex z satisfying $|z| \leq 1$.

Proof.   Suppose that Eq. (4.8) holds. Then by Theorem 11.3.1 in Brockwell and Davis (1991), Eq. (4.2) has the unique T-variate stationary causal solution

$$
\mathbf{Y}_n = \sum_{j=0}^{\infty} \Psi_j \varepsilon_{n-j}, \tag{4.9}
$$

where $\{\Psi_j, j \geq 0\}$ is a sequence of $T \times T$ matrices whose components are absolutely summable in j. By the equivalence between periodic correlation and multivariate stationarity discussed in Sec. 2, it follows that $\{X_n\}$ is PC with period T.

The seasonal causality in Eq. (2.6) follows from Eq. (4.9). Isolating the Tth component of $\mathbf{Y}_n$ in Eq. (4.9) gives

$$X_{nT+T} = \sum_{j=0}^{\infty} (0, 0, \ldots, 0, 1) \Psi_j \varepsilon_{n-j}, \tag{4.10}$$

which shows that Eq. (2.6) holds for $\nu = T$ when the absolute summability of the Ψ_js in j is used. To show that Eq. (2.6) holds for $1 \leq \nu \leq T - 1$, one recurses Eq. (4.1) ν times into the form

$$X_{nT+\nu} = \sum_{k=0}^{p**} \alpha_k(\nu) X_{nT-k} + \sum_{k=0}^{q**} \beta_k(\nu) Z_{nT+\nu-k} \tag{4.11}.$$

The explicit values of $\alpha_k(\nu)$, $\beta_k(\nu)$, and the finite orders p^{**} and q^{**} are not important here and will not be elaborated upon. Equation (4.9) can be used to show that $X_{nT}, X_{nT-1}, \ldots, X_{nT-p**}$ are each causal linear combinations of $Z_{nT}, Z_{nT-1}, \ldots$. Using Eq. (4.11) and the componentwise absolute summability of the Ψ_js in j proves that Eq. (2.6) holds for $1 \leq \nu \leq T - 1$ as well.

In a similar manner, one can obtain invertibility conditions for PARMA models. The PARMA invertibility condition is

$$\det\left(\Theta_0 + \sum_{k=1}^{q*} \Theta_k z^k \right) \neq 0 \tag{4.12}$$

for all $|z| \leq 1$. If Eq. (4.12) holds, the invertible representation

$$Z_{nT+\nu} = \sum_{k=0}^{\infty} \pi_k(\nu) X_{nT+\nu-k} \tag{4.13}$$

can be made where

$$\sum_{k=0}^{\infty} |\pi_k(\nu)| < \infty \tag{4.14}$$

for $1 \leq \nu \leq T$. As in the ARMA case, model invertibility is needed for parameter identifiability, see Bentarzi and Hallin (1994).

Bentarzi and Hallin (1994) give necessary and sufficient PARMA causality and invertibility conditions phrased more directly in terms of the $\phi_k(\nu)$s and $\theta_k(\nu)$s; however, these conditions are considerably more complex than the classical univariate ARMA causality and invertibility conditions.

Under any conditions such that the solution to Eq. (4.1) can be expressed in the causal form in Eq. (2.6) the $\psi_k(\nu)$s can be determined recursively.

Plugging Eq. (2.6) into Eq. (4.1) and equating white noise coefficients gives $\psi_0(\nu) = \theta_0(\nu)$ and the recursion

$$\psi_k(\nu) = \sum_{j=1}^{\min[k, p(\nu)]} \phi_j(\nu)\psi_{k-j}(\nu - j) + \theta_k(\nu)1_{[k \leq q(\nu)]} \tag{4.15}$$

for $k \geq 1$; in Eq. (4.15), $\psi_k(j)$ is interpreted periodically in j with period T. When PARMA parameters and model orders do not depend on the season ν, Eq. (4.15) reduces to its well-known stationary counterpart, see (3.3.3) and (3.3.4) in Brockwell and Davis (1991). The autocovariance function of the PARMA model can be computed from the $\psi_k(\nu)$s through Eq. (2.8). In practice, the infinite sums in Eq. (2.8) will have to be truncated. When Eq. (4.8) holds, one can justify the geometric convergence of the $\psi_k(\nu)$s to zero as $k \to \infty$ uniformly in ν.

EXAMPLE 4.2 Consider the first order periodic autoregressive model (PAR(1))

$$X_{nT+\nu} - \phi_1(\nu)X_{nT+\nu-1} = \theta_0(\nu)Z_{nT+\nu}. \tag{4.16}$$

Then Eq. (4.4) gives $p^* = 1$ and $q^* = 0$. Further computations show that Eq. (4.8) reduces to

$$\det(\Phi_0 - \Phi_1 z) = 1 - z\prod_{\nu=1}^{T} \phi_1(\nu) \neq 0. \tag{4.17}$$

Hence, a unique causal PC solution with period T exists to Eq. (4.16) whenever

$$\left|\prod_{\nu=1}^{T} \phi_1(\nu)\right| < 1 \tag{4.18}$$

which we henceforth assume. From a stability standpoint, Eq. (4.18) shows that the PAR(1) model can "expand" during some seasons ($|\phi_1(\nu)| > 1$ for some ν) as long as there is an overall yearly "contraction".

The $\psi_k(\nu)$s in Eq. (2.6) are obtained by solving Eq. (4.15):

$$\psi_k(\nu) = \theta_0(\nu - k)\prod_{i=0}^{k-1} \phi_1(\nu - i), \tag{4.19}$$

where $\theta_0(j)$ and $\phi_1(j)$ are interpreted periodically in j with period T; a product over an empty set of indices is taken as unity.

To get the covariance structure of the PAR(1) model, we first derive its seasonal variance. Taking a variance on both sides of Eq. (4.16) and using the notation $\sigma^2(\nu) = \mathrm{Var}[X_{nT+\nu}]$ gives

$$\sigma^2(\nu) = \phi_1^2(\nu)\sigma^2(\nu - 1) + \theta_0^2(\nu). \tag{4.20}$$

Recursing Eq. (4.20) T times and requiring $\sigma^2(0) = \sigma^2(T)$ gives

$$\sigma^2(T) = \frac{\sum_{j=1}^{T}[\prod_{k=j+1}^{T}\phi_1^2(k)]\theta_0^2(j)}{1 - \prod_{j=1}^{T}\phi_1^2(j)}. \tag{4.21}$$

Further manipulations in Eq. (4.20) provide the desired variance:

$$\sigma^2(\nu) = \left(\prod_{k=1}^{\nu}\phi_1^2(k)\right)\frac{\sum_{j=1}^{T}[\prod_{m=j+1}^{T}\phi_1^2(m)]\theta_0^2(j)}{1 - \prod_{j=1}^{T}\phi_1^2(j)}$$
$$+ \sum_{k=1}^{\nu}\left[\prod_{j=k+1}^{\nu}\phi_1^2(j)\right]\theta_0^2(k). \tag{4.22}$$

The covariance and correlation structures of the PAR(1) model can be obtained by recursing Eq. (4.16) and taking expectations:

$$\gamma_\nu(h) = \sigma^2(\nu - h)\prod_{k=0}^{h-1}\phi_1(\nu - k); \tag{4.23}$$

$$\rho_\nu(h) = \sqrt{\frac{\sigma^2(\nu - h)}{\sigma^2(\nu)}}\prod_{k=0}^{h-1}\phi_1(\nu - k), \tag{4.24}$$

for $h \geq 0$. In Eqns. (4.23) and (4.24), $\sigma^2(j)$ and $\phi_1(j)$ are interpreted periodically in j with period T.

When compared to the autocovariance and autocorrelation functions of causal AR(1) models, Eqns. (4.23) and (4.24) are quite complex. One can verify that the expressions in Eqns. (4.23) and (4.24) reduce to the classical AR(1) forms when the PAR(1) parameters are constant in ν. In general, calculations in PARMA models are very tedious.

A class of models that is frequently considered for periodic series is the seasonal ARMA (SARMA) or seasonal Box–Jenkins model class. Basically, SARMA models for $\{X_n\}$ are solutions to ARMA difference equations with all lags taken at multiples of the period T:

$$X_{nT+\nu} - \sum_{k=1}^{p}\phi_k X_{n(T-k)+\nu} = U_{nT+\nu} + \sum_{k=1}^{q}\theta_k U_{n(T-k)+\nu}, \tag{4.25}$$

where $\{U_n\}$ is mean zero white noise with variance σ^2_U. Contrary to its name, the SARMA model in Eq. (4.25) is a stationary model with an auto-covariance function that is zero unless the lag is a multiple of the period T. Nonzero correlations between different seasons in a SARMA model can be induced by considering an ARMA $\{U_n\}$ instead of white noise:

$$U_n - \sum_{k=1}^{\tilde{p}} \alpha_k U_{n-k} = V_n + \sum_{k=1}^{\tilde{q}} \beta_k V_{n-k}. \tag{4.26}$$

In Eq. (4.26), $\{V_n\}$ is mean zero white noise with variance σ^2_V.

The model obtained for $\{X_n\}$ by combining Eqns. (4.25) and (4.26) is again stationary. Of course, one would need to let model coefficients vary periodically in Eq. (4.25) and/or Eq. (4.26) to get a PC model. The upshot is that SARMA models should not be expected to capture periodic correlation.

The difference between SARMA and PARMA models can be seen intuitively by considering a monthly averaged temperature series. In a regression sense, a periodic autoregressive model (PAR) with $p(\nu) \equiv 2$, for instance, would use the most recent February and March temperatures to explain the coming April's temperature. In contrast, the seasonal autoregression with $p = 2$ in Eq. (4.25) would use April's temperature during the two most recent *years* to explain the coming April's temperature. Such a SAR model ignores the two most recent months in the series, arguably the most important regressors. If one considers a SARMA model by combining Eqns. (4.25) and (4.26), then a forecast of this coming April's temperature may depend on, among other things, the most recent February and March temperatures; however, such a SARMA forecast is independent of the month being forecasted since the SARMA model is stationary. If the covariance structure of the series is truly periodic, then a SARMA model should not be expected to provide accurate forecasts. McLeod (1993) demonstrates the inadequacies of SARMA models in forecasting PC streamflow series.

A practical problem with PARMA models is their large number of parameters. The mean zero PARMA model in Eq. (4.1) has $T + \sum_{\nu=1}^{T}(p(\nu) + q(\nu))$ parameters—an additional T are added when considering non-zero seasonal means. This large number of parameters makes parameter estimation problems difficult. Consequently, much of the literature is devoted to fitting parsimonious PARMA models to PC series. Thompstone et al. (1985) suggest grouping similar seasons into blocks to reduce the number of PARMA parameters. Hannan (1955), Jones and Brelsford (1967), and Lund et al. (1995) consider reducing the number of model parameters with short Fourier series. For example, in a PAR(1) model for a monthly series, one could attempt to consolidate the 12 autoregressive parameters into three via

$$\phi_1(\nu) = C_0 + C_1 \cos(2\pi(\nu - C_2)/12). \tag{4.27}$$

Fourier reductions are particularly useful when seasonal changes in the correlation structure of the series are not abrupt.

Layered modeling is another way of fitting a parsimonious PARMA model. In a layered approach, one identifies a first layer that removes the periodic correlation in the series. For example, Bloomfield et al. (1994) found that a PAR(1) model fit in Eq. (4.16) to a series of monthly stratospheric ozone concentrations left residuals that were stationary but nonwhite. Hence, Eq. (4.16) and a "second layer" where $\{Z_n\}$ satisfies the ARMA difference equation

$$Z_n - \sum_{k=1}^{p} \alpha_k Z_{n-k} = \omega_n - \sum_{k=1}^{q} \beta_k \omega_{n-k} \tag{4.28}$$

were considered. In Eq. (4.28) $\{\omega_n\}$ is mean zero white noise with variance σ_ω^2.

Combining Eqns. (4.28) and (4.16) yields a PARMA model with $p(\nu) \equiv p + 1$, $q(\nu) \equiv q$, and $2T + p + q$ parameters. Note that σ_ω^2 is not a free parameter as the constraint $\text{Var}[Z_n] = 1$ must be imposed.

Returning to the general case, if $\{Z_n\}$ admits a stationary causal solution in the form

$$Z_n = \sum_{k=0}^{\infty} \eta_k \omega_{n-k}, \tag{4.29}$$

where the η_ks are absolutely summable in k, then combining Eqns. (2.6), (4.19), and (4.29) gives

$$X_{nT+\nu} = \sum_{k=0}^{\infty} \tilde{\psi}_k(\nu) \omega_{nT+\nu-k}, \tag{4.30}$$

where

$$\tilde{\psi}_k(\nu) = \sum_{l=0}^{k} \eta_{k-l} \theta_0(\nu - l) \prod_{i=0}^{l-1} \phi_1(\nu - i). \tag{4.31}$$

In Eq. (4.31), $\phi_1(j)$ and $\theta_0(j)$ are interpreted periodically in j with period T.

5. YULE-WALKER ESTIMATION IN PAR MODELS

This section considers parameter estimation in periodic autoregressive (PAR) models. Suppose that we have a sample $(X_1, X_2, \ldots, X_N)'$ from a zero mean causal Gaussian periodic autoregressive series with known period T:

$$X_{nT+\nu} - \sum_{k=1}^{p(\nu)} \phi_k(\nu) X_{nT+\nu-k} = \theta_0(\nu) Z_{nT+\nu}, \tag{5.1}$$

where $\{Z_n\}$ is mean zero, unit variance, Gaussian white noise. Our goal is to estimate the parameters $\phi_k(\nu)$ for $1 \leq \nu \leq T$ and $1 \leq k \leq p(\nu)$, and $\theta_0(\nu)$ for $1 \leq \nu \leq T$.

It is convenient to work with the autocovariance function $\gamma(\nu, l) = E[X_{nT+\nu} X_{nT+l}]$ and its sample estimate

$$\hat{\gamma}_N(\nu, l) = d^{-1} \sum_{n=0}^{M} X_{nT+\nu} X_{nT+l}. \tag{5.2}$$

In Eq. (5.2), $d = N/T$ represents the number of years of data and is assumed to be a positive integer for simplicity. The indices in Eq. (5.2) satisfy $M = \lfloor d - \max(\nu, l)/T \rfloor$, $1 \leq \nu \leq T$, and $1 \leq l \leq N - MT$.

Exploiting the relationship between PAR models and vector autoregressions in Eq. (4.2), Pagano (1978) establishes the following result.

LEMMA 5.1 In a causal Gaussian PAR model,

  (i) $\hat{\gamma}_N(\nu, l) \to \gamma(\nu, l)$ almost surely as $N \to \infty$;

  (ii) $E[(\hat{\gamma}_N(\nu, l) - \gamma(\nu, l))^2] \to 0$ as $N \to \infty$;

  (iii) $d\,\mathrm{Cov}[\hat{\gamma}_N(\nu, l), \hat{\gamma}_N(\nu', l')] \to \sum_{i=-\infty}^{\infty} [\gamma(\nu, \nu'+iT)\gamma(l, l'+iT) + \gamma(\nu, l'+iT)\gamma(l, \nu'+iT)]$ as $N \to \infty$.

The asymptotic joint normality of $\hat{\gamma}_N(\nu, l)$ can also be deduced as in Hannan (1970):

THEOREM 5.2 In a causal Gaussian PAR model, the random variables

$$\Delta(\nu, l) = d^{1/2}[\hat{\gamma}_N(\nu, l) - \gamma(\nu, l)] \tag{5.3}$$

for $1 \leq \nu \leq T$ and $0 \leq l \leq q$, where q is some fixed positive integer, are asymptotically jointly normal as $N \to \infty$ with mean zero and an asymptotic covariance as given in the right hand side of Lemma 5.1 (iii).

The PAR parameters can be related to $\gamma(\nu, l)$ via the Yule–Walker equations. Multiplying Eq. (5.1) by $X_{nT+\nu-l}$ and taking expectations gives

$$\gamma(\nu, \nu - l) - \sum_{k=1}^{p(\nu)} \phi_k(\nu)\gamma(\nu - k, \nu - l) = \theta_0^2(\nu)1_{[l=0]}, \qquad (5.4)$$

for $1 \leq \nu \leq T$ and $l \geq 0$.

The Yule–Walker estimates $\hat{\phi}_k(\nu)$ are defined as solutions to the moment-type estimating equations

$$\hat{\gamma}_N(\nu, \nu - l) - \sum_{k=1}^{p(\nu)} \hat{\phi}_k(\nu)\hat{\gamma}_N(\nu - k, \nu - l) = \hat{\theta}_0^2(\nu)1_{[l=0]}, \qquad (5.5)$$

where $1 \leq \nu \leq T$ and $0 \leq l \leq p(\nu)$.

Let $\boldsymbol{\phi}(\nu) = (\phi_1(\nu), \dots, \phi_{p(\nu)}(\nu))'$ and $\hat{\boldsymbol{\phi}}(\nu) = (\hat{\phi}_1(\nu), \dots, \hat{\phi}_{p(\nu)}(\nu))'$ denote actual and Yule–Walker estimated autoregressive parameters during season ν. The joint limit distribution of $d^{1/2}(\hat{\boldsymbol{\phi}}(\nu) - \boldsymbol{\phi}(\nu))$ for $1 \leq \nu \leq T$ can be deduced from Eqns. (5.4) and (5.5), and Theorem 5.2 (see Pagano, 1978). The result is stated below.

THEOREM 5.3   For causal Gaussian PAR series,

$$d^{1/2}[\hat{\boldsymbol{\phi}}(\nu) - \boldsymbol{\phi}(\nu)] \xrightarrow{d} N(0, F_\nu^{-1}) \qquad (5.6)$$

for each season ν. Furthermore, for any $\nu \neq \nu'$, $d^{1/2}[\hat{\boldsymbol{\phi}}(\nu)) - \boldsymbol{\phi}(\nu)]$ is asymptotically independent of $d^{1/2}[\hat{\boldsymbol{\phi}}(\nu') - \boldsymbol{\phi}(\nu')]$. Here, F_ν is the Fisher information matrix whose components are given by

$$F_\nu(j, l) = \theta_0(\nu)^{-2}\gamma(\nu - j, \nu - l) \qquad \text{for} \qquad 1 \leq j, l \leq p(\nu). \qquad (5.7)$$

The Yule–Walker estimates are asymptotically most efficient in a Gaussian PAR model because their asymptotic covariance matrices are the inverses of the corresponding Fisher information matrices. We refer the interested reader to Pagano (1978) for the details.

Finally, the $\theta_0(\nu)$s are estimated with the $l = 0$ in Eq. (5.5):

$$\hat{\theta}_0^2(\nu) = \hat{\gamma}_N(\nu, \nu) - \sum_{k=1}^{p(\nu)} \hat{\phi}_k(\nu)\hat{\gamma}_N(\nu - k, \nu). \qquad (5.8)$$

6. MAXIMUM LIKELIHOOD ESTIMATION

This section considers maximum likelihood estimation of PARMA parameters from the sample $\mathbf{X} = (X_1, X_2, \ldots, X_N)'$ taken from a mean zero causal invertible Gaussian PARMA model. The likelihood estimates are found by maximizing the PARMA likelihood; hence, one needs to be able to compute the likelihood function in terms of the model parameters in Eq. (4.1). For notation, let ϕ and θ denote PARMA autoregressive and moving average parameters respectively, and $L(\phi, \theta; \mathbf{X})$ the PARMA likelihood function.

The likelihood L has the familiar multivariate normal form

$$L(\phi, \theta; \mathbf{X}) = (2\pi)^{-N/2}(\det \Sigma)^{-1/2}\exp(-\tfrac{1}{2}\mathbf{X}'\Sigma^{-1}\mathbf{X}), \tag{6.1}$$

where Σ is the covariance matrix of $\mathbf{X}$. Evaluation of L has proven difficult due to the problem of inverting Σ; nonetheless, some work has been done in this direction. Vecchia (1985a, 1985b) computes an approximate L that is useful when N is large. Li and Hui (1988) compute an exact L, but their methods require Cholesky decompositions of large matrices. Recently, Adams and Goodwin (1995) presented a recursive control-type algorithm that converges to the maximum likelihood parameter estimates.

The approach that we take for likelihood evaluation involves the innovations algorithm, see Brockwell and Davis, Chapter 5 (1991). The innovations method of likelihood evaluation computes a sequence of one-step ahead predictors and their mean squared errors. Let

$$\hat{X}_{n+1} = E[X_{n+1}|X_1, \ldots, X_n] \tag{6.2}$$

and

$$v_n = E[(X_{n+1} - \hat{X}_{n+1})^2] \tag{6.3}$$

be the best predictors and their mean squared errors for $0 \leq n \leq N - 1$. We take $\hat{X}_1 = 0$ and note that, due to normality, $\hat{X}_{n+1}$ is linear in $X_1, \ldots, X_n$ and

$$E[(\hat{X}_{n+1} - X_{n+1})^2] = E[(\hat{X}_{n+1} - X_{n+1})^2|X_1, \ldots, X_n]. \tag{6.4}$$

Whenever Σ is invertible, the likelihood in Eq. (6.1) has the equivalent innovations representation

$$L(\phi, \theta; \mathbf{X}) = (2\pi)^{-N/2}\left(\prod_{j=0}^{N-1} v_j\right)^{-1/2} \exp\left(-\frac{1}{2}\sum_{j=1}^{N} \frac{(X_j - \hat{X}_j)^2}{v_{j-1}}\right). \tag{6.5}$$

Hence, the objective reduces to computing the one step ahead predictors and their mean squared errors. This can be accomplished without matrix inversions with the innovational recursions, see Brockwell and Davis (1991)

$$\hat{X}_{n+1} = \sum_{j=1}^{n} \Delta_{nj}(X_{n+1-j} - \hat{X}_{n+1-j}),$$ (6.6)

for $n \geq 1$, where the Δ_{ij}s and v_ns are computed recursively with $v_0 = \kappa(1,1)$,

$$\Delta_{n,n-k} = v_k^{-1}\left[\kappa(n+1,k+1) - \sum_{j=0}^{k-1} \Delta_{k,k-j}\Delta_{n,n-j}v_j\right],$$ (6.7)

and

$$v_n = \kappa(n+1,n+1) - \sum_{j=0}^{n-1} \Delta_{n,n-j}^2 v_j.$$ (6.8)

In Eqns. (6.7) and (6.8), $\kappa(i,j) = E[X_i X_j]$. κ can be computed from $\gamma_\nu(h)$ which can be evaluated with Eqns. (2.8) and (4.15). Equations (6.7) and (6.8) can be solved in the order $v_0, \Delta_{1,1}, v_1, \Delta_{2,2}, \Delta_{2,1}, v_2, \Delta_{3,3},$ $\Delta_{3,2}, \Delta_{3,1}, v_3,$ etc.

The above innovational computations can be demanding in general; however, the methods simplify considerably for periodic autoregressions. Specifically, one can easily show that

$$\hat{X}_{nT+\nu} = \sum_{k=1}^{p(\nu)} \phi_k(\nu)X_{nT+\nu-k} \quad \text{and} \quad v_{nT+\nu-1} = \theta_0(\nu)^2$$ (6.9)

when $nT + \nu - p(\nu) \geq 1$. Hence, evaluating the exact likelihood for PAR models is not difficult.

EXAMPLE 6.1   Consider the causal PAR(1) model in Eq. (4.16). Then $\hat{X}_1 = 0$ and Eq. (6.9) gives $\hat{X}_{nT+\nu} = \phi_1(\nu)X_{nT+\nu-1}$ and $v_{nT+\nu-1} = \theta_0^2(\nu)$. Using this in Eq. (6.5) gives

$$-2 \ln[L(\phi, \theta; X)] = N \ln(2\pi) + \ln(\sigma^2(1)) + \frac{X_1^2}{\sigma^2(1)}$$

$$+ \sum_{j=1}^{N-1} \ln(\theta_0^2(j)) + \sum_{j=2}^{N} \frac{[X_j - \phi_1(j)X_{j-1}]^2}{\theta_0^2(j)}.$$ (6.10)

In Eq. (6.10), $\sigma^2(1)$ is computed from Eq. (4.22) and $\theta_0(j)$ and $\phi_1(j)$ are interpreted periodically in j with period T.

Let $\hat{\phi}$ and $\hat{\theta}$ denote the maximum likelihood (ML) estimators obtained by maximizing the likelihood function in Eq. (6.5). The ML estimators are equivalently obtained by minimizing

$$\sum_{j=0}^{N-1} \ln(v_j) + \sum_{j=1}^{N} \frac{(X_j - \hat{X}_j)^2}{v_{j-1}} \tag{6.11}$$

in the PARMA parameters.

We now consider the asymptotic properties of the ML estimators for PAR models; more general PARMA models will be considered elsewhere. For the PAR model, Eqns. (6.9) and (6.11) show that, asymptotically, the ML estimates of $\phi_k(\nu)$ and $\theta_0(\nu)$ can be obtained by minimizing

$$d \sum_{\nu=1}^{T} \ln(\theta_0(\nu)^2) + \sum_{n=0}^{d-1} \sum_{\nu=1}^{T} \frac{(X_{nT+\nu} - \hat{X}_{nT+\nu})^2}{\theta_0^2(\nu)}, \tag{6.12}$$

where $\hat{X}_{nT+\nu}$ is obtained from Eq. (6.9) with the slight modification that X_j is taken as zero whenever $j \le 0$. The likelihood equations for estimating $\phi_k(\nu)$ corresponding to Eq. (6.12) are

$$\sum_{n=0}^{d-1} \left(X_{nT+\nu} - \sum_{k=1}^{p(\nu)} \hat{\phi}_k(\nu) X_{nT+\nu-k} \right) X_{nT+\nu-l} = 0, \tag{6.13}$$

for $1 \le \nu \le T$ and $1 \le l \le p(\nu)$. The ML estimate for $\theta_0^2(\nu)$ is then given by

$$\hat{\theta}_0^2(\nu) = d^{-1} \sum_{n=0}^{d-1} \left(X_{nT+\nu} - \hat{X}_{nT+\nu} \right)^2 \tag{6.14}$$

for $1 \le \nu \le T$.

Note that Eq. (6.13) is identical to Eq. (5.4) for $1 \le \nu \le T$ and $1 \le l \le p(\nu)$. Consequently, the ML and Yule-Walker estimates of $\phi_k(\nu)$ are asymptotically equivalent and the limit distribution of $\{\hat{\phi}_k(\nu)\}$ is as given in Theorem 5.3.

One advantage of likelihood estimation over Yule–Walker estimation in PAR models occurs when there are constraints on the model parameters such as those in Eq. (4.27). In this case, the likelihood of Eq. (6.11) can always be numerically minimized as a function of the "constrained model" parameters; however, it is not obvious how the Yule–Walker equations should be modified, if even possible, to obtain such constrained estimates. In the next Section, we will consider an estimating equation parameter estimation approach that easily accounts for parametric constraints.

7. AN ESTIMATING EQUATION APPROACH

This section studies an estimating equation approach to PARMA parameter estimation. Throughout, we assume that the PARMA parameters ϕ and θ are known functions of a parameter vector β; hence, we allow for the possibility of parametric constraints such as those in Eq. (4.27).

The quasilikelihood estimating equation, Godambe, (1985); Godambe and Heyde (1987), for estimating β is

$$\sum_{j=1}^{N} \frac{[X_j - \hat{X}_j(\beta)]}{v_{j-1}(\beta)} \frac{d\hat{X}_j(\beta)}{d\beta} = 0, \tag{7.1}$$

where $\hat{X}_j(\beta)$ and $v_{j-1}(\beta)$ are the one-step ahead linear predictors and their mean squared errors as discussed in the last section. We use the notation $\hat{X}_j(\beta)$ to emphasize that $\hat{X}_j$ depends on β. The quasilikelihood estimate $\hat{\beta}$ of β is obtained as a solution to the vector valued equation Eq. (7.1).

Throughout, we are assuming that $\{X_n\}$ is a causal invertible Gaussian PARMA series; further, we tacitly assume that all PARMA parameters are identifiable. When $p(\nu)$ and $q(\nu)$ are such that $p^* \geq 1$ and $q^* \geq 1$ in Eq. (4.2), it is known that causality and invertibility may not be sufficient for parameter identifiability, see Dunsmuir and Hannan (1976). In the case of a periodic autoregression ($q^* = 0$) or a periodic moving-average ($p^* = 0$), causality and invertibility is sufficient for parameter identifiability.

Let $S(\beta)$ denote the left hand side of Eq. (7.1). Using normality and the innovations orthogonality property that $\hat{X}_j(\beta) - X_j$ is uncorrelated with $X_1, \ldots, X_{j-1}$, it is straightforward to show that $S(\beta)$ is a mean zero martingale. Suppose further that one can establish that

$$N^{-1} \sum_{j=1}^{N} v_{j-1}^{-1}(\beta) \left(\frac{d\hat{X}_j(\beta)}{d\beta}\right) \left(\frac{d\hat{X}_j(\beta)}{d\beta}\right)' \to V(\beta), \tag{7.2}$$

where the convergence in Eq. (7.2) is in probability and $V(\beta)$ is some positive definite matrix. Then $V(\beta)$ is called a quasilikelihood information matrix. Under reasonably general regularity conditions (see Hutton and Nelson (1986), one can prove the following result.

THEOREM 7.1   As $N \to \infty$,

$$N^{1/2}(\hat{\beta} - \beta) \to N(0, V^{-1}(\beta)), \tag{7.3}$$

in distribution where $\hat{\beta}$ is a consistent solution to Eq. (7.1).

Proof. We outline the proof only; full details will be pursued elsewhere. A Taylor expansion gives

$$S(\hat{\boldsymbol{\beta}}) = S(\boldsymbol{\beta}) + \left(\frac{dS(\boldsymbol{\beta})}{d\boldsymbol{\beta}}\right)_{\boldsymbol{\beta}=\boldsymbol{\beta}^*} (\hat{\boldsymbol{\beta}} - \boldsymbol{\beta}), \tag{7.4}$$

where $\boldsymbol{\beta}^* = \boldsymbol{\beta} + \gamma(\hat{\boldsymbol{\beta}} - \boldsymbol{\beta})$ and $|\gamma| < 1$. Since $S(\hat{\boldsymbol{\beta}}) = 0$, we obtain

$$N^{1/2}(\hat{\boldsymbol{\beta}} - \boldsymbol{\beta}) = -N^{-1/2}S(\boldsymbol{\beta})\left(N^{-1}\frac{dS(\boldsymbol{\beta})}{d\boldsymbol{\beta}}\right)^{-1}_{\boldsymbol{\beta}=\boldsymbol{\beta}^*}. \tag{7.5}$$

Under appropriate regularity conditions, see Hall and Heyde (1980), one has

$$N^{-1/2}S(\boldsymbol{\beta}) \rightarrow N(0, V(\boldsymbol{\beta})) \tag{7.6}$$

in distribution and

$$-N^{-1}\left(\frac{dS(\boldsymbol{\beta})}{d\boldsymbol{\beta}}\right)_{\boldsymbol{\beta}=\boldsymbol{\beta}^*} \rightarrow V(\boldsymbol{\beta}) \tag{7.7}$$

in probability. The result in the theorem follows by applying Slutsky's Theorem.

The quasilikelihood information matrix $V(\boldsymbol{\beta})$ can be estimated by

$$\widehat{V(\boldsymbol{\beta})} = N^{-1}\sum_{j=1}^{N} v_{j-1}^{-1}(\hat{\boldsymbol{\beta}})\left(\frac{d\hat{X}_j(\boldsymbol{\beta})}{d\boldsymbol{\beta}}\right)_{\boldsymbol{\beta}=\hat{\boldsymbol{\beta}}}\left(\frac{d\hat{X}_j(\boldsymbol{\beta})}{d\boldsymbol{\beta}}\right)'_{\boldsymbol{\beta}=\hat{\boldsymbol{\beta}}}. \tag{7.8}$$

As a corollary to Theorem 7.1, one obtains

$$N^{1/2}[\widehat{V(\boldsymbol{\beta})}]^{1/2}(\hat{\boldsymbol{\beta}} - \boldsymbol{\beta}) \rightarrow N(0, I) \tag{7.9}$$

in distribution, where I denotes the identity matrix.

Considering a PAR model, Eq. (7.1) reduces asymptotically to

$$\sum_{n=0}^{d-1}\sum_{\nu=1}^{T}\frac{(X_{nT+\nu} - \sum_{k=1}^{p(\nu)}\phi_k(\nu)X_{nT+\nu-k})}{\theta_0^2(\nu)}D_n(\nu) = 0, \tag{7.10}$$

where

$$D_n(\nu) = \sum_{k=1}^{p(\nu)}\frac{d\phi_k(\nu)}{d\boldsymbol{\beta}}X_{nT+\nu-k}. \tag{7.11}$$

In particular, for the PAR(1) model in Eq. (4.16), the quasilikelihood equation is

$$\sum_{n=0}^{d-1}\sum_{\nu=1}^{T}\frac{[X_{nT+\nu} - \phi_1(\nu)X_{nT+\nu-1}]}{\theta_0^2(\nu)}\frac{d\phi_1(\nu)}{d\boldsymbol{\beta}}X_{nT+\nu-1} = 0. \tag{7.12}$$

The quasilikelihood information matrix for the PAR(1) model is given by

$$V(\boldsymbol{\beta}) = T^{-1} \sum_{\nu=1}^{T} \theta_0^{-2}(\nu)\gamma_{\nu-1}(0) \frac{d\phi_1(\nu)}{d\boldsymbol{\beta}} \frac{d\phi_1(\nu)'}{d\boldsymbol{\beta}}. \tag{7.13}$$

In Eq. (7.13), $\gamma_{\nu-1}(0) = \text{Var}[X_{nT+\nu-1}]$.

Acknowledgments

Robert Lund's research was supported by NSF Grant DMS 9703838. Ishwar Basawa's research was supported by NSF and ONR Grants. We thank the referee for helpful comments on the first draft of this manuscript.

REFERENCES

Adams, G. J. and G. C. Goodwin, Parameter estimation for periodic ARMA models, *J. Time Ser. Anal., 16*: 127–146 (1995).

Anderson, P. L. and A. V. Vecchia, Asymptotic results for periodic autoregressive moving-average processes, *J. Time Ser. Anal., 14*: 1–18 (1993).

Anderson, P. L. and M. M. Meerschaert, Periodic moving averages of random variables with regularly varying tails, *Ann. Statist., 25*: 771–785, (1997).

Bentarzi, M. and M. Hallin, On the invertibility of periodic moving-average models, *J. Time Ser. Anal., 15*: 263–268 (1994).

Bloomfield, P., H. L. Hurd, and R. B. Lund, Periodic correlation in stratospheric ozone data, *J. Time Ser. Anal., 15*: 127–150 (1994).

Brockwell, P. J. and R. A. Davis. *Time Series: Theory and Methods*, Second Edition, Springer-Verlag, New York, 1991.

Chu, K., *Analysis of a test for periodic correlation*, Master's Thesis, Department of Statistics, Colorado State University, 1996.

Dunsmuir, W. and E. J. Hannan, Vector linear time series models, *Adv. Appl. Prob., 8*: 339–364, 1976.

Gardner, W. and L. E. Franks, Characterization of cyclostationary random signal processes, *IEEE Trans. Infn. Theory 21*: 4–14 (1975).

Godambe, V. P., The foundations of finite sample estimation in stochastic processes, *Biometrika, 72*: 419–428 (1985).

Godambe, V. P. and C. C. Heyde, Quasilikelihood and optimal estimation, *Inter. Stat. Rev., 55*: 231–244 (1987).

Gladyšhev, E. G., Periodically correlated random sequences, *Soviet Math, 2*: 385–388 (1961).

Hall, P. and C. C. Heyde, *Martingale Limit Theory and its Applications*, Academic Press, New York, 1980.

Hannan, E. J., A test for singularities in Sydney rainfall, *Australian J. Phys. 8*: 289–297 (1955).

Hannan, E. J., *Multiple time series*, John Wiley and Sons, New York, 1970.

Hurd, H. L., Representation of strongly harmonizable periodically correlated processes and their covariances, *J. Multivariate Anal., 29*: 53–67 (1989).

Hurd, H. L. and N. L. Gerr, Graphical methods for determining the presence of periodic correlation, *J. Time Ser. Anal., 12*: 337–350, (1991).

Hurd, H. L. and J. Leskow, Estimation of the Fourier coefficient functions and their spectral densities for ϕ-mixing almost periodically correlated processes, *Statist. Prob. Letters, 14*: 299–306 (1992).

Hutton, J. E. and P. I. Nelson, Quasilikelihood estimation for semimartingales, *Stoch. Proc. and Their Applications, 22*: 245-257 (1986).

Jones, R. H. and W. M. Brelsford, Time series with periodic structure, *Biometrika, 54*: 403–408 (1967).

Lawrance, A. J. and N. T. Kottegoda, Stochastic modeling of river flow time series, Journal of *Roy. Statist. Soc., A, 140*: 1–31 (1977).

Leskow, J., Analysis of time series stationarity with applications, Preprint, 1997.

Li, W. K. and Y. V. Hui., An algorithm for the exact likelihood of periodic autoregressive moving average models, Comm. Statist. *Simula. Comput. 17*: 1483–1494, (1988).

Lund, R. B., H. L. Hurd, P. Bloomfield, and R. L. Smith, Climatological time series with periodic correlation, *J. Climate, 11*: 2787–2809 (1995).

Monin, A. S., Stationary and periodic time series in the general circulation of the atmosphere in *Proceedings Symposium on Time Series Analysis*, (ed. M. Rosenblatt), 1963, 144–151.

McLeod, A. I., Parsimony, model adequacy and periodic correlation in time series forecasting, *Inter. Statisti. Rev., 61*: 387-393 (1993).

Pagano, M. On periodic and multiple autoregressions, *Ann. Statist. 6*: 1310–1317 (1978).

Parzen, E. and M. Pagano. An approach to modeling seasonally stationary time series, *J. Econometrics, 9*: 137–153 (1979).

Thompstone, R. M., K. W. Hipel and A. I. McLeod, Grouping of Periodic Autoregressive Models, in *Time Series Analysis: Theory and Practice 6*: 35–49, (eds O. D. Anderson, J. K. Ord and E. A. Robinson), Elsiver Science, North Holland, 1985.

Vecchia, A. V., Periodic autoregressive-moving average (PARMA) modeling with applications to water resources, *Water Resources Bulletin, 21*: 721–730 (1985a).

Vecchia, A. V., Maximum likelihood estimation for periodic autoregressive moving average models, *Technometrics, 27*: 375–384 (1985b).

Vecchia, A. V. and R. Ballerini, Testing for periodic autocorrelations in seasonal time series data, *Biometrika, 78*: 53–63 (1991).

3
Modeling Time Series of Count Data

RICHARD A. DAVIS and YING WANG   Colorado State University, Fort Collins, Colorado

WILLIAM T. M. DUNSMUIR   University of New South Wales, Sydney, Australia

1. INTRODUCTION

The focus of this chapter is on the similarities and differences in model building and interpretation for two types of conditional mean processes in Poisson regression models for time series of counts: parameter-driven and observation-driven specifications. For a parameter-driven model, it is shown that under general conditions, the Poisson maximum likelihood estimator of the regression parameter based on a model without serial correlation is consistent and asymptotically normal with an easily computable covariance matrix. This covariance matrix depends on the covariance structure of the latent process. A method of testing for the existence of a latent process is developed and compared to existing test statistics via simulation. Once the existence of a latent process has been detected, a simple and easily implementable method for estimating the autocorrelation of the latent process is given. The resulting estimates are shown to be consistent and the standard errors of the estimates are also provided. A test statistic for testing serial dependence in the latent process is proposed based on the study of the distribution of autocorrelation estimates. Estimation of the regression and latent process parameters in a parameter-driven model is developed using an approximation to the likelihood. New and existing formulations of observation-driven models are also discussed. Properties of these observation-driven models are studied and likelihood based estimation procedures are presented. Issues surrounding model identification and estimation for both parameter and observation driven models are illustrated with two data sets.

1.1 Overview

In recent years there has been a growing literature on models and methods for analyzing time series of counts. The need for such a development is clearly demonstrated in applications such as modeling of disease incidence, as for example in the modeling of polio counts in U.S., see Zeger (1988). Another area that is increasingly important is the modeling of disease counts and their relationship to putative causal factors such as weather and atmospheric pollution. Numerous articles have appeared in this vein, for example Campbell (1994) Jorgensen et al. (1995) Saldiva et al. (1995) and Schwartz et al. (1996). Another potential application of these methods is to time series of counts obtained from defect measurements on the output of some commercial or manufacturing process in which, as is the case for continuous output measurements, there is interest in relating process control variables to the outcome defect rate which could be a binomial count or a Poisson count. Other recent applications are to micro time scale financial data in which the rate at which transactions form during the day might be the variable of interest or to road traffic injuries or fatalities—see Brannas and Johansson (1994) or Harvey and Fernandes (1989), for example.

In the setting where the researcher wishes to relate disease outcomes to pollution, count measurements arise naturally because often the disease is relatively rare and because the time scale or geographical scale on which pollution measurements fluctuate are necessarily small. For example, in a study of the relationship between temperature and the incidence of sudden infant death syndrome (SIDS) Campbell worked with daily measurements of temperature and SIDS counts. If the disease is even rarer (e.g., polio) the time scale that is suitable might be one month. Aggregation over longer time periods in both of these examples would obliterate or at least blur the link between the putative risk factor and the disease outcome.

As in all time series modeling applications, the possibility of serial dependence within the data has to be investigated. In a linear model in which the response variable is continuous and possibly even Gaussian, the effects of serial correlation in the regression error structure have been well understood for about half a century. Likewise the various forms of model structure (serially correlated errors, lagged dependent variables, for example) are very well understood as is the theory of state-space models.

Only in the last 5 to 10 years has there been serious attention paid to the analogous results and models in the area of regression modeling for count data. The recent literature has seen various models and methods proposed and we will review some of these here.

However what is lacking, in our view, is a comprehensive comparison of these various models and methods in terms of their statistical properties and their performance on a variety of real data sets. Also, there is not yet a structured and statistically well founded approach to model building and diagnosis as is readily found in the literature for linear time series analysis. This chapter is aimed at making modest progress towards the goal of filling in these details or at the very least, identifying the need for additional theory and practice.

Analysis of time series of counts is concerned with the usual issues that arise in the standard linear modeling paradigm, namely: model fitting, hypothesis testing or relationship building, and forecasting. The use to which the model will be applied is important in the selection of model as we discuss below. In anticipation of that discussion we mention that models which are based on a presumption of a latent stochastic process governing the (conditional) mean function in the distribution of the observed count process are difficult to forecast with since the latent process is the means by which serial dependence is modeled and yet this is not directly observable. Furthermore, this is a non-Gaussian setting so that predictors are typically nonlinear. On the other hand models which incorporate lagged values of the dependent count process directly into their mean function are easy to forecast with.

Reviews of the models considered in this paper can also be found in Brockwell and Davis (1996), and Fahrmeir and Tutz (1994). While the literature contains many suggestions, little comparative and analytical work has been carried out. Often asymptotic distributions, so necessary for valid inference (i.e., in which the effect of serial dependence on regression standard errors), are only loosely argued. For example, if a regression on time is performed (e.g., to build a linear trend term in the linear predictor) without normalization by the number of time periods the usual asymptotics do not apply when the trend is negative simply because the information matrix does not grow without bound.

A proper treatment of estimation of serial dependence has not yet been given. We provide some contributions to this below. Typically in Poisson regression the use of the residuals from the fit (based on observed counts minus the fitted values) will seriously underestimate the true serial dependence. We summarize existing methods and propose new ones for avoiding this underestimation. In addition we compute standard errors and the analogue of the Box–Pierce portmanteau test for serial correlation widely used in linear time series models.

1.2   Desiderata of Models

Zeger and Qaqish (1988) implicitly offer desiderata that observation driven models should satisfy. For specificity, we assume that the count data, denoted by Y_t, follow a Poisson distribution with mean, conditional upon the r-dimensional regression variable $\mathbf{x}_t$ and a random process ν_t, given by

$$\mu_t = \exp(\mathbf{x}_t^T \boldsymbol{\beta} + \nu_t)$$

in which the process ν_t may depend on lagged values of the observed counts Y_{t-l}. It is assumed that the joint distribution of the ν_t depends on a vector of parameters γ. If $\nu_t \equiv 0$, then this is the familiar Poisson regression model. The three desiderata described in Zeger and Qaqish (1988) are:

D.1.   The marginal mean of Y_t should be approximated as

$$E(Y_t) = E(\mu_t) \approx \exp(\mathbf{x}_t^T \boldsymbol{\beta})$$

so that the regression coefficients $\boldsymbol{\beta}$ can be interpreted as the proportional change- in the marginal expectation of Y_t given a unit change in the regressor variable. D.1 is useful for interpretation.

D.2.   Both positive and negative serial dependence should be possible in the model.

D.3.   The estimates of $\boldsymbol{\beta}$ and γ should be approximately orthogonal making their estimation easier (presumably because a two stage procedure could be used).

Condition D.3 is met for linear regression models with time series errors. However, for modeling count data, D.3 may be overly restrictive since the conditional mean and variance of Y_t are linked. Also with rapid computing now widely available requirement D.3 is no longer necessary.

In addition to these three criteria, we offer three more which fill out the modeling paradigm for count data. These are:

D.4.   The ability to easily forecast with the model. This is often the primary goal in many time series applications.

D.5.   A method for model fitting and inference should be reasonably straightforward to implement and control.

D.6.   Diagnostic tools should be available for identification of a class of models, for the assessment of model adequacy, and for detection of outliers, etc.

1.3 Generalized State-Space Models

Linear state-space models and the associated Kalman recursions have had a profound impact on time series analysis and related areas. The techniques were originally developed in connection with the control of linear systems (for accounts of this subject see Davis and Vinter, 1985, and Hannan and Deistler, 1988). In recent years, non-linear state-space models have been developed to handle a wide range of situations not easily covered under the linear framework. In this subsection, we provide a brief overview of generalized state-space models.

A state-space model for a time series $\{Y_t, t = 1, 2, \ldots\}$ consists of two equations referred to as the observation and state equations. Generalized state-space models can be loosely characterized as either "parameter driven" or "observation driven." The observation equation is the same for both models, but the state vectors of a parameter driven model evolve independently of the past history of the observation process, while the state vectors of an observation driven model depend on past observations.

The *observation equation* specifies the distribution of Y_t given a state variable S_t. For ease of presentation, we assume the state-variable is univariate, although it is often taken to be vector valued. Specifically, if $\mathbf{Y}^{(t)}$ and $\mathbf{S}^{(t)}$ denote the t–dimensional column vectors $\mathbf{Y}^{(t)} = (Y_1, Y_2, \ldots, Y_t)^T$ and $\mathbf{S}^{(t)} = (S_1, S_2, \ldots, S_t)^T$, respectively, then it is assumed that Y_t given $(S_t, \mathbf{S}^{(t-1)}, \mathbf{Y}^{(t-1)})$ is independent of $(\mathbf{S}^{(t-1)}, \mathbf{Y}^{(t-1)})$ with conditional probability density,

$$p(y_t|s_t, \mathbf{s}^{(t-1)}, \mathbf{y}^{(t-1)}) = p(y_t|s_t), \quad t = 1, 2, \ldots. \tag{1.1}$$

For the parameter-driven model, S_{t+1} given $(S_t, \mathbf{S}^{(t-1)}, \mathbf{Y}^{(t)})$ is assumed to be independent of $(\mathbf{S}^{(t-1)}, \mathbf{Y}^{(t)})$ with conditional density function,

$$p(s_{t+1}|s_t, \mathbf{s}^{(t-1)}, \mathbf{y}^{(t)}) := p(s_{t+1}|s_t), \quad t = 1, 2, \ldots. \tag{1.2}$$

This latter equation is known as the *state-equation* for the parameter driven model.

The joint density of the observation and state variables can be computed directly from Eqns (1.1) and (1.2) as

$$
\begin{aligned}
p(y_1, \ldots, y_n, s_1, \ldots, s_n) &= p(y_n|s_n, \mathbf{s}^{(n-1)}, \mathbf{y}^{(n-1)}) p(s_n, \mathbf{s}^{(n-1)}, \mathbf{y}^{(n-1)}) \\
&= p(y_n|s_n) p(s_n|\mathbf{s}^{(n-1)}, \mathbf{y}^{(n-1)}) p(\mathbf{y}^{(n-1)}, \mathbf{s}^{(n-1)}) \\
&= p(y_n|s_n) p(s_n|s_{n-1}) p(\mathbf{y}^{(n-1)}, \mathbf{s}^{(n-1)}) \\
&= \cdots \\
&= \left(\prod_{j=1}^{n} p(y_j|s_j)\right)\left(\prod_{j=2}^{n} p(s_j|s_{j-1})\right) p_1(s_1),
\end{aligned}
\tag{1.3}
$$

and since Eq. (1.2) implies that $\{S_t\}$ is Markov

$$p(y_1, \ldots, y_n | s_1, \ldots, s_n) = \left(\prod_{j=1}^{n} p(y_j | s_j) \right). \tag{1.4}$$

We conclude that $Y_1, \ldots, Y_n$ are conditionally independent given the state variables $S_1, \ldots, S_n$, so that the dependence structure of $\{Y_t\}$ is inherited from that of the state-process $\{S_t\}$. The sequence of state-variables $\{S_t\}$ is often referred to as the *hidden* or *latent* generating process associated with the observed process.

In an observation-driven model specification, it is again assumed that Y_t, conditional on the vector $(S_t, \mathbf{S}^{(t-1)}, \mathbf{Y}^{(t-1)})$, is independent of $(\mathbf{S}^{(t-1)}, \mathbf{Y}^{(t-1)})$. The model is specified by the conditional densities

$$p(y_t | \mathbf{s}^{(t)}, \mathbf{y}^{(t-1)}) = p(y_t | s_t), \quad t = 1, 2, \ldots, \tag{1.5}$$

$$p(s_{t+1} | \mathbf{y}^{(t)}) = p_{S_{t+1} | \mathbf{Y}^{(t)}}(s_{t+1} | \mathbf{y}^{(t)}), t = 0, 1, \ldots, \tag{1.6}$$

where $p(s_1 | y^{(0)}) := p_1(s_1)$ for some prespecified initial density $p_1(s_1)$.

The advantage of the observation driven state equation (1.6) is that the forecast density $p(y_{t+1} | \mathbf{y}^{(t)})$ is easy to compute from the relation

$$p(y_{t+1} | \mathbf{y}^{(t)}) = \int p(y_{t+1} | s_{t+1}) p(s_{t+1} | \mathbf{y}^{(t)}) \, d\mu(s_t). \tag{1.7}$$

In other words computing the best predictor of Y_{t+1} in terms of $\mathbf{Y}^{(t)}$ and calculating the joint density function of $\mathbf{y}_n$ given by

$$p(y_1, \ldots, y_n) = \prod_{t=1}^{n} p(y_t | \mathbf{y}^{(t-1)}) \tag{1.8}$$

are relatively simple tasks. Calculation of the joint density function is particularly important for estimation and making inferences about the parameters of the model. On the other hand, for a parameter driven model, computation of the forecast density function and hence the joint density function, is difficult requiring recursive updating of the densities $p(s_t | \mathbf{y}^{(t)})$ and $p(s_{t+1} | \mathbf{y}^{(t)})$ via Bayes's theorem, see Brockwell and Davis (1996).

While the observation-driven model formulation has a decided edge for forecasting future values of the series and for calculating the joint density, the evolutionary properties of the time series are more difficult to characterize than for the parameter-driven model. Specifically, ergodic and asymptotic stationarity of a time series based on an observation-driven model can be quite difficult to establish and this is the subject of ongoing research. However, for a process based on a parameter-driven model, such properties are typically inherited by those assumed for the underlying state-variables.

1.4 Two Examples

To illustrate the issues, theory and methods we will use the following two examples throughout our discussion.

1.4.1 Asthma Presentations at a Sydney Hospital

The data arose from a single hospital (at Campbelltown) as part of a larger (ongoing) study into the relationship between atmospheric pollution and the number of asthma cases presenting themselves to various emergency departments in local hospitals in the South West region of Sydney, Australia. It is not our purpose here to examine this issue since the data required to do this properly is not yet fully compiled. In addition, for much of the record that we examine, measurements on other variables, such as pollen counts, which could influence the asthma attacks that people experience, are not available. However, the analysis we provide here would be required as a first step in constructing a model relating the putative causes and the outcome counts. Regional measurements are also required because the weather patterns in Sydney are such as to move pollution around substantially from day to day.

Figure 1 shows the daily number of asthma presentations from January 1, 1990–December 31, 1993. It is apparent from this figure that there is some

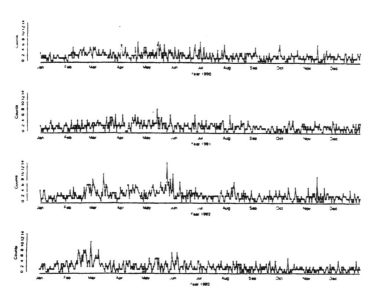

Figure 1 Asthma presentations at a Sydney hospital.

form of seasonal pattern with higher activity occurring in the Fall (March–May). There is also a suggestion of an increasing trend in time. Note that this is a distinct possibility regardless of whether asthma is more prevalent in the region simply because it is an area which has a young and growing population. Of interest would be to test the significance of any trend, suitably adjusted for serial dependence. There are also patterns of departure from a seasonal cycle in the data (e.g., in 1992) which exhibit clear signs of positive serial dependence.

Not apparent, but consistent with behavioral aspects of the population, is the possibility of a day of the week effect (which also could be confounded with weekly pattern in pollution). General practitioners (private physicians) are less available on Sundays and people tend to rely more on the emergency departments of hospitals should an asthma attack occur. As it turns out that the day of the week effect can be characterized by being similar on Tuesday through Saturday and elevated incidence rate occurring on Sunday and Monday with roughly similar size effects on these two days. The Monday rise might be explained in terms of patients who hold off through Sunday hoping for improvement but who then require hospital attention on Monday. In any case testing the significance of day of week effects is important in daily data of these type.

1.4.2   Polio Incidence

The second data set that we will consider consists of the monthly number of cases of poliomyelitis in the U.S. for the year 1970–1983 as reported by the Center for Disease Control. These data, which were originally presented in Zeger (1988), have become a standard example in the field. The data, which are graphed in Fig. 2 reveal some seasonality and the possibility of a slight decreasing trend. The main objective in modelling this data is the detection of a decreasing trend. Although some researchers have considered the observed count of 14 for November, 1972 to be an outlier, there is no corroborating evidence—we did not remove it in order to compare with existing analyses reported here which also did not remove or adjust it. Chan and Ledolter (1995) found a slight change in the slope and intercept terms of the model when the analysis is adjusted for the outlier.

1.5   Outline of Chapter

The focus of this chapter is on the similarities and differences in model building and interpretation for the two types of conditional mean processes, the parameter driven and observation driven specifications. We will concentrate exclusively on the case where the observations, conditional upon a

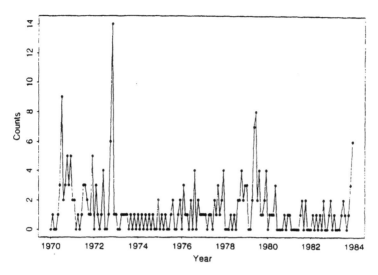

Figure 2 Monthly number of cases of poliomyelitis in the U.S. (1970–1983).

specified mean process, are independent Poisson observations. This class of models is useful in a wide range of applications in their own right. Focusing on the Poisson case allows the exposition of the main ideas without the distraction of the more general class of exponential family models.

Section 2 is concerned with defining the parameter driven model in terms of a latent process specification. The probabilistic properties of such models are developed. Inferential methods are reviewed and illustrated on the time series of polio data. Section 3 provides a parallel exposition for observation driven models. Section 4 gives data analysis results on three data sets including the polio data, UK SIDS, and the asthma data.

2. PARAMETER DRIVEN MODELS

In this Section, we consider a class of parameter-driven or latent process models for modeling time series of counts. We will focus on the situation when the "observation equation" is governed by a Poisson distribution. Denote the time series of counts by $Y_1, \ldots, Y_n$ and suppose that for each t, $\mathbf{x}_t$ is a r-dimensional regression vector whose first component is 1. It will also be convenient to introduce the notation $\mathbf{Y}^{(t)} = (Y_1, \ldots, Y_t)^T$, $X^{(t)} = (\mathbf{x}_1, \ldots, \mathbf{x}_t)$, and $\mathbf{\varepsilon}^{(t)} = (\varepsilon_1, \ldots, \varepsilon_t)^T$, where $\{\varepsilon_t\}$ is the latent process.

In this setting, the state-variable S_t described in Sec. 1.3, can be set to either the multivariate vector consisting of $(\mathbf{x}_t^T, \varepsilon_t)^T$ or just ε_t.

The conditional distribution of Y_t given $\mathbf{x}_t$ and ε_t corresponding to the observation equation of Eq. (1.1) is assumed to be Poisson with mean $u_t = \varepsilon_t \exp\{\mathbf{x}_t^T \boldsymbol{\beta}\}$ denoted by,

$$Y_t | \varepsilon_t, \mathbf{x}_t \sim \mathcal{P}(\varepsilon_t \exp\{\mathbf{x}_t^T \boldsymbol{\beta}\}), \tag{2.1}$$

where $\boldsymbol{\beta} = (\beta_1, \dots, \beta_r)^T$. The analogue of the state-equation in this context is a specification of the distributional properties of the latent process $\{\varepsilon_t\}$. Unlike the standard state-space formulation considered in Sec. 1.3, it is not necessary for the ε_t to have a Markov structure. At this point, we only assume that $\{\varepsilon_t\}$ is a non-negative strictly stationary time series with mean 1 and autocovariance function (ACVF)

$$\gamma_\varepsilon(h) = E(\varepsilon_{t+h} - 1)(\varepsilon_t - 1). \tag{2.2}$$

A Markov structure could be imposed on the state-variables by replacing the ε_t with high-dimensional random vectors consisting of the lagged components of the ε_t. However, for our purposes, this reformulation offers no real advantage.

The assumption of non-negativity of the ε_t is clear in order to ensure that the conditional mean of Y_t is non-negative. The condition that $E(\varepsilon_t) = 1$ is imposed for identifiability reasons; otherwise, if $c = E(\varepsilon_t) \neq 1$, then c can be absorbed into the intercept term in the exponent of μ_t. (That is one would replace ε_t with ε_t/c and β_1 with $\beta_1 + \ln c$.)

In order to meet the non-negativity constraint on the ε_t, it is often convenient to model the logarithms of the ε_t. Letting $\delta_t = \ln \varepsilon_t$, then the conditional mean of Y_t can be written as

$$u_t = \exp\{\mathbf{x}_t^T \boldsymbol{\beta} + \delta_t\}.$$

Of course, in order for the corresponding ε_t to have mean 1, we must assume $Ee^{\delta_t} = 1$. Unless the $\{\delta_t\}$ is a stationary Gaussian process, there is not an explicit relationship between the ACVF's of $\{\varepsilon_t\}$ and $\{\delta_t\}$. In the case when $\{\varepsilon_t\}$ is a stationary log-normal process, i.e., $\{\delta_t\}$ is a stationary Gaussian process with ACVF $\gamma_\delta(\cdot)$, then there is a nice connection between the ACVF's. First, in order to satisfy the identifiability requirement that $E(\varepsilon_t) = E[\exp(\delta_t)] = 1$ it is required that $\delta_t \sim N(-\sigma_\delta^2/2, \sigma_\delta^2)$ $[\sigma_\delta^2 := \gamma_\delta(0)]$, so that the mean is -0.5 times the variance. Then, with this choice of mean and variance in the log-normal distribution, $\gamma_\varepsilon(h) = E(\exp\{\delta_{t+h} + \delta_t\} - 1) = e^{\gamma_\delta(h)} - 1$ for all h.

2.1 Means, Variances, and Autocovariances of Y_t

In this Section various key facts about the moments of the observed count process, Y_t, are derived and relationships between the first and second moments of the latent process, ε_t, are provided. While most of these results are available elsewhere (Zeger (1988) for example) it is useful to have these collected together for easy reference. Throughout, expectations, variances, and covariances are conditional upon the regressors x_t (and this will not be explicitly noted) but not on the latent process unless otherwise indicated in the usual way. Recall that $E(\varepsilon_t) = 1$.

Mean of Y_t:

$$\mu_t = E(Y_t) = E[E(Y_t|\varepsilon_t)] = \exp(x_t^T \beta).$$

Variance of Y_t:

$$\text{Var}(Y_t) = E[\text{Var}(Y_t|\varepsilon_t)] + \text{Var}[E(Y_t|\varepsilon_t)] = \mu_t + \sigma_\varepsilon^2 \mu_t^2.$$

Autocovariance function of Y_t:

$$\text{Cov}(Y_{t+h}, Y_t) = \mu_t \mu_{t+h}[E(\varepsilon_t \varepsilon_{t+h}) - 1] = \mu_t \mu_{t+h} \gamma_\varepsilon(h).$$

Autocorrelation function of Y_t:

$$\text{Cor}(Y_s, Y_t) = \frac{\mu_s \mu_t \gamma_\varepsilon(s-t)}{\sqrt{[\mu_s + \mu_s^2 \sigma_\varepsilon^2][\mu_t + \mu_t^2 \sigma_\varepsilon^2]}}$$

$$= \frac{\rho_\varepsilon(s-t)}{\sqrt{[1 + (\sigma_\varepsilon^2 \mu_s)^{-1}][1 + (\sigma_\varepsilon^2 \mu_t)^{-1}]}}$$

and this is not free of the regressors x_t, as is to be expected. In the case when there are no regression terms other than a constant, i.e., $r = 1$, the process $\{Y_t\}$ is then stationary with autocorrelation function (ACF) given by

$$\rho_Y(h) := \text{Cor}(Y_{t+h}, Y_t) = \frac{\gamma_\varepsilon(h)}{\mu^{-1} + \sigma_\varepsilon^2},$$

where $\mu = e^{\beta_1}$. Since $\mu > 0$ we see that

$$|\rho_Y(h)| \le |\rho_\varepsilon(h)|,$$

where $\rho_\varepsilon(h) := \text{Cor}(\varepsilon_{t+h}, \varepsilon_t)$. To illustrate this last observation, consider Fig. 3 in which the ACF for the $\{Y_t\}$ process is shown along with that for the latent process. Clearly, even when the mean of Y_t is stationary, the ACF of the observed count process tends to underestimate that of the latent process. Because of this, methods are required to estimate the underlying correlation and to test if it is zero or not. Such methods also need to be applicable when the regression terms are present.

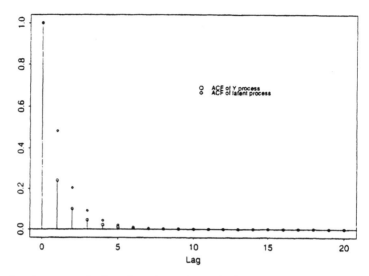

Figure 3 Autocorrelation function of the $\{Y_t\}$ and latent processes.

2.2 Preliminary Estimates and Diagnostics

In a linear model with time series errors, for example,

$$Y_t = \mathbf{x}_t^T \boldsymbol{\beta} + \varepsilon_t,$$

the first step in fitting such models, is to determine the autocovariance structure of the time series of errors $\{\varepsilon_t\}$. Assuming that $\{\varepsilon_t\}$ is a linear process such as an ARMA, the parameter $\boldsymbol{\beta}$ is estimated using ordinary least squares (OLS) by regressing the data vector $(Y_1, \ldots, Y_n)^T$ onto the $\mathbf{x}_t$. While this estimate ignores the dependence structure of the $\{\varepsilon_t\}$, the OLS estimate has the same asymptotic efficiency as the MLE of $\boldsymbol{\beta}$ under a wide class of models for the ε_t process (see for example Hannan (1970)). The asymptotic covariance matrix of the OLS (and MLE) estimate does depend on the covariance structure of the ε_t. Once a consistent estimator of $\boldsymbol{\beta}$ has been found, then the ACVF of the ε_t can be consistently estimated from the sample ACVF of the residuals defined by, $\hat{\varepsilon}_t = Y_t - \mathbf{x}_t^T \hat{\boldsymbol{\beta}}_{\text{OLS}}$. A model is then selected for the regression parameter $\boldsymbol{\beta}$ and the parameters of the model for the ε_t can be re-estimated using MLE.

In this Section we consider carrying out an analogous procedure applied to our parameter-driven model. The first step is to estimate the $\boldsymbol{\beta}$ vector using generalized linear models (GLM) or Poisson regression.

The GLM estimate $\hat{\boldsymbol{\beta}}$ of $\boldsymbol{\beta}$ is obtained by maximizing the function

$$l(\boldsymbol{\beta}) = -\sum_{t=1}^{n} \exp(\mathbf{x}_t^T \boldsymbol{\beta}) + \sum_{t=1}^{n} Y_t \mathbf{x}_t^T \beta.$$

Note that $l(\boldsymbol{\beta})$ is the log-likelihood (apart from a constant) corresponding to a Poisson regression model without a latent process. In order to derive the asymptotic behavior of $\hat{\boldsymbol{\beta}}$ we assume there exists a sequence of nonsingular matrices M_n such that the regressors obey the following conditions:

$$M_n \left[\sum_{t=1}^{n} \mathbf{x}_t \mathbf{x}_t^T \exp(\mathbf{x}_t^T \boldsymbol{\beta}) \right] M_n^T \rightarrow \Omega_l(\boldsymbol{\beta}), \tag{2.3}$$

$$M_n \left\{ \sum_{t=1}^{n} \mathbf{x}_t \mathbf{x}_{t+h}^T \exp[(\mathbf{x}_t^T + \mathbf{x}_{t+h}^T)\boldsymbol{\beta}] \right\} M_n^T \rightarrow \Omega_h(\boldsymbol{\beta}) \text{ (uniformly in } h), \tag{2.4}$$

$$\max_{1 \leq t \leq n} |M_n^T \mathbf{x}_t| \exp(\mathbf{x}_t^T \boldsymbol{\beta}) \rightarrow 0, \tag{2.5}$$

and for $h \leq 0$,

$$M_n \sum_{t=1}^{1-h} \mathbf{x}_t \mathbf{x}_{t+h}^T \exp[(\mathbf{x}_t^T + \mathbf{x}_{t+h}^T)\boldsymbol{\beta}] M_n^T \rightarrow 0, \tag{2.6}$$

and is uniformly bounded in h as $n \rightarrow \infty$. Similarly, for $h > 0$,

$$M_n \sum_{t=n-h}^{n} \mathbf{x}_t \mathbf{x}_{t+h}^T \exp[(\mathbf{x}_t^T + \mathbf{x}_{t+h}^T)\boldsymbol{\beta}] M_n^T \rightarrow 0 \tag{2.7}$$

and is uniformly bounded in h as $n \rightarrow \infty$. Following the theorem below, we show that these conditions are met for a wide range of regression variables. Before stating the theorem, define

$$\Omega_n := \text{Cov} \left[M_n \sum_{t=1}^{n} \mathbf{x}_t (Y_t - \mu_t) \right]$$

$$= M_n \sum_{t=1}^{n} \sum_{s=1}^{n} \mathbf{x}_t E[(Y_t - \mu_t)(Y_s - \mu_s)] \mathbf{x}_s^T M_n^T.$$

THEOREM 2.1   Let $\hat{\boldsymbol{\beta}}$ be the GLM estimate of $\boldsymbol{\beta}$ obtained by maximizing $l(\boldsymbol{\beta})$ (that is, assuming that the latent process $\{\varepsilon_t\}$ is not present in the model for the observed mean) for the parameter-driven model specified in Eqns (1.1)–(1.4). Further assume that the $\{\mathbf{x}_t\}$ satisfy Eqns (2.3)–(2.7), and $\sum_{h=0}^{\infty} |\gamma_\varepsilon(h)| < \infty$. Then,

$$\Omega_n \to \Omega_I(\boldsymbol{\beta}) + \Omega_{II}(\boldsymbol{\beta}) \tag{2.8}$$

and

$$\hat{\boldsymbol{\beta}} \to \boldsymbol{\beta}, \tag{2.9}$$

where

$$\Omega_{II}(\boldsymbol{\beta}) = \sum_{h=-\infty}^{\infty} \Omega_h(\boldsymbol{\beta})\gamma_\varepsilon(h).$$

Moreover, if

$$M_n \sum_{t=1}^{n} \mathbf{x}_t \, \exp(\mathbf{x}_t^T \boldsymbol{\beta})(\varepsilon_t - 1) \overset{d}{\to} N[\mathbf{0}, \Omega_{II}(\boldsymbol{\beta})], \tag{2.10}$$

then

$$M_n^{-1}(\hat{\boldsymbol{\beta}} - \boldsymbol{\beta}) \to N[\mathbf{0}, \Omega_I^{-1}(\boldsymbol{\beta}) + \Omega_I^{-1}(\boldsymbol{\beta})\Omega_{II}(\boldsymbol{\beta})\Omega_I^{-1}(\boldsymbol{\beta})]$$

as $n \to \infty$.

Remark 2.1.   The matrix, $M_n\Omega_I^{-1}(\boldsymbol{\beta})M_n^T$, is the asymptotic covariance matrix associated with the usual GLM estimate, while $M_n\Omega_I^{-1}(\boldsymbol{\beta})\Omega_{II}(\boldsymbol{\beta})\Omega_I^{-1}(\boldsymbol{\beta})M_n^T$ is the additional contribution to the asymptotic covariance due to the existence of the latent process. Clearly if the latent process has small covariance then this second term will not contribute very much.

Remark 2.2.   In the following subsection, conditions under which the various covariance matrices converge to their asymptotic limits are described and in some cases the form of the limit is given explicitly in terms of integrals.

Remark 2.3.   The convergence in Eq. (2.10) holds under a variety of conditions on the latent process. For example, if $\{\varepsilon_t\}$ is strongly mixing with a suitable rate or if $\{\delta_t = \ln \varepsilon_t\}$ is a linear process, then the central limit theorem in Eq. (2.10) holds.

Brannas and Johansson (1994) state that results of Judge (1985) can be used to establish the consistency and asymptotic normality of the GLM estimates even when autocorrelation is present. The form of the covariance matrix given above shows explicitly how autocorrelation inflates the true asymptotic covariance matrix.

2.2.1 Calculating the Asymptotic Covariance Matrix

Equations (2.3)–(2.7) hold for a variety of regression functions including:

1. Trend type terms in which the regression function depends on n and t and has the form

   $$\mathbf{x}_t = \mathbf{f}(t/n),$$

   for some vector-valued continuous function $\mathbf{f}(\cdot)$ defined on the unit interval $[0, 1]$. For example $\mathbf{f}$ could be a vector of polynomials. In particular, when $\mathbf{f}^T(x) = (1, x)$ this would specify a linear trend term in the regression. On the other hand, the choice of regression variables $\mathbf{x}_t^T = (1, t)$ would not give rise to a consistent estimate of $\boldsymbol{\beta}$ for $\beta_2 \leq 0$. Two attractive features about using regression variables of the form $\mathbf{x}_t = \mathbf{f}(t/n)$ are that the asymptotic covariance matrix can be computed in closed form and the normalizing matrix M_n is given by $n^{-1/2}I_r$, where I_r is the identity matrix. To see this, note that

   $$\Omega_n := \frac{1}{n}\sum_{t=1}^{n}\sum_{s=1}^{n}\mathbf{x}_t\mathbf{x}_s^T E[(Y_t - \mu_t)(Y_s - \mu_s)]$$

   $$= \frac{1}{n}\sum_{t=1}^{n}\mathbf{x}_t\mathbf{x}_t^T \mu_t + \frac{1}{n}\sum_{t=1}^{n}\sum_{s=1}^{n}\mathbf{x}_t\mathbf{x}_s^T \mu_t\mu_s\gamma_\varepsilon(s - t)$$

   $$= \frac{1}{n}\sum_{t=1}^{n}\mathbf{f}(t/n)\mathbf{f}^T(t/n)e^{\mathbf{f}^T(t/n)\boldsymbol{\beta}}$$

   $$+ \frac{1}{n}\sum_{t=1}^{n}\sum_{s=1}^{n}\mathbf{f}(t/n)\mathbf{f}^T(s/n)e^{(\mathbf{f}^T(t/n)+\mathbf{f}^T(s/n))\boldsymbol{\beta}}\gamma_\varepsilon(s - t)$$

   $$\to \Omega_I(\boldsymbol{\beta}) + \Omega_{II}(\boldsymbol{\beta})$$

   where

   $$\Omega_I(\boldsymbol{\beta}) = \int_0^1 \mathbf{f}(x)\mathbf{f}^T(x)e^{\mathbf{f}^T(x)\boldsymbol{\beta}}\,dx$$

   and

   $$\Omega_{II}(\boldsymbol{\beta}) = \left(\int_0^1 \mathbf{f}(x)\mathbf{f}^T(x)e^{2\mathbf{f}^T(x)\boldsymbol{\beta}}\,dx\right)\sum_{h=-\infty}^{\infty}\gamma_\varepsilon(h).$$

2. Harmonic terms to specify annual effects or day of the week effects. For example $x_t = \cos(2\pi t/12)$. The convergence of Ω_n can be established using elementary properties of trigonometric functions.
3. Stationary processes as might arise from a seasonally adjusted temperature series. (See Campbell (1994) for an example of this.) Ergodic properties of the process can be used to establish convergence of the covariance matrix Ω_n.

2.2.2 Application to the Polio Data

We now apply the results of Theorem 2.1 to the polio data example from Zeger (1988). Here we use the same regression variables as in Zeger (1988) consisting of an intercept term, a linear trend, and harmonics at periods of 6 and 12 months. Specifically,

$$\mathbf{x}_t = (1, t'/1000, \cos(2\pi t'/12), \sin(2\pi t'/12), \cos(2\pi t'/6), \sin(2\pi t'/6))^T,$$

where $t' = (t - 73)$ to locate the intercept term at January, 1976 as in Zeger's analysis.

	Zeger		GLM fit		Asym	Simulations	
	$\hat{\beta}_Z$	s.e. $(\hat{\beta}_Z)$	$\hat{\beta}_{GLM}$	s.e $(\hat{\beta}_{GLM})$	s.e. $(\hat{\beta}_{GLM})$	mean $\hat{\beta}_{GLM}$	s.d. $\hat{\beta}_{GLM}$
Intercept	0.17	0.13	0.207	0.075	0.205	0.150	0.213
Trend $\times 10^{-3}$	−4.35	2.68	−4.799	1.399	4.115	−4.887	3.937
$\cos(2\pi t/12)$	−0.11	0.16	−0.149	0.097	0.157	−0.145	0.144
$\sin(2\pi t/12)$	−0.48	0.17	−0.532	0.109	0.168	−0.531	0.168
$\cos(4\pi t/12)$	0.20	0.14	0.169	0.098	0.122	0.167	0.123
$\sin(4\pi t/12)$	−0.41	0.14	−0.432	0.101	0.125	−0.440	0.125

The asymptotic standard errors for the GLM estimates given in the above theorem are estimated using the values of $\hat{\sigma}_\varepsilon^2 = 0.77$ and $\hat{\rho}_\varepsilon(h) = (0.77)^h$ reported in Zeger (1988, Table 3). These were obtained from the formulae

$$\Omega_t = \frac{1}{n} \sum_{t=1}^{n} \mathbf{x}_t \mathbf{x}_t^T \exp(\mathbf{x}_t^T \boldsymbol{\beta})$$

and

$$\Omega_{II} = \frac{1}{n} \sum_{t=1}^{n} \sum_{s=1}^{n} \mathbf{x}_t \mathbf{x}_s^T \exp[(\mathbf{x}_t^T + \mathbf{x}_s^T)\boldsymbol{\beta}] \gamma_\varepsilon (s - t).$$

Use of the correct standard errors for the trend term would lead to the conclusion that the trend is not significant whereas use of the standard errors produced by the GLM analysis would lead to declaring the trend to be significant.

The final two columns of the above table report the results of 1000 simulations of time series of length $n = 168$ using the GLM fitted values, $\hat{\boldsymbol{\beta}}_{GLM}$, as the true values. The latent process was assumed to be a log-normal autoregression of order 1 with $\phi = 0.82$ and $\sigma_\delta^2 = 0.57$. The mean of this autoregression was chosen as $-\sigma_\delta^2/2 = -0.285$ in order to satisfy the conditions of the above theorem. The average over the 1000 simulated values of $\hat{\beta}_1$ is observed as 0.150 which is significantly biased downwards from the true value of 0.207 used in the simulations. The other parameters appear to be estimated without substantial bias. The standard deviation of the GLM estimates observed over the 1000 replications are reported in the last column of the above table. These are in good agreement with the standard errors obtained from the asymptotic theory (column 6).

2.3   Testing for the Existence of a Latent Process

Prior to the estimation of autocovariances it is reasonable to test for the existence of a latent process. Brannas and Johansson (1994) review the following statistic

$$S = \frac{\sum_{t=1}^{n}[(Y_t - \hat{\mu}_t)^2 - Y_t]}{[2 \sum_{t=1}^{n} \hat{\mu}_t^2]^{1/2}}$$

derived by several authors and based on a local alternative hypothesis or the Lagrange multiplier test of the Poisson distribution against a negative binomial or more general Katz distribution. A variant, introduced by Dean and Lawless (1989) "to improve on the small sample performance of the test", also considered by them is

$$S_a = \frac{\sum_{t=1}^{n}[(Y_t - \hat{\mu}_t)^2 - Y_t + \hat{h}_t \hat{\mu}_t]}{[2 \sum_{t=1}^{n} \hat{\mu}_t^2]^{1/2}},$$

where h_t is the tth diagonal element of the "hat" matrix. The "hat" matrix for generalized linear models extends that for linear regression and is defined in Fahrmeir and Tutz (1994, p.127), for example, as

$$H = \Lambda^{1/2} X (X^T \Lambda X)^{-1} X^T \Lambda^{1/2},$$

where

$$\Lambda = \mathrm{diag}(\mu_1, \cdots, \mu_n)$$

and

$$X = (\mathbf{x}_1, \mathbf{x}_2, \ldots, \mathbf{x}_n)^T$$

is the design matrix. Note that use of the $\hat{h}_t$ adjusts for the first part, $\hat{\Omega}_I = (X^T \Lambda X)$, of the asymptotic covariance matrix in Theorem 2.1. Under the hypothesis of no latent process this is appropriate. Later, when we examine the bias in autocovariance estimates when there is a latent process, additional adjustments are required to account for the complete asymptotic variance of $\hat{\beta}$.

Both of these statistics are asymptotically distributed as a $N(0, 1)$ variate under the null hypothesis of no latent process. They are used in a one-sided test. Monte Carlo work reviewed in Brannas and Johansson and corroborated by us, suggests that S_a has better size properties in small samples. For this reason, we will consider S_a from here on.

We also introduce an alternative test specifically designed for overdispersion due to the existence of a latent process in a Poisson observation process. Since this test uses higher moment properties of Poisson observations it was considered as a possibly more powerful test statistic than S_a. Also it might be more appropriate for the lagged regression models introduced in Sec. 3 since for these the distributional theory of S_a is not clearly appropriate. Under the null hypothesis that there is no latent process (i.e. $\varepsilon_t \equiv 1$) the Pearson residuals

$$e_t = \frac{Y_t - \hat{\mu}_t}{\sqrt{\hat{\mu}_t}}$$

have approximately zero mean and unit variance. Hence the statistic

$$Q = \left[\frac{1}{n} \sum_{t=1}^{n} e_t^2 - 1 \right] \Big/ \hat{\sigma}_Q,$$

where

$$\hat{\sigma}_Q^2 = \frac{1}{n} \left(\frac{1}{n} \sum_{t=1}^{n} \frac{1}{\hat{\mu}_t} + 2 \right),$$

could be used to test for a latent process. The expression for $\hat{\sigma}_Q^2$ is easily derived using the fact that a Poisson random variable Y_t with mean μ_t has fourth central moment $E(Y_t - \mu_t)^4 = \mu_t + 3\mu_t^2$. Under the hypothesis that the variance of the latent process is zero

$$Q \sim N(0, 1)$$

approximately.

Using 1000 replications of a time series of length $n = 168$ obtained from simulating independent Poisson variates (i.e., with no latent process present) with mean $\mu_t = \hat{\mu}_t$, the GLM fit to the polio data considered in Zeger (1988), gave simulated type I errors for Q which are severely lower than the nominal values. The observed mean and standard deviation of Q are -0.23 and 0.788, respectively, explaining the low coverage type I errors rates observed. To adjust for the negative bias alternate estimates of residuals which adjust for the effect of fitted values $\hat{\mu}_t$ could be used. For example one could use the divisor $n - p$ instead of n in the numerator of Q. The resulting statistic had appreciably better performance than Q but was still on the conservative side. A second reasonably simple approach would be to use standardized Pearson residuals. Standardized Pearson residuals are

$$\tilde{e}_t = e_t / (1 - h_t)^{0.5},$$

where h_t is the tth value of the "hat" matrix. The sum of these h_t values is r (the dimension of β). Using these standardized residuals we define

$$\tilde{Q} = \left(\frac{1}{n} \sum_{t=1}^{n} \tilde{e}_t^2 - 1 \right) \Big/ \hat{\sigma}_Q.$$

With this definition we get based on the same 1000 replications as used for Q above, mean and standard deviation of $\tilde{Q}$ as 0.011 and 0.826, respectively, and clearly improved, although still low, type I errors as

α	0.100	0.050	0.025	0.010
Empirical $P(\tilde{Q} > z_{1-\alpha})$	0.073	0.037	0.022	0.004

with corresponding significance points

α	0.100	0.050	0.025	0.010
$z_{1-\alpha}^*$ s.t. $P(\tilde{Q} > z_{1-\alpha}^*)$	1.10	1.48	1.89	2.21

On this basis, $\tilde{Q}$ is preferred to the unadjusted version Q.

We next compare the size and size adjusted power of the two statistics $\tilde{Q}$ and S_a. First the size properties are as follows for a simulation of 1000 replicates assuming no latent process is present:

Case 1: Linear Regression $1 + 1t/100$ for $t = 1, \ldots, 100$.

α	0.100	0.050	0.025	0.010
Empirical $P(\tilde{Q} > z_{1-\alpha})$	0.089	0.048	0.026	0.003
Empirical $P(S_a > z_{1-\alpha})$	0.085	0.045	0.027	0.011

Case 2: Cosine regression $1 + \cos(2\pi t/12)$ for $t = 1, \ldots, 100$.

α	0.100	0.050	0.025	0.010
Empirical $P(\tilde{Q} > z_{1-\alpha})$	0.077	0.038	0.021	0.013
Empirical $P(S_a > z_{1-\alpha})$	0.099	0.056	0.025	0.009

In addition the power of the test to detect departures from the null hypothesis were investigated using 1000 replicates with the same regression models. The latent process was generated using a log-normal distribution. The latent process had variance $\sigma_\varepsilon^2 = 0.05$ chosen to give a small deviation from the null hypothesis. The autocovariance was simulated using an autoregressive process with $\phi = 0$ and $\phi = 0.9$. The results, again based on 1000 replications are for a size 0.05 test using empirical significance points obtained under the null hypothesis.

	Linear regression $1 + 1t/100$		Cosine regression $1 + 1\cos(2\pi t/12)$	
	$\phi = 0$	$\phi = 0.9$	$\phi = 0$	$\phi = 0.9$
Power of $\tilde{Q}$	0.460	0.306	0.290	0.230
Power of S_a	0.491	0.316	0.410	0.342

On the basis of this limited simulation it would appear as though the S_a statistic has better type I error rates (i.e. closer to normal distribution) and larger power especially for a regressor which is seasonally varying. Note also that if the latent process has positive autocorrelation ($\phi = 0.9$) the power of S_a and $\tilde{Q}$ are reduced relative to the case of a white noise latent process ($\phi = 0$). Further research is required to demonstrate that S_a is preferred in all plausible circumstances.

2.4 Estimating the Variance of the Latent Process

We now assume that the test proposed in the last section rejects the hypothesis of no latent process. Zeger proposed the estimate

$$\hat{\sigma}_{\varepsilon.Z}^2 = \frac{\sum_{t=1}^{n}[(Y_t - \hat{\mu}_t)^2 - \hat{\mu}_t]}{\sum_{t=1}^{n} \hat{\mu}_t^2}$$

of σ_{ε}^2 which is approximately unbiased for σ_{ε}^2 and would be exactly unbiased if the $\hat{\mu}_t$ were replaced by the true values μ_t. Brannas and Johansson (1994) suggested the estimate

$$\hat{\sigma}_{\varepsilon.BJ}^2 = \frac{\sum_{t=1}^{n} \hat{\mu}_t^2[(Y_t - \hat{\mu}_t)^2 - \hat{\mu}_t]}{\sum_{t=1}^{n} \hat{\mu}_t^4}$$

which is derived by using ordinary least squares (OLS) regression of $(Y_t - \hat{\mu}_t)^2 - \hat{\mu}_t$ on its approximate expected value of $\sigma_{\varepsilon}^2 \hat{\mu}_t^2$. (Recall $\text{Var}(Y_t) = \mu_t + \sigma_{\varepsilon}^2 \mu_t^2$.)

The above estimates are not necessarily optimal in any sense other than being approximately unbiased. More generally, consider weighted estimates of the form:

$$\hat{\sigma}_{\varepsilon.W}^2 = \left(\sum_{t=1}^{n} W_t^2\right)^{-1} \sum_{t=1}^{n} W_t^2 E_t,$$

where

$$E_t = \hat{\mu}_t^{-2}[(Y_t - \hat{\mu}_t)^2 - \hat{\mu}_t].$$

Note that because $E(\hat{\sigma}_{\varepsilon.W}^2) \approx \sigma_{\varepsilon}^2$ the weighted estimate is approximately unbiased and would be exactly so if $\hat{\mu}_t$ was replaced by μ_t. Zeger's estimate corresponds to choosing weights $W_t^2 = \hat{\mu}_t^2$ and Brannas and Johansson's to choosing $W_t^2 = \hat{\mu}_t^4$. Note also that these weighted estimates are not guaranteed to be positive. However, it is unlikely that a negative estimate will be produced from these methods if the test Sec. 2.3 is applied to determine if the latent process is present.

The optimal weights for minimizing the variance of $\hat{\sigma}_{\varepsilon.W}^2$ are given by

$$W_t^{*2} = 1/\text{Var}(E_t).$$

Calculation of the variances required for this are complicated since they depend on moments up to order 4 for the latent process when it is a sequence of independent random variables. For latent processes with autocorrelation the calculation is further complicated. In addition these higher moments and autocovariances must be estimated.

Under the assumption that the latent process is a sequence of independent and identically distributed random variables, the optimal weights are

$$W_t^{*2} = 1/\text{Var}(E_t) \approx \mu_t^4/B_t,$$

where

$$B_t = E[(Y_t - \mu_t)^4] - (\mu_t + \sigma_\varepsilon^2 \mu_t^2)^2.$$

Calculation of these optimal weights would require an iterative approach starting with an initial estimate of σ_ε^2.

Note that the variance of the optimally weighted estimator is

$$\text{Var}(\hat{\sigma}_{\varepsilon,W^*}^2) \approx \frac{1}{(\sum_{t=1}^{n} \mu_t^4/B_t)}$$

and that of Zeger's estimator is

$$\text{Var}(\hat{\sigma}_{\varepsilon,Z}^2) \approx \frac{\sum_{t=1}^{n} B_t}{(\sum_{t=1}^{n} \mu_t^2)^2}.$$

For the polio data, using the GLM fit to obtain $\hat{\mu}_t$ and using the value of $\sigma_\delta^2 = 0.57$ the values of these variances are approximately $\text{Var}(\hat{\sigma}_{\varepsilon,W}^2) \approx 0.46^2$ and $\text{Var}(\hat{\sigma}_{\varepsilon,Z}^2) \approx 0.53^2$ indicating a modest improvement in estimation of variance using the optimal weighting based on the assumption (incorrect in this case) that the latent process is independent.

In the non-independent case, the calculation of optimal weights will be complicated by their dependence on unknown covariances. From now on we will use the "optimal" weights derived above under the assumption of independence and evaluate their performance in the correlated case using simulations. To implement the above optimal weighting scheme an initial estimate of the variance is required. One possibility is to use the weights based on the assumption that the latent process has zero variance and then use the resulting estimate to obtain the optimal weights.

2.5 Methods for Estimating the Autocovariances of the Latent Process

Zeger (1988) suggested method of moment estimators for the autocorrelations of the latent process $\{\varepsilon_t\}$. These are as follows:

$$\hat{\rho}_{\varepsilon,Z}(\tau) = \hat{\gamma}_{\varepsilon,Z}(\tau)/\hat{\sigma}_{\varepsilon,Z}^2$$

with autocovariance estimates given by

$$\hat{\gamma}_{\varepsilon,Z}(\tau) = \sum_{t=\tau+1}^{n} \{(Y_t - \hat{\mu}_t)(Y_{t-\tau} - \hat{\mu}_{t-\tau})\} \Big/ \sum_{t=\tau+1}^{n} \hat{\mu}_t \hat{\mu}_{t-\tau}.$$

The estimates of variance and autocovariances are not guaranteed to form a non-negative definite sequence and therefore the autocorrelations are not guaranteed to be less than one in absolute value.

Brannas and Johansson (1994) suggest OLS type estimators derived by letting $\tilde{Y}_t = (Y_t - \hat{\mu}_t)$ and noting that from the autocovariance function of Y_t in Sec. 2.1,

$$E(\tilde{Y}_t \tilde{Y}_{t+\tau}) \approx \rho_\varepsilon(\tau) \sigma_\varepsilon^2 \mu_t \mu_{t+\tau},$$

so that regressing $\tilde{Y}_t \tilde{Y}_{t+\tau}$ on $\hat{\mu}_t \hat{\mu}_{t+\tau}$ leads to estimates

$$\hat{\gamma}_{\varepsilon,\mathrm{BJ}}(\tau) = \sum_{t=\tau+1}^{n} \hat{\mu}_t \hat{\mu}_{t+\tau} \{(Y_t - \hat{\mu}_t)(Y_{t-\tau} - \hat{\mu}_{t-\tau})\} \Big/ \sum_{t=\tau+1}^{n} \hat{\mu}_t^2 \hat{\mu}_{t-\tau}^2.$$

In both the Zeger definition and the least squares definition of Brannas and Johansson there is no attempt to standardize the individual terms in the summations for unequal variances. By analogy with the use of weights in forming estimates of the variance of the latent process some form of weighting could be useful in forming autocovariance estimates. We consider weighted estimates of autocovariances which are required to be unbiased for any underlying stationary latent process. Calculation of the variance of any such estimates will require estimation and knowledge of the autocovariances. However, in the case where the latent process is a sequence of independent random variables the variance of these weighted estimates is readily computable as we will show. Since the hypothesis of independence is of primary interest, obtaining minimum variance estimates is desirable.

Consider the following weighted estimates:

$$\hat{\gamma}_{\varepsilon,W}(h) = \frac{1}{\sum_{t=1}^{n-h} W_t W_{t+h}} \sum_{t=1}^{n-h} W_t W_{t+h} \hat{V}_t \hat{V}_{t+h},$$

where $\hat{V}_t = e_t / \sqrt{\hat{\mu}_t}$. These estimates are approximately unbiased for the true covariance since the individual terms satisfy

$$E(\hat{V}_t \hat{V}_{t+h}) \approx (\mu_t \mu_{t+h})^{-1} \mathrm{Cov}(Y_t, Y_{t+h})$$

$$= \gamma_\varepsilon(h).$$

Because of this unbiasedness, use of V_t as the basis for constructing weighted estimates seems reasonable. The approximate variance of these estimates under the assumption that the latent process is white noise is:

$$\mathrm{Var}(\hat{\gamma}_{\varepsilon,W}(h)) \approx \frac{1}{\left(\sum_{t=1}^{n-h} W_t W_{t+h}\right)^2} \sum_{t=1}^{n-h} W_t^2 W_{t+h}^2 (\sigma_\varepsilon^2 + 1/\mu_t)(\sigma_\varepsilon^2 + 1/\mu_{t+h}).$$

Alternatively the following weights can be derived directly from the finite sample approximation as

$$W_t^* = (\hat{\sigma}_\varepsilon^2 + 1/\hat{\mu}_t)^{-1}.$$

Zeger's estimates use weights $W_t^Z = \hat{\mu}_t$ which leads to an approximate variance

$$\text{Var}(\hat{\gamma}_{\varepsilon,W^Z}(h)) \approx \frac{1}{(\sum_{t=1}^{n-h} \hat{\mu}_t\hat{\mu}_{t+h})^2} \sum_{t=1}^{n-h} \hat{\mu}_t^2 \hat{\mu}_{t+h}^2 (\hat{\sigma}_\varepsilon^2 + 1/\hat{\mu}_t)(\hat{\sigma}_\varepsilon^2 + 1/\hat{\mu}_{t+h}).$$

A similar expression can be derived for the Brannas-Johansson estimates in which the weights are $W_t^{BJ} = \hat{\mu}_t^2$.

Whatever the weighting scheme used the standard errors of the estimated covariances can be estimated, under the assumption that the latent process is an independent sequence, by

$$s.e.(\hat{\gamma}_{\varepsilon,W}(h)) \approx \left[\frac{1}{(\sum_{t=1}^{n-h} \hat{W}_t\hat{W}_{t+h})^2} \sum_{t=1}^{n-h} \hat{W}_t^2 \hat{W}_{t+h}^2 (\hat{\sigma}_\varepsilon^2 + 1/\hat{\mu}_t)(\hat{\sigma}_\varepsilon^2 + 1/\hat{\mu}_{t+h})\right]^{1/2}.$$

Integral approximations reported in Davis, Dunsmuir, and Wang (1998) and simulations (not reported in detail here) lead to the following observations.

Remark 2.4. The optimal estimates outperform the Zeger estimates which in turn outperform the BJ estimates. The performance advantages decrease as the degree of autocorrelation increases positively.

Remark 2.5. When there is substantial autocorrelation present there is also substantial bias in the estimates of the $\gamma_\varepsilon(h)$—this occurs in all estimates to a similar degree. This would impact the estimation of the correct asymptotic variance in the GLM estimates of β of the above theorem and would tend to lead to underestimating the correct standard errors of the GLM estimates when there is substantial positive serial correlation present.

Remark 2.6. However for the estimation of $\hat{\rho}_\varepsilon(1)$ the bias is not as severe. For some purposes such as construction of a suitable correlation model for use in an efficient estimation procedure this reduction in bias is a good property. However (see Sec. 2.7) all biases depend on the form of the regressor with worse bias for the linear trend regression than the cosine regression. This indicates that any bias adjustment procedure should account for the form of the regression function—see Section 2.7 for some proposals.

Remark 2.7.   The asymptotic formulae for $s.e.(\hat{\gamma}_{\varepsilon,W}(h))$ give unbiased estimates of the simulated standard deviations when there is no serial dependence, i.e., in the situation that they are derived under. This means that this formula will be useful in calculating an overall test of autocovariance (see Sec. 2.6 below). Further the asymptotic formulae provide reasonable estimates of standard deviations even in the cases where the null hypothesis is not true.

2.6   Tests for Zero Autocorrelation in the Latent Process

In Brannas and Johansson (1994) the performance of the standard Box–Pierce and Ljung–Box portmanteau statistics is investigated. Three types of residuals (Pearson, Anscombe and their own proposal) are used to define the autocorrelations in the standard fashion. However there will be some inaccuracies in any such procedures unless the standard errors are properly calculated. The problem with the correlated Poisson model is that the variance and covariances have different forms of dependence on the mean function μ_t and there is no single normalization of residuals which will simultaneously eliminate this dependence from the variance and from the covariance terms required to construct autocorrelations. Hence the usual normalization in the Box–Pierce and Ljung–Box portmanteau statistics will be incorrect.

Brannas and Johansson (1994) observe that in the portmanteau statistics constructed in this fashion "the sizes are significantly too high". One explanation of their results might be that the Box–Pierce and Ljung–Box statistics need to be adjusted to account for the correct standard errors. To illustrate this consider the use of Pearson residuals in the standard formulas for autocorrelation estimates which are, because the sample mean of the Pearson residuals e_t is near zero, equal to

$$\hat{\rho}_{\varepsilon.P}(h) = \frac{\sum_{t=1}^{n-h} e_t e_{t+h}}{\sum_{t=1}^{n} e_t^2}.$$

Using the types of formulas that are given above it can be shown that, under the assumption that the latent process is an IID sequence with mean 1 and variance σ_ε^2,

$$\mathrm{Var}(\hat{\rho}_{\varepsilon.P}(h)) \approx \left[\frac{1}{n}\sum_{t=1}^{n}(1 + \mu_t\sigma_\varepsilon^2)\right]^{-2}$$

$$\times \left[\frac{1}{n^2}\sum_{t=1}^{n-h}(1 + \mu_t\sigma_\varepsilon^2)(1 + \mu_{t+h}\sigma_\varepsilon^2)\right] =: V_h/n.$$

These can be readily estimated using the fitted mean and estimated variance. Also, for understanding the theoretical aspects of these estimated autocorrelations the true values of the mean function μ_t and variance σ_ε^2 can be used. For the cosine regression introduced above the value of $V_h/n = 1.19/n$ for $h = 1$ compared with the Ljung–Box value of $(n - h)/[n(n + 2)] = 0.97/n$ and $V_h/n = 0.72/n$ for $h = 6$ compared with the Ljung–Box value of $(n - h)/[n(n + 2)] = 0.92/n$. While these individual differences are substantial their effect will not be so large in a test statistic constructed using a number of autocorrelation estimates. In particular for this cosine regression case the mean of the V_h/n over the first 13 lags is $0.94/n$ while that of $(n - h)/[n(n + 2)]$ is $0.91/n$ indicating very little average difference in the correct variances for the autocorrelations based on Pearson residuals and the incorrect variances implicit in the Ljung–Box statistic. Indeed, in this case, the use of Pearson residuals in standard estimates of autocorrelations leads to a Ljung–Box statistic with good nominal coverage. In other situations the average of the first several values of $n\text{Var}[\hat{\rho}_{\varepsilon,P}(h)]$ may not be so close to unity and there will be a need to properly calculate the asymptotic variances. However, based on the results presented above this is not difficult to do computationally, and, because it is always the correct procedure, we would recommend its universal use in these models.

Although these effects are small there will be circumstances where they could be large enough to badly distort the inference about serial autocorrelations using the classical (unadjusted) procedures from time series analysis of stationary sequences. To be accurate under a wide range of regression functions the classical statistical procedures for testing for the existence of autocorrelation should really be modified in the present situation. Under the assumption that the latent process is white noise with positive variance the $\hat{\gamma}_{\varepsilon,W}(h)$ are asymptotically distributed as independent normal random variables with standard deviations estimated by $s.e.[\hat{\gamma}_{\varepsilon,W}(h)]$ derived above. Hence a test statistic for zero autocovariance can be constructed based on the first several autocovariance estimates. The following statistic (analogous to the Box–Jenkins's portmanteau statistic) is proposed for testing for serial dependence in the mean of the observed count time series. For a given maximum lag length L define

$$H^2 = \sum_{h=1}^{L}[\hat{\gamma}_\varepsilon(h)/s.e.(\hat{\gamma}_\varepsilon(h))]^2.$$

Under the hypothesis of independence H^2 will have an approximate χ^2 distribution on L df. The following simulations indicate that this is a reasonable approximation.

Simulation results comparing the performance of H^2 using the estimates of Zeger, Brannas and Johansson and the optimally weighted estimates (i.e., those using W_t^*) are presented below. In the following cases a white noise latent process is assumed to be present. The simulation results are based on 1000 replications of time series with sample size 100.

Case 1: Type I Errors. Linear trend, latent process is IID log-normal with $\text{Var}(\varepsilon_t) = 1$.

$\alpha = 0.05$	Observed type I error	Observed 95%-ile	Power against		
			$\phi = 0.2$	$\phi = 0.4$	$\phi = 0.6$
$P(H^2_{\text{OPT}} > \chi^2_{10.(1-\alpha)})$	0.033	16.66	0.122	0.379	0.731
$P(H^2_{Z} > \chi^2_{10.(1-\alpha)})$	0.027	16.31	0.119	0.317	0.664
$P(H^2_{\text{BJ}} > \chi^2_{10.(1-\alpha)})$	0.030	14.24	0.078	0.221	0.495
Ljung–Box using e_t	0.044	17.88	0.126	0.370	0.721

Case 2: Type I Errors. Cosine trend, latent process is IID log-normal with $\text{Var}(\varepsilon_t) = 1$.

$\alpha = 0.05$	Observed type I error	Observed 95%-ile	Power against		
			$\phi = 0.2$	$\phi = 0.4$	$\phi = 0.6$
$P(H^2_{\text{OPT}} > \chi^2_{10.(1-\alpha)})$	0.046	17.91	0.069	0.267	0.648
$P(H^2_{Z} > \chi^2_{10.(1-\alpha)})$	0.063	19.51	0.058	0.117	0.484
$P(H^2_{\text{BJ}} > \chi^2_{10.(1-\alpha)})$	0.139	25.85	0.052	0.094	0.209
Ljung–Box using e_t	0.042	17.20	0.090	0.296	0.699

The "optimal" estimates used in the test statistic provide approximately similar performance to the Ljung-Box statistic. Both of these are superior to the use of Zeger's estimates which in turn is superior to the BJ estimates. Since H^2_{OPT} is based on the correct variances of autocovariances while the Ljung–Box statistic based on Pearson residuals does not we would recommend the use of H^2_{OPT} universally.

2.7 Bias Adjustment for Estimation of Autocovariances

Note that the use of Pearson residuals for estimation of autocorrelations in the non-null case (i.e., where there is serial correlation) is biased in general because

$$E\left(\frac{1}{n}\sum_{t=1}^{n-h}e_t e_{t+h}\right) \approx \left[\frac{1}{n}\sum_{t=1}^{n-h}\mu_t^{1/2}\mu_{t+h}^{1/2}\right]\gamma_\varepsilon(h)$$

while

$$E\left(\frac{1}{n}\sum_{t=1}^{n}e_t^2\right) \approx \frac{1}{n}\sum_{t=1}^{n}\mu_t^{-1}[\mu_t + \sigma_\varepsilon^2\mu_t^2]$$

$$= 1 + \left(\frac{1}{n}\sum_{t=1}^{n}\mu_t\right)\sigma_\varepsilon^2.$$

In the case when $\mu_t \equiv \mu$, we have

$$E\left(\frac{1}{n}\sum_{t=1}^{n-h}e_t e_{t+h}\right) \approx \mu\gamma_\varepsilon(h) \qquad \text{and} \qquad E\left(\frac{1}{n}\sum_{t=1}^{n}e_t^2\right) \approx 1 + \mu\sigma_\varepsilon^2$$

giving

$$\hat{\rho}_{\varepsilon,P}(h) \approx \frac{\mu\sigma_\varepsilon^2}{(1 + \mu\sigma_\varepsilon^2)}\rho_\varepsilon(h),$$

which is as observed for the autocorrelation of the Y_t process in Sec. 2.1. For the cosine regression model used in Sec. 2.3, the bias factors for autocorrelation estimation are 0.75 at $h = 1$ and 0.57 at $h = 6$.

While this theory suggest we can expect bias in the standard ACF estimates using Pearson residuals, the theoretically asymptotically unbiased estimates $\hat{\gamma}_Z(h)$ and $\hat{\gamma}_{OPT}(h)$ are in fact also substantially biased towards zero when the true autocovariances were positive as arises for example in the autoregression with positive parameter. For the purposes of good estimates of asymptotic covariance matrix for the GLM estimates of the regression parameters removal of bias from the estimates of variance and autocovariance is warranted.

We now consider bias-correction strategies for the estimators of autocovariance. Standard approaches to adjusting for bias include use of $n - p$ or use of the 'hats' h_t to adjust for the use of fitted means in place of true means when calculating residuals. We attempted this, but similarly to their use in the Pearson residual based ACF they had little effect on the bias. Also different types of residuals, as suggested by Brannas and Johansson could be used. But we doubt this will overcome the problem.

The Zeger and optimally weighted estimates would be unbiased if the true mean μ_t was used in their definition. However as we will demonstrate the effect of variability in the estimates of β can have a substantial impact on bias. The traditional adjustment procedures do not allow for the impact of substantial positive autocorrelation in the latent process on the standard errors of regression estimates, which in turn can effect bias because the estimates mean for these Poisson models is an exponential function of a linear function in the estimated β. Accordingly it is the *variance* of $\hat{\beta}$ that is important to the *bias* in $\hat{\mu}$. Fortunately, simple expansions to adjust for use of $\hat{\mu}_t$ in place of μ_t in the above definitions can be derived in a relatively straightforward way using the asymptotic distribution for β derived in Sec. 2.2.

Bias adjusted Zeger and the optimally weighted estimates, proposed in Davis, Dunsmuir, and Wang (1998), are defined as follows:

$$
\hat{\sigma}^2_{Z.UB} = \frac{\sum_{t=1}^{n}\{(Y_t - \hat{\mu}_t)^2 + \hat{\mu}_t^2 \exp(-2\mathbf{x}_t^T \hat{G}_n \mathbf{x}_t) \\ \times [\exp(2\mathbf{x}_t^T \hat{G}_n \mathbf{x}_t) - 2\exp(\mathbf{x}_t^T \hat{G}_n \mathbf{x}_t/2) + 1]) - \hat{\mu}_t\}}{\sum_{t=1}^{n} \hat{\mu}_t^2 \exp(-2\mathbf{x}_t^T \hat{G}_n \mathbf{x}_t)},
$$

$$
\hat{\sigma}^2_{OPT.UB} = \frac{\sum_{t=1}^{n}\left(\frac{W_t^*}{\hat{\mu}_t}\right)^2 \{(Y_t - \hat{\mu}_t)^2 + \hat{\mu}_t^2 \exp(-2\mathbf{x}_t^T \hat{G}_n \mathbf{x}_t) \\ \times [\exp(2\mathbf{x}_t^T \hat{G}_n \mathbf{x}_t) - 2\exp(\mathbf{x}_t^T \hat{G}_n \mathbf{x}_t/2) + 1] - \hat{\mu}_t\}}{\sum_{t=1}^{n} W_t^{*2} \exp(-2\mathbf{x}_t^T \hat{G}_n \mathbf{x}_t)},
$$

in which the estimate of the asymptotic covariance matrix is given by

$$
\hat{G}_n = \hat{\Omega}_{I.n}^{-1} + \hat{\Omega}_{I.n}^{-1}\hat{\Omega}_{II.n}\hat{\Omega}_{I.n}^{-1},
$$

where

$$
\hat{\Omega}_{I.n} = \sum_{t=1}^{n} \mathbf{x}_t \mathbf{x}_t^T \hat{\mu}_t,
$$

$$
\hat{\Omega}_{II.n} = \sum_{h=-L}^{L} \sum_{t=\max(1-h.1)}^{\min(n-h.n)} \mathbf{x}_t \mathbf{x}_{t+h}^T \hat{\mu}_t \hat{\mu}_{t+h} \hat{\gamma}_{\varepsilon.Z}(h) \tag{2.11}
$$

for some maximum lag L specified somewhat arbitrarily, and

$$
W_t^{*2} = 1/\mathrm{Var}(E_t) = \mu_t^4/B_t.
$$

If ε_t is log-normal,

$$B_t = \mu_t + \mu_t^2[7\exp(\sigma_\delta^2) - 5] + \mu_t^3[6\exp(3\sigma_\delta^2) - 14\exp(\sigma_\delta^2) + 8]$$
$$+ \mu_t^4[\exp(6\sigma_\delta^2) - 4\exp(3\sigma_\delta^2) - \exp(2\sigma_\delta^2) + 8\exp(\sigma_\delta^2) - 4].$$

The bias adjusted estimates for autocovariances are given by

$$\hat{\gamma}_{Z.UB}(h) = \left(\sum_{t=1}^{n-h} \left[(Y_t - \hat{\mu}_t)(Y_{t+h} - \hat{\mu}_{t+h}) + \hat{\mu}_t\hat{\mu}_{t+h}e^{-(\mathbf{x}_t+\mathbf{x}_{t+h})^T\hat{G}_n(\mathbf{x}_t+\mathbf{x}_{t+h})/2} \right.\right.$$

$$\left.\left. \times (e^{(\mathbf{x}_t+\mathbf{x}_{t+h})^T\hat{G}_n(\mathbf{x}_t+\mathbf{x}_{t+h})/2} - e^{\mathbf{x}_t^T\hat{G}_n\mathbf{x}_t/2} - e^{\mathbf{x}_{t+h}^T\hat{G}_n\mathbf{x}_{t+h}/2} + 1) \right] \right)$$

$$\left/ \left(\sum_{t=1}^{n-h} \hat{\mu}_t\hat{\mu}_{t+h}e^{-(\mathbf{x}_t+\mathbf{x}_{t+h})^T\hat{G}_n(\mathbf{x}_t+\mathbf{x}_{t+h})/2} \right),\right.$$

and

$$\hat{\gamma}_{OPT.UB}(h) = \left(\sum_{t=1}^{n-h} \frac{W_t W_{t+h}}{\hat{\mu}_t\hat{\mu}_{t+h}} \left[(Y_t - \hat{\mu}_t)(Y_{t+h} - \hat{\mu}_{t+h}) \right.\right.$$

$$+ \hat{\mu}_t\hat{\mu}_{t+h}e^{-(\mathbf{x}_t+\mathbf{x}_{t+h})^T\hat{G}_n(\mathbf{x}_t+\mathbf{x}_{t+h})/2}$$

$$\left.\left. \times (e^{(\mathbf{x}_t+\mathbf{x}_{t+h})^T\hat{G}_n(\mathbf{x}_t+\mathbf{x}_{t-h})/2} - e^{\mathbf{x}_t^T\hat{G}_n\mathbf{x}_t/2} - e^{\mathbf{x}_{t+h}^T\hat{G}_n\mathbf{x}_{t+h}/2} + 1) \right] \right)$$

$$\left/ \left(\sum_{t=1}^{n-h} W_t W_{t+h}e^{-(\mathbf{x}_t+\mathbf{x}_{t+h})^T\hat{G}_n(\mathbf{x}_t+\mathbf{x}_{t-h})/2} \right),\right.$$

where

$$W_t = (\hat{\sigma}_\varepsilon^2 + 1/\hat{\mu}_t)^{-1}.$$

The corresponding autocorrelation estimates are

$$\hat{\rho}_{Z.UB}(h) = \hat{\gamma}_{Z.UB}(h)/\hat{\sigma}_{Z.UB}^2 \qquad (2.12)$$

and

$$\hat{\rho}_{OPT.UB}(h) = \hat{\gamma}_{OPT.UB}(h)/\hat{\sigma}_{OPT.UB}^2. \qquad (2.13)$$

It is straightforward to show that these adjusted estimates are consistent. In particular $\hat{G}_n$ will converge to zero as $n \to \infty$. This implies that the adjustment to the denominator will tend to unity while that of the numerator will tend to zero as is required for the adjustments to be asymptotically negligible. The following tables compare the Zeger and bias-adjusted Zeger estimates, along with the optimal and bias-adjusted optimal estimates of covariance and correlation for two Poisson regression models. The sample size is 100 and the latent process in both cases is assumed to be a log-normal AR(1) process with $\phi = 0.9$ and $\sigma_\delta^2 = 0.6931$. The summarized results are based on 1000 replications. In the computation of the bias-adjustment factors, we used $L = 10$ in Eq. (2.11) for all the realizations. For the autocorrelation estimates, the simulation results reported below are conditional on $S_a > 1.645$ so that the null hypothesis of no latent process is rejected at the 5 per cent level. This was done to eliminate values of autocorrelations badly effected by zero or near zero variance estimates.

Case 1: Linear regression $1 + t/100$ for $t = 1, \ldots, 100$.
Simulation results for the autocovariances:

Lag	True		Means					S.D.		
		Z	Z.UB	OPT	OPT.UB	Z	Z.UB	OPT	OPT.UB
0	1.00	0.49	0.71	0.52	0.73	0.29	0.59	0.31	0.56
1	0.87	0.39	0.59	0.41	0.59	0.26	0.53	0.26	0.50
2	0.75	0.31	0.49	0.33	0.49	0.23	0.47	0.23	0.45
3	0.66	0.24	0.41	0.26	0.41	0.20	0.42	0.20	0.40
4	0.58	0.19	0.34	0.20	0.34	0.18	0.38	0.19	0.36
5	0.51	0.14	0.27	0.16	0.28	0.16	0.34	0.17	0.33
6	0.45	0.10	0.22	0.12	0.23	0.15	0.31	0.16	0.30

Simulation results for the autocorrelations. In 3 out of 1000 replications S_a was less than 1.645.

Lag	True		Means					S.D.		
		Z	Z.UB	OPT	OPT.UB	Z	Z.UB	OPT	OPT.UB
1	0.87	0.79	0.81	0.79	0.81	0.18	0.16	0.18	0.16
2	0.75	0.61	0.65	0.61	0.65	0.20	0.19	0.20	0.18
3	0.66	0.46	0.52	0.48	0.52	0.22	0.21	0.21	0.20
4	0.58	0.35	0.42	0.37	0.42	0.23	0.22	0.22	0.21
5	0.51	0.25	0.32	0.26	0.33	0.24	0.24	0.23	0.23
6	0.45	0.17	0.25	0.19	0.26	0.26	0.25	0.24	0.24

Case 2: Cosine regression $1 + \cos(2\pi t/12)$ for $t = 1, \ldots, 100$.
Simulation results for the autocovariances:

Lag	True	Z	Z.UB	OPT	OPT.UB	Z	Z.UB	OPT	OPT.UB
			Means				S.D.		
0	1.00	0.60	0.80	0.61	0.80	0.38	0.62	0.35	0.56
1	0.87	0.49	0.67	0.50	0.67	0.35	0.57	0.32	0.52
2	0.75	0.42	0.58	0.42	0.58	0.31	0.50	0.29	0.47
3	0.66	0.36	0.50	0.35	0.49	0.28	0.46	0.27	0.43
4	0.58	0.30	0.42	0.29	0.42	0.26	0.42	0.25	0.40
5	0.51	0.25	0.37	0.25	0.36	0.23	0.38	0.23	0.37
6	0.45	0.21	0.31	0.21	0.31	0.21	0.34	0.21	0.34

Simulation results for the autocorrelations. In 5 out of 1000 replications S_a was less than 1.645.

Lag	True	Z	Z.UB	OPT	OPT.UB	Z	Z.UB	OPT	OPT.UB
			Means				S.D.		
1	0.87	0.81	0.82	0.81	0.82	0.21	0.19	0.21	0.19
2	0.75	0.68	0.70	0.68	0.69	0.22	0.21	0.22	0.20
3	0.66	0.57	0.59	0.56	0.58	0.26	0.25	0.24	0.23
4	0.58	0.48	0.50	0.46	0.48	0.30	0.29	0.27	0.27
5	0.51	0.41	0.44	0.39	0.42	0.31	0.31	0.28	0.28
6	0.45	0.34	0.37	0.32	0.35	0.31	0.31	0.28	0.28

Note that the bias improved versions of the estimates of the autocovariances do indeed have better bias properties but at the expense of higher variance. The bias adjusted estimates for the autocovariances have the same magnitude of bias, but the optimal ones have smaller variances. The estimates of autocorrelations also have better bias properties, although the unadjusted estimates perform reasonably well, and have comparable variances. This means that for correlation estimation and identification purposes, the bias adjusted estimates are preferable to the unadjusted versions. However for purposes in which an unbiased estimate of scale is required even the bias adjusted estimates of are biased towards 0. This will impact the estimation of standard errors of $\hat{\beta}_{GLM}$ and the method of Zeger for example.

2.8 Estimation

Zeger's treatment of the estimation of the latent process model is based on a quasi-likelihood approach used to correct for serial correlation in the latent process $\{\varepsilon_t\}$. Assumptions on the distributional properties of this process are not explicitly stated but for much of his treatment these are not required. However for the alternative specification in terms of the $\{\delta_t\}$ process the requirement that these be normally distributed is made quite explicitly in the treatment given in Chan and Ledolter (1995). Indeed as we will demonstrate later, it is difficult to obtain the exact and large sample statistical properties required for inference in this log-normal specification without the assumption of normality.

Detection of autocorrelation using the observed count process, Y_t, is not straightforward as we showed in Sec. 2.1. Typically use of the auto-correlations for Y_t will lead to underestimates of the true degree of auto-correlation in the latent process ε_t. This is noted in Zeger (1988). Also, in practice, the regression on $\mathbf{x}_t$ will need to be performed before attempting to estimate this autocorrelation. However such estimation should adjust for autocorrelation in order to arrive at correct inferences. Estimation of the parameters θ in the model by direct numerical maximization of the like-lihood function is difficult since the likelihood cannot be written down in closed form. (For model in Eq. (2.1), from Eq. (1.3) the likelihood is the n-fold integral,

$$\int_{-\infty}^{\infty} \cdots \int_{-\infty}^{\infty} \exp\left\{ \sum_{t=1}^{n} (\mathbf{x}_t^T \beta y_t - \varepsilon_t \exp(\mathbf{x}_t^T \beta)) \right\}$$

$$\times \left(\prod_{t=1}^{n} \varepsilon_t^{y_t} \right) L(\theta; \varepsilon^{(n)})(d\varepsilon_1 \cdots d\varepsilon_n) \Big/ \prod_{t=1}^{n} (y_t!),$$

where $L(\theta; \varepsilon^{(n)})$ is the likelihood based on $\varepsilon_1, \ldots, \varepsilon_n$.) To overcome this difficulty, Chan and Ledolter (1995) proposed an algorithm, called Monte Carlo EM (MCEM), whose iterates $\theta^{(i)}$ converge to the maximum likelihood estimate. The E-step of the algorithm requires calculation of the conditional expectation of the log-likelihood of the complete data:

$$Q(\theta|\theta^{(i)}) = E_{\theta^{(i)}}(\ln L(\theta; \varepsilon^{(n)}, \mathbf{Y}^{(n)})|\mathbf{Y})$$

$$= \int \ln L(\theta; \varepsilon^{(n)}, \mathbf{y}^{(n)}) f(\varepsilon^{(n)}|\mathbf{y}^{(n)}; \theta^{(i)}) \, d\varepsilon^{(n)}. \tag{2.14}$$

In the M-step, $Q(\theta|\theta^{(i)})$ is maximized as a function of θ to obtain the updated estimate $\theta^{(i+1)}$. Direct calculation of Q is intractable. Suppose for the moment that it is possible to generate replicates of $\varepsilon^{(n)}$ from the conditional distribution of $\varepsilon^{(n)}$ given $\mathbf{Y}^{(n)}$ when $\theta = \theta^{(i)}$. If we denote m separate replicates of $\varepsilon^{(n)}$ by $\varepsilon_1^{(n)}, \ldots, \varepsilon_m^{(n)}$, then a Monte Carlo approximation to Q in Eq. (2.14) is given by

$$Q_m(\theta|\theta^{(i)}) = \frac{1}{m} \sum_{j=1}^{m} \ln L(\theta; \varepsilon_j^{(n)}, \mathbf{Y}^{(n)}).$$

The M-step is easy to carry out using Q_m in place of Q. The difficult steps in the algorithm are the generation of replicates of $\varepsilon^{(n)}$ given $\mathbf{Y}^{(n)}$ and the choice of m. Chan and Ledolter (1995) discuss the use of the Gibb's sampler for generating the desired replicates and give some guidelines on the choice of m.

In their analyses of the polio data, Zeger (1988) and Chan and Ledolter (1995) included as regression components an intercept, a slope, and harmonics at periods of 6 and 12 months. Specifically, they took $\varepsilon_t = \exp(\delta_t), \delta_t = \phi\delta_{t-1} + z_t, z_t$ is IID $N(0, \sigma^2)$, and

$$\mathbf{x}_t = (1, t/1000, \cos(2\pi t/12), \sin(2\pi t/12), \cos(2\pi t/6), \sin(2\pi t/6))^T.$$

The implementation of Chan and Ledolter's MCEM method by Kuk and Cheng (1997) gave estimates $\hat{\beta} = (0.247, -3.871, 0.162, -0.482, 0.414, -0.011)^T$, $\hat{\phi} = 0.648$, and $\hat{\sigma}^2 = 0.281$. The negative coefficient of $t/1000$ indicates a slight downward trend in the monthly number of polio cases.

2.8.1 The Likelihood Function

Suppose $\{\varepsilon_t\}$ is the latent process in a parameter-driven model

$$Y_t|\varepsilon_t, \mathbf{x}_t \sim \mathcal{P}(\varepsilon_t \exp\{\mathbf{x}_t^T\beta\}),$$

where $\varepsilon_t = \exp(\delta_t)$ and $\{\delta_t\}$ is an autoregressive process of order p [AR(p)]. That is, $\{\delta_t\}$ satisfies the recursions

$$\delta_t = \phi_1\delta_{t-1} + \cdots + \phi_p\delta_{t-p} + z_t, \qquad z_t \sim \text{IID}(0, \sigma^2). \tag{2.15}$$

The likelihood of the complete data $(\mathbf{y}, \delta) = (y_1, \ldots, y_n; \delta_1, \ldots, \delta_n)$ is given by

$$f(\mathbf{y};\boldsymbol{\delta}) = f(\mathbf{y}|\boldsymbol{\delta})f(\boldsymbol{\delta})$$

$$= \prod_{i=1}^{n} f(y_i|\delta_i)f(\boldsymbol{\delta})$$

$$= \left[\prod_{i=1}^{n} \frac{\exp\{-\exp(\delta_i + \mathbf{x}_i^T\boldsymbol{\beta})\}\exp\{(\delta_i + \mathbf{x}_i^T\boldsymbol{\beta})y_i\}}{y_i!}\right] \frac{\exp\{-\frac{1}{2}\boldsymbol{\delta}^T V\boldsymbol{\delta}\}}{(2\pi)^{n/2}|V^{-1}|^{1/2}}$$

$$= \frac{|V|^{1/2}}{C_1}\exp\left\{-\sum_{i=1}^{n}\exp(\delta_i + \mathbf{x}_i^T\boldsymbol{\beta}) + \sum_{i=1}^{n}(\delta_i + \mathbf{x}_i^T\boldsymbol{\beta})y_i - \frac{1}{2}\boldsymbol{\delta}^T V\boldsymbol{\delta}\right\}$$

$$= \frac{|V|^{1/2}}{C_1}\exp\{-(\mathbf{e}^{\boldsymbol{\delta}})^T\mathbf{e}^{X\boldsymbol{\beta}} + \mathbf{y}^T\boldsymbol{\delta} + \mathbf{y}^T X\boldsymbol{\beta} - \frac{1}{2}\boldsymbol{\delta}^T V\boldsymbol{\delta}\}, \qquad (2.16)$$

where

$$C_1 = (2\pi)^{n/2}\left(\prod_{i=1}^{n} y_i!\right),$$

$\mathbf{e}^{\boldsymbol{\delta}} = (e^{\delta_1}, \ldots, e^{\delta_n})^T$, $\mathbf{e}^{X\boldsymbol{\beta}} = [\exp(\mathbf{x}_1^T\boldsymbol{\beta}), \ldots, \exp(\mathbf{x}_n^T\boldsymbol{\beta})]^T$ and V^{-1} is the covariance matrix of $\boldsymbol{\delta}$.

Our objective is to estimate the model parameters $\boldsymbol{\theta} = (\boldsymbol{\beta}^T, \boldsymbol{\phi}^T, \sigma^2)^T$, where $\boldsymbol{\phi} = (\phi_1, \ldots, \phi_p)^T$. As noted above, the likelihood of the observed data $\mathbf{y}$ does not have a simple closed form and maximum likelihood estimation is intractable. Instead, we approximate the likelihood of the complete data $(\mathbf{y}, \boldsymbol{\delta})$ by a distribution for which the $\boldsymbol{\delta}$ are easily integrated in order to compute the marginal distribution of $\mathbf{y}$. We then estimate the model parameters by maximizing the approximate likelihood.

2.8.2 The Approximate Likelihood

A Taylor expansion at $\boldsymbol{\delta}_0 = (\delta_1^{(0)}, \ldots, \delta_n^{(0)})$ on the term $(\mathbf{e}^{\boldsymbol{\delta}})^T\mathbf{e}^{X\boldsymbol{\beta}}$ in Eq. (2.16) gives

$$(\mathbf{e}^{\boldsymbol{\delta}})^T\mathbf{e}^{X\boldsymbol{\beta}} = \mathbf{b}_0^T\mathbf{e}^{X\boldsymbol{\beta}} + (\boldsymbol{\delta} - \boldsymbol{\delta}_0)^T K\mathbf{b}_0 + \frac{1}{2}(\boldsymbol{\delta} - \boldsymbol{\delta}_0)^T BK(\boldsymbol{\delta} - \boldsymbol{\delta}_0) \qquad (2.17)$$

where $\mathbf{b}_0 = [\exp(\delta_1^{(0)}), \ldots, \exp(\delta_n^{(0)})]^T$, $B = \text{diag}[(\exp(\delta_1^{(0)}), \ldots, \exp(\delta_n^{(0)})$, $K = \text{diag}[\exp(\mathbf{x}_1^T\boldsymbol{\beta}), \ldots, \exp(\mathbf{x}_n^T\boldsymbol{\beta})]$. Let

$$\tilde{\mathbf{y}} = \mathbf{y} - K\mathbf{b}_0 + BK\boldsymbol{\delta}_0, \qquad (2.18)$$

the approximate likelihood of the complete data $(\mathbf{y}, \boldsymbol{\delta})$ is then given by

$$f_a(\mathbf{y}, \boldsymbol{\delta}) = \frac{|V|^{1/2}}{C_1} \exp\{\mathbf{y}^T \boldsymbol{\delta} + \mathbf{y}^T X \boldsymbol{\beta} - \tfrac{1}{2} \boldsymbol{\delta}^T V \boldsymbol{\delta}$$

$$- [\mathbf{b}_0^T e^{X\boldsymbol{\beta}} + (\boldsymbol{\delta} - \boldsymbol{\delta}_0)^T K \mathbf{b}_0 + \tfrac{1}{2}(\boldsymbol{\delta} - \boldsymbol{\delta}_0)^T BK(\boldsymbol{\delta} - \boldsymbol{\delta}_0)]\}$$

$$= \frac{|V|^{1/2}}{C_1} \exp\{(\mathbf{y} - K\mathbf{b}_0 + BK\boldsymbol{\delta}_0)^T \boldsymbol{\delta} - \tfrac{1}{2}\boldsymbol{\delta}^T (BK + V)\boldsymbol{\delta}$$

$$- \tfrac{1}{2}\boldsymbol{\delta}_0^T BK\boldsymbol{\delta}_0 + \boldsymbol{\delta}_0^T K\mathbf{b}_0 + \mathbf{y}^T X\boldsymbol{\beta} - \mathbf{b}_0^T e^{X\boldsymbol{\beta}}\}$$

$$= \frac{|V|^{1/2}}{C_1} \exp\{\tilde{\mathbf{y}}^T \boldsymbol{\delta} - \tfrac{1}{2}\boldsymbol{\delta}^T (BK + V)\boldsymbol{\delta}$$

$$- \tfrac{1}{2}\boldsymbol{\delta}_0^T BK\boldsymbol{\delta}_0 + \boldsymbol{\delta}_0^T K\mathbf{b}_0 + \mathbf{y}^T X\boldsymbol{\beta} - \mathbf{b}_0^T e^{X\boldsymbol{\beta}}\}$$

$$= \frac{|V|^{1/2}}{C_1} \exp\{-\tfrac{1}{2}[\boldsymbol{\delta} - (BK + V)^{-1}\tilde{\mathbf{y}}]^T (BK + V)[\boldsymbol{\delta} - (BK + V)^{-1}\tilde{\mathbf{y}}]$$

$$+ \tfrac{1}{2}\tilde{\mathbf{y}}^T (BK + V)^{-1}\tilde{\mathbf{y}} - \tfrac{1}{2}\boldsymbol{\delta}_0^T BK\boldsymbol{\delta}_0 + \boldsymbol{\delta}_0^T K\mathbf{b}_0 + \mathbf{y}^T X\boldsymbol{\beta} - \mathbf{b}_0^T e^{X\boldsymbol{\beta}}\}.$$

So the conditional distribution of $\boldsymbol{\delta}$ given $\mathbf{y}$ is

$$\boldsymbol{\delta}|\mathbf{y} \sim N[(BK + V)^{-1}\tilde{\mathbf{y}}, (BK + V)^{-1}] \qquad (2.19)$$

and the approximate distribution of $\mathbf{y}$ is

$$f_a(\mathbf{y}) = \frac{|V|^{1/2}}{|BK + V|^{1/2}(\prod_{i=1}^{n} y_i!)} \exp\{\tfrac{1}{2}\tilde{\mathbf{y}}^T (BK + V)^{-1}\tilde{\mathbf{y}}$$

$$- \tfrac{1}{2}\boldsymbol{\delta}_0^T BK\boldsymbol{\delta}_0 + \boldsymbol{\delta}_0^T K\mathbf{b}_0 + \mathbf{y}^T X\boldsymbol{\beta} - \mathbf{b}_0^T e^{X\boldsymbol{\beta}}\} \qquad (2.20)$$

The calculation of $|BK + V|$ and $(BK + V)^{-1}\tilde{\mathbf{y}}$ are based on the innovations algorithm (Brockwell and Davis (1991)) that avoids inverting the matrix $(BK + V)$ directly. We treat $(BK + V)$ as a covariance matrix of $\tilde{\mathbf{y}}$ and decompose it as

$$BK + V = CDC^T,$$

where C is a lower triangular matrix with all the diagonal elements 1, $D = \text{diag}(v_0, v_1, \cdots, v_{n-1})$, and v_i is the mean squared error of the one-step predictor of $\tilde{y}_i$. Then the determinant of $(BK + V)$ is given by

$$|BK + V| = |C||D||C^T| = \prod_{i=0}^{n-1} v_i.$$

Moreover the innovations algorithm can easily be adapted to compute the conditional mean, $E(\boldsymbol{\delta}|\mathbf{y}) = (BK + V)^{-1}\tilde{\mathbf{y}}$ specified in Eq. (2.19).

The parameters of the model are then estimated as follows.

1. Fix initial values of $\boldsymbol{\delta} = \boldsymbol{\delta}^{(0)}, \boldsymbol{\phi} = \boldsymbol{\phi}^{(0)}$, and $\sigma^2 = \sigma^{2(0)}$;

2. For fixed $\delta^{(j)}$, $\phi^{(j)}$ and $\sigma^{2(j)}$, maximize $\mathbf{y}^T X \boldsymbol{\beta} - \mathbf{b}_0^T e^{X\boldsymbol{\beta}}$ to get $\boldsymbol{\beta}^{(j+1)}$; this is comparable to Poisson regression.

3. For fixed $\delta^{(j)}$ and $\boldsymbol{\beta}^{(j+1)}$, maximize

$$\log \frac{|V|}{|BK + V|} + \tilde{\mathbf{y}}^T (BK + V)^{-1} \tilde{\mathbf{y}}$$

to find $\phi^{(j+1)}$ and $\sigma^{2(j+1)}$, the estimates of ϕ and σ^2 respectively;

4. For fixed $\boldsymbol{\beta}^{(j+1)}$, $\phi^{(j+1)}$ and $\sigma^{2(j+1)}$, use $\delta^{(j+1)} = (BK + V)^{-1}\tilde{\mathbf{y}}$ iteratively to get the estimates of δ. After convergence, set $\delta^{(j+1)} = \delta^{(\infty)}$ and $\delta_0 = \delta^{(\infty)}$;

5. Increment j, go to step 2 and continue to convergence.

We applied this algorithm to the polio data. Using the regression terms

$$\mathbf{x}_t = [1, t/1000, \cos(2\pi t/12), \sin(2\pi t/12), \cos(2\pi t/6), \sin(2\pi t/6)]^T.$$

and an AR(1) model for the latent process $\{\delta_t\}$ we obtain

These values are in reasonably close agreement with those cited earlier by

$\hat{\beta}_1$	$\hat{\beta}_2$	$\hat{\beta}_3$	$\hat{\beta}_4$	$\hat{\beta}_5$	$\hat{\beta}_6$	$\hat{\phi}$	$\hat{\sigma}^2$
0.407	−4.236	0.153	−0.466	0.402	−0.008	0.664	0.244

Kuk and Cheng (1997) and require a great deal less computation. Note that these estimates are not directly comparable with the earlier analyses because the regression terms are not centered at january, 1976 in order to match the analysis of Kuk and Cheng.

Durbin and Koopman (1997) consider a Gaussian state equation and a non-Gaussian observation process. They approximate the density $p(y_t|\varepsilon_t)$ by a Gaussian conditional density $g(y_t|\varepsilon_t)$ to arrive at a Gaussian approximation to the likelihood. Simulation techniques are then used to compute the adjustment from the approximate likelihood to the correct likelihood. The conditional mode and expected value of ε_t are also calculated efficiently using their method. An interesting future research question is to compare the performance of their methods with that proposed here and with the use of full simulation of the exact likelihood based on the Markov chain Monte Carlo ideas described above.

3. OBSERVATION DRIVEN MODELS

3.1 Review of Existing Models

In this Section we review various observation driven models for count data. In addition we offer some alternatives that have been considered in the literature to date.

In an early contribution, Harvey and Fernandes (1989) considered observations Y_t which have a Poisson distribution conditional upon a mean process ε_t. They specify the conditional density of ε_t conditional upon information up to time $t-1$ as that of a gamma distribution with parameters $a_{t|t-1}$ and $b_{t|t-1}$. This density corresponds to a conjugate prior for the Poisson distribution. The posterior density of ε_t given observations up to time t is also gamma with parameters a_t and b_t. The two sets of parameters are assumed to be linked through the relations $a_{t|t-1} = \omega a_t$ and $b_{t|t-1} = \omega b_t$ for some $0 < \omega \leq 1$. Using this formulation the stochastic mechanism for the transition from ε_{t-1} to ε_t is defined implicitly. The log-likelihood for ω is defined and the forecast function $E(Y_{T+1}|\mathbf{Y}^{(\mathbf{T})})$ is shown to follow an exponentially weighted moving average of past Y_t's. Similar models, based on appropriate conjugate distributions, are developed for the binomial, multinomial and negative exponential observation distributions. Harvey and Fernandes also extend their methods to incorporate estimation of explanatory variables by modifying, in the Poisson case, the mean function to $\varepsilon_t \exp(\mathbf{x}_t^T \boldsymbol{\beta})$ and apply this model to various time series of counts.

In a different direction, and one that we pursue in more detail, Zeger and Qaqish (1988) considered the following model for Poisson counts. Let

$$\mathbf{H}_t = (\mathbf{Y}^{(t-1)}, \mathbf{X}^{(t)})$$

be the past of the observed count process and the past and present of the regressor variables. Zeger and Qaqish (1988) assume that the conditional distribution of $Y_t|\mathbf{H}_t$ is Poisson with mean given by

$$\mu_t = \exp(\mathbf{x}_t^T \boldsymbol{\beta}) \prod_{i=1}^{p} \left[\frac{\max(Y_{t-i}, c)}{\exp(\mathbf{x}_{t-i}^T \boldsymbol{\beta})} \right]^{\gamma_i}, \tag{3.1}$$

where $0 < c < 1$ is a constant which prevents $Y_{t-1} = 0$ becoming an absorbing state. An alternative form considered by them is

$$\mu_t = \exp(\mathbf{x}_t^T \boldsymbol{\beta}) \prod_{i=1}^{p} \left[\frac{Y_{t-i} + c}{\exp(\mathbf{x}_{t-i}^T \boldsymbol{\beta}) + c} \right]^{\gamma_i},$$

in which, in the case $p = 1$, c is interpreted as an immigration rate adding to counts at every time point. In both of these forms estimation of c could be problematic and the inferential theory for its estimation does not seem to have been worked out; because it represents an end point parameter this might be problematic. Thus the parameter c seems to be somewhat arbitrary and, apart from the interpretations just given for the case $p = 1$, its true meaning in more general cases is not clear.

An alternative considered in Zeger and Qaqish (1988) is

$$\mu_t = \exp(\mathbf{x}_t^T \boldsymbol{\beta}) \exp\left(\sum_{i=1}^{p} \gamma_i Y_{t-i} \right). \tag{3.2}$$

Note that the Y_{t-i} enter without any form of mean correction or centering.

Zeger and Qaqish (1988) argue that the model in Eq. (3.1) is preferred on three criteria. In particular the model in Eq. (3.2) cannot be stationary unless, at least in the case $p = 1$, $\gamma_1 \leq 0$ thereby excluding the possibility of positive dependence.

The model in Eq. (3.2) is applied to data in Fahrmeir and Tutz (1994). They offer some comments on the problems of stationarity (p. 195) but appear to ignore this issue when it comes to actually fitting an observation driven model to real data. When their form of the model is used with the asthma data the estimated long run mean and seasonal effects are badly distorted.

It might be thought that the difficulties with the model in Eq. (3.2) could be overcome by subtracting the deterministic mean from the Y_t to arrive at

$$\mu_t = \exp(\mathbf{x}_t^T \boldsymbol{\beta}) \exp\left\{ \sum_{i=1}^{p} \gamma_i [Y_{t-i} - \exp(\mathbf{x}_{t-i}^T \boldsymbol{\beta})] \right\}, \tag{3.3}$$

but in fact this will not lead to a stationary process as can be seen by using an argument similar to the one given in Zeger and Qaqish.

3.2   Observation Driven Models in Standardized Errors

A different form of mean correction is proposed here which could result in the generation of stationary processes. However, as we will see, this is at the expense of a Markov (finite lag) structure for the Y_t process.

To introduce the model let, for $\lambda \geq 0$,

$$e_t = (Y_t - \mu_t)/\mu_t^{\lambda}$$

and

$$W_t := \log(\mu_t) = \mathbf{x}_t^T \boldsymbol{\beta} + \sum_{i=1}^{p} \gamma_i e_{t-i},$$

and assume that

$$Y_t | \mathbf{Y}^{(t-1)} \sim \mathcal{P}(\mu_t).$$

Note that the conditional mean, μ_t, is based on the whole past and so the process $\{Y_t\}$ is no longer Markov. However it could lead to stationary solutions although establishing this with rigor seems to be difficult. Note that $\{\mu_t\}$ is a pth order Markov chain. We discuss some of its properties below. However for the purposes of inference that discussion is not required.

Extensions to autoregressive moving average filters applied to past values of e_t can also be made as follows. Let

$$W_t = \mathbf{x}_t^T \boldsymbol{\beta} + \sum_{i=1}^{\infty} \tau_i e_{t-i},$$

where

$$\sum_{i=1}^{\infty} \tau_i z^i = \left(1 - \sum_{i=1}^{q} \phi_i z^i\right)^{-1} \left(1 + \sum_{i=1}^{p} \gamma_i z^i\right) - 1$$

$$= \phi(z)^{-1} \gamma(z) - 1$$

and note that $\sum_{i=1}^{\infty} \tau_i e_{t-i}$ is the one step ahead predictor of the e_t based on an ARMA(q, p) model. In such a specification the infinite past is required but in practice this will not be available. Since the joint distribution of the e_t is not known initial conditions which conform to the distribution of values for $t \leq 0$ cannot be specified. The simplest proposal for practical applications might be to begin the recursions required for computing $\sum_{i=1}^{\infty} \tau_i e_{t-i}$ be setting $e_t = 0$ for $t \leq 0$.

Shephard (1995, unpublished) suggested a model quite similar to that proposed above. He presents an argument, based on a Taylor series linearization of the link function, for using $\lambda = 1$ in the definition of e_t at least for the Poisson case. We assume that τ_t is determined by information available at time t, i.e., τ_t depends on past observations $(y_{t-1}, y_{t-2}, \ldots, y_{t-p})$ and covariates up to time t. Define

$$\log(\mu_t) = \mathbf{x}_t^T \boldsymbol{\beta} + \sum_{i=1}^{q} \phi_i z_{t-i}$$

where

$$z_t = \mathbf{x}_t^T \boldsymbol{\beta} + \sum_{i=1}^{q} \phi_i z_{t-i} + e_t + \sum_{k=1}^{p} \gamma_k e_{t-k}.$$

This model is called the generalized linear ARMA (or GLARMA) model by Shephard (1995). Note that

$$z_t = \phi(B)^{-1} \mathbf{x}_t^T \boldsymbol{\beta} + \phi(B)^{-1} \gamma(B) e_t,$$

so that

$$
\begin{aligned}
\log(\mu_t) &= \mathbf{x}_t^T \boldsymbol{\beta} + \sum_{i=1}^{q} \phi_i z_{t-i} \\
&= \mathbf{x}_t^T \boldsymbol{\beta} + [1 - \phi(B)][\phi(B)^{-1} \mathbf{x}_t^T \boldsymbol{\beta} + \phi(B)^{-1} \gamma(B) e_t] \\
&= \phi(B)^{-1} \mathbf{x}_t^T \boldsymbol{\beta} + [\phi(B)^{-1} - 1] \gamma(B) e_t.
\end{aligned}
$$

This formulation differs from that proposed by us above in the following ways:

1. It uses $\lambda = 1$. This means that the e_t do not have unit variance but instead have variance that depends on the mean function μ_t. For $\lambda = 0.5$, $\{e_t\}$ is a weakly stationary martingale difference sequence whereas this is not the case for $\lambda = 1$. Nevertheless the μ_t sequence may still, at least in the case where $q = 0$ and $p = 1$, be a stationary first order Markov process.
2. It defines the mean as a distributed lag in the previous regression terms. The expected value of the mean function will not be the current deterministic part of the process mean, even approximately, which might make interpretation of the coefficients $\boldsymbol{\beta}$ difficult for practical use.
3. The model implies that there must always be at least one non-trivial lag term in the autoregressive component. Our model does not require this and allows pure MA, pure AR and mixed ARMA models.

3.3 Properties of the Process

Note that $\{e_t : t \le s - 1\}$, given initial conditions $\{\mu_t : -p + 1 \le t \le 0\}$ is equivalent to $\{Y_t : t \le s - 1\}$ or $\{\mu_t : t \le s - 1\}$ from which it follows that the e_t form a martingale difference sequence since

$$E(e_s | \mathcal{F}_{s-1}^c) = 0$$

where $\mathcal{F}_{s-1}^c$ is the σ-algebra generated by $\{e_t : t \le s - 1\}$. Hence the e_t have zero mean and variance

$$E(e_t^2) = E[E(e_t^2 | \mu_t)] = \mu_t^{1-2\lambda}$$

which, for $\lambda = 0.5$, is unity. Also, from the martingale difference property, the covariance, for $s \ne t$, is

$$E(e_t e_s) = 0.$$

From the above properties we have, for any λ,

$$E(W_t) = \mathbf{x}_t^T \boldsymbol{\beta}$$

which is a desirable property for the log mean (compare with the first desiderata listed in Sec. 1.2). Also

$$\text{Var}(W_t) = \sum_{i=1}^{p} \gamma_i^2 \mu_{t-i}^{1-2\lambda},$$

and for $l > 0$

$$\text{Cov}(W_t, W_{t+l}) = \sum_{i=1}^{p-l} \gamma_i \gamma_{i+l} \mu_{t-i}^{1-2\lambda},$$

and again, if $\lambda = 0.5$, the covariances do not depend on time t. Because the process $\{W_t\}$ possesses a number of nice properties in the case $\lambda = 0.5$, we consider this case in more depth.

Although we know that the W_t have mean exactly equal to the linear predictor, it does not follow that the μ_t have mean equal to the $e^{\mathbf{x}_t \boldsymbol{\beta}}$. We have

$$W_t = \mathbf{x}_t^T \boldsymbol{\beta} + U_t \approx \mathbf{x}_t^T \boldsymbol{\beta} + U_t'$$

in the sense that the distributions will be similar, where U_t' is a Gaussian stationary sequence with mean zero and variances and covariances matched to those for U_t. Roughly speaking U_t' is a latent process. Hence, using the results obtained for latent processes we have, again in the case $\lambda = 0.5$,

$$E(\mu_t) = E(e^{W_t}) \approx \exp[\mathbf{x}_t^T \boldsymbol{\beta} + \tfrac{1}{2} \text{Var}(U_t)] = \exp\left(\mathbf{x}_t^T \boldsymbol{\beta} + \tfrac{1}{2} \sum_{i=1}^{p} \gamma_i^2 \right),$$

so that, in practice the bias of $E(\mu_t)$ as an estimate of $\exp(\mathbf{x}_t^T \boldsymbol{\beta})$ can be approximately adjusted for and, perhaps most importantly the interpretation of the regression coefficients, other than the intercept or constant term, are (at least to the level of approximation) interpretable as the amount by which the mean of Y_t would change for a unit change in the regressors.

While the distribution of e_t is not normally distributed the linear combination $U_t = \sum_{i=1}^{p} \gamma_i e_{t-i}$ will have a distribution more closely approximated by a sequence of correlated normal random variables. The extent to which the joint distribution of the sequence $\{e_t\}$ differs from a process of independent Gaussian random variables with zero mean and unit variance will govern the extent to which the approximation

$$E(e^{W_t}) = \exp\left(\mathbf{x}_t^T \boldsymbol{\beta} + \frac{1}{2} \sum_{i=1}^{p} \gamma_i^2 \right)$$

holds.

Overall, the $\lambda = 0.5$ case has a number of desirable properties:

1. It is easily interpretable on the linear predictor scale *and* on the scale of the mean μ_t. The regression parameters can be roughly interpreted as the amount by which the mean of the count process at time t will change for a unit change in the regressor variable.

2. An approximately unbiased plot of the μ_t can be generated by

$$\hat{\mu}_t = \exp\left(\hat{W}_t - \frac{1}{2}\sum_{i=1}^{p} \hat{\gamma}_i^2 \right),$$

where parameter estimates have been used throughout.

3. The model is easy to use for prediction. In fact $\hat{\mu}_t$ should be used as the one step ahead forecast of Y_t, given a value for $\mathbf{x}_t$ or a reliable forecast of it.

4. The model provides a mechanism for adjusting the inference about the regression parameter $\boldsymbol{\beta}$ for a form of serial dependence.

5. The model is generalizable to "autoregressive" type lag structure and mixed autoregressive-moving average structure.

3.4 Estimation and Inference

3.4.1 Estimation

In this Section we will only consider estimation and inference properties for the observation-driven model of Sec. 3.2 when $\lambda = 0.5$. It might be thought to be reasonable to use an iterated GLM procedure in which at each stage, the values of the sequence $\{e_t\}$ are used in a standard GLM procedure to obtain estimates of $\boldsymbol{\beta}$ and $\boldsymbol{\xi} = (\boldsymbol{\phi}^T, \boldsymbol{\gamma}^T)^T$, which are then used to redefine $\{e_t\}$ and so on iteratively. This does not appear to converge to the values which maximize the likelihood. Furthermore, and perhaps more importantly, the standard errors obtained will not be correct for $\boldsymbol{\beta}$ and $\boldsymbol{\xi}$ using this iterated method. However computation of the likelihood and its first and second derivative is not difficult and can be done recursively using a simple Newton-Raphson update procedure. Correct standard errors are then also available.

Ignoring terms which do not involve the parameters, the log-likelihood, as a function of $\boldsymbol{\theta} = (\boldsymbol{\beta}^T, \boldsymbol{\xi}^T)^T$, is given by

$$L(\boldsymbol{\theta}) = \sum_{t=1}^{n}[Y_t W_t(\boldsymbol{\theta}) - e^{W_t(\boldsymbol{\theta})}],$$

where

$$W_t(\boldsymbol{\theta}) = \log(\mu_t) = \mathbf{x}_t^T \boldsymbol{\beta} + \sum_{i=1}^{\infty} \tau_i(\boldsymbol{\xi})e_{t-i}(\boldsymbol{\theta}) \tag{3.4}$$

and

$$e_t = (Y_t - \mu_t)/\sqrt{\mu_t}.$$

First and second derivatives are given by the following expressions

$$\frac{\partial L}{\partial \boldsymbol{\theta}} = \sum_{t=1}^{n}(Y_t - \mu_t)\frac{\partial W_t}{\partial \boldsymbol{\theta}} = \sum_{t=1}^{n} e_t \mu_t^{1/2}\frac{\partial W_t}{\partial \boldsymbol{\theta}}$$

and

$$\frac{\partial^2 L}{\partial \boldsymbol{\theta}\partial \boldsymbol{\theta}^T} = \sum_{t=1}^{n}\left[(Y_t - \mu_t)\frac{\partial^2 W_t}{\partial \boldsymbol{\theta}\partial \boldsymbol{\theta}^T} - \mu_t\frac{\partial W_t}{\partial \boldsymbol{\theta}}\frac{\partial W_t}{\partial \boldsymbol{\theta}^T}\right]$$

$$= \sum_{t=1}^{n}\left[e_t \mu_t^{1/2}\frac{\partial^2 W_t}{\partial \boldsymbol{\theta}\partial \boldsymbol{\theta}^T} - \mu_t\frac{\partial W_t}{\partial \boldsymbol{\theta}}\frac{\partial W_t}{\partial \boldsymbol{\theta}^T}\right].$$

In order to calculate these, recursive expressions for $\partial W_t/\partial \boldsymbol{\theta}$ and $\partial^2 W_t/\partial \boldsymbol{\theta}\partial \boldsymbol{\theta}^T$ can readily be derived and programmed. Thus calculation of the likelihood and its first and second derivatives (from which the standard errors are derived—see below) can be implemented in a straightforward fashion. We have found that implementation in *Splus* on a PC provides reasonably fast computation of estimates based on the Newton–Raphson method.

3.4.2 Initial Estimates

To initialize the Newton–Raphson recursions we have found that using the GLM estimates without terms for the autoregressive moving average terms together with zero initial values for the ARMA terms gives reasonable starting values. Convergence in all cases reported below (in which the first derivatives were less than 10^{-8}) occurred within at most six iterations from these starting conditions.

3.4.3 Asymptotics

It would appear that the asymptotic properties of the maximum likelihood estimates considered in the previous section are straightforward to establish due primarily to the martingale difference properties of $\{e_t\}$. Nevertheless, a formal proof of these asymptotic properties is difficult without a more detailed understanding on the underlying distributional structure of the $\{e_t\}$ sequence. Assuming that the regressor sequences $x_{j,t}$ satisfy the conditions stated in Sec. 2.2 and that

$$\Omega^{-1} = \lim_{n \to \infty} \frac{1}{n}\sum_{t=1}^{n} e^{W_t(\boldsymbol{\theta}_0)} W_t'(\boldsymbol{\theta}_0) W_t'(\boldsymbol{\theta}_0)^T$$

exists and is non-singular, we conjecture that

$$\sqrt{n}(\hat{\boldsymbol{\theta}} - \boldsymbol{\theta}_0) \xrightarrow{d} N(0, \Omega^{-1}).$$

Asymptotic standard errors of the estimates are then easily obtained from

$$\hat{\Omega} = -\left(\frac{1}{n}\frac{\partial^2 L(\hat{\boldsymbol{\theta}})}{\partial \boldsymbol{\theta}\partial \boldsymbol{\theta}^T}\right)^{-1}$$

as

$$\hat{\sigma}_{\hat{\boldsymbol{\theta}},j} = \sqrt{\frac{\hat{\Omega}_{jj}}{n}}.$$

4. EXAMPLES

4.1 Analysis of the Polio Data

We provide an analysis of the polio data. These data have been much analyzed (list of references) but all these analyses appear to provide different inferences especially about the important issue of trend. A summary is provided for reference in Table 1.

Clearly there is lack of agreement on the values of the parameter estimates and the standard errors that should be attached to them. Some of these standard errors appear to be unreasonably small in the light of the other results and those obtained below using the observation driven model.

We fit the model in Eq. (3.4) to the data with the same regression terms as used in Zeger (1988) (and centered in an identical way as described there). Two models were considered. The first started with $p = 6$ (the first 6 moving

Table 1 Summary of Analyses for the Trend Coefficient for the Polio Data

Study	Trend ($\hat{\beta}$)	s.e.($\hat{\beta}$)	t-statistic
GLM Estimate	−4.80	1.40	−3.43
Zeger (1988)	−4.35	2.68	−1.62
Chan & Ledolter (1995)	−4.62	1.38	−3.35
Kuk & Cheng (1997) MCNR	−3.79	2.95	−1.28
Jorgensen et al. (1995)	−1.61	0.018	−89.7
Fahrmeir and Tutz (1994) 5 lags in Y	−3.33	2.00	−1.67

average terms) and $q = 0$ (no autoregressive components). The second started with $p = 0$ and $q = 6$. In both cases any autoregressive or moving average terms which were not significant at the 5 per cent level were dropped from the model and the model refitted to arrive at the results shown in Table 2.

For the fitted AR(1,5) model, the autocorrelations of the e_t using Zeger's estimates and the optimal estimation method do not appear to be significant at any lag up to lag 12 (the largest values estimated). Their mean is 0.0186 and variance is 1.5015. When the largest five values of the e_t, all of which exceed 3 are excluded, the mean is -0.096 with variance 1.099. It would seem as though the values of the process at times $t = 7, 34, 35, 74, 113$ corresponding to the large Pearson (predictive) residuals e_t could be considered as outliers.

Use of the Q statistic to determine if a latent process is present gives a value of 1.64 (1.99 adjusted for bias due to fitting parameters) which is significant at the 5 per cent level on a one sided test (at the 2.3 per cent level using the bias adjusted version). Note that the distributional properties of the alternative statistics S_a and $\tilde{Q}$ are not established for these lagged regression models. It is likely that the simpler statistic Q will be conservative, however, so the marginally significant result obtained here indicates that not all the additional variation in the observed counts is accounted for by the lagged regression model.

Table 2   Observation-Driven Models Fitted to the Polio Data

Model term	AR $\hat{\theta}$	$\hat{\sigma}_{\theta}^{(1)}$	MA $\hat{\theta}$	$\hat{\sigma}_{\theta}^{(2)}$
$\hat{\beta}_0$ (intercept)	0.138	0.117	0.130	0.114
$\hat{\beta}_1$ (trend)	−3.83	2.26	−3.93	2.18
$\hat{\beta}_2$ (annual cosine)	−0.099	0.105	−0.099	0.118
$\hat{\beta}_3$ (annual sine)	−0.506	0.128	−0.531	0.141
$\hat{\beta}_4$ (semi-annual cosine)	0.230	0.127	0.211	0.117
$\hat{\beta}_5$ (semi-annual sine)	−0.397	0.123	−0.393	0.116
$\hat{\phi}_1$ or $\hat{\gamma}_1$	0.227	0.053	0.218	0.056
$\hat{\gamma}_2$	—	—	0.127	0.046
$\hat{\phi}_5$ or $\hat{\gamma}_5$	0.105	0.050	0.087	0.043
$L(\hat{\beta}, \hat{\xi})$ (log-likelihood)	−119.6	—	−118.9	—
$\hat{\beta}_0 + \frac{1}{2}\sum \hat{\gamma}_i^2$ adjusted intercept	—	—	0.166	—

4.2 Analysis of the UK SIDS Data

Campbell (1994) considers the time series of sudden infant death syndrome cases in the U.K. with the objective of understanding the relationship between temperature fluctuations and the number of daily SIDS deaths. On the basis of a model which does not include temperature he reasons that the method of Zeger is required to adjust for serial dependence in the observed counts. He then applies Zeger's (1988) method to a model in which temperature effects are included.

In our view there is no need to adjust for serial dependence in the full model (which includes temperature effects). In fact when we analyze these data we find that evidence for serial dependence through a latent process is non-existent. Using the model numbers that Campbell uses we find, upon application of the tests for the existence of a latent process the test statistic values shown in Table 3.

Note that when a linear trend effect is included (model 4) there is no evidence of the existence of a latent process.

We also examined the evidence for serial dependence using an observation driven model. Several ARMA terms were added to model 4 including an ARMA(1,0), ARMA((1,7),0), ARMA(0,1). In none of these models was there any evidence of significant effects, all estimated ARMA coefficients being not significant at anywhere near the 5 per cent level.

In conclusion, application of methods for detecting a latent process or for fitting an observation driven model leads to the conclusion that, provided the full set of regression terms (including temperature most importantly), there is no evidence of serial dependence additional to that in the regressor terms.

4.3 Analysis of the Asthma Data

Exploratory analysis of these data suggested the need to include model terms for a Sunday effect, a Monday effect, a possible increasing linear trend in time and Fourier series terms to model the seasonal pattern in the data. It was found necessary to include terms for $\cos(2\pi kt/365)$ and $\sin(2\pi kt/365)$ for $k = 1, 2, 3, 4$. Initial estimates of regression parameters were obtained using a Poisson regression. The log-likelihood was -796.48. The residuals from this fit indicated that there was significant serial dependence in the observed asthma counts. Indeed application of the test statistics described above gave the following results. The Q statistic was observed to be 2.41 and the variance of the residuals was 0.72 with standard error of 0.024 indicating significant additional variance than that explained by the Poisson regression alone. The portmanteau

Table 3   Test Statistics S_a and H^2_{OPT} for the SIDS Data

Model	S_a test	ACF test H^2_{OPT}, $L = 20$
Model 1	3.07	241.2
Model 2	1.60	109.1
Model 3	1.44	79.0
Model 4	0.58	12.30

test for serial dependence was observed to be 46.7 on 20 df which is highly significant (p = 0.00064) also indicating serial dependence. Significant values in the autocorrelation estimates (based on Zeger's method) were observed at lags 1, 2, 3, 5, 7, 10. The trend term was not significant in the Poisson regression and also in the observation driven model to be described next. Accordingly details of its estimation will not be given and it was omitted from the model. Thus the asthma presentations observed at Campbelltown appear not to have a significant increasing trend over the 4 years of observations. All Fourier terms had at least one significant coefficient of each pair. These were retained in the model to be described next.

To model the observed serial dependence we used the GLMARMA model in Eq. (3.4) with lags 1, 2, 3, 5, 7, 10 for the AR component and no moving average component, at least initially. The resulting estimates are given in Table 4. The log-likelihood value is -776.22 for this model indicating substantial improvement in the fit. Note that the AR coefficients at lags 2 and 5 are not significant. The model was refit dropping the AR coefficients at these two lags. The results are also shown in Table 4 for ease of comparison with the previous model. Note that the log-likelihood is now -778.24 which shows a non-significant change from the model with the additional AR terms at lags 2 and 5.

Application of the various test statistics to the residuals (although probably not valid for the lagged model) are as follows. The Q statistic was now observed to be 1.75 and the variance of the residuals was 0.052 with standard error of 0.024 still indicating significant additional variance than that explained by the subset GLMARMA model. The portmanteau test for serial dependence was observed to be 18.0 on 20 df which is not significant. The lag 2 autocorrelation is significantly different from zero at about the 1 per cent level. No other significant values in the autocorrelation estimates (based on Zeger's method) were observed.

Table 4 Summary of Observation-Driven Model Fits to the Asthma Data

Term	AR (1, 2, 3, 5, 7, 10)		AR (1, 3, 7, 10)	
	Parameter	*s.e.*	Parameter	*s.e.*
(Intercept)	0.553	0.031	0.532	0.030
Sunday effect	0.233	0.054	0.240	0.054
Monday effect	0.245	0.054	0.244	0.054
$\cos(2\pi t/365)$	−0.163	0.039	−0.163	0.037
$\sin(2\pi t/365)$	0.360	0.039	0.362	0.036
$\cos(4\pi t/365)$	−0.066	0.039	−0.067	0.038
$\sin(4\pi t/365)$	0.021	0.039	0.021	0.035
$\cos(6\pi t/365)$	−0.080	0.038	−0.080	0.036
$\sin(6\pi t/365)$	0.008	0.038	0.009	0.036
$\cos(8\pi t/365)$	−0.148	0.038	−0.152	0.036
$\sin(8\pi t/365)$	−0.057	0.038	−0.057	0.035
ϕ_1	0.044	0.017	0.047	0.017
ϕ_2	0.026	0.018	—	—
ϕ_3	0.046	0.018	0.049	0.017
ϕ_5	0.023	0.018	—	—
ϕ_7	0.058	0.017	0.059	0.017
ϕ_{10}	0.038	0.018	0.041	0.018

In summary there remains some slight overdispersion not explained by the lagged AR component of this observation driven model.

An alternative is to model the lag dependence using a moving average model. The most parsimonious MA is MA(1, 3) with log-likelihood value −786.87, but this is not an improvement over the AR. On the other hand the MA model is easier to use in practice involving a finite linear combination of past "errors" so that it might be preferred even if the model does not result in a noticeable improvement.

Acknowledgments

We gratefully acknowledge the assistance of Louise Helby for providing the daily asthma admissions data at Sydney hospitals and to Professor Mike Campbell for providing us with his UK SIDS and regression series used in our analysis. This research supported in part by NSF DMS Grant No. 9504596.

REFERENCES

Brannas, K. and P. Johansson, Time series count data regression, *Commun. Statist. Theory and Meth.*, *23* (*10*): 2907-2925 (1994).

Brockwell, P. J. and R. A. Davis, *Time Series: Theory and Methods* (*2nd edition*), Springer-Verlag, New York, 1991.

Brockwell, P. J. and R. A. Davis, *Introduction to Time Series and Forecasting*, Springer-Verlag, New York, 1996, Sec. 8.8.

Campbell, M. J., Time series regression for counts: an investigation into the relationship between sudden infant death syndrome and environmental temperature, *J. R. Statist. Soc. A*, *157*: 191–208 (1994).

Chan, K. S. and J. Ledolter, Monte Carlo EM estimation for time series models involving counts, *J. Amer. Statist. Assoc.*, *90*: 242–252 (1995).

Davis, M. H. A. and R. Vinter, *Stochastic Modeling and Control*, Chapman & Hall, New York, 1985.

Davis, R. A., W. T. M. Dunsmuir and Y. Wang, On autocorrelation in a Poisson regression model. Preprint 1998.

Dean, C. and J. F. Lawless, Tests for detecting overdispersion in Poisson regression models, *J. Amer. Statist. Assoc.*, *84*: 467–472 (1989).

Durbin, J. and S. J. Koopman, Monte Carlo maximum likelihood estimation for non-Gaussian state space models, *Biometrika*, *84*: 669–684 (1997).

Fahrmeir, L. and G. Tutz, *Multivariate Statistical Modeling Based on Generalized Linear Models*, Springer-Verlag, New York, 1994, Chapters 6–8.

Hannan, E. J., *Multiple Time Series*, John Wiley & Sons, New York, 1970.

Hannan, E. J. and M. Deistler, *The Statistical Theory of Linear Systems*, John Wiley & Sons, New York, 1988.

Harvey, A. C. and C. Fernandes, Time series models for count or qualitative observations, *J. Business and Economic Statistics*, *7*: 407–417 (1989).

Jorgensen, B., S. Lundbye-Christensen, X.-K. Song and L. Sun, A state space model for multivariate longitudinal count data, *Technique Report 148*, Department of Statistics, University of British Columbia, 1995.

Judge, G. G., *The Theory and Practice of Econometrics* (*2nd edition*), John Wiley & Sons, New York, 1985.

Kuk, A. Y. C. and Y. W. Cheng, The Monte Carlo Newton Raphson algorithm, *J. Statist. Comput. Simul.*, *59*: 233–250 (1997).

Saldiva, P. H., C. A. Pope, 3rd, J. Schwartz and D. W. Dockery, Air pollution and mortality in elderly people: a time-series study in Sao Paulo, Brazil, *Archives of Environmental Health*, *50*: 159–163 (1995).

Schwartz, J., C. Spix and G. Touloumi, Methodological issues in studies of air pollution and daily counts of deaths or hospital admissions, *J. Epidemiology and Community Health: 50* (*Supplement*), S3-S11 (1996).

Shephard, N., Generalized linear autoregressions, *Working Paper*, Nuffield College, Oxford, 1995.

Zeger, S. L. A regression model for time series of counts, *Biometrika*, *75*: 621–629 (1988).

Zeger, S. L. and B. Qaqish, Markov regression models for time series: a quasi-likelihood approach, *Biometrics*, *44*: 1019–1031 (1988).

4

Seasonal and Cyclical Long Memory

JOSU ARTECHE   University of the Basque Country, Bilbao, Spain

PETER M. ROBINSON   London School of Economics and Political Science, London, England

1. INTRODUCTION

There has recently been great interest in time series with long memory, namely series whose dependence decays slowly in the sense that autocovariances are not summable and the spectral density is unbounded. This concept has been extended to SCLM (Seasonal/Cyclical Long Memory) where the dependence between seasonal or cyclic observations decays similarly slowly. We discuss issues related to SCLM processes such as modeling, estimation, statistical inference, applications and extensions.

Long memory of a covariance stationary series x_t, $t = 0, \pm1, \pm2, \ldots$, may be modeled in the frequency domain by the spectral distribution function, $F(\lambda)$, or spectral density $f(\lambda) = dF(\lambda)/d\lambda$, satisfying

$$\gamma_j = \int_{-\pi}^{\pi} f(\lambda) \cos(j\lambda)\, d\lambda \tag{1.1}$$

where $\gamma_j = E(x_t - Ex_0)(x_{t+j} - Ex_0)$ is the lag-j autocovariance of x_t. In a semiparametric setup $f(\lambda)$ is typically assumed to behave as

$$f(\lambda) \sim C|\lambda|^{-2d} \quad \text{as} \quad \lambda \to 0 \tag{1.2}$$

where $0 < C < \infty$ and the memory or persistence parameter, d, satisfies $d < 1/2$ for stationarity and $d > -1/2$ for invertibility. x_t is said to have long memory if $d > 0$, short memory if $d = 0$ and negative memory if $d < 0$. For reviews see Beran (1994a) or Robinson (1994c). Under additional assumptions, see Yong (1974),

$$\gamma_j \sim K j^{2d-1} \qquad \text{as} \qquad j \to \infty, \tag{1.3}$$

where K is a positive constant when $0 < d < 1/2$. Equation (1.3) implies that the autocovariances decay at a slow hyperbolic rate rather than the exponential one typical of stationary ARMA processes, and they are eventually positive.

Many time series move in a regular or quasi-regular manner showing a cyclical evolution that produces oscillating autocorrelations and peaks in the spectral density whose locations define the cycles, a spectral peak at frequency ω reflecting a cycle of period $2\pi/\omega$. A particular case occurs when the spectral density has peaks at seasonal frequencies $\omega_h = 2\pi h/s$, $h = 1, 2, \ldots, [s/2]$, where s is the number of observations per year ($s = 4$ for quarterly data, $s = 12$ for monthly data) and $[s/2]$ denotes the integer part of $s/2$, that is $s/2$ if s is even and $(s-1)/2$ if s is odd. In this case we say that x_t is a seasonal process. Nerlove (1964) described seasonality as "that characteristic of a time series that gives rise to spectral peaks at seasonal frequencies". In this sense we consider seasonality a special case of cyclical behavior.

In this paper we focus on processes whose spectral density has a singularity or a zero at any frequency ω, $0 < \omega \leq \pi$, such that

$$f(\omega + \lambda) \sim C|\lambda|^{-2d} \qquad \text{as} \qquad \lambda \to 0, \ |d| < 1/2 \tag{1.4}$$

where C is a positive constant. Thus $f(\lambda)$ has a pole at $\lambda = \omega$ if $d > 0$ and a zero if $d < 0$. When $f(\lambda)$ satisfies Eq. (1.4) for every seasonal frequency $\omega = \omega_h$, $h = 1, 2, \ldots, [s/2]$, possibly with the memory parameter, d, varying across h, we say that the process has "seasonal long memory". However, for non-seasonal time series, perhaps of annual data, we can have cyclic behavior such that Eq. (1.4) holds for a single ω or for a single $\omega \in (0, \pi]$ as well as $\omega = 0$. We thus use the terminology SCLM (Seasonal/Cyclical Long Memory) for processes satisfying Eq. (1.4) for one or more $\omega \in (0, \pi]$ (though strictly $-1/2 < d < 0$ entails "negative dependence", not long memory).

SCLM processes might be described in terms of their autocovariances just as mentioned in Eq. (1.3) for standard long memory processes. A characteristic of autocovariances of SCLM processes is oscillating slow decay such that often $\gamma_j = O(j^{2d-1})$ as $j \to \infty$ but with oscillations whose amplitude depends on ω instead of the eventual monotonic decay in Eq. (1.3) of standard long memory processes at frequency zero.

The models traditionally used for seasonal and cyclical time series are stationary short memory processes on the one hand, or nonstationary processes due to a deterministic component such as seasonal dummies or to a stochastic trend such as seasonal unit roots. This work is reviewed in Sec. 2

in order to place SCLM in some perspective. The modeling of SCLM is described in more detail in Sec. 3. Section 4 discusses several parametric and semiparametric methods of estimation in SCLM processes. Tests of seasonal integration and cointegration are reviewed in Sec. 5. All this work assumes knowledge of the location of the poles/zeros in $f(\lambda)$, as is reasonable in a seasonal setting, but not necessary in a cyclic one. Section 6 describes approaches for estimating ω in parametric and semiparametric SCLM processes. Section 7 concludes the chapter with some mention of extensions and applications.

2. MODELLING SEASONALITY AND CYCLES

Seasonality has traditionally been considered a nuisance that obscures the more important components of time series (e.g. growth and cyclical components), and several seasonal adjustment procedures have been proposed. They are typically based on the idea that a time series, possibly after logarithmic transformation, is additively composed of three different components, the trend-cycle, T_t, the seasonal, S_t, and the irregular component, I_t,

$$x_t = T_t + S_t + I_t. \tag{2.1}$$

Traditionally T_t includes also the possibility of a cyclical component, considering the cycle as a periodic component with period larger than the number of observations per year. This implies a spectral peak at some frequency between zero and $2\pi/s$ which may be indistinguishable from a stochastic trend, characterized by a spectral pole at the origin. However, there may be cycles of period different from the seasonal ones, s/j, for $j = 1, 2, \ldots, [s/2]$. To allow for this behavior we can include a cyclic component, C_t, in Eq. (2.1),

$$x_t = T_t + C_t + S_t + I_t. \tag{2.2}$$

The additive form in Eqns (2.1) and (2.2) is often known as Unobserved Component (UC) or Structural Time Series model. The seasonally adjusted series is obtained by subtracting an estimate of S_t. We group the different methods of estimation of S_t and adjustment of x_t in two classes, "model-free" and "model-based" adjusting procedures. The "model-free" techniques ignore the seasonal and other structure of the series. They are based on the application of a succession of moving averages, perhaps the most widely used being the US Bureau of the Census X-11 procedure, Shiskin et al. (1967) and the X-11 ARIMA, Dagum (1980) which apply two-sided filters. The "model-based" seasonal adjustment procedures adapt to the characteristics of each series by estimation of parametric models. Some of these models are described below.

Seasonal adjustment procedures have been criticized for causing undesirable effects such as spectral dips at seasonal frequencies or distortion of the spectral density at other frequencies, see Nerlove (1964) or Bell and Hillmer (1984). Furthermore, the UC models in Eqns (2.1) and (2.2) suppose that each component in x_t can be specified separately and independently of the remainder, whereas the same model can include two or more components (for example the stochastic seasonal processes classified as (b), (c) and (d) below include an irregular component). Such factors have encouraged the use of seasonally unadjusted data.

Most of the processes described in this Section are seasonal, modeling a specific cyclical behavior. However, other cyclic patterns can be modelled similarly by suitably choosing the dummy variables, cosinusoids or lag operators in the models described below.

One of the earliest models for seasonality is the deterministic, strictly periodic form

$$x_t = \sum_{k=1}^{s} a_k D_{kt} \tag{2.3}$$

where $D_{kt} = 1$ if $t - k$ is a multiple of s (the number of observations per year) and 0 otherwise and

$$\sum_{k=1}^{s} a_k = 0,$$

which may be achieved by subtracting a constant from the original series. We can rewrite Eq. (2.3) as a function of sine and cosine waves,

$$x_t = \sum_{h=1}^{[s/2]} \Psi_{h,t}, \tag{2.4}$$

where

$$\Psi_{h,t} = \alpha_h \cos(\omega_h t) + \beta_h \sin(\omega_h t), \quad \omega_h = \frac{2\pi h}{s}, \tag{2.5}$$

$$\alpha_h = \frac{2}{s} \sum_{k=1}^{s} a_k \cos(k\omega_h),$$

$$\beta_h = \frac{2}{s} \sum_{k=1}^{s} a_k \sin(k\omega_h),$$

for $1 \leq h < s/2$, and if s is even $\beta_{s/2} \sin(k\omega_{s/2})$ is zero and

$$\alpha_{s/2} = \frac{1}{s} \sum_{k=1}^{s} a_k \cos(k\omega_{s/2}),$$

see Hannan (1963). x_t in Eq. (2.4) can equivalently be written $x_t = \sum_{h=1}^{[s/2]} r_h \cos(\omega_h t - \theta_h)$ where $r_h = \sqrt{\alpha_h^2 + \beta_h^2}$ is the h-th amplitude and $\theta_h = \arctan(\beta_h/\alpha_h)$ is the h-th phase. It is rarely plausible that time series have such a rigid deterministic behavior as Eqns (2.3) or (2.4) impose, so a stochastic error term is often added. If this irregular component is well behaved and the frequencies ω_h are known, then α_h and β_h in Eq. (2.5) or a_k in Eq. (2.3) can be estimated through simple regression methods. In fact least squares estimates have desirable orthogonality properties under uncorrelated errors, and are Gauss-Markov efficient under quite general (albeit short memory) autocorrelated errors.

The processes in Eqns (2.3) and (2.4) are completely deterministic, and if α_h, β_h are fixed parameters they are non-stationary so that it does not make sense to speak of a spectral distribution function or spectral density. However the spectral behavior of stochastic seasonal time series will give us relevant information on the characteristics of the process. According to spectral characteristics we distinguish four classes of stochastic seasonal/cyclical processes:

(a) Stationary with spectral distribution function with jumps and thus not absolutely continuous.
(b) Stationary with absolutely continuous spectral distribution function everywhere and smooth, positive, spectral density.
(c) Stationary with absolutely continuous spectral distribution function but spectral density with one or more singularities or zeros.
(d) Non-stationary so that no spectral distribution function exists.

(a) *Stationary process with jumping spectral distribution.* This kind of process is defined by Eqns (2.4) and (2.5) but $\Psi_{h,t}$ is made stochastic by allowing α_h and β_h to be random variables satisfying

$$E[\alpha_h] = E[\beta_h] = 0, \qquad E[\alpha_h^2] = E[\beta_h^2] = \sigma_h^2 \qquad \text{for all } h$$
$$E[\alpha_h \alpha_i] = E[\beta_h \beta_i] = 0 \quad h \neq i, \qquad E[\alpha_h \beta_i] = 0 \quad \text{for all } h, i. \tag{2.6}$$

Under Eq. (2.6), x_t is covariance stationary with lag-j autocovariance

$$\gamma_j = E(x_t x_{t-j}) = \sum_{h=1}^{[s/2]} \sigma_h^2 \cos(\omega_h j) = \int_{-\pi}^{\pi} \cos(j\lambda)\, dF(\lambda) \quad j = 0, \pm 1 \ldots .$$

Although α_h and β_h are random variables, they are fixed in a particular realization. Thus, although $\Psi_{h,t}$ is stationary, the model is still deterministic, only two observations are necessary to determine α_h and β_h, and once this has been done the remainder of the series can be forecast with zero mean squared error. The spectral distribution function, $F(\lambda)$, is a step function consisting of jumps of magnitude $\sigma_h^2/2$ at frequencies $-\omega_h$ and ω_h, for $h = 1, \ldots, [s/2]$. Since $F(\lambda)$ is not continuous the spectral density does not exist. However, in a similar manner as Stieltjes integration is carried out, we can define the so-called line or discrete spectrum, that is a discrete function with values $\sigma_h^2/2$ at frequencies $-\omega_h$ and ω_h for $h = 1, \ldots, [s/2]$. The line spectrum at ω_h gives the relative importance of a cycle of period s/h in the variance of x_t.

(b) *Stationary process with absolutely continuous spectral distribution and smooth spectral density.* The models in Eqns (2.3) and (2.4) assume that the cyclic behavior in x_t is constant across time and does not change its form. However, in many time series the seasonal/cyclical behavior is likely to change across time. Of course the variation must be slow (otherwise we cannot speak of seasonality or cycle) in such a way that the periodical structure seems to persist and the series has a quasi-periodic behavior. Hannan (1964) allows for this behavior in the model

$$x_t = \sum_{h=1}^{[s/2]} \Psi_{h,t}, \quad \Psi_{h,t} = \alpha_{h,t}\cos(\omega_h t) + \beta_{h,t}\sin(\omega_h t), \tag{2.7}$$

where $\omega_h = 2\pi h/s$ are seasonal frequencies and $\alpha_{h,t}$ and $\beta_{h,t}$ are not constant but evolving with time. Hannan (1964) assumed

$$E[\alpha_{h,t}] = E[\beta_{h,t}] = 0 \qquad \text{for all } h \text{ and all } t$$

$$E[\alpha_{h,t}\alpha_{h,t-j}] = E[\beta_{h,t}\beta_{h,t-j}] = c_h\rho_h^j$$

$$E[\alpha_{h,t}\alpha_{i,s}] = E[\beta_{h,t}\beta_{i,s}] = 0 \qquad \text{for } h \neq i \text{ and all } t, s \tag{2.8}$$

$$E[\alpha_{h,t}\beta_{i,s}] = 0 \qquad \text{for all} \qquad h, i \text{ and all } t, s$$

Thus the lag-j autocovariance of $\Psi_{h,t}$ is

$$E[\Psi_{h,t}\Psi_{h,t-j}] = c_h\rho_h^j\cos(\omega_h j) \tag{2.9}$$

Stationarity of $\Psi_{h,t}$ entails $|\rho_h| < 1$. However, ρ_h has to be close to 1 to avoid a fast changing behavior of $\Psi_{h,t}$. When $|\rho_h| < 1$, $\Psi_{h,t}$ is stationary and non deterministic with absolutely continuous spectral distribution and smooth spectral density,

$$f_h(\lambda) = \frac{c_h}{2\pi} \sum_{j=-\infty}^{\infty} \rho_h^j \cos(\omega_h j)\cos(\lambda j)$$

$$= \frac{c_h}{4\pi}\left\{\frac{1-\rho_h^2}{1+\rho_h^2-2\rho_h\cos(\lambda-\omega_h)} + \frac{1-\rho_h^2}{1+\rho_h^2-2\rho_h\cos(\lambda+\omega_h)}\right\} \tag{2.10}$$

which, for ρ_h near to unity, will concentrate around $\lambda = \omega_h$. Hannan et al. (1970) considered a parameterization of $\alpha_{h,t}$ and $\beta_{h,t}$ obeying Eq. (2.8),

$$\alpha_{h,t} = \rho_h \alpha_{h,t-1} + \varepsilon_{h,t}, \quad \beta_{h,t} = \rho_h \beta_{h,t-1} + \varepsilon_{h,t}^\dagger, \quad |\rho_h| < 1, \tag{2.11}$$

where $\varepsilon_{h,t}$ and $\varepsilon_{h,t}^\dagger$ have zero mean and common variance σ_h^2, and all correlations between ε, $\varepsilon^\dagger$ and between two time points and for differing values of h vanish. Substituting Eq. (2.11) in $\Psi_{h,t}$ in Eq. (2.7), we find that $\Psi_{h,t}$ is an ARMA(2, 1) process

$$[1 - 2\rho_h \cos(\omega_h)L + \rho_h^2 L^2]\Psi_{h,t} = \eta_{h,t} - \rho_h \cos(\omega_h)\eta_{h,t-1} - \rho_h \sin(\omega_h)\eta_{h,t-1}^\dagger,$$
$$\tag{2.12}$$

where

$$\eta_{h,t} = \varepsilon_{h,t} \cos(\omega_h t) + \varepsilon_{h,t}^\dagger \sin(\omega_h t)$$

$$\eta_{h,t}^\dagger = \varepsilon_{h,t} \sin(\omega_h t) - \varepsilon_{h,t}^\dagger \cos(\omega_h t)$$

are thus zero mean random variables with variance σ_h^2 and inherit the uncorrelatedness properties of $\varepsilon_{h,t}$ and $\varepsilon_{h,t}^\dagger$. The lag-$j$ autocovariance and spectral density of $\Psi_{h,t}$ are Eqns (2.9) and (2.10) with $c_h = \sigma_h^2/(1 - \rho_h^2)$. Consequently the spectrum of x_t is a smooth function

$$f(\lambda) = \sum_{h=1}^{[s/2]} f_h(\lambda) \tag{2.13}$$

which shows peaks (the sharper the closer ρ_h is to 1) around seasonal frequencies ω_h, $h = 1, 2, mathinner..., [s/2]$.

In addition to the specific ARMA in Eq. (2.12) we can use many other ARMA processes to model a changing cyclical behavior. In particular, if the spectrum of an AR(2), $(1 - \phi_1 L - \phi_2 L^2)x_t = \varepsilon_t$, contains a peak at frequency λ^* within the range $0 < \lambda^* < \pi$, its exact position is

$$\lambda^* = \cos^{-1}\left[\frac{-\phi_1(1 - \phi_2)}{4\phi_2}\right].$$

For example the spectrum of the AR part in Eq. (2.12) has a peak at

$$\lambda^* = \cos^{-1}\left[\frac{(1 + \rho_h^2)\cos\omega_h}{2\rho_h}\right]$$

so that λ^* is closer to ω_h the closer ρ_h is to 1. We can also use the seasonal lag operator, L^s, ($L^s x_t = x_{t-s}$) to define the seasonal ARMA(1, 1) model

$$(1 - \phi_s L^s)x_t = (1 + \theta_s L^s)\varepsilon_t \tag{2.14}$$

where ε_t is white noise with variance σ^2. When ϕ_s and θ_s are inside the unit circle, x_t is stationary and invertible with smooth spectral density

$$f(\lambda) = \frac{\sigma^2}{2\pi} \frac{1 + \theta_s^2 + 2\theta_s \cos(\lambda s)}{1 + \phi_s^2 - 2\phi_s \cos(\lambda s)}$$

If $\phi_s > 0$ and $\theta_s > 0$, $f(\lambda)$ exhibits peaks at the seasonal harmonic frequencies, $\omega_h = 2\pi h/s$, $h = 1, 2, \ldots, [s/2]$, as well as at zero. More general seasonal ARMA processes can be defined as

$$\Phi_s(L^s)x_t = \Theta_s(L^s)\varepsilon_t \tag{2.15}$$

where $\Phi_s(L^s)$ and $\Theta_s(L^s)$ are polynomials in the seasonal lag operator with zeros outside the unit circle, see Box and Jenkins (1976).

(c) *Stationary process with absolutely continuous spectral distribution and singularities or zeros in its spectral density.* The structure of α_h and β_h in Eq. (2.11) may generate a relatively rapid change in the seasonal pattern, whereas the definition of seasonality implies a regular or quasi-regular behavior. The closer ρ_h is to 1 the more regular the movement of $\Psi_{h,t}$. In fact we can choose $\rho_h = 1$, but in this case $\Psi_{h,t}$ ceases to be stationary. Instead we can assume that $\alpha_{h,t}$ and $\beta_{h,t}$ evolve as

$$(1 - L)^{d_h}\alpha_{h,t} = \varepsilon_{h,t}, \qquad (1 - L)^{d_h}\beta_{h,t} = \varepsilon_{h,t}^\dagger, \tag{2.16}$$

where $\varepsilon_{h,t}$ and $\varepsilon_{h,t}^\dagger$ are defined as in Eq. (2.11). Thus $\alpha_{h,t}$ and $\beta_{h,t}$ are fractional ARIMA$(0, d_h, 0)$ processes and they are stationary if $d_h < 1/2$ and invertible if $d_h > -1/2$, see Hosking (1981). The slowly changing behavior necessary for seasonality requires $d_h > 0$ and stationarity entails $d_h < 1/2$. Under these circumstances the spectral density of $\alpha_{h,t}$ and $\beta_{h,t}$, $f_0(\lambda)$, satisfies Eq. (1.4) for $\omega = 0$. Their lag-j autocovariance is

$$\gamma_{h,j}^\dagger = E[\alpha_{h,t}\alpha_{h,t-j}] = E[\beta_{h,t}\beta_{h,t-j}] = \sigma_h^2 \frac{\Gamma(1 - 2d_h)\Gamma(j + d_h)}{\Gamma(d_h)\Gamma(1 - d_h)\Gamma(j + 1 - d_h)}.$$

Thus the lag-j autocovariance of $\Psi_{h,t}$ is

$$E[\Psi_{h,t}\Psi_{h,t-j}] = \gamma_{h,j}^\dagger \cos(j\omega_h)$$

and its spectral density is

$$f_h(\lambda) = \frac{1}{2\pi} \sum_{j=-\infty}^{\infty} \gamma_{h,j}^\dagger \cos(j\omega_h)e^{-i\lambda j} = \frac{1}{4\pi} \sum_{j=-\infty}^{\infty} \gamma_{h,j}^\dagger e^{-i\lambda j}(e^{ij\omega_h} + e^{-ij\omega_h})$$

$$= \tfrac{1}{2}f_0(\lambda - \omega_h) + \tfrac{1}{2}f_0(\lambda + \omega_h).$$

The multiplication of $\alpha_{h,t}$ by $\cos(\omega_h t)$ and $\beta_{h,t}$ by $\sin(\omega_h t)$ produces a phase shift such that the spectral pole moves from zero in $\alpha_{h,t}$ and $\beta_{h,t}$ to ω_h in $\Psi_{h,t}$. Thus, the process defined by Eqns (2.7) and (2.16) has an absolutely continuous spectral distribution but its spectral density is not smooth, but goes to ∞ (if $d_h > 0$) or is zero (if $d_h < 0$), at frequencies $\pm\omega_h$ as described in Eq. (1.4). This is the SCLM property that characterizes the processes we focus attention on in this paper. A more detailed description of models with this property is given in the next section.

(d) *Non-stationary and non-deterministic stochastic seasonal process.* If $\alpha_{h,t}$ and $\beta_{h,t}$ are determined by the fractional ARIMAs in Eq. (2.16) but with $d_h \geq 1/2$, then they, and thus $\Psi_{h,t}$ in Eq. (2.7), are non-stationary. In this case there does not exist a spectral distribution. Nevertheless, the frequency domain is still an adequate framework to detect seasonality using the pseudospectrum. If $u_t = \tau(L)x_t$ is stationary with spectrum $f_u(\lambda)$, the pseudospectrum of x_t is $f(\lambda) = |\tau(e^{i\lambda})|^{-2}f_u(\lambda)$. For example, if $d_h = 1$ in Eq. (2.16) or equivalently $\rho_h = 1$ in Eq. (2.11), then $\Psi_{h,t}$ is a non-stationary ARMA(2, 1) process

$$\tau_h(L)\Psi_{h,t} = \eta_{h,t} - \cos(\omega_h)\eta_{h,t-1} - \sin(\omega_h)\eta_{h,t-1}^{\dagger}$$

where $\tau_h(L) = 1 - 2\cos(\omega_h)L + L^2$. The non-stationarity comes from the fact that the AR polynomial, $\tau_h(L)$, has zeros at $\cos\omega_h \pm \sqrt{\cos^2\omega_h - 1}$, with modulus one. However $\tau_h(L)\Psi_{h,t}$ is a stationary MA(1). Since $|\tau_h(e^{i\lambda})|^{-2} = [2(\cos\omega_h - \cos\lambda)]^{-2}$ diverges at $\lambda = \pm\omega_h$, then the pseudospectrum of $\Psi_{h,t}$ goes to infinity at frequencies $\pm\omega_h$, reflecting a strong cyclical pattern with period $2\pi/\omega_h = s/h$. Hannan et al. (1970) estimated this model using optimal signal extraction methods, see also Hannan (1967).

In the Box–Jenkins framework we can define the seasonal ARIMA(P, D, Q) time series

$$\Phi_s(L^s)(1 - L^s)^D x_t = \Theta_s(L^s)\varepsilon_t \tag{2.17}$$

where the ε_t are white noise $(0, \sigma^2)$, $\Phi_s(L^s)$ and $\Theta_s(L^s)$ are polynomials in the lag operator with zeros outside the unit circle, and D is a positive integer in Box and Jenkins (1976) but could instead be fractional, see Hosking (1984). Then Eq. (2.17) defines the fractional seasonal ARIMA(P, D, Q), that is stationary if $D < 1/2$ and non-stationary if $D \geq 1/2$. The spectrum ($D < 1/2$) or pseudospectrum (if $D \geq 1/2$) of x_t is

$$f(\lambda) = \frac{\sigma^2}{2\pi} \frac{|\Theta_s(e^{i\lambda s})|^2}{|\Phi_s(e^{i\lambda s})|^2} \left(2\sin\frac{\lambda s}{2}\right)^{-2D} \tag{2.18}$$

and diverges if $D > 0$ or is zero if $D < 0$ at frequencies $\omega_h = 2\pi h/s$, $h = 0, \ldots, [s/2]$, that is at the origin and seasonal frequencies. The seasonal difference operator, $(1 - L^s)$, can be written as the product of the difference operator, $(1 - L)$, and the seasonal summation operator, $S(L) = (1 + L + \cdots + L^{s-1})$, such that the pole in Eq. (2.18) at the origin corresponds to the operator $(1 - L)$, and the spectral poles at seasonal frequencies are due to $S(L)$. Thus $(1 - L^s)$ includes a stochastic trend in addition to the seasonal factor. This is why sometimes, e.g. Harvey (1989), $S(L)$ is used instead of $(1 - L^s)$ to model the seasonal component of the UC models in Eqns (2.1) and (2.2).

Another type of non-stationarity may be described by a different data generating process for each season. This phenomenon is often modeled via the Periodic ARIMA process, e.g. Troutman (1979), Tiao and Grupe (1980), Osborn (1991), Franses and Ooms (1995),

$$\Phi_q(L)(1 - L^s)^{d_q} x_T^q = \Theta_q(L)\varepsilon_T^q \qquad q = 1,\ldots,s, \qquad T = 1,2,\ldots \quad (2.19)$$

where ε_T^q is white noise with variance σ_q^2, the index q indicates the season or situation of the observation in the cycle (for example different months) and T represents the year such that $x_T^q = x_{(T-1)s+q}$. Thus, Eq. (2.19) allows for s different models, one per season. When the zeros of $\Phi_q(L)$ and $\Theta_q(L)$ lie outside the unit circle, and $d_q < 1/2$, then Eq. (2.19) is stationary for every $q = 1, 2, \ldots, s$. Although x_t^q may be stationary, x_t is non-stationary if some parameters vary with q. In this case the autocovariances of x_t depend on q and therefore are not time invariant and we cannot use frequency domain techniques. This kind of process is usually analysed in a multivariate set-up using a vector ARMA representation. Define the $s \times 1$ vector $z_T = (x_T^1, \ldots, x_T^s)' = (x_{(T-1)s+1}, \ldots, x_{Ts})'$. The periodic process in Eq. (2.19) can be written in vector ARMA form as

$$A(L^*)C(L^*)z_T = B(L^*)u_T \qquad T = 1,2,\ldots, \quad (2.20)$$

where $u_T = (\varepsilon_T^1, \ldots, \varepsilon_T^q)'$, $C(L^*) = \text{diag}\{(1 - L^*)^{d_q}\}$, $A(L^*)$ and $B(L^*)$ are matrix polynomials in L^*, and the operator L^* is the lag operator for the index T, $L^* z_T = z_{T-1}$. This implies seasonal difference in the elements of z_T, $L^* x_T^q = L^s x_{(T-1)s+q} = x_{(T-2)s+q} = x_{T-1}^q$. The vector z_T is stationary if $d_q < 1/2$ for $q = 1, \ldots, s$, and $|A(z)|$ has zeros outside the unit circle, and is invertible if $d_q > -1/2$ for $q = 1, \ldots, s$, whereas the zeros of $|B(z)|$ lie outside the unit circle. Under stationarity z_T has a spectral density matrix $f_z(\lambda)$. Although x_t is non-stationary the expectation of the sample autocovariances of x_t converges to the autocovariances of a stationary process with spectral density function

$$f(\lambda) = \frac{1}{s} R(e^{i\lambda})' f_z(s\lambda) R(e^{-i\lambda}) \quad (2.21)$$

where $R(r)$ is a $s \times 1$ vector with k-th element r^k, see Tiao and Grupe (1980). Thus asymptotically we can use Eq. (2.21) to classify periodic processes in the same way as non-periodic seasonal models.

3. SCLM PROCESSES

This Section describes commonly used parametric models of the class (c) of stochastic seasonal processes introduced in the previous section, that is processes whose spectral density satisfies Eq. (1.4). We say that such processes have SCLM, and using the notation in Engle et al. (1989) we denote them by $I_\omega(d)$ (integrated of order d at ω).

Though Eq. (1.4) is a semiparametric condition, only imposing knowledge of $f(\lambda)$ around ω, it is interesting to describe parametric processes satisfying Eq. (1.4), specifying short memory as well as long memory components of x_t, for example for the purpose of Monte Carlo simulation. Some examples have been introduced in the previous Section, for example Eqns (2.7) and (2.16) or (2.17). In case of Gaussianity it suffices to specify the mean, μ, and $f(\lambda)$ for all $\lambda \in (-\pi, \pi]$, or equivalently γ_j, for all j. Autocovariances of SCLM processes have a slow decay typical of long memory along with oscillations depending on the frequency ω such that, for $d > 0$, $\sum |\gamma_j| = \infty$, although this is consistent with $\sum \gamma_j$, and thus $f(0)$, being finite. These observations apply to non-Gaussian (finite variance) series as well as Gaussian ones, though there remains the possibility that x_t may not exhibit long memory in second moments but in some other way (for example x_t^2 could have long memory), as briefly discussed in Section 7.

Two SCLM models have been stressed in the literature, being natural extensions of models for long memory at zero frequency, namely the fractional noise and the fractional ARIMA.

3.1 Seasonal Fractional Noise

This kind of stationary process is characterized by a spectral density

$$f(\lambda) = c|1 - \cos(s\lambda)| \sum_{j=-\infty}^{\infty} \left| \lambda + \frac{2\pi}{s}j \right|^{-2(1+d)} \tag{3.1}$$

and lag-j autocovariance

$$\gamma_j = \frac{V(x_1)}{2} \left(\left| \frac{j}{s} + 1 \right|^{2d+1} - 2 \left| \frac{j}{s} \right|^{2d+1} + \left| \frac{j}{s} - 1 \right|^{2d+1} \right) \tag{3.2}$$

where s is the number of observations per year, c is a positive constant and $d < 1/2$, see Jonas (1983), Carlin and Dempster (1989) or Ooms (1995). The spectrum in Eq. (3.1) satisfies Eq. (1.4) for $\omega = 2\pi h/s$, $h = 0, 1, \ldots, [s/2]$, the γ_j in Eq. (3.2) have slow and oscillating decay as $j \to \infty$, and if $d > 0$ they are not absolutely summable. This kind of process generalizes the fractional noise described by Mandelbrot and Van Ness (1968), characterized by Eqns (3.1) or (3.2) with $s = 1$, and having typical long memory behavior at frequency zero.

3.2 SCLM in the Box-Jenkins Set-up

Andel (1986), and later and in more depth Gray et al. (1989, 1994), analysed the so-called Gegenbauer process

$$(1 - 2L\cos\omega + L^2)^d x_t = u_t \tag{3.3}$$

where u_t has positive and continuous spectrum, $f_u(\lambda)$, and d can be any real number. For example when u_t is a stationary and invertible ARMA(p, q) Eq. (3.3) is called GARMA (Gegenbauer ARMA). The spectral density of x_t in Eq. (3.3) is

$$f(\lambda) = (2(\cos\omega - \cos\lambda))^{-2d} f_u(\lambda) \tag{3.4}$$

and satisfies Eq. (1.4), so x_t has SCLM at frequency ω for $|d| < 1/2$ when $\omega \neq 0, \pi$. When $\omega = 0$ and u_t is an ARMA(p, q), then Eq. (3.3) is the fractional ARIMA$(p, 2d, q)$, $(1 - L)^{2d} x_t = u_t$, so that x_t is stationary if $d < 1/4$ and invertible when $d > -1/4$. If $\omega = \pi$, x_t is stationary if $d < 1/4$ and invertible when $d > -1/4$. When u_t is iid$(0, \sigma^2)$, and $d < 1/2$, the auto-covariances of x_t are

$$\gamma_j = \frac{\sigma^2}{2\sqrt{\pi}} \Gamma(1 - 2d)(2\sin\omega)^{(1/2)-2d} [P_{j-(1/2)}^{2d-(1/2)}$$

$$\times (\cos\omega) + (-1)^j P_{j-(1/2)}^{2d-(1/2)}(-\cos\omega)] \tag{3.5}$$

where $P_a^b(z)$ are associated Legendre functions, see Chung (1996a). The asymptotic behavior of γ_j in Eq. (3.5) is

$$\gamma_j \sim K\cos(j\omega)j^{2d-1} \qquad \text{as} \qquad j \to \infty \tag{3.6}$$

where K is a finite constant that depends on d but not on j, see Gray et al. (1989) or Chung (1996a), so γ_j has the slow and oscillating decay typical of SCLM.

Hosking (1984), Porter-Hudak (1990) and Ray (1993) among others, proposed use of the fractional seasonal difference operator, $(1 - L^s)^d$, where d can be any real number. Porter-Hudak (1990) used the operator $(1 - L^{12})^d$ in monthly monetary U.S.A. aggregates and Ray (1993) used

$(1 - L^3)^{d_3}(1 - L^{12})^{d_{12}}$ for monthly IBM revenue data. Note that for even s, $(1 - L^s)^d$ can be decomposed into the product of operators of type $(1 - 2L\cos\omega + L^2)^d$. For instance if $s = 4$,

$$(1 - L^4)^d = (1 - 2L\cos\omega_0 + L^2)^{d/2}(1 - 2L\cos\omega_1 + L^2)^d$$

$$\times (1 - 2L\cos\omega_2 + L^2)^{d/2} \tag{3.7}$$

for $\omega_0 = 0$, $\omega_1 = \pi/2$ and $\omega_2 = \pi$. Thus x_t in $(1 - L^4)^d x_t = u_t$ is $I_0(d)$, $I_{\pi/2}(d)$ and $I_\pi(d)$.

In order to allow for different persistence parameters across different frequencies, Chan and Terrin (1995), Chan and Wei (1988), Giraitis and Leipus (1995) and Robinson (1994a) used the model

$$(1 - L)^{d_0}\left\{\prod_{j=1}^{h-1}(1 - 2L\cos\omega_j + L^2)^{d_j}\right\}(1 + L)^{d_h}x_t = u_t \tag{3.8}$$

where the ω_j can be any frequencies between 0 and π and u_t has continuous and positive spectrum. Thus x_t in Eq. (3.8) is $I_{\omega_j}(d_j)$ for $j = 0, 1, 2, \ldots, h$, where $\omega_0 = 0$ and $\omega_h = \pi$. When u_t is a stationary and invertible ARMA, Giraitis and Leipus (1995) used the terminology ARUMA for such x_t. When $|d_j| < 1/2$ for $j = 0, 1, \ldots, h$, Eq. (3.8) can be expressed as

$$\sum_{j=0}^{\infty} \pi_j x_{t-j} = u_t$$

or

$$x_t = \sum_{j=0}^{\infty} \psi_j u_{t-j}$$

where $\pi_0 = \psi_0 = 1$ and

$$\pi_j = \sum_{\substack{0 \le k_0, \ldots, k_h \le j \\ k_0 + \cdots + k_h = j}} C_{k_0}^{(-d_0/2)}(\eta_0)C_{k_1}^{(-d_1)}(\eta_1)\ldots C_{k_{h-1}}^{(-d_{h-1})}(\eta_{h-1})C_{k_h}^{(-d_h/2)}(\eta_h) \tag{3.9}$$

for $j = 1, \ldots$, where $\eta_i = \cos\omega_i$, $i = 0, 1, \ldots, h$, and $C_k^{(d)}(x)$ are orthogonal Gegenbauer polynomials. Similarly ψ_j is Eq. (3.9) with $d_0, \ldots, d_h$ instead of $-d_0, \ldots, -d_h$, see Giraitis and Leipus (1995). The weights π_j in Eq. (3.9) have the asymptotic behavior

$$\pi_j \sim K\left[j^{-1-d_0} + (-1)^j j^{-1-d_h} + \sum_{k=1}^{h-1} j^{-d_k-1}(\cos(\omega_k j) + v_k)\right] \tag{3.10}$$

where K is a finite constant and v_k is a constant depending on $d_0, \ldots, d_h$ and ω_k. Similarly the ψ_j behave asymptotically as in Eq. (3.10) with $d_0, \ldots, d_h$ instead of $-d_0, \ldots, -d_h$.

The complicated form of Eq. (3.8) impedes calculation of explicit formulae for autocovariances, which have only been obtained for the Gegenbauer process in Eq. (3.3), see Eq. (3.5). If there is more than one spectral pole/zero, only asymptotic behavior has been established. Giraitis and Leipus (1995) showed that the autocovariances of Eq. (3.8) satisfy

$$\gamma_j \sim K \sum_{k=0}^{h} j^{2d_k - 1} \cos(j\omega_k) \qquad \text{as } j \to \infty.$$

Thus π_j, ψ_j and γ_j have slow decay with oscillations depending on the different ω_k. Eventually it is the largest persistence parameter which governs the behavior of π_j, ψ_j and γ_j.

The model in Eq. (3.8) allows for spectral poles/zeros at any frequency $\omega_j \in [0, \pi]$. One particular case occurs when ω_j are seasonal frequencies, $\omega_j = 2\pi j/s$, $j = 1, 2, \ldots, [s/2]$. Then Eq. (3.8) has been called "flexible ARFISMA," Hassler (1994) or "flexible (seasonal) ARMA$(p, d, q)_s$," Ooms (1995).

4. ESTIMATION IN SCLM PROCESSES

Hurst (1951) introduced the rescaled range statistic (R/S) to measure long memory in the flows of the river Nile, and R/S has been analysed, applied and extended by a number of subsequent authors. However R/S does not extend readily to the SCLM context, so we explore alternative approaches.

Statistical inference in long memory processes can be parametric or semiparametric. Parametric methods are generally more efficient if they are based on a correct and complete specification of $f(\lambda)$, but even estimates of the persistence parameter, d in Eq. (1.4), can be inconsistent if $f(\lambda)$ is misspecified at frequencies far from ω. Semiparametric techniques, that only assume partial knowledge of $f(\lambda)$ around a known frequency (like in Eq. (1.4)) are less efficient but guarantee consistency under much more general circumstances.

4.1 Parametric Estimation

Consider the covariance stationary process, x_t, satisfying

$$\phi(L)(x_t - \mu_0) = \varepsilon_t \tag{4.1}$$

where

$$\phi(z) = 1 - \sum_{j=1}^{\infty} \phi_j z^j, \qquad \sum_{j=1}^{\infty} \phi_j^2 < \infty, \tag{4.2}$$

$\mu_0 = Ex_t$ and the ε_t have zero mean and are uncorrelated with variance σ_0^2, for all t. All the stationary and invertible processes described in previous sections can be written as Eq. (4.1) satisfying Eq. (4.2). Suppose that the ϕ_j and σ_0^2, as well as μ_0, are unknown, but we know a function

$$\phi(z; \theta) = 1 - \sum_{j=1}^{\infty} \phi_j(\theta) z^j$$

where θ is a $k \times 1$ vector such that there exists an unknown θ_0 for which $\phi_j(\theta_0) = \phi_j$ for all j, and therefore $\phi(z; \theta_0) = \phi(z)$. The spectral density of x_t is given by

$$f(\lambda) = \frac{\sigma_0^2}{2\pi} |\phi(e^{i\lambda})|^{-2}, \qquad -\pi < \lambda \leq \pi, \tag{4.3}$$

and the lag-j autocovariance by

$$\gamma_j = \int_{-\pi}^{\pi} f(\lambda) \cos(j\lambda)\, d\lambda.$$

For any admissible θ we introduce

$$\gamma_j(\theta) = \frac{1}{2\pi} \int_{-\pi}^{\pi} h(\lambda; \theta) \cos(j\lambda)\, d\lambda,$$

$$h(z; \theta) = |\phi(e^{iz}; \theta)|^{-2}.$$

In this Section we consider so-called Gaussian estimates, although Gaussianity is not required to achieve good asymptotic properties. Denote by $\Delta(\theta)$ the $n \times n$ Toeplitz matrix with (i,j)-th element $\gamma_{i-j}(\theta)$, by $\mathbf{1}$ the $n \times 1$ vector of ones and by x the $n \times 1$ vector of observations $(x_1, x_2, \ldots, x_n)'$. For

$$L_a(\theta, \mu, \sigma^2) = \tfrac{1}{2}\log \sigma^2 + \tfrac{1}{2}\log |\Delta(\theta)| + \frac{1}{2\sigma^2}(x - \mu\mathbf{1})' \Delta(\theta)^{-1}(x - \mu\mathbf{1}) \tag{4.4}$$

define

$$(\hat{\theta}_a, \hat{\mu}_a, \hat{\sigma}_a^2) = \arg\min_{\theta, \mu, \sigma^2} L_a(\theta, \mu, \sigma^2)$$

where the minimization is carried out over an appropriate set. In case the ε_t in Eq. (4.1) (and therefore x_t) are Gaussian, $\hat{\theta}_a$ is a maximum likelihood estimate of θ_0.

As in other optimization problems introduced below, σ_0^2 and μ_0 can be estimated in closed form and the nonlinear optimization carried out only with respect to θ. Under regularity conditions $\hat{\theta}_a$ is consistent and

$$\sqrt{n}(\hat{\theta}_a - \theta_0) \xrightarrow{d} N_k(0, \Omega^{-1}) \tag{4.5}$$

where $\xrightarrow{d}$ means convergence in distribution, $N_k(\cdot, \cdot)$ is a k-variate normal and

$$\Omega = \frac{1}{4\pi} \int_{-\pi}^{\pi} \frac{\partial}{\partial \theta} \log h(\lambda; \theta_0) \frac{\partial}{\partial \theta'} \log h(\lambda; \theta_0) \, d\lambda. \tag{4.6}$$

Since the function $h(z; \theta)$ is known, Ω can be consistently estimated by, for example, substituting θ_0 in Eq. (4.6) by a consistent estimate of it (e.g. $\hat{\theta}_a$). These asymptotic properties do not rely on x_t being Gaussian, though under Gaussianity $\hat{\theta}_a$ is also asymptotically efficient.

We can approximate $L_a(\theta, \mu, \sigma^2)$ by

$$L_b(\theta, \mu, \sigma^2) = \tfrac{1}{2} \log \sigma^2 + \frac{1}{2\sigma^2} \sum_{t=1}^{n} \varepsilon_t^2(\theta, \mu) \tag{4.7}$$

where $\varepsilon_t(\theta, \mu) = \phi(L; \theta)(x_t - \mu)$ and $x_t = 0$ for $t \leq 0$. We call

$$(\hat{\theta}_b, \hat{\mu}_b, \hat{\sigma}_b^2) = \arg\min_{\theta, \mu, \sigma^2} L_b(\theta, \mu, \sigma^2)$$

a (nonlinear) least squares estimate. Under regularity conditions, $\hat{\theta}_b$ has the same asymptotic properties as $\hat{\theta}_a$.

Next define the centered periodogram

$$I_n(\lambda; \mu) = \frac{1}{2\pi n} \left| \sum_{t=1}^{n} (x_t - \mu) e^{it\lambda} \right|^2. \tag{4.8}$$

Whittle (1953) proposed to approximate $L_a(\theta, \mu, \sigma^2)$ by

$$L_c(\theta, \mu, \sigma^2) = \frac{1}{2\pi} \int_{-\pi}^{\pi} \left\{ \log \sigma^2 h(\lambda; \theta) + \frac{I_n(\lambda; \mu)}{\sigma^2 h(\lambda; \theta)} \right\} d\lambda \tag{4.9}$$

and the estimates

$$(\hat{\theta}_c, \hat{\mu}_c, \hat{\sigma}_c^2) = \arg\min_{\theta, \mu, \sigma^2} L_c(\theta, \mu, \sigma^2).$$

Under regularity conditions, $\hat{\theta}_c$ has the same asymptotic properties as $\hat{\theta}_a$ and $\hat{\theta}_b$.

Finally define the (uncentered) periodogram

$$I_n(\lambda) = \frac{1}{2\pi n} \left| \sum_{t=1}^{n} x_t e^{it\lambda} \right|^2. \tag{4.10}$$

Define the Fourier or harmonic frequencies $\lambda_j = 2\pi j/n$, and the discrete approximation to $L_c(\theta, \mu, \sigma^2)$, see Hannan (1973b)

$$L_d(\theta, \sigma^2) = \frac{1}{n} \sum_j' \left\{ \log \sigma^2 h(\lambda_j; \theta) + \frac{I_n(\lambda_j)}{\sigma^2 h(\lambda_j; \theta)} \right\} \qquad (4.11)$$

where $\sum_j'$ runs over all $j = 1, \ldots, n - 1$, such that $0 < h(\lambda_j; \theta) < \infty$ for all admissible θ. By omitting $j = 0$ and n we avoid the need to estimate μ_0. Let

$$(\hat{\theta}_d, \hat{\sigma}_d^2) = \arg\min_{\theta, \sigma^2} L_d(\theta, \sigma^2)$$

where the minimization is over a compact subset of R^{k+1}. Then $\hat{\theta}_d$ typically has the same asymptotic properties as $\hat{\theta}_a$, $\hat{\theta}_b$ and $\hat{\theta}_c$ described above.

The relative computational needs of $\hat{\theta}_a$, $\hat{\theta}_b$, $\hat{\theta}_c$ and $\hat{\theta}_d$, which we call Gaussian estimates, depend on the parameterization we impose. In general, $\hat{\theta}_b$ is more easily calculated than $\hat{\theta}_a$ since it avoids the matrix inversion in Eq. (4.4) and $\hat{\theta}_d$ more easily calculated than $\hat{\theta}_b$ and $\hat{\theta}_a$ because $h(\lambda; \theta)$ is typically of simpler form than $\varepsilon_t(\theta, \mu)$ and $\gamma_j(\theta)$. Moreover $\hat{\theta}_d$ makes especially convenient use of the fast Fourier transform.

The above discussion has made no reference to long memory or SCLM models, and in fact $\hat{\theta}_a$, $\hat{\theta}_b$, $\hat{\theta}_c$ and $\hat{\theta}_d$ and their asymptotic properties were originally obtained for short memory time series models such as stationary and invertible ARMA's, see for example Whittle (1953) or Hannan (1973b). However the discussion also seems relevant to SCLM models. In fact, for long memory models with a spectral pole/zero only at the origin, Fox and Taqqu (1986), Dahlhaus (1989), Giraitis and Surgailis (1990), Heyde and Gay (1993) and Hosoya (1997) provide asymptotic properties for $\hat{\theta}_c$ which are identical to those earlier obtained for short memory processes, see for example Eq. (4.5). Li and McLeod (1986) and Sowell (1986, 1992) discuss computational aspects of $\hat{\theta}_a$ for fractional ARIMA processes

$$\Phi(L)(1 - L)^d (x_t - \mu_0) = \Theta(L)\varepsilon_t \qquad (4.12)$$

where the zeros of $\Phi(z)$ and $\Theta(z)$ lie outside the unit circle. $\hat{\theta}_b$ for invertible, possibly non-stationary fractional ARIMA processes has been analysed by Beran (1995). Beran (1994b) proposed a modified version of $\hat{\theta}_b$ for long memory processes that is robust against outliers. Asymptotic theory for $\hat{\theta}_d$ has not been considered explicitly for long memory models with a spectral pole at zero frequency but it can be done by avoiding the spectral singularity with the omission of frequencies close to the origin in $L_d(\theta, \sigma^2)$. In case of long memory at frequency zero $\hat{\theta}_d$ has an extra advantage over $\hat{\theta}_a$, $\hat{\theta}_b$ and $\hat{\theta}_c$, because these are affected by $\hat{\mu}_a$, $\hat{\mu}_b$ and $\hat{\mu}_c$ which converge more slowly than $\sqrt{n}$, see Vitale (1973), Adenstedt (1974) and Samarov and Taqqu (1988), as discussed by Cheung and Diebold (1994) via Monte Carlo analysis.

The discussion of Gaussian estimates also seems relevant to SCLM models with spectral poles/zeros at known frequencies different from zero. Consider

$$\Phi(L)\prod_{j=0}^{h}(1 - 2L\cos\omega_j + L^2)^{d_j}(x_t - \mu_0) = \Theta(L)\varepsilon_t \qquad (4.13)$$

where $d_j > 0$ for all j, and $d_j < 1/2$ if $\omega_j \neq 0, \pi$, and $d_j < 1/4$ if $\omega_j = 0, \pi$, $\Theta(z)$ and $\Phi(z)$ have their roots outside the unit circle and the ε_t are as before. In this case

$$h(\lambda;\theta) = \left|\frac{\Theta(e^{i\lambda})}{\Phi(e^{i\lambda})}\right|^2 \prod_{j=0}^{h}[2(\cos\lambda - \cos\omega_j)]^{-2d_j}$$

where $\theta = (\Phi_1,\ldots,\Phi_p,\Theta_1,\ldots,\Theta_q,d_0,\ldots,d_h)'$. Giraitis and Leipus (1995) obtain consistency of $\hat{\theta}_c$ but they do not establish the asymptotic distribution, although a non-Gaussian limit distribution is conjectured. For vector x_t in Eq. (4.1) Hosoya (1996, 1997) considered a multivariate extension of $L_c(\theta,\mu,\sigma^2)$ and obtained an analogous result to Eqns (4.5) and (4.6).

Following Kashyap and Eom (1988) we can also proceed by regressing $\log I_n(\lambda_j)$ on $\log h(\lambda_j;\theta)$ over $j = 1,\ldots,n-1$, though this approach leads to less efficient than Gaussian estimates. In fact Ray (1993) used this technique to estimate d_3 and d_{12} in the SCLM process

$$\phi_0(L)\phi_3(L^3)\phi_{12}(L^{12})(1 - L^3)^{d_3}(1 - L^{12})^{d_{12}}x_t = \theta_0(L)\theta_3(L^3)\theta_{12}(L^{12})\varepsilon_t$$
$$(4.14)$$

where the ε_t are white noise. Ray (1993) used these estimates as a first step in the estimation of the complete model in Eq. (4.14) for monthly IBM revenues.

4.2   Semiparametric Estimation

When we are interested only in estimation of the persistence parameter, d in Eq. (1.4), we only need to specify $f(\lambda)$ around ω in order to obtain consistent estimates that we call semiparametric. This is a clear advantage with respect to parametric estimates that need a complete and correct specification of $f(\lambda)$ over the whole band of Nyqvist frequencies for consistency, though in the event of such specification the parametric estimates have the competing advantage of converging faster.

Due to their simplicity, perhaps the most popular semiparametric procedures are variants of the log-periodogram estimate introduced by Geweke and Porter-Hudak (1983). Consider a least squares regression of $\log I_n(\omega + \lambda_j)$ on $-2\log \lambda_j$ and an intercept, where $I_n(\lambda)$ is the periodogram defined in Eq. (4.10) and $\lambda_j = 2\pi j/n$ are Fourier frequencies. The regression is carried out for $j = 1, \ldots, m$, where the "bandwidth" m is an integer between 1 and $n/2$ and in practice is much less than n, and for asymptotic theory satisfies at least

$$\frac{1}{m} + \frac{m}{n} \to 0 \qquad \text{as } n \to \infty. \tag{4.15}$$

The original version, due to Geweke and Porter-Hudak (1983), uses instead of $-2\log \lambda_j$ the regressor $-\log\{4\sin^2(\lambda_j/2)\}$, but as indicated by Robinson (1995a), use of the simpler $-2\log \lambda_j$, which corresponds more naturally to Eq. (1.4), leads to equivalent asymptotic properties. These authors assumed $\omega = 0$, when, because $I_n(\lambda)$ is an even function, regression of $\log I_n(\lambda_j)$ on $-2\log|\lambda_j|$ for $j = \pm 1, \ldots, \pm m$ is equivalent to using frequencies for $j = 1, \ldots, m$. When $\omega \neq 0, \pi$, $I_n(\omega + \lambda)$ is not necessary symmetric about ω and information on both sides of the pole/zero can make a substantial difference. Thus a log-periodogram estimate for such ω is

$$\hat{d} = -\frac{1}{2}\frac{\sum_{j=\pm 1}^{\pm m} v_j \log I_n(\omega + \lambda_j)}{\sum_{j=\pm 1}^{\pm m} v_j^2} \tag{4.16}$$

where $v_j = \log|j| - \frac{1}{m}\sum_1^m \log l$. Work on estimating Eq. (1.4) with $\omega = 0$ suggests two possible modifications to this scheme. Due to anomalous behavior of the periodogram very close to a spectral pole/zero, see Robinson (1995a), Kunsch (1986) and Hurvich and Beltrao (1993,1994), Kunsch (1986) and Robinson (1995a) trimmed out some frequencies close to ω. The second type of modification is an efficiency improvement suggested by Robinson (1995a) and based on pooling adjacent periodogram ordinates. Incorporating these two suggestions we have the estimate

$$\hat{d}^{(J)} = -\frac{1}{4}\frac{\sum_k' v_k[\log \hat{I}_{\omega Jk} + \log \tilde{I}_{\omega Jk}]}{\sum_k' v_k^2} \tag{4.17}$$

where $\hat{I}_{\omega Jk} = \sum_{j=1}^J I_n(\omega + \lambda_{k+j-J})$, $\tilde{I}_{\omega Jk} = \sum_{j=1}^J I_n(\omega - \lambda_{k+j-J})$, J is a positive integer (the pooling number) and $\sum_k'$ is a sum over $k = l + J, l + 2J, \ldots, m$. When the pooling number, $J = 1$, and the trimming number, $l = 0$, then Eq. (4.17) reduces to Eq. (4.16). When $\omega = 0$ Robinson (1995a) proved that under Gaussianity

$$\sqrt{m}(\hat{d}^{(J)} - d) \xrightarrow{d} N\left(0, \frac{J\psi'(J)}{4}\right) \qquad \text{as } n \to \infty$$

where $\psi'(z) = (d/dz)\psi(z)$ and $\psi(z)$ is the digamma function defined as $(d/dz)\log\Gamma(z)$ where $\Gamma(z)$ is the Gamma function. The same asymptotics follow for $\omega \neq 0$ in Eq. (1.4), see Arteche (1998). For $\omega = 0$ Velasco (1998c) relaxes the assumption of Gaussianity and only imposes boundness of the fourth moments of the ε_t in Eq. (4.1) to obtain consistency and asymptotic normality (using a suitably tapered periodogram) with variance $3J\psi'(J)/4$. Note that tapering increases the variance. Still for $\omega = 0$, and assuming Gaussianity, Velasco (1998a) proves consistency of $\hat{d}^{(J)}$ for the non-stationary case $d \in [1/2, 1)$ and also shows that $\hat{d}^{(J)}$ is asymptotically normal with variance $J\psi'(J)/4$ for the non-tapered estimate if $d \in [1/2, 3/4)$, and $3J\psi'(J)/4$ for $d \in [1/2, 3/2)$ in the tapered case. The good properties in finite samples of $\hat{d}^{(1)}$ for $d \in [1/2, 1)$ are shown in Hurvich and Ray (1995). These results seem to extend straightforwardly to the case $\omega \neq 0$.

Related with the parametric Gaussian estimates described in the previous section, Kunsch (1987) and Robinson (1995b) considered a semiparametric approximation of $L_d(\theta, \sigma^2)$ in Eq. (4.11). The estimate, $\tilde{d}$, is the argument that minimizes

$$Q(C, d) = \frac{1}{2m} \sum_{j=\pm 1}^{\pm m} \left\{ \log C|\lambda_j|^{-2d} + \frac{|\lambda_j|^{2d}}{C} I_n(\omega + \lambda_j) \right\} \tag{4.18}$$

where m satisfies at least Eq. (4.15). The estimate $\tilde{d}$ has been called the Gaussian semiparametric or local Whittle estimate. When $\omega = 0$ only frequencies on one side of ω are used, due to the symmetry of $I_n(\lambda)$ at the origin. Without requiring Gaussianity, Robinson (1995b) obtained consistency and asymptotic normality for the case $\omega = 0$ such that

$$\sqrt{m}(\tilde{d} - d) \xrightarrow{d} N(0, 1/4).$$

Note that $\tilde{d}$ is asymptotically more efficient than $\hat{d}^{(J)}$ because $J\psi'(J) \downarrow 1$ as $J \to \infty$. The same asymptotics hold for $\omega \neq 0$, see Arteche (1998). Velasco (1998b) extended Robinson's results to non-stationary processes obtaining consistency for $d \in [1/2, 1)$ and asymptotic normality when $d \in [1/2, 2/3)$ ($d \in [1/2, 3/4)$ under Gaussianity). A multivariate extension of this estimator is studied by Lobato (1995).

Robinson (1994b) proposed an alternative technique to estimate d in case

$$f(\omega + \lambda) \sim L\left(\frac{1}{|\lambda|}\right)|\lambda|^{-2d} \quad \text{as } \lambda \to 0 \tag{4.19}$$

where $L(z)$ is a slowly varying function, that is a positive measurable function satisfying

$$\frac{L(tz)}{L(z)} \to 1 \quad \text{as } z \to \infty \quad \text{for all } t > 0.$$

Note that Eq. (4.19) specializes to Eq. (1.4) when $L(z)$ is a constant. The proposed "averaged periodogram" estimate is

$$\hat{d}_{qm\omega} = \frac{1}{2} - \frac{\log\{\hat{F}(q\lambda_m)/\hat{F}(\lambda_m)\}}{2\log q} \tag{4.20}$$

where

$$\hat{F}(\lambda) = \frac{2\pi}{n} \sum_{j=\pm 1}^{\pm[\lambda n/2\pi]} I_n(\omega + \lambda_j), \tag{4.21}$$

and $q \in (0,1)$ is a user chosen number and m again satisfies at least Eq. (4.15). With only second moment restrictions and without requiring Gaussianity, Robinson (1994b) showed the consistency of $\hat{d}_{qm\omega}$ for $\omega = 0$. Assuming Gaussianity, Lobato and Robinson (1996a) obtained the asymptotic distribution of $\hat{d}_{qm\omega}$ for $\omega = 0$. This is normal for $d \in (0, 1/4)$ and non-normal (related to Rosenblatt process) for $d \in (1/4, 1/2)$. The same properties are likely to hold for $\omega \neq 0$.

Janacek (1982) introduced an alternative method to estimate d through estimation of the Fourier coefficients of $\log f(\lambda)$ using the log-periodogram. Although originally this estimate was proposed for long memory at frequency zero, Janacek claimed that this method can be naturally extended to SCLM time series.

A number of other semiparametric estimates have been proposed for the $\omega = 0$ case that seem capable of extending to general ω, such as the time domain ones of Robinson (1994c), and the one of Parzen (1986) and Hidalgo and Yajima (1997) that achieves an efficiency improvement over the estimates described above.

5. TESTING SEASONAL/CYCLICAL INTEGRATION AND COINTEGRATION

The characteristics of the process generating the series depend strongly on the value of the persistence parameter, d. In particular, d determines if the process has long memory (stationary or non-stationary), short memory or negative memory (invertible or non-invertible). Some interesting situations that may require a rigorous test are

1. $d = 0$ (short memory) against $d > 0$ (long memory) or $d < 0$ (negative memory),
2. $d = 1/2$ ("just" non-stationarity) against $d > 1/2$ (non-stationarity) or $d < 1/2$ (stationarity),
3. $d = -1/2$ ("just" non-invertibility) against $d > -1/2$ (invertibility) or $d < -1/2$ (non-invertibility).

The hypotheses involved in (1) can be tested using simple t tests based on the estimates and their asymptotic distributions described in Sec. 4 or by Lagrange Multiplier tests as those proposed in the parametric case by Robinson (1994a) or in a semiparametric setting by Lobato and Robinson (1996b). t-tests on (2) and (3) can be carried out using those estimates whose asymptotic properties hold for non-stationary or non-invertible processes.

Traditionally, interest has focused on testing the possibility of unit roots where d in Eq. (1.4) is an integer. Some early work is due to Dickey, Hasza and Fuller (1984) who test the possibility of a seasonal unit root of the form

$$(1 - L^s)x_t = \varepsilon_t \qquad t = 1, 2, \ldots$$

where the ε_t are iid $(0, \sigma^2)$ random variables, against the alternative

$$x_t = \alpha x_{t-s} + \varepsilon_t$$

with $|\alpha| < 1$. They provide percentiles for the proposed test statistic. One of the limitations of this procedure is that it is a joint test for unit roots at the origin and seasonal frequencies, $\omega_h = 2\pi h/s$, $h = 1, 2, \ldots, [s/2]$, see Eq. (3.7) for the case $s = 4$. Furthermore the alternative is a specified form of s-th order autoregressive process. Hylleberg et al. (1990), using quarterly data, extended this procedure allowing for an individual test at zero and at every seasonal frequency that is robust to behavior at other frequencies. Some extensions of this procedure to monthly data are Beaulieu and Miron (1993) and Franses (1991). The null hypothesis in each case is pure integrability $(I_\omega(1))$ and the alternative is pure stationarity or short memory $(I_\omega(0))$. Canova and Hansen (1995) extended the test of Kwiatkowsky et al. (1992) to the seasonal case, testing the null of stationarity $(I_\omega(0))$ against the alternative of pure integration $(I_\omega(1))$. Bearing in mind the properties of these two types of test, that basically differ in the specification of the null and alternative, the simultaneous use of both procedures has been advised in order to test for pure integrability. The same conclusion of both types of test (that is one rejects and the other does not reject the null) provides strong evidence in favour of the result implied by both procedures. If one test contradicts the other, then we need a more thorough analysis. In this case we may have fractional integration.

A general test, based on the parametric model in Eq. (3.8) and allowing for fractional and integer $I_\omega(d)$ as null and alternative, has been proposed by Robinson (1994a) and applied to quarterly macroeconomic data by Gil-Alaña and Robinson (1997). Suppose

$$\phi(L)(x_t - \mu) = u_t \qquad t = 1, 2, \ldots$$
$$x_t = 0 \qquad t \leq 0$$

where u_t is a short memory covariance stationary sequence with zero mean, and $\phi(z)$ is a known function. Consider the function $\phi(z;\vartheta)$ where ϑ is a p-dimensional vector of real valued parameters such that $\phi(z;\vartheta) = \phi(z)$ if and only if

$$H_0: \vartheta = 0.$$

The hypotheses of principle interest entail ϕ of the form

$$\phi(L;\vartheta) = (1-L)^{d_0+\vartheta_{i_0}}\left\{ \prod_{j=1}^{h-1}(1-2L\cos\omega_j + L^2)^{d_j+\vartheta_{i_j}}\right\}(1+L)^{d_h+\vartheta_{i_h}} \quad (5.2)$$

where for each j, $\vartheta_{i_j} = \vartheta_l$ for some l and for each l there is at least one j such that $\vartheta_{i_j} = \vartheta_l$. The null hypothesis is that the $p \times 1$ vector $(p \le h+1)$ $\vartheta = (\vartheta_1, \vartheta_2, \ldots, \vartheta_p)'$ is a vector of zeros. Thus fractional seasonal and cyclical integration is allowed in the null and alternative in contrast with the focus on testing for a unit root against autoregressive alternatives in much of the literature. To avoid estimation of the persistence parameters, Robinson (1994a) used a score test although undoubtedly the same asymptotic behavior can be expected of Wald and likelihood ratio tests. When u_t is white noise the proposed test statistic is

$$R = \frac{n}{\tilde{\sigma}^4}\tilde{a}'\tilde{A}^{-1}\tilde{a}$$

where

$$\tilde{\sigma}^2 = \frac{1}{n}\sum_1^n u_t^2, \qquad u_t = \phi(L;0)x_t, \qquad \tilde{a} = -\frac{2\pi}{n}\sum_j{}'\Psi(\lambda_j)I_u(\lambda_j),$$

$I_u(\lambda)$ is the periodogram of u_t defined in Eq. (4.10), $\Psi(\lambda) = Re\{(\partial/\partial\vartheta)\log\phi(e^{i\lambda};0)\}$ and $\tilde{A} = (2/n)\sum_j \Psi(\lambda_j)\Psi(\lambda_j)'$ where the primed sum is over

$$\lambda_j \in M = \{\lambda: -\pi < \lambda < \pi, \ \lambda \notin (\omega_l - \lambda_1, \omega_l + \lambda_1), \ l = 0, 1, \ldots, h\}$$

and ω_l are the distinct poles of $\Psi(\lambda)$ on $(-\pi, \pi]$. Asymptotically equivalent expressions for $\tilde{a}$ and $\tilde{A}$ can be found in Robinson (1994a), as well as a time domain test statistic. Robinson (1994a) also proposed a modification of R that allows for parametric weak correlation in u_t so long as its spectrum is bounded and bounded away from zero and of known parametric form. Unlike the techniques earlier described these procedures have the advantage of being standard in the sense that the test statistic has a χ_p^2 limit distribution under the null and a limiting non-central χ_p^2 distribution against Pitman or local alternatives, and are asymptotically locally most powerful.

Hylleberg et al. (1990) considered the possibility of seasonal cointegration, which they defined as follows.

A pair of series each of which are integrated at frequency ω are said to be cointegrated at that frequency if a linear combination of the series is not integrated at ω.

Hylleberg et al. (1990) pointed out that in case of several spectral poles, as for example x_t in Eq. (3.8) the procedure in Engle and Granger (1987) to test for cointegration at zero frequency is invalid, so that prior to any test for cointegration we have to filter the data in such a way that only the pole at the frequency where we suspect the cointegration occurs remains. For instance, if we want to test for cointegration at the origin, we have first to remove seasonal roots, for example by applying the seasonal summation operator, $S(L) = (1 + L + \cdots + L^{s-1})$, to the original series and then performing a standard cointegration test such as those discussed in Engle and Granger (1987).

Engle and Granger (1987) and Hylleberg et al. (1990) consider only the possibility that a linear combination of $I_\omega(1)$ processes is $I_\omega(0)$. But our definition of SCLM or $I_\omega(d)$ processes allows for the possibility of fractional integration and cointegration. In this sense Engle et al. (1989) define cyclical cointegration in the following manner,

A vector of series x_t, each component $I_\omega(d)$ (integrated of order d at frequency ω), may be said to be cointegrated at that frequency if there exists a vector α_ω such that $z_t^\omega = \alpha_\omega' x_t$ is integrated of lower order at ω.

As in the definition of SCLM, the case of cointegration at every seasonal frequency is known as seasonal cointegration.

6.   ESTIMATION OF THE FREQUENCY ω

Most analyses of SCLM models assume that the frequency ω where the spectral pole occurs is known. Of course, seasonal frequencies are known, but in cyclical time series, ω may well be unknown.

The literature on estimating ω in cyclical long memory is of recent date and it is of interest to consider first earlier work on estimating frequency in an alternative model, namely the deterministic periodic time series

$$x_t = \alpha_0 \sin \omega t + \beta_0 \cos \omega t + u_t \tag{6.1}$$

where u_t is stationary with mean zero and spectral density, $f_u(\lambda)$, continuous and positive at ω. Whittle (1952) found that the least squares estimate of ω in Eq. (6.1), $\hat{\omega}$, is the periodogram maximizer and has a variance $O(n^{-3})$.

Walker (1971) (for u_t white noise) and Hannan (1971, 1973a) extended Whittle's work and, without assuming Gaussianity, found that for $\omega \neq 0, \pi$,

$$n^{3/2}(\hat{\omega} - \omega) \xrightarrow{d} N\left(0, \frac{48\pi f_u(\omega)}{\alpha_0^2 + \beta_0^2}\right). \tag{6.2}$$

In case $\omega = 0, \pi$, Hannan (1973a) showed that there exists an integer valued random variable, n_0, with $P(n_0 < \infty) = 1$ such that $\hat{\omega} = \omega$ for $n > n_0$, so that $\hat{\omega}$ will be equal to the value it estimates for a large enough sample size. Mackisack and Poskitt (1989) proposed a different technique based on the maximization of the transfer function calculated by fitting high order autoregressions to x_t. Only $\sqrt{n}$-consistency for $\omega \in (0, \pi)$ is rigorously proved (although it is claimed that the variance of the estimate is $O(n^{-(5/2)})$ when the order of the autoregression is $O(n^{1/2})$), and their method is computationally intensive. A different approach has been suggested by Quinn and Fernandes (1991). The technique is based on fitting ARMA(2, 2) models in an iterative way and they propose a simple algorithm that converges rapidly. The same asymptotic distribution, Eq. (6.2), as the maximizer of the periodogram is obtained. A similar procedure with the same asymptotic distribution is described in Truong-Van (1990).

In Eq. (6.1) only one sinusoidal component is assumed. However a multiple finite number of components can describe seasonal or cyclical movement,

$$x_t = \sum_{j=1}^{r} \{\alpha_j \cos(\omega_j t) + \beta_j \sin(\omega_j t)\} + u_t. \tag{6.3}$$

In this context estimation of r, the number of cosinusoids, has been treated by Quinn (1989), Kavalieris and Hannan (1994), Hannan (1993) and Wang (1993) among others. Estimation of the ω_j has been analysed in Chen (1988a, b), Walker (1971) and Kavalieris and Hannan (1994).

The estimation of ω in cyclical long memory models may be necessary to determine the periodicity of the cycle and as a first step prior to estimation of remaining parameters. Yajima (1995) considered the model

$$f(\lambda; \omega, \theta) = g(\lambda; \omega, \theta)|\lambda - \omega|^{-2d} \quad \omega \in [0, \pi] \quad \text{and } 0 < d < 1/2, d \in \theta \tag{6.4}$$

where θ is a parameter vector including d, and the function g obeys some regularity conditions, such that the GARMA process is a special case of Eq. (6.4). The estimate of ω considered by Yajima is the periodogram maximizer. He obtains n^{α}-consistency under Gaussianity for any $\alpha \in (0, 1)$ and shows that the Whittle estimates of θ obtained by minimizing

$$U_n(\hat{\omega}, \theta) = \int_{-\pi}^{\pi} \left\{ \log f(\lambda; \hat{\omega}, \theta) + \frac{I_n(\lambda)}{2\pi f(\lambda; \hat{\omega}, \theta)} \right\} d\lambda \tag{6.5}$$

are $\sqrt{n}$-consistent and asymptotically normal. Yajima does not provide any distribution theory for his estimate of ω, but a non-normal distribution is conjectured.

Chung (1996a, b) obtained an estimate of $\eta = \cos \omega$ in Gegenbauer processes,

$$\phi(L)(1 - 2L\eta + L^2)^d (x_t - \mu) = \theta(l)\varepsilon_t$$

and claimed asymptotic properties for conditional sum of squares estimates, compare Eq. (4.7), including a non-normal limit distribution for the estimate of ω but a normal limit distribution for the estimates of the remaining parameters.

A joint estimation of all the frequencies $\omega_j, j = 0, 1, \ldots, h$, and the rest of the long and short memory parameters in the model in Eq. (4.13) is proposed by Giraitis and Leipus (1995). They obtain consistency of the Whittle estimates obtained minimizing $U_n(\omega, \theta)$ defined in Eq. (6.5), but no asymptotic distribution.

Hidalgo (1998) proposes an alternative semiparametric technique to estimate ω in a process satisfying Eq. (1.4) with $d \in (0, 1/2)$. The estimate $\hat{\omega}_H$ is the argument that maximizes the estimate of d proposed by Hidalgo and Yajima (1997),

$$\hat{d}^* = \frac{1}{m} \sum_{p=1}^{m} \hat{d}_p \tag{6.6}$$

where $\hat{d}_p = a_1/a_2$ and

$$a_1 = \frac{1}{p} \sum_{l=1}^{p} w(l) \log \hat{f}_p(\lambda_l) - \left(\frac{1}{p} \sum_{l=1}^{p} w(l) \right) \log \hat{f}_p(\lambda_{p+1}),$$

$$a_2 = -2 \int_0^1 w(u) \log u \, du,$$

where $w(l) = (l/p)^{1/c} - (l/p)^{(1/c+1)}$, $c > 1$ and $\hat{f}_p(\lambda_l)$ is a particular moving average of periodogram ordinates at frequencies close to ω. Without assuming Gaussianity Hidalgo (1998) shows that $nk^{-(1/2)}(\hat{\omega}_H - \omega)$, for $k \to \infty$ suitably slowly with n, has a normal limit distribution.

7. CONCLUSION AND EXTENSIONS

This paper has discussed modeling and inference in SCLM processes having spectral density satisfying Eq. (1.4). The combination of seasonal or cyclic behavior and long memory can lead to several extensions:

1. The autoregressive coefficients π_j in Eq. (3.9) of the SCLM model in Eq. (3.8) can be useful for forecasting. Although obtaining $C_k^{(d)}$ from a given estimate of d, $\hat{d}$, can be done recursively, the accurate generation of the π_j's gets more difficult as the number of spectral poles and the sample size increase, and deserves attention.

2. The definition in Eq. (1.4) of SCLM imposes an asymptotically symmetric behavior in $f(\lambda)$ around ω. Nevertheless, if $\omega \neq 0$, f need not actually be symmetric. We can generalize from SCLM to SCALM (Seasonal Cyclical Asymmetric Long Memory) processes by defining

$$
\begin{aligned}
f(\omega + \lambda) &\sim C_1 \lambda^{-2d_1} \\
f(\omega - \lambda) &\sim C_2 \lambda^{-2d_2} \qquad \text{as } \lambda \to 0^+
\end{aligned}
\tag{7.1}
$$

where C_1, d_1 can be different from C_2, d_2. Then Eq. (1.4) is a restriction of Eq. (7.1) that happens when $C_1 = C_2 = C$ and $d_1 = d_2 = d$. Discussion of Eq. (7.1) has began in Arteche (1998) and Arteche and Robinson (1998).

3. Some financial series such as asset returns appear to be approximately serially uncorrelated. However, there are nonlinear transformations, such as squares, that can exhibit autocorrelation as modeled in the extensive ARCH and stochastic volatility literature, following Engle (1982) or Taylor (1986, 1994). Moreover there is evidence of long memory autocorrelation in the squares of some series. The first model that causes this effect is the general GARCH process proposed by Robinson (1991), who uses it as an alternative in testing for no-ARCH. His model is sufficiently general to describe SCLM behavior in the squares, as may be appropriate in financial data.

Acknowledgments

Research supported by ESRC Grant R000235892. Josu Arteche also acknowledges financial support from the Bank of Spain and UPV Grant 038.321-HB039/97. The referees are thanked for their comments.

REFERENCES

Adenstedt, R. K., On large sample estimation for the mean of a stationary random sequence, *Ann. Statist.*, 2,6: 1095–1107 (1974).

Andel, J., Long-memory time series models, *Kybernetica, 22*: 105–123 (1986).

Arteche, J., *Seasonal and cyclical long memory in time series*. Ph.D. Thesis, LSE, University of London, 1998.

Arteche, J. and P. M. Robinson, Semiparametric inference in seasonal and cyclical long memory processes. Preprint, 1998.

Beaulieu, J. J. and J. A. Miron, Seasonal unit roots in aggregated U.S. data, *J. Econometrics, 55*: 305–328 (1993).

Bell, W. R. and S. C. Hillmer, Issues involved with the seasonal adjustment of economic time series. *J. Business and Econ. Statist.* 2: 291–320 (1984).

Beran, J., *Statistics for Long-Memory Processes*. Monographs on statistics and applied probability 61, Chapman and Hall, New York 1994a.

Beran, J., On a class of M-estimators for gaussian long-memory models, *Biometrika, 81, 4*: 755–766 (1994b).

Beran, J., Maximum likelihood estimation of the differencing parameter for invertible short and long memory autoregressive integrated moving average models, *J. R. Statist. Soc. B, 57, 4*: 659–672 (1995).

Box, G. E. P. and M. Jenkins, *Time Series Analysis: Forecasting and Control*, Holden-Day, San Francisco, 1976.

Canova, F. and B. E. Hansen, Are seasonal patterns constant over time? A test for seasonal stability, *J. Business and Econ. Statist., 13, 3*: 237–252 (1995).

Carlin, J. B. and A. P. Dempster, Sensitivity analysis of seasonal adjustments: Empirical case studies, *J. Amer. Statist. Assoc, 84*: 6–20 (1989).

Chan, N. H. and N. Terrin, Inference for unstable long-memory processes with applications to fractional unit root autoregressions, *Ann. Statist. 23, 5*: 1662–1683 (1995).

Chan, N. H. and C. Z. Wei, Limiting distributions of least squares estimates of unstable autoregressive processes, *Ann. Statist. 16, 1*: 367–401 (1988).

Chen, Z. G., An alternative consistent procedure for detecting hidden frequencies, *J. Time Ser. Anals. 9, 3*: 301–317 (1988a).

Chen, Z. G., Consistent estimate for hidden frequencies in a linear process, *Adv. Probab, 20*: 295–314 (1988b).

Cheung, Y. W. and F. X. Diebold, On maximum likelihood estimation of the differencing parameter of fractionally integrated noise with unknown mean, *J. Econometrics 62*: 301–316 (1994).

Chung, C. F., Estimating a generalized long-memory process, *J. Econometrics 73*: 237–259 (1996a).

Chung, C. F., A generalized fractionally integrated autoregressive moving-average process, *J. Time Ser. Anal. 17, 2*: 111-140 (1996b).

Dagum, E. B., The X-11 ARIMA seasonal adjustment method, *Statistics Canada, Catalogue 12-564E* (1980).

Dahlhaus, R., Efficient parameter estimation for self-similar processes, *Ann. Statist., 17*: 1749–1766 (1989).

Dickey, D. A., D. P. Hasza and W. A. Fuller, Testing for unit roots in seasonal time series, *J. Amer. Statis. Assoc. 79*: 355–367 (1984).

Engle, R. F., Autoregressive conditional heteroscedasticity with estimates of the variance of United Kingdom inflation, *Econometrica, 50, 4*: 987–1007 (1982).

Engle, R. F. and W. J. Granger, Co-integration and error correction: Representation, estimation, and testing, *Econometrica, 55, 2*: 251–276 (1987).

Engle, R. F., C. W. J. Granger and J. J. Hallman, Merging short and long-run forecasts: An application of seasonal cointegration to monthly electrical sales forecasting, *J. Econometrics 40*: 45–62 (1989).

Fox, R. and M. S. Taqqu, Large sample properties of parameter estimates for strongly dependent stationary Gaussian time series, *Ann. Statist. 14*: 517–532 (1986).

Franses, P. H., Model selection and seasonality in time series, *Tinbergen Institute Series, 18* (1991).

Franses, P. H. and M. Ooms, A periodic long-memory ARFIMA$(0, D_s, 0)$ model for quarterly UK inflation. Report 9511/A, Erasmus University Rotterdam, The Netherlands, 1995.

Geweke, J. and S. Porter-Hudak, The estimation and application of long memory time series models, *J. Time Ser. Anal. 4*: 221–238 (1983).

Gil-Alaña, L. A. and P. M. Robinson, Testing of unit root and other non-stationary hypothesis in macroeconomic time series, *J. Econometrics, 80, 2*: 241–268 (1997).

Giraitis, L. and R. Leipus, A generalized fractionally differencing approach in long memory modeling, *Liet. Matem. Rink. 35, 1*: 65–81 (1995).

Giraitis, L. and D. Surgailis, A central limit theorem for quadratic forms in strongly dependent linear variables and its application to asymptotic normality of Whittle's estimate, *Probab. Theory and Related Fields,* *86*: 87–104 (1990).

Gray, H. L., N. F. Zhang and W. A. Woodward, On generalized fractional processes, *J. Time Ser. Anal. 10*: 233–257 (1989).

Gray, H. L., N. F. Zhang and W. A. Woodward, On generalized fractional processes—a correction, *J. Time Ser. Anal. 15, 5*: 561–562 (1994).

Hannan, E. J., The estimation of seasonal variation in economic time series, *J. Amer. Statist. Assoc. 58*: 31–44 (1963).

Hannan, E. J., The estimation of a changing seasonal pattern, *J. Amer. Statis. Assoc., 59*: 1063–1077 (1964).

Hannan, E. J., Measurement of a wandering signal amid noise, *J. Appl. Probab. 4*: 90–102 (1967).

Hannan, E. J., Non-linear time series regression, *J. Appl. Probab. 8*: 767–780 (1971).

Hannan, E. J., The estimation of frequency, *J. Appl. Probab. 10*: 510–519 (1973a).

Hannan, E. J., The asymptotic theory of linear time series models, *J. Appl. Probab. 10*: 130–145 (1973b).

Hannan, E. J., Determining the number of jumps in a spectrum, Developments in Time Series Analysis (ed. Subba Rao) Chapman & Hall, London, 1993, 127–138.

Hannan, E. J., R. D. Terrel and N. E. Tuckwell, The seasonal adjustment of economic time series, *Int. Econ. Rev. 2, 1*: 24–52 (1970).

Harvey, A. C., *Forecasting structural Time Series Models and the Kalman Filter*, Cambridge University Press, U.K. 1989.

Hassler, U., (Mis)specification of long-memory in seasonal time series, *J. Time Ser. Anal. 15*: 19–30 (1994).

Heyde, C. C. and R. Gay, Smoothed periodogram asymptotics and estimation for processes and fields with possible long-range dependence, *Stochastic Process. Appl. 45*: 169–182 (1993).

Hidalgo, J., Estimation of the pole of long-range processes, preprint, 1998.

Hidalgo, J. and Y. Yajima, Semiparametric estimation of the long-range parameter, preprint, 1997.

Hosking, J. R. M., Fractional differencing, *Biometrika, 68*: 165–176 (1981).

Hosking, J. R. M., Modelling persistence in hydrological time series using fractional differencing, *Water Resour. Res. 20*: 1898–1908 (1984).

Hosoya, Y., The quasi-likelihood approach to statistical inference on multiple time series with long-range dependence, *J. Econometrics, 73*: 217–236 (1996).

Hosoya, Y., A limit theory for long-range dependence and statistical inference in related models, *Ann. Statist. 25*: 105–137 (1997).

Hurst, H. E., Long-term storage capacity of reservoirs, *Trans. Amer. Soc. Civil Engineers 1*: 519–543 (1951).

Hurvich, C. M. and K. I. Beltrao, Asymptotics for the low-frequency ordinates of the periodogram of long-memory time series. *J. Time Ser. Anal., 14, 5*: 455–472 (1993).

Hurvich, C. M. and K. I. Beltrao, Acknowledgment of priority for "asymptotics for the low-frequency ordinates of the periodogram of long-memory time series", *J. Time Ser. Anal., 15*: 64 (1994).

Hurvich, C. M. and B. K. Ray, Estimation of the memory parameter for nonstationary or noninvertible fractionally integrated processes, *J. Time Ser. Anal., 16, 1*: 17–42 (1995).

Hylleberg, S., R. F. Engle, C. W. J. Granger and B. S. Yoo, Seasonal integration and cointegration. *J. Econometrics 44*: 215–238 (1990).

Janacek, G. J., Determining the degree of differencing for time series via the log-spectrum, *J. Time Ser. Anal., 3, 3*: 177–183 (1982).

Jonas, A. J., *Persistent Memory Random processes*, Ph.D. Thesis, Department of Statistics, Harvard University, 1983.

Kashyap, R. L. and K. B. Eom, Estimation in long-memory time series model, *J. Time Ser. Anal., 9, 1*: 35–41 (1988).

Kavalieris, L. and E. J. Hannan, Determining the number of terms in a trigonometric regression, *J. Time Ser. Anal., 15, 6*: 613–625 (1994).

Kunsch, H. R., Discrimination between monotonic trends and long-range dependence, *J. Appl. Probab., 23*: 1025–1030 (1986).

Kunsch, H. R., Statistical aspects of self-similar processes, *Proc. First World Congress Bernoulli Soc.*, (eds. Yu. Prohorov and V. V. Sazanov) VNU Science Press, Utrecht, *1*: 67–74, 1987.

Kwiatkowsky, D., P. C. B., Phillips, P. Schmidt and Y. Shin, Testing the null hypothesis of stationarity against the alternative of a unit root, *J. Econometrics, 53*: 159–178, (1992).

Li, W. K. and A. I. McLeod, Fractional time series modelling. *Biometrika 73*: 217–221 (1986).

Lobato, I. N., *Multivariate Analysis of Long-Memory Time Series in the Frequency Domain*. Ph.D. Thesis, University of London, 1995.

Lobato, I. N. and P. M. Robinson, Averaged periodogram estimation of long-memory, *J. Econometrics, 73*: 303–324 (1996a).

Lobato, I. N. and P. M. Robinson, A nonparametric test for I(0). Forthcoming, *Review of Economic Studies* (1996b).

Mackisack, M. S. and D. S. Poskitt, Autoregressive frequency estimation, *Biometrika, 76, 3*: 565–575 (1989).

Mandelbrot, B. B. and J. N. Van Ness, Fractional Brownian motions, fractional noises and applications, *SIAM Review 10*: 422–437 (1968).

Nerlove, M., Spectral analysis of seasonal adjustment procedures, *Econometrica, 32, 3*: 241–286 (1964).

Ooms, M., Flexible seasonal long-memory and economic time series, preprint 1995.

Osborn, D. R., The implications of periodically varying coefficients for seasonal time series processes, *J. Econometrics, 48*: 373–384 (1991).

Parzwen, E., Quantile spectral analysis and long-memory time series, *J. Appl. Probab. 23A*: 41–54 (1986).

Porter-Hudak, S., An application of seasonal fractionally differenced model to the monetary aggregates, *J. Amer. Statist. Assoc., 85, 410*: 338–344 (1990).

Quinn, B. G., Estimating the number of terms in a sinusoidal regression, *J. Time Ser. Anal., 10, 1*: 71–75 (1989).

Quinn, B. G. and J. M. Fernandes, A fast efficient technique for the estimation of frequency, *Biometrika, 78, 3*: 489–497 (1991).

Ray, B. K., Long-range forecasting of IBM product revenues using a seasonal fractionally differenced ARMA model, *Int. J. Forecasting 9*: 255–269 (1993).

Robinson, P. M., Testing for strong serial correlation and dynamic conditional heteroskedasticity in multiple regression, *J. Econometrics, 47*: 67–84 (1991).

Robinson, P. M., Efficient tests of non-stationary hypotheses, *J. Amer. Statist. Assoc. 89*: 1420–1437 (1994a).

Robinson, P. M., Semiparametric analysis of long-memory time series, *Ann. Statist. 22*: 515–539 (1994b).

Robinson, P. M., Time series with strong dependence, *Advances in Econometrics*, Cambridge University Press, 1994c, *1*: 47–95.

Robinson, P. M., Log-periodogram regression of time series with long-range dependence, *Ann. Statist., 23, 3*: 1048–1072 (1995a).

Robinson, P. M., Gaussian semiparametric estimation of long-range dependence, *Ann. Statist., 23, 5*: 1630–1661 (1995b).

Samarov, A. and M. Taqqu., On the efficiency of the sample mean in long memory noise, *J. Time Ser. Anal., 9, 2*: 191–200 (1988).

Shiskin, J., A. H. Young and J. C. Musgrave., The X-11 variant of the census method II seasonal adjustment program. *Technical Paper 15*, Washington DC: Bureau of the Census, US Department of Commerce, 1967.

Sowell, F., *Fractionally Integrated Vector Time Series*. Ph.D. Dissertation, Duke University, Durham, N.C., 1986.

Sowell, F., Maximum likelihood estimation of stationary univariate fractionally integrated time series models, *J. Econometrics 53*: 105–188 (1992).

Taylor, S. J., *Modelling Financial Time Series*, John Wiley & Sons, New York, 1986.

Taylor, S. J., Modelling stochastic volatility: A review and comparative study, *Math. Finance, 4, 2*: 183–204 (1994).

Tiao, G. C., and M. R. Grupe, Hidden periodic autoregressive-moving average models in time series data, *Biometrika, 67, 2*: 365–373 (1980).

Troutman, B. M., Some results in periodic autoregression, *Biometrika, 66, 2*: 219–228 (1979).

Truong-van, B., A new approach to frequency analysis with amplified harmonics, *J. Roy. Stat. Soc. B, 52, 1*: 203–221 (1990).

Velasco, C., Non-stationary log-periodogram regression, *J. Econometrics*, Forthcoming, (1998a).

Velasco, C., Gaussian semiparametric estimation of non-stationary time series, *J. Time Ser. Anal.,* Forthcoming, (1998b).

Velasco, C., Non-gaussian log-periodogram regression, preprint, 1998c.

Vitale, R. A., An asymptotically efficient estimate in time series analysis, *Quart. Appl. Math., XXX*: 421–440 (1973).

Walker, A. M., On the estimation of a harmonic component in a time series with stationary independent residuals, *Biometrika, 58, 1*: 21–36 (1971).

Wang, X., An AIC type estimator for the number of cosinusoids, *J. Time Ser. Anal., 14*: 431–440 (1993).

Whittle, P., The simultaneous estimation of a time series harmonic components and covariance structure, *Trabajos de Estadística, 3*: 43–57 (1952).

Whittle, P., Estimation and information in stationary time series, *Ark. Mat., 2*: 423–434 (1953).

Yajima, Y., Estimation of the frequency of unbounded spectral densities, preprint, 1995. University of Tokyo, working paper.

Yong, C. H., *Asymptotic Behaviour of Trigonometric Series,* Chinese Univ., Hong Kong, 1974.

5
Nonparametric Specification Procedures for Time Series

DAG TJØSTHEIM   University of Bergen, Bergen, Norway

1.   INTRODUCTION

Interactive modeling of time series data requires formal or informal specification procedures at various stages of the model building. For example, since linearity represents a great simplification, a linearity test ought to be implemented at an early stage, as described in Sec. 2. Irrespective of whether linearity is rejected or not, a specification of the most significant lags may be considered and is surveyed in Sec. 3.

If linearity is rejected, one possibility is to search for a parametric model among such classsical models as the threshold, the exponential autoregressive, the bilinear, or the more flexible class of smooth transition models, (cf. Granger and Teräsvirta (1993). If instead a nonparametric approach is followed, some simplification would usually be required, among other things, to avoid the curse of dimensionality. The class of additive models represents one such simplification, and there are at least two ways of estimating such models, as will be indicated in Sec. 4. In that Section we also describe additivity tests.

Finally, the aim of much of time series analysis is to build a model where the residuals of the model are independent identically distributed (iid) random variables. If the residuals are not iid, this may give cause to re-evaluating and re-estimating the model. Nonparametric tests of independence are treated in Sec. 5.

All of the specification tests emphasized in this paper can be put into a common nonparametric framework, and demonstrating and clarifying this framework in a variety of different situations is the main purpose of the paper. The key idea of our procedure is to construct two statistics, which estimate the same quantity under H_0—for example linearity or independence—but different quantities under the alternative hypothesis H_A. Moreover, a distance function is introduced measuring the distance between the statistics. The null hypothesis is rejected if a large value of the distance functional occurs. The critical value is derived from the distribution of the distance functional under H_0. This distribution can either be constructed from asymptotic theory, often using a U-statistic argument, or from a randomization device. For small and moderate sample sizes it is our experience that the randomization argument gives a much better approximation than the asymptotics.

The two statistics can both be constructed nonparametrically (independence and additivity tests), or one nonparametric and one parametric statistic (linearity tests). In the latter case our construction is similar to that of Härdle and Mammen (1993), who limit themselves to the iid case, but treat more general parametric models than the linear one.

In some cases it is possible to transform the test situation to a situation where only a single test statistic is involved, as in testing of linearity and additivity, by looking at local second order derivatives. This requires analysis of functionals of nonparametric derivatives, and it is advantageous to use the technique of local polynomials, which has been quite heavily promoted recently by Fan and Gijbels (1995a).

It should be made clear at the outset that our tests are designed to be used more as exploratory tools than as formal tests. Therefore we are not so concerned with formal power properties and guarding against pathological worst case alternatives. Such problems are difficult and challenging but they would completely change the tenor of the present paper were they to be explored. Our goal has rather been to construct exploratory devices which would be of direct use in a number of commonly encountered situations in model building. This in turn generates quite a host of open theoretical problems, and we will try to point out some as we proceed. In this sense the present survey could be seen as a continuation of Tjøstheim (1994, 1996).

Finally, it should be mentioned that all of our nonparametric techniques are based on the assumption of stationarity. In applications in economics, for example, series are often thought to be nonstationary, and a first step consists in transforming them, using first differences or other devices, to obtain stationarity. Very recently an attempt has been made, Karlsen and Tjøstheim (1997) to extend nonparametrics to the class of null recurrent processes (which includes the random walk case).

2. LINEARITY TESTS

Estimating and analysing linear models are simple tasks compared to the same for nonlinear models. It is therefore important to decide early in the modeling process whether a linear or a nonlinear model should be entertained. This fact constitutes the motivation for linearity tests. Such tests can roughly be divided into the classes of parametric and nonparametric tests. Both types of test have as a prerequisite that we work with a stationary series $\{X_t\}$. It is perhaps worthwhile to stress that in practice it is sometimes hard to distinguish between nonlinearity and nonstationarity. If nonstationarity is suspected, there is an extra reason for concern, as it is well-known, (cf. Teräsvirta et al. (1994)) that nonlinear models are more vulnerable to changes over time than linear ones.

2.1 Parametric Tests

The principle of these tests is most easily explained for a lag-one time series

$$X_t = f(X_{t-1}, \boldsymbol{\theta}, e_{t-1}) + e_t.$$

Here $\{X_t, t \geq 0\}$ is the time series of interest and $\{e_t, t \geq 0\}$ is the series of innovations generating the series; usually assumed to consist of iid random variables. The function f is assumed to be known and could for example be a threshold or an exponential function. A flexible class of models is obtained by letting f belong to the smooth transition class; i.e. f could be exponential autoregressive,

$$f(x, \boldsymbol{\theta}) = \{\theta_1 + \theta_2 \exp(-\theta_3 x^2)\} x,$$

or logistic autoregressive,

$$f(x, \boldsymbol{\theta}) = \{\theta_1 x + \theta_2 x \{[1 + \exp(-\theta_3(x - \theta_4))]^{-1} - 1/2\}.$$

Usually, but not always, there will be a single value such that if $\boldsymbol{\theta}' = \boldsymbol{\theta}_0'$, then $f(x, \boldsymbol{\theta})$ is linear. Here $\boldsymbol{\theta}'$ is a subvector of $\boldsymbol{\theta}$. In the exponential case $\boldsymbol{\theta}' = [\theta_2, \theta_3] = [0, 0]$ gives linearity, but so does $\theta_2 = 0$ or $\theta_3 = 0$ separately, which may create an identification problem. The same is true for the logistic model with $\boldsymbol{\theta}' = [\theta_2, \theta_3] = [0, 0]$.

The test of $\theta' = \theta_0'$ is carried out using a Lagrange multiplier test. The procedure is described in much more detail with an outline of optimality properties in Luukkonen et al. (1988a and b), Saikkonen and Luukkonen (1988), Granger and Teräsvirta (1993) and Teräsvirta (1994). This way of testing is optimal if it is used against a specific alternative (specific f) and it turns out that this particular alternative is in fact true. This seldom (never) happens in practice of course, and one could resort to a battery of tests for various alternatives. In this class of alternatives it must also be taken into account that the lag structure is unknown, and therefore it is important that such tests are fairly robust not only to a wrongly chosen f, but also to a wrongly specified lag structure.

The Lagrange multiplier tests do possess some degree of robustness, and they are used by many econometricians. Some other parametric tests for example Keenan (1985), Tsay (1986) come out as special cases of the Lagrange formalism by choosing f appropriately, whereas for example the tests based on neural networks, Teräsvirta et al. (1993) are not covered by it.

2.2 Nonparametric Tests—the Spectral Domain

The first nonparametric tests were proposed in the spectral domain, and I will give a brief presentation of them before I turn to tests based on the conditional mean and the conditional variance, which will be our primary concern in this section.

These tests originated with Subba Rao and Gabr (1980) and were improved by Hinich (1982). In particular Hinich has continued to work with them, and a number of results have been obtained by him and his co-workers, Ashley et al. (1986), Brockett et al. (1988).

The spectral density of a stationary process with an absolutely summable covariance function K is given by

$$f(\omega) = (2\pi)^{-1} \sum_{t=-\infty}^{\infty} K(t) \exp(-i\omega t)$$

where

$$K(t) = \text{cov}(X_t, X_0) = \int_{-\pi}^{\pi} \exp(i\omega t) f(\omega) \, d\omega.$$

Similarly, for a zero-mean process with an absolutely summable third moment function $K(s, t) = E(X_s X_t X_0)$ the bispectrum is defined by

$$f_B(\omega_1, \omega_2) = (2\pi)^{-1} \sum_{s=-\infty}^{\infty} \sum_{t=-\infty}^{\infty} K(s, t) \exp(-i\omega_1 s - i\omega_2 t)$$

so that

$$K(s, t) = \int_{-\pi}^{\pi} \int_{-\pi}^{\pi} \exp(i\omega_1 s + i\omega_2 t) f_B(\omega_1, \omega_2) \, d\omega_1 \, d\omega_2.$$

It is well known that for a Gaussian process

$$f_B(\omega_1, \omega_2) \equiv 0. \tag{2.1}$$

As noted by Subba Rao and Gabr (1980) and Hinich (1982) for a linear process

$$X_t = \sum_{i=0}^{\infty} \alpha_i e_{t-i}$$

we have the identity

$$B(\omega_1, \omega_2) = \frac{|f_B(\omega_1, \omega_2)|^2}{f(\omega_1) f(\omega_2) f(\omega_1 + \omega_2)} \equiv c, \tag{2.2}$$

where c is a constant, and Eqns (2.1) and (2.2) can now be taken as starting points for tests of Gaussianity and linearity.

It was mentioned already by Subba Rao and Gabr (1980) that Eqns (2.1) and (2.2) may hold in non-Gaussian and non-linear cases, and the main criticism leveled against the bispectrum test has been that it fails to detect certain symmetric nonlinear patterns and that it needs considerably more observations to match the power of the best parametric test for a given alternative.

The so-called BDS test (Brock, Dechert and Scheinkman (1987)) exploits the correlation structure in an entirely different way, using concepts from chaos theory. It could be thought of as a linearity test, but I have stressed the independence aspect of this test, and the reader is referred to Sec. 5.1 for more details.

2.3 Testing Linearity in the Conditional Mean and Conditional Variance

An informal and much used exploratory technique in regression and time series analysis is to construct plots of the conditional mean and the conditional variance of the dependent variable Y given an explanatory variable X_k. This corresponds to a nonparametric regression of Y on X_k, or in a time series context, of X_t on X_{t-k}. In the time series case whole series of plots are displayed in the book by Tong (1990). Looking at the plot of a single series in this way it is not always easy to determine whether an apparent nonlinearity is due to a genuine nonlinearity or is due to statistical fluctuations inherent in the nonparametric estimation method. The results to be presented in the remainder of this section could be seen as an attempt to quantify and formalize this much used looking–at–plot procedure. Essentially it amounts to computing the statistical fluctuations of the plots that can be expected under the null hypothesis of linearity. The presentation will be based almost exclusively on Hjellvik and Tjøstheim (1995,1996) and Hjellvik et al. (1998).

Before we start we need to clarify the concept of linearity. Ideally we would like to measure and test deviations from linear autoregressive models

$$X_t = \sum_{i=1}^{p} a_i X_{t-i} + e_t, \tag{2.3}$$

where $\{e_t\}$ consists of iid random variables (or from the more general ARMA models), and where the characteristic polynomial $z^p - \sum_{i=1}^{p} a_i z^{p-i}$ has its zeros inside the unit circle in the complex plane. For some of the theoretical/asymptotical derivations a martingale difference assumption on $\{e_t\}$ could be natural, but much of the practical implementations are based on bootstrapping of residuals, for which the iid assumption would be needed.

For an autoregressive model the conditional mean of X_t given $X_{t-1}, \ldots, X_{t-p}$ is linear and is given by

$$M(x_1, \ldots, x_p) = E(X_t | X_{t-1} = x_1, \ldots, X_{t-p} = x_p) = \sum_{i=1}^{p} a_i x_i. \tag{2.4}$$

Similarly, the conditional variance is constant and given by

$$V(x_1, \ldots, x_p) = \operatorname{var}(X_t | X_{t-1} = x_1, \ldots, X_{t-p} = x_p) \equiv \sigma_e^2, \tag{2.5}$$

where $\sigma_e^2 = \operatorname{var}(e_t) = E(e_t^2)$. Hence, eventually we would like a linear model to mean an AR (ARMA) model with linear conditional mean and constant conditional variance. However, estimating $M(x_1, \ldots, x_p)$ nonparametrically is difficult in practice due to the curse of dimensionality for p moderate or large. Moreover, a graphical display would be difficult to interpret for $p > 2$.

For these reasons we have resorted to looking at one lag at a time; i.e. we estimate $M_k(x) = E(X_t | X_{t-k} = x)$ and $V_k(x) = \operatorname{var}(X_t | X_{t-k} = x)$ nonparametrically and compare them with the linear regression of X_t on X_{t-k}. The difficulty which then surfaces is that it is possible to construct examples, see also Tong (1990), where $M(x_1, \ldots, x_p)$ is linear and $M_k(x)$ is nonlinear. This means that to obtain a test of linearity in the sense of Eqns (2.4) and (2.5), if we are basing ourselves on $M_k(x)$ and $V_k(x)$, we have to find the null distribution of our test statistics under the null hypothesis of Eq. (2.3). This is exactly what we do in the randomization or bootstrap version of our test, whereas the asymptotic theory presented in Sec. 2.5 is worked out under more restrictive conditions.

We are now ready to introduce our test functionals. These functionals are all essentially based on measuring the deviation from linearity of the quantities $M_k(x)$ and $V_k(x)$.

If, with no restriction, we assume that $\{X_t\}$ is zero-mean, then the linear regression of X_t on X_{t-k} is given by $\rho_k X_{t-k}$ where $\rho_k = \text{corr}(X_t, X_{t-k})$, and the squared difference linearity test functional (cf. Hjellvik and Tjøstheim (1995, 1996)) is defined by

$$L(M_k) = \int (M_k(x) - \rho_k x)^2 w(x)\, dF(x) \tag{2.6}$$

where w is a weight function and F is the cumulative distribution function of X_t. The weight function can be chosen to have compact support or to be smooth, and it is included to screen out outliers and to make the asymptotic analysis easier (cf. Sec. 2.5). Large values of L_k would be taken as an indication of nonlinearity.

The conditional variance can be treated likewise using the functional

$$L(V_k) = \int (V_k(x) - \sigma_k^2)^2 w(x)\, dF(x)$$

where $\sigma_k^2 = (1 - \rho_k^2)\,\text{var}(X_t)$ is the residual variance in a linear regression of X_t on X_{t-k}. Actually, following the practice of ARCH modeling, Engle (1982), Bollerslev et al. (1994) the $L(V_k)$-functional will be applied to the residuals e_t of a linear AR (or ARMA) model fit. Under the hypothesis of the model in Eq. (2.3) (where $\{e_t, t \geq 0\}$ no longer have to be independent) we have

$$L(V_k) = \int (V_k(e) - \sigma_e^2)^2 w(e)\, dF_e(e) \tag{2.7}$$

where $\sigma_e^2 = \text{var}(e_t)$. In practice (cf. Sec. 2.4), the residuals must be estimated, but this does not alter the first order asymptotics of the test.

The functionals in Eqns (2.6) and (2.7) were introduced in Hjellvik and Tjøstheim (1995). Clearly one can also look at aggregated functionals,

$$L_{\text{ave}}(M_k) = \frac{1}{k} \sum_{i=1}^{k} L(M_i),$$

$$L_{\text{sup}}(M_k) = \max\{L(M_i), i = 1, \ldots, k\},$$

$$L_{\text{ave}}(V_k) = \frac{1}{k} \sum_{i=1}^{k} L(V_i),$$

$$L_{\text{sup}}(V_k) = \max\{L(V_i), i = 1, \ldots, k\},$$

measuring the deviation from linearity up to and including lag k.

Alternatively, one could use (Hjellvik et al. (1997)) the trivial fact that the derivative of a linear function is a constant, and the second order derivative is zero to introduce the functionals

$$L'(M_k) = \int \{M_k'(x) - \rho_k\}^2 w(x) \, dF(x),$$

$$L''(M_k) = \int \{M_k''(x)\}^2 w(x) \, dF(x),$$

and

$$L'(V_k) = \int \{V_k'(x)\}^2 w(x) \, dF(x).$$

This raises the issue of estimating the derivatives $M_k'(x)$, $M_k''(x)$ and $V_k'(x)$, and we focus our attention on the general problem of estimation in Sec. 2.4.

2.4 Estimation

There are two issues involved; i.e. that of estimating individual functions such as $M_k(x), V_k(x)$ and that of estimating the integrals over these functions. The integrals can be estimated numerically; i.e.

$$\hat{L}(M_k) = \int \{\hat{M}_k(x) - \hat{\rho}_k x\}^2 w(x) \hat{p}(x) \, dx$$

assuming that there is a density function p such that $dF(x) = p(x) \, dx$. However, we have found it more convenient to use the empirical mean, so that

$$\hat{L}(M_k) = \frac{1}{n} \sum \{\hat{M}_k(X_t) - \hat{\rho}_k X_t\}^2 w(X_t)$$

and

$$\hat{L}(V_k) = \frac{1}{n} \sum \{\hat{V}_k(\hat{e}_t) - \hat{\sigma}_e^2\}^2 w(\hat{e}_t)$$

where $\hat{e}_t$ are the residuals from a linear AR($\hat{p}$) fitting; i.e.

$$\hat{e}_t = X_t - \hat{a}_1 X_{t-1} - \cdots - \hat{a}_p X_{t-\hat{p}}.$$

The other functionals are estimated in an entirely analogous fashion. It should also be noted that in all applications in Hjellvik and Tjøstheim (1995, 1996) and Hjellvik et al. (1998) we have normalized the $\{X_t\}$-process so that it has mean zero and standard deviation 1.

We next turn our attention to the estimates of M_k, V_k. In Hjellvik and Tjøstheim (1995, 1996) ordinary kernel estimates were used; i.e.

$$\hat{M}_k(x) = \frac{\sum X_{t+1} K_h(X_t - x)}{\sum K_h(X_t - x)}$$

with $K(\cdot)$ being a kernel function; e.g. the standard Gaussian kernel or Epanechnikov, and where $K_h(\cdot) = h^{-1} K\{h^{-1}(\cdot)\}$. Similarly,

$$\hat{V}_k(x) = \frac{\sum X_{t+1}^2 K_h(X_t - x)}{\sum K_h(X_t - x)} - \{\hat{M}_k(x)\}^2.$$

Recently local polynomial estimation has been increasingly popular (cf. Fan and Gijbels (1995a)). In some situations this method has an advantage over the kernel method. Locally at a point x we approximate $M_k(x)$ by a polynomial of order T, so that

$$M_k(z) \approx \sum_{i=0}^{T} \frac{M_k^{(i)}(x)}{i!} (z - x)^i$$

with z lying in a neighborhood of x. Here $M_k^{(i)}(x)$ denotes the ith derivative of $M_k(x)$. Consider the following least squares problem: let $\hat{\gamma}_i, i = 1, \ldots, T$ minimize

$$\left\{ \sum_{t=k+1}^{n} X_t - \sum_{i=1}^{T} \frac{\gamma_i}{i!} (X_{t-k} - x)^i \right\}^2 K\{h^{-1}(X_{t-k} - x)\}$$

where K is again a non-negative function, serving as a kernel function, and h is the bandwidth controlling the size of the local neighborhood. Then $\hat{\gamma}_i$ estimates $M_k^{(i)}(x)$, $i = 0, \ldots, T$. Let $\gamma = \{M_k(x), M_k^{(1)}(x), \ldots, M_k^{(T)}(x)\}^\tau$, where τ denotes transposed. The least squares theory provides the solution

$$\hat{\gamma} = \{\hat{M}_{k.T}(x), \hat{M}_{k.T}^{(1)}, \ldots, \hat{M}_{k.T}^{(T)}(x)\}^\tau = (X^\tau K X)^{-1} X^\tau K Y$$

where $Y = (X_{k+1}, \ldots, X_n)^\tau$, K is a $(n-k) \times (n-k)$ diagonal matrix with $K_h(X_{i+k-1} - x)$ along the diagonal and X is a $(n-k) \times (T+1)$ matrix with the ith row $[1, (X_{i+k} - x), \ldots, (X_{i+k} - x)^T / T!]$. The special case $T = 0$ corresponds to ordinary kernel estimation. The theory of local polynomial regression has been developed in a number of papers, see e.g. Stone (1977, 1985), Tsybakov (1986), Fan (1992, 1993), Ruppert and Wand (1994), Fan and Gijbels (1995a and 1995b), Masry and Fan (1997).

It should be noted that $\hat{\gamma}_i = \hat{\gamma}_{i,T}$ is a function of T, and thus for each T we get a new functional

$$\hat{L}_T(M_k) = \frac{1}{n} \sum \{\hat{M}_{k.T}(X_t) - \hat{\rho}_k X_t\}^2 w(X_t)$$

and similarly for $\hat{L}_T(M_k')$ and $\hat{L}_T(M_k'')$.

The conditional variance is estimated by approximating $E(X_t^2|X_{t-k})$ by a local polynomial, which is achieved by letting δ_i, $i = 1, \ldots, T$ be the minimizers of

$$\sum_{t=k+1}^{n} \left\{ X_t^2 - \sum_{i=0}^{T} \frac{\delta_i}{i!}(X_{t-k} - x)^i \right\}^2 K\{h^{-1}(X_{t-k} - x)\}$$

and then subtracting $\hat{M}_k^2(x)$.

The main advantage of using local polynomial estimation from our point of view is larger robustness when it comes to choosing the bandwidth. When $h \to \infty$, then the kernel estimate $\hat{M}_{k,0}(x) \to \bar{X}$, the sample mean, whereas $\hat{M}_{k,1}(x)$, the intercept of the local linear least squares estimation, can be shown to tend to a straight line $\hat{\alpha} + \hat{\beta}x$, which is the global regression line in a linear regression of X_t on X_{t-k}. Moreover $\hat{M}_{k,2}(x)$ tends to a parabola resulting from a global regression of X_t on X_{t-k} and X_{t-k}^2. This means that for large h, $\hat{L}_T(M)$, $T \geq 1$ will still be able to reveal nonlinearity and give a serviceable null distribution for the linear case, whereas $\hat{L}_0(M)$ will not. We refer to Hjellvik et al. (1998) for a closer discussion and more details.

Let us close Sec. 2.4 by tying the functionals to the general principles outlined in the introduction. The two statistics mentioned there, in the present context, clearly are $\hat{M}_k(x)$ and $\hat{\rho}_k x$. They both estimate the same quantity $\rho_k x$ under a linearity hypothesis of $M_k(x) = \rho_k x$. (But it should be noted that in the randomization/bootstrap approach we will rather evaluate L_k-functionals under the null hypothesis in Eq. (2.3)). In the case of $L_1(M_k')$, $\hat{M}_k'(x)$ and $\hat{\rho}_k$ play the same role, whereas in the case of $L_T(M'')$ only the statistic $\hat{M}_k''(x)$ remains.

In a fundamental paper Härdle and Mammen (1993) suggested a comparison of nonparametric and parametric estimates in an iid regression situation. They also introduced a scaling factor so that, in our notation,

$$L^0(M) = \frac{1}{n}\sum\{M_k(x) - \mathcal{K}_{h,n}\rho_k x\}^2,$$

where $\mathcal{K}_{h,n}$ is a (random) smoothing operator whose purpose it is to eliminate smoothing bias terms that occur under the null hypothesis. Using $\mathcal{K}_{h,n}$ one obtains optimal convergence under a Pitman alternative. In practice such a factor does not play a large role, but it is advantageous if local asymptotic theory or an asymptotic theory of the bootstrap is to be developed.

2.5   Asymptotic Theory

Consistency can be proved relatively easily by an addition–subtraction argument. To illustrate, let $F_n(x) = n^{-1} \sum_t 1(X_t \leq x)$ denote the empirical distribution function. Then $L(M_k) - \hat{L}_T(M_k)$ can be written

$$
L(M_k) - \hat{L}_T(M_k) = \int \{M_k(x) - \rho_k x\}^2 w(x)\, dF(x)
$$

$$
- \int \{\hat{M}_{k,T}(x) - \hat{\rho}_k x\}^2 w(x)\, dF_n(x)
$$

$$
= \int \{\hat{M}_{k,T}(x) - \hat{\rho}_k x\}^2 w(x)\{dF(x) - dF_n(x)\}
$$

$$
+ \int [\{M_k(x) - \rho_k x\}^2 - \{\hat{M}_{k,T}(x) - \hat{\rho}_k x\}^2]
$$

$$
\times w(x)\, dF(x).
$$

If w has compact support, the second integral can be made small if $\hat{\rho}_k \overset{\text{a.s.}}{\to} \rho_k$ and $\hat{M}_{k,T}(x) \overset{\text{a.s.}}{\to} M_k(x)$ uniformly on the support of w. Assumptions such that this holds, are not difficult to find. If $M_k(x)$ is bounded on the support of w, then the integrand in the first integral will be bounded for n large enough, and the desired result will follow from convergence results on the empirical distribution function. The other functionals can be treated likewise. A proof of weak consistency is presented in Hjellvik et al. (1998). The consistency results, see e.g. Theorem 3.1 in Hjellvik et al. (1998) show that the test statistics will converge to zero in probability or almost surely if the model is linear. Therefore, large values of the statistics will indicate departure from the linearity hypothesis.

It is much more difficult to prove distributional results. Such proofs are based on U-statistic arguments and are quite intricate as is demonstrated in Hjellvik et al. (1998). To give a feeling for what is involved I list the regularity conditions for a typical result. These are stated for a general AR(1) regression model

$$
Y_t = m(X_t) + e_t, \tag{2.8}
$$

where $\{X_t\}$ is a stationary series having a marginal density $p(x)$.

A1.   The kernel function K is a symmetric density function with a bounded support in R^1, and $|K(x_1) - K(x_2)| \leq c|x_1 - x_2|$ for all x_1 and x_2 in its support. The weight function $w(\cdot)$ is continuous and with compact support contained in $\{p(x) > 0\}$.

A2. For all t, $E\{e_t|X_t, X_{t-1}, \ldots ; Y_{t-1}, Y_{t-2}, \ldots\} = E\{e_t|X_t\} = 0$, $E\{X_t^8\} < \infty$ and $E\{Y_t^8\} < \infty$. Further, $E\{Y_t^2|X_s = x\}$ is a bounded function of x for arbitrary values of t and s.

A3. The joint density of distinct elements of $(X_1, Y_1, X_s, Y_s, X_t, Y_t)$ $(t > s > 1)$ is continuous and bounded by a constant independent of s and t.

A4. The process $\{(X_t, Y_t)\}$ is absolutely regular, i.e.

$$\beta(j) \equiv \sup_{i \geq 1} E\Big\{ \sup_{A \in \mathcal{F}_{i+j}^\infty} |P(A|\mathcal{F}_1^i) - P(A)| \Big\} \to 0 \qquad \text{as} \quad j \to \infty,$$

where $\mathcal{F}_i^j$ is the σ-field generated by $\{(X_k, Y_k): k = i, \ldots, j\}$, $(j \geq i)$. Further, for a constant $\delta \in (0, 0.5)$, $\sum_{k=1}^\infty k^2 \beta^{\delta/(1+\delta)}(k) < \infty$.

A5. As $n \to \infty$, then $h \to 0$ and $nh^{(2+4\delta)/(1+\delta)} / \log n \to \infty$.

These conditions lead to asymptotic normality results (they are formulated for $T = 2$, but similar results can be formulated for a general T) as given in Hjellvik et al. (1998):

Theorem A: Under the hypothesis of linearity ($m(x) = \rho x$), the conditions (A1)–(A5) imply asymptotic normality with the following rates:

(i) $\hat{L}_2(m) \sim \mathcal{N}[a_1/(nh), \sigma_1^2/(n^2h)]$,

(ii) $\hat{L}_2(m') \sim \mathcal{N}[a_2/(nh^3), \sigma_2^2/(n^2h^5)]$,

(iii) $\hat{L}_2(m'') \sim \mathcal{N}[a_3/(nh^5), \sigma_3^2/(n^2h^9)]$.

The constants a_1, a_2, a_3, σ_1^2, σ_2^2, σ_3^2 are defined in Hjellvik et al. (1997).

Theorem B: For a general $m(\cdot)$ in Eq. (2.8), under the conditions (A1)–(A5)

(i) $\hat{L}_2(m) \sim \mathcal{N}(a_4, \sigma_4^2/n)$,

(ii) $\hat{L}_2(m') \sim \mathcal{N}(a_5, \sigma_5^2/n)$,

(iii) $\hat{L}_2(m'') \sim \mathcal{N}(a_6, \sigma_6^2/n)$.

Again the relevant constants a_4, a_5, a_6, σ_4^2, σ_5^2, σ_6^2, are given in Hjellvik et al. (1997).

The main idea in the proof of the asymptotic distributional results is to perform the Hoeffding decomposition on the test statistic and then apply asymptotic results on U-statistics, cf. Yoshihara (1976), Denker and Keller (1983), Hall (1984), de Jong (1987), and Hjellvik et al. (1998). The results of Theorem B are in a sense trivial. One could instead assume that the alternative hypothesis H_A deviates from H_0 at a certain rate related to the sample size, cf. Härdle and Mammen (1993). Since it is arguable whether such a hypothesis has any practical implications in the current context, we do not pursue this any further.

It should be noted that for a fixed bandwidth h the asymptotic variances and biases of $\hat{L}(m')$ and $\hat{L}(m'')$ are larger than for $\hat{L}(m)$. However, for a given n, typically a larger bandwidth would be optimal for $\hat{L}(m')$ and $\hat{L}(m'')$ as compared to $\hat{L}(m)$. Based on finite sample evidence, cf. Hjellvik et al. (1998) none of the tests dominates the other two.

2.6   Finite Sample Properties and Use of the Asymptotics

Unfortunately the finite sample distribution is not close to that predicted by asymptotic theory unless n is very large. This is shown in Table 1, which is taken from Hjellvik et al. (1998). It shows the relationship between estimated and simulated means of $\hat{L}_2(M_1)$, $\hat{L}_2(M_1')$ and $\hat{L}_2(M_1'')$ for $\{X_t\} = \{e_t\}$, the latter consisting of iid random variables. A trapezoidal weight function has been used and a bandwidth proportional to $n^{-1/9}$, see below and also Hjellvik et al. (1998). Table 1 also shows the simulated significance level for a nominal level of 0.05. It is seen that even for large sample sizes, the approximation is not very good. Actually, a rectangular weight function on $[-1, 1]$ gave somewhat better results, the empirical sizes of $\hat{L}_2(M_1)$, $\hat{L}(M_1')$ and $\hat{L}(M_1'')$ being 0.032, 0.062 and 0.066, respectively, for a sample size of $n = 2^{21} = 2\,097\,152$. Note that the tabulated empirical sizes give a direct indication of the convergence in distribution implied by Theorem A. We believe that the slow convergence, which occurs for a large set of reasonable bandwidths, is typical, as is also indicated by similar experiments in Hjellvik and Tjøstheim (1995, 1996). In fact, it is believed to be a general feature of the test functionals used not only here, but also in independence testing, say, and we will come back to this in Sec. 5. We think that the reason for the bad approximation is that, unlike a standard parametric setting, the next order terms in the Edgeworth expansion for these kind of functionals are very close to the leading normal approximation term given in Theorem A. It cannot be ruled out that for each case there *exists* a bandwidth giving a fairly accurate approximation for a fixed moderate sample size, but in practice, when, unlike a simulation experiment, the truth is not known, it would be difficult to find such a bandwidth.

It should be noted that very recently better approximation to the asymptotic distribution has been reported in a *fixed* design case by Poggi and Portier (1997). Alternatively, a better approximation can be obtained by considering a fixed x, Matsudo and Shigeno (1996).

Even though the asymptotic theory requires very large sample sizes to work satisfactorily, it is useful in giving rough ideas as to which bandwidth should be chosen. For example, cf. Hjellvik et al. (1998), by considering

Table 1 The ratio between the asymptotic values given by Theorem A and simulated values for the mean and the standard deviation of $\hat{L}_2(M_1)$, $\hat{L}_2(M_1')$ and $\hat{L}_2(M_1'')$, and the empirical sizes for these statistics when they have been centered by the asymptotic mean and scaled by the asymptotic standard deviation of Theorem A. A critical value of 1.645 corresponding to the nominal size of 0.05 for the standard normal distribution has been used. The model is $X_t = e_t$, and the number of realizations are 500.

	128	256	512	2^{10}	2^{11}	2^{12}	...	2^{15}	2^{16}	2^{17}	2^{18}	2^{19}	2^{20}	2^{21}
$\hat{L}_2(M_1)$														
mean	1.76	1.64	1.69	1.59	1.56	1.46	...	1.38	1.34	1.38	1.31	1.27	1.25	1.23
sd	1.75	1.50	1.64	1.57	1.45	1.49	...	1.40	1.28	1.30	1.27	1.17	1.17	1.20
size	0.004	0.020	0.012	0.010	0.020	0.016	...	0.012	0.022	0.020	0.024	0.024	0.026	0.030
$\hat{L}_2(M_1')$														
mean	0.480	0.516	0.590	0.620	0.656	0.668	...	0.777	0.803	0.874	0.886	0.887	0.889	0.912
sd	0.520	0.537	0.641	0.684	0.672	0.748	...	0.870	0.840	0.919	0.937	0.919	0.895	0.968
size	0.294	0.248	0.236	0.214	0.188	0.202	...	0.140	0.138	0.100	0.092	0.100	0.122	0.080
$\hat{L}_2(M_1'')$														
mean	0.423	0.462	0.527	0.572	0.616	0.646	...	0.760	0.794	0.863	0.902	0.901	0.880	0.929
sd	0.492	0.481	0.575	0.618	0.632	0.732	...	0.775	0.811	0.945	0.951	0.962	0.883	0.998
size	0.356	0.300	0.258	0.256	0.214	0.222	...	0.158	0.140	0.106	0.092	0.098	0.122	0.084

mean square errors for $\hat{M}_T(x)$ for $T = 0, 1, 2$, optimal bandwidth for $T = 2$ is proportional asymptotically to $n^{-1/9}$, whereas for $T = 0$ and $T = 1$, the optimal bandwidths are given by

$$h_0(x) = \frac{1}{n^{1/5}} \left(\frac{\sigma^2(x)J}{p(x)\mu_2^2\{m''(x) + 4m'(x)p'(x)/p(x)\}^2} \right)^{1/5}$$

and

$$h_1(x) = \frac{1}{n^{1/5}} \left(\frac{\sigma^2(x)J}{p(x)\mu_2^2\{m''(x)\}^2} \right)^{1/5},$$

respectively. Here $\sigma(x) = \mathrm{Var}(Y|X = x)$, $\mu_2 = \int t^2 K(t)\,dt$ and $J = \int K^2(t)\,dt$. Qualitatively these values for the bandwidths are confirmed by experiments in Hjellvik et al. (1998).

2.7   A Randomization/Bootstrap Approach to Testing

The outcome of the experiment described in Sec. 2.6 means that for small and moderate sample sizes the asymptotic distribution cannot be used to construct the null distribution of the functionals. An alternative is to create the null distribution by randomizing (permuting) or by bootstrapping the residuals

$$\hat{e}_t = X_t - \sum_{i=1}^{\hat{p}} \hat{a}_i X_{t-i}$$

from the best linear autoregressive (or ARMA) fit to $\{X_t\}$. Randomized/ bootstrapped values $\hat{L}^*(\cdot)$ of the functional in the null situation are created by inserting in the expression for $\hat{L}(\cdot)$ randomized/bootstrapped linear versions

$$X_t^* = \sum_{i=1}^{\hat{p}} \hat{a}_i X_{t-i}^* + e_t^*$$

of $\{X_t\}$. By taking a sufficiently large number of replicas $\{e_t^*\}$ of $\{e_t\}$, in this way we can construct a null distribution for $\hat{L}(\cdot)$. Both the conditional mean and the conditional variance functionals can be treated in this way, and for more details I refer to Hjellvik and Tjøstheim (1995, 1996) and Hjellvik et al. (1998).

    In these papers, as a standard, 40 replicas were used to create the null distribution. The distribution was smoothed by a nonparametric integrated kernel type estimate and the critical region corresponding to an α-level test is then obtained by selecting the $(1 - \alpha)$-quantile c_α^* of the bootstrap distribution, and the hypothesis of linearity is rejected if $\hat{L}(\cdot) \geq c^*_\alpha$, where $\hat{L}(\cdot)$ is the value of $\hat{L}$ as computed from the original data series $\{X_t\}$. Such an approach produces very dramatic improvements of the level of the test, and even for $n = 100$ the level is quite accurately determined.

Ideally one would like to analyse the randomization/bootstrap approach theoretically. Asymptotic theory is lacking for the bootstrap in this situation, but work is now in progress, Neumann and Kreiss (1996). It is a highly nontrivial task, and large sample sizes may again be needed to check the validity of the asymptotic theory.

Another approach is to look at the randomization(permutation) aspect and treat this as a finite sample problem. Actually, if the residuals e_t were known exactly, the permutations would provide a test with exact levels (cf. also Sec. 5.3). On the other hand if the order p is correct the difference between $\hat{e}_t$ and e_t is of order $O_p(n^{-1/2})$, i.e. a relatively quick convergence, and from this point of view the good results obtained by randomization/ bootstrapping is not so surprising.

The randomization/bootstrap approach applied to the test functionals $L(\cdot)$ has been evaluated on a large group of examples, both simulated and real data. It has also been compared to parametric tests and the bispectrum test. It performs well even for small sample sizes. We refer to Hjellvik et al. (1997). Instead of going into details here we will rather mention one example where it fares poorly to show the limitations of our approach. For the bilinear model

$$X_t = (-0.9 - 0.1e_{t-1})X_{t-1} + e_t + 2.0$$

our test fails completely. The reason is, Hjellvik and Tjøstheim (1995) that $\hat{M}_k(x)$ is very close to being linear and $\hat{V}_k(x)$ very close to being a constant for this process. Thus the nonlinearity is not manifesting itself in terms of our test statistics, and cannot be picked up. The Lagrange multiplier test against a bilinear alternative (perhaps not a very realistic situation in practice, because it presupposes that it is known that a bilinear model of order 1 would be the appropriate alternative) works well in this case.

3. SELECTING SIGNIFICANT LAGS

Selecting the order of a linear autoregressive model has been a research topic that has been extensively worked on over the years starting from Akaike's (1969) path-breaking paper. If the data at hand fail to pass a linearity test of the kind described in Sec. 2.7, then this issue takes on new importance, since if nonparametric analysis should be used subsequently, the curse of dimensionality would effectively preclude having many lags in the model. Actually the emphasis of the problem shifts a bit. It is not so much determining the order of the model as determining a few main lags which capture the essential features of the data. For example, it may be desirable to include one or two higher order lags to model seasonality.

Both the problem of determining the order of the model and that of selecting significant lags are handled by essentially the same approach. As far as we know the order problem was first attacked in a fully nonparametric context in Auestad and Tjøstheim (1990). This has been followed up, extended and deepened in subsequent work by Cheng and Tong (1992, 1993), Granger and Lin (1994), Tjøstheim and Auestad (1994b), Yao and Tong (1994) and very recently by Tschernig (1996), Tschernig and Yang (1996) and Yang and Tschernig (1996). See also the work by Vieu (1994).

The gist of the problem is as follows: For a given collection of data $\{X_1, \ldots, X_n\}$ select a set of lags $i_1, \ldots, i_p$ where $i_p \leq L$, some upper limit, such that the mean square prediction error

$$E\{X_t - M(X_{t-i_1}, \ldots, X_{t-i_p})\}^2$$

is minimized. Here

$$M(X_{t-i_1}, \ldots, X_{t-i_p}) = E(X_t | X_{t-i_1}, \ldots, X_{t-i_p})$$

is the optimal least squares nonlinear predictor of X_t based on $X_{t-i_1}, \ldots, X_{t-i_p}$. Clearly, it is reasonable to let L and p increase as the number of observations n gets large. Even though the curse of dimensionality puts a relative strong restriction on the choice of p, it should be realized that the main objective is not to estimate $M(x_1, \ldots, x_p)$ for all possible sets of points $(x_1, \ldots, x_p)$ but rather to estimate $i_1, \ldots, i_p$ based on $M(x_1, \ldots, x_p)$ evaluated at the observations themselves. It is conceivable that good estimates of $i_1, \ldots, i_p$ can be obtained even if the estimates of $M(X_{t-1}, \ldots, X_{t-i_p})$ are quite poor. This argument has been formalized and elaborated upon (in the order case) in a paper by Cheng and Tong (1992).

In practice $E\{X_t - M(X_{t-i_1}, \ldots, X_{t-i_j})\}^2$ must be estimated for each subset of indices $\{i_1, \ldots, i_j\}$ and there are several difficulties. The obvious estimate is

$$\frac{1}{n} \sum_t \{X_t - \hat{M}(X_{t-i_1}, \ldots, X_{t-i_j})\}^2 \tag{3.1}$$

where M could be estimated by, say,

$$\hat{M}(x_1, \ldots, x_j) = \frac{\sum_t X_t K_h(X_{t-i_1} - x_1) \cdots K_h(X_{t-i_j} - x_j)}{\sum_t K_h(X_{t-i_1} - x_1) \cdots K_h(X_{t-i_j} - x_j)} \tag{3.2}$$

where a product kernel has been used. Obviously, a local polynomial estimator could be used, Tschernig and Yang (1996). However inserting Eq. (3.2) in Eq. (3.1), the prediction error expressed in Eq. (3.1) can be made arbitrarily small by taking the bandwidth h small enough.

There are at least two ways of avoiding this difficulty. In Tjøstheim and Auestad (1994b) $\hat{M}(\cdot)$ is computed as in Eq. (3.2) but a penalty factor is added using an analogy to the FPE-criterion in linear analysis; i.e. we let $\{Y_t\}$ be a process independent of $\{X_t\}$ but having identical properties, and, following Akaike (1969), we consider the effect of using the predictor $\hat{M}_X$ estimated from the X–process on the Y–process. We also introduce a weight function w as in Sec. 2. It is found in Tjøstheim and Auestad (1994b) that asymptotically the final prediction error (FPE) (modulo an order in probability argument) is given by

$$\text{FPE}(i_1, \ldots, i_j)$$

$$= E\left[n^{-1} \sum_t \{ Y_t - \hat{M}_X(Y_{t-i_1}, \ldots, Y_{t-i_j}) \}^2 w^2(Y_{t-i_1}, \ldots, Y_{t-i_j}) \right]$$

$$\sim E\{ V(X_{t-i_1}, \ldots, X_{t-i_j}) w^2(X_{t-i_1}, \ldots, X_{t-i_j}) \}$$

$$+ (nh^j)^{-1} J^j \int V(x_{i_1}, \ldots, x_{i_j}) w^2(x_{i_1}, \ldots, x_{i_j}) \, dx_{i_1} \cdots dx_{i_j}. \qquad (3.3)$$

where $J = \int K^2(z) \, dz$ and where

$$V(x_{i_1}, \ldots, x_{i_j}) = \text{Var}(X_t | X_{t-i_1} = x_{i_1}, \ldots, X_{t-i_j} = x_{i_j}).$$

It is seen that if $h \to 0$ faster than $n^{-1/j}$, then the penalty term tends to infinity, and such choices of bandwidth would therefore be prohibited. To be able to use Eq. (3.3) in practice we must estimate the quantities involved. In Tjøstheim and Auestad (1994b) a simplified version of the *estimated* FPE is given by

$$\widehat{\text{FPE}}(i_1, \ldots, i_j) = \sum R_t \frac{1 + (nh^j)^{-1} J^j B_j}{1 - (nh^j)^{-1} [2\{K(0)\}^j - J^j] B_j} \qquad (3.4)$$

where $R_t = \{ X_t - \hat{M}(X_{t-i_1}, \ldots, X_{t-i_j}) \}^2 w^2(X_{t-i_1}, \ldots, X_{t-i_j})$ and

$$B_j = n^{-1} \sum_t \frac{w^2(X_{t-i_1}, \ldots, X_{t-i_j})}{\hat{p}(X_{t-i_1}, \ldots, X_{t-i_j})}$$

with $\hat{p}(\cdot)$ being the estimated joint density of $X_{t-i_1}, \ldots, X_{t-i_j}$. A more general version of Eq. (3.4) is given in formula (14) of Tjøstheim and Auestad (1994b), where details and recommendations are also given for the estimation of p, the choice of h etc.

The above estimation scheme is very time consuming, especially if many h-values are tried out for each set of potential lags $\{i_1, \ldots, i_j\}$. Therefore, instead of searching exhaustively for all combinations $\{i_1, \ldots, i_j\}$, $i_j \leq L$, $j \leq p$, we have done a stepwise procedure where the first lag is found by taking argmin of $\widehat{\mathrm{FPE}}(i_1)$. If i_1^* is the lag obtained, then i_1^* is retained in all subsequent sets of lags such that the "optimal" two-set of lags is determined by letting i_2 range between 1 and L, $i_2 \neq i_1^*$, and finding the combination $\{i_1^*, i_2\}$ minimizing $\widehat{\mathrm{FPE}}(i_1^*, i_2)$, and so on.

It is indicated in Tjøstheim and Auestad (1994b) how a nonparametric analogy to the Akike information criterion (AIC) could be set up, and how the procedure could be applied to the conditional variance in addition to the conditional mean. The procedure works fairly well on the chosen examples (simulated and real), but clearly there are many unsolved theoretical and practical problems, and the procedure must be used with care. For example, there is a clear tendency of overestimating the number of lags needed in the model.

An alternative way of doing the penalization is to use cross-validation. This is in many ways simpler to use in practice, and it has been demonstrated by Cheng and Tong (1993) that it is equivalent asymptotically to using a penalization of the type given in Eq. (3.3). We refer to Cheng and Tong (1993) and Yao and Tong (1994) for an outline of the theory and practical examples for the cross-validation procedure. Especially if the lag structure of the conditional variance is to be determined, cross-validation presents itself as a particularly interesting alternative.

The curse of dimensionality has already been mentioned as an obstacle in nonparametric order selection. If there are reasons to believe, for example, based on tests described in Sec. 4, that the data are well approximated by an additive model, one could try a selection–of–lags procedure based on minimizing

$$E\{X_t - f_1(X_{t-i_1}) - \cdots - f_p(X_{t-i_p})\}^2$$

where the functions f_i could be estimated by, say, the marginal integration procedure outlined in Sec. 4 or by back-fitting, Hastie and Tibshirani (1990). Again one could use an approach based on the FPE-criterion or on cross-validation. In this case the penalty factor would be of order $(nh)^{-1}$ as $n \to \infty$ as compared to $(nh^j)^{-1}$ for Eq. (3.3).

3.1   A Modified FPE-Criterion

In a recent paper Tschernig and Yang (1996) make several significant contributions to this field: they base their development on the FPE approach but they try both local linear estimators and the Nadaraya–Watson kernel estimators used in Tjøstheim and Auestad (1994b). They also highlight the difference between the two methods when it comes to computation of bias, and they are able to find optimal—in the sense of balancing bias squared and variance—bandwidths in both cases.

The concept of consistency has played a certain role in developing the various versions of of the linear parametric criteria. It is well-known that in the linear case neither FPE nor AIC are consistent, i.e. if the true model is AR(p), then the order estimate $\hat{p}$ will not converge to p with probability one. Actually, there is a positive probability of overestimating the order asymptotically. Alternative criteria like BIC and LIL penalizes harder and yield consistency, see e.g. Koreisha and Yoshimoto (1991) for a review.

In the nonparametric case both the FPE and the cross-validated criteria are consistent as pointed out by Cheng and Tong (1993) and by Vieu (1994), and more recently by Tschernig and Yang (1996) for the FPE case. However, the latter authors go further. They show that the probability of overestimation goes to zero much more slowly than the corresponding error of underestimation, and they use this to suggest a corrected (still consistent) criterion where higher lags are penalized harder. This criterion does quite a lot better than the other versions, incuding those of Tjøstheim and Auestad (1994b), in a quite extensive simulation study conducted by the authors, and it appears that the modified criterion definitely deserves further attention.

4. ADDITIVE MODELING AND ADDITIVITY TESTS

If the data do not pass a linearity test, large simplifications can still be obtained if the data can be modeled additively; i.e. if X_t with a good approximation can be represented as

$$X_t + \sum_{j=1}^{p} f_j(X_{t-i_i}) + e_t.$$

With no loss of generality we will write $f_i(X_{t-i})$ in this Section. The traditional way of estimating the f's nonparametrically has been by backfitting, Hastie and Tibshirani (1990). This is a widely available and useful method. A disadvantage is that asymptotic theory is difficult to work out. In the independent regression case various theoretical optimality results have been obtained by Stone (1977, 1985) using spline methods. In this chapter we will put the emphasis on a new method, which is now being known as marginal integration. This method was to my knowledge first proposed in a paper by Auestad and Tjøstheim (1991). It was extended and treated more formally in Tjøstheim and Auestad (1994a). Independently the technique was discovered by Newey (1994), Linton and Perch Nilsen (1995). Subsequently the method has been applied to a number of new areas, and especially Oliver Linton has sought to demonstrate its potential, Chen and Haïdle (1994), Severance-Lossin and Sperlich (1995), Chen et al. (1996), Linton (1997a, 1997b). A rigorous account in the time series case is given in Masry and Tjøstheim (1997). See also Fan et al. (1995).

The idea behind marginal integration is very simple. In the model

$$X_t = c + \sum f_i(X_{t-i}) + e_t, \tag{4.1}$$

it is well-known that in general $E(X_t|X_{t-i}) \neq f_i(X_{t-i})$. However,

$$E(X_t|X_{t-1}, \ldots, X_{t-i}) = c + \sum f_i(X_{t-i}),$$

and by integrating over the joint marginal distribution $F_{(k)}$ of $X_{t-1}, \ldots, X_{t-k-1}, X_{t-k+1}, \ldots, X_{t-p}$; i.e. by taking the expectation (not conditional!) with respect to these variables, we get due to the additivity

$$
\begin{aligned}
P(x_k) &= \int E(X_t|X_{t-1} = x_1, \ldots, X_{t-p} = x_p) \\
&\quad \times dF_{(k)}(x_1, \ldots, x_{k-1}, x_{k+1}, \ldots, x_p) \\
&= c + c_1 + f_k(x_k)
\end{aligned}
$$

where $c_1 = \sum_{i \neq k} E\{f_i(X_i)\}$. Going back to the original equation it easily follows that

$$P(x_k) = f_k(x_k) - E\{f_k(X_t)\} + E(X_t).$$

In Auestad and Tjøstheim (1991) and Tjøstheim and Auestad (1994a) the appellation "projection" was used for $P(x_k)$. Strictly speaking this is misleading since mathematically $P(x_k)$ is not a projection.

The quantity $P(x_k)$ can be estimated in essentially two ways: One possibility is to use numerical integration and put

$$
\begin{aligned}
\tilde{P}(x_k) &= \int \hat{M}(x_1, \ldots, x_p)\hat{p}_{(k)}(x_1, \ldots, x_{k-1}, x_{k+1}, \ldots, x_p) \\
&\quad \times dx_1 \cdots dx_{k-1} dx_{k+1} \cdots dx_p.
\end{aligned}
$$

Alternatively, and this is the estimator used in Tjøstheim and Auestad (1994a),

$$\hat{P}(x_k) = \frac{1}{n} \sum_t \hat{M}(X_{t-1}, \ldots, X_{t-k+1}, x_k, X_{t-k-1}, \ldots, X_{t-p})$$

where again $M(\cdot)$ is estimated using one of the methods mentioned in Sec. 3. The estimator $\hat{P}(\cdot)$ has been analysed with an increasing degree of rigour (in the kernel case) in Tjøstheim and Auestad (1994a) and in Masry and Tjøstheim (1997). In the regression case with an iid set of explanatory variables we refer for example to Linton and Perch Nielsen (1995) and to Fan et al. (1998), where also local polynomial estimation has been used for $M(\cdot)$.

The projectors $P(X_k)$ were originally intended, Auestad and Tjøstheim (1991), Tjøstheim and Auestad (1994a) mainly as a means of looking for functional shapes of the additive components, so that subsequently parametric models could be built for these. Some examples are shown in Tjøstheim and Auestad (1994a) and in the time series regression case in Masry and Tjøstheim (1997) for the first order system

$$Y_{t+1} = 0.5Y_t + 0.5X_{t+1}^2 + 0.5Z_{t+1} + 0.5Z_{t+1}$$

$$\times [\{1 + \exp(-Z_{t+1}\}^{-1} - 0.5] + e_{t+1},$$

$$X_{t+1} = 0.5X_t + \varepsilon_{t+1},$$

$$Z_{t+1} = 0.5Z_t + \eta_{t+1},$$

where $\{e_t\}$, $\{\varepsilon_t\}$ and $\{\eta_t\}$ are generated as independent processes consisting of Gaussian iid random variables with zero mean and variance one. The $\{X_t\}$, $\{Y_t\}$ and $\{Z_t\}$ processes were subsequently adjusted so that they have zero mean and unit variance. The projection estimates of Y_{t+1} on Y_t, X_{t+1} and Z_{t+1} based on the kernel estimate of

$$m(x, y, z) = E(Y_{t+1} \mid X_{t+1} = x, Y_t = y, Z_{t+1} = z)$$

were computed for the scaled processes on the set $[-3, 3] \times [-3, 3] \times [-3, 3]$. Plots of $\hat{P}_X(x)$, $\hat{P}_Y(y)$ and $\hat{P}_Z(z)$ for 10 independent realizations and for $n = 500$ and $n = 2000$ are given in Fig. 1, which is taken from Masry and Tjøstheim (1997), to which we refer for more details. The plots clearly reveal the linear dependence on Y_t and the character of the nonlinear dependence on X_{t+1} and Z_{t+1}.

If there are not too many components, fairly accurate estimates of the components $f_k(\cdot)$ in Eq. (4.1) can be obtained and they could be used as an end result of the analysis. However, if many components are involved Tjøstheim and Auestad (1994a) found using ordinary kernel methods that the estimates are severely biased. The curse of dimensionality is implicitly making itself felt through the estimation of $M(x_1, \ldots, x_p)$, but the functional shapes can still be used for identification purposes including checking whether certain lags are linear or missing (flat projectors). Moreover, the estimates could be used as input estimates in a combined backfitting/marginal integration scheme. In two interesting papers Linton (1997a,

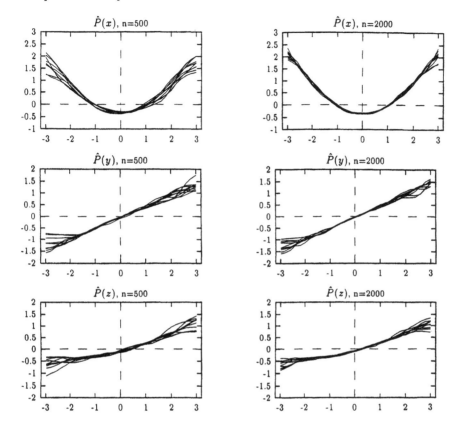

Figure 1

1997b) has obtained a certain type of optimality for this case. Hengartner (1996) treats a similar problem in the case of many covariates, and still another method has very recently been presented by Kim et al. (1997). Also conditional variance structures and generalized link models have been considered.

Thus the marginal integration method has been steadily and significantly improved by using local polynomials, combination with back-fitting and by computational devices to speed up the procedure, but the methodology still needs to be more extensively tested to determine its true potential.

As is the case for backfitting, the marginal integration method can be extended to a hierarchy of models allowing higher order interactions. This was indicated in Tjøstheim and Auestad (1994a). But first note that non-additive models cannot in general be precisely characterized by one-dimensional projections. For example, the model

$$X_t = X_{t-1} f(X_{t-2}) + e_t$$

has projections $P_1(x_1) = k_1 x_1$ and $P_2(x_2) = k_2 f(x_2)$, where k_2 is zero if $E(X_t) = 0$. In such a case, if one relies only on one-dimensional projectors, the process will be wrongly identified as a linear autoregressive process.

Higher dimensional projectors could be introduced to tackle models that allow interaction between subsets of lagged variables $\{X_{t-i}\}$. Two-dimensional projectors can be defined by

$$P_{jk}(x_j, x_k) = E\{M(X_{t-i_1}, \ldots, X_{t-i_j} = x_j, \ldots, X_{t-i_k} = x_k, \ldots, X_{t-i_p})\}.$$

If we have a model which is additive in the component variables and their second order interactions; i.e.

$$M(x_1, \ldots, x_p) = c + \sum_j f_j(x_j) + \sum_{j<k} f_{j,k}(x_j, x_k), \tag{4.2}$$

and if we assume

$$\int f_j(x_j) p(x_j) \, dx_j = 0$$

$$\int f_{i,j}(x_i, x_j) p(x_i) dx_i = \int f_{i,j}(x_i, x_j) p(x_j) \, dx_j = 0, \tag{4.3}$$

where the marginal density $p(\cdot)$ is assumed to exist, then the components $f_i(x_i)$ and $f_{i,j}(x_i, x_j)$ can be properly identified by a combination of two-dimensional and one-dimensional projectors, cf. Tjøstheim and Auestad (1994a) and Sperlich et al. (1997). The set of Eq. (4.3) do not imply restriction of generality; they merely ensures uniqueness.

The possibility of using representations such as Eq. (4.2) for testing of additivity was indicated in Tjøstheim and Auestad (1994a) and has been developed in Sperlich et al. (1997). We will return to this point in Sec. 4.2 after a brief look at asymptotics.

4.1 Asymptotics

One attractive feature of marginal integration, as compared to backfitting, is that an asymptotic theory can be worked out. It is illustrative to compare with the asymptotic theory of the conventional nonparametric estimate of the conditional mean (kernel or local polynomial). The conditional mean estimator $\hat{M}(\mathbf{x}) = \hat{M}(x_1, \ldots, x_j)$ is under a set of relatively weak regularity conditions, Robinson (1983), Masry and Tjøstheim (1995) asymptotically normally distributed

$$\hat{M}(\mathbf{x}) \sim \mathcal{N}\left(M(\mathbf{x}) + B_{n,h}(\mathbf{x}), (nh^j)^{-1} \frac{\{\int K^2(t)\, dt\} V(\mathbf{x})}{p(\mathbf{x})} \right),$$

where an expression for the bias $B_{n,h}(\mathbf{x})$ can be found in e.g. Auestad and Tjøstheim (1990) or Masry and Tjøstheim (1995). For j moderate and large the curse of dimensionality acts in conjunction with the slow convergence rate of $O_p\{(nh^j)\}^{-1/2}$. The quantities $\hat{P}_k(x)$ are derived from the j-dimensional quantity $\hat{M}(\mathbf{x})$, but are one-dimensional, and intuitively one might expect that they follow a one-dimensional rate $O_p(nh)^{-1/2}$. Indeed, under a string of regularity conditions which are stated in Masry and Tjøstheim (1997) we have for the weighted (typically with a 0–1 weight function w) projection estimator

$$\hat{P}_{k,w} = n^{-1} \sum_t \hat{M}(X_{t-1}, \ldots, X_{t-k+1}, x_k, X_{t-k-1}, \ldots, X_{t-p})$$

$$\times w_{(k)}(X_{t-1}, \ldots, X_{t-p})$$

that it is asymptotically normal

$$\hat{P}_{k,w}(x_k) \sim \mathcal{N}(P_{k,w}(x_k), (nh)^{-1} \int \frac{V_{[k]}(\mathbf{y}, x_k) w_{(k)}(\mathbf{y}) p_{(k)}^2(\mathbf{y})}{p_{[k]}(\mathbf{y}, x_k)} d_{(k)}\mathbf{y} \qquad (4.4)$$

with

$$P_{k,w}(x_k) = \int M(y_1, \ldots, y_{k-1}, x_k, y_{k+1}, \ldots, y_p) w_{(k)}(\mathbf{y}) p_{(k)}(\mathbf{y}) d_{(k)}\mathbf{y},$$

$$V_{[k]}(\mathbf{y}, x_k) = V(y_1, \ldots, y_{k-1}, y_k = x_k, y_{k+1}, \ldots, y_p)$$

and similarly for $p_{[k]}(\mathbf{y}, x_k)$, where $w_{(k)}(\mathbf{y})$ and $p_{(k)}(\mathbf{y})$ mean that the k-th coordinate has been removed in the weight function and the $(p-1)$-dimensional density function, respectively, and where finally

$$d_{(k)}\mathbf{y} = dy_{i_1} \cdots dy_{i_{k-1}} dy_{i_{k+1}} \cdots dy_{i_p}.$$

It is seen that it is important to include the weight function, because otherwise the integral may not converge.

A variance formula is implicit in Eq. (4.4). A similar formula can be found for the bias.

Analogous formulas have been developed for the marginal integration regression case including local polynomial estimation. There, a much more refined treatment of the bandwidth parameter has been undertaken, allowing for a bandwidth h_k in the direction k of interest which is different from the bandwidth parameters in the other directions, see Fan. et al. (1998).

To my knowledge little has been done to investigate the finite sample approximation to the asymptotic formulae. In a sense quite severe trouble could be expected as the $\hat{P}(x)$-functional is roughly of the same type as the linearity functional treated in Sec. 2. The simulation experiments in Tjøstheim and Auestad suggest a much stronger bias than that explained by first order asymptotics for $n = 500$ and p moderately large, and this has to do with the curse of dimensionality which operates through higher order terms. It should be noted that more recent estimation methods, Linton (1997a), Hengartner (1996) and Kim et al. (1997) seem to have improved considerably on the estimation scheme used in Tjøstheim and Auestad (1994a).

4.2 Tests of Additivity

In principle additivity tests can be built up quite analogously to the linearity test presented in Sec. 2, as mentioned already in Tjøstheim and Auestad (1994a). One can compare the additive model estimate with the full non-parametric estimate. Work along these lines has just started and so far has only been set up in a quite general regression setting with independent sets of explanatory variables, cf. Linton and Gozalo (1996b) and Sperlich et al. (1997). Linton and Gozalo (1996b) look at a link function formulation where for a vector of explanatory variables $\mathbf{X} = [X_1, \ldots, X_p]$ and a response variable Y

$$G\{m(\mathbf{x})\} = c + \sum_{j=1}^{p} f_j(x_j)$$

with $m(\mathbf{x}) = E(Y|\mathbf{X} = \mathbf{x})$. The observations $\mathbf{X}_1, \ldots, \mathbf{X}_n$ are assumed to be iid. To simplify let $G \equiv 1$. Then Linton and Gozalo (1996b) and Sperlich et al. (1997) essentially propose estimating $m(\mathbf{x})$ nonparametrically, via the Nadaraya–Watson estimator or using local polynomials, and then the components f_j are estimated by marginal integration in an additive model fit to obtain

$$\tilde{m}(\mathbf{x}) = \hat{c} + \sum \hat{f}_i(x_i).$$

The two estimates are then compared using e.g. a functional of type

$$\hat{A}(m) = n^{-1} \sum_{j=1}^{n} \{\hat{m}(\mathbf{X}_j) - \tilde{m}(\mathbf{X}_j)\}^2 w(\mathbf{X}_j).$$

If p is moderate or large the full nonparametric estimate $\hat{m}$ will be hampered by the curse of dimensionality. In analogy to the linearity test functional, very large number of observations are probably required to use asymptotics effectively, and a bootstrap approach analogous to that of Sec. 2 could be tried as an alternative.

In an attempt to avoid the curse of dimensionality and to obtain a more finely tuned test, in Sperlich et al. (1997) the alternative is more restricted. In the model

$$Y = m(\mathbf{X}) + \sigma(\mathbf{X})e_t$$

it is assumed that $m(\mathbf{x})$ is either additive or, if not, can at most have second order interactions, so that

$$m(\mathbf{x}) = c + \sum_i f_i(x_i) + \sum_{j<i} f_{ij}(x_i, x_j),$$

and where the normalization conditions in Eq. (4.3) are imposed. It was explained in the beginning of this Section how these quantities can be estimated. In analogy to the linearity test we now also have the advantage that we can concentrate at two lags at a time, and in this way get an indication as to which pairs of components are pairwise additive and which are not. Suitably normalized, cf. Sperlich et al. (1997) we get a test where pairwise additivity in X_j and X_i is rejected if

$$\frac{1}{n} \sum_t \{\hat{f}_i(X_{i,t}) + \hat{f}_j(X_{j,t}) - \hat{f}_{i,j}(X_{i,t}, X_{j,t})\}^2 w(X_{i,t}, X_{j,t})$$

is large enough. An asymptotic analysis, based on U-statistic theory, is given in Sperlich et al. (1997), where it is also explained how a bootstrap version can be set up.

In the linearity case an alternative way of testing was obtained by checking whether $\hat{M}_k''(\cdot)$ is significantly different from zero. Clearly if we have additivity in X_i and X_j, so that either

(1) $m(x_1, \ldots, x_p) = f_i(x_i) + f_j(x_j) + f_{(i,j)}(x_1, \ldots, x_p)$

where $f_{(i,j)}(\cdot)$ means that x_i, x_j is missing, or

(2) $m(x_1, \ldots, x_p) = f_i(x_i) + f_j(x_j) + \sum_{\ell,k} f_{\ell k}(x_\ell, x_k)$

where neither i nor j are allowed to enter in the summation over ℓ, k, then

$$\frac{\partial^2 m(x_1, \ldots, x_p)}{\partial x_i \partial x_j} = 0.$$

This second order partial derivative may be estimated by local polynomial estimation. If p is moderately large, the curse of dimensionality would be troublesome. Alternatively, one could first project over x_i, x_j and then test whether $\partial^2 \hat{P}_{i,j}(x_i, x_j)/\partial x_i \partial x_j$ is significantly different from zero using a test functional of type:

$$\frac{1}{n} \sum_t \left\{ \frac{\partial^2 \hat{P}_{i,j}(X_i, X_j)}{\partial X_{i,t} \partial X_{j,t}} \right\}^2 .$$

There are several other tests of additivity of quite different nature, both in the time series case and in the regression case. We refer to Chen et al. (1995) for three time series tests.

5. TESTS OF INDEPENDENCE

Much of this section is taken from the survey paper Tjøstheim (1996).

Specification tests for independence are important for two reasons. Independence may be part of the model definition; in economic theory, for example, models have been proposed where $\log X_t - \log X_{t-1}$ should be iid random variables. More generally, certain phenomena can be assumed to be described by a renewal process, such that the increments $\{X_t - X_{t-1}\}$ should form a series of independent variables. Clearly there is a need for testing such hypotheses.

The second, and perhaps more important reason, consists in examining whether the residuals from a model fitting procedure are independent or not. The goal of the model fitting process should be (most would agree) to incorporate every structural feature of the data into the model, so that the residuals are left with no structure; i.e. they should be iid.

Traditionally the Box–Ljung–Pierce (Box and Pierce, (1970), Ljung and Box (1978) see also Dufour and Roy (1986)) test has been used for testing of independence of residuals from a linear ARMA time series model. If we denote by $\{\hat{e}_t\}$ the estimated residuals from a model fit, then the Box–Ljung–Pierce statistic tests for independence by testing for uncorrelatedness among the residuals $\{\hat{e}_t\}$. Although the test is nonparametric in nature, it does not have the wide applicability usually associated with nonparametric tests. This is due to the strong connections between autocorrelations and linear (parametric) Gaussian models. We illustrate this by a very simple example, the pure ARCH model, where the residuals $\{e_t\}$ are modeled by

$$e_t = \sqrt{a + be_{t-1}^2}\,\eta_t$$

with a and b being non-negative constants, and $\{\eta_t\}$ is a series of iid zero-mean random variables with variance 1. Then

$$\mathrm{var}(e_t|e_{t-1} = x) = a + bx^2,$$

and hence if large fluctuations (large e_{t-1}'s) have occurred, then large fluctuations can be expected in the future. In financial time series terminology the fluctuations of $\{e_t\}$ measures the risk, and ARCH models and the corresponding generalized GARCH models are much used and empirically quite well substantiated in finance. The point in this context is that the Box–Ljung–Pierce test fails to detect the very structure that modelers of financial data are looking for. It is trivial to see that for the ARCH residuals $\{e_t\}$ we have $\mathrm{corr}(e_t, e_{t-i}) = 0$ for all i, and this has brought the Box–Ljung–Pierce test into disrepute among econometricians and finance analysts, and it has brought on quite an intensive quest for alternative tests. A natural modification is to replace the Box–Ljung–Pierce test by other traditional nonparametric tests like the rank correlation test, the van der Waerden test or permutation type tests. A very systematic investigation of rank statistics and their theory and use in a time series situation has been undertaken by M. Hallin and M. Puri and their coworkers, see e.g. Hallin et al. (1985) and Hallin and Mélard (1988). Most of this work has been carried through in a linear framework, but Benghabrit and Hallin (1992) is an exception. The full potential of these methods in a nonlinear situation remains to be established. However, as is documented by Skaug and Tjøstheim (1996), neither the rank correlation nor the van der Waerden test work particularly well in a nonlinear environment.

An ad hoc procedure for avoiding the above problems is simply to square the residuals and compute the correlations of the squares,

$$\hat{\rho}_{e^2}(k) = \frac{\sum (\hat{e}_t^2 - \bar{\hat{e}}^2)(\hat{e}_{t-k}^2 - \bar{\hat{e}}^2)}{\sum (\hat{e}_t^2 - \bar{\hat{e}}^2)^2}.$$

If fourth order moments exist and if the $\hat{e}_t$'s are estimated residuals from an ARMA model of known order, then it can be shown, McLeod and Li (1983) that the asymptotic distribution of $\hat{\rho}_{e^2}(k)$ is normal, and a χ^2-statistic analogous to the Box–Ljung–Pierce statistic can be constructed. For the ARCH example $e_t^2 = (a + be_{t-1}^2)\eta_t^2$. The e_t^2's are correlated variables, and hence the McLeod–Li test will be able to pick up this dependence, which is one reason why this test has gained popularity among econometricians.

Lawrance and Lewis (1987) argue that in case one has residuals from a nonlinear model, the autocorrelations of the squared residuals may be difficult to handle, since it involves fourth order moments, and they suggest computing correlations between e_s and e_t^2, which should be "one-sided" for the ARCH model, or between e_t and $(X_t - \mu)^2$ or e_s^2 and $X_t - \mu$.

5.1 The Correlation Integral of Chaos Theory

Quite a different approach to independence testing has recently originated from the correlation integral of chaos theory. In Grossberger and Procaccia (1983) the correlation integral was introduced as a means of measuring the fractal dimension of deterministic data. It measures serial dependence patterns in the sense that it keeps track of the frequency with which temporal patterns are repeated in a data sequence. Let $\{x_1, \ldots, x_n\}$ be a sequence of numbers and let

$$\mathbf{x}_t^{(k)} = [x_t, x_{t-1}, \ldots, x_{t-k+1}], \quad k \le t \le n.$$

Then the correlation integral for embedding dimension k is given by

$$C_{k,n}(\varepsilon) = \frac{2}{n(n-1)} \sum_{1 \le s < t \le n} 1(\|\mathbf{x}_t^{(k)} - \mathbf{x}_s^{(k)}\| < \varepsilon)$$

where $\|\mathbf{x}\| = \max_{1 \le i \le m} |x_i|$, where $1(\cdot)$ is the indicator function and $\varepsilon > 0$ is a cut-off threshold which could be a multiple of the standard deviation in the case of a stationary process. Let

$$C_k(\varepsilon) = \lim_{n \to \infty} C_{k,n}(\varepsilon).$$

If $\{X_t\}$ is an absolutely regular, (Bradley (1986), or see condition (A4) in Sec. 2.5) stationary process, the above limit exists and is given by

$$C_k(\varepsilon) = \int 1(\|\mathbf{x}^{(k)} - \mathbf{y}^{(k)}\| < \varepsilon) \, dF_k(\mathbf{x}^{(k)}) \, dF_k(\mathbf{y}^{(k)}),$$

where F_k is the joint cumulative distribution function of $\mathbf{X}_t^{(k)}$. Since $1(\|\mathbf{x}^{(k)} - \mathbf{y}^{(k)}\| < \varepsilon) = \Pi_{i=1}^k 1(|x_i - y_i| < \varepsilon)$ it is easily seen that if $\{X_t\}$ consists of iid random variables, then

$$C_k(\varepsilon) = \{C_1(\varepsilon)\}^k,$$

and this expression can be used as a basis for a test of independence. Note that no moments of $\{X_t\}$ need exist.

The sampling theory of $C_{k,n}$ can be derived, Brock et al. (1991) by exploiting that $C_{k,n}$ is a generalized U-statistic, Serfling (1980), and Denker and Keller (1983) with a symmetric kernel. Under the hypothesis of independence, and excluding the case of uniformly distributed random variables, Brock et al. (1991) have established that

$$\frac{\sqrt{n} \, [C_{k,n}(\varepsilon) - \{C_{1,n}(\varepsilon)\}^k]}{V_{k,n}} \to \mathcal{N}(0, 1),$$

where $V_{k,n}$ is a normalizing constant, whose numerical value is given in Tjøstheim (1996).

The test is called the BDS test after its originators, Brock, Dechert and Scheinkman (1987), and it is discussed in considerable detail in Brock et al. (1992). It is now reasonably well-tried. It is effective against a wide range of alternatives including ARCH and its generalizations GARCH. There are problems with obtaining the right level though. For small, moderate and even quite large sample sizes cases of overestimation of the level have been reported giving the test a high false alarm rate, Brock et al. (1991) and Mizrach (1991). In power experiments on simulated data as a general rule the level has been determined by simulations, so that it does not influence the power comparison of those experiments. For real data this cannot be done, and one has to be careful in applying the test in the form given above. Its small sample properties have been discussed in Hsieh and LeBaron (1988). In the light of the results to be reported later in this Section, the bootstrap should be a natural alternative to using the asymptotic distributions in determining critical values.

Howell Tong and his coworkers have recently worked intensively on various aspects of chaos theory and nonlinear time series. The reader is referred to Tong (1995) and references therein. A recent development is a bootstrap type test of chaotic behavior, Yao and Tong (1996).

5.2 Tests Based on the Empirical Distribution Function

Independence between stochastic variables is defined in terms of distribution functions or—when they exist—density functions. Both distribution functions and density functions can be estimated nonparametrically, and it is therefore natural to consider independence tests in terms of these quantities. The idea goes back at least to a paper by Hoeffding (1948), who treated distribution functions. In this section and in Sec. 5.3 we will give an account of recent developments in this area.

To test independence we need an indicator of dependence so that the null hypothesis of independence can be rejected for large values of this dependence measure. The problem of measuring dependence between two random variables X and Y can be formulated as a problem of measuring the distance between the two bivariate distributions $F_{X,Y}$ and $F_X \otimes F_Y$, where $F_{X,Y}$, F_X and F_Y are the joint and marginal distribution functions, respectively, and where $F_X \otimes F_Y(x, y) = F_X(x) F_Y(y)$. An alternative would be to use characteristic functions, Pinkse (1993). Let $\Delta(\cdot, \cdot)$ be a candidate for such a distance functional. It will not be assumed that Δ is a metric, but it is natural to require, Skaug and Tjøstheim (1996) that

$$\Delta(F_{X,Y}, F_X \otimes F_Y) \geq 0 \quad \text{and} \quad \Delta(F_{X,Y}, F_X \otimes F_Y) = 0$$

$$\text{iff} \quad F_{X,Y} = F_X \otimes F_Y. \tag{5.1}$$

Moreover one may require invariance under transformations, or more precisely

$$\Delta(F_{X,Y}^*, F_X^* \otimes F_Y^*) = \Delta(F_{X,Y}, F_X \otimes F_Y) \tag{5.2}$$

where $F_X^*(x) = F_X\{g^{-1}(x)\}$ and $F_{X,Y}^*(x, y) = F_{X,Y}\{g^{-1}(x), g^{-1}(y)\}$. Here g is an increasing function, and F_X^*, F_Y^*, $F_{X,Y}^*$ are the marginal and bivariate distribution functions of the random variables $g(X)$, $g(Y)$ and $\{g(X, Y)\}$, respectively.

For distance functionals not satisfying Eq. (5.2) we can at least obtain scale and location invariance by standardizing such that $E(X) = E(Y) = 0$ and $\text{var}(X) = \text{var}(Y) = 1$, assuming that the second moments exist. We have used such a standardization for all of our functionals.

Conventional distance measures between two distribution functions F_1 and F_2 are the Kolmogorov–Smirnov distance

$$\Delta_1(F_1, F_2) = \sup_{\mathbf{x}} |F_1(\mathbf{x}) - F_2(\mathbf{x})|$$

and the Cramer–von Mises type distance

$$\Delta_2(F_1, F_2) = \int \{F_1(\mathbf{x}) - F_2(\mathbf{x})\}^2 \, dF_1(\mathbf{x}). \tag{5.3}$$

Here Δ_1 satisfies Eqns (5.1) and (5.2), whereas Δ_2 satisfies Eq. (5.1) but not Eq. (5.2).

By letting $F_1 = F_{X,Y}$ and $F_2 = F_X \otimes F_Y$ these distance functions can be taken as measures of dependence, and it is easily seen that they fit into the general framework of specification testing outlined in the introduction. An and Cheng (1990) have used the Kolmogorov–Smirnov distance in connection with their linearity test. It could be converted into an independence test, but apart from this, as far as I know, all the work pertaining to measuring dependence and testing of independence has been done in terms of the Cramer–von Mises distance, Hoeffding (1948), Blum et al. (1961), Rosenblatt (1975), Rosenblatt and Wahlen (1992), Skaug and Tjøstheim (1993b, 1996), Delgado (1996) and Hong (1998).

A natural estimator $\hat{\Delta}$ of a distance functional Δ is obtained by putting

$$\hat{\Delta}(F_{X,Y}, F_X \otimes F_Y) = \Delta(\hat{F}_{X,Y}, \hat{F}_X \otimes \hat{F}_Y),$$

where $\hat{F}$ is the empirical distribution function given by

$$\hat{F}_X(x) = \frac{1}{n} \sum_{i=1}^{n} 1(X_i \leq x) \qquad \hat{F}_{X,Y}(x, y) = \frac{1}{n} \sum_{i=1}^{n} 1(X_i \leq x)1(Y_i \leq y)$$

for given observations $\{(X_1, Y_1), \ldots, (X_n, Y_n)\}$. Similarly for a stationary time series $\{X_t\}$,

$$\hat{F}_k(x, y) \doteq \hat{F}_{X_t, X_{t-k}}(x, y) = \frac{1}{n-k} \sum_{t=k+1}^{n} 1(X_t \le x) 1(X_{t-k} \le y).$$

The work centered around the Cramer–von Mises type statistic in Eq. (5.3) was already started by Hoeffding (1948), who studied finite sample distributions in a non-time series situation in some special cases. A asymptotic theory was provided by Blum, Kiefer and Rosenblatt (1961) in the case of having observations (X_i, Y_i) where the pairs $\{(X_1, Y_1), \ldots, (X_n, Y_n)\}$ are iid. This was extended to the time series case with a resulting test of serial independence in Skaug and Tjøstheim (1993b).

In the time series case the Cramer–von Mises distance at lag k is given by

$$D_{F_k} = \int \{F_k(x, y) - F(x)F(y)\}^2 \, dF_k(x, y) \tag{5.4}$$

where F_k and F are the joint and marginal distributions of (X_t, X_{t-k}) and X_t, respectively. Replacing theoretical distributions by empirical ones leads to the estimate

$$\hat{D}_k = \frac{1}{n-k} \sum_{t=k+1}^{n} \{\hat{F}_k(X_t, X_{t-k}) - \hat{F}(X_t)\hat{F}(X_{t-k})\}^2.$$

Assuming $\{X_t\}$ to be ergodic, we have, Skaug and Tjøstheim (1993b), that $\hat{D}_{F_k} \overset{a.s}{\to} D_{F_k}$ as $n \to \infty$. If we recall the general test principles stated in the introduction, we see that we now have a test where $\hat{F}_k$ and $\hat{F} \otimes \hat{F}$ estimates the same quantity under the null hypothesis but different quantities under the alternative hypothesis in accordance with the general principles outlined there.

The asymptotic distribution of $\hat{D}_{F_k}$ is not normal. Rather, using U-statistic theory, cf. also Denker and Keller (1983), Carlstein (1988) and Skaug (1993) we have, Skaug and Tjøstheim (1993b), the convergence in distribution

$$n\hat{D}_{F_k} \overset{\mathcal{L}}{\to} \sum_{i,j=1}^{\infty} \eta_i \eta_j W_{ij}^2 \qquad \text{as} \qquad n \to \infty, \tag{5.5}$$

where $\{W_{ij}\}$ is a sequence of independent identically distributed $\mathcal{N}(0, 1)$ variables, and where the η_m's are the eigenvalues of the eigenvalue problem

$$\int g(x, y)h(y) \, dF(x, y) = \eta h(x) \tag{5.6}$$

with $g(x, y) = \int \{1(x \le w) - F(w)\}\{1(y \le w) - F(w)\} \, dF(w)$.

A test of the null hypothesis of independence, or rather pairwise independence at lag k can now be constructed. In the light of our previous results it is reasonable to reject H_0 if large values of $\hat{D}_{F_k}$ is observed. Thus a test of level ε is:

$$\text{reject} \quad H_0 \quad \text{if} \quad n\hat{D}_{F_k} > u_{n,\varepsilon}$$

where $u_{n,\varepsilon}$ is the upper ε-point in the null distribution of $n\hat{D}_{F_k}$. Since the exact distribution of $\hat{D}_{F_k}$ is unknown, we can use the asymptotic approximation furnished by Eq. (5.5). For $n = 50$, 100 and k small this works well. However, as k increases in general, Skaug and Tjøstheim (1993b) it was found that the level was seriously overestimated. The test statistic in Eq. (5.4) has recently been modified and improved by Hong (1998); this modification also giving a good approximation for a much wider choice of k's.

Under the hypothesis of $\{X_t\}$ being iid, permutations or the bootstrap is a natural tool for constructing the null distribution and critical values. Actually, the permutations give exact levels, and in a sense removes the need for a very accurate asymptotic theory.

Under the alternative hypothesis that X_t and X_{t-k} are dependent, the test statistic $\hat{D}_{F_k}$ will in general be asymptotically normal, but the power function will be very complicated, and we have not tried to obtain an asymptotic expression for it. A more careful asymptotic analysis with several extensions and new insights are provided by Hong (1998).

To extend the scope to testing of serial independence among $[X_t, \ldots, X_{t-k}]$ or alternatively between several random variables for which there are iid observations for each of them, one might use a functional

$$\hat{D}_{1,k} = \frac{1}{n} \sum_{t=k+1}^{n} \{\hat{F}_{1,k}(X_t, X_{t-1}, \ldots, X_{t-k}) - \hat{F}(X_t)\hat{F}(X_{t-1}) \cdots \hat{F}(X_{t-k})\}^2.$$

The asymptotic theory under the null hypothesis of independence for such a test has been examined by Delgado (1996), but due to the curse of dimensionality problems can be expected for moderately large k's. As an alternative Skaug and Tjøstheim (1993b) used a "Box–Ljung–Pierce analogy", testing for pairwise independence cumulatively in the pairs (X_t, X_{t-1}), $(X_t, X_{t-2}), \ldots, (X_t, X_{t-k})$ using the statistic

$$\hat{D}_F^{(k)} = \sum_{i=1}^{k} \hat{D}_{F_i}.$$

The asymptotic theory of such a test is given in Skaug (1993), and corresponding to Eq. (5.5), under the null hypothesis of independence, we obtain

$$n\hat{D}_F^{(k)} \rightarrow \sum_{i,j=1}^{\infty} \eta_i \eta_j W_{i,j}(k) \qquad \text{as} \qquad n \rightarrow \infty,$$

where $\{\eta_m,\ m \geq 1\}$ are as in Eq. (5.6) and $\{W_{ij}(k)\}$ is a sequence of iid χ^2-variables with k degrees of freedom. Exact levels can again be obtained by permutations.

Since our test is an omnibus test against a non-specified alternative of dependence, it will obviously have lower power for a fully specified alternative than a test designed explicitly for that alternative. The test has been examined in a number of simulation experiments. It performs surprisingly well in a linear Gaussian situation with very little power loss compared to a correlation type test, which is optimal in this situation. However, it has a very disappointing behavior for the important class of ARCH models, although it is improved in Hong's (1998) modified version. He has also included a larger class of experiments than those treated in Skaug and Tjøstheim (1993b). It remains to explain these properties theoretically.

It is clear that this way of testing independence can be extended to testing, say, conditional independence, since if X and Y are conditionally independent given Z, then the conditional distributions are related by the relationship

$$F_{X,Y|Z}(x,y|z) = F_{X|Z}(x|z)F_{Y|Z}(y|z)$$

which means that the associated distribution functions must satisfy $F_{X,Y,Z} \otimes F_Z = F_{X,Z} \otimes F_{Y,Z}$ and a test of conditional independence can be constructed by considering the distance functional $\Delta(F_{X,Y,Z} \otimes F_Z, F_{X,Z} \otimes F_{Y,Z})$. Linton and Gozalo (1996a) has explored this possibility.

5.3 Measures and Tests Based on Density Functions

In this Section density functions of all variables will be assumed to exist. Most of the material is taken from Skaug and Tjøstheim (1993a, 1996) and Tjøstheim (1996), to which we refer for more details.

For two random variables X and Y having a joint density $p_{X,Y}$ and marginals p_X and p_Y we measure the degree of dependence by $\Delta(p_{X,Y}, p_X \otimes p_Y)$ where Δ now is a distance measure between two bivariate density functions. The requirements of Eqns (5.1) and (5.2) discussed in Sec. 5.2 will be natural to consider here too. All of the functionals that I consider will be of the type

$$\Delta = \int B\{p_{X,Y}(x,y), p_X(x), p_Y(y)\} p_{X,Y}(x,y)\,dx\,dy \tag{5.7}$$

where B is a real-valued function. If B is of the form $B(z_1, z_2, z_3) = D(z_1/z_2 z_3)$ for some function D, we have

$$\Delta = \int D\left\{ \frac{p_{X,Y}(x,y)}{p_X(x)p_Y(y)} \right\} p_{X,Y}(x,y)\,dx\,dy \tag{5.8}$$

which by a change of variable formula for integrals is seen to have the property shown in Eq. (5.2). Moreover, if $D(u) \geq 0$ and $D(u) = 0$ iff $u = 1$, then Eq. (5.1) is fulfilled. Several well known distance measures are of this type. For instance letting $D(u) = (1 - u^{-1/2})$ we obtain the Hellinger distance

$$
H = \int \{ \sqrt{p_{X,Y}(x,y)} - \sqrt{p_X(x)p_Y(y)} \}^2 \, dx \, dy
$$

$$
= \int 2 \left\{ 1 - \sqrt{\frac{p_X(x)p_Y(y)}{p_{X,Y}(x,y)}} \right\} p_{X,Y}(x,y) \, dx \, dy \tag{5.9}
$$

between $p_{X,Y}$ and $p_X \otimes p_Y$. The Hellinger distance is a metric and thus satisfies Eq. (5.1). It is seen to be a special case of the so-called directed divergence measure (cf. Tjøstheim (1996)).

The familiar Kullback–Leibler information (entropy) distance is obtainable as a limiting case of a divergence measure, and it is given as:

$$
I = \int \log \left\{ \frac{p_{X,Y}(x,y)}{p_X(x)p_Y(y)} \right\} p_{X,Y}(x,y) \, dx \, dy. \tag{5.10}
$$

Since it is of type shown in Eq. (5.8), it satisfies Eq. (5.2), and it can be shown to satisfy Eq. (5.1), see Robinson (1991). Other distance measures, not of the form in Eq. (5.8), have been suggested by Chan and Tran (1992) and Rosenblatt and Wahlen (1992). A weighted difference functional was introduced in Skaug and Tjøstheim (1993a). It is related to a series expansion of I, and it is given by

$$
J = \int \{ p_{X,Y}(x,y) - p_X(x)p_Y(y) \} p_{X,Y}(x,y) \, dx \, dy. \tag{5.11}
$$

The distance measure J has neither of the properties shown in Eqns (5.1) and (5.2). Thus there are obvious arguments for dismissing it offhand. The reason this has not been done, is that it has worked consistently well on examples used in the literature for comparing serial independence tests. Moreover, its simple structure is ideal for exemplifying the asymptotic structure of the estimated functionals.

All of the above measures are conceptually based on the distance between two arbitrary multivariate densities $p_1(\mathbf{x})$ and $p_2(\mathbf{x})$. It is therefore, as was the case for the distribution functions, obvious how one can extend the above formulas to the case of measuring distance between $p_{X_1,\ldots,X_k}$ and $p_{X_1} \otimes \cdots \otimes p_{X_k}$. However, estimating such a measure would involve estimating higher dimensional densities which is difficult due to the curse of dimensionality. Therefore, again we look at functionals built up from measuring pairwise independence. For a stationary time series and serial independence we introduce

$$\Delta^{(k)} = \sum_{i=1}^{k} \Delta(p_i, p \otimes p)$$

with $p_i = p_{X_t, X_{t-i}}$ and $p = p_{X_t}$. For a given functional $\Delta = \Delta(p, q)$ depending on two densities p and q we estimate Δ by $\hat{\Delta} = \Delta(\hat{p}, \hat{q})$. There are several ways of estimating the densities p and q, but I will only consider nonparametric kernel estimates; i.e.

$$\hat{p}(\mathbf{x}) = \frac{1}{n} \sum_{t=1}^{n} K_h(\mathbf{x} - \mathbf{X}_t)$$

for given observations $\{\mathbf{X}, \ldots, \mathbf{X}_n\}$. As in Sec. 2.4, $K_h(\mathbf{x} - \mathbf{X}_t) = h^{-d} K\{h^{-1}(\mathbf{x} - \mathbf{X}_t)\}$ where $h = h_n$ is the bandwidth, K is the kernel function, and d is the dimension of $\mathbf{X}$. The kernel function was taken to be a product of one-dimensional kernels in Skaug and Tjøstheim (1993a, 1996). This is convenient but not necessary.

Once estimates for $p_{X,Y}, p_X$ and p_Y have been obtained in the integral in Eq. (5.7) for Δ, the integral could have been computed by numerical integration, but again we have opted for empirical averages. Then in effect we replace $p(x, y)\, dx\, dy = dF(x, y)$ by $d\hat{F}(x, y)$ so that corresponding to Eq. (5.7),

$$\hat{\Delta}_k = \frac{1}{n - k} \sum_{t=k+1}^{n} B\{\hat{p}_k(X_t, X_{t-k}), \hat{p}(X_t)\hat{p}(X_{t-k})\}$$

in case we are considering serial dependence for a stationary process with observations $\{X_1, \ldots, X_n\}$. We have found it convenient to introduce a weight function $w(X_t, X_{t-k})$ such that

$$\hat{\Delta}_k = \hat{\Delta}_{k,w} = \frac{1}{n - k} \sum_{t=k+1}^{n} B\{\hat{p}_k(X_t, X_{t-k}), \hat{p}(X_t)\hat{p}(X_{t-k})\} w(X_t, X_{t-k}).$$

Typically $w(x, y) = 1\{|x| \leq \lambda \operatorname{sd}(X)\} 1\{|y| \leq \lambda \operatorname{sd}(Y)\}$ where λ usually is between 2 and 3. The purpose of the weighing is twofold: We want to screen off outliers, and the asymptotic theory simplifies with this device. In effect it means that we are only measuring the dependence within the support of w. We could let the support of w tend to infinity as $n \to \infty$, but such an n-dependence would lead to complications in the asymptotic analysis (cf. Robinson (1991) for a related case). Moreover, it is not clear how the n-dependence should be chosen in practice for a given finite sample.

When a weight function w is included, then the theoretical counterpart is

$$\Delta_{k,w} = \int B\{p_k(x, y), p(x)p(y)\} w(x, y) p_k(x, y)\, dx\, dy$$

which is what the estimate $\hat{\Delta}_k$ should be measured against. In Skaug and Tjøstheim (1993a, 1996) an asymptotic theory is derived for the weighted functionals $\hat{I}_{k,w}$, $\hat{J}_{k,w}$ and $\hat{H}_{k,w}$ corresponding to the functionals in Eqns (5.9), (5.10) and (5.11), the weighing being as above. Then Eqns (5.1) and (5.2) are only approximately fulfilled. Similar derivations can be done for more general functionals, and it is indicated how an asymptotic theory can be developed in the more general cases in the Appendix of Tjøstheim (1996). Asymptotic normality can be obtained under a weak set of regularity conditions, Skaug and Tjøstheim (1996)

It should be noted that the leading term in an asymptotic expansion of the standard deviation is of order $n^{-1/2}$ for all of the three functionals $\hat{H}^{(k)}$, $\hat{I}^{(k)}$ and $\hat{J}^{(k)}$. This is the same as the standard deviation of parametric estimates in a regular parametric estimation problem, whereas the standard deviation for the estimators $\hat{p}(x)$ and $\hat{p}_k(x,y)$ are of order $(nh)^{-1/2}$ or $(nh^2)^{-1/2}$, respectively. In the regular parametric case the next order term of the Edgeworth expansion is of order n^{-1}, and for moderately large values of n the first order term $n^{-1/2}$ will dominate. Similarly, Hall, (1992) for $\hat{p}(x)$ and $\hat{p}_k(x,y)$ the next terms are of order $(nh)^{-1}$ and $(nh^2)^{-1}$. However, for the functionals constructed above the next order terms are generally much closer in order. For $\hat{J}$ it is shown in Skaug and Tjøstheim (1993a) that the next order terms are of order $n^{-1/2}h$ and $(nh)^{-1}$, and since $h = h_n = O(n^{-1/6})$ or $O(n^{-1/5})$, the first order asymptotic theory based on the normal approximation cannot be expected to be accurate unless n is very large. Basing a test on such a theory may be hazardous as the real level will typically deviate substantially from the nominal level. This is amply demonstrated in Skaug and Tjøstheim (1993a, 1996) where sometimes the level was twice that of the nominal level for a sample size of 100 observations. This is very similar to the problems with the linearity test, and it is not surprising, since the source of the trouble is the same in both cases, and it is our belief that such problems are likely to occur when one deals with estimation of functionals defined by an integral over a quantity where smoothing enters.

An obvious remedy is to include higher order terms in the asymptotic expansion, but this is problematic, as they are difficult to compute analytically and would involve complicated expressions which have to be estimated. This again suggests use of the bootstrap or permutations as an alternative for constructing the null distribution. One may anticipate that it picks up higher order terms of the Edgeworth expansion (cf. Hall (1992), although no rigorous analysis to confirm this has been carried through for the present functionals.

Actually, it is sufficient—if not better—to attack this problem via permutations: let $X^{(\cdot)} = (X^{(1)}, \ldots, X^{(n)})$ be the order statistic of our data sample. The fact that $X^{(\cdot)}$ is a sufficient statistic under H_0, Lehmann (1959), makes the construction of a permutation test possible. Conditioned on $X^{(\cdot)} = (x^{(1)}, \ldots, x^{(n)})$ the distribution of $(X_1, \ldots, X_n)$ is constructed by putting equal probability $1/n!$ on each of the $n!$ permutations of $x^{(1)}, \ldots, x^{(n)}$. The distribution of the test statistic $\hat{\Delta}$ conditioned on the observed value of $X^{(\cdot)}$, the permutation distribution, is then constructed at each of these permutations. Define the ε-quantile in the permutation distribution

$$c_\varepsilon(X^{(\cdot)}) = \inf_c \left\{ c : P(\hat{\Delta} \geq c|x^{(\cdot)}) \leq \varepsilon \right\}.$$

The permutation test has an exact level. In practice for all but very small sample sizes it is impossible to compute $c_\varepsilon(X^{(\cdot)})$ exactly, so Monte Carlo methods are used to obtain an approximation. This does not affect the level of the test, but has influence on the power properties.

The fact that the permutation test yields an exact level underscores one of the main points of this paper: Instead of obtaining the null distribution by an approximate asymptotic expression, it is generally better to construct it via randomization. A correct level is then guaranteed for the independence test and approximately so for the linearity test. We believe that very little power is lost using this approach, but this claim should be substantiated by theoretical analysis. An empirical result in this direction can be found in Table 6 of Skaug and Tjøstheim (1996) and similar evidence is given by Hong (1998). It is also worth noting that permutation tests have been considered in a time series situation by Dufour and Hallin (1991) and by Dufour and Roy (1985, 1986).

As for the linearity test it is quite difficult to conduct asymptotic power studies. Such studies would also be unreliable unless n is rather large for the reasons just mentioned. For a finite sample size ($n = 100$), the test functionals $\hat{H}$, $\hat{I}$ and $\hat{J}$ are inferior to the correlation functional and to the functional derived from the empirical distribution function for first order AR and MA models, Skaug and Tjøstheim (1993b, 1996). However, for a number of nonlinear examples, including the ARCH model, the functionals $\hat{H}$, $\hat{I}$ and $\hat{J}$ are far superior to the correlation functional, and comparable and in some cases better than the BDS functional. This is illustrated in Tables 2 and 3 (taken from Skaug and Tjøstheim (1996), and where $\hat{C}_1^{(q)}$, $\hat{C}_2^{(q)}$ and $\hat{C}_3^{(q)}$ denote the correlation test, the Spearman rank correlation test and the van der Waerden rank test, respectively) for the ARCH process

$$X_t = e_t\sqrt{1 + 0.5X_{t-1}^2} \tag{5.12a}$$

and the GARCH process

$$X_t = e_t\sqrt{h_t}, \qquad h_t = 1 + 0.1X_{t-1}^2 + 0.8h_{t-1}. \tag{5.12b}$$

Table 2   Power of the Tests Against the ARCH Alternative in Eq. (5.12a) with $n = 250$ and $\varepsilon = 0.05$

	$q=1$	$q=2$	$q=3$	$q=4$	$q=5$	$q=6$	$q=7$	$q=8$	$q=9$	$q=10$
$\hat{H}^{(q)}$	0.874	0.813	0.740	0.669	0.607	0.549	0.505	0.461	0.417	0.381
$\hat{I}^{(q)}$	0.861	0.802	0.721	0.655	0.592	0.535	0.490	0.444	0.405	0.375
$\hat{J}^{(q)}$	0.976	0.948	0.896	0.827	0.756	0.692	0.622	0.563	0.513	0.469
$\hat{C}_1^{(q)}$	0.213	0.218	0.210	0.194	0.174	0.166	0.153	0.152	0.145	0.143
$\hat{C}_2^{(q)}$	0.074	0.068	0.064	0.068	0.058	0.052	0.049	0.050	0.048	0.050
$\hat{C}_3^{(q)}$	0.973	0.937	0.923	0.910	0.889	0.871	0.831	0.813	0.809	0.800
BDS	0.950	0.880	0.740	*	*	*	*	*	*	*

Table 3   Power of the Tests Against the GARCH Alternative in Eq. (5.12b) with $n = 250$ and $\varepsilon = 0.05$

	$q=1$	$q=2$	$q=3$	$q=4$	$q=5$	$q=6$	$q=7$	$q=8$	$q=9$	$q=10$
$\hat{H}^{(q)}$	0.180	0.252	0.277	0.294	0.297	0.291	0.288	0.280	0.268	0.254
$\hat{I}^{(q)}$	0.179	0.232	0.264	0.287	0.288	0.289	0.281	0.278	0.268	0.261
$\hat{J}^{(q)}$	0.331	0.434	0.479	0.510	0.514	0.506	0.494	0.478	0.462	0.444
$\hat{C}_1^{(q)}$	0.085	0.092	0.095	0.099	0.103	0.105	0.105	0.105	0.106	0.103
$\hat{C}_2^{(q)}$	0.052	0.053	0.055	0.061	0.059	0.054	0.053	0.055	0.055	0.057
$\hat{C}_3^{(q)}$	0.277	0.285	0.336	0.357	0.355	0.352	0.321	0.320	0.329	0.323
BDS	0.210	0.230	0.220	*	*	*	*	*	*	*

In both cases $\{e_t, t \geq 1\}$ is a sequence of iid Gaussian variables with mean zero and variance 1.

A more thorough documentation and a real data example involving exchange rates are given in Skaug and Tjøstheim (1996). Hong (1998) put forward the interesting hypothesis that empirical distribution functions are generally advantageous for the conditional mean, whereas densities seem to be better for nonlinearities in the conditional variance. So far there is no theory to verify such a hypothesis. Actually, an asymptotic optimality theory is largely lacking. It would be quite difficult to establish and possibly of somewhat restricted practical applicability due to the higher order asymptotics involved, especially for density based tests. So far comparative studies have mainly been based on evidence brought forward by simulations; clearly not an entirely satisfactory situation. This remains an important area for further research.

5.4 An Independence Test Based on the Conditional Mean and the Conditional Variance

The tests described in Sections 5.2 and 5.3 are general purpose tests, and for special classes of alternatives it is possible to find alternative tests with better properties. For the threshold process for example, one would think that an independence test based on the conditional mean would have an advantage, whereas a test based on the conditional variance should have good properties in the ARCH case. In fact such tests were proposed in Hjellvik and Tjøstheim (1996) as "special cases" of the linearity test, and we thus have a possibility of building a synthesis between linearity and independence tests.

To be specific, consider the test functionals

$$I_M(k) = \int M_k^2(x) w(x) \, dF(x)$$

and

$$I_V(k) = \int \{V_k(x) - \sigma_X^2\}^2 w(x) \, dF(x)$$

where we use essentially the same notation as in Sec. 2.3, and where $\sigma_X^2 = E(X_t^2)$. If $\{X_t\}$ consists of iid random variables, then $M_k(x) \equiv E(X_t) = 0$ in the zero-mean case, and $V_k(x) \equiv \sigma_X^2$, so that the functionals are appropriate for tests of independence.

The corresponding estimated functionals are

$$\hat{I}_M(k) = \frac{1}{n} \sum \hat{M}_k^2(X_t) w(X_t)$$

and

$$\hat{I}_V(k) = \frac{1}{n} \sum \{\hat{V}(X_t) - \hat{\sigma}_X^2\}^2 w(X_t).$$

An asymptotic theory analogous to that of the linearity tests can be built, but again there will be trouble in getting accurate levels, unless the number of observations are huge, due to the combined effect of the smoothing and integration operation.

A permutation (or bootstrap) approach works much better in building the null distribution, and in this case, as in Sec. 5.3, the permutation test has

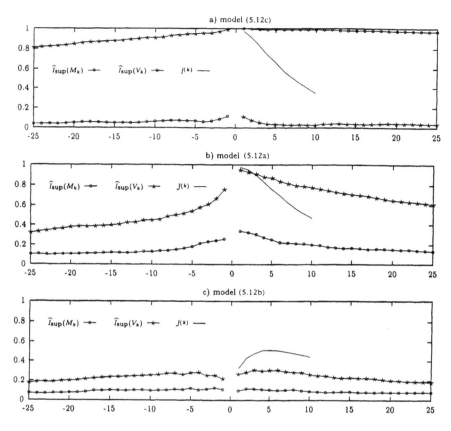

Figure 2 Power of tests as function of k

exact levels. In Fig. 2, taken from Hjellvik and Tjøstheim (1996), we have compared this way of testing with the general tests of independence of Tables 2 and 3 for the processes defined in Eqns (5.12a), (5.12b) and the threshold model defined by

$$X_t = \begin{cases} -0.5X_{t-1} + e_t, & X_{t-1} \leq 1 \\ 0.4X_{t-1} + e_t, & X_{t-1} > 1 \end{cases}. \tag{5.12c}$$

It is seen, as predicted, that $\hat{I}_M(k)$ and $\hat{I}_V(k)$ perform better for the models which are explicitly defined in terms of the conditional mean or the conditional variance, respectively.

Acknowledgment

I am grateful to an anonymous referee for a number of useful remarks.

REFERENCES

Akaike, H., Fitting autoregressions for predictions, *Ann. Inst. Statistist. Math., 21*: 243-247 (1969).

An, H.-Z. and B. Cheng, A Kolmogorov–Smirnov type statistic with application to test for normality in time series, *Int. Stat. Rev. 59*: 287–307 (1990).

Ashley, R. A., D. M. Patterson and M. J. Hinich, A diagnostic test for nonlinear serial dependence in time series fitting errors, *J. Time Ser. Anal. 7*: 165–178 (1986).

Auestad, B. and D. Tjøstheim, Identification of nonlinear time series: first order characterization and order determination. *Biometrika 77*: 669-687 (1990).

Auestad, B. and D. Tjøstheim, *Functional identification in nonlinear time series* in *Nonparametric Functional Estimation and Related Topics* (G. Roussas ed.), Kluwer Academic, Amsterdam, 493-507, 1991.

Benghabrit, Y. and M. Hallin, Optimal rank-based tests against first-order superdiagonal bilinear dependence, *J. Stat. Plan. and Inf. 32*: 45–61 (1992).

Blum, J. R., J. Kiefer, and M. Rosenblatt, Distribution free tests of independence based on the sample distribution function, *Ann. Math. Stat. 32*: 485–498 (1961).

Bollerslev, T., Generalized autoregressive conditional heteroscedasticity, *J. Econometrics, 31*: 307–327 (1986).

Bollerslev, T., R. F. Engle and D. B. Nelson, *Arch models* in *Handbook of Econometrics* (R. F. Engle and D. L. McFadden eds.), Elsevier, Amsterdam, 1994, *4*: 2959–3038.

Box, G. E. P. and D. Pierce, Distribution of residual autocorrelation in autoregressive integrated moving average time series models. *J. Amer. Statist. Assoc., 65*: 1509–1526 (1970).

Bradley, R., *Basic properties of strong mixing conditions* in Dependence in Probability and Statistics. A Survey of Recent Results (E. Eberlein and M. S. Taqqu, eds.), Birkhäuser, Boston, 1984, p. 169.

Brock, W. A., W. D. Dechert and J. A. Scheinkman, *A test for independence based on the correlation dimension.* Preprint, Department of Economics, University of Wisconsin, 1987.

Brock, W. A., W. D. Dechert, J. A. Scheinkman and B. LeBaron, *A test of independence based on the correlation dimension,* Department of Economics, University of Wisconsin, 1991.

Brock, W. A., D. A. Hsieh and B. LeBaron, *Nonlinear Dynamics, Chaos and Instability: Statistical Theory and Economic Evidence,* MIT Press, Cambridge, Massachusetts, 1992.

Brockett, P. L., M. J. Hinich and D. Patterson, Bispectral based tests for the detection of Gaussianity and linearity in time series. *J. Amer. Statist. Assoc., 83*: 657–664 (1988).

Carlstein, E., Degenerate U-statistics based on non-independent observations. *Calcutta Stat. Assoc. Bulletin, 37*: 55–65 (1988).

Chan, N. H. and L. T. Tran, Nonparametric tests for serial independence, *J. Time Ser. Anal., 13*: 19–28 (1992).

Chen, R. and W. Härdle, *Estimation and variable selection in additive nonparametric regression models.* Preprint, Sonderforschungsbereich 373, Humboldt-Universität zu Berlin, Wirtschaftswissenschaftliche Fakultät, 1994.

Chen, R., W. Härdle, O. Linton and E. Severance-Lossin, *Estimation in additive nonparametric regression.* Proceedings of the COMPSTAT conference SEMMERING (W. Härdle and M. Schimek eds.), Physika Verlag, Heidelberg, 1996.

Chen, R., J. S. Liu and R. S. Tsay, Additivity tests for nonlinear autoregression. *Biometrika, 82*: 369–383 (1995).

Cheng, B. and H. Tong, On consistent nonparametric order determination and chaos (with discussion), *J. Roy. Stat. Soc. Ser. B, 54*: 427–474 (1992).

Cheng, B. and H. Tong, H. On residual sums of squares in nonparametric autoregression. *Stochastic Process. Appl., 48*: 157–174 (1993).

De Jong, P., A central limit theorem for generalized quadratic forms, *Probab. Theory and Related Fields, 75*: 261–277 (1987).

Delgado, M., Testing serial independence using the sample distribution function, *J. Time Ser. Anal., 17*: 271–286 (1996).

Denker, M. and G. L. Keller, On U-statistics and von Mises' statistics for weakly dependent processes, *Z.* Wahrscheinlichkeitstheorie verw. *Gebiete, 64*: 505–522 (1983).

Dufour, J. M. and M. Hallin, An exponential bound for the permutational distribution of a first order autocorrelation coefficient, *Statistique et Analyse des Données, 15*: 45-56 (1991).

Dufour, J. M. and R. Roy, Some robust exact results on sample auto correlations and tests of randomness, *J. Econometrics, 29*: 257, *41*: 279–281 (1985).

Dufour, J. M. and R. Roy, Generalized portmanteau statistics and tests of randomness, *Comm. Statist. A. Theory and Methods, 15*: 2953-2972 (1986).

Engle, R. F., Autoregressive conditional heteroscedasticity with estimates of variance of U.K. inflation, *Econometrica, 50*: 987–1008 (1982).

Fan, J., Design-adaptive nonparametric regression. *J. Amer. Statist. Assoc., 87*: 998–1004 (1992).

Fan, J., Local linear regression smoothing and their minimax efficiencies. *Ann. Statist., 21*: 196–216 (1993).

Fan, J. and I. Gijbels, *Local Polynomial Modeling and Its Application. Theory and Methodologies.* Chapman and Hall, New York, 1995a.

Fan, J. and I. Gijbels, Data-driven bandwidth selection in local polynomial fitting: variable bandwidth and spatial adaption. *J. Roy. Statist. Soc. Ser. B, 57*: 371–394 (1995b).

Fan, J., W. Härdle and E. Mammen, *Direct estimation of low dimensional components in additive models.* Preprint, Sonderforschungsbereich 373, Humboldt-Universität zu Berlin, Wirtschaftswissenschaftliche Fakultät, Ann. Statist. 1998.

Granger, C. W. J. and J.-L. Lin, Using the mutual information coefficient to identify lags in nonlinear time series models, *J. Time Ser. Anal., 15*: 371–384 (1994).

Granger, C. W. J. and T. Teräsvirta, *Modelling Nonlinear Economic Relationships,.* Oxford University Press, Oxford, 1993.

Grossberger, P. and I. Procaccia, Measuring the strangeness of strange attractors, *Physica, 9D*: 189 (1983).

Hall, P., Central limit theorem for integrated square error of multivariate nonparametric density estimators. *J. Mult. Analysis, 14*: 1–16 (1984).

Hall, P., *The Bootstrap and Edgeworth Expansion.* Springer, New York, 1992, Chapters 3 and 4.

Hallin, M. and G. Mélard, Rank based tests for randomness against first order serial independence, *J. Amer. Statist. Assoc., 83*: 1117–1129 (1988).

Hallin, M., J. F. Ingenbleek and M. L. Puri, Linar serial rank tests for randomness against ARMA alternatives, *Ann Statist. 13*: 1156–1181 (1985).

Härdle, W. and E. Mammen, Comparing nonparametric versus parametric regression fits, *Ann. Statist., 21*: 1926–1947 (1993).

Hastie, T. and R. Tibshirani, *Generalized Additive Models*, Chapman and Hall, London, 1990.

Hengartner, N. W., *Rate optimal estimation of additive regression via the integration method in the presence of many covariates.* Preprint, Department of Statistics, Yale University, New Haven CT, 1996.

Hinich, M., Testing for Gaussianity and linearity of a stationary time series, *J. Time Ser. Anal., 3*: 169–176 (1982).

Hjellvik, V. and D. Tjøstheim, Nonparametric tests of linearity for time series. *Biometrika, 82*: 351–368 (1995).

Hjellvik, V. and D. Tjøstheim, Nonparametric statistics for testing linearity and serial independence. *J. Nonparametric Stat., 6*: 223–251 (1996).

Hjellvik, V., Q. Yao and D. Tjøstheim, Linearity testing using polynomial approximation. *J. Stat. Plan. and Inf., 68*: 295–321, 1998.

Hoeffding, W., A non-parametric test of independence. *Ann. Math. Stat, 19*: 546-557 (1948).

Hong, Y., Testing for pairwise serial independence via the empirical distribution function. *J. Roy. Stat. Ser. B. 60*: 429–454, 1998.

Hsieh, D. and B. LeBaron, *Small sample properties of the BDS statistics I and II.* Preprint, Graduate School of Business, University of Chicago and University of Wisconsin, 1988.

Karlsen, H. and D. Tjøstheim, *Nonparametric estimation in null recurrent time series models.* Preprint, Department of Mathematics, University of Bergen, 1997.

Keenan, D. M., A Tukey nonadditivity-type test for time series non-linearities, *Biometrika, 72*: 39–44 (1985).

Kim, W., O. B. Linton and N. W. Hengartner, *A nimble method of estimating additive nonparametric regression*, Preprint, Cowles Foundation, Yale University, New Haven, Connecticut, 1997.

Koreisha, S. and G. Yoshimoto, A comparison among identification procedures for autoregressive moving average models, *Inter. Stat. Rev.*, *51*: 37-57 (1991).

Lawrance, A. J. and P. A. W. Lewis, Higher order residual analysis for nonlinear time series with autoregressive correlation structures. *Inter. Stat. Rev.*, *55*: 21-35 (1987).

Lehmann, E. L, Testing Statistical Hypothesis. Wiley, New York, 1959, p. 133.

Linton, O., Efficient estimation of additive nonparametric regression models. *Biometrika, 84*: 469– 473 (1997a).

Linton, O., *Efficient estimation of generalized additive nonparametric regression models.* Preprint, Cowles Foundation, Yale University, New Haven, Connecticut, 1997b.

Linton, O. and P. Gozalo, *Conditional independence restrictions: testing and estimation.* Preprint, Cowles Foundation, Yale University, New Haven, Connecticut, 1996a.

Linton, O. and P. Gozalo, *Testing additivity in generalized nonparametric regression models.* Preprint, Cowles Foundation, Yale University, New Haven, Connecticut, 1996b.

Linton, O. and W. Härdle, Estimation of additive regression models with known links, *Biometrika, 83*: 529–540 (1995).

Linton, O. and J. P. Nielsen, A kernel method of estimating structured nonparametric regression based on marginal integration, *Biometrika, 82*: 93–101 (1995).

Ljung, G. M. and G. E. P. Box, On a measure of lack of fit in time series models, *Biometrika, 65*: 297–303 (1978).

Luukkonen, R., P. Saikkonen and T. Teräsvirta, Testing linearity in univariate time series, *Scand. J. Statist., 15*: 161–175 (1998a).

Luukkonen, R., P. Saikkonen and T. Teräsvirta, Testing linearity against smooth transition autoregression, *Biometrika. 75*: 491–499 (1988b).

Masry, E. and J. Fan, Local polynomial estimation of regression functions for mixing processes. *Scand. J. Statist., 24*: 165–180 (1997).

Masry, E. and D. Tjøstheim, Nonparametric estimation and identification of ARCH non-linear time series: strong convergence and asymptotic normality. *Econometric Theory, 11*: 258–289 (1995).

Masry, E. and D. Tjøstheim, Additive nonlinear ARX time series and projection estimates. *Econometric Theory, 13*: 214–252 (1997).

Matsuda, Y. and H. Shigeno, *A nonparametric test for nonlinearity by the weighted least squares method.* Preprint, Department of Mathematics and Computing Sciences, Tokyo Institute of Technology, 1996.

McLeod, A. I. and W. K. Li, Diagnostic checking ARMA time series models using squared-residuals and autocorrelations, *J. Time Ser. Anal., 4*: 269–273 (1983).

Mizrach, B., *A simple nonparametric test of independence of order* (p). Preprint, Department of Finance. The Wharton School, Philadelphia, 1991.

Newey, W. K., Kernel estimation of partial means. *Econometric Theory, 10*: 233–253 (1994).

Pinkse, C. A. P., *A general characteristic function measure applied to serial independence testing.* Preprint, London School of Economics, 1993.

Poggi, J.-M. and B. Portier, A test of linearity for functional autoregressive models, *J. Time Ser. Anal., 18*: 615–640 (1997).

Rios, R., Towards a nonparametric test of linearity for time series, *J. Stat. Plan. and Infer.*, to appear (1998).

Robinson, P., Nonparametric estimators for time series, *J. Time Ser. Anal., 4*: 185–207 (1983).

Robinson, P., Consistent nonparametric entropy-based testing, *Rev. Econ. Stud., 58*: 437–453 (1991).

Rosenblatt, M., A quadratic measure of deviation of two-dimensional density estimates and a test of independence. *Ann. Statist., 3*: 1–14 (1975).

Rosenblatt, M. and B. Wahlen, A nonparametric measure of independence under a hypothesis of independent components. *Statistics and Probab. Letters, 15*: 245–252 (1992).

Ruppert, D. and M.-P. Wand, Multivariate locally weighted least squares regression, *Ann. Statist., 22*: 1346–1370 (1994).

Saikkonen, P. and R. Luukkonen, Lagrange multiplier tests for testing non-linearities in time series models. *Scand. J. Statist.*, *15*: 55-68 (1988).

Serfling, R. J., *Approximations Theorems of Mathematical Statistics*, Wiley, New York, 1980, Chap. 5.

Severance-Lossin, E. and S. Sperlich, *Estimation of derivatives for additive separable models*. Preprint, Sonderforschungsbereich 373, Humboldt-Universität zu Berlin, Wirtschaftswissenschaftliche Fakultät, 1995.

Skaug, H. J., *The limit distribution of the Hoeffding statistic for tests of serial independence*. Preprint, Department of Mathematics, University of Bergen, 1993.

Skaug, H. J. and D. Tjøstheim, *Nonparametric Tests of Serial Independence* in *Developments in Time Series Analysis*. The Priestley Birthday Volume (T. Subba Rao, ed.). Chapman and Hall. London, 1993a, 207-230.

Skaug, H. J. and D. Tjøstheim, A nonparametric test of serial independence based on the empirical distribution function, Theorems 1 and 2, *Biometrika, 80*: 591-602 (1993b).

Skaug, H. J. and D. Tjøstheim, *Testing for serial independence using measures of distance between densities* in *Athens Conference on Applied Probability and Time Series, Vol. II, in memory of E. J. Hannan. Springer Lecture Notes in Statistics 115* (P. M. Robinson and M. Rosenblatt, eds.), Springer, Berlin, 1996, 363-377.

Sperlich, S., L. Yang and D. Tjøstheim, *Nonparametric estimation and testing of interaction*. Preprint, Sonderforschungbereich 373, Humboldt-Universität zu Berlin, Wirtschaftswissenschaftliche Fakultät, 1997.

Stone, C. J., Consistent nonparametric regression, *Ann. Statist.*, *5*: 595-645 (1977).

Stone, C. J., Additive regression and other nonparametric models. *Ann. Statist. 13*: 689-705 (1985).

Subba Rao, T. and M. M. Gabr, A test for linearity of stationary time series, *J. Time Ser. Anal., 1*: 145-158 (1980).

Teräsvirta, T., Specification, estimation and evaluation of smooth transition autoregressive models. *J. Amer. Statist. Assoc., 89*: 208-218 (1994).

Teräsvirta, T., C.-E. Lin and G. W. J. Granger, Power of the neural network linearity test, *J. Time Ser. Anal., 7*: 51-72 (1993).

Teräsvirta, T., D. Tjøstheim and C. W. J. Granger, *Aspects of modelling nonlinear time series* in *Handbook of Econometrics* (R. F. Engle and D. L. McFadden eds.) Elsevier, Amsterdam, 1994, *4*: 2919–2960.

Tjøstheim, D., Non-linear time series: a selective review, *Scand. J. Statist.*, *21*: 97–130 (1994).

Tjøstheim, D., Measures of dependence and tests of independence, *Statistics*, *28*: 249–284 (1996).

Tjøstheim, D. and B. Auestad, Nonparametric identification of nonlinear time series: Projections. *J. Amer. Statist. Assoc.*, *89*: 1398–1409 (1994a).

Tjøstheim, D. and B. Auestad, Nonparametric identification of nonlinear time series: selecting significant lags, *J. Amer. Statist. Assoc.*, *89*: 1410–1419 (1994b).

Tong, H., *Non-linear Time Series. A Dynamical System Approach.* Oxford University Press, Oxford, 1990, p. 13.

Tong, H., A personal overview of nonlinear time series from a chaos perspective (with discussion). *Scand. J. Stat.*, *22*: 399–446 (1994).

Tsay, R., Nonlinearity tests for time series, *Biometrika 73*: 461–466 (1986).

Tschernig, R., *Nonlinearities in German Unemployment rates: a nonparametric analysis.* Preprint, Sonderforschungsbereich 373, Humboldt-Universität zu Berlin, Wirtschaftswissenschaftliche Fakultät, 1996.

Tschernig, R. and L. Yang, *Nonparametric lag selection in time series.* Report Humboldt Sonderforschungsbereich 373, Humboldt-Universität zu Berlin, Wirtschaftswissenschaftliche Fakultät, 1996.

Tsybakov, A. B., Robust reconstruction of functions by the local-approximation method. *Problems of Information Transmission,* *22*: 133–146 (1986).

Vieu, P., Choice of regressors in nonparametric estimation, *Comput. Statist. and Data Anal.,* *17*: 575–594 (1994).

Yang, L. and R. Tschernig, *Nonparametric lag selection in seasonally nonstationary time series.* Preprint, Sonderforschungsbereich 373, Humboldt-Universität zu Berlin, Wirtschaftswissenschaftliche Fakultät, 1996.

Yao, Q. and H. Tong, On subset selection in non-parametric stochastic regression. *Statistica Sinica,* *4*: 51–70 (1994).

Yao, Q. and H. Tong, *A bandwidth selector and an informal test for operational determinism.* Preprint, Institute of Mathematics, University of Kent at Canterbury, 1996.

Yoshihira, K., Limiting behaviour of U-statistics for stationary absolutely regular processes. Z. Wahrscheinlichkeitstheorie verw. *Gebiete, 35*: 237–252 (1976).

6

Parameter Estimation and Model Selection for Multistep Prediction of a Time Series: A Review

R. J. BHANSALI University of Liverpool, Liverpool, England

1. INTRODUCTION

A direct method for multistep prediction of a time series consists of either estimating the parameters of a pre-specified model separately for each lead time, h, or selecting and fitting a new model for each h. A more standard "plug-in" method, by contrast, involves identifying and fitting an initial model to the time series by maximum likelihood; the multistep forecasts are then constructed from the model equation, but with the unknown future values replaced by their own forecasts. A review of some of the recent developments in implementing the direct method for multistep prediction of both stationary and non-stationary time series and for time series regression models is given from a parametric and non-parametric point of view and by explicitly recognizing that the model fitted for one-step prediction may not necessarily coincide with the actual stochastic structure generating the time series.

An important question in empirical time series analysis is how to predict the future values of an observed time series on the basis of its recorded past. A fundamental contribution towards answering this question, due to Kolmogorov (1939) and Wiener (1949), is commonly known as the Wiener–Kolmogorov prediction theory, and it helps also to unify many of the recent developments in Time Series Analysis, especially those based on linear models.

To gain an appreciation of the elements of the Wiener–Kolmogorov approach, assume that the observed time series is a realization of a discrete-time stationary process, $\{x_t\}$, with mean 0, covariance function $R(s) = E(x_t x_{t+s})$, correlation function $c(s) = R(s)/R(0)$ and spectral density function $f(\mu)$. Further suppose that $R(s)$ is absolutely summable and $f(\mu)$ is non-vanishing, i.e.,

$$\sum_{s=-\infty}^{\infty} |R(s)| < \infty, \tag{1.1}$$

$$f(\mu) = (2\pi)^{-1} \sum_{s=-\infty}^{\infty} R(s) \exp(-is\mu) \neq 0, \qquad |\mu| \leq \pi. \tag{1.2}$$

Under these assumptions, $\{x_t\}$ admits moving average and autoregressive representations, see Brillinger (1975):

$$x_t = \sum_{j=0}^{\infty} b(j)\varepsilon_{t-j}, \qquad \sum_{j=0}^{\infty} a(j)x_{t-j} = \varepsilon_t, \tag{1.3}$$

in which $a(0) = b(0) = 1$ and $\{\varepsilon_t\}$ is a sequence of uncorrelated random variables each with mean 0 and variance σ^2 and the $b(j)$ and $a(j)$ are absolutely summable coefficients such that if

$$B(z) = \sum_{j=0}^{\infty} b(j)z^j, \qquad A(z) = \sum_{j=0}^{\infty} a(j)z^j$$

$B(z) = [A(z)]^{-1}, B(z) \neq 0, A(z) \neq 0, |z| \leq 1$. Moreover, if $f(\mu)$ is treated as known, the $b(j)$, $a(j)$ and σ^2 can be found by the spectral factorization procedure, see, e.g., Bhansali (1974).

Suppose that $\{x_t, t \leq n\}$ is known and prediction of $\{x_{n+h}, h = 1, \ldots, J\}$ is required, where $J \geq 1$ and n are pre-specified. Denote the linear least-squares predictor of x_{n+h} by

$$\hat{x}_n(h) = -\sum_{j=1}^{\infty} \phi_h(j)x_{n+1-j}$$

and the mean squared of prediction by $V(h) = E[\{x_{n+h} - \hat{x}_n(h)\}^2]$. As is well-known, see Whittle (1963), if

$$G_h(z) = -\sum_{j=1}^{\infty} \phi_h(j)z^j$$

denotes the generating function of the $-\phi_h(j)$, $G_h(z)$ and $V(h)$ may be expressed in terms of the $A(z)$, $B(z)$ and σ^2 as follows:

$$G_h(z) = z^{-h+1} \left[1 - \left\{ A(z) \sum_{j=0}^{h-1} b(j) z^j \right\} \right],$$ (1.4)

$$V(h) = \sigma^2 \sum_{j=0}^{h-1} b^2(j).$$ (1.5)

A "parametric' and a "non-parametric" approach is available for implementing the Wiener–Kolmogorov prediction theory with an observed time series of length T, assuming that after a suitable transformation, if necessary, it is stationary with 0 mean.

In the parametric approach, $\{x_t\}$ is postulated to follow a linear finite parameter model and $f(\mu)$ is assumed to be a function of $s \geq 1$ parameters. A popular class of models is the autoregressive-moving average, ARMA, model of order (p, q) for which $s = p + q$; see Box and Jenkins (1970), whose work has been influential in the use of differencing for transforming a possibly non-stationary time series to one that is stationary. If the value of s is known, a maximum likelihood procedure may be employed for estimating the spectral parameters from the data; the resulting estimates are known to be consistent, asymptotically normal and efficient, see Hannan and Deistler (1988).

An autoregressive model fitting approach is commonly adopted for implementing the non-parametric approach, especially if only the class of linear models is being considered. The observed series is treated as a realization of an infinite-order autoregressive process, but an autoregressive model of order k is fitted, usually by least-squares or an equivalent procedure, where $k = k(T)$ say is a function of T such that $k(T) \to \infty$ as $T \to \infty$, but sufficiently slowly; see Parzen (1969) and Berk (1974).

It should be stressed that conceptually the "non-parametric" autoregressive model fitting approach is different from the fitting of a "parametric" autoregressive process of order p to the observed time series. The main distinction between the two is that in the "non-parametric' approach, an autoregressive approximation is constructed for the unknown generating structure of the process, while in the "parametric" approach the fitted autoregressive process is interpreted as a suitable model for this generating structure and, from the point of view of subsequent statistical inference, it is treated as being equivalent to the "true" stochastic process generating the observed time series. Thus, whereas the value of k in the former approach is allowed to grow with the series length, T, that of p in the "parametric" approach is treated as a fixed constant and invariably set equal to the actual order of the generating autoregressive process.

A conceptual difficulty in adopting the parametric approach is that for an observed time series, there may not be a "true" finite parameter model; see Hannan (1987) for a discussion of this point. An additional practical difficulty in adopting either the parametric or the non-parametric approach is how to select the order of the model to fit. In the last thirty years or so, there has been a considerable development on this question, see Bhansali (1993b) for a review, and the use of a model selection criterion, like the Akaike (1973) information criterion (AIC), or a Bayesian criterion (BIC) of Schwarz (1978) is recommended.

Having chosen a model to fit using either the parametric or the non-parametric approach, the one-step prediction is obtained from the model equation itself, but the multistep prediction by a plug in method in which the model equation is iterated repeatedly and the unknown future values are replaced by their own forecasts.

If the selected model coincides with that generating the observed time series, the plug-in method yields the maximum likelihood estimators of the Wiener–Kolmogorov prediction constants for all $h \geq 1$. If, however, the selected model differs from the generating model, the plug-in method still estimates the one-step linear least-squares prediction constants, but restricted to only the class of postulated models; for multistep prediction, by contrast, this optimality property does not hold and the theoretical parameters estimated by the method do not coincide with the h-step linear least-squares prediction constants, not even those restricted to the class of the selected model.

The stochastic structure generating an observed time series is invariably unknown in practice and there is no certainty that the model chosen by a selection criterion would necessarily coincide with the generating model, especially if it has an infinite number of parameters, or if an underparametrized model, like exponential smoothing, is deliberately employed because of its convenience and ease of use. This observation suggests that for $h > 1$, the plug-in method when used either in conjunction with the parametric or the non-parametric approach may be improved upon by adopting a different procedure.

There has recently been considerable development in an alternative direct method of lead-time dependent model selection and/or parameter estimation for multistep prediction. A main objective of this chapter is to review these developments, placing particular emphasis on the qualitative differences in the asymptotic sampling properties of the direct method if the parametric approach is used for studying these properties, and if, instead, the non-parametric approach is used.

For multistep prediction of a stationary process from only a finite amount of its past, Grenander and Rosenblatt (1954) find the relevant linear least-squares prediction constants separately for each lead time; they do not, however, consider the statistical questions of parameter estimation and model selection.

2. PARAMETER ESTIMATION FOR LINEAR STATIONARY MODELS

Suppose that $\{x_t\}$ is a discrete-time autoregressive process of order m,

$$\sum_{u=0}^{m} a_m(u)x_{t-u} = \varepsilon_t, \qquad a_m(0) = 1, \tag{2.1}$$

where $m \geq 0$ is finite or infinite but unknown and $\{\varepsilon_t\}$ is a sequence of independent random variables each with mean 0, variance σ^2 and finite fourth moment, the $a_m(u)$ are absolutely summable real coefficients such that

$$\sum_{j=0}^{m} a_m(j)z^j \neq 0, \qquad |z| \leq 1. \tag{2.2}$$

2.1   Direct and Plug-in Methods for Fitting Autoregressions

Having observed $x_1, \ldots, x_T$, let K denote a known upper bound on the amount of past to be used for h-step prediction, $h \geq 1$.

Denote the kth order (direct) linear least-squares predictor of x_{n+h} based on the past $\{x_n, x_{n-1}, \ldots, x_{n-k+1}\}$, $k \geq 1$, by

$$\bar{x}_{Dhk}(n) = -\sum_{j=1}^{k} \phi_{Dhk}(j)x_{n+1-j}, \tag{2.3}$$

the corresponding prediction error by $z_{Dhk}(n) = x_{n+h} - \bar{x}_{Dhk}(n)$ and let $\Phi_{Dh}(k) = [\phi_{Dhk}(1), \ldots, \phi_{Dhk}(k)]'$, $r_h(k) = [R(h), \ R(h+1), \ldots, R(h+k-1)]'$ and $\Gamma(k) = [R(u-v)](u, v, = 1, \ldots, k)$. We have

$$\Phi_{Dh}(k) = -\Gamma(k)^{-1}r_h(k), \tag{2.4}$$

$$V_D(h, k) = E[\{z_{Dhk}(n)\}^2] = R(0) + r_h(k)'\Phi_{Dh}(k). \tag{2.5}$$

As in Shibata (1980) and Bhansali (1996), let $\hat{\Phi}_{Dh}(k) = [\hat{\phi}_{Dhk}(1), \ldots, \hat{\phi}_{Dhk}(k)]'$ be the kth order direct estimate of the h-step prediction constant, $\Phi_{Dh}(k)$, obtained by regressing x_{t+h} on $x_t, x_{t-1}, \ldots, x_{t-k+1}$, $t = K, K+1, \ldots, T-h$, and let $\hat{V}_{Dh}(k)$ denote the residual error variance in this regression. We have,

$$\hat{\Phi}_{Dh}(k) = -\hat{\Gamma}_h(k)^{-1}\hat{r}_h(k), \tag{2.6}$$

$$\hat{V}_{Dh}(k) = \hat{d}_h(0) + \hat{r}_h(k)'\hat{\Phi}_{Dh}(k), \tag{2.7}$$

where $\hat{\Gamma}_h(k) = [C_{hT}(u,v)](u,v = 1,\ldots,k)$, $\hat{r}_h(k) = [C_{hT}(-h+1,1),\ldots,$ $C_{hT}(-h+1,k)]'$, $\hat{d}_h(0) = C_{hT}(-h+1,-h+1)$, the subscript D stands for the direct method and, with $N = T - h - K + 1$,

$$C_{hT}(u,v) = N^{-1}\sum_{t+K}^{T-h} x_{t-u+1}x_{t-v+1}.$$

For $h = 1$, we write $a(k) \equiv \Phi_{D1}(k)$, $\sigma^2(k) \equiv V_D(1,k)$, $\hat{\Phi}_{Dh}(k) = \hat{a}(k)$, $\hat{V}_{Dh}(k) = \hat{\sigma}^2(k)$ and now $\hat{a}(k)$ and $\hat{\sigma}^2(k)$ provide the usual least-squares estimators of the autoregressive parameters; furthermore, as discussed in Sec. 1, the direct and plug-in estimators coincide for this value of h.

For $h > 1$, however, the direct and plug-in estimators do not coincide. As in Yamamoto (1976), the kth order plug-in estimator, $\hat{\Phi}_{Ph}(k) = [\hat{\phi}_{Phk}(1),\ldots,\hat{\phi}_{Phk}(k)]'$ of the h-step prediction constants is given by, with $e_k = [1,0,\ldots,0]'$,

$$\hat{\Phi}_{Ph}(k) = -e_k'\hat{\Omega}(k)^h, \tag{2.8}$$

and the underlying theoretical parameter by $\Phi_{Ph}(k) = -e_k'\Omega(k)^h$. Here, $\Omega(k)$ is a $k \times k$ companion matrix for $a(k)$ and it has $-a(k)'$ in its first row, an identity matrix of dimension $k - 1$ in its bottom left hand $(k-1) \times (k-1)$ corner and a vector of zeroes of dimension $k - 1$ in its last column and $\hat{\Omega}(k)$ is defined analogously but with $\hat{a}(k)$ replacing $a(k)$.

From Eq. (1.5), a kth order plug-in estimate of $V(h)$ is given by

$$\hat{V}_{Ph}(k) = \hat{\sigma}^2(k)\sum_{j=0}^{h-1}\{\hat{b}_{Pk}(j)\}^2,$$

where $\hat{b}_{Pk}(0) = 1$, and if $\hat{b}_{Pk.h-1} = [\hat{b}_{Pk}(1),\ldots,\hat{p}_{Pk}(h-1)]'$, $\hat{H}_k(h-1) = [\hat{a}_k(u-v)](u,v = 1,\ldots,h-1)$, $\hat{a}_k(j) = 0$, $j > 0$, $\hat{a}_{k.h-1} = [\hat{a}_k(1),\ldots,\hat{a}_k(h-1)]'$,

$$\hat{b}_{Pk.h-1} = -\hat{H}_k(h-1)^{-1}\hat{a}_{k.h-1},$$

see Bhansali (1989, 1993a).

Let $\bar{x}_{Phk}(n)$ denote the theoretical kth order h-step plug-in predictor, which is defined analogously to, $\bar{x}_{Dhk}(n)$ in Eq. (2.3) but with the $\phi_{Phk}(j)$ replacing the $\phi_{Phk}(n)$ and let $V_P(h,k)$ denote the corresponding mean squared error of $\bar{x}_{Phk}(n)$. A closed-form expression for $V_P(h,k)$ is given by Bhansali (1996) and as discussed there, if m is finite and $k \geq m$,

$\Phi_{Dh}(k) = \Phi_{Ph}(k) = [\phi_h(1), \ldots, \phi_h(k)]' = \Phi(k, h)$,   say,   and $V_P(h, k) = V_D(h, k) = V(h)$, where $\phi_h(j) = 0, j > m$.

By contrast, if $k < m$, m finite or infinite, then for $h = 1$, $V_P(1, k) = V_D(1, k) = \sigma^2(k)$, but for $h > 1$

$$V_p(h, k) \geq V_D(h, k) \geq V(h) \tag{2.13}$$

and in this sense for multistep prediction with an incorrect model, the plug in method does not estimate the coefficients of the linear least-squares predictor, not even that restricted to the class of incorrect models, see also Findley (1983) for a discussion of this point.

2.2 The Prediction Mean Squared Error Criterion

It should be observed that in the preceding discussion the criterion of minimum mean squared error of prediction has been used for a comparison of the direct and plug-in methods. Despite some of the known difficulties in using the least-squares approach, for example its arbitrariness and sensitivity to outliers, the theory of prediction based on this criterion is both elegant and useful in practice. Clements and Hendry (1993) show, however, that for a comparison of multistep predictions across alternative methods and/or models this criterion is inadequate because mean squared error of h-step prediction, $h \geq 1$, is not invariant to non-singular, scale-preserving linear transforms, even for a model class that is itself invariant to such transforms. An implication of this non-invariance, according to these authors, is that rankings based on the mean squared error of prediction criterion can vary by choosing alternative representations of a process and an approach can, for example, dominate all others for a comparison in terms of the first differences of a time series yet loose to one of the others in terms of comparisons based on the actual values of the process.

Thus, consider an autoregressive process of order 1:

$$x_t = \delta x_{t-1} + \varepsilon_t, \qquad |\delta| < 1, \tag{2.10}$$

where $\{\varepsilon_t\}$ is as in Eq. (2.1). The two-step linear least-squares predictor of x_{n+2} based on $\{x_{n-u}, u \leq 0\}$, has a mean squared prediction error of $\sigma^2(1 + \delta^2)$. If, however, the first differences, $w_t = x_t - x_{t-1}$, of x_t are considered then the two-step linear least-squares predictor of $\{w_t\}$ has a mean squared error of $\sigma^2\{1 + (\delta - 1)^2\}$. If $\delta > 0.5$, the prediction based on w_t has a smaller mean squared error than that based on x_t, but the converse holds if $\delta < 0.5$ and now the prediction based on x_t has a smaller mean squared error than that on w_t. The authors show that this result holds also for multivariate processes and that the inclusion of parameter estimation effect could also alter the rankings based on the prediction mean squared error criterion.

The main assumption made, however, for comparing the mean squared errors of the direct and plug-in methods is that the process to be predicted, $\{x_t\}$, is stationary and that both these methods are applied to the same process; in this situation, provided the generating structure of $\{x_t\}$ is unknown or misspecified, Eq. (2.9) holds.

As an example, suppose that $k = 1$ and $h = 2$, that is, an autoregression of order 1, Eq. (2.14), is fitted for two-step prediction but $\{x_t\}$ in fact does not follow this model. We now have, from Eqns. (2.4), (2.5) and (2.8),

$$\Phi_{D2}(1) = -c(2), \qquad \Phi_{P2}(1) = -c^2(1);$$

$$V_D(2, 1) = \{1 - c^2(2)\}R(0),$$

$$V_P(2, 1) = \{1 - 2c^2(1)c(2) + c^4(1)\}R(0)$$

Thus,

$$\{V_P(2, 1)/V_D(2, 1)\} = 1 + [\{c(2) - c^2(1)\}^2\}/\{1 - c^2(2)\}] \geq 1, \qquad (2.11)$$

with the equality on the right-hand-side of Eq, (2.11) occurring if $\{x_t\}$ actually follows the model in Eq. (2.10).

Suppose, however, that $\{x_t\}$ in fact follows a moving average model of order 1,

$$x_t = \varepsilon_t + \beta\varepsilon_{t-1}, \qquad |\beta| < 1.$$

Now, since the correlation function, $c_x(k)$, say, of $\{x_t\}$ vanishes for all $k > 1$, $\{V_P(2, 1)/V_D(2, 1) - 1\} = \{c_x(1)\}^4 = \{\beta/(1 + \beta^2)\}^4 > 0$. Also, the first differences, $w_t = x_t - x_{t-1}$, of $\{x_t\}$ follow a moving average process of order 2,

$$w_t = \varepsilon_t + (\beta - 1)\varepsilon_{t-1} - \beta\varepsilon_{t-2},$$

with correlation function, $c_w(1) = (1 - \beta)^2/\{2(1 - \beta + \beta^2)\}$, $c_w(2) = -\beta/\{2(1 - \beta + \beta^2)\}$, $c_w(k) = 0$, $k > 2$. Hence, from Eq. (2.11), we have, $V_P(2, 1) > V_D(2, 1)$.

2.3 Asymptotic Properties of the Direct and Plug-in Autoregressive Estimates

If m is finite, $k = m$ and $h = 1$, the asymptotic distribution of the estimated autoregressive parameters, $\hat{a}(k)$ and $\hat{\sigma}^2(k)$ was derived in a classic paper by Mann and Wald (1943). For $h > 1$ and $k = m$, Yamamoto (1976) gives the asymptotic distribution of $\hat{\Phi}_{ph}(m)$ and Stine (1987) that of $\hat{V}_{Ph}(m)$. These authors show that for a Gaussian process, $\hat{\Phi}_{Ph}(m)$ and $\hat{V}_{Ph}(m)$ are asymptotically equivalent to the maximum likelihood estimators of $\Phi_{Ph}(m)$ and $V(h)$ and they are asymptotically efficient in the sense that their asymptotic covariance matrix attains the Cramer–Rao lower bound applicable to this situation. The case $k \neq m$, m finite or infinite, is considered by Kromer (1969), Ogata (1980), Bhansali (1981) and Kunitomo and Yamamoto (1985).

The direct estimators $\hat{\Phi}_{Dh}(m)$ and $\hat{V}_{Dh}(m)$ by contrast are asymptotically inefficient for all $h \geq 1$, if $m < \infty$ and $k = m$; see Kabaila (1981), Hosoya and Taniguchi (1982) and Bhansali (1997), who derive the asymptotic distribution of these estimators. The last author also gives explicit expressions for evaluating the loss in parameter estimation efficiency in this situation. Although this result partially explains the relative lack of popularity of the direct method in practice, it is probably not surprising: whereas the plug-in method corresponds to applying the method of maximum likelihood for estimating $\Phi(h)$ and $V(h)$, the direct method corresponds to applying the least-squares method.

As discussed in Sec. I, however, if a "non-parametric" approach is adopted for deriving the asymptotic distributions of the direct and plug-in estimators, the two methods are asymptotically equivalent in the sense that the asymptotic distributions of the two sets of estimators coincide. This result has been derived by Bhansali (1993a) under Berk's (1974) regularity conditions and requiring that m in Eq. (2.1) is infinite, k is a function of T such that $k \to \infty$, $k^3/T \to 0$ as $T \to \infty$ and certain additional regularity conditions hold; it may also be gleaned as an iterated limit from the results of Bhansali (1997).

An intuitive explanation of why the two methods are now asymptotically equivalent has also been given by Bhansali (1993a) in the context of estimation of $V(h)$, who shows that although the h-step prediction error, z_{n+h}, has not been observed, the asymptotic distributions of $\hat{V}_{Ph}(k)$ and $\hat{V}_{Dh}(k)$ is the same as that of a standard non-parametric estimator of the variance, $V(h)$, of this process [see Anderson (1971), for a definition of the latter] and thus the effect of estimating the autoregressive coefficients for the estimation of $V(h)$ by either of these two methods is asymptotically negligible as $k \to \infty$.

The difference highlighted above between the relative asymptotic behavior of the direct and plug-in methods accords with the main intuitive motivation for considering the direct method, namely that the stochastic structure generating an observed time series is usually unknown and increasingly more complex parametrizations may be needed, increasing with the series length, T, for obtaining a realistic approximation to this generating structure. Nevertheless, with a finite T, the direct method may be expected to be less biased but more variable than the plug-in method in approximating the unknown generating structure.

2.4   Direct and Plug-in Estimates for ARMA Models

A number of results described above for the fitting of autoregressive models extend also to the fitting of autoregressive moving average models. Below we illustrate the nature of these results for the situation in which an autoregressive moving average model of order $(1, 1)$, ARMA$(1, 1)$, is fitted but the process generating the observed time series may or may not follow this model. Suppose that the former holds and that

$$x_t - \delta x_{t-1} = \varepsilon_t - \beta \varepsilon_{t-1}, |\delta| < 1, |\beta| < 1, \tag{2.12}$$

where $\delta \neq \beta$ and $\{\varepsilon_t\}$ is a sequence of independent normal variables, each with mean 0 and variance σ^2. For ease of exposition suppose also that $J = 2$, i.e., prediction only up to two steps ahead is being considered.

As is well-known, $\{x_t\}$ now admits representations Eq. (1.3) with $b(j) = (\delta - \beta)\delta^{j-1}$, $a(j) = (\beta - \delta)\beta^{j-1}(j \geq 1)$, and based on a knowledge of $\{x_t, t \leq n\}$, the linear least-squares predictors of x_{n+h}, $h = 1$, 2, and the corresponding mean squared errors of prediction are given by:

$$\hat{x}_n(1) = -\sum_{j=1}^{\infty}(\beta - \delta)\beta^{j-1}x_{n+1-j}, \qquad V(1) = \sigma^2, \tag{2.13}$$

$$\hat{x}_n(2) = -\sum_{j=1}^{\infty}\delta(\beta - \delta)\beta^{j-1}x_{n+1-j}, \qquad V(2) = \sigma^2\{1 + (\delta - \beta)^2\delta^2\}. \tag{2.14}$$

Having observed $x_1, \ldots, x_T$, δ and β are estimated by a likelihood procedure and an estimator of $\hat{x}_n(1)$ and $\hat{x}_n(2)$ constructed by replacing δ and β in Eqns. (2.13) and (2.14) by their estimates. Bloomfield (1972) and Yamamoto (1981) give expressions for the increase in the mean squared error of prediction due to the estimation of δ and β; also, Ansley and Newbold (1981) consider the estimation of $V(h)$, $h \geq 1$.

Suppose now that $\{x_t\}$, admitting representations in Eq. (1.3), does not actually follow the ARMA$(1, 1)$ model in Eq. (2.12), but this model has nevertheless been fitted to the observed time series by a likelihood procedure as discussed above.

The question of what theoretical parameter, $\Theta^0 = [\delta^0, \beta^0]'$, say, the likelihood estimator $\hat{\Theta} = [\hat{\delta}, \hat{\beta}]'$, is now estimating has attracted some attention; see Kabaila (1983), Hannan (1979) and Grenander and Rosenblatt (1957). It follows that the relevant parameter minimizes the one-step mean squared error of prediction within the class of the incorrect ARMA $(1, 1)$ models:

Since for any choice of $\Theta = [\delta, \beta]'$, $\phi_1(j) = -(\beta - \delta)\beta^{j-1}$, the one-step mean squared error of prediction is given by

$$V_\Theta(1) = E\left[\left\{x_{n+1} + \sum_{j=1}^{\infty}(\beta - \delta)\beta^{j-1}x_{n+1-j}\right\}^2\right]$$

$$= \int_{-\pi}^{\pi} |1 - \delta\exp(-i\mu)|^2 |1 - \beta\exp(-i\mu)|^{-2} f_0(\mu)\, d\mu,$$

where $f_0(\mu)$ denotes the true spectral density of the observed process. The theoretical parameter corresponding to the maximum likelihood estimator is given by Θ^0, where Θ^0 is the value of Θ that minimizes $V_\Theta(1)$, i.e.,

$$\Theta^0 = \operatorname*{arginf}_\Theta V_\Theta(1).$$

Dahlhaus and Wefelmayer (1996) have proved a result establishing asymptotic efficiency of $\hat{\Theta}$ for Θ^0 within the class of the incorrect ARMA(1, 1) models.

To understand the role of the direct method in this situation, suppose that $h = 2$ and that $\{x_t\}$ in fact follows an ARMA(1, 1) model. Now, since $x_{n+2} = \hat{x}_n(2) + z_{n+2}$, where $\hat{x}_n(2)$ is given by Eq. (2.14), we may write

$$x_{n+2} - \tau(1 - \beta B)^{-1}x_n = z_{n+2} \tag{2.15}$$

where $\tau = (\beta - \delta)\delta$, $z_{n+2} = \varepsilon_{n+2} + (\delta - \beta)\varepsilon_{n+1}$ is the two-step prediction error and B denotes the backward shift operator, $B^j x_t = x_{t-j}$. It will be convenient to put $\Theta(2) = [\tau, \beta]'$.

If only $x_1, \ldots, x_T$ are observed, the direct estimate of $\Theta(2)$ for two-step prediction is obtained by minimizing

$$Q(2) = \sum_{t=1}^{T} z_{t+2}^2 = \sum_{t=1}^{T}\{x_{t+2} - \tau(1 - \beta B)^{-1}x_t\}^2 \tag{2.16}$$

with respect to $\Theta(2)$ by a least-squares method. Stoica and Soderstrom (1984) show that this estimate is unique; but, since, by our hypothesis, x_t actually follows an ARMA(1, 1) model, it is asymptotically inefficient as compared with the maximum likelihood estimate.

Suppose now that $\{x_t\}$ does not in fact follow an ARMA(1, 1) model but τ and β are estimated by minimizing $Q(2)$. As in Stoica and Soderstrom (1984), these direct estimates are still unique. Moreover, Dahlhaus and Wefelmeyer (1996) suggest that within the class of the incorrect ARMA(1, 1) models, the direct estimate is optimal in the sense of being asymptotically efficient. An additional property of the direct estimate is that its theoretical parameter, $\Theta^0(2)$, say, minimizes the two-step mean squared error of prediction within the class of ARMA(1, 1) models:

For a given value of $\Theta(2) = [\tau, \beta]'$, the two-step mean squared error of prediction is given by

$$V^{(2)}_{\Theta(2)} = \int_{-\pi}^{\pi} |1 - \tau\{1 - \beta \exp(-i\mu)^{-1}|^2 f_0(\mu) \, d\mu. \tag{2.17}$$

The theoretical parameter corresponding to the direct estimate when the fitted ARMA(1, 1) model is incorrect is the value, $\Theta^0(2)$, say, of $\Theta(2)$ that minimizes Eq. (2.17), while that corresponding to the maximum likelihood estimator does not minimize this quantity, and in this sense the direct method may be preferred to the maximum likelihood method when the fitted model may be incorrect.

As an alternative to estimating the two-step prediction constants, τ and β, by minimizing $Q(2)$, one could instead consider estimating the original coefficients, δ and β, of the ARMA(1, 1) model, Eq. (2.12), by minimizing $Q(2)$. This approach may not be recommended, however, due to two difficulties: first, there is no certainty that the resulting estimate is unique, see Stoica and Soderstrom (1984); secondly, as discussed above, whereas the maximum likelihood estimator is asymptotically optimal whether or not the fitted ARMA(1, 1) model is correct, this is not so for the estimate so obtained.

A related reference is Stoica and Nehorai (1989), who consider estimation of the parameters of an ARMA model by minimizing a weighted sum of the mean squared errors of prediction up to J steps ahead, and study possible merits of this approach by a simulation study. However, as the estimation of the original coefficients of the ARMA model, e.g., δ and β in Eq, (2.12), rather than that of the multistep prediction constants is being considered, the authors' failure to discern an improvement in the multistep forecasts by the use of their techniques is probably not surprising, but it may not be viewed as a proper assessment of the direct method considered in this chapter.

Although for exposition purposes, the review above is restricted to the class of ARMA(1, 1) models, the results discussed for this model continue to hold also for a general ARMA model, irrespective of whether or not it is the correct model. As discussed by Hannan (1987), however, there are various technical and conceptual difficulties in adopting a "non-parametric" approach to the fitting of an ARMA model. Moreover, often a principal motivation for fitting an ARMA model is to obtain a parsimonious description of the correlation structure of a relatively short time series and for such series it is usual to consider only small values of the order and the adoption of a "non-parametric" approach may accordingly seem counter-intuitive.

These reasons explain why results paralleling those described above for the "non-parametric" autoregressive model fitting approach are as yet not available for ARMA models; see, however, Potscher (1990) who considers order selection for an ARMA model when the upper bound (P, Q), say, for the order has not been pre-specified, but does not explicitly consider the question of multistep prediction.

3. PARAMETER ESTIMATION FOR NON-STATIONARY MODELS

It will be convenient throughout this Section to denote the observed time series by $Y_1, \ldots, Y_T$ and to assume that it is a realization of a discrete-time process $\{y_t\}$ such that a process, $\{x_t\}$, obtained from $\{y_t\}$ after a suitable transformation is a 0–mean stationary process with covariance and spectral density functions satisfying Eqns. (1.1) and (1.2).

A standard method for forecasting a (possibly) non-stationary time series is Exponentially Weighted Moving Average, for which the forecast of $y_{n+h}(h \geq 1)$ based on a knowledge of $\{y_{n-j}, j \leq 0\}$ is given by

$$\hat{y}_n(h) = (1 - \beta) \sum_{j=1}^{\infty} \beta^j y_{n+1-j} \tag{3.1}$$

where $0 < \beta < 1$ and the right-hand-side of Eq. (3.1) does not depend upon the prediction lead time, h. For an observed time series, this procedure is usually implemented by replacing β by its maximum likelihood estimator, $\hat{\beta}$, for all lead times, $h \geq 1$.

This procedure is known to be linear least-squares for one-step prediction when the first difference, x_t, say, of y_t follows a first-order moving average process:

$$y_t - y_{t-1} = x_t = \varepsilon_t - \beta \varepsilon_{t-1}. \tag{3.2}$$

On the other hand, since Eq. (3.1) involves only one parameter and it is convenient to employ, this procedure is used routinely in practice, even when Eq. (3.2) may not be an appropriate model. Thus, as in Sec. 1 and Sec. 2, the use of direct method for parameter estimation could be appropriate in this situation.

Tiao and Xu (1993) have investigated this last possibility. In their approach, β is estimated separately for each lead time by a least-squares procedure in which the sum of squares of the within sample h-step forecast errors is minimized. Thus, their procedure for estimating β is similar to that described in Sec. 2, see Eq. (2.16), for the direct estimation of multistep prediction constants for a stationary time series model, and they also establish for the direct estimator of β, results parallel to those described in Sec. 2 for the latter set of estimates. In particular, they show that if the dth difference of y_t, $d = 0$, 1, follows a standard ARMA process and β is estimated separately for each h, the estimator converges in probability to the value, β^0, say, of β that minimizes the h-step mean squared error of prediction when the model in Eq. (3.2) is erroneously used for forecasting instead of the correct model. Moreover, for the situation in which the model in Eq. (3.2) is correct, they derive the joint asymptotic distribution of the direct estimates, $\hat{\beta}_h$, $h = 1, \ldots, J$, and introduce diagnostic tools for assessing whether this model holds for an observed time series; see Bhansali (1983, 1993a) for similar results for a stationary time series.

A related reference is Cox (1961) who examines the behavior of the exponentially-weighted moving average when the actual process follows a stationary AR(1) model or an AR(1) process with superimposed error. An explicit expression for the one-step mean squared error of prediction and for the value of β minimizing this error is given.

Haywood and Tunnicliffe–Wilson (1997) extend the work of Tiao and Xu (1993) to the class of non-stationary time series, y_t, such that an appropriately transformed series, x_t, say, is stationary with 0 mean and a spectral density function $f(\mu)$ postulated to follow a linear model:

$$f(\mu) = \sum_{i=1}^{K} g_i f_i(\mu), \tag{3.3}$$

where the $f_i(\mu)$ are known spectral components and the g_i are unknown parameters. It should be noted that although this model holds for the exponential smoothing scheme, Eq. (3.1), the spectral density function of $x_t = y_t - y_{t-1}$ is not a linear function of β and a parameter related to β is considered instead. More generally, however, the structural models of Harvey (1989) do follow the specification in Eq. (3.3). Direct estimation of the g_i by minimizing the sum of squares of the h-step prediction errors is considered. However, a frequency domain approximation for this sum of squares, which reduces to the standard Whittle likelihood for $h = 1$, is obtained. A numerical algorithm for finding the minimum of their frequency domain estimation criterion is developed and an illustration of its behavior with observed and simulated series is given. An advantage of these authors' approach is that they are able to characterize in the frequency domain the manner in which the direct method may help in improving the fit of the model in Eq. (3.3) to the periodogram of the data.

Chen (1995) considers a two-parameter exponential smoothing scheme and gives theoretical results for evaluating the h-step prediction interval when the second difference of the observed process, y_t, is a linear stationary process satisfying Eq. (1.3). For an extension of these results to the seasonal Holt–Winters forecasting method see Chen (1996); also, Chen (1993) compares the forecasts provided by the simple exponential smoothing scheme in Eq. (3.1) with that provided by autoregressive model fitting. Although this author does not explicitly study the use of the direct method for parameter estimation, his results are interesting because of the emphasis on studying these *ad hoc* procedures from a theoretical point of view.

Clements and Hendry (1996) investigate the role of a multistep method of parameter estimation, called the dynamic estimation method in their terminology, for forecasting economic time series. These authors recognise that model misspecification is a necessary but not always a sufficient condition for sustaining the dynamic estimation method, and carry out an extensive Monte Carlo investigation to identify features of a time series that might favor this method. In their simulation study, observations are generated from a non-seasonal Holt model such that its second differences follow a second-order moving average model, but an autoregressive model with and without a constant "drift" term is fitted either to the original undifferenced data or to observations obtained after taking first differences. Thus their simulations focus on potential biases in estimating unit root processes with neglected moving average error. Although conditions which might favor the dynamic estimation method have been identified, it should be observed that the actual parameter estimation method used is different from the direct method considered in this paper. We illustrate differences between these two methods in terms of the ARMA(1,1) model, Eq. (2.12), but with $\beta \equiv 0$ and consider only forecasting two steps ahead. In this example, the dynamic estimation method corresponds to estimating the parameter δ by minimizing $Q(2)$ defined by the first equality in Eq. (2.16); the direct method by contrast corresponds to estimating the two-step prediction constant $\tau = \delta^2$ by minimizing this quantity; see also Eq. (2.11) and the preceding discussion in Sec. 2.3. From a statistical point of view, as discussed there, the former procedure may not be optimal even when a mis-specified model is being fit and it may not be recommended. Secondly, while the authors are correct in surmising that reasonable alternatives for dealing with the particular types of model misspecification they and other authors have investigated may be available, it should be recognised that a main motivation for considering the direct method is that frequently the precise nature of this misspecification would be unknown and in this situation the direct method could lead to improved forecasts in a mean squared sense.

4. MODEL SELECTION

It is now standard practice to use an order selection criterion for selecting the order of an autoregressive model. As is well-known, Bhansali (1993b), a number of currently available model selection criteria may be written as special cases of a general criterion of the following form:

$$\text{AIC}_\alpha(p) = T \ln \hat{\sigma}^2(p) + \alpha p (p = 0, 1, \ldots, K) \tag{4.1}$$

in which $\alpha > 0$ is either a fixed constant or a function of T, and the order is selected by minimizing this criterion. Thus, the Akaike information criterion (AIC) is a special case of Eq. (4.1), with $\alpha = 2$ and so is a Bayesian criterion (BIC), say, of Schwarz (1978), with $\alpha = \ln T$.

If m in Eq. (4.1) is finite, the order selected by minimizing the AIC_α criterion in Eq. (4.1) with a fixed α is known, Shibata (1976), not to provide a consistent estimator of the order. If, however, $\alpha = \alpha(T)$, a function of T, such that $\alpha(T) \to \infty$ as $T \to \infty$ then the selected order is consistent for m. Hannan and Quinn (1979) give a lower bound on the rate of growth of $\alpha(T)$ with T and show that any choice of $\alpha(T) > 2 \log \log T$ yields consistent order selection.

Shibata (1980, 1981), on the other hand, has shown that if m in Eq. (2.1) is infinite, the order selection by AIC and by a criterion of the form in Eq. (4.1) with a fixed $\alpha > 1$ is asymptotically efficient from the point-of-view of autoregressive spectral estimation and one-step prediction, but the order selection by BIC and any other consistent criterion by contrast is not asymptotically efficient in the sense defined by this author.

Although the results discussed above are important and contribute towards clarifying the relationship between the objectives of consistency and predictive efficiency in model selection, they do not apply when the objective is prediction more than one step ahead. For this problem, Shibata (1980) suggested estimating the prediction constants separately for each lead time by the direct method, in particular by solving Eqns (2.6) and (2.7), and also selecting the order to be fitted separately for each h. As Shibata considered only the fitting of autoregressive models and assumed explicitly that m in Eq. (2.1) is infinite, he suggested the following criterion, with $\alpha = 2$ and $N = T - h - K + 1$, for selecting the order:

$$Sh_\alpha(k) = \hat{V}_{Dh}(k)(N + \alpha k). \tag{4.2}$$

It is clear, however, that α in Eq. (4.2) need not be restricted to 2 and that other values could also be considered and that one may also let $\alpha = \alpha(T)$, a function of T, such that $\alpha(T) \to \infty$ as $T \to \infty$. Moreover, the following h-step generalizations of the criterion in Eq. (4.1) and of the FPE $_\alpha$ criterion of

Bhansali and Downham (1977) may also be considered as an alternative to Eq. (4.2):

$$\text{AICh}_\alpha(k) = T \ln \hat{V}_{Dh}(k) + \alpha k, \tag{4.3}$$

$$\text{FPEh}_\alpha(k) = \hat{V}_{Dh}(k)(1 + \alpha k/T). \tag{4.4}$$

Gersch and Kitagawa (1983) and Findley (1983) were early in recognizing the statistical methodological importance of Shibata's suggestion of selecting a new model for each lead time, and they have illustrated by various examples the wide applicability of this suggestion for many problems involving multistep prediction.

Thus, Gersch and Kitagawa (1983) compare the one-month and twelve-month ahead forecasts produced by two different models, called M1 and M2, for forecasting an economic time series containing trend and seasonal components. Here both models contain a local polynomial trend component, a seasonal component and a purely random disturbance term, but M2 in addition contains a globally stochastic stationary component, postulated to be an autoregression. These authors report that for one-step prediction, the standard Akaike information criterion, Eq. (4.1) with $\alpha = 2$, selects the model M2 and its overall predictive performance is superior to that of M1. The converse holds, however, for forecasting twelve steps ahead; the use of an optimum twelve-step ahead AIC criterion, analogous to Eq. (4.3) above with $\alpha = 2$, now selects the model M1 and its predictions are closer to the actual observations than those of M2.

Also, Findley (1983, 1985, 1991) in a series of papers has advanced theoretical arguments in favor of lead-time dependent model selection and parameter estimation for multistep prediction. His work has contributed immensely in generating wide-spread recognition of the importance of using the direct method for forecasting, especially when the model selected for one-step prediction may be interpreted only as an approximation to the generating structure of an observed time series.

Bhansali (1996) has recently established a theoretical optimality result for the direct method of selecting and fitting a separate autoregressive model for each lead time. His result generalizes to $h > 1$ the work of Shibata (1980) discussed above establishing optimality of AIC and related criteria for one-step prediction. An asymptotic lower bound for the h-step mean squared error of prediction when m in Eq. (2.1) is infinite and the h-step prediction constants are estimated by solving Eq. (2.6), but with $k = \tilde{k}_h$, say, treated as a random variable, whose value is selected anew for each h by a selection criterion, is developed. The orders selected by the h-step criteria in Eqns (4.2)–(4.4) with a fixed $\alpha > 1$ are shown to be asymptotically efficient in the sense that the bound is attained in the limit if $\hat{k}_h$ is selected by any of these criteria. A lower bound for the h-step mean squared error of prediction of the plug-in method, but with the initial order treated as a random variable and selected from the data by an order selection criterion is also derived. The results point to a two-fold advantage of the direct method for multistep prediction: first, the asymptotic lower bound on its mean squared error is smaller than that for the plug-in method; secondly, whereas the former bound is attainable asymptotically, that for the plug-in method is not even asymptotically attainable.

For a finite autoregressive process, Bhansali (1997) has studied the asymptotic behavior of the order selected by the criteria in Eqns (4.2)–(4.4). If α remains fixed as $T \to \infty$, the selected order is shown not to be consistent, and in this sense the clash between the objectives of efficiency and consistency continues to hold also when selecting a model for multistep prediction.

Hurvich and Tsai (1997) give a conceptual motivation for the criterion in Eq. (4.3) in terms a generalized Kullback–Leibler information measure and suggest a corrected version of this criterion, suitable for small values of T; the correction here is analogous to that introduced earlier by Hurvich and Tsai (1989) and it is especially useful when the ratio K/T is not small.

Bhansali (1998) also gives a motivation for the criterion in Eq. (4.4) and implicitly for the criteria in Eqns (4.3) and (4.2) in a least-squares framework by introducing this criterion as an asymptotically unbiased estimator of the prediction mean squared error with prediction constants estimated by the direct method.

A characteristic feature of the model selection procedures discussed above is that forecasts at different lead times may use different models. Liu (1996) has suggested a simultaneous procedure for constructing joint multiperiod forecasts. In his approach, m in Eq. (2.1) is assumed to be finite and forecasts up to the lead time J are constructed by fitting a multivariate regression model and (single) order, q, say, for constructing the simultaneous forecasts is selected from the data. The author develops suitable versions of the BIC and related criteria for achieving this objective. Although in the numerical examples given by the author, the procedure appears to work well, no theoretical result establishing its optimality has been given and, under the assumption of a finite m, it is not clear why the author's procedure need necessarily outperform the plug-in method with the value of m selected by a consistent criterion.

Lin and Granger (1994) and Hurvich (1987) are two additional references in which lead-time dependent model selection has been advocated. The former authors consider multistep prediction for a nonlinear time series and report that the procedure of selecting a new model for each lead time gives satisfactory results. Also, by contrast with the least-squares procedure considered in this Section and in Sec. 2 and Sec. 3, Hurvich (1987) considers estimation of the autoregressive coefficients in the frequency domain by a variant of the Yule–Walker method in which instead of the standard "positive definite" estimate of the auto-covariances, obtained by Fourier transforming the raw periodogram, an estimate obtained by Fourier transforming a window spectral estimate is used. It is not clear, however, why this procedure should be optimal.

5. TIME SERIES REGRESSION MODELS

In many applications, especially in economics and control engineering, it is common to fit regression models in which the error term is postulated to follow a linear time series model. Hannan and Deistler (1988) provide an exposition of the current state of the art for fitting such models, which is largely centered on fitting these models by maximum likelihood and using AIC, BIC, or a related criterion for determining the dimension of the model to fit. There has been interest, however, in using variants of the direct method for parameter estimation.

Weiss (1991) considers "multistep" estimation for a regression model involving lagged dependent variables. The parameters are estimated by a least-squares approach in which a weighted average of the sum of squares of forecast errors, averaged over lead times $h = 1, \dots, J$, is minimized. The accuracy of the forecasts is then assessed by recognising explicitly that the fitted model may be misspecified. Several analytical results establishing consistency of the estimators are given. The author also provides useful references to earlier work in the econometric literature in which a similar approach to parameter estimation was taken. It should be noted, however, that whereas in Sec. 2 and Sec. 3 the parameters are estimated separately for each lead time, this is not necessarily the case in this author's work and in this sense his approach is different from the direct method discussed here. This author's approach would be relevant when an objective is to minimize a pre-specified weighted sum of the multistep prediction errors while the direct method is suitable when the h-step mean squared error of prediction is to be minimized separately for each h.

Greco et al. (1984) and Milanese and Tempo (1985) are additional references on applications of variants of the direct method in control engineering; see also Pillai et al. (1992).

6. CONCLUDING REMARKS

An objective of the present paper is to provide a review of the current state-of-the-art for implementing the direct method of lead-time dependent model selection and/or parameter estimation for multistep prediction. This method is interpreted as an alternative to the widely-used plug-in method in which the multistep forecasts are based on a single model fitted to the observed time series. However, whereas the plug-in method is implicitly based on the premise that the correct model has fitted, the direct method does not make this assumption and it is suitable when the actual fitted model may not necessarily coincide with the stochastic structure generating the observed time series. Two well-defined contexts in which the latter situation may arise are the following:

1. When an underparametrized model is fitted, often deliberately, for multistep prediction;
2. When the generating structure of the observed time series is complex and it may involve an infinite number of parameters.

The lead-time dependent parameter estimation is likely to be particularly suitable in the former situation, and in Sec.2 and Sec. 3 we have given a review of the current literature in which this method has been successfully implemented.

By contrast, the lead-time dependent model selection, including the parameter estimation, is likely to be more suitable in the latter situation, especially when the non-parametric autoregressive model fitting is considered. In Sec. 4, we have given a review of the literature where this possibility has been investigated.

An established current practice for multistep prediction of an observed time series, after it has been suitably transformed to stationarity, is to fit a single finite parameter model, such as an ARMA model. For this situation, whether or not the direct method would improve the forecasts depends upon how closely the fitted model describes the stochastic structure generating the observed time series. If the two are not close, then the direct method is likely to be useful and it may be recommended.

Conversely, if the two agree exactly, the direct method is likely to be inefficient and it may not be recommended.

It should be emphasised that the use of standard diagnostic checks is not always sufficient to ensure that a suitable model has been fitted for multistep prediction, since, as has been discussed by Bhansali (1996), these diagnostic checks are designed to test whether the residuals of a fitted time series model are "white noise" while an objective of fitting a time series model for h-step prediction, $h > 1$, is to ensure that the residuals of the chosen model follow a moving average process of order $h - 1$. As a consequence, it is possible for a fitted model to pass some of the commonly-used diagnostic tests and yet to find that its multistep predictions could be improved by fitting a different model. In the references already cited in the chapter, empirical evidence has been provided to suggest that even when the model to fit has been determined by a model selection criterion such as the BIC, the direct method can lead to an improvement in the multistep forecasts, even though the extent of this improvement may occasionally be small.

REFERENCES

Akaike, H., Information theory and an extension of the maximum likelihood principle in *2nd International Symp. on Info. Theory* (B. N. Petrov and F. Csaki, eds.). Akademia Kiado, Budapest, 1973.

Anderson, T. W., *The Statistical Analysis of Time Series*. Wiley, New York, 1971, p. 467.

Ansley, C. F. and P. Newbold, On the bias in the estimates of forecast mean square error. *J. Amer. Statist. Assoc.*, 76: 569–578 (1981).

Berk, K. N., Consistent autoregressive spectral estimate. *Ann. Statist.*, 2: 489–502 (1974).

Bhansali, R. J., Asymptotic properties of the Wiener–Kolmogorov predictor—I, *J. Roy. Statist. Soc.*, B36: 61–73 (1974).

Bhansali, R. J., Effects of not knowing the order of an autoregressive process on the mean squared error of prediction—I, *J. Amer. Statist. Assoc.*, 76: 588–597 (1981).

Bhansali, R. J., The inverse partial correlation function of a time series and its applications, *J. Mult. Analysis, 13*: 310–327 (1983).

Bhansali, R. J., Estimation of the moving average representation of a stationary process by autoregressive model fitting, *J. Time Ser. Anal., 10*: 215–232 (1989).

Bhansali, R. J., Estimation of the prediction error variance and an R^2 measure by autoregressive model fitting, *J. Time Ser. Anal., 14*: 125–146 (1993a).

Bhansali, R. J., Order selection for linear time series models: a review. In *Developments in Time Series Analysis* (T. Subba Rao, ed.), Chapman and Hall, London, 1993b, pp. 50–66.

Bhansali, R. J., Asymptotically efficient autoregressive model selection for multistep prediction, *Ann. Inst. Statist. Math., 48*: 577–602 (1996).

Bhansali, R. J., Direct autoregressive predictors for multistep prediction: order selection and performance relative to the plug in predictors, *Statistica Sinica, 7*: 425–449 (1997).

Bhansali, R. J., Autoregressive model selection for multistep prediction. *J. Statistical Planning and Inference* (forthcoming) 1998.

Bhansali, R. J. and D. Y. Downham, Some properties of the order of an autoregressive model selected by a generalization of Akaike's FPE criterion, *Biometrika, 64*: 547–551 (1977).

Bloomfield, P., On the error of prediction of a time series, *Biometrika, 59*: 501–507 (1972).

Box, G. E. P. and G. M. Jenkins, *Time Series Analysis: Forecasting and Control.* Holden Day, New York, 1970.

Brillinger, D. R., *Time Series: Data Analysis and Theory,* Holt, New York, 1975, p. 78.

Chen, C., Some robustness properties of the simple exponential smoothing predictor, *J. Japan Statist. Soc., 23*: 201–214 (1993).

Chen, C., Theoretical results and empirical studies of the forecasting accuracy of the Holt's linear exponential smoothing method, *Ryukyu Math. J., 8*: 1–25 (1995)

Chen, C., Some statistical properties of the Holt–Winters seasonal forecasting method, *J. Japan Statist. Soc., 26*: 173–187 (1996).

Clements, M. P. and D. F. Hendry, On the limitations of comparing mean square forecast errors. *J. Forecasting, 12*: 617–637 (1993).

Clements, M. P. and D. F. Hendry, Multi-Step estimation for forecasting, *Oxford Bulletin of Econ. and Statist., 58*: 657–684 (1996).

Cox, D. R., Prediction by exponentially weighted moving averages and related methods. *J. Roy. Statist. Soc., B23*: 414–422 (1961).

Dahlhaus, R. and W. Wefelmeyer, Asymptotically optimal estimation misspecified time series models, *Ann. Statist., 24*: 952–974 (1996).

Findley, D. F., On the use of multiple models for multi-period forecasting. In *Proc. Bus. Econ. Statist. Sect.,* Amer. Statist. Assoc., Washington, D.C., 1983, pp. 528–531.

Findley, D. F., Model selection for multi-step ahead forecasting. In *Proc. 7th Symp. Identification and System Parameter Estimation* (H. A. Baker and P. C. Young, eds.). Pergamon, Oxford, 1985, pp. 1039–1044.

Findley, D. F., Convergence of finite multi-step predictors from incorrect models and its role in model selection, *Note Mat.*, *11*: 145–155 (1991).

Gersch, W. and G. Kitagawa, The prediction of time series with trends and seasonalities. *J. Bus. Econ. Statist.*, *1*: 253–264 (1983).

Greco, C., G. Menga, E. Mosca and G. Zappa, Performance improvement of self-tuning controllers by multistep horizons: the musmar approach, *Automatica, 20*: 681–699 (1984).

Grenander, U. and M. Rosenblatt, An extension of the theorem of G. Szego and its application to the study of stochastic processes. *Trans. Amer. Math. Soc., 76*: 112–126 (1954).

Grenander, U. and M. Rosenblatt, *Statistical Analysis of Stationary Time Series. Wiley:* New York, 1957.

Hannan, E. J. (1979). The Statistical theory of linear systems. In *Developments in Statistics* (ed. P. R. Krishnaiah), pp. 83–121, New York: Academic Press.

Hannan, E. J., Rational transfer function approximation. *Statistical Science, 2*; 135–161 (1987).

Hannan, E. J. and M. Deistler, *The Statistical Theory of Linear Systems,* Wiley, New York, 1988, p.32.

Hannan, E. J. and B. G. Quinn, The determination of the order of an autoregression, *J. Roy. Statist. Soc., B41*: 190–195 (1979).

Harvey, A. C., *Forecasting Structural Time Series Models and the Kalman Filter,* Cambridge University Press, Cambridge, 1989.

Haywood, J. and G. Tunnicliffe Wilson, Fitting time series models by minimizing multistep-ahead errors: a freqency domain approach, *J. Roy. Statist. Soc., 59*: 237–254 (1997).

Hosoya, Y. and M. Taniguchi, A central limit theorem for stationary processes and the parameter estimation of linear processes, *Ann. Statist., 10*: 132–153 (1982).

Hurvich, C. M., Automatic selection of a linear predictor through frequency domain cross-validation, *Comm. Statist. Theory and Methods, 16*: 3199–3234 (1987).

Hurvich, C. M. and C.-L. Tsai, Regression and time series model selection in small samples, *Biometrika, 76*: 297–307 (1989).

Hurvich, C. M. and C.-L. Tsai, Selection of a multistep linear predictor for short time series, *Statistica Sinica, 7*: 395–406 (1997).

Kabaila, P. V. Estimation based on one step ahead prediction versus estimation based on multi-step ahead prediction, *Stochastics, 6*: 43–55 (1981).

Kabaila, P. V., Parameter values of ARMA models minimizing the one-step ahead prediction error when the true system is not in the model set, *J. Appl. Prob., 20*: 405–408 (1983).

Kolmogorov, A. N., Sur l'interpolation et l'extrapolation des suites stationnaires, *C. R. Acad. Sci. Paris, 208*: 2043–2045 (1939).

Kromer, R. E. Asymptotic properties of the autoregressive spectral estimator. Technical Report No. 13, Stanford University, Dept. of Statistics, USA, 1969.

Kunitomo, N. and T. Yamamoto, Properties of predictors in misspecified autoregressive models. *J. Amer. Statist. Assoc., 80*: 941–950 (1985).

Lin, J.-L. and C. W. J. Granger, Forecasting from non-linear models in practice, *J. Forecasting, 13*: 1–9 (1994).

Liu, S.-I., Model selection for multiperiod forecasts, *Biometrika, 83* 861–873 (1996).

Mann, H. B. and A. Wald, On the statistical treatment of stochastic difference equations, *Econometrica, 11*: 173–220 (1943).

Milanese, M. and R. Tempo, Optimal algorithms theory for robust estimation and prediction, *IEEE Trans. Auto. Control, AC-30*: 730–738 (1985).

Ogata, Y., Maximum likelihood estimates of incorrect Markov models for time series and the derivation of AIC, *J. Appl. Prob., 17*: 59–72 (1980).

Parzen, E. Multiple time series modelling. In *Multivariate Analysis II* (P. R. Krishnaiah, ed.), Academic Press, New York, 1969.

Pillai, S. U., Shim, T. I. and M. H. Benteftifa, A new spectrum extension method that maximizes the multistep minimum prediction error—generalization of the maximum entropy concept, *IEEE Trans. Signal Proc., 40*: 142–158 (1992).

Potscher, B. M., Estimation of autoregressive moving average order given an infinite number of models and approximation of spectral densities, *J. Time Ser. Anal., 11*: 165–179 (1990).

Schwarz, G., Estimation of the dimension of a model, *Ann. Statist., 6*: 461-464 (1978).

Shibata, R., Selection of the order of an autoregressive model by Akaike's information criterion, *Biometrika, 63*: 117–126 (1976).

Shibata, R., Asymptotically efficient selection of the order of the model for estimating the parameters of a linear process, *Ann. Statist., 8*: 147–164 (1980).

Shibata, R., An optimal autoregressive spectral estimate, *Ann. Statist., 9*: 300–306 (1981).

Stine, R. A., Estimating properties of autoregressive forecasts, *J. Amer. Statist. Assoc., 82*: 1072–1078 (1987).

Stoica, P. and A. Nehorai, On multistep prediction error methods or time series models, *J. Forecasting, 8*: 357–368 (1989).

Stoica, P. and T. Soderstrom, Uniqueness of estimated k-step prediction models of ARMA processes, *Syst. Control Letters, 4*: 325–331 (1984).

Tiao, G. C. and D. Xu, Robustness of MLE for multi-step predictions: the exponential smoothing case, *Biometrika, 80*: 623–641 (1993).

Weiss, A. A., Multi-step estimation and forecasting in dynamic models, *J. Econometrics, 48*: 135–149 (1991).

Whittle, P., *Prediction and Regulation by Linear Least Squares Methods.* English University Press, London, 1963, p. 32.

Wiener, N., *Extrapolation, Interpolation and Smoothing of Stationary Time Series,* Wiley, New York, 1949.

Yamamoto, T., Asymptotic mean square prediction error for an autoregressive model with estimated coefficients, *Appl. Statist., 25*: 123–127 (1976).

Yamamoto, T., Predictions of multivariate autoregressive-moving average models. *Biometrika, 68*: 485–492 (1981).

7

Nonlinear Estimation for Time Series Observed on Arrays

ROBERT H. SHUMWAY and SUNG-EUN KIM
University of California, Davis, California

ROBERT R. BLANDFORD Center for Monitoring Research, Arlington, Virginia

1. INTRODUCTION

Recently, rekindled interest in isolating directional signals propagating across arrays has led to reconsidering detection and parameter estimation for certain complex valued nonlinear regression models. We consider maximum likelihood estimation of directional parameters for such models under various stochastic signal assumptions, where the signal and noise spectral structures are estimated via analysis of variance or the ECM algorithm. We use asymptotic distributions for variance components and F-Statistics for detecting the signal and large sample normal theory for estimating the velocities and azimuths. Applications to designing infrasonic arrays for monitoring a nuclear test ban treaty are considered. In particular, we treat estimation and detection for infrasonic signals under various correlation scenarios, using optimal and suboptimal versions of beam forming as objective functions.

An interesting class of statistical problems arises when analyzing commonly occurring practical situations involving signals generated by various physical processes. The questions of interest, in these situations, relate to the characterization of special propagation patterns that these signals may exhibit as they are observed on a vector time series, composed of single

227

recording instruments or *sensors* arranged in a spatial pattern called an *array*. A signal from a fixed source may be assumed to propagate across this array as a *plane wave*, for example, which allows one to predict exactly when the signal will arrive at each sensor in an array, as a function of the velocity and angle of approach or *azimuth*. Knowing the time delays, it is implicitly possible to invert to obtain the velocities and azimuths corresponding to the propagating wave. This, in turn, leads to an estimator for the location of the source that generated the signal. The preceding discussion becomes more concrete when it is related to an actual practical situation and we shall use a set of data of contemporary interest as an example.

We consider using a small infrasonic array to detect a surface explosion that may originate from mining activities or from a nuclear test, see for example Simons, (1996), Blandford (1996). Infrasound in the atmosphere, i.e. sound waves whose frequencies are less than the lowest audible frequency, propagate great distances without substantial loss of energy. Sounds at these frequencies can be caused by volcanic explosions, earthquakes and nuclear or mining explosions. Experience has shown that infrasonic waves are approximately plane. The problem of optimal configuring of such arrays for purposes of monitoring a comprehensive nuclear test ban treaty is presently under consideration. For example, Fig. 1 shows time series from three microbarographic instruments in a small triangular array, called Small Fry, with approximately one kilometer sides on Palmyra Island in the Pacific that recorded an event called Tanana which was detonated about 25 km South of Kiritimati Island (Christmas Island at the time of the explosion). For a description of infrasonic recording instruments, see Cook and Bedard (1971). Although the coordinates of the array generating the three series shown in Fig. 1 were not available, it appears that the strongest part of the signal begins shortly after point 1410, first at series B, followed by A, followed by C as can be seen from the negative pulse 4230. We can imagine the sensors being located at various coordinates, (x_j, y_j) in the plane, as in Fig. 2, where we show a single location (x_1, y_1) and a plane wave arriving from the southwest (225°) with a frequency proportional to the distance between the lines. Problems of interest for this infrasonic array are (1) detecting the plane wave signal and (2) estimating the unknown parameters, mainly the velocity and azimuth of the propagating signal, which map directly into an estimator for the location of the explosion. An additional problem is determining the optimum geometry, i.e. the number and locations of the recording instruments, for a new array, for monitoring purposes that will tend to produce the smallest location error.

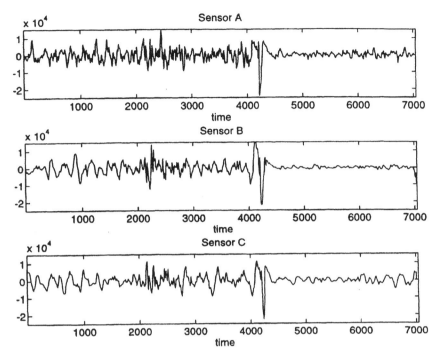

Figure 1.   Three series for the event Small Fry at a small triangular array on Palmyra Island showing arrival shortly after point 4343 (sampling rate = 10 pts/s).

In the preceding situation, it is convenient to regard the signal as a three-dimensional stationary space-time process, say $s(t, x_j, y_j) = s(t, \mathbf{r}_j)$, where $\mathbf{r}_j = (x_j, y_j)'$ denotes the location of the jth sensor in the x, y plane. A fixed plane wave with frequency ν_0, in cycles per unit time, and wavenumber coordinates $\boldsymbol{\theta}_0 = (\theta_{10}, \theta_{20})'$, in cycles per unit distance, can be modeled as $\cos[2\pi(\nu_0 t + \mathbf{r}_j'\boldsymbol{\theta}_0)]$. One might then assume a fixed or random signal appearing at time delay

$$T_j(\boldsymbol{\theta}_0) = -\frac{\mathbf{r}_j'\boldsymbol{\theta}_0}{\nu_0} \tag{1}$$

at location $j, j = 1, 2, \ldots, N$. Generally, the spatial locations are irregularly spaced and can be quite sparse, as in the above infrasonics example. In the case of an array, we assume the coordinates $\mathbf{r}_j = (x_j, y_j)', j = 1, \ldots, N$ for the N time series that are observed in the array and the corresponding time

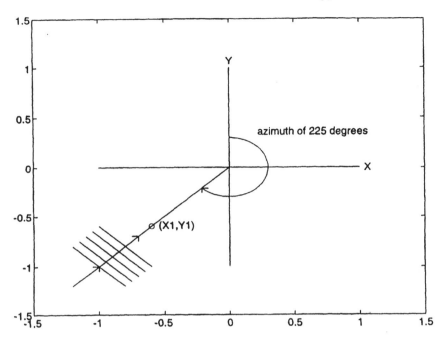

Figure 2. Schematic diagram showing a plane wave arriving at location (x_1, y_1) at Azimuth 225°. Parallel lines denote the frequency of the plane wave.

delays, represented by $T_j(\theta_0)$ from Eq. (1). For the case of sparsely observed two-dimensional series, it is convenient to write the series actually observed at the jth sensor as

$$y_j(t) = s(t - T_j(\theta_0)) + n_j(t) \tag{2}$$

for $t = 0, 1, \ldots, T - 1, j = 1, 2, \ldots, N$, where $n_j(t)$ is a stationary noise process. Later, we will discuss various possible assumptions for the signal and noise processes and will settle on a particular set of assumptions that seem to lead to tractable computational results. Note that in cases where the signal cannot be assumed to be identical on each sensor (or perfectly correlated), one might make the assumption

$$y_j(t) = s_j(t - T_j(\theta_0)) + n_j(t), \tag{3}$$

where the signal is a general process that differs from sensor to sensor and has some correlation structure between series. In particular, the infrasonics problem is a case in point, where the design is influenced by a correlation, declining with distance, between the signals observed at different recording instruments. The wavenumber vector $\boldsymbol{\theta}_0 = (\theta_{10}, \theta_{20})'$ is nonlinearly related to the velocity and azimuth parameters, which are of primary interest. The velocity c_0 and azimuth α_0 are expressed as

$$c_0 = \frac{\nu_0}{\|\boldsymbol{\theta}_0\|} \qquad (4)$$

and

$$\alpha_0 = \tan^{-1}\left(\frac{\theta_{10}}{\theta_{20}}\right), \qquad (5)$$

where α_0 is measured clockwise in radians. Generally we convert to degrees by multiplying by $180/\pi$ and interpret the angle obtained, beginning with $0°$ north. $\|A\|^2 = \text{tr } AA^*$ will denote the usual squared norm of the matrix A, where * denotes the complex conjugate transpose of the matrix.

For the model, given by Eq. (2), we may assume either that the signal $s(t)$ is an unknown deterministic function or that it is random with some unknown autocovariance function and associated spectrum. Random signal approaches have been given by Thomson (1982) and in the engineering literature by Wax (1991). Böhme (1996) considers both the random and deterministic signal models for more general models. Stochastic signal approaches involving the multivariate signal model in Eq. (3) have mainly assumed general correlation structures for both the signal vector $\mathbf{s}(t; \boldsymbol{\theta}) = (s_1(t - T_1(\boldsymbol{\theta})), \ldots, s_N(t - T_N(\boldsymbol{\theta})))'$ and for the noise vector $\mathbf{n}(t) = (n_1(t), n_2(t), \ldots, n_N(t))'$. In the stationary case, this means that the signal and noise vectors can be characterized by $N \times N$ spectral matrices, say $f_s(\nu)$ and $f_n(\nu)$, where ν is frequency in cycles per unit time.

It should be noted that a substantial engineering literature is available that considers the case where there are multiple signals present on each channel. For such cases the MUSIC (Multiple Signal Characterization Algorithm) is quite popular and has been compared to maximum likelihood by Stoica and Nehorai (1989). The Haykin (1991) edited volume contains a number of excellent papers that approach array detection and estimation from either of the above two points of view. For the multiple deterministic signal case, an approach analogous to stepwise multiple regression has been proposed by Shumway (1983). The optimization problem and resulting resolution of multiple propagating signals is quite difficult and we do not consider this problem here. Rather, we feel that in our application it is more important to concentrate on the case where each sensor may contain a different signal such as would be described by the model in Eq. (3). The influence on signal decorrelation as a function of distance has not been considered before in detail and it is this case that we concentrate on here.

Approaches to detection and maximum likelihood estimation based on the more usual deterministic signal interpretation can be found, for example, in Hinich and Shaman (1972), Hinich (1981), Shumway (1983), Stoica and Nehorai (1990) or Brillinger (1985). The high resolution estimator of Capon (1969) is based on fixed signal arguments. Applications to seismology have been discussed in Shumway (1983, 1988) and, recently, by Harris (1990), where signal detection and azimuth estimation are of interest for seismic signals originating from earthquakes and various kinds of explosions. In the deterministic signal context, the noises are assumed to have some spectral matrix, say $f_n(\nu)$, and the signal is a fixed unknown function. A special case that is extremely tractable is the uncorrelated noise case, where

$$f_n(\nu) = P_n(\nu)I, \tag{6}$$

with identical spectra at each channel and I denoting the $N \times N$ identity matrix here and in the sequel. Fixed signal approaches, that assume uncorrelated noise components, are in Shumway (1983), whereas the correlated case is covered by Brillinger (1985).

We can envision two separate problems of interest for the signal models discussed in the previous Section. The first is detecting the presence of the signal and amounts to testing $s(t) = 0$ for the fixed signal model or $f_s(\nu) = 0$ for the random signal models. The second problem is in estimating the velocity and azimuth or wavenumber θ of the signal in the presence of the nuisance parameters of the signal and noise processes. We develop first an approach to detecting the signal and estimating velocity and azimuth under the model in Eq. (2) using both the random signal and fixed signal assumptions. Both estimators are shown to be based on the well known procedure of *beam forming*. The asymptotic variance of the maximum likelihood estimators, which is similar under both conventions, is still a function of the signal and noise spectra but we use a random effects spectral analysis of variance approach to give simple estimators for these two nuisance parameters. We then derive optimal estimators for the velocity and azimuth under the general signal correlation structure implied by the multivariate signal model in Eq. (3). Eventually, we apply the simple estimator to the infrasonic data discussed above.

2. MAXIMUM LIKELIHOOD ESTIMATION

It is convenient to consider first the problems of detecting a signal and of estimating the parameters for wavenumber vector θ, that are involved nonlinearly as delays in the observed signal through Eq. (1). Note that the velocity c_0 and azimuth α_0 in Eqns (4) and (5) are further nonlinear functions of θ_0. We consider first the univariate signal case, represented by Eq. (2) and then the multivariate signal case modeled in Eq. (3).

In the random signal case, we restrict attention to the case of uncorrelated noise components, which leads to particularly tractable forms for the likelihood ratio detector and for the maximum likelihood estimators for the wavenumber parameters. In this case, it is easier to handle the problem of estimating the spectral nuisance parameters, as in the analysis of variance approach given in Sec. 3. The multivariate signal case is also treated because of its importance in modeling situations where the signal is different on each channel and may have a correlation or coherence that decreases with distance. This has implications for array design in that adding more sensors at larger spacing may be of little incremental value in improving the precision of velocity and azimuth estimators.

Fixed signal models are treated for the model implied by Eq. (3), with a treatment of cases involving both uncorrelated and correlated noise components. Again, the case where noise components are uncorrelated, as specified in Eq. (6), is of importance because it leads to the tractable and well known beam formed estimators and simple large sample variances of the wavenumber parameters.

2.1 Random Signal Models

To begin, consider a random signal model based on Eq. (3), rewritten in vector notation as

$$\mathbf{y}(t) = \mathbf{s}(t; \theta_0) + \mathbf{n}(t), \tag{7}$$

and assume that the signal and noise series $\mathbf{n}(t)$ are linear processes, i.e., can be written as linearly filtered combinations of white noise vectors with square summable coefficients and components with finite fourth order moments. These assumptions seem to be reasonable for many applications and can be used to verify that conditions in Robinson (1972) hold, leading to consistency and asymptotic normality of the estimators for velocity and azimuth. If we denote the $N \times N$ spectral matrices of the processes by $f_y(\nu; \theta_0), f_s(\nu)$ and $f_n(\nu)$, we have

$$f_y(\nu; \theta_0) = Z(\nu; \theta_0) f_s(\nu) Z^*(\nu; \theta_0) + f_n(\nu) \tag{8}$$

where $Z(\nu; \theta_0) = \mathrm{diag}\{z_1(\nu; \theta_0), z_2(\nu; \theta_0), \ldots, z_N(\nu; \theta_0)\}$ denotes a diagonal matrix with diagonal elements defined by

$$z_j(\nu; \theta_0) = \exp\{-2\pi i \nu T_j(\theta_0)\} \tag{9}$$

Note that we are using $*$ to denote the conjugate transpose of a vector or matrix.

We consider first the estimation of the directional parameters under the assumptions of the univariate signal model in Eq. (2) and then look at the more involved model in Eq. (3). In general, assume that the signal and noises are both $N \times 1$ vector linear processes, with covariance functions satisfying $\sum_h |hR^s_{jk}(h)| < \infty$ and $\sum_h |hR''_{jk}(h)| < \infty$ for $j, k = 1, \ldots, N$. Then, for frequencies of the form $\nu^T_k = k_T/T, k_T = 0, 1, \ldots, T - 1$, define the discrete Fourier transform (DFT) of the observed data by

$$Y_j(k_T) = T^{-1/2} \sum_{t=0}^{T-1} y_j(t) \exp\{-2\pi i \nu^T_k t\}. \tag{10}$$

We index the frequency arguments by T to indicate that they may be chosen to be close to some fixed target frequency ν_0 as T increases. Now, for a fixed target frequency ν_0, choose L neighboring frequencies of the form $\nu_\ell = \nu_0 + \ell/T, \ell = -(L-1)/2, \ldots, 0 \ldots, (L-1)/2$, i.e. a bandwidth of approximately L/T of frequencies in the neighborhood of ν_0, where L is arbitrarily taken to be odd. Let $Y_{j\ell}, \ell = 1, 2, \ldots, L$ be the sample of DFT's in the neighborhood of the frequency ν_0. If the signal and noise spectra are reasonably smooth in the narrow band about ν_0 and if the plane wave is at some frequency in this interval, satisfying Eq. (1), then the vectors $\mathbf{Y}_\ell = (Y_{1\ell}, Y_{2\ell}, \ldots, Y_{N\ell})'$ will be approximately complex Gaussian and uncorrelated $O(T^{-1})$, see, for example, Hannan (1970), with covariance matrix $f_y(\nu_0; \boldsymbol{\theta}_0)$, as given in Eq. (8). Note that the special case where the noise and signal spectral matrices are diagonal with identical elements in the diagonals is of interest. This means taking the noise spectral matrix as Eq. (6) and the signal spectral matrix as $P_s(\nu_0)I$, where $P_s(\nu_0)$ and $P_n(\nu_0)$ will denote the signal and noise spectra respectively in the sequel. In this special case, it is clear that

$$f_y(\nu_0; \boldsymbol{\theta}_0) = P_s(\nu_0)\mathbf{z}(\boldsymbol{\theta}_0)\mathbf{z}^*(\boldsymbol{\theta}_0) + P_n(\nu_0)I \tag{11}$$

where $\mathbf{z}(\boldsymbol{\theta}_0) = (z_1(\boldsymbol{\theta}_0), z_2(\boldsymbol{\theta}_0), \ldots, z_N(\boldsymbol{\theta}_0))'$ with

$$z_j(\boldsymbol{\theta}_0) = \exp\{2\pi i \mathbf{r}'_j \boldsymbol{\theta}_0\}, \tag{12}$$

because of Eq. (1) and (8). Note that we can suppress the dependence on frequency in the vectors $\mathbf{z}(\boldsymbol{\theta}_0)$ and in the spectra P_s, P_n in the band near ν_0.

The previous assumptions imply that we can approximate the narrow band log likelihood as a function of the parameters $\boldsymbol{\theta}, P_s, P_n$ as

$$\log L(\boldsymbol{\theta}, P_s, P_n) \propto -L \log |f_y(\nu_0; \boldsymbol{\theta})| - \sum_{\ell=1}^{L} \mathbf{Y}_\ell^* [f_y(\nu_0; \boldsymbol{\theta}]^{-1} \mathbf{Y}_\ell$$

$$= -NL \log P_n - L \log \left(1 + \frac{N}{r}\right)$$

$$- \frac{1}{P_n} \sum_{\ell=1}^{L} \mathbf{Y}_\ell^* \mathbf{Y}_\ell + \frac{1}{P_n} \frac{\sum_{\ell=1}^{L} |\mathbf{z}^*(\boldsymbol{\theta})\mathbf{Y}_\ell|^2}{N+r} \tag{13}$$

where we have used

$$r = \frac{P_n}{P_s} \tag{14}$$

for the inverse of the *signal-to-noise ratio*.

If we ignore the nuisance parameters P_s and P_n, it is clear that the log likelihood is a monotone function of the statistic

$$B(\boldsymbol{\theta}) = \sum_{\ell=1}^{L} |\mathbf{z}^*(\boldsymbol{\theta})\mathbf{Y}_\ell|^2$$

$$= \sum_{l=1}^{L} \left| \sum_{j=1}^{N} \exp\{-2\pi i \mathbf{r}_j' \boldsymbol{\theta}\} Y_{j\ell} \right|^2, \tag{15}$$

which is the observed power along a *beam* corresponding to the wavenumber $\boldsymbol{\theta}$. The *beam power* can be searched over $\boldsymbol{\theta}$ to get a maximum likelihood estimator, say

$$\hat{\boldsymbol{\theta}} = \text{argmax}_{\boldsymbol{\theta}} \ B(\boldsymbol{\theta})$$

for the wavenumber coordinates. This is the classical estimator for the wavenumber and it is satisfying that it is the maximum likelihood estimator as well.

For a fixed $\boldsymbol{\theta}$, the likelihood function in Eq. (13) is maximized by differentiating the log likelihood in Eq. (13) with respect to P_n and $r = P_n/P_s$ and solving. We obtain

$$\hat{P}_n(\boldsymbol{\theta}) = \frac{SSE(\boldsymbol{\theta})}{(N-1)L} \tag{16}$$

for the noise, where the sum of squares for error $SSE(\boldsymbol{\theta})$ is

$$SSE(\boldsymbol{\theta}) = \sum_{\ell=1}^{L} \left(\|\mathbf{Y}_\ell\|^2 - \frac{|\mathbf{z}^*(\boldsymbol{\theta})\mathbf{Y}_\ell|^2}{N} \right)$$

$$= \left(\sum_{\ell=1}^{L} \|\mathbf{Y}_\ell\|^2 - \frac{B(\boldsymbol{\theta})}{N} \right)$$

The maximum likelihood estimator for the signal spectrum is

$$\hat{P}_s(\boldsymbol{\theta}) = \frac{1}{N^2 L}\left(B(\boldsymbol{\theta}) - NL\hat{P}_n(\boldsymbol{\theta})\right) \tag{18}$$

For the random signal case, a large L theory with decreasing bandwidth can be developed, using arguments similar to those of Robinson (1972), that express $B(\boldsymbol{\theta})$ and its derivatives in terms of the sample autocovariance function. We get consistency for $\hat{\boldsymbol{\theta}}$ and asymptotic normality from the large sample behavior of the autocovariance function. Large sample theory tied to maximizing the approximate log likelihood in Eq. (13) implies that, for a true value of $\boldsymbol{\theta}_0$ and fixed values for P_n and P_s, the large-sample distribution of $\hat{\boldsymbol{\theta}}$ is approximately normal with mean $\boldsymbol{\theta}_0$ and covariance matrix

$$\text{cov } \hat{\boldsymbol{\theta}} = \left\{-E\frac{\partial^2 \log L(\boldsymbol{\theta}_0)}{\partial \boldsymbol{\theta}_0 \partial \boldsymbol{\theta}_0'}\right\}^{-1} \tag{19}$$

Having done this, note that the velocity and azimuth are nonlinear functions, say $\mathbf{h}(\boldsymbol{\theta}) = (c, \alpha)' = (h_1(\boldsymbol{\theta}), h_2(\boldsymbol{\theta}))'$, and the delta method implies that the function $\mathbf{h}(\hat{\boldsymbol{\theta}})$ is approximately normal with mean $\mathbf{h}(\boldsymbol{\theta}_0)$ and

$$\text{cov } \mathbf{h}(\hat{\boldsymbol{\theta}}) = \left(\frac{\partial \mathbf{h}}{\partial \boldsymbol{\theta}_0}\right)\text{cov } \hat{\boldsymbol{\theta}}\left(\frac{\partial \mathbf{h}}{\partial \boldsymbol{\theta}_0}\right)'. \tag{20}$$

For the velocity and azimuth functions, note that

$$\frac{\partial \mathbf{h}}{\partial \boldsymbol{\theta}} = \frac{1}{\|\boldsymbol{\theta}\|^3}\begin{pmatrix} -\nu\theta_1 & -\nu\theta_2 \\ \|\boldsymbol{\theta}\|\theta_2 & -\|\boldsymbol{\theta}\|\theta_1 \end{pmatrix}. \tag{21}$$

Applying Eq. (19) to the log likelihood in Eq. (13) gives

$$\frac{\partial^2 \log L(\boldsymbol{\theta})}{\partial \boldsymbol{\theta}\partial \boldsymbol{\theta}'} = -\frac{(2\pi)^2}{P_n(N+r)}\sum_{j,k}\sum_{\ell=1}^{L}\exp\{-2\pi i(\mathbf{r}_j - \mathbf{r}_k)'\boldsymbol{\theta}\}$$

$$\times Y_{j\ell}Y_{k\ell}^*(\mathbf{r}_j - \mathbf{r}_k)(\mathbf{r}_j - \mathbf{r}_k)'$$

Hence,

$$-E\left\{\frac{\partial^2 \log L(\boldsymbol{\theta})}{\partial \boldsymbol{\theta}\partial \boldsymbol{\theta}'}\right\} = (2\pi)^2\frac{L}{(N+r)}\frac{P_s}{P_n}\sum_{j,k}(\mathbf{r}_j\mathbf{r}_j' - \mathbf{r}_j\mathbf{r}_k' - \mathbf{r}_k\mathbf{r}_j' + \mathbf{r}_k\mathbf{r}_k')$$

$$= \frac{2(2\pi)^2 N^2 L}{r(N+r)}\ R, \tag{22}$$

where

$$R = \frac{1}{N} \sum_{j=1}^{N} (\mathbf{r}_j - \bar{\mathbf{r}})(\mathbf{r}_j - \bar{\mathbf{r}})' \tag{23}$$

is the sample covariance matrix of the array coordinates. It follows that $\hat{\boldsymbol{\theta}}$ will be approximately normal with mean $\boldsymbol{\theta}_0$ and approximate covariance matrix

$$\text{cov}(\hat{\boldsymbol{\theta}}) \approx \frac{1}{2(2\pi)^2} \frac{1}{L} \frac{r}{N} \left(1 + \frac{r}{N}\right) R^{-1}. \tag{24}$$

Then, defining the vectors $\boldsymbol{\theta} = (\theta_1, \theta_2)'$ and $\tilde{\boldsymbol{\theta}} = (\theta_2, -\theta_1)'$, we obtain

$$
\text{cov}\begin{pmatrix} \hat{c} \\ \hat{\alpha} \end{pmatrix} \approx \frac{1}{2(2\pi)^2} \frac{1}{L} \frac{r}{N} \left(1 + \frac{r}{N}\right) \frac{1}{\|\boldsymbol{\theta}_0\|^6}
$$
$$
\times \begin{pmatrix} \nu^2 \boldsymbol{\theta}_0' R^{-1} \boldsymbol{\theta}_0 & -\nu \|\boldsymbol{\theta}_0\| \boldsymbol{\theta}' R^{-1} \tilde{\boldsymbol{\theta}}_0 \\ -\nu \|\boldsymbol{\theta}_0\| \boldsymbol{\theta}_0' R^{-1} \tilde{\boldsymbol{\theta}}_0 & \|\boldsymbol{\theta}_0\|^2 \tilde{\boldsymbol{\theta}}_0' R^{-1} \tilde{\boldsymbol{\theta}}_0 \end{pmatrix} \tag{25}
$$

for the covariance matrix of the velocity and azimuth. This exhibits nicely the dependence of the variances of azimuth and velocity on the geometry of the array, as embodied in the covariance matrix of locations R. Note the variances also depend on the noise-to-signal ratio r, so that we still need an estimator for this quantity. In the next Section, we consider estimating the signal and noise spectra using a spectral analysis of variance.

A generalization of the model given above to the case where there is a general signal correlation structure seems to be useful in cases where the signal cannot assumed to be identical on each channel. In such cases, it is reasonable to keep the noise uncorrelated, so that we are interested in the case where the spectral matrix is given by $f_y(\nu_0; \boldsymbol{\theta}_0)$ in Eq. (8) with the noise spectral matrix in diagonal form, namely $f_n(\nu_0) = P_n(\nu_0)I$. In the sequel, we will often suppress the frequency argument and write $P_n(\nu_0) = P_n$. Hence, the assumed form is

$$f_y(\nu_0; \boldsymbol{\theta}_0) = Z(\boldsymbol{\theta}_0) f_s(\nu_0) Z^*(\boldsymbol{\theta}_0) + P_n(\nu_0) I_N, \tag{26}$$

with the matrix $Z(\boldsymbol{\theta}_0) = \text{diag}\{z_1(\boldsymbol{\theta}_0), z_2(\boldsymbol{\theta}_0), \dots, z_N(\boldsymbol{\theta}_0)\}$ is a diagonal matrix with the indicated components in the diagonal. The log likelihood in this case has the form

$$\log L(\theta) \propto -L \log |f_y(\nu_0; \theta)| - \sum_\ell \mathbf{Y}_\ell^* [f_y(\nu_0; \theta)]^{-1} \mathbf{Y}_\ell$$

$$= -LN \log P_n - L \log \left| I + \frac{1}{P_n} f_s \right|$$

$$-\frac{1}{P_n} \sum_\ell \mathbf{Y}_\ell^* \mathbf{Y}_\ell + \frac{1}{P_n} \sum_l \mathbf{Y}_\ell^* Z(\theta) \left(I + P_n f_s^{-1} \right)^{-1} Z^*(\theta) \mathbf{Y}_\ell, \quad (27)$$

using standard matrix identities, see, for example, Jazwinski (1970). If we define

$$C = (I + P_n f_s^{-1})^{-1}, \tag{28}$$

it is clear that, in the case where the noise spectrum P_n and the signal spectral matrix f_s are given, the optimum detector in Eq. (24) has the form

$$GB(\theta) = \frac{1}{P_n} \sum_{j,k} \sum_\ell Y_{j\ell}^* Y_{k\ell} c_{jk} \exp\{2\pi i (\mathbf{r}_j - \mathbf{r}_k)' \theta\}, \tag{29}$$

that, on comparison with Eq. (15), can be regarded as a *generalized beam forming* statistic. A likelihood ratio test of no signal, i.e. $f_s = 0$, would require maximum likelihood estimators for θ, P_n and f_s, as discussed in Sec. 3.1 in the full space for P_n only under $f_s = 0$.

Assuming that signal and noise spectra are known, one can just maximize Eq. (29) to obtain the estimated wavenumber $\hat{\theta}$ and the associated velocity and azimuth. Following through the usual likelihood theory, we obtain the variance covariance matrix of $\hat{\theta}$ as

$$\text{cov } \hat{\theta} \approx \frac{1}{(2\pi)^2} \frac{P_n}{L} D^{-1} \tag{30}$$

where D is the matrix

$$D = \sum_{j,k} c_{jk} f_{kj}^s (\mathbf{r}_j - \mathbf{r}_k)(\mathbf{r}_j - \mathbf{r}_k)', \tag{31}$$

leading to the covariance matrix

$$\text{cov}\begin{pmatrix} \hat{c} \\ \hat{\alpha} \end{pmatrix} \approx \frac{1}{(2\pi)^2} \frac{P_n}{L} \frac{1}{\|\theta_0\|^6} \begin{pmatrix} \nu^2 \theta_0' D^{-1} \theta_0 & -\nu \|\theta_0\| \theta_0' D^{-1} \tilde{\theta}_0 \\ -\nu \|\theta_0\| \theta_0' D^{-1} \tilde{\theta}_0 & \|\theta_0\|^2 \tilde{\theta}_0' D^{-1} \tilde{\theta}_0 \end{pmatrix}. \tag{32}$$

The above covariance matrix allows one to assess the effect of a known signal correlation, as a function of the cross-spectral matrix of the signal process f_s, when the optimum detector is of the form of Eq. (24). This requires prior knowledge of the signal spectral matrix and of the noise spectrum and again, we defer this problem to the next section. In general, we can characterize the effect of declining coherence by expressing the spectral matrix of the signal in the form

$$f_{jk}^s(\nu) = P_s(\nu)|\rho_{jk}| \tag{33}$$

where $|\rho_{jk}|$ will be the magnitude of the coherence, that will be parameterized later using a model of Mack and Flinn (1971).

In many cases, the use of the simple beam forming estimator in Eq. (15) will be preferred if the coherence is reasonably strong over the frequencies and distances occurring within the array. Hence, the asymptotic variance covariance matrix of estimators maximizing Eq. (15) will be of interest. In order for the result to hold we must have

$$E\left[\frac{\partial B(\boldsymbol{\theta}_0)}{\partial \boldsymbol{\theta}_0}\right] = \sum_{l=1}^{L} \sum_{j,k} (\mathbf{r}_j - \mathbf{r}_k) f_{jk}^s = 0$$

which occurs if f_{jk}^s is purely real and we make this assumption for the variance computation. Note that any phase differences are assumed to be contained in the function $\mathbf{Z}(\boldsymbol{\theta})$ anyway so that the assumption does not seem too rigid. Note if we call the estimator $\hat{\boldsymbol{\theta}}_b$, the new covariance matrix will be

$$\operatorname{cov} \hat{\boldsymbol{\theta}}_b = V^{-1} W V^{-1} \tag{34}$$

where

$$W = E\left[\frac{\partial B(\boldsymbol{\theta}_0)}{\partial \boldsymbol{\theta}_0} \frac{\partial B(\boldsymbol{\theta}_0)}{\partial \boldsymbol{\theta}_0}^*\right]$$

and V is just Eq. (19) evaluated for $B(\boldsymbol{\theta}_0)$. Evaluating the above yields $V/(2\pi)^2$ as Eq. (31) with $c_{jk} = 1$ and

$$\frac{W}{(2\pi)^2} = \sum_{j,k,j',k'} (\mathbf{r}_j - \mathbf{r}_k)(\mathbf{r}_{j'} - \mathbf{r}_{k'})' f_{jj'}^s f_{k'k}^s$$
$$+ 2P_n \sum_{j,k,k'} (\mathbf{r}_j - \mathbf{r}_k)(\mathbf{r}_j - \mathbf{r}_{k'})' f_{kk'}^s + 2P_n^2 N^2 R. \tag{35}$$

Finally, the covariance matrix corresponding to (32) becomes

$$\operatorname{cov}\begin{pmatrix} \hat{c}_b \\ \hat{\alpha}_b \end{pmatrix} \approx \frac{1}{(2\pi)^2} \frac{P_n}{L} \frac{1}{\|\boldsymbol{\theta}_0\|^6}$$
$$\times \begin{pmatrix} \nu^2 \boldsymbol{\theta}_0' V^{-1} W V^{-1} \boldsymbol{\theta}_0 & -\nu\|\boldsymbol{\theta}_0\|\boldsymbol{\theta}_0' V^{-1} W V^{-1} \tilde{\boldsymbol{\theta}}_0 \\ -\nu\|\boldsymbol{\theta}_0\|\boldsymbol{\theta}_0' V^{-1} W V^{-1} \tilde{\boldsymbol{\theta}}_0 & \|\boldsymbol{\theta}_0\|^2 \tilde{\boldsymbol{\theta}}_0' V^{-1} W V^{-1} \tilde{\boldsymbol{\theta}}_0 \end{pmatrix}. \tag{36}$$

The preceding material allows us to consider both the classic beam formed estimator $B(\hat{\boldsymbol{\theta}})$, obtained by maximizing Eq. (15), and the optimal weighted beam $GB(\boldsymbol{\theta})$, obtained by maximizing Eq. (29) under various assumed correlation structures. This enables a variance to be computed for both estimators under various correlation configurations. If a reasonable model can be developed for correlations expected under various spacings, the variance properties of any particular array geometry can be computed for each estimator. This enables the comparison of various array geometries which is done in Sec. 4.2. Of course, in practice, we may either assume a correlation structure or let the data determine it using the ECM algorithm to be discussed in Sec. 3.1.

2.2 Deterministic Signal Models

An alternate approach to that given in the previous Sections, Shumway (1983) assumes that, on transformation, as in Eq. (10), we may write a model for the discrete Fourier transforms of the data as

$$\mathbf{Y}_\ell = \mathbf{z}(\boldsymbol{\theta})S_\ell + \mathbf{N}_\ell, \tag{37}$$

where the signal is a fixed unknown function with discrete Fourier transform S_ℓ and the noise vector $\mathbf{n}(t)$ has the transform $\mathbf{N}_\ell$, for $\ell = 1,\ldots,L$ frequencies in the neighborhood of ν_0. In the general case, one assumes that the spectral matrix of the noise is $f_n(\nu_0) = E(\mathbf{N}_\ell \mathbf{N}_\ell^*)$. If we assume that the noise is uncorrelated from sensor to sensor, i.e., *spatially white*, with $f_n(\nu_0) = P_n(\nu_0)I$, as in Shumway (1983), we may write down an approximate log likelihood in the form

$$\log L(\boldsymbol{\theta}, S_1, S_2,\ldots, S_L, P_n) \approx -NL \log P_n - \frac{1}{P_n} \sum_{\ell=1}^{L} \|\mathbf{Y}_\ell - \mathbf{z}(\boldsymbol{\theta})S_\ell\|^2 \tag{38}$$

and we obtain the beam formed estimator

$$\hat{S}_\ell(\boldsymbol{\theta}) = N^{-1}\mathbf{z}^*(\boldsymbol{\theta})\mathbf{Y}_\ell \tag{39}$$

for the signal and

$$
\begin{aligned}
\hat{P}_n(\boldsymbol{\theta}) &= (NL)^{-1} \sum_{\ell=1}^{L} \|\mathbf{Y}_\ell - \mathbf{z}(\boldsymbol{\theta})\hat{S}_\ell(\boldsymbol{\theta})\|^2 \\
&= (NL)^{-1} \sum_{\ell=1}^{L} \left(\|\mathbf{Y}_\ell\|^2 - \frac{|\mathbf{z}^*(\boldsymbol{\theta})\mathbf{Y}_\ell|^2}{N} \right) \\
&= (NL)^{-1} \left(\sum_{\ell=1}^{L} \|\mathbf{Y}_\ell\|^2 - \frac{B(\boldsymbol{\theta})}{N} \right) \\
&= \frac{SSE(\boldsymbol{\theta})}{NL},
\end{aligned}
\tag{40}
$$

for the noise spectrum, where $SSE(\theta)$ is the sum of squares for error that appeared previously in Eq. (17).

This exhibits the log likelihood as a monotone function of the beam power, as before, and we obtain the maximum likelihood estimator as the $\hat{\theta}$ maximizing the power in the beam over the L frequencies in the neighborhood of ν_0. Of course, this assumes that the noise processes are uncorrelated from channel to channel. A likelihood ratio test of the hypothesis $S_1 = S_2 = \cdots = S_L = 0$ leads to using a detector proportional to

$$F(\hat{\theta}) = \frac{(N-1)}{N} \frac{B(\hat{\theta})}{SSE(\hat{\theta})}, \tag{41}$$

which has asymptotically an F distribution, as given in the next section. The detection statistic is exhibited in Eq. (41) as a function of the power in the beam divided by the noise power, when they are both evaluated at $\hat{\theta}$, the wavenumber maximizing the log likelihood in Eq. (38).

For the above deterministic signal case, the large L argument doesn't work but for large N, large T and fixed L, with the covariance matrix of the locations R converging to a limit, Wu (1982) obtained the large-sample distributions of the joint estimators for the velocity and azimuth, conditional on the signals, as a slightly modified version of Eq. (25), namely

$$\operatorname{cov}\begin{pmatrix} \hat{c} \\ \hat{\alpha} \end{pmatrix} \approx \frac{1}{2(2\pi)^2} \frac{1}{L} \frac{r}{N} \frac{1}{\|\theta_0\|^6} \begin{pmatrix} \nu^2 \theta_0' R^{-1} \theta_0 & -\nu\|\theta_0\|\theta' R^{-1} \bar{\theta}_0 \\ -\nu\|\theta_0\|\theta_0' R^{-1} \bar{\theta}_0 & \|\theta_0\|^2 \bar{\theta}_0' R^{-1} \bar{\theta}_0 \end{pmatrix} \tag{42}$$

with the signal power part of the noise-to-signal ratio, in this case, defined by

$$P_s = L^{-1} \sum_{\ell=1}^{L} |S_\ell|^2, \tag{43}$$

which is interpreted as the average signal power over the L frequencies.

Note that one might also write Eq. (37) in the form

$$Y_{j\ell} = \mu_{j\ell}(\theta) + V_{j\ell}$$

where

$$\log \mu_{j\ell}(\theta) = 2\pi i r_j' \theta + \log|A_\ell| + i\phi_\ell,$$

where A_ℓ and ϕ_ℓ are the real and imaginary parts of the signal. The mean is expressed as a linear function of the unknown parameters and Brillinger (1985) notes that this is in the form of a generalized linear model with link function given by the logarithm. The covariance matrix of the wavenumber vector $\hat{\boldsymbol{\theta}}$ can be computed from the usual generalized least squares approach but we do not pursue that direction here.

A more general model, in the fixed signal case, assumes that the noise is correlated from channel to channel, with $N \times N$ spectral matrix f_n. Replacing the spectrum P_n by f_n gives the modified log likelihood

$$\log L(\boldsymbol{\theta}, S_1, S_2, \ldots, S_L, P_n) \propto -NL \log |f_n|$$

$$- \sum_{\ell=1}^{L} (\mathbf{Y}_\ell - \mathbf{z}(\boldsymbol{\theta}) S_\ell)^* f_n^{-1} (\mathbf{Y}_\ell - \mathbf{z}(\boldsymbol{\theta}) S_\ell) \quad (44)$$

and we obtain the maximum likelihood estimator

$$\hat{S}_\ell = [\mathbf{z}^*(\boldsymbol{\theta}) f_n^{-1} \mathbf{z}(\boldsymbol{\theta})]^{-1} \mathbf{z}^*(\boldsymbol{\theta}) f_n^{-1} \mathbf{Y}_\ell \quad (45)$$

for the signal. For this case, we may obtain some classical estimators in the literature. Capon (1969) interprets Eq. (45) as the minimum variance estimator that passes the signal undistorted (BLUE) and proposes maximizing a sample version of the output power or variance of the signal estimator Eq. (45), namely

$$C(\boldsymbol{\theta}) = [\mathbf{z}^*(\boldsymbol{\theta}) M^{-1} \mathbf{z}(\boldsymbol{\theta})]^{-1} \quad (46)$$

where

$$M = \sum_{\ell=1}^{L} \mathbf{Y}_\ell \mathbf{Y}_\ell^* \quad (47)$$

is the observed spectral matrix. In cases where M is not of full rank $L < N$, one can replace M by $M + cI$, where c is a small regularization constant.

Brillinger (1985) has considered inference for the case where the signal is assumed to be present at ν_0, say, for $\ell = \ell'$ in the re-indexed frequencies and not present for $\ell \neq \ell'$. Performing the likelihood ratio test of no signal, i.e. $S_{\ell'} = 0$ leads to a test statistic of the form

$$D(\boldsymbol{\theta}) = \frac{\mathbf{z}^*(\boldsymbol{\theta}) S^{-1} \mathbf{z}(\boldsymbol{\theta})}{\mathbf{z}^*(\boldsymbol{\theta}) M^{-1} \mathbf{z}(\boldsymbol{\theta})} - 1, \quad (48)$$

where $S = M - \mathbf{Y}_{\ell'} \mathbf{Y}_{\ell'}^*$ denotes the sum over the noise frequencies. The test statistic in Eq. (48) is maximized over $\boldsymbol{\theta}$. Note that the test statistic is almost that of Capon (1970), where we scale $C(\boldsymbol{\theta})$ by a quadratic form with the estimated noise spectral matrix. Brillinger (1985) compares the performance of the above test statistics on contrived data with a noise correlation structure and a periodic signal at frequency ℓ'. Qualitative performances of $F(\boldsymbol{\theta}), C(\boldsymbol{\theta})$ and $D(\boldsymbol{\theta})$ improve over the beam power $B(\boldsymbol{\theta})$; the beam power and F-Statistic give the same maximum but the three aforementioned statistics have the advantages that their null distributions (see next Section) do not depend on any nuisance parameters, when the signal is absent.

There is a fundamental difference between the points of view taken by the models used by Brillinger (1985) and the models that are used in Shumway (1983) and advocated here. The Brillinger statistic $D(\theta)$ is derived as the outcome of a likelihood ratio test of signal or no-signal at the pure frequency $\ell'(\nu_0)$ and assumes that there will be no signal present at adjacent frequencies under either the null or alternative hypotheses. Such statistics will work extremely well when the signal is a pure frequency, such as a sine wave, but may not work quite as well when the signal frequencies are distributed over a band, say $\nu_0 \pm (L - 1)/2T$, which is the assumption made here. There is always the difficulty of deciding how broad a frequency band should be taken in order to capture significant signal power. Noise power estimators such as Eq. (40) are derived under the assumption that the signal is present at all frequencies in the narrow band and not under the assumption that it is restricted to a single frequency component. Of course, it is easy to modify the Brillinger statistic to encompass a broader frequency band by deducting additional adjacent frequencies from M, when computing S. In the next Section, we derive distributions under the stochastic signal assumptions and compare with those of the fixed signal case.

3. SIGNAL DETECTION

At the beginning of Sec. 2.1, we considered maximum likelihood estimation of the wavenumber vector θ and the associated velocity and azimuth coordinates c and α. The likelihood function in Eq. (13) was a monotone function of the beam power $B(\theta)$, so that the value θ maximizing the beam power is the maximum likelihood estimator. Under regularity conditions, $\hat{\theta} \to \theta_0$ almost surely as $L \to \infty$, and $L/T \to 0$ the asymptotic normality and covariance matrix in Eq. (24) for the wavenumber vector $\hat{\theta}$ followed.

However, a simpler *analysis of variance* method seems more reasonable in the special case where $f_s = P_s I, f_n = P_n I$ and we take that approach first. We summarize in the Sec. 3.2 a number of results that have been obtained under the assumption that the signal is an unknown fixed function.

3.1 Random Signal Models

We may apply the usual approach employed in random effects analysis of variance developments to estimate the signal and noise spectra under the random signal model that implies the spectral matrix in Eq. (11). This form contains the signal and noise spectra as unknown parameters, in addition to the wavenumber parameter vector $\boldsymbol{\theta}$. We apply the argument to the *power components* suggested by the terms in Eq. (41) in the the deterministic signal analysis. Arranging the numerator and denominator as in Table 1 below, we see that we need only verify the performance of the two components under the random signal assumptions.

Now, we have $SSE(\hat{\boldsymbol{\theta}}) \rightarrow SSE(\boldsymbol{\theta}_0)$ in probability and

$$SSE(\boldsymbol{\theta}_0) = \sum_{\ell=1}^{L} \| \mathbf{Y}_\ell - \mathbf{z}(\boldsymbol{\theta}_0)\frac{1}{N}\mathbf{z}^*(\boldsymbol{\theta}_0)\mathbf{Y}_\ell \|^2$$

$$= \sum_{\ell=1}^{L} \mathbf{N}_\ell^* \left(I - \frac{\mathbf{z}(\boldsymbol{\theta}_0)z(\boldsymbol{\theta}_0)^*}{N} \right) \mathbf{N}_\ell,$$

which is a sum of Hermitian forms in idempotent matrices of rank $(N-1)$, and we obtain

$$SSE(\hat{\boldsymbol{\theta}}) \rightarrow P_n \frac{\chi^2_{2L(N-1)}}{2}$$

in distribution. To handle the signal sum of squares component, we note that $B(\hat{\boldsymbol{\theta}}) \rightarrow B(\boldsymbol{\theta}_0)$, which is a sum of squares with components $\mathbf{z}^*(\boldsymbol{\theta}_0)(z(\boldsymbol{\theta}_0)S_\ell + \mathbf{N}_\ell)$, that are uncorrelated complex normals with mean zero and variances $N^2 P_s + N P_n$, implying that

$$\frac{B(\hat{\boldsymbol{\theta}})}{N} \rightarrow (P_n + NP_s)\frac{\chi^2_{2L}}{2}.$$

The two components are approximately independent because the linear forms $\mathbf{z}^*(\boldsymbol{\theta}_0)\mathbf{N}_\ell$ and $(I - \mathbf{z}(\boldsymbol{\theta}_0)\mathbf{z}^*(\boldsymbol{\theta}_0)/N)\mathbf{N}_\ell$ are uncorrelated.

The above computations imply that the statistic $F(\hat{\boldsymbol{\theta}})$ in Eq. (41), derived under the fixed signal assumption, has a limiting that is proportional to an F-distribution with $2L$ and $2L(N-1)$ degrees of freedom under the assumption that the signal is random. Note also that the F-Statistic can also be

Table 1   ANOVA

Source	Sum of squares	Approx. E(SS)	df
Signal	$N^{-1}B(\hat{\boldsymbol{\theta}}) = \dfrac{1}{N}\sum_{\ell} |\mathbf{z}^*(\hat{\boldsymbol{\theta}})\mathbf{Y}_\ell|^2$	$LP_n(\nu_0) + LNP_s(\nu_0)$	$2L$
Error	$SSE(\hat{\boldsymbol{\theta}}) = \sum_{\ell} \| \mathbf{Y}_\ell - \mathbf{z}(\hat{\boldsymbol{\theta}})\hat{S}_\ell \|^2$	$L(N-1)P_n(\nu_0)$	$2L(N-1)$

expressed as the ratio of the mean-square errors in Table 1 and is distributed as $cF_{2L,2L(N-1)}$ where the constant

$$c = \left(1 + N\frac{P_s}{P_n}\right) \tag{49}$$

is proportional to the number of sensors and the signal to noise ratio $1/r$. The constant becomes 1 when the signal power $P_s = 0$. We may also derive the approximate expected mean squares given in Table 1 through similar arguments, suggesting that the usual estimators in Eqns (16) and (18) are approximately unbiased for the noise and signal spectra respectively.

For the multivariate signal model with signal spectral matrix f_s, there are no simple ANOVA results but we may still derive a procedure for maximizing the log likelihood in Eq. (27) using either a Newton-Raphson approach and the vector scores and information matrix Eq. (30) or the EM algorithm on the complete data log likelihood. Now, regarding the generating model as

$$\mathbf{Y}_\ell = Z(\boldsymbol{\theta})\mathbf{S}_\ell + \mathbf{N}_\ell \tag{50}$$

where $Z(\boldsymbol{\theta})$ is the diagonal matrix defined below Eq. (26), and $\mathbf{S}_\ell = (S_{1\ell}, S_{2\ell}, \ldots, S_{L\ell})'$ is the random signal vector with spectral matrix f_s. Considering the complete data as the signal vector above plus the noise vectors $\mathbf{N}_1, \mathbf{N}_2, \ldots, \mathbf{N}_L$, we obtain

$$\log \tilde{L}(\boldsymbol{\theta}, f_s, P_n) \propto -L\log|f_s| - \sum_{\ell=1}^{L} \mathbf{S}_\ell^* f_s^{-1} \mathbf{S}_\ell$$
$$- LN\log P_n - \frac{1}{P_n}\sum_{\ell=1}^{L} \|\mathbf{N}_\ell\|^2$$

for the log likelihood. To maximize the incomplete data log likelihood in Eq. (27), we consider applying the Expected Conditional Maximization (ECM) algorithm of Meng and Rubin (1993). In this version of the EM algorithm, we apply the expectation and maximization steps for a given $\hat{\boldsymbol{\theta}}$, where $\hat{\boldsymbol{\theta}}$ is the maximizer for a fixed f_s, P_n. Note that the EM algorithm of Dempster Laird and Rubin (1977) is a rather convenient way of handling maximum likelihood estimation in multidimensional random signal context, see Shumway, (1988), Feder and Weinstein (1988), Böhme (1996). Now applying the E step of the EM algorithm to the complete data log likelihood above for a fixed $\boldsymbol{\theta}$ gives

$$\hat{f}_s = L^{-1}\sum_\ell E(\mathbf{S}_\ell \mathbf{S}_\ell^* | \mathbf{Y}_1, \ldots, \mathbf{Y}_L)$$
$$= L^{-1}\sum_\ell (\tilde{\mathbf{S}}_\ell(\boldsymbol{\theta})\tilde{\mathbf{S}}_\ell^*(\boldsymbol{\theta}) + \tilde{\Sigma}(\boldsymbol{\theta})) \tag{51}$$

for the next iterate of the signal spectral matrix, where

$$\tilde{S}_\ell(\theta) = E(S_\ell | Y_1, \ldots, Y_L)$$
$$= f_s Z^*(\theta)(Z(\theta)f_s Z^*(\theta) + P_n I)^{-1} Y_\ell$$
$$= (I + P_n f_s^{-1})^{-1} Z^*(\theta) Y_\ell \tag{52}$$

and

$$\tilde{\Sigma}(\theta) = \text{cov}(S_\ell | Y_1, \ldots, Y_L)$$
$$= f_s - f_s Z^*(\theta)(Z(\theta)f_s Z^*(\theta) + P_n I)^{-1} Z(\theta)f_s$$
$$= P_n(I + P_n f_s^{-1})^{-1}, \tag{53}$$

using standard matrix identities from Jazwinski (1970). The estimator for the noise spectrum will be

$$\hat{P}_n = (NL)^{-1} \sum_\ell (\|Y_\ell - Z(\theta)\tilde{S}_\ell(\theta)\|^2 + \text{tr}(Z(\theta)\tilde{\Sigma}(\theta)Z^*(\theta))). \tag{54}$$

To use the iterative ECM algorithm, one might use the following sequence of steps.

1. Determine an initial estimator for $\hat{\theta}$ as the maximizer of the beam power $B(\theta)$ in Eq. (15).
2. Compute initial estimators for $\hat{P}_n$ and $\hat{f}_s = \hat{P}_s I$, assuming no signal correlation, using Eqns (16) and (18).
3. Update $\tilde{S}_\ell(\hat{\theta})$ and $\tilde{\Sigma}(\hat{\theta})$ from Eqns (52) and (53).
4. Update $\hat{f}_s$ and $\hat{P}_n$ using Eqns (51) and (54).
5. Repeat steps 3 and 4 to convergence.
6. Update θ by finding the maximizer of $GB(\theta)$ in Eq. (29).
7. Return to 3.

The above procedures is rather involved and it is often the case that one can estimate the signal spectral matrix from prior data, assuming that the signal spectral matrix in known in the generalized beam formed estimator $GB(\theta)$. This is illustrated in Sec. 4 where we assign a functional form to $|\rho_{jk}|$, where $f_{jk}^s = P_s |\rho_{jk}^s|$ based on a theoretical relation between correlation and distance.

3.2   Deterministic Signal Models

Much attention has been focused on the fixed signal model in Eq. (37) under various assumptions on the noise vector N_ℓ. First, we note that Shumway (1983) has considered the fixed signal case under the assumption under the case that the noise is spatially white, i.e. $f_n = P_n I$. In this case, the likelihood ratio detector is of the form in Eq. (41) and consistency of the least squares estimators $\hat{\theta} \to \theta_0$ can be established by appealing to Wu (1982), where the signal now is bounded. It follows that the analysis of variance components in Table 1 have the same approximate expected sums of squares, except that P_s is now given by Eq. (43). Also,

$$SSE(\theta_0) = \sum_{\ell=1}^{L} \mathbf{Y}_{\ell}^{*} \left(I - \mathbf{z}(\theta_0) \frac{1}{N} \mathbf{z}^{*}(\theta_0) \right) \mathbf{Y}_{\ell}$$

is approximately $P_n \, \chi^2_{2L(N-1)}/2$ and

$$\frac{B(\theta_0)}{N} = \sum_{\ell=1}^{L} \mathbf{Y}_{\ell}^{*} \mathbf{z}(\theta_0) \frac{1}{N} \mathbf{z}^{*}(\theta_0) \mathbf{Y}_{\ell}$$

is approximately $P_n \, \chi^2_{2L}(\delta^2)/2$ where the non-centrality parameter is

$$\delta^2 = \frac{2NL}{r}, \tag{55}$$

which is clearly directly proportional to the number of sensors, N, the number of frequencies (bandwidth), L, and the noise-to-signal ratio. Under the hypothesis of no signal, the detector $F(\hat{\theta})$ has approximately the F distribution with $2L$ and $2L(N-1)$ degrees of freedom. Note that, under the spatially white assumption, the distribution of the statistic does not involve any nuisance parameters.

When there is a general noise correlation structure, say an $N \times N$ spectral matrix f_n. We note first that the conventional beam power $B(\theta)$, given by Eq. (15), is $\mathbf{z}^{*}(\theta) f_n^{-1} \mathbf{z}(\theta) \chi^2_{2L}/2$, which exhibits the approximate distribution as being proportional to a chi-squared distribution, where the proportionality constant depends on the unknown parameters, f_n and θ. Even in the case of spatially uncorrelated noise, the distribution will be $P_n \chi^2_{2L}/2$, again, depending on an unknown parameter. The same holds for Capon's high resolution estimator in Eq. (46), which is approximately $\chi^2_{2(L-N+1)}/\mathbf{z}^{*}(\theta) f_n \mathbf{z}(\theta)$ when the signal is absent for $\ell \neq \ell'$, see Capon and Goodman (1970). $C(\theta)$ has an additional problem, namely, that the distribution is singular unless $L > N$, i.e. there are more frequencies than sensors. Brillinger (1985) derives the statistic $D(\theta)$ in Eq. (48) as the outcome of a likelihood ratio test of no signal at frequency ℓ' and shows that the null distribution is $F_{2,2(N-L)}/(N-L)$ which does not depend on unknown parameters when the signal is absent. Again, it has the restriction that the number of series must be greater than the number of frequencies.

To summarize the case of an unknown deterministic signal, if the noise is spatially white, the F-Statistic $F(\theta)$ of Shumway (1983) has a known limiting distribution under both the null and alternative hypotheses. Under the correlated noise case, the Brillinger (1985) test statistic has a known limiting distribution under the null hypothesis. Neither of the two statistics has a null distribution depending on the values of unknown parameters so they will both have known critical values for setting false alarm probabilities.

4. INFRASONIC MONITORING OF NUCLEAR TESTING

The monitoring of nuclear testing using infrasonic arrays can provide useful information in addition to that provided by seismic arrays. Data consisting of accurate azimuthal estimates from a number of arrays can be combined with data from seismic arrays to give accurate confidence regions for the locations of the measured event. We focus here on the initial problem of detection and estimating azimuth and velocity from purely infrasonic signals observed at small arrays.

Of particularly crucial interest is the secondary problem of developing an array design for monitoring nuclear testing. In general, the main problems of interest are detecting an infrasonic signal and estimating the velocity and azimuth of the arrival, so that an event can be located. We seek optimal configurations for the number of stations and for their locations, i.e., choices for $\mathbf{r}_j = (x_j, y_j)'$, $j = 1, \ldots, N$ that produce enhanced detection performance and accurate resolution of the velocity and azimuth. The Infrasound Experts Group of the Geneva Conference on Disarmament has recommended an infrasound array design consisting of four elements, with three elements forming an equilateral triangle and the fourth at the center of the triangle. The Experts recommended that the sides of the triangle be in the range 1 to 3 km and these are the configurations that we focus on here. Complications that influence the optimal design are loss of signal correlation as the separation between the sensors increases.

4.1 Detection and Estimation

We look first at a simple example involving the three sensors in Fig. 1, that recorded the event Tanana at a small triangular array on Palmyra Island. To provide a baseline example, we took 2048 points (about 200 seconds) of data and re-aligned the signal to correspond with an approximate velocity of 0.3 km/sec and an approximate azimuth of 225°. The array geometry, shown later in Fig. 6, was a triangle with 1 km sides at locations $\mathbf{r}_1 = (0, 0.577)'$, $r_2 = (0.5, -0.289)'$ and $r_3 = (-0.5, -0.289)'$, leading to a diagonal covariance matrix R in Eq. (23) that has $1/6$ on the diagonals. The re-aligned signals at sensors A, B and C are shown in Fig. 3; the signal spectrum was centered at 0.044 Hz (cycles per second), i.e., at a period of

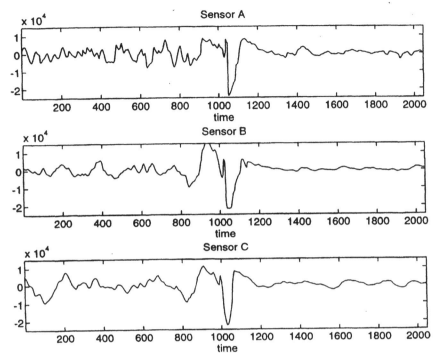

Figure 3. Small Fry series of Fig. 1 after aligning and delaying to correspond with an approximate velocity of 0.3 km/s. and an Azimuth of 225°. The array geometry is a triangle with 1 km sides at locations $A = (0, 0.577)$, $B = (0.5, -0.289)$, $C = (-0.5, -0.289)$.

about 23 s. The coherence between sensors was reasonably high, in this case, with values 0.86, 0.83 and 0.82 for the inter-sensor pairs **AB**, **AC** and **BC** respectively. Hence, there is reason to believe that random univariate signal model in Eq. (2) will work reasonably well.

In order for the large L asymptotics to have a chance, we first chose a band broad enough to include the entire signal, with $L = 17$ frequency ordinates in the band running from 0.005 to 0.08 Hz. The estimated signal-to-noise ratio $P_s/P_n = 1/r$, in this band, using the maximum likelihood estimators in Eqns (16) and (18) was 3.16. For the broad band, the estimator for the velocity was 0.26(0.03) km/sec, using Eq. (15) to get the estimator and Eq. (25) to get the standard deviation (in parentheses). This compares to the velocity of 0.30 km/sec that we had to input by lining up the largest minima of the signal on each sensor with this velocity. The azimuth estimator is 225(8)°, which is right on the known azimuth. The standard deviation implies an approximate 95 per cent confidence interval from 209 to 241°. For comparison, we tried a narrow band estimator, based on $L = 3$ frequency ordinates, in the band running from approximately 0.040 to 0.048 Hz. The signal-to-noise ratio in this band was 7.36 and the estimated velocity of 0.24(0.05) km/sec is still slightly off. Again, the azimuth estimator was 229(11)° with a slightly larger uncertainty. The increase in the signal-to-noise ratio in the narrower band does not completely compensate for the loss in bandwidth.

Figure 4 shows the relief plots (a) and (c) and contour plots (b) and (d) of the F-detector given by the F-Statistic in Eq. (41) which holds for both the deterministic and random signal models. The detector is plotted as a

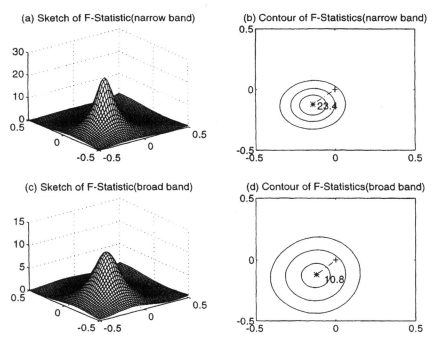

Figure 4. Relief (a) and (c) and contour (b) and (d) plots of F-Statistics for detecting infrasonic signal at 0.044 Hz and two bandwidths. Wavenumber coordinates range from −0.5 to 0.5 cycles per km. (b) Narrow band F shows peak at $\alpha = 229(11)°$ and velocity $c = 0.24(0.05)$ km/s. Contours from the peak are $F = 15, 9, 5$. (d) Broad band F shows peak at $\alpha = 225(8)°$ and velocity $c = 0.26(0.03)$ km/s. Contours from the peak are $F = 8, 4, 2$.

function of the wavenumber coordinates $-0.5 \leq \theta_1, \theta_2 \leq 0.5$, where the slowest velocity in the plot corresponds to the velocity on one of the edges, e.g. $\theta_1 = \theta_2 = 0.5$, or 0.062 km/sec, using Eq. (4). Very high velocities occur near the center $\theta_1 = \theta_2 = 0$, which corresponds to an infinite velocity. The azimuth can be seen visually as the angle made by a line drawn from the center to the maximum; here, it is 225°. Note that the asymptotics for this will be less sensitive to large L results, although one will still need to have the sample length T large. The plotted F-Statistic, for the narrow band case has $2L = 2(3) = 6$ and $2L(N - 1) = 6(2) = 12$ degrees of freedom. Note that $F_{6,12}(0.01) = 4.82$ and $F_{6,12}(0.001) = 8.38$ so that the 0.01 and 0.001 significance points are exceeded by all velocities and azimuths within the first two contours of the top plot. The plotted F-Statistic, for the broad band case has $2L = 2(17) = 34$ and $2L(N - 1) = 34(2) = 68$ degrees of freedom. Note that $F_{34.68}(0.01) = 1.95$ and $F_{34.68}(0.001) = 2.42$ so that the 0.01 and 0.001 significance points are exceeded by all velocities and azimuths within the first two contours of the bottom plot.

It is also of interest to determine the estimated velocities and azimuths and their uncertainties under the coherence structure estimated at the array. This requires estimating the variance of the beam formed estimator maximizing $B(\theta)$ under the assumption that there is less than perfect coherence using Eq. (36). This can be compared to the variance of generalized beam maximizing $GB(\theta)$, given by Eq. (32). Both equations require estimating the $N \times N$ signal spectral matrix $f_s(\nu)$ and the noise spectrum $P_n(\nu)$. In the case of the $\hat{\theta}$ maximizing the beam forming estimator, we used Eqns (51)–(53) and the EM steps 1 to 5 to obtain the maximum likelihood estimators for $f_x(\nu)$ and $P_n(\nu)$, with fixed at $\hat{\theta}$. For the case of $\hat{\theta}_b$, the maximizer of the general beam, we simply used the ECM algorithm, following Steps 1–7 as given. Figure 5 shows the results using a slightly broader band (L = 19) and we note that the estimators and their standard errors are nearly the same, a reflection, no doubt of the relatively high empirical coherence between the three sensors in this case.

The high coherence obtained for the case above will not always be the rule and it is of interest to consider the consequences of declining correlations that would be present if the sensors were placed further apart and if the signal to be detected appeared in different frequency bands. The next Section considers the effects of changing these input parameters using a theoretical model for coherence as a function of distance.

4.2 Array Design

It is of interest to determine the array configuration that might be nearly optimal for detection and azimuth estimation of surface and underground

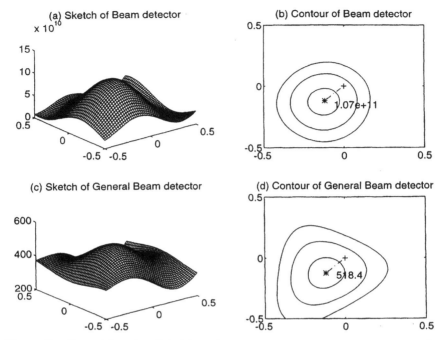

Figure 5. Broad band relief (a) and (c) and contour (b) and (d) plots of beam power and correlated beam power detectors. Wavenumber coordinate range from -0.5 to 0.5 cycles/km. (b) Beam power shows peak at $\alpha = 225(7)^\circ$ and velocity $c = 0.26(0.03)$ km/s. (d) Correlated detector shows peaks at $\alpha = 225(6)^\circ$ and velocity $c = 0.26(0.03)$ km/s.

explosions. We note that the analysis is complicated by the fact that the signal correlation will decrease as the separation between the sensors increases. Blandford (1996) has made extrapolations for coherence as a function of distance, using a model of Mack and Flinn (1971), for the distance ranges implied by an array that will have three elements arranged in a triangle with vertices separated by d kilometers and a center element. Figure 6 shows the locations of the sensors in the examples of the previous Section and the theoretical coherence as a function of separation and frequency assuming the Mack-Flinn model. It can be seen that coherence is reasonably high at low frequencies that are less than, say 0.5 Hz., but that predicted coherence goes down significantly at higher frequencies, particularly for the separation of

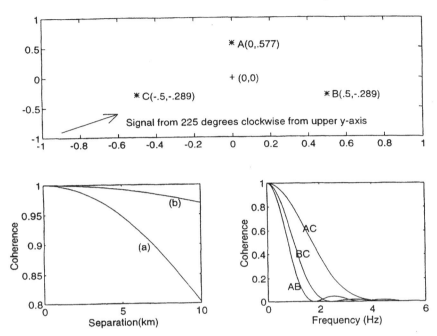

Figure 6. Sensor location (top) with coherence as a function of separation at (0.044 Hz) and frequency (bottom). Note the difference between coherence assumed for (a) parallel to wavefronts within 5° azimuth range and (b) normal to wavefronts within 0.01 km/s velocity range.

around 1 km that predominates for the triangular array under consideration.

It is interesting to compute the azimuthal standard deviations implied by the Mack-Flinn coherence model for various intersensor separations on the simple triangular array. The approach taken assumes that either the values maximizing the beam or the generalized beam are used, with the variances computed from the appropriate asymptotic expressions in Eqns (36) and (32) respectively. Figure 7 shows the results at various distance ranges and periods. It is clear that the beam and generalized beam behave similarly, except at the longer distance ranges. The signal to noise ratio $1/r$ was taken as 0.75^2, which is regarded as sufficient for an analyst to declare a detection if three channels with this signal to noise ratio are beamformed, resulting in a beam amplitude $1/r$ or 1.5. Bandwidths were $0.01, 0.02, 0.04$ and 0.2 Hz respectively for cases (a)–(d) in Fig. 7. Azimuthal uncertainties are larger for longer periods and actually get quite small for the higher frequencies. One

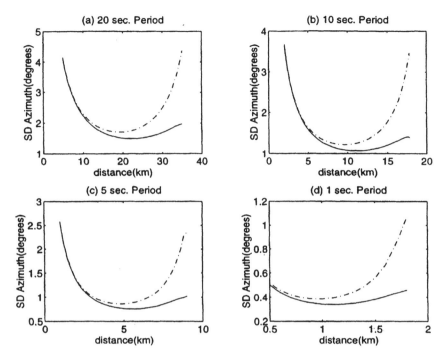

Figure 7. Estimated large sample standard deviation of azimuth estimators as a function of distance using the theoretical coherence model and triangular array. The solid line is for the weighted beam forming estimator and the dotted line is for the simple beam forming estimator.

second period standard deviations remain below 1° for intersensor distances less than one km. It is interesting to note that the minimum variance is obtained for separation distances that are exactly equal to the period, i.e. 20 km for a 20 s period, 10 km for a 10 s period, 5 km for a 5 s period and 1 km for a 1 s period.

Blandford (1996) also increases the number of elements, considering an $N = 7$ element array consisting of the array above plus a small inverted interior triangle, with 0.2 km on a side. With a signal-to-noise ratio of $(0.75)^2$, standard deviations were in the 0.5-2.0° range for periods less than 5 s. However, there is little improvement for 10 and 20 s periods. Blandford (1996) summarizes the computation by noting that for the 5 s period signal, which is expected from a 1 kt atmospheric nuclear explosion, the 1 Km array would have an azimuth estimation error approximately equal to the best historically observed residual error. The conclusions are somewhat preliminary because the underlying data is available only for much long periods and small inter-sensor distances. Analysis of additional signal data at more appropriate periods and spacings is of critical importance.

5. DISCUSSION

The approach of this chapter has been to concentrate on the case where it is known that either a common single signal is present on each sensor or that different signals appear with some unknown correlation structure. The signal is propagating at an unknown velocity from a generally unknown direction. The unknown parameters are related to the wavenumber vector parameter θ and these parameters appear nonlinearly in a random coefficient regression model. The random coefficients correspond to the Fourier transform of the unknown signals which have some known or unknown spectral matrix. For a fixed value of the parameter, say $\theta = \theta_0$, the model is exactly regression with a design matrix that depends on θ_0.

Hence, for the single perfectly correlated signal case, we were able to exploit analogies with the usual random effects analysis of variance problem in statistics to derive maximum likelihood estimators for the signal and noise spectra and to derive the distribution of an analog of the usual F-Statistic as the optimum likelihood ratio detector. Then, depending on the usual asymptotics, we argued that the maximizer of the log likelihood, or equivalently the maximizer of $B(\theta)$ the beam power yielded an estimator $\hat{\theta}$ converging in probability to θ_0, the true value, with an asymptotic variance covariance matrix depending on θ_0 and the signal and noise spectra P_s and P_n.

For the case where a slightly different signal is observed at each channel, with a coherence that may decline as a function of the array configuration, we derived the optimum likelihood ratio detector as the generalized beam $G(\theta)$. In this case, the signal spectral matrix f_s and noise spectrum P_n were estimated by mixing EM steps for a fixed $\hat{\theta}_0$ with conditional maximization steps to get updated $\hat{\theta}$ values. The asymptotic covariance matrix was a function here of θ_0, f_s and f_n. The popularity of beam forming $B(\hat{\theta})$ as an estimation method even when there may be non-identical signals led to deriving the asymptotic covariances of $\hat{\theta}$ under the assumption of a general spectral signal matrix f_s.

The application of the above theory to the infrasonic data showed the value of being able to compute the variance of the azimuth and velocity estimators under differing array configurations so that the design of such arrays under various correlation assumptions became feasible.

An additional problem of some interest that we do not consider in detail is the case where one may have more than a single signal observed at each sensor, for example, as in the model

$$y_j(t) = \sum_{k=1}^{p} s_k(t - T_j(\boldsymbol{\theta}_k)) + n_j(t), \tag{56}$$

where $T_j(\boldsymbol{\theta}_k)$ is the time delay experienced by the kth signal at the jth sensor. In the frequency domain, this implies that

$$\mathbf{Y}_\ell = \sum_{k=1}^{p} \mathbf{z}(\boldsymbol{\theta}_k) S_{k\ell} + \mathbf{N}_\ell. \tag{57}$$

In the deterministic signal case, Shumway (1983) proposed a stepwise approach for determining the number of signals and for estimating the wavenumber coordinates of each signal and gave an example of resolving two seismic signals. Such a stepwise approach is also feasible for the stochastic signal case using sequential likelihood ratio tests. In the stochastic signal case, the covariance matrix can be written in the form

$$f_y(\boldsymbol{\theta}) = Z(\boldsymbol{\theta}) f_s Z^*(\boldsymbol{\theta}) + P_n I \tag{58}$$

where we have suppressed the dependence on frequency and $Z(\boldsymbol{\theta}) = (\mathbf{z}(\boldsymbol{\theta}_1), \ldots, \mathbf{z}(\boldsymbol{\theta}_p))$. Another approach, developed in the engineering literature, notes that the above special structure implies that sum of the inner products of the last $N - p$ eigenvectors of $f_y(\boldsymbol{\theta})$ with any column of $Z(\boldsymbol{\theta})$ should be zero. Finding the values of $\boldsymbol{\theta}$ that minimize the resulting inner products gives the MUSIC estimators for $\boldsymbol{\theta}_1, \ldots, \boldsymbol{\theta}_p$. Wax (1991) has given a solution that involves a multiparameter search and the use of a model selection criterion on a log likelihood based on the above setup, see also Stoica and Nehorai (1989), Böhme (1996).

Acknowledgments

The authors are pleased to acknowledge the comments of two reviewers that led to improvements and clarifications of the original exposition. The research of the first two authors was supported, in part, by the Nuclear Treaty Programs Office under Contract DSWA01-97-C-0150, issued by the Defense Special Weapons Agency.

REFERENCES

Blandford, R.R., Design of infrasonic arrays. Air Force Technical Applications Commission Technical Report, Center for Monitoring Research, 1300 N. 17th St., Arlington, VA 22209, 1996.

Böhme, J. F., Retrieving signals from array data. ES Husebye AM Dainty ed. Monitoring a Comprehensive Test Ban Treaty. Boston: Kluwer Academic, 1996, pp 587–610.

Brillinger, D. R., A maximum likelihood approach to frequency-wavenumber analysis. *IEEE Trans. on Acoustics, Speech and Signal Processing. 33*: 1076–1085, 1985.

Capon, J., High resolution frequency wavenumber spectrum analysis. *Proc. IEEE 57*: 1408–1418, 1969.

Capon, J. N. and R. Goodman, Probability distributions for estimators of the frequency wavenumber spectrum. *Proc. IEEE 58*: 1785–1786, 1970.

Cook, R. K. and A. G. Bedard, On the measurement of infrasound. *Geophys. J. R. Astr. Soc. 26*: 5–11, 1971.

Dempster, A. P., N. M. Laird and D. B. Rubin (1977). Maximum likelihood from incomplete data via the EM algorithm. *J. Royal Statist. Soc., B, 39*: 1–38, 1977.

Feder, M. and E. Weinstein, Parameter estimation of superimposed signals using the EM Algorithm. *Trans. on Acoust. Speech and Sig. Proc. 36*: 477–489, 1988.

Hannan, E. J., Multiple Time Series. New York: Wiley, 1970, p. 249.

Harris, D. B., Comparison of the direction estimation performance of high-frequency seismic arrays and three-component stations. *Bull. Seismolog. Soc. Amer. 80*:1951–1968, 1990.

S. Haykin ed., Advances in Spectrum Analysis and Array Processing, Vol. II. Englewood Cliffs: Prentice Hall, 1991.

Hinich, M. J., Frequency-wavenumber array processing. *J. Acoust. Soc. Amer. 69*: 732–737, 1981

Hinich, M. J. and P. Shaman, Parameter estimation for an R-dimensional plane wave observed with additive independent Gaussian errors. *Ann. Math. Statist. 43*:153–169, 1972.

Jazwinski, A, H., Stochastic Processes and Filtering Theory. New York: Academic Press, 1970, pp. 261–262.

Mack, H. and Flinn, E. A., Analysis of the spatial coherence of short period acoustic gravity waves in the atmosphere. *Geophys. J.R. Astr. Soc. 26*: 255–269.

Meng, X. L. and D. B. Rubin, Maximum likelihood estimation via the ECM algorithm: A general framework. *Biometrika 80*: 267–278, 1993.

Robinson, P. M., Non-linear regression for multiple time series. *J. Appl. Prob. 9*: 758–768, 1972.

Shumway. R. H., Replicated time series regression: An approach to signal estimation and detection. Brillinger, Krishnaiah ed. Handbook of Statistics, Vol. 3, Time Series in the Frequency Domain. Amsterdam: North Holland, 1983, pp 383–408.

Shumway, R. H., Applied Time Series Analysis. Englewood Cliffs: Prentice Hall, 1988.

Simons, D. J., Atmospheric methods for nuclear test monitoring. E. S. Husebye and A. M. Dainty ed. Monitoring a Comprehensive Test Ban Treaty. Boston: Kluwer Academic, 1996, pp 135–141.

Stoica, P. and A. Nehorai, MUSIC, maximum likelihood and Cramer-Rao lower bound. *IEEE Trans. Acoust. Speech and Sig. Proc. 37*: 720–751, 1989.

Stoica, P. and A. Nehorai, Performance study of conditional and unconditional direction of arrival estimation. *IEEE Trans. Acoust. Speech and Sig. Proc. 38*: 1783–1795, 1990.

Thomson, P. J., Signal estimation using an array of recorders. *Stoch. Proc. Appl. 13*: 201–214, 1982.

Wax, M., Detection and localization of multiple sources via the stochastic signal model. *IEEE Trans. Sig. Proc. 39*: 2450–2456, 1991.

Wu. J. S.-H., Asymptotic properties of nonlinear least squares estimators in a replicated time series model. Ph.D Dissertation, The George Washington University, Washington, DC, 1982

8

Some Contributions to Multivariate Nonlinear Time Series and to Bilinear Models

T. SUBBA RAO and W. K. WONG* University of Manchester Institute of Science and Technology, Manchester, England

1. INTRODUCTION

We define higher-order cumulants for random vectors and matrices, and use these definitions for evaluating higher-order cumulants of stationary vector time series admitting a linear representation. These expressions are useful for obtaining tests for Gaussianity and linearity of vector time series. We consider multivariate extensions of well-known bilinear models and show that they satisfy Yule–Walker type equations in high-order cumulants. The estimation of the parameters (using frequency domain approach) of multivariate bilinear models is considered. The methods are illustrated with real time series.

It is well-known that many time series encountered in practice are nonlinear and nonGaussian. For example, the Wolfer's sunspot numbers and Canadian Lynx data have been shown to be nonlinear and nonGaussian, see Subba Rao and Gabr (1980), Tong (1990) and Priestley (1988). Both frequency domain and time domain approaches have been developed for the analysis of nonlinear time series. Higher order spectra (cumulant spectra) play an important role in the frequency domain analysis of nonGaussian time series, and in particular statistical tests for linearity and Gaussianity, see Subba Rao and Gabr (1980). So far the analysis

*Now at the University of Stirling, Scotland

is restricted mainly to the univariate time series. In recent years, attempts have been made to extend these methods to multivariate case. Though not very extensive and exhaustive as in the univariate case, the research literature on multivariate Gaussian case is quite extensive and well documented, see the books by Hannan (1970), Lütkepohl (1991) and Reinsel (1993).

Our object here is to describe new results in the analysis of multivariate (nonGaussian) time series and in particular bilinear models. In Sec. 2, we obtain expressions for cumulants of random matrices which are extensions of the results well known in univariate case, Leonov and Shiryaev (1959). Using the results obtained in Sec. 2, we derive expressions for higher-order cumulant spectra of vector time series satisfying a linear process representation. This is analogous to the well-known relation given by Bartlett, Brillinger and Rosenblatt in the univariate case. This relation can be used for constructing a test for linearity of vector time series similar to the Subba Rao and Gabr (1980) test. In Sec. 4, symmetries of cumulants of random vectors are discussed. Multivariate bilinear models are introduced in Sec. 5, and the higher order properties of bilinear processes are studied in Sec. 6 and we show that they satisfy Yule–Walker type difference equations. A test for Gaussianity is briefly discussed in Sec. 7. The estimation of the parameters of the bilinear models is considered in Sec. 8 and the methods are illustrated with real data in Sec. 9.

2. CUMULANTS OF RANDOM MATRICES

To study statistical properties of random vectors and random matrices, we need to define and evaluate higher order moments and cumulants. The results obtained here will be of general interest. In later Sections of this paper, we will be considering properties of multivariate bilinear time series models. The second-order properties of bilinear models are similar to linear ARMA models, and hence there is a need to calculate higher order cumulants. Though we have not considered here, if we wish to extend the nonlinear canonical correlation technique derived by Subba Rao and da Silva (1992) to multivariate case, the higher-order moments derived here will be very useful.

DEFINITION 2.1   Let $\mathbf{X} = (x_1, x_2, \ldots, x_k)'$ be a $k \times 1$ random vector, (where x_i's are real) with distribution function F_x and moment generating function $\Phi_x(t)$ given by

$$\Phi_x(t) = E(e^{t'\mathbf{X}})$$

where $t' = (t_1, t_2, \ldots, t_k)$. The kth order joint cumulant is defined as

$$\text{cum}(x_1, x_2, \ldots, x_k) = k! \frac{\partial^k}{\partial t_1 \partial t_2 \ldots \partial t_k} \{\ln \Phi_x(t)\}_{t=0} \tag{2.1}$$

In other words, the kth order cumulant is the kth coefficient of $t_1 t_2, .., t_k/k!$ in the Taylor series expansion of $\ln \Phi_x(t)$ about the origin. One can show, Leonov and Shiryaev (1959), Brillinger (1965) that

$$\text{cum}(x_1, x_2, \ldots, x_k) = \Sigma(-1)^{p-1}(p-1)! \left(E \prod_{j \in \nu_1} x_j \right) \ldots E\left(\prod_{j \in \nu_p} x_j \right) \tag{2.2}$$

where the summation extends over all partitions $\{\nu_1, \nu_2, \ldots, \nu_p\}$, $p = 1, 2, \ldots, k$ of $\{1, 2, \ldots, \ldots k\}$.

It is well-known that cumulants are semi-invariants, that is, the change of origin only affects the mean (or the first cumulant) not the higher order cumulants. We observe

$$\text{cum}(x_i) = E(x_i), \text{cum}(x_i, x_i) = \text{var}(x_i), \text{cum}(x_i, x_j) = \text{cov}(x_i, x_j)$$

Further, the following properties, see Rosenblatt (1966), Brillinger and Rosenblatt (1967) are often used in later sections.

Let $\mathbf{X} = (x_1, x_2, \ldots, x_k)'$ and $\mathbf{Y} = (y_1, y_2, \ldots, y_k)'$ be two random vectors whose cumulants exist. Then

1. If $a_1, a_2, \ldots, a_k$ are constants, then

$$\text{cum}(a_1 x_1, a_2 x_2, \ldots, a_k x_k) = a_1 a_2, \ldots, a_k \text{cum}(x_1, x_2, \ldots, x_k)$$

2. $\text{cum}(x_1, x_2, .., x_k)$ is symmetric in its arguments
3. If any group of the $\mathbf{X}$'s are independent of the remaining X's, then

$$\text{cum}(x_1, x_2, \ldots, x_k) = 0$$

4. $\text{cum}(x_1 + y_1, x_2, \ldots, x_k) = \text{cum}(x_1, x_2, \ldots, x_k) + \text{cum}(y_1, x_2, \ldots, x_k)$
5. If $\mathbf{X}$ and $\mathbf{Y}$ are independent, then

$$\text{cum}(x_1 + y_1, x_2 + y_2, \ldots, x_k + y_k)$$
$$= \text{cum}(x_1, x_2, \ldots, x_k) + \text{cum}(y_1, y_2, \ldots, y_k).$$

Now consider the L random variables

$$y_l = \prod_{k=1}^{k_l} x_{lk}, \qquad l = 1, 2, \ldots, L. \tag{2.3}$$

If all the necessary cumulants of x_{lk} exist, then (see Brillinger, 1975).

$$\text{cum}(y_1, y_2, \ldots, y_k) = \sum_\nu \text{cum}(x_{ij}; ij \in \nu_1) \ldots \text{cum}(x_{ij}, ij \in \nu_p) \tag{2.4}$$

where the summation is over all indecomposable partitions of $\{11, 12, \ldots, 1k_1 | \ldots | L1, L2 \ldots, Lk_L\}$.

We are interested here in the extension of the above results, in particular Eq. (2.4), to random matrices which are Kronecker products of another set of random matrices.

DEFINITION 2.2   Let $A = (a_{ij})$ be an $m \times n$ matrix, and let $B = (b_{ij})$ be a $p \times q$ matrix. Then the Kronecker product of A and B is defined as

$$A \otimes B = \begin{bmatrix} Ab_{11} & \cdots & Ab_{1q} \\ & \vdots & \\ Ab_{p1} & \cdots & Ab_{pq} \end{bmatrix} \tag{2.5}$$

The matrix Eq. (2.5) which is of order $mp \times nq$ can be written as

$$A \otimes B = [a_{i_1 j_1} b_{i_2 j_2}] \tag{2.6}$$

where $i_1 j_1$ and $i_2 j_2$ are in lexicographic order with i_2 or j_2 increasing by one when i_1 or j_1 completes a cycle.

The position of the element $a_{i_1 j_1} b_{i_2 j_2}$ in the matrix $A \otimes B$ given by Eq. (2.6) corresponds to the coordinate $(i_1 i_2, j_1 j_2)$ where $i_1 i_2$ and $j_1 j_2$ refer, respectively, to the $((i_2 - 1)m + i_1)$ th row and $((j_2 - 1)n + j_1)$ th column. Now we can define the joint cumulants of random matrices.

DEFINITION 2.3   Let $X^{(l)} = [x_{ij}^{(l)}]$ be a random matrix of order $m_l \times n_l$, $(l = 1, 2, \ldots, k)$. If the kth order scalar joint cumulants $\text{cum}(x_{i_1 j_1}^{(l_1)}, x_{i_2 j_2}^{(l_2)} \ldots x_{i_k j_k}^{(l_k)})$ is denoted by $\kappa_{i_1 i_2 \ldots i_k, j_1 j_2 \ldots j_k}^{(l_1, l_2, \ldots, l_k)}$ (and if it exists, then we define the kth order joint cumulant matrix of order $\prod_{j=1}^{k} m_{lj} \times \prod_{j=1}^{k} n_{lj}$

$$\text{cum}\{X^{(l_1)}, X^{(l_2)}, \ldots, X^{(l_k)}\} = [\kappa_{i_1 i_2 \ldots i_k, j_1 j_2 \ldots j_k}^{(l_1, l_2, \ldots, l_k)}] \tag{2.7}$$

where the subscripts $i_1 i_2 \ldots i_k$ and $j_1 j_2 \ldots, j_k$ corresponds to the element of the $(i_1 + \sum_{l=2}^{k}(i_l - 1)m_l)$th row, $(j_1 + \sum_{l=2}^{k}(j_l - 1)n_l)$ th column.

THEOREM 2.1   Let $X^{1\ell_1}, X^{2\ell_2} \ldots, X^{k\ell_k}$ be a sequence of random matrices, where each $X^{j\ell_j}$ is of dimension $p_j \ell_j \times q_j \ell_j$. Let $A^{j\ell_j}$ and $B^{j\ell_j}$ be matrices of constants and of dimensions $r_j \ell_j \times p_j \ell_j$ and $q_j \ell_j \times s_j \ell_j$ respectively. Then

$$\text{cum}\left\{\sum_{l_1=1}^{r_1} A^{1l_1} X^{1l_1} B^{1l_1}, \ldots, \sum_{l_k=1}^{r_k} A^{kl_k} X^{kl_k} B^{kl_k}\right\}$$

$$= \sum_{\ell_1=1}^{r_1} \cdots \sum_{\ell_k=1}^{r_k} (A^{(1l_1)} \otimes \cdots \otimes A^{(kl_k)}) \operatorname{cum}\{X^{1l_1}, \ldots, X^{kl_k}\}$$

$$\times \{B^{1l_1} \otimes \cdots \otimes B^{kl_k}\} \tag{2.8}$$

where X^{jl_j} is a $p_j l_j \times q_j l_j$ random matrix, $A^{(jl_j)}$ and $B^{(jl_j)}$ are constant matrices of dimensions $r_j l_j \times p_j l_j$ and $q_j l_j \times s_j l_j$ respectively.

The proof of the above Theorem is straightforward, and is given in Wong's PhD thesis submitted to UMIST (1993).

In the univariate case, $\operatorname{cum}(x_1, x_2, \ldots, x_k)$ is symmetric about its arguments, i.e. the cumulant remains the same regardless of the order of its arguments. Permutation matrices, see MacRae (1974), Magnus and Neudecker (1988) are widely used to change the order of the Kronecker products of matrices.

Let $I_{(p,q)}$ be a $pq \times pq$ matrix partitioned into $p \times q$ submatrices such that (ij)th submatrix has 1 in position (ji) and 0's elsewhere. $I_{(p,q)}$ is called the permuted identity matrix and can be used to change the order of a Kronecker product. Then the following results hold.

Let A be an $m \times n$ matrix, B be a $p \times q$ matrix, b is a $p \times 1$ column vector. Then

$$I_{(m,p)}(A \otimes B)I_{(q,n)} = B \otimes A$$

$$I_{(m,p)}(A \otimes b) = (b \otimes A)$$

$$I_{(p,m)}(b \otimes A) = (A \otimes b) \tag{2.9}$$

For proof of the above see Magnus and Neudecker (1988).

Also we have $I_{(p,q)}^{-1} = I_{(q,p)}$ and $I_{(p,q)} \operatorname{vec}(\mathcal{B}) = \operatorname{vec}(\mathcal{B}')$, where vec operator transforms a matrix into a vector by stacking the columns of the matrix one underneath the other.

An immediate application of the above is that if A and B are two matrices defined above but random then

$$I_{(m,p)} \operatorname{cum}(A, B)I_{(q,n)} = \operatorname{cum}(B, A) \tag{2.10}$$

In general one can show that if $\nu = \{1, 2, \ldots, k\}$ and ν' contains the same elements as ν but in a different order, then

$$\operatorname{cum}\{X^{(l)}; l \in \nu'\} = P \operatorname{cum}\{X^{(l)}; l \in \nu\} Q \tag{2.11}$$

where P and Q are suitable permutation matrices. These results have been used by Jou and Shaman (1988) to obtain higher order moments of Bilinear models. The following result is a generalisation of Eq. (2.3) given by Leonov and Shiryaev (1959), for details see Wong (1993).

THEOREM 2.2   Let $\mathbf{Y}^{(l)} = \mathbf{X}^{(l_1)} \otimes \cdots \otimes \mathbf{X}^{(l_l)}$ $(l = 1, 2, \ldots, L)$ where $\mathbf{X}^{(m_n)}$ are random matrices of arbitrary finite dimensions. Then

$$\text{cum}\{\mathbf{Y}^{(1)}, \mathbf{Y}^{(2)}, \ldots, \mathbf{Y}^{(L)}\}$$

$$= \sum_{r=1}^{m} P_r[\text{cum}\{\mathbf{X}^{(m_n)}, m_n \in \nu^r\}$$

$$\otimes \cdots \otimes \text{cum}\{\mathbf{X}^{(m_n)}, m_n \in \nu_{p_r}^{(m)}\}]Q_r \tag{2.12}$$

where P_r and Q_r are permutation matrices, and the ordered set $\nu^{(r)} = \{\nu_1^{(r)}, \ldots, \nu_{p_r}^{(r)}\}$ are the indecomposable partitions with respect to the original partitions

$$\nu^* = \{l_1, l_2, \ldots, l_{k_1} | \ldots . | l_1, l_2, \ldots, l_{k_L}\}$$

Proof.   See Wong (1993, p35).

    To illustrate the result of Eq. (2.12) consider the evaluation of $\text{cum}\{X^{(1)} \otimes X^{(2)}, X^{(3)}\}$, where the random matrices $X^{(l)}, l = 1, 2$ and 3 are all of dimensions $d \times d$. The original partition is $\{12|3\}$ and the indecomposable partitions are $\{123\}$, $\{1|23\}$ and $\{13|2\}$. If $I_{d \times d}$ is a $d \times d$ identity matrix and $I_{(d \times d)}$ is the $d^2 \times d^2$ permuted identity matrix such that $I_{(d.d)}\{X^{(3)} \otimes X^{(2)}\}I_{(d.d)} = X^{(2)} \otimes X^{(3)}$, then from Eq. (2.12), we get

$$\text{cum}\{X^{(1)} \otimes X^{(2)}, X^{(3)}\} = \text{cum}\{X^{(1)}, X^{(2)}, X^{(3)}\}$$

$$+ \text{cum}\{X^{(1)}\} \otimes \text{cum}\{X^{(2)}, X^{(3)}\}$$

$$+ P[\text{cum}\{X^{(1)}, X^{(3)}\} \otimes \text{cum}\{X^{(2)}\}]Q \tag{2.13}$$

where $P = Q = I_{d \times d} \otimes I_{(d.d)}$.

    We now use the above results to derive expressions for the kth order cumulants of d dimensional stationary time series $\{X_t\}$.

    Let $\{\mathbf{X}_t\}$ be a $d \times 1$ stationary (up to kth order) vector time series. Let us denote the kth order scalar cumulant $\text{cum}\{X_{i_1 t_1}, X_{i_2 t_2}, \ldots, X_{i_k t_k}\}$ by $k_{i_1 i_2 \ldots i_k}$ $(t_1 - t_k, \ldots, t_{k-1} - t_k)$. Then following Eq. (2.7), we have the kth order cumulant vector (of order $d^k \times 1$), as

$$\text{cum}(\mathbf{X}_{t_1}, \mathbf{X}_{t_2}, \ldots, \mathbf{X}_{t_k}) = \mathcal{K}(t_1 - t_k, t_2 - t_k, \ldots, t_{k-1} - t_k)$$

$$= [k_{i_1 i_2 \ldots i_k}(t_1 - t_k, \ldots, t_{k-1} - t_k)] \tag{2.14}$$

where the subscripts, $i_1 i_2 \ldots i_k$ indicate the position in the column vector. For example if $d = 2$ and $k = 3$, then writing $k_{i_1 i_2 i_3}(s_1, s_2)$ as $k_{i_1 i_2 i_3}$, we have the vector

$$\mathcal{K}(s_1, s_2) = (k_{111}, k_{112}, k_{121}, k_{122}, k_{211}, k_{212}, k_{221}, k_{222}) \tag{2.15}$$

Now we can define the kth order cumulant spectral vector of dimension $d^k \times 1$ as the Fourier transform of Eq. (2.14), and is given by (if it exists)

$$f(\omega_1, \omega_2, \ldots, \omega_{k-1}) = \frac{1}{(2\pi)^{k-1}} \cdot \sum_{s_1, s_2, \ldots, s_{k-1} = -\infty}^{\infty} \mathcal{K}(s_1, s_2, \ldots, s_{k-1})$$

$$\times \exp\left[-i \sum_{j=1}^{k-1} s_j \omega_j\right] \tag{2.16}$$

where $\sum_{j=1}^{k} \omega_j = 0 \pmod{2\pi}$, see Brillinger and Rosenblatt (1967).

Suppose $d = 2$, and $\gamma(s) = \mathrm{cum}(\mathbf{X}_t, \mathbf{X}_{t+s})$ and the second-order spectral matrix $g(\omega)$ of order 2×2 is given by

$$g(\omega) = \frac{1}{2\pi} \sum_{-\infty}^{\infty} \gamma(s) e^{-is\omega} \tag{2.17}$$

Then the second order spectral vector $(2^2 \times 1)$ is related to $g(\omega)$ by the relation

$$f(\omega) = \mathrm{vec}(g(\omega)) \tag{2.18}$$

3. CUMULANT PROPERTIES OF LINEAR PROCESSES

Let $\{X_t\}$ be a d dimensional stationary time series, and we say X_t is linear if it can be represented as

$$X_t = \sum_{j=-\infty}^{\infty} A_j \mathbf{e}_{t-j} \tag{3.1}$$

where $\{e_t\}$ is a sequence of independent, identically distributed random vectors such that

$$\mathrm{cum}(e_t) = E(e_t) = \mathbf{0}$$

$$\mathrm{cum}(e_{t_1}, e_{t_2}, \ldots, e_{t_n}) = \begin{cases} \mathbf{C}_n & \text{if } t_1 = t_2 = \ldots = t_n \\ 0 & \text{otherwise} \end{cases}$$

where $\mathbf{C}_n$ is a column vector of order $d^n \times 1$ (see Eq. (2.14)). Since X_t and e_t are stationary, they have the spectral representations:

$$X_t = \int_{-\pi}^{\pi} e^{it\omega} \, dZ_x(\omega), e_t = \int_{-\pi}^{\pi} e^{it\omega} \, dZ_e(\omega) \tag{3.2}$$

where $Z_x(\omega)$ and $Z_e(\omega)$ are respectively the vector processes of orthogonal increments of X_t and $\mathbf{e}_t$. From Eqns (3.1) and (3.2) we obtain

$$dZ_x(\omega) = H(\omega) \, dZ_e(\omega) \tag{3.3}$$

where $H(\omega) = \Sigma A_j e^{-ij\omega}$. From Eq. (3.1) we have the covariance matrix

$$\gamma(s) = E(X_{t+s} - \mu)(X_t - \mu)' = \sum_{j=-\infty}^{\infty} A_{j+s} \Sigma A_j' \qquad (3.4)$$

where

$$\Sigma = E(e_t e_t') \text{ and } C_2 = \text{vec}(\Sigma).$$

Substituting for $\gamma(s)$ from Eq. (3.4) into Eq. (2.17), we obtain,

$$g(\omega) = \frac{1}{2\pi} H(\omega) \Sigma H^*(\omega),$$

$$f(\omega) = \text{vec}(g(\omega)) = \frac{1}{2\pi} (H(-\omega) \otimes H(\omega)) \text{vec}(\Sigma). \qquad (3.5)$$

The above second-order properties of the linear processes can be generalised to higher-order, and they are given in the following lemma.

LEMMA 3.1   Let $\{X_t\}$ be a stationary process satisfying Eq. (3.1) Then

(a)   the kth order cumulant vector is

$$\mathcal{K}(s_1, s_2, \ldots, s_{k-1}) = \Sigma \{A_{j+s_1} \otimes A_{j+s_2} \ldots \otimes A_j\} C_k \qquad (3.6)$$

(b)   the kth order spectral vector is

$$f_k(\omega_1, \omega_2, \ldots, \omega_{k-1}) = \frac{1}{(2\pi)^{k-1}} \{H(\omega_1) \otimes \ldots \otimes H(\omega_k)\} C_k \qquad (3.7)$$

where $\omega_k = -\sum_{j=1}^{k-1} \omega_j$

Proof.   From Eq. (3.1), we have, the kth order cumulant vector

$$\mathcal{K}(s_1, s_2, \ldots, s_{k-1}) = \text{cum}\{X_{t+s}, X_{t+s_2}, \ldots, X_{t+s_{k-1}}, X_t\}$$

$$= \text{cum}\left\{\sum_{j_1} A_{j_1} e_{t+s_1-j_1}, \ldots, \sum_{j_k} A_{j_k} e_{t-j_k}\right\}.$$

Using Theorem 2.1 and observing that X_t is a column vector and $\mathbf{B}$ is a scalar of unit value, we can show that the above cumulant is equal to

$$= \sum_{j_1, j_2, \ldots, j_n} \{A_{j1} \otimes A_{j_2} \ldots \otimes A_{j_k}\} \text{cum}\{e_{t+s_1-j_1}, \ldots, e_{t+s_{k-1}-j_{k-1}}, e_{t-j_k}\}$$

$$= \sum \{A_{j+s_1} \otimes \mathbf{A}_{j+s_2} \ldots \otimes A_j\} C_k$$

We can obtain Eq. (3.7) by substituting for $\mathcal{K}(s_1, s_2 \ldots, s_{k-1})$ from Eq. (3.6) into Eq. (2.16). Alternatively, we can obtain the relation Eq. (3.7) as follows:

From Eq. (3.3) we have

$$\text{cum}\{dZ_x(\omega_1), dZ_x(\omega_1), \ldots, dZ_x(\omega_k)\}$$

$$= \text{cum}\{H(\omega_1)dZ_e(\omega_1), \ldots, H(\omega_k)dZ_e(\omega_k)\} \tag{3.8}$$

Now using Theorem (2.1), we can show that the right-hand-side expression of Eq. (3.8) is equal to

$$= \{H(\omega_1) \otimes H(\omega_2) \ldots \otimes H(\omega_k)\} f_{k.e}(\omega_1, \omega_2 \ldots, \omega_k) \, d\omega_1 d\omega_2 \ldots d\omega_k$$

where $\omega_k = -\sum_{j-1}^{k-1} \omega_j$. Since $\{e_t\}$ are iid vectors, we have $f_{k.e}(\omega_1, \omega_2, \ldots, \omega_k) = (2\pi)^{1-k}.C_k$ and hence the result Eq. (3.7). Using Eqns (3.5) and (3.7), we arrive at the following useful result.

THEOREM 3.1 Let the time series $\{X_t\}$ satisfy Eq. (3.1). Then

$$f^*(\omega_1, \omega_2, \ldots, \omega_{k-1})\{g(\omega_1) \otimes g(\omega_2) \ldots \otimes g(\omega_k)\}^{-1} f(\omega_1, \omega_2, \ldots, \omega_{k-1})$$

$$= \frac{1}{(2\pi)^{k-2}}.C_k'\{\Sigma \otimes \Sigma \ldots \otimes \Sigma\}^{-1} C_k \tag{3.9}$$

where $\omega_k = -\sum_{j=1}^{k-1}$.

Proof. Consider the relation in Eq. (3.5)

$$\mathbf{g}(\omega) = \frac{1}{2\pi} H(\omega)\Sigma H^*(\omega),$$

and consider the Kronecker product of the matrices

$$(2\pi)^k(g(\omega_1) \otimes g(\omega_2) \ldots \otimes g(\omega_k))$$

$$= \{H(\omega_1)\Sigma H^*(\omega_1)\} \otimes \ldots \otimes \{H(\omega_k)\Sigma H(\omega_k)\}$$

$$= \{H(\omega_1) \otimes H(\omega_2) \otimes \ldots H(\omega_k)\}\{\Sigma \otimes \Sigma \otimes \ldots \otimes \Sigma\}$$

$$\times H^*(\omega_1) \otimes H^*(\omega_2) \ldots \otimes H^*(\omega_k)\}$$

Hence

$$\{g(\omega_1) \otimes g(\omega_2) \otimes \ldots \otimes g(\omega_k)\}^{-1} = (2\pi \mathcal{H}_2^{-1}\{\Sigma^{[k]}\}^{-1} \mathcal{H}_1^{-1} \tag{3.10}$$

where $\mathcal{H}_1 = \mathcal{H}(\omega_1) \otimes H(\omega_2) \ldots \otimes H(\omega_k)$, $\mathcal{H}_2 = \mathcal{H}_1^*$ and $\Sigma^{[k]} = \Sigma \otimes \Sigma \ldots \otimes \Sigma$. From Eq. (3.7) we have

$$f_k(\omega_1, \omega_2, \ldots, \omega_k) = \frac{1}{(2\pi)^{k-1}} \mathcal{H}_1 C_k \quad \text{and}$$

$$f_k^*(\omega_1, \omega_2, \ldots, \omega_k) = \frac{1}{(2\pi)^{k-1}} C_k' \mathcal{H}_2 \tag{3.11}$$

The result in Eq. (3.9) follows from Eqns (3.10) and (3.11). It is interesting to consider the univariate case, *i.e* $d = 1$.

SPECIAL CASE   Consider the case when $d = 1$ and $k = 3$. We have

$$g(\omega) = \frac{\sigma_e^2}{2\pi} |H(\omega)|^2,$$

$$f_2(\omega_1, \omega_2) = \frac{1}{2\pi} H(\omega_1)H(\omega_2)H(\omega_3)C_3 \qquad (3.12)$$

where $\omega_3 = -\omega_1 - \omega_2$. From Eq. (3.12), we obtain

$$\frac{|f_2(\omega_1, \omega_2)|^2}{g(\omega_1)g(\omega_2)g(-\omega_1 - \omega_2)} = \text{constant.} \qquad (3.13)$$

The result in Eq. (3.13) is well known and is used for testing linearity of the stationary time series by Subba Rao and Gabr (1980). The result is true for any $k > 3$.

What the result in Eq. (3.9) tells us is that if the vector time series $\{X_t\}$ is linear in the sense that it has the representation in Eq. (3.1), then the normalised function defined by the left-hand-side expression Eq. (3.9) is always a constant. This can be considered as a linear characterisation in terms of higher order spectra for vector time series. Recently Wong (1997) has purposed frequency domain tests for Gaussianity and linearity.

Now let us define a scalar random process $y_t = \alpha' X_t$, where α is a column vector of constants and X_t is defined as in Eq. (3.1). Let $g_y(\omega)$ be the second-order spectral density function of $\{y_t\}$ and let $f_y(\omega_1, \omega_2, \ldots, \omega_{k-1})$ be the kth order cumulant spectral density function. Then we have

$$g_y(\omega) = \alpha' g_x(\omega)\alpha$$

$$f_y(\omega_1, \omega_2, \ldots, \omega_{k-1}) = (\alpha^{[k]})' f_{k,x}(\omega_1, \omega_2, \ldots, \omega_{k-1}) \qquad (3.14)$$

where $g_x(\omega)$ is given by Eq. (3.5) and $f_{k,x}(\omega_1, \omega_2, \ldots, \omega_{k-1})$ is a column vector given by Eq. (3.7) and $\alpha^{[k]} = \alpha \otimes \alpha \ldots \otimes \alpha$

From Eq. (3.14), it is clear that

$$\frac{|f_y(\omega_1, \omega_2, \ldots \omega_{k-1})|^2}{g_y(\omega_1)g_y(\omega_2) \ldots g(\omega_k)}$$

is not a constant (i.e. not independent of $\omega_1, \omega_2, \ldots, \omega_k$) showing that a linear combination of a vector time series $\{X_t\}$ satisfying Eq. (3.1) is not linear. In otherwords for testing linearity of vector time series linear combinations cannot be used (in contrast to Gaussianity).

4. SYMMETRIES OF CUMULANTS OF RANDOM VECTORS

It is well-known that in the case of a univariate time series $(d = 1)$, $\text{cum}(x_{t_1}, x_{t_2}, \ldots, x_{t_n})$ is the same as $\text{cum}(x_{t_1}, x_{t_3}, x_{t_2}, \ldots, x_{t_n})$ or any other permutation. This is not the case when $d \neq 1$. For example $\mathcal{K}(s_1, s_2)$ as given by Eq. (2.15) is not the same as $\mathcal{K}(s_2, s_1)$. However, we can obtain one from the other using permutation matrices.

Let $I_{(p.q)}$ be a $pq \times pq$ permuted identity matrix, see MacRae (1974) as considered in Sec. 2. Let $J_{1k} = I_{d^k \times d^k}$ and define

$$J_{ik} = I_1 \otimes I_2 \otimes \ldots \otimes I_{i-2} \otimes I_{(d.d)} \otimes I_{i=1} \otimes \ldots I_k \tag{4.1}$$

where $I_1, I_2 \ldots I_{i-2}, I_{i+1}, \ldots, I_k$ are all $d \times d$ identity matrices. J_{ik} is a $d^k \times d^k$ matrix and it has the property that

$$J_{ik} \text{cum}(X_1, X_2, \ldots, X_{i-1}, X_i, \ldots X_k) = \text{cum}(X_1, \ldots, X_i, X_{i-1}, \ldots, X_k) \tag{4.2}$$

where $X_1, X_2, \ldots, X_k$ are stationary random vectors of order $d \times 1$.

In the case $k = 3$, we have $J_{13} = I_{d^3 \times d^3}, J_{23} = I_{(d.d)} \otimes I$ and $J_{33} = I \otimes I_{(d.d)}$, see also Jou and Shaman (1988). Let

$$J^{(k)}_{i_1 i_2 \ldots i_j} = J_{i_1 k} J_{i_2 k} \ldots J_{i_j k} \tag{4.3}$$

where $J_2^{(3)} = J_{23}, J_3^{(3)} = J_{33}, J_{23}^{(3)} = J_{23} J_{33}$ and so on. Note that $J_{232}^{(3)} = J_{323}^{(3)}$.

Let $\mathcal{K}(s_1, s_2, \ldots, s_{k-1})$ be the kth order cumulant vector (of order $d^k \times 1$) and let us define the kth order cumulant spectral vector as

$$f(\omega_1, \omega_2, \ldots, \omega_{k-1}) = \frac{1}{(2\pi)^{k-1}} \sum_{s_1, s_2, \ldots s_{k-1} = -\infty}^{\infty} \mathcal{K}(s_1, s_2, \ldots, s_{k-1}) \ldots$$

$$\times \exp\left(-\sum_{j=1}^{k-1} s_j \omega_j\right) \tag{4.4}$$

when it exists.

Using the above permutation matrices we can obtain one set of cumulant vectors from the others by proper multiplication. In Table 1 we give these matrices when $k = 3$.

For example $\mathcal{K}(-s_2, s_1 - s_2) = J_{23}^{(3)} \mathcal{K}(s_1, s_2)$ and $f(\omega_3, \omega_1) = J_{23}^{(3)} f(\omega_1, \omega_2)$ where $\omega_1 + \omega_2 + \omega_3 = 0 \pmod{2\pi}$. We can find similar multiplication matrices for $k = 4$, see Wong (1993), p. 39.

We can use the above properties to derive the cumulant properties of multivariate bilinear processes.

Table 1. Symmetrics of Third Order Cumulants

$\mathcal{K}_3$	f_3	$J^{(3)}_{i_1\ldots i_j}$	$\mathcal{K}_3$	f_3	$J^{(3)}_{i_1\ldots i_j}$
$\mathcal{K}(s_1, s_2)$	$f(\omega_1, \omega_2)$	$J^{(3)}_1$	$\mathcal{K}(-s_2, s_1 - s_1)$	$f(\omega_3, \omega_1)$	$J^{(3)}_{23}$
$\mathcal{K}(s_2, s_1)$	$f(\omega_2, \omega_1)$	$J^{(3)}_2$	$\mathcal{K}(s_2 - s_1, -s_1)$	$f(\omega_2, \omega_3)$	$J^{(3)}_{32}$
$\mathcal{K}(s_1 - s_2, -s_2)$	$f(\omega_1, \omega_3)$	$J^{(3)}_3$	$\mathcal{K}(-s_2, s_2 - s_1)$	$f(\omega_3, \omega_2)$	$J^{(3)}_{23}$

5. MULTIVARIATE BILINEAR MODELS

Bilinear models have been introduced by Granger and Andersen (1978) and Subba Rao (1977) to describe many nonlinear time series. A systematic study of the bilinear models for the univariate case has been made by Subba Rao (1981) and the details were given in the book by Subba Rao and Gabr (1984). A univariate time series x_t is said to be a bilinear process if it satisfies the difference equation

$$x_t = \sum_{l=1}^{p_1} a_l x_{t-l} + \sum_{l=1}^{q_1} c_l e_{t-l} + \sum_{m=1}^{p_2} \sum_{n=1}^{q_2} b_{mn} x_{t-m} e_{t-n} + e_t \tag{5.1}$$

where $\{\dot{e}_t\}$ is a sequence of independent, identically distributed random variables. The model in Eq. (5.1) is referred to as $BL(p_1, q_1, p_2, q_2)$, see Subba Rao (1981). The properties of the above model have been extensively studied by many, for example, Terdik (1985), Guegan (1994), Liu (1989) and Pham Dinh Tuan and Tran (1981).

An extension of the univariate bilinear models to the multivariate case has been first considered by Subba Rao (1985), Stensholt and Tjøstheim (1987), Stensholt (1989), Liu (1989). Subba Rao (1985) gave a Volterra expansion for the mulivariate bilinear models.

We now consider an extension of the model in Eq. (5.1) to the multivariate time series case.

Let X_t be a d-dimensional random vector with elements $(x_{1t}, x_{2t}, \ldots, x_{dt})$. Let $e'_t = (e_{1t}, e_{2t}, \ldots, e_{dt})$, where $\{e_t\}$ is a sequence of independent random vectors with $E(e_t) = \mathbf{0}$, $E(e_t e'_t) = \mathbf{\Sigma}$. Let the ith component x_{it} satisfy the difference equation,

$$x_{it} = \sum_{l=1}^{p_1}\sum_{j=1}^{d} a_{ijl} x_{j,t-\ell} + \sum_{l=1}^{q_1}\sum_{j=1}^{d} c_{ijl} e_{j,t-l} + \sum_{m=1}^{p_2}\sum_{n=1}^{q_2}\sum_{j,v=1}^{d} b_{ijmnv} x_{j,t-m} e_{v,t-n} + e_{it}$$

$$(i = 1, 2\ldots, d) \tag{5.2}$$

The model in Eq. (5.2) can be written in the matrix form as follows:

$$X_t = \sum_{l=1}^{p_1} A_l X_{t-l} + \sum_{l=1}^{q_1} c_{l\nu} e_{t-\ell}^\nu + \sum_{m=1}^{p_2} \sum_{n=1}^{q_2} B_{mn\nu} X_{t-m} e_{t-n} + D_\nu e_t^\nu \tag{5.3}$$

where $A_l = [a_{ijl}]$, $B_{mn\nu} = [b_{ijmn\nu}]$ are $d \times d$ matrices, $C_{l\nu} = [c_{i\nu l}]$ is a $d \times 1$ vector and D_ν be the νth column of $d \times d$ identity matrix. Here we used $\dot{e}_t^\nu$ to denote the component $e_{\nu t}$ to introduce Einstein summation convention. For example, $c_{l\nu} e_{t-l}^\nu$ stands for $\sum_\nu c_{l\nu} e_{\nu,t-\ell}$, $D_\nu e_t^\nu$ for $\sum D_\nu e_{\nu,t}$ and $B_{mn\nu} X_{t-m} e_{t-n}^\nu = \sum_\nu B_{mn\nu} X_{t-m} e_{\nu,t-n}$. This convention is used to avoid too many summations, see McCullagh (1987), p. 2.

We define the model in Eq. (5.3) as $MDL(d; p_1, q_1, p_2, q_2)$. In the univariate case, statistical properties of the model $BL(p, o, p, 1)$ have been extensively studied, and these properties have been extended to lower triangular bilinear models. Here we consider multivariate extensions, for details, see Wong (1993) and also Subba Rao (1985), Liu (1989), Stensholt and Tjøstheim (1997).

Consider the model $MBL(d; p, o, p, 1)$. Define the vector $\chi_t = (X_t', X_{t-1}', \ldots, X_{t-p+1}')$, and let D_ν be the νth column of the $dp \times d$ matrix $[I_{p \times p}, \mathbf{0}, \ldots, \mathbf{0}]'$. Further let

$$\mathcal{A} = \begin{bmatrix} \mathbf{A}_1 & \mathbf{A}_2 & \ldots\ldots & \mathbf{A}_p \\ \mathbf{I} & \mathbf{0} & \ldots\ldots & \mathbf{0} \\ & \vdots & & \\ \mathbf{0} & \mathbf{0} & \ldots & \mathbf{I} \quad \mathbf{0} \end{bmatrix}, \quad \mathcal{B}_\nu = \begin{bmatrix} \mathbf{B}_{11\nu} & \mathbf{B}_{21\nu} & \ldots & \mathbf{B}_{p1\nu} \\ \mathbf{0} & \mathbf{0} & \ldots & \mathbf{0} \\ & \vdots & & \\ \mathbf{0} & \ldots\ldots\ldots & & \mathbf{0} \end{bmatrix}. \tag{5.4}$$

We can now write the model $MBL(d; p, o, p, 1)$ in the form

$$\chi_t = \mathcal{A}\chi_{t-1} + \mathcal{B}_\nu \chi_{t-1} e_t^\nu + \mathcal{D}_\nu e_t^\nu \tag{5.5}$$

where $\chi_t, \mathcal{D}_\nu$ are $dp \times 1$ vectors and $\mathcal{A}, \mathcal{B}_\nu$ are $dp \times dp$ matrices. As before $\mathcal{D}_\nu e_t^\nu = \sum_{\nu=1}^d \mathcal{D}_\nu e_{\nu,t}$ etc.

The above form of Eq. (5.5) is not Markovian, but can be written in the Markovian form using the approach by Pham Dinh and Tran (1981). Define $\mathcal{Z}_t = (\mathcal{A} + \mathcal{B}_\nu e_t^\nu)\chi_t$. Then (5.5) can be written in the form

$$Z_t = (\mathcal{A} + \mathcal{B}_\nu e_t^\nu)Z_{t-1} + (\mathcal{A} + \mathcal{B}_\nu e_t^\nu)(\mathcal{D}_\nu e_t^\nu)$$

$$\chi_t = Z_{t-1} + \mathcal{D}_\nu e_t^\nu \tag{5.6}$$

We see the model $MBL(d; p, o, p, 1)$ has only the error term e_t, e_{t-1} and not lower order lag terms. This restricts the usefulness of this model in practical analysis of real data. In order to make the models more useful, we need to include lower order terms. Therefore, we define the lower triangular bilinear model $LBL(d; p_1, q_1, p_2, q_2)$ as follows: we say the random vector X_t satisfies the lower triangular bilinear model if it satisfies the difference equation.

$$X_t = \sum_{l=1}^{p_1} A_l X_{t-\ell} + \sum_{l=1}^{q_1} c_{l\nu} e_{t-l}^{\nu} + \sum_{n=1}^{q_2} \sum_{m \geq n}^{p_2} B_{mn\nu} X_{t-m} e_{t-n}^{\nu} + \mathcal{D}_{\nu} e_t^{\nu} \tag{5.7}$$

Using a method introduced by Tang (1987), Wong (1993) has shown that Eq. (5.7) can be written in the form

$$\chi_t = A \chi_{t-1} + \mathcal{B}_{\nu} \chi_{t-1} e_{t-1}^{\nu} + C_{\nu} e_{t-1}^{\nu} + D_{\nu} e_t^{\nu} \tag{5.8}$$

The two representations in Eqns (5.5) and (5.8) look deceptively similar. The fundamental difference is that in the representation in Eq. (5.5), the state vector χ_t contains only the lagged values $(X_t, X_{t-1}, \ldots, X_{t-p+1})$, where χ_t in Eq. (5.8) contains not only $(X_t, X_{t-1}, \ldots, X_{t-p-1})$ but also variables which contain the lagged variables $e_{t-2}, e_{t-3}, \ldots, e_{t-q_2}$. From now onwards we consider Eq. (5.8) for our statistical analysis.

6. HIGHER ORDER CUMULANTS OF BILINEAR MODELS

In the previous section we have shown that lower triangular bilinear model $LBL(d, p)$ where $p = \max(p_1, q_1, p_2, q_2)$, given by Eq. (5.7) can be written in a compact form of Eq. (5.8). We will use this form for obtaining the second order and higher order cumulants of the process $\{X_t\}$ satisfying the model in Eq. (5.7). Consider the model in Eq. (5.8)

$$\chi_t = A \chi_{t-1} + \mathcal{B}_{\nu} \chi_{t-1} e_{t-1}^{\nu} + C_{\nu} e_{t-1}^{\nu} + D_{\nu} e_t^{\nu} \tag{6.1}$$

For convenience, we write $\langle \chi_t, \chi_t, \ldots, \chi_t \rangle$ for the cumulant of $\{\chi_t, \chi_t, \ldots \chi_t\}$. We will assume that χ_t is independent of e_s, for $t < s$. We note that the joint cumulant of three or more random vectors or variables which contain χ_t or e_s^{ν} will be zero for $s \geq t$ if one of the elements is e_s^{ν} $(s \geq t)$.

Now consider Eq. (6.1). We have

$$\langle \chi_{t+1}, e_{t+1}^{\nu} \rangle = D_{\nu_1} \sigma^{\nu \nu_1} \tag{6.2}$$

where $\sigma^{\nu \nu_1} = \text{cov}(e_{\nu,t}, e_{\nu_1,t})$ is the (ν, ν_1)th element of the matrix Σ. From Eqns (6.1) and (6.2), we have the mean

$$\mathcal{K}_{t+1} = E(\chi_{t+1}) = \langle \chi_{t+1} \rangle$$
$$= A \mathcal{K}_t + \mathcal{B}_{\nu_1} D_{\nu_2} \sigma^{\nu_1 \nu_2} \tag{6.3}$$

Let $\rho(\mathbf{A}) = \max_i \{|\lambda_i(\mathbf{A})|\}$, where $\lambda_i(\mathbf{A})$ is the ith eigenvalue of $\mathbf{A}$. $\rho(\mathbf{A})$ is also known as the spectral radius of $\mathbf{A}$. From Eq. (6.3), we see that if $\rho(\mathcal{A}) < 1$, then as $t \to \infty$,

$$\mathcal{K}_t \to \mathcal{K} = (\mathbf{I} - \mathcal{A})^{-1} \mathcal{B}_{\nu_1} D_{\nu_2} \sigma^{\nu_1 \nu_2} \tag{6.4}$$

$\rho(\mathcal{A}) < 1$ is a sufficient condition for the asymptotic first-order stationarity of the time series $\{X_t\}$. Let $\mathcal{K}_t(0) = \text{cum}(\chi_t, \chi_t) = \text{vec}[E(\chi_t - \mathcal{K})(\chi_t - \mathcal{K})']$.

From Eqns (6.1) and (6.3), we have

$$
\begin{aligned}
\mathcal{K}_{t+1}(0) &= \langle \chi_{t+1}, \chi_{t+1} \rangle \\
&= \langle \mathcal{A}\chi_t, \mathcal{A}\chi_t \rangle + \langle \mathcal{B}_\nu \chi_t e_t^\nu, \mathcal{B}_\nu \chi_t e_t^\nu \rangle \\
&\quad + \langle \mathcal{C}_\nu e_t^\nu, \mathcal{C}_\nu e_t^\nu \rangle + \langle \mathcal{D}_\nu e_{t+1}^\nu, \mathcal{D}_\nu e_{t+1}^\nu \rangle \\
&\quad + (I + J_{22})\{ \langle \mathbf{A}\chi_t, \mathbf{B}_\nu \chi_t \mathbf{e}_t^\nu \rangle \\
&\quad + \langle \mathbf{A}\chi_t, \mathbf{C}_\nu e_t^\nu \rangle + \langle \mathbf{B}_\nu \chi_t e_t^\nu, \mathbf{C}_\nu e_t^\nu \rangle \}
\end{aligned}
\tag{6.5}
$$

where J_{22} is the permutation matrix defined earlier.

We can show $\langle \mathcal{A}\chi_t, \mathcal{A}\chi_t \rangle = (\mathcal{A} \otimes \mathcal{A})\mathcal{K}_t(0)$,

$$
\langle C_{\nu_1} e_t^{\nu_1}, C_{\nu_2} e_t^{\nu_2} \rangle = (C_{\nu_1} \otimes C_{\nu_2})\sigma^{\nu_1 \cdot \nu_2},
$$

$$
\langle D_{\nu_1} e_{t+1}^{\nu_1}, D_{\nu_2} e_{t+1}^{\nu_2} \rangle = (D_{\nu_1} \otimes D_{\nu_2})\sigma^{\nu_1 \nu_2}.
$$

If we assume that $\rho[\mathcal{A} \otimes \mathcal{A} + \mathcal{B}_{\nu_1} \otimes \mathcal{B}_{\nu_2}\sigma^{\nu_1 \nu_2}] < 1$ (this is a sufficient condition for the second-order stationarity of $\{X_t\}$) we can show, see Wong (1993), that as $t \to \infty$,

$$
\begin{aligned}
\mathcal{K}(o) &= \text{cum}(\boldsymbol{\chi}_t, \boldsymbol{\chi}_t) \\
&= [\mathcal{A} \otimes \mathcal{A} + (\mathcal{B}_{\nu_1} \otimes \mathcal{B}_{\nu_2})\sigma^{\nu_1 \nu_2}]^{-1} \\
&\quad \times \{ (\mathcal{B}_{\nu_1} \otimes \mathcal{B}_{\nu_2})[(\mathcal{K} \otimes \mathcal{K})\sigma^{\nu_1 \nu_2} + (\mathcal{D}_{\nu_3} \otimes \mathcal{D}_{\nu_4})\sigma^{\nu_1 \nu_4}\sigma^{\nu_2 \nu_3}] \\
&\quad + [C_{\nu_1} \otimes C_{\nu_2} + D_{\nu_1} \otimes D_{\nu_2}]\sigma^{\nu_1 \nu_2} \\
&\quad + (J + J_{22})[(AD_{\nu_2}) \otimes (\mathcal{B}_{\nu_1}\mathcal{K}) + (AD_{\nu_2}) \otimes C_{\nu_1} \\
&\quad + [B_{\nu_1}\mathcal{K}) \otimes C_{\nu_2}]\sigma^{\nu_1 \nu_2} \}
\end{aligned}
\tag{6.6}
$$

For the lagged moments, it is sufficient to consider the case $\mathcal{K}(s)$, $s \geq 1$ since $\mathcal{K}(-s) = J_{22}\mathcal{K}(s)$. When $s = 1$ we can show

$$
\begin{aligned}
\mathcal{K}(1) &= \langle \chi_{t+1}, \chi_t \rangle \\
&= (\mathcal{A} \otimes I)\mathcal{K}(o) + [(\mathcal{B}_{\nu_1}\mathcal{K}) \otimes D_{\nu_2} + C_{\nu_1} \otimes D_{\nu_2}]\sigma^{\nu_1 \nu_2}
\end{aligned}
\tag{6.7}
$$

and for all $s \geq 2$,

$$
\mathcal{K}(s) = (\mathcal{A} \otimes I)\mathcal{K}(s - 1) = (\mathcal{A}^{s-1} \otimes I)\mathcal{K}(1)
\tag{6.8}
$$

The above reduces to the form given by Subba Rao (1985) for the bivariate case when $q_1 = 0, p_1 = p_2 = q_2 = 1$. The difference equation for $\mathcal{K}(s)$ given by Eq. (6.8) shows that the second-order covariance structure for the lower triangular Bilinear model $LBL(d; p)$ is similar to the linear multivariate

ARMA $(p, 1)$ model. In view of this it is important to calculate higher-order cumulants which possibly can distinguish between linear ARMA models and the Bilinear models. Sesay and Subba Rao (1988, 1991) have shown that for the univariate Bilinear models high order cumulants do satisfy Yule–Walker type difference equations. The derivation of these equations for multivariate case, though straightforward, are lengthy (and can be found in the PhD Thesis of Wong (1993)). Here we just give the final results. These are as follows:

$$\mathcal{K}(s, 0) = (\mathcal{A} \otimes I \otimes I)\mathcal{K}(s - 1, 0), s \geq 2$$

$$\mathcal{K}(s + \tau, s) = (\mathcal{A} \otimes I \otimes I)\mathcal{K}(s + \tau - 1, s), \tau \geq 2 \qquad (6.9)$$

Similarly for the fourth-order cumulants

$$\mathcal{K}(s, 0, 0) = (\mathcal{A} \otimes I \otimes I)\mathcal{K}(s - 1, 0, 0), s \geq 2$$

$$
\begin{aligned}
\mathcal{K}(s, s, 0) = &[\mathcal{A} \otimes \mathcal{A} \otimes I \otimes I + (\mathcal{B}_{\nu_1} \otimes \mathcal{B}_{\nu_2} \otimes I \otimes I)\sigma^{\nu_1 \nu_2}]\mathcal{K}(s - 1, s - 1, 0) \\
&+ (I + J_2^{(4)})\{(\mathcal{B}_{\nu_1} \otimes \mathcal{B}_{\nu_2} \otimes I \otimes I)[\mathcal{K} \otimes \mathcal{K}(s - 1, 0) \\
&+ J_3^{(4)}\mathcal{K}(s - 1) \otimes \mathcal{K}(s - 1)] \\
&+ (\mathcal{A} \otimes \mathcal{B}_{\nu_1} \otimes I \otimes I)[D_{\nu_2} \otimes \mathcal{K}(s - 1, 0)] \\
&+ (I \otimes \mathcal{B}_{\nu_1} \otimes I \otimes I)[C_{\nu_2} \otimes \mathcal{K}(s - 1, 0)]\sigma^{\nu_1 \nu_2} \qquad (6.10)
\end{aligned}
$$

Similarly we can obtain difference equations for the fourth-order cumulants in higher lags.

We pointed out earlier that higher-order cumulants can distinguish Bilinear models and linear vector ARMA models. The discrimination between Bilinear models and nonGaussian linear models may not be that easy using higher-order cumulants as can be shown as follows: see Wong (1993), p. 135. Consider the vector ARMA (p, q) model

$$X_t = A_1 X_{t-1} + \cdots + A_p X_{t-p} + C_1 e_{t-1} + C_2 e_{t-2} + \cdots + C_q e_{t-q} + e_t \qquad (6.11)$$

where $\{e_t\}$ is a iid nonGaussian sequence of independent vectors but with finite kth order cumulants. Assume $\rho(\mathcal{A}) < 1$, where $\mathcal{A}$ is defined as before. Then for $s > q$,

$$\mathcal{K}(s) = (\mathcal{A} \otimes I)\mathcal{K}(s - 1),$$

$$\mathcal{K}(s, s) = (\mathcal{A} \otimes \mathcal{A} \otimes \mathbf{I})\mathcal{K}(s - 1, s - 1)$$

$$\mathcal{K}(s, s, s) = (\mathcal{A} \otimes \mathcal{A} \otimes \mathcal{A} \otimes \mathbf{I})\mathcal{K}(s - 1, s - 1, s - 1) \qquad (6.12)$$

These equations may look like the equations we obtained for Bilinear models. In the case of linear models, the second-order covariances and higher-order moments decay to zero at an exponential rate. In the case of bilinear models, the rate of covergence (to zero) is very slow depending on the nonlinear behavior of the models. In this context it may be worth pointing out that the behavior of the (sample) covariances in the case of nonlinear models resembles the linear long range memory processes, see PhD Thesis of Maria Augusta da Silva (1994). There is often the possibility of confusion between these two processes.

7. TESTS FOR GAUSSIANITY OF VECTOR TIME SERIES

Recently Subba Rao and Wong (1997) have proposed tests based on measures of skewness and kurtosis for testing Gaussianity under the assumption that the series X_t is stationary. We use the measures defined by Mardia (1970) in the classical case. Let X be a d dimensional random vector with mean μ and variance covariance matrix Σ. Mardia (1970) defined skewness measure as

$$\beta_{1.d} = [E(Y \otimes Y \otimes Y)]'(\Sigma \otimes \Sigma \otimes \Sigma)^{-1}[E(Y \otimes Y \otimes Y)] \tag{7.1}$$

and the kurtosis measure as

$$\beta_{2.d} = E[Y'\Sigma^{-1}Y]^2 = E[\Sigma\Sigma\sigma^{ij}Y_iY_j]^2 \tag{7.2}$$

where $Y = X - \mu$. If X is Gaussian, then $\beta_{1.d} = 0$ and $\beta_{2.d} = d(d+2)$.
An alternative way of writing Eq. (7.1) is to write it in a scalar form

$$\beta_{1.d} = \sum_{r,s,t=1}^{d} \sum_{r,s',d'=1}^{d} \sigma^{rr'}\sigma^{ss'}\sigma^{tt'}\mu_{111}^{rst}\mu_{111}^{r's't'} \tag{7.3}$$

where $\mu_{111}^{rst} = E(Y_rY_sY_t)$. The form of Eq. (7.3) is used for testing the hypotheses. Here we describe briefly the test and consider the statistical power of the test. For details of the test, we refer to the paper of Subba Rao and Wong (1998).

The simulation results described here are contained in a MSc dissertation submitted by Perez (1997), and the computer programs written by Jose Perez of the tests can be obtained from the first author.

Let $(X_1, X_2, \ldots, X_n)$ be a sample from the stationary time series $\{X_t\}$. We define the sample moments as follows:

$$m_i = \frac{1}{n} \sum_{t=1}^{n} x_{i,t},$$

$$m_{i_1 i_2 \dots i_r} = \frac{1}{n} \sum_{t=1}^{n} x_{i_1 t} x_{i_2 t} \dots x_{i_r t} \tag{7.4}$$

$$M_{i.i_2 \dots i_r} = \frac{1}{n} \sum_{t=1}^{n} (x_{i,t} - m_{i_1})(x_{i_2 t} - m_{i_2}) \dots (x_{i_r t} - m_{ir})$$

The sample variance covariance matrix is defined as

$$S = \frac{1}{n} \sum_{t=1}^{n} (X_t - \bar{X})(X_t - \bar{X})' \tag{7.5}$$

We define the sample measures of skewness and kurtosis as

$$b_{1,d} = \sum_{i_1 j_1 k_1} \sum_{i_2 j_2 k_2} S^{i_1 i_2} S^{j_1 j_2} S^{k_1 k_2} M_{i_1 j_1 k_1} M_{i_2 j_2 k_2}$$

$$b_{2,d} = \sum_{i,j,k,l=1}^{d} S^{ij} S^{kl} M_{ijkl} \tag{7.6}$$

Under the null hypothesis we can show that $b_{1,d}$ converges (in distribution) to $\sum [M_{rst}]^2$. Subba Rao and Wong (1998) have shown that for large n, nb_{1d} is a weighted χ^2 with $q = d(d+1)(d+2)/6$ degrees of freedom.

Similarly, under the null hypothesis b_{2d} for large n is asymptotically distributed as normal with mean $d(d+2)$ and variance V/n, where

$$V = \sum_{ijkl=1}^{d} \sum_{q=-\infty}^{\infty} \left[8\zeta_{ij}^2(q)\zeta_{kl}^2(q) + 16\zeta_{ij}(q)\zeta_{kl}(q)\zeta_{il}(q)\zeta_{kl}(q) \right] \tag{7.7}$$

where

$$\zeta(q) = \Sigma^{-1/2} \gamma(q) \Sigma^{-1/2}, \Sigma = E[(X_t - \mu)(X_t - \mu)'], \gamma(q)$$
$$= E[(X_{t+q} - \mu)(X_t - \mu)'].$$

We accept the null hypothesis at 100α per cent significance level when

$$T_1 = nb_{1,d} < \chi^2_{q,\alpha} \quad \text{and} \quad T_2 = \left| \frac{b_{2,d} - d(d+2)}{\sqrt{V/n}} \right| < Z_{\alpha/2}.$$

To illustrate the power of the test, 200 samples from bivariate models have been generated with different sample sizes ($n = 150, 250, 350$ and 450). The series are generated from the following models, for details, see Jose Perez (1997).

Series A $X_t = AX_{t-1} + e_t,$

where $A = \begin{bmatrix} 0.3 & -0.1 \\ 0.7 & 0.2 \end{bmatrix}$

Series B $X_t = A_1 X_{t-1} + A_2 X_{t-2} + e_t,$

where $A_i = \begin{bmatrix} -0.5 & 0.2 \\ 0.3 & 0.4 \end{bmatrix}, \qquad A_2 = \begin{bmatrix} 0.10 & 0.30 \\ 0.15 & 0.01 \end{bmatrix}$

Series C $X_t = AX_{t-1} + B_1 X_{t-1} e_{1,t-1} + B_2 X_{t-1} e_{2,t-1} + \mathbf{e_e}$

where   A is the same as in Series A, and

$$B_1 = \begin{bmatrix} 0.03 & 0.1 \\ 0.02 & -0.02 \end{bmatrix}, \qquad B_2 = \begin{bmatrix} -0.2 & 0.12 \\ 0.07 & 0.15 \end{bmatrix}.$$

In all the above simulations, we assume that $\{e_t\}$ are iid random vectors, and each distributed normally with mean $\mu' = (0,0)$ and variance co-variance matrix

$$\Sigma = \begin{bmatrix} 6.0 & -2.0 \\ -2.0 & 3.0 \end{bmatrix}.$$

In evaluating Eq. (7.7), we have to choose some truncation point when summing over q; and this depends on the rate of decay of $\zeta_{ij}(q)$. We have chosen here truncation point as 20, i.e. $\zeta_{ij}(q), -20 \leq q \leq 20$.

In Table 2, we give the percentage of times the null hypothesis has been accepted in each case when the tests were applied. We see from the results in Table 2, that the probability of rejecting the null hypothesis when the alternative is true is increasing as the sample size is increasing. We have also applied the above tests to real data, and the results were described in the paper by Subba Rao and Wong (1998).

Table 2   Percentage of Times the Null Hypothesis Has Been Accepted

		Number of observations			
Series	Test	150	250	350	450
A	Skewness	95.5%	98%	94%	96.5%
	Kurtosis	98%	98%	97%	97%
B	Skewness	94%	94.5%	93.5%	93.5%
	Kurtosis	98%	97%	96%	97%
C	Skewness	29%	11.5%	5%	1%
	Kurtosis	57%	40%	25.5%	8%

8. ESTIMATION OF LOWER TRIANGULAR BILINEAR MODELS

In this Section we consider frequency domain approach to estimation of the parameters, and this is an extension of the method described by Sesay and Subba Rao (1992) for the univariate case. An alternative time domain approach was considered by Stensholt (1989) in her PhD Thesis submitted to UMIST. For estimation purposes, we consider the model $LBL(d; p_1, 0, p_2, q_2)$

$$X_t = \mu + \sum_{l=1}^{p_1} A_l X_{t-l} + \sum_{n=1}^{q_2} \sum_{m \geq n}^{p_2} B_{mn\nu} X_{t-m} e_{t-n}^{\nu} + e_t \tag{8.1}$$

where $\{e_t\}$ are iid random vectors each distributed $N(0, \Sigma)$ and $p_1 \geq p_2 \geq q_2$. The difference between the two models in Eqns (5.3) and (8.1) is that we included the constant term μ here and also the present model does not have any moving average terms.

Let $(X_1, X_2, \ldots, X_n)$ be a sample from the d-dimensional stationary time series $\{X_t\}$ and let

$$d(\lambda) = \frac{1}{\sqrt{2\pi n}} \sum_{t=1}^{n} X_t e^{it\lambda} \tag{8.2}$$

and $I(\lambda) = d(\lambda)d^*(\lambda)$ be the periodogram matrix. Let $\mathbf{f}(\omega)$ be the second-order spectral density matrix of the process $\{X_t\}$, and it is a function of the parameters $\boldsymbol{\theta}$. From now onwards, we write $f(\omega, \theta)$ for $f(\omega)$ to identify its dependence on the parameters. We minimise

$$L(\theta) = \frac{n}{2\pi} \int_{-\pi}^{\pi} [\ln |f(\omega, \theta)| + Tr(f^{-1}(\omega, \theta)I(\omega))] \, d\omega \tag{8.3}$$

which is widely known as Whittle's criterion. The original derivation of this criterion was based on the Gaussian likelihood principles. However, one can justify using Eq. (8.3), see Dzhaparidze (1986) using the finite Fourier transforms $d(\omega_j)$, $\omega_j = 2\pi j/n, j = 1, 2, \ldots, [n/2]$ which, under suitable conditions, are distributed as independent complex normal vectors, see Brillinger, (1975). For minimising purposes, we consider the discrete version of Eq. (8.3), namely

$$L(\theta) \simeq \sum_{j=1}^{[n/2]} [\ln |f(\omega_j, \theta)| + Tr(f^{-1}(\omega_j, \theta)I(\omega_j))] \tag{8.4}$$

Let $\hat{\theta}$ be the value of θ which minimises $L(\theta)$. Under Rosenblatt's strong mixing condition, Dzhaparidze (1986) obtained the asymptotic distribution of $\hat{\theta}$. He has shown that, under suitable conditions, $\sqrt{n}(\hat{\theta} - \theta)$ is

asymptotically distributed as multivariate normal with mean 0 and vari-
ance- covariance matrix

$$\Gamma^{-1}(\Gamma + C)^{-1} \tag{8.5}$$

where

$$\Gamma_{ij} = \frac{1}{4\pi} \int T_r \left\{ f^{-1}(\lambda, \theta) \frac{\partial f(\lambda, \theta)}{\partial \theta_i} f^{-1}(\lambda, \theta) \frac{\partial f(\lambda, \theta)}{\partial \theta_j} \right\} d\lambda \tag{8.6}$$

$$C_{ij} = \frac{1}{\pi} \int_{-\pi}^{\pi} \int_{-\pi}^{\pi} T_r \left\{ \left[\frac{\partial f(\mu, \theta)}{\partial \theta_j} \otimes \frac{\partial f(\lambda, \theta)}{\partial \theta_i} \right] f_4(\lambda, -\lambda, \mu, -\mu, \theta) \, d\lambda \, d\mu \right.$$

Note that $f_4(\lambda_1, \lambda_2, \lambda_3, \lambda_4)$ is a $d^2 \times d^2$ matrix with the fourth order cumulant spectrum $f_{ijkl}(\lambda_1, \lambda_2, \lambda_3, \lambda_4)$ as an entry at $((i-1)d + k)$th row and
$(j-1)d + l)$th column, for details see Wong (1993).

In order to minimise the criterion in Eq. (8.4), we need the spectral
density matrix of X_t satisfying the model in Eq. (8.1). The second-order
covariance matrices obtained in Sec. 6 are still valid even though an extra
constant term is included in the model. Only the mean of X_t is affected. Let
$\mathbf{M} = (\boldsymbol{\mu}', 0, \ldots, 0)'$, included in the model in Eq. (8.1). Let $\mathcal{K} = E(X_t)$. Then,

$$\mathcal{K} = (I - A)^{-1}[M + (\mathcal{B}_{\nu_1} D_{\nu_2})\sigma^{\nu_1 \nu_2}] \tag{8.7}$$

Let

$$R = I - \mathcal{A} \otimes \mathcal{A} - (\mathcal{B}_{\nu_1} \otimes \mathcal{B}_{\nu_2})\sigma^{\nu_1 \nu_2}$$

$$S = (\mathcal{B}_{\nu_1} D_{\nu_3} \otimes \mathcal{B}_{\nu_2} D_{\nu_4})\sigma^{\nu_1 \nu_4}\sigma^{\nu_2 \nu_3}$$

$$\times \{ \mathcal{B}_{\nu_1}\mathcal{K} \otimes \mathcal{B}_{\nu_2}\mathcal{K} + D_{\nu_1} \otimes D_{\nu_2} + (I + J_{22})(\mathcal{B}_{\nu_1}\mathcal{K} \otimes \mathcal{A}D_{\nu_2}) \}\sigma^{\nu_1 \nu_2}$$

$$T = (\mathcal{B}_{o_1}\mathcal{K} \otimes D_{\nu_2})\sigma^{\nu_1 \nu_2}$$

Then the second-order cumulant vector is given by

$$\mathcal{K}(s) = \begin{cases} R^{-1}S & \text{if } s = 0 \\ (\mathcal{A}^s \otimes I)\mathcal{K}(o) + (\mathcal{A}^{s-1} \otimes I)T & \text{if } s \geq 1 \end{cases} \tag{8.8}$$

The second-order spectral density vector is given by $(z = e^{-i\omega})$

$$2\pi f(\omega) = \sum_{-\infty}^{\infty} \mathcal{K}(s)z^s$$

$$= \mathcal{K}(o) + (I + J_{22}^c) \left[\left[\sum_{s=1}^{\infty} \mathcal{K}(s)z^s \right] \right.$$

$$= \mathcal{K}(o) + (I + J_{22}^c)\{(H^{-1}(z)z \otimes I)[(\mathcal{A} \otimes I)\mathcal{K}(o) + T] \tag{8.9}$$

where $H(z) = I - Az$, and J_{22}^c is an operator defined on any complex matrix V such that $J_{22}^c V = J_{22} \bar{V}$, where $\bar{V}$ is the complex conjugate of V. We minimise Eq. (8.4) using Newton-Raphson algorithm (as modified by Marquardt).

Let θ denote the q-dimensional vector of parameters, the elements of which are $\mu, \{A_l\}$ and $\{B_{mn\nu}\}$. Though the variance covariance matrix Σ is also a parameter, we first treat it as if it is known and estimate the rest of the parameters.

Newton–Raphson method requires the first and second derivatives of the function to be minimised. For convenience, we denote $f(\omega_k)$ by f_k, $I(\omega_k)$ by I_k and the derivatives $D_i g$ which stands for the first derivative of the function g with respect to the parameter θ_i etc. ... Then we can show

$$D_i l(\theta) = \sum_{k=1}^{\lfloor n/2 \rfloor} \{ T_r[f_k^{-1}.D_i f_k] - T_r[f_k^{-1} D_i f_k f_k^{-1} I_k] \}$$

$$= \sum_{k=1}^{\lfloor n/2 \rfloor} T_r \{ [f_k^{-1} D_i f_k . f_k^{-1}][f_k - I_k] \} \tag{8.10}$$

and the second derivative

$$D_{ij}(\theta) = \frac{\partial l(\theta)}{\partial \theta_i \partial \theta_j}$$

$$= \sum T_r \{ [\mathcal{D}_j (f_k^{-1}.D_i f_k . f_k^{-1})][f_k - I_k] + f_k^{-1} D_i f_k . f_k^{-1} . f_k^{-1} . D_j f_k \}$$

$$\simeq \sum_{k=1}^{\lfloor n/2 \rfloor} T_r \{ f_k^{-1}.D_i f_k . f_k^{-1} . D_j f_k \} \tag{8.11}$$

where $i \leq i, j \leq q$.

Let $G(\theta) = [D_i l(\theta)]$ be a q dimensional column vector and $H(\theta) = [D_{ij}(\theta)]$ be a $q \times q$ matrix. The Newton–Raphson iterative equation is given by

$$\theta^{(i)} = \theta^{(i-1)} - H^{-1}(\theta^{(i-1)})G(\theta^{(i-1)}) \tag{8.12}$$

where $\theta^{(o)}$ is the initial value of the parameters, for example, obtained by Repeated Residual Method (RRM) introduced by Subba Rao (1977).

Instead of considering Eq. (8.12), we use Marquardt algorithm given by

$$\theta^{(i)} = \theta^{(i-1)} - [\tilde{H}(\theta^{(i-1)}) + \lambda I]^{-1} \tilde{G}(\theta^{(i-1)}) \tag{8.13}$$

where

$$\tilde{G}_i(\theta) = \frac{G_i(\theta)}{\sqrt{H_{ii}(\theta)}}$$

$$\tilde{H}_{ij}(\theta) = \frac{H_{ij}(\theta)}{\sqrt{H_i(\theta)H_j(\theta)}} \tag{8.14}$$

and λ is chosen such that the convergence is very fast. We iterate until

$$\frac{|l(\theta^{(i)}) - l(\theta^{(i-1)})|}{1 + |l(\theta^{(i-1)})|} < 10^{-6} \tag{8.15}$$

Having estimated the parameters θ, we can estimate Σ as follows;
Let

$$\hat{e}_t^{(i)} = \begin{cases} 0 & \text{if} \quad t \leq \max(p_1, p_2, q_2) \\ X_t - \mu^{(i)} - \Sigma_l A_l^{(i)} X_{t-l} & - \sum_{m,n} B_{mnv}^{(i)} X_{t-m} \hat{e}_{t-n} \end{cases}$$

The variance covariance matrix Σ is estimated by

$$\Sigma^{(i)} = \frac{1}{n-p} \sum_{t=p+1}^{n} \hat{e}_t^{(i)} \hat{e}_t^{(i)'} \tag{8.16}$$

where $p = \max(p_1, p_2, q_2)$. To start the iteration, we set $\lambda = 0.01$ and $\theta^{(o)}$ obtained by the RRM method. The above iterative equations need explicitly the first order derivatives of the spectral density matrix with respect to the parameters. The derivatives are given as follows: (for details see Wong (1993).

$$2\pi \frac{\partial f(\omega)}{\partial \mu_i} = DK(o) + (I + J_{22}^c)$$

$$\times \{(H^{-1}(z)z \otimes I)[(A \otimes I)DK(o) + DT]\}$$

$$2\pi \frac{\partial f(\omega)}{\partial a_{ijkl}} = DK(o) + (I + J_{22}^c)$$

$$\times \{[(H^{-1}(z)E^{4-1}(z)z^2) \otimes I][(A \otimes I)K(o) + T]$$

$$+ [H^{-1}(z)z \otimes I]$$

$$\times [(E \otimes I)K(o) + (A \otimes I)DK(o) + DT]\} \tag{8.17}$$

$$2\pi \frac{\partial f(\omega)}{\partial b_{ijmnv}} = DK(o) + (\mathbf{I} + J_{22}^c)$$

$$\times \{(H^{-1}(z)z \otimes I)[(A \otimes I)DK(o) + DT]\}$$

We use the following criteria for the order determination.

$$AIC = \ln|\hat{\Sigma}| + \frac{2q}{n}$$

$$BIC = \ln|\hat{\Sigma}| + \frac{\ln(n)q}{n} \tag{8.18}$$

where q is the total number of parameters in the model. The procedure involves estimating the best AR model, i.e. estimating the parameter vector $(\mu, A_1, A_2, \ldots, A_p)$ and then using these to estimate nonlinear parameters, either using the Repeated Residual Method where we use the estimated residuals obtained from the AR model and then fit the entire model using the observations $(X_1, X_2, \ldots, X_n)$ and the fitted residuals from the AR model. Alternatively, one can use iterate equations given above using the AR parameters as starting values and the rest of the parameters set equal to zero. Setting so many parameters to zero can lead to many numerical problems. It may also be pointed out here that we use the criteria in Eq. (8.18) as only guidelines to determine the orders. We are not aware of any studies carried out to determine the suitability of these criteria for determining the orders for nonlinear models (in particular Bilinear models). We believe it is worth investigating these. In view of the large number of parameters involved in fitting these models, it may be useful to set some parameters equal to zero (or some known value) which is based on a *a priori* knowledge. These models we call Restricted models.

As in the linear time series situations we can test the residual vectors. The hypothesis we test are that $\{e_t\}$ are a sequence of independent vector (white noise). Let $(e_1, e_2, \ldots, e_n)$ be the residuals and let $\bar{e} = (1/n)\Sigma e_t$ and $S = (1/n)\Sigma(e_t - \bar{e})(e_t - \bar{e})'$. Define $\varepsilon_t = S^{-1/2}(e_t - \bar{e})$. $(t = 1, 2, \ldots, n)$. Let

$$\mathbf{R}_s^{(1)} = \frac{1}{n} \sum_{t=1}^{n-s} \varepsilon_{t+s} \otimes \varepsilon_t, \, s > 1 \tag{8.19}$$

For large n, the vector $R_s^{(1)}$ is asymptotically normal with mean zero. Under the null hypothesis the statistic $T_1 = n \sum_{s=1}^{m} \mathbf{R}_s^{(1)} \mathbf{R}_s^{(1)}$ is approximately a χ^2 with md^2 degrees of freedom. One of the referees rightly pointed out that the degrees of freedom need to be corrected for estimation of the parameters. We did not investigate this here. Reject the hypothesis if $T_1 > \chi^2_{\alpha,f}$, where $f = md^2$.

9. REAL DATA

In this section we fit the above models to real data.

9.1   West German Income and Consumption data

For our first illustration we consider the quarterly seasonally adjusted West German disposable income and expenditure from 1960 to 1978 giving us 92 observation. The data was used by Lütkepohl (1991) who assumed the series is Gaussian and fitted a VAR model to the data. The data analysed here is the first differences of the natural logarithms of the series. We performed the Gaussianity test described earlier on the data set, see for details Subba Rao and Wong (1998). The test has shown that the income data is nonGaussian and expenditure data is Gaussian, and jointly they are nonGaussian. The data and first differences of logarithms are given in Figs. 1a and 1b. We denoted the income data by $\{x_{1t}\}$ and the expenditure by $\{x_{2t}\}$. Following Lütkepohl (1990), we fitted the models to the first 73 observations and then used the fitted models to forecast the observations $(x_{77}, \ldots, x_{80})$. It must be pointed out that the dataset is very short for nonlinear analysis, and therefore one must be careful in drawing any serious conclusions from these models. Also it looks that there is a single outlier in the income series and this possibility is not investigated here. We do not know whether this single observation has any affect on the conclusions drawn here.

The AIC criterion suggested that AR(2) model is the best linear model. The restricted repeated residual method suggested inclusion of two bilinear

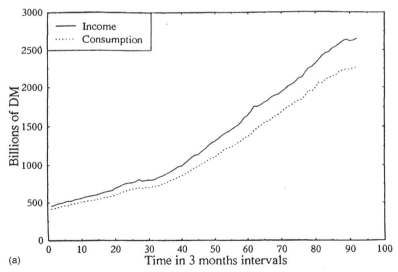

(a)

Figure 1a.   West German quarterly seasonally adjusted economic data.

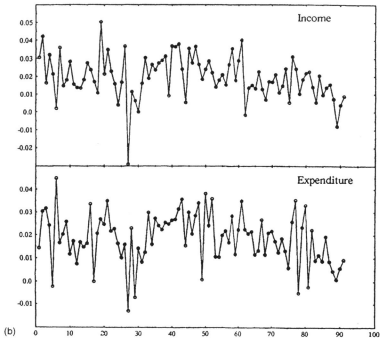

Figure 1b. First difference logarithms of income and consumption

parameters b_{21211} and b_{21212} (see the model in Eq. (5.2) in the scaler form). The estimates of the parameters are given in Table 3.

From Table 3, we see that a rise (or fall) in the disposable income will cause the future expenditure to rise (or fall) and vice versa. This is consistent with the well-known observation that consumer's spending, at least in the short run, cause economy to grow. It is interesting to see that both bilinear terms consist of the income component affecting the expenditure component. One may interpret this as an indication that the nonlinearity in consumption expenditures is not inherent but caused by external factors, such as disposable income. This seems to be consistent with the behavior of the consumers. One of the features in modern economic systems is the loosening of credit policy. This factor enables people to spend a lot more money even though their income rise is relatively small. However, once they fall into the trap of overspending, or worse still if there is a recession, the spending behavior tends to swing to the other extreme of not buying anything even though they could afford to.

Table 3 West German Data

Parameter	VAR(2)	Bilinear model
μ	$\begin{bmatrix} 0.021 \\ 0.020 \end{bmatrix}$	$\begin{bmatrix} 0.015 \\ 0.007 \end{bmatrix}$
A_1	$\begin{bmatrix} -0.088 & 0.284 \\ 0.272 & -0.316 \end{bmatrix}$	$\begin{bmatrix} -0.110 & 0.323 \\ 0.108 & -0.206 \end{bmatrix}$
A_2	$\begin{bmatrix} 0.005 & 0.082 \\ 0.345 & 0.004 \end{bmatrix}$	$\begin{bmatrix} 0.021 & 0.085 \\ 0.214 & 0.120 \end{bmatrix}$
$b_{21211} x_{1,t-2} e_{1,t-1}$	—	9.802
$b_{21211} x_{1,t-2} e_{2,t-1}$	—	−30.369
$\hat{\Sigma} \times 10^{-4}$	$\begin{bmatrix} 1.371 & 0.576 \\ 0.576 & 0.812 \end{bmatrix}$	$\begin{bmatrix} 1.331 & 0.515 \\ 0.515 & 0.707 \end{bmatrix}$

In Tables 4 and 5, we give forecasts obtained by the linear models and the bilinear models.

For the disposable income, we see that the forecasts obtained by bilinear model are no way better than linear model, whereas, for the expenditure they are better. This is consistent with our earlier conclusion obtained by using the Gaussianity test that income data is nearly Gaussian, whereas expenditure data is skewed and hence not consistent with the Gaussianity assumption.

Table 4 Disposable Income Forecasts in $\times 10^{-2}$

t	True values	VAR(2)	Bilinear
77	0.2426	0.2025	0.2052
78	0.1016	0.2530	0.2658
79	0.1829	0.1581	0.1571
80	0.2176	0.1980	0.2030
$MSE \times 10^{-4}$		0.6383	0.7309

Table 5 Consumption Expenditure Forecast in $\times 10^{-2}$

t	True values	VAR(2)	Bilinear
77	0.3535	0.1565	0.1720
78	−0.0512	0.1976	0.1041
79	0.2334	0.2642	0.2375
80	0.3305	0.1464	0.1544
MSE $\times 10^{-4}$		0.3388	0.2202

9.2   U.S. Treasury Bills Discount Rate and Bonds Yield

For our second illustration we consider the Quarterly U.S. treasury bills discount rate (x_{1t}) and yield on treasury bonds (x_{2t}). These two data sets are given in Figs 2a and 2b. To make it stationary and reduce the variation in the data, we took logarithms of the data and then differenced the data twice. In each of the two series, there seems to be a single outlier. Since we did not have any statistical tests to conclude that they are outliers, we did not discard them. The models have been fitted to the first ninety observations and the forecasts was done for the rest of the observations $(x_{91}, \ldots, x_{134})$. In

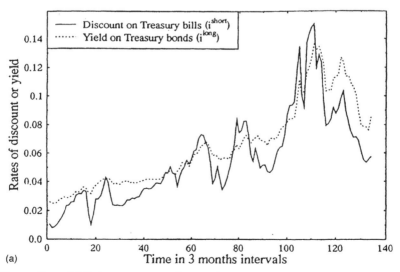

Figure 2a.   U.S. interest rate time series.

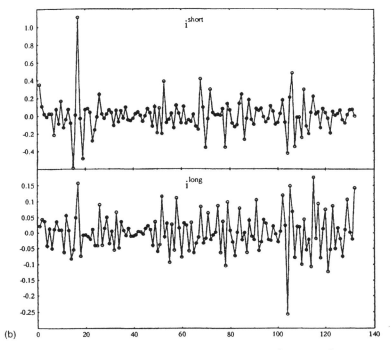

Figure 2b. Transformed U.S. interest rates data (i^{short} and i^{long}).

the 1980's, due to the need to finance very large government budget deficit and other political reasons, the US Government has raised the interest rates.

To study the effects of high interest rates policy in the 1980's (corresponding to $t = 101$ onwards), we divided the prediction period into two parts $t = 91$ to 100 and $t = 101$ to 134. The AIC criterion has led to the choice of AR(4) model. The fitted parameters are given in Table 6.

Since this data is transformed twice (after logs) it is not that easy to interpret the coefficients. Large coefficients (1.624 and 0.956) indicate that the rate of change in the bond yields has a linear positive effect on the treasury bill discount rate. In Table 7, we give the mean square error of forecasts for both linear and bilinear models separately for the periods 91–100 and 101–134.

For the first period, the forecasts obtained by the bilinear model are better than the bilinear model, whereas the linear model has coped well for the second period. It is also clear that there is a structural change during the two periods.

Table 6 US Treasury Bills

Parameter	VAR(4)	Bilinear
μ	$\begin{bmatrix} 0 \\ 0 \end{bmatrix}$	$\begin{bmatrix} 0.005 \\ 0.012 \end{bmatrix}$
A_1	$\begin{bmatrix} -0.519 & 1.624 \\ 0.020 & -0.566 \end{bmatrix}$	$\begin{bmatrix} -0.483 & 1.316 \\ 0.003 & -0.484 \end{bmatrix}$
A_2	$\begin{bmatrix} -0.663 & 0.956 \\ -0.019 & -0.462 \end{bmatrix}$	$\begin{bmatrix} -0.638 & 0.850 \\ -0.034 & -0.322 \end{bmatrix}$
A_3	$\begin{bmatrix} -0.249 & -0.174 \\ 0.071 & -0.577 \end{bmatrix}$	$\begin{bmatrix} -0.206 & -0.260 \\ 0.076 & -0.507 \end{bmatrix}$
A_4	$\begin{bmatrix} -0.180 & 0.460 \\ -0.031 & -0.227 \end{bmatrix}$	$\begin{bmatrix} -0.140 & 0.313 \\ -0.024 & -0.229 \end{bmatrix}$
$b_1 x_{2,t-3} e_{2,t-2}$	—	28.197
$b_1 x_{2,5-3} e_{2,t-3}$	—	9.254
$b_2 x_{2,t-3} e_{2,t-2}$	—	6.106
$b_2 x_{1,t-3} e_{2,t-2}$	—	−1.445
$\hat{\Sigma} \times 10^{-2}$	$\begin{bmatrix} 2.209 & 0.374 \\ 0.374 & 0.157 \end{bmatrix}$	$\begin{bmatrix} 1.774 & 0.298 \\ 0.298 & 0.163 \end{bmatrix}$

Table 7 Mean Square Error of Forecast

Period	VAR(4) MSE	Bilinear MSE
91–100	0.747	0.475
101–134	2.324	3.897

9.3 Canadian Muskrat and Mink data

For our last illustration, we consider the well-known bivariate time series considered by several authors before, see Tong (1990). There are sixty four observations on muskrat (x_{1t}) and mink (x_{2t}). The data are plotted in Figs 3a and 3b. Following previous authors we first took natural logarithms and difference only the muskrat data and the mean 10.8 is subtracted from the

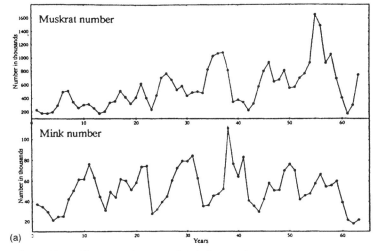

Figure 3a. Canadian muskrat and mink numbers.

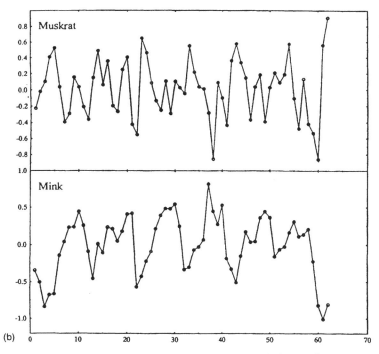

Figure 3b. Transformed Canadian muskrat and mink numbers.

Mink data. We fitted models to the first 61 observations and then used the fitted model to forecast the next two observations. The data length is very short to draw any valid conclusions. The AIC criteria has chosen VAR(4) model and the nonlinear fitted is LBL (2, 4, 0, 4, 2). The coefficients are given in Table 8.

The prey-predator relationship is well illustrated by the coefficients in the models. Large negative value (-0.715) indicates how an increase in number of mink is followed by a decrease in muskrat growth rate in the following year. The first distinctive feature of the selected bilinear model is that all the eight bilinear terms are caused by the first component residuals, that is the innovations of the muskrat growth rate. The second feature is that the

Table 8 Fitted Models for Muskrat and Mink Data

Parameter	VAR(4)	Bilinear
μ	$\begin{bmatrix} -0.002 \\ -0.002 \end{bmatrix}$	$\begin{bmatrix} 0.117 \\ 0.162 \end{bmatrix}$
A_1	$\begin{bmatrix} 0.263 & -0.715 \\ 0.503 & 0.611 \end{bmatrix}$	$\begin{bmatrix} -0.483 & 1.316 \\ 0.003 & -0.484 \end{bmatrix}$
A_2	$\begin{bmatrix} -0.176 & 0.330 \\ -0.104 & 0.340 \end{bmatrix}$	$\begin{bmatrix} -0.638 & 0.850 \\ -0.034 & -0.322 \end{bmatrix}$
A_3	$\begin{bmatrix} 0.155 & -0.027 \\ 0.050 & -0.273 \end{bmatrix}$	$\begin{bmatrix} -0.206 & -0.260 \\ 0.076 & -0.507 \end{bmatrix}$
A_4	$\begin{bmatrix} 0.039 & 0.169 \\ 0.188 & -0.179 \end{bmatrix}$	$\begin{bmatrix} -0.140 & 0.313 \\ -0.024 & -0.229 \end{bmatrix}$
$b_1 x_{1,t-4} e_{1,t-2}$	—	-1.279
$b_1 x_{1,t-2} e_{1,t-2}$	—	-0.597
$b_1 x_{1,t-3} e_{1,t-2}$	—	0.184
$b_1 x_{1,t-4} e_{1,t-1}$	—	0.118
$b_2 x_{2,t-2} e_{1,t-2}$	—	-0.858
$b_2 x_{1,t-1} e_{1,t-1}$	—	-0.528
$b_1 x_{2,t-2} e_{1,t-1}$	—	0.119
$b_2 x_{2,t-4} e_{1,t-2}$	—	0.262
$\Sigma \times 10^{-2}$	$\begin{bmatrix} 6.367 & 1.901 \\ 1.901 & 5.078 \end{bmatrix}$	$\begin{bmatrix} 5.458 & 1.940 \\ 1.940 & 6.914 \end{bmatrix}$

nonlinearity in the muskrat growth rate is strongly inherent, as indicated by the four bilinear terms in the tables. These observations are difficult to interpret biologically.

In this case the forecast obtained for the years 1910 and 1911 by the linear model are slightly better than the bilinear model. In view of the small sample size, it is advisable not to attach too much significance to the forecasts obtained here.

CONCLUSIONS

The foregoing study on the three data sets are not exhaustive. If the sample sizes are not too small, the number of parameters in the bilinear models are small, the estimates obtained by Whittle's estimate are good in terms of AIC and forecasting ability. Except for muskrat and mink data (here the data set is too small), the forecasts obtained by bilinear models are usually good. We believe nonlinear models are very useful to describe nonGaussian series. However, it must be pointed out that it is better to entertain the nonlinear models only after testing for Gaussianity and linearity, see for example the test by Subba Rao and Wong, (1998). One of the problems encountered in multivariate nonlinear modeling is the large number of parameters compared to the length of the data. It is here we have to explore the parsimonious relationships that may exist before embarking on nonlinear model building. The things to consider are (i) whether the series are separately nonlinear and their joint relationship is linear or (ii) alternatively their joint relationship is nonlinear whereas they are individually linear. We did not explore these options here, but hope to pursue these in the near future. We did not discuss the possibility of the presence of outliers here. However, it may be worth pointing out to an observation made by Subba Rao (1979), that bilinear models can produce bursts which can look like outliers.

Acknowledgments

We wish to thank the two referees for reading the paper very carefully and making many constructive suggestions.

REFERENCES

Brillinger, D. R. (1963) An Introduction to Polyspectra. *Ann. Maths. Stat.* *36*: 1351-1374.

Brillinger, D. R. (1975) Time Series: Data Analysis and Theory. Holt. Reinehart and Winston. New York.

Brillinger, D. R. and Rosenblatt, M. (1967) Asymptotic Theory of Estimates of kth Order Spectra. Spectral Analysis of Time Series. Ed. by B. Harris, pp. 153–188. John Wiley, New York.

Da Silva, Maria Eduarda Augusta (1994) Some contributions to the analysis of Bilinear Time Series Models. Unpublished PhD thesis submitted to UMIST.

Dzhaparidze, K. O. (1986) Parameter Estimation and Hypothesis Testing in Spectral Analysis of Stationary Time Series. Springer-Verlag, New York.

Granger, C. W. J. and Andersen, A. P. (1978) An Introduction to Bilinear Time Series Analysis. Vandenhoeck and Ruprecht. Gottingen.

Guegan, D. (1994) Series Chronologiques nonlineaires a' temps descret. Economica.

Hannan, E. J. (1970) Multiple Time Series. John Wiley, New York.

Jou, Y. and Shaman, P. (1988) The Third Order Cumulant Function and Bispectral Density of the Bilinear Time Series Model $BL(p, o, p, 1)$. Technical Report, Department of Statistics, University of Pennsylvania, Philadelphia, Pa 19104, U.S.A.

Leonov, V. P. and Shiryaev, A. N. (1959) On a method of calculation of semi-invariants. *Theory of Probability and its Applications.* 4: 319–329.

Liu, J. (1989) On the Existence of a General Multiple Bilinear Time Series. *J. Time Ser. Anal. 10,* 4: 341–355.

Lütkepohl, H. (1991) Introduction to Multiple Time Series Analysis. Springer-Verlag. New York.

MacRae, E. C. (1974) Matrix Derivatives with an Application to an Adaptive Linear Decision Problem. *Ann. Statist. 2*: 337–346.

Magnus, J. R. and Neudecker, H. (1988) Matrix Differential Calculus with Applications to Statistics and Econometrics. John Wiley, New York.

Mardia, K. V (1970) Measures of multivariate skewness and kurtosis with application. *Biometrika 57*: 519–530.

McCullagh, P (1987) Tensor Methods in Statistics, Chapman and Hall, London.

Perez Jose A. (1997) Some Contributions to Test of Gaussianity of Multivariate Time Series. MSc dissertation submitted to UMIST.

Pham Dinh T. and Tet Tran L. (1981) On the First Order Bilinear Time Series Models. *J. Appl. Prob. 18*: 617–627.

Priestley, M. B. (1988) Nonlinear and Nonstationary Time Series Analysis. Academic Press. London.

Reinsel, G. C. (1993) Element of Multivariate Time Series Analysis. Springer-Verlag, New York.

Rosenblatt, M. (1966) Remarks on Higher Order Spectra. "Multivariate Analysis". Ed. by P. R. Krishnaiah, 383–389. Academic Press.

Sesay, S. A. O. and Subba Rao, T. (1988) Yule–Walker Difference Equations for Higher Order Moments and Cumulants for Bilinear Time Series Models. *J. Time Ser. Anal. 9*: 385–401.

Sesay, S. A. O. and Subba Rao, T. (1991) Difference Equations for Higher order Moments and Cumulants for bilinear Time Series Models $BL(p,0,p,1)$ *J. Time Ser. Anal. 11*: 385–401.

Sesay, S. A. O. and Subba Rao, T. (1992) Frequency domain estimation of bilinear time series models. *J. Time Ser. Anal. 13*: 521–545.

Stensholt, B. K. and Tjøstheim, D. (1987) Multiple Bilinear Time Series Models. *J. Time Ser. Anal. 8, 2*: 221–233.

Stensholt, B. K. (1989) Statistical Analysis of Multivariate Bilinear Models. Unpublished PhD Thesis submitted to UMIST.

Subba Rao, T. (1977) On the Estimation of Bilinear Time Series Models. *Bull. Int. Statist. Inst., 41*. Paper presented at the 41st Session of ISI Meeting, New Delhi.

Subba Rao, T. (1981) On the Theory of Bilinear Time Series Models. *J. Roy. Statist. Soc. B. 43*: 244-255.

Subba Rao, T. (1979) Discussion on a paper on "Robust estimation of Power spectra" presented to the Royal Statistical Society. *J. Roy. Statist. Soc. 41*: 313–353, p. 347.

Subba Rao, T. (1985) Statistical Analysis of Bivariate Bilinear Time Series Models. Paper presented at the International Symposium on Advances in Multivariate Statistical Analysis. Indian Statistical Institute, Calcutta. 16–20 Dec, 1985.

Subba Rao, T. and Gabr, M. M. (1980) A Test for Linearity of Stationary Time Series. *J. Time Ser. Anal. 1, 2*: 145–158.

Subba Rao, T. and Gabr, M. M. (1984) An Introduction to Bispectral Analysis and Bilinear Time Series Models. Lecture notes in Statistics, Vol. 24. Springer-Verlag, New York.

Subba Rao, T. and Maria Eduarda A. da Silva (1992) Identification of bilinear time series models $BL(p, o, p, 1)$. *Statistica Sinica, 2*: 465–478.

Subba Rao, T. and Wong, W. K. (1998) Tests for Gaussianity and Linearity of Multivariate Stationary Time Series. *J. Stat. Plan. and Inf. 68*: 373–386.

Tang, Z (1987) Bilinear Stochastic Processes and Time Series, Unpublished PhD Thesis, Oregon State University, U.S.A.

Terdik, Gy. (1985) Transfer Functions and Conditions for Stationarity of Bilinear Models with Gaussian Residuals. *Proc. Roy. Soc. (London) A 400*: 315–330.

Tong, H. (1990) Nonlinear Time Series. Oxford University Press, Oxford.

Wong, W. K. (1993) Some contributions to Multivariate Stationary Nonlinear Time Series. Unpublished PhD Thesis submitted to UMIST.

Wong, W. K. (1997) Frequency domain tests of multivariate Gaussianity and Linearity. *J. Time Ser. Anal. 18, 2*: 181–194.

9

Optimal Testing for Semi-Parametric AR Models—From Gaussian Lagrange Multipliers to Autoregression Rank Scores and Adaptive Tests

MARC HALLIN* Institute of Statistics and Operations Research, European Center for Advanced Research in Economics, and Université Libre de Bruxelles, Brussels, Belgium

BAS J. M. WERKER Université Libre de Bruxelles, Brussels, Belgium

1. INTRODUCTION

1.1 Hypothesis Testing and Time Series

Despite its importance in many aspects of the analysis, hypothesis testing is a less frequently addressed subject than several others in the area of time series. Whereas a huge literature deals with the many facets of point estimation problems, several aspects of hypothesis testing remain largely unexplored, even in the simple context of finite-order autoregressive (AR) models. This situation is all the more surprising in view of the basic role of hypothesis testing techniques in a variety of problems connected with the analysis of time series data: model identification, order selection, correlogram inspection, diagnostic checking, etc.

*Research supported by an A.R.C. contract of the Communauté française de Belgique, the European Human Capital contract ERB CGRX-CT 940693 and the Fonds d'Encouragement à la Recherche de l'Université Libre de Bruxelles.

In this chapter we deliberately confine ourselves to the simplest of all time series models, the traditional AR(p) one, under the usual *causality* (*stationarity*) assumptions—the asymptotics of explosive or unit-root AR models being slightly more complicated, see Jeganathan (1995 and 1997). The innovation density, however, will be considered unspecified—a semi-parametric point of view which is by far more realistic than the classical Gaussian one. In this restricted and assumedly well-known context, we will consider the problem of testing arbitrary linear restrictions on the vector of autoregressive parameters. A survey of possible methods—most of which are optimal in a certain sense—is provided. The autoregressive parameter thus will be the only parameter of interest, though the innovation density, which is treated here as a nuisance, could be of equal interest, especially for prediction.

The autoregressive model has been selected because of its simplicity and pedagogical virtues. We nevertheless believe that several of the procedures described below will be new to many readers—despite the widespread use of AR models in everyday practice. And, the results described here can be expected to hold, with adequate modification, in much more general semi-parametric situations: mixed ARMA models, random coefficient AR models, models with ARCH or GARCH innovation processes, nonlinear models, etc.

1.2 Definitions and Main Notation

Let $(X_{-p+1}, \ldots, X_0, X_1, \ldots, X_t, \ldots, X_n)'$ be an observed series, of length $n + p$. Throughout the paper, we assume that $\mathbf{X}^{(n)} = (X_1, \ldots, X_n)'$ satisfies the stochastic difference equation (AR(p) model)

$$X_t - \sum_{i=1}^{p} \theta_i X_{t-i} = \varepsilon_t, \qquad t = 0, \pm 1, \pm 2, \ldots, \tag{1.1}$$

where $\{\varepsilon_t\}, t = 0, \pm 1, \pm 2, \ldots$ is an iid sequence with mean zero and probability density f (the *innovation* density). We denote by $P_{f;\boldsymbol{\theta}}^{(n)}$ the probability distribution of $\mathbf{X}^{(n)}$, conditional on $(X_{-p+1}, \ldots, X_0)$.

As usual, it is assumed that the parameter $\boldsymbol{\theta} = (\theta_1, \ldots, \theta_p)' \in \mathbb{R}^p$ (the *autoregressive* parameter) is such that all the roots of the *characteristic* polynomial

$$\theta(z) = 1 - \sum_{i=1}^{p} \theta_i z^i, \qquad z \in \mathbb{C} \tag{1.2}$$

lie outside the unit disk. The set of all parameter values $\theta \in \mathbb{R}^p$ satisfying this condition is denoted as Θ; for $\theta \in \Theta$, the stochastic equation in Eq. (1.1) admits a stationary solution.

We do not assume here that $\mathbf{X}^{(n)}$ is a realization of that stationary solution. Indeed, irrespective of the starting values $(X_{-p+1}, \ldots, X_0)$, $\mathbf{X}^{(n)}$ is asymptotically stationary, in the sense that, denoting by λ the root of Eq. (1.2) which is closest to the unit circle, there exists a stationary solution $\{X_t^*\}$ of Eq. (1.1) such that the difference $X_t - X_t^*$ is $O_P(|\lambda|^{-t})$ as $t \to \infty$; the influence on $\mathbf{X}^{(n)}$ and on the conditional distribution $P_{f;\theta}^{(n)}$ of the starting values $X_{-p+1}, \ldots, X_0$ thus is asymptotically negligible; see Sec. 1.4 for further discussion.

The null hypotheses we are interested in are the *linear* hypotheses, under which θ belongs to some linear restriction of Θ or, equivalently, satisfies some given set of linear constraints. Such hypotheses are characterized by a $p \times r$ matrix Ω, of full rank $r \leq p$, and by an element θ_0 of $\mathbb{R}^p$: denoting by $\mathcal{M}(\Omega)$ the r-dimensional linear subspace of $\mathbb{R}^p$ spanned by the columns of Ω, we consider the hypothesis under which $\theta - \theta_0$ belongs to $\Theta \cap \mathcal{M}(\Omega)$, and thus satisfies a set of $p - r$ linearly independent linear constraints. Of course, for $\theta_0 \in \mathcal{M}(\Omega)$, this notation makes little sense ($\theta_0 = 0$ would characterize the same hypothesis); therefore, we tacitly assume that either $\theta_0 = 0$, or $\theta_0 \in \mathbb{R}^p \backslash \mathcal{M}(\Omega)$.

In the parametric approach to hypothesis testing, the innovation density f is specified up to a scale factor—typically, the standard error σ_f—and we denote by $H_f^{(n)}(\theta_0; \Omega)$ the linear hypothesis characterized by f, θ_0 and Ω. More specifically,

$$H_f^{(n)}(\theta_0; \Omega) \overset{\text{def}}{=} \{P_{f;\theta}^{(n)} | \theta - \theta_0 \in \Theta \cap \mathcal{M}(\Omega)\}. \tag{1.3}$$

Such a heavy notation is necessary since, in many places below, precise statements have to be made, which require an exact specification of n, f, θ_0 and Ω.

In the semi-parametric approach, except maybe for some technical assumptions, the innovation density remains unspecified (within the family $\mathcal{F}$ of densities satisfying the required technical assumption; these assumptions depend upon the context, and the definition of $\mathcal{F}$ accordingly may vary from one section to another), and plays the role of a *nuisance*; we then use the following notation

$$H^{(n)}(\theta_0; \Omega) \overset{\text{def}}{=} \{P_{f;\theta}^{(n)} | f \in \mathcal{F}; \theta - \theta_0 \in \Theta \cap \mathcal{M}(\Omega)\}$$
$$= \bigcup_{f \in \mathcal{F}} H_f^{(n)}(\theta_0; \Omega). \tag{1.4}$$

The notation $H_f^{(n)}(\theta_0)$, $H^{(n)}(\theta_0)$ and $H_f^{(n)}(\Omega)$, $H^{(n)}(\Omega)$ will be used instead of $H_f^{(n)}(\theta_0; \mathbf{O})$, $H^{(n)}(\theta_0; \mathbf{O})$ and of $H_f^{(n)}(\mathbf{0}; \Omega)$, $H^{(n)}(\mathbf{0}; \Omega)$, respectively. Clearly, under $H_f^{(n)}(\theta_0)$, the (conditional) distribution of $\mathbf{X}^{(n)}$ is entirely specified as $P_{f;\theta_0}^{(n)}$; under $H^{(n)}(\theta_0)$, the value θ_0 of the autoregressive parameter θ is specified, but not the innovation density f.

An important particular case is the one considered in several order-identification methods, where null hypotheses of the form $\theta_p = 0$ (i.e., $AR(p-1)$ dependence) are tested against $\theta_p \neq 0$ ($AR(p)$ dependence) for $p = 1, \ldots$. In our notation, this corresponds to null hypotheses of the form $H_f^{(n)}(\Omega)$ (parametric case) or $H^{(n)}(\Omega)$, (semi-parametric case), with

$$\Omega = \begin{pmatrix} \mathbf{I}_{(p-1)\times(p-1)} \\ 0 \quad \cdots \quad 0 \end{pmatrix} \tag{1.5}$$

($\mathbf{I}_{p\times p}$, here and in the sequel, stands for the $p \times p$ unit matrix).

The function $\mathbf{Z}^{(n)} : \Theta \times \mathbb{R}^{n+p} \mapsto \mathbb{R}^n$ mapping $(\theta, (X_{-p+1}, \ldots, X_n)')$ to $\mathbf{Z}^{(n)}(\theta) = (Z_1(\theta), \ldots, Z_n(\theta))'$, where

$$Z_t(\theta) \overset{\text{def}}{=} X_t - \sum_{i=1}^{p} \theta_i X_{t-i} \tag{1.6}$$

is called the residual function; $\mathbf{Z}^{(n)}(\theta)$ is the residual series associated with the parameter value θ. Obviously, the semi-parametric hypothesis $H^{(n)}(\theta_0)$ holds iff $Z_t^{(n)}(\theta_0)$ coincides with ε_t for $t = 1, \ldots, n$.

Using the above notation, the likelihood function for $P_{f;\theta}^{(n)}$—recall that, for fixed n, we are working here with conditional likelihoods—takes the very simple form

$$L_{f;\theta}^{(n)}(\mathbf{X}^{(n)}) = \prod_{t=1}^{n} f(Z_t(\theta)). \tag{1.7}$$

1.3 Outline of the Chapter

Section 2 starts with the most traditional optimal parametric testing procedure: the (Gaussian) Lagrange multiplier method, also known as the Rao score test. After a brief description of the method, we show, in Sec. 2.2, how the Lagrange multiplier test statistic actually reduces to a quadratic form in the residual autocorrelations resulting from a (Gaussian) maximum likelihood estimation of θ under the null. Though completely parametric in its spirit, this method however also can be considered as a semi-parametric one, since it remains valid under a broad range of innovation densities.

In Sec. 3, non-Gaussian versions of the same Lagrange multiplier method are described, suggesting a generalized concept of residual autocorrelation, adapted to the density f under consideration, but reducing to the classical concept in the case of a Gaussian f. The non-Gaussian Lagrange multiplier test statistic then also takes the form of a quadratic function of these generalized residual autocorrelations.

The most adequate framework for a study of asymptotically optimal testing methods, however, is not that of Lagrange multipliers, and Sec. 4.1–4.3 are devoted to an elementary presentation of some elements of Le Cam's theory of locally asymptotically normal (LAN) experiments. Under very general conditions on the innovation density f, AR(p) experiments indeed are LAN, with central sequences (playing, locally and asymptotically, the role of a sufficient statistic) generalizing the vector of scores on which the Lagrange multiplier methods were based in Sec. 2. For fixed f, the Le Cam theory then automatically provides (Sec. 4.4) the form of the locally asymptotically optimal tests, along with (Sec. 4.5) explicit values for their local powers and asymptotic relative efficiencies (AREs).

Now, the Le Cam approach still is of a parametric nature. If the semiparametric aspects of the problem are taken into account, the innovation f plays the role of a nuisance parameter, and deserves a special treatment. Two main attitudes are possible in this respect: either the nuisance is "eliminated" in some appropriate way, or it is estimated. These two attitudes are considered in Sec. 5 (permutation and rank tests) and Sec. 6 (adaptive tests), respectively.

The classical theory of hypothesis testing, as developed, for instance, in Lehmann (1986), traditionally eliminates nuisance parameters by applying either the unbiasedness, or the invariance principle. We briefly show, in Sec. 5.1 and 5.2, how unbiasedness in the present context leads to the consideration of permutation tests. Similarly (Sec. 5.3), the invariance argument yields tests based on a maximal invariant: the rank tests.

Despite the restriction to rank-based test statistics, optimality will be preserved if rank-based versions of the central sequences can be obtained. Such versions are provided in Sec. 5.4, where we also show how they correspond to rank-based versions of the generalized correlograms defined in Sec. 3.2. Depending on the choice of a score function, autocorrelations of the van der Waerden, Wilcoxon, Laplace, or Spearman types can be constructed, leading to tests which are optimal under the normal, logistic, or double exponential densities, respectively. Some details are given in Sec. 5.5 about the alignment problem, i.e., the problem of substituting an estimate for the actual value, under the null, of the parameter. Asymptotic relative efficiencies are also derived, showing that rank tests are not only more robust and more reliable under the null: they also can be more powerful than their parametric competitors.

Section 5.6 is devoted to a variant of the rank-based methods described in the previous sections: the so-called ranked residual tests, inspired from

Kreiss (1987a and 1990a). These tests rely on a somewhat hybrid class of statistics, where both the ranks of estimated residuals and the observations themselves are involved. Section 5.7 describes the application, in the AR(p) context, of a concept first developed by Gutenbrunner and Jurečková (1992) for linear models with independent observations: (auto)regression rank scores. This method, quite remarkably, avoids the estimation of any nuisance parameters, thus offering an interesting alternative to the alignment approach. The particular problem of testing AR($p - 1$) against AR(p) dependence is considered as an illustration, demonstrating the power and striking simplicity of the method.

Section 6 reviews the problem of estimating the underlying innovation density from the data, and to obtain adaptive tests, which uniformly achieve optimal asymptotic performances and perform as well, asymptotically, as if the "true" innovation density were known. Associated with these tests, we propose in Sec. 6.4 a concept of adaptive residual correlogram. Finally, in Sec. 6.5, we show how the benefits of adaptivity and those of rank tests can be combined, and introduce adaptive rank tests which, in some sense, are a form of aligned permutation tests.

Section 7 provides some general comments and conclusions.

1.4 The Initial Value Problem

The initial value problem is present in all time series models involving difference equations. Specifying f and θ in Eq. (1.1) indeed is not sufficient if a likelihood function is to be written for the observed series $(X_{-p+1}, \ldots, X_0, \mathbf{X}^{(n)})$: a specification of the distribution of the initial values $(X_{-p+1}, \ldots, X_0)$ is also required. Several solutions can be, and have been adopted, which all are asymptotically equivalent; see Billingsley (1961, pp. 4–5), for a general discussion of this fact.

Assuming independence between $(X_{-p+1}, \ldots, X_0)$ and $\{\varepsilon_t, t \geq 1\}$, and a joint probability density g_θ for $(X_{-p+1}, \ldots, X_0)$, the general form of the unconditional likelihood is

$$L_{f:\theta}^{(n)}(X_{-p+1}, \ldots, X_0, \mathbf{X}^{(n)}) = g_\theta(X_{-p+1}, \ldots, X_0) \prod_{t=1}^{n} f(Z_t(\theta)). \qquad (1.8)$$

An implicit specification of g_θ is often achieved by imposing stationarity; this is the attitude adopted, for instance, by Kreiss (1987a). The assumption that the observed series is a finite realization of the stationary solution of Eq. (1.1) under innovation density f indeed characterizes g_θ as the p-dimensional marginal density of this uniquely defined stationary distribution. But, even so, and with the notable exception of Gaussian innovations, this stationary density is quite a complicated function of f and θ. While unnecessarily restricting the generality of the approach, the stationarity assumption thus does not bring any practical tractable solution to this specification problem—except, perhaps, under Gaussian conditions.

On the other hand, if the stationarity assumption is to be relaxed, some plausible and intuitively interpretable density g_θ should be proposed for the starting values; but no convincing candidate seems to qualify. Specifying the density of $(X_{-p+1}, \ldots, X_0)$ in a possibly non-Gaussian context thus seems highly unrealistic, and g_θ, typically, should be treated as a nuisance parameter.

Fortunately, all this discussion is somewhat futile, since the influence on $L_{f;\theta}^{(n)}$ of g_θ is typically negligible. If it is assumed, for instance, that

$$\log(g_\theta(X_{-p+1}, \ldots, X_0)) - \log(g_{\theta+n^{-1/2}\tau}(X_{-p+1}, \ldots X_0)) = o_P(1), \qquad (1.9)$$

under Eq. (1.8), as $n \to \infty$, for any $\tau \in \mathbb{R}^p$, it is easy to see—cf., e.g., Swensen (1985), who also refers to Billingsley (1961)—that, under very general assumptions (those required, for instance, in Sec. 4), the influence of g_θ is, locally and asymptotically, nil. More precisely, in the LAN context of Sec. 4, the local experiments converge weakly to a limit which does not depend on g_θ; the two forms of Eq. (1.7) and (1.8) of the likelihood $L_{f;\theta}^{(n)}$ are thus equivalent for our purpose.

In a more general, nonlinear context, where the starting value problem is considerably more complex than in the present linear autoregression case, Koul and Schick (1997) replace Eq. (1.9) with the L_1 continuity condition

$$\lim_{t \to \theta} \left[\int |g_t(x_1, \ldots, x_p) - g_\theta(x_1, \ldots, x_p)| \, dx \right] = 0 \qquad (1.10)$$

(actually, their initial value density g_θ also depends on the innovation density f, and, in an adaptive context, they even require a stronger continuity property with respect to both θ and f), and provide a very careful treatment of this problem.

Equation (1.10) is satisfied in the linear AR case, and clearly implies Eq. (1.9). However, the conditional approach adopted here allows for skipping this starting value controversy. This is why it is adopted here, as the easiest way to obtain the very simple likelihood of Eq. (1.7), without implying any loss of generality under large-sample situations: the reader willing to assume the stationarity of $\mathbf{X}_n$ can do so safely, without any modification, in the subsequent sections.

2. GAUSSIAN LAGRANGE MULTIPLIERS

2.1 The Gaussian Lagrange Multiplier Test

One of the most classical ways of obtaining asymptotic optimal tests is the Lagrange multiplier method, also known as the Rao score method, first proposed by Rao (1947). For its implementation in the AR context, the reader is referred to Godfrey (1979), Hosking (1980), Pötscher (1983), or Pham (1987). As we shall see in Sec. 4, Le Cam's concept of weak convergence of statistical experiments will provide the adequate framework for a precise description of the optimality properties of the method.

Denote by $L_{\mathscr{G};\theta}^{(n)}$ the Gaussian likelihood obtained by substituting a $\mathscr{N}(0,\sigma^2)$ density for f in Eq. (1.7), and consider the derivatives, with respect to θ, of the resulting Gaussian log-likelihood

$$n^{-1/2}\mathrm{grad}_{\theta}\log L_{\mathscr{G};\theta}^{(n)}(\mathbf{X}^{(n)}) = -\frac{n^{-1/2}}{\sigma^2}\sum_{t=1}^{n}Z_t(\theta)\,\mathrm{grad}_{\theta}(Z_t(\theta))$$

$$= -\frac{n^{-1/2}}{\hat{\sigma}_n^2}\sum_{t=1}^{n}Z_t(\theta)\,\mathrm{grad}_{\theta}(Z_t(\theta)) + o_P(1)$$

$$\stackrel{\mathrm{def}}{=} \Delta_{\mathscr{G};\theta}^{(n)}, \tag{2.1}$$

where $\hat{\sigma}_n^2 \stackrel{\mathrm{def}}{=} n^{-1}\sum_{t=1}^{n}Z_t^2(\theta)$. In the present case, Eq. (1.6) yields

$$\mathrm{grad}_{\theta}(Z_t(\theta)) = \mathrm{grad}_{\theta}\left(X_t - \sum_{i=1}^{p}\theta_i X_{t-i}\right)$$

$$= -(X_{t-1},\ldots,X_{t-p})', \tag{2.2}$$

so that

$$\Delta_{\mathscr{G};\theta}^{(n)} = \frac{n^{-1/2}}{\hat{\sigma}_n^2}\sum_{t=1}^{n}Z_t(\theta)\begin{pmatrix}X_{t-1}\\ \vdots\\ X_{t-p}\end{pmatrix}$$

$$\xrightarrow{\mathscr{L}} \mathscr{N}(\mathbf{0},\Gamma_{\mathscr{G}}(\theta)) \tag{2.3}$$

under $H_f^{(n)}(\theta)$ as $n \to \infty$ where f is any density such that

$$\int xf(x)\,dx = 0 \quad \text{and} \quad \int x^2 f(x)\,dx \stackrel{\mathrm{def}}{=} \sigma^2 < \infty. \tag{2.4}$$

The asymptotic covariance matrix $\Gamma_{\mathscr{G}}(\theta)$ is the autocovariance matrix of order p of the stationary AR(p) solution of Eq. (1.1) under standard normal innovations; namely, denoting by $g_u(\theta)$ the Green's functions associated with (1.1), characterized by

$$\left(1 - \sum_{i=1}^{p} \theta_i z^i\right)^{-1} \overset{\text{def}}{=} \sum_{u=0}^{\infty} g_u(\boldsymbol{\theta}) z^u \qquad |z| < 1,$$

we have

$$\Gamma_{\mathscr{G}}(\boldsymbol{\theta}) = \left(\sum_{u=0}^{\infty} g_u(\boldsymbol{\theta}) g_{u+|i-j|}(\boldsymbol{\theta})\right). \tag{2.5}$$

It is easily seen that the g_u's are equicontinuous functions of $\boldsymbol{\theta} \in \Theta$, so that $\Gamma_{\mathscr{G}}(\boldsymbol{\theta})$ itself is continuous with respect to $\boldsymbol{\theta}$.

Denote by $\hat{\boldsymbol{\theta}}^{(n)}$ the Gaussian maximum likelihood estimator of $\boldsymbol{\theta}$, constrained by the condition $\boldsymbol{\theta} - \boldsymbol{\theta}_0 \in \mathscr{M}(\Omega)$, and define

$$Q_{\mathscr{G}}^{(n)}(\boldsymbol{\theta}) \overset{\text{def}}{=} \Delta_{\mathscr{G};\boldsymbol{\theta}}^{(n)'} [\Gamma_{\mathscr{G}}^{-1}(\boldsymbol{\theta}) - \Omega(\Omega' \Gamma_{\mathscr{G}}(\boldsymbol{\theta})\Omega)^{-1}\Omega'] \Delta_{\mathscr{G};\boldsymbol{\theta}}^{(n)}$$

$$= \Delta_{\mathscr{G};\boldsymbol{\theta}}^{(n)'} \Gamma_{\mathscr{G}}^{-1/2}(\boldsymbol{\theta}) [\mathbf{I}_{p \times p} - \Pi_{\Gamma_{\mathscr{G}}^{1/2}(\boldsymbol{\theta})\Omega}] \Gamma_{\mathscr{G}}^{-1/2}(\boldsymbol{\theta}) \Delta_{\mathscr{G};\boldsymbol{\theta}}^{(n)}, \tag{2.6}$$

where $\Pi_{\mathbf{N}} \overset{\text{def}}{=} \mathbf{N}(\mathbf{N}'\mathbf{N})^{-1}\mathbf{N}'$ stands for the orthogonal (Euclidean) projection matrix onto $\mathscr{M}(\mathbf{N})$, and $\mathbf{M}^{1/2}$ for the symmetric square root of the positive definite matrix $\mathbf{M}$.

At (asymptotic) probability level α, the Gaussian Lagrange multiplier test then consists in rejecting the null hypothesis $H_{\mathscr{G}}^{(n)}(\boldsymbol{\theta}_0; \Omega)$ (the subscript $\mathscr{G}$ in $H_{\mathscr{G}}^{(n)}$ refers to a Gaussian innovation density) whenever the quadratic test statistic $Q_{\mathscr{G}}^{(n)}(\hat{\boldsymbol{\theta}}^{(n)})$ exceeds the $(1 - \alpha)$-quantile $\chi^2_{p-r;1-\alpha}$ of a chi-square distribution with $p - r$ degrees of freedom— thus a typical parametric procedure. However, since Eq. (2.3) holds under any innovation density $g \in \mathscr{F} \overset{\text{def}}{=} \{f : \text{Eq. (2.4) holds}\}$, the same test can be performed for the semiparametric null hypothesis $H^{(n)}(\boldsymbol{\theta}_0; \Omega)$.

2.2 A Correlogram-Based Interpretation

The Lagrange multiplier test statistic in Eq. (2.6) can also be expressed as a quadratic form in the residual autocorrelation coefficients. Let

$$r_u^{(n)}(\boldsymbol{\theta}) \overset{\text{def}}{=} (n - u)^{-1} \sum_{t=u+1}^{n} \frac{Z_t(\boldsymbol{\theta}) Z_{t-u}(\boldsymbol{\theta})}{\hat{\sigma}_n^2}; \tag{2.7}$$

since it has been assumed that the innovation, hence the observed process, is centered at the origin, $r_u^{(n)}(\hat{\boldsymbol{\theta}}^{(n)})$, $u = 1, \ldots$ is the residual correlogram associated with the maximum likelihood estimate $\hat{\boldsymbol{\theta}}^{(n)}$. From the identity $X_t = \sum_{u=0}^{t-1} g_u(\boldsymbol{\theta}) Z_{t-u}(\boldsymbol{\theta})$, one easily obtains

$$
\Delta_{\mathscr{G}:\theta}^{(n)} = n^{1/2} \begin{pmatrix} \sum_{u=1}^{n-1} g_{u-1}(\theta) r_u^{(n)}(\theta) \\ \vdots \\ \sum_{u=p}^{n-1} g_{u-p}(\theta) r_u^{(n)}(\theta) \end{pmatrix} + o_P(1) \tag{2.8}
$$

under $H^{(n)}(\theta)$, as $n \to \infty$. Substituting this correlogram-based version of $\Delta_{\mathscr{G}:\hat{\theta}^{(n)}}^{(n)}$ into the test statistic in Eq. (2.6) thus provides an interpretation of the Lagrange multiplier test statistic in terms of the estimated residual correlogram obtained after constrained Gaussian maximum likelihood AR fitting.

Such an interpretation allows for connecting the Lagrange multiplier test with residual correlogram inspection. Just as the more popular Box–Pierce portmanteau test statistic, the Lagrange multiplier test statistic appears as a quadratic summary of residual autocorrelations. An important difference is that Eq. (2.6) is a weighted quadratic form, whereas the Box–Pierce statistic is an unweighted sum of squares; the weights are such that the resulting test enjoys local asymptotic optimality properties, the exact nature of which will appear more clearly in Sec. 4 below.

In the particular case of testing $\mathrm{AR}(p-1)$ against $\mathrm{AR}(p)$ dependence, Eq. (2.6) can be expressed, Garel and Hallin (1997), as

$$
Q_{p-1}^{(n)} = n(\mathbf{T}_{p-1}^{(n)})' \left[\begin{pmatrix} 1 & 0 & \cdots & 0 \\ 0 & & & \\ \vdots & & \mathbf{w}_{p-1}^2 & \\ 0 & & & \end{pmatrix} \right.
$$

$$
\left. - \begin{pmatrix} 1 & 0 & \cdots & 0 \\ & \mathbf{I}_{(p-1)\times(p-1)} & \end{pmatrix} \mathbf{W}_{p-1}^{-2} \begin{pmatrix} 1 \\ 0 \\ \vdots \\ 0 \end{pmatrix} \mathbf{I}_{(p-1)\times(p-1)} \right] \mathbf{T}_{p-1}^{(n)}, \tag{2.9}
$$

with (letting $\hat{r}_{p-1:i}$ and $\hat{g}_{p-1:u}$ denote the residual autocorrelations and Green's functions resulting from fitting an $\mathrm{AR}(p-1)$ model to the observed series)

$$
n^{1/2}\mathbf{T}_{p-1} = \begin{pmatrix} (n-1)^{1/2}\hat{r}_{p-1;1} \\ \sum_{i=2}^{n-1} \hat{g}_{p-1;i-1}(n-i)^{1/2}\hat{r}_{p-1;i} \\ \sum_{i=2}^{n-1} \hat{g}_{p-1;i-2}(n-i)^{1/2}\hat{r}_{p-1;i} \\ \vdots \\ \sum_{i=2}^{n-1} \hat{g}_{p-1;i-p+1}(n-i)^{1/2}\hat{r}_{p-1;i} \end{pmatrix}, \tag{2.10}
$$

$$
\mathbf{w}_{p-1}^2 = \begin{pmatrix} \sum_{i=1}^{n-1}(\hat{g}_{p-1;i})^2 & \sum_{i=1}^{n-1}\hat{g}_{p-1;i}\hat{g}_{p-1;i-1} & \cdots & \sum_{i=1}^{n-1}\hat{g}_{p-1;i}\hat{g}_{p-1;i-p+2} \\ \sum_{i=1}^{n-1}\hat{g}_{p-1;i}\hat{g}_{p-1;i-1} & \sum_{i=1}^{n-1}(\hat{g}_{p-1;i})^2 & \cdots & \sum_{i=1}^{n-1}\hat{g}_{p-1;i}\hat{g}_{p-1;i-p+3} \\ \vdots & \vdots & \ddots & \vdots \\ \sum_{i=1}^{n-1}\hat{g}_{p-1;i}\hat{g}_{p-1;i-p+2} & \sum_{i=1}^{n-1}\hat{g}_{p-1;i}\hat{g}_{p-1;i-p+3} & \cdots & \sum_{i=1}^{n-1}(\hat{g}_{p-1;i})^2 \end{pmatrix} \tag{2.11}
$$

(under $\theta = (\theta_1, \ldots, \theta_{p-1}, 0)'$, $g_{-1} = \ldots = g_{-p+2} = 0$), and

$$
\mathbf{W}_{p-1}^2 = \begin{pmatrix} 1 & 0 & \cdots & 0 \\ 0 & & & \\ \vdots & & \mathbf{0} & \\ 0 & & & \end{pmatrix} + \mathbf{w}_{p-1}^2. \tag{2.12}
$$

3. NON-GAUSSIAN LAGRANGE MULTIPLIERS

3.1 Non-Gaussian Lagrange Multiplier Tests

Though generally limited, in practice, to the Gaussian version just described, Lagrange multiplier tests theoretically also exist for general innovation densities f. Assume that Eq. (2.4) holds, and that f admits a derivative f'; letting $\phi_f \overset{\text{def}}{=} -f'/f$, assume furthermore that

$$
\mathscr{I}_f \overset{\text{def}}{=} \int \phi_f^2(x)f(x)\,dx < \infty. \tag{3.1}
$$

Using the same notation, and the same definitions as above, but replacing the Gaussian scores

$$\text{grad}_\theta \log(\exp\{-[Z_t(\theta)]^2/2\sigma^2\}) = -\frac{Z_t(\theta)}{\sigma^2}\text{grad}_\theta Z_t(\theta)$$

with the general ones

$$\text{grad}_\theta \log f(Z_t(\theta)) = \frac{f'(Z_t(\theta))}{f(Z_t(\theta))}\text{grad}_\theta Z_t(\theta) = -\phi_f(Z_t(\theta))\text{grad}_\theta Z_t(\theta),$$

(3.1)

the vector in Eq. (2.1) takes the form

$$\Delta_{f;\theta}^{(n)} \overset{\text{def}}{=} n^{-1/2}\text{grad}_\theta \log L_{f;\theta}^{(n)}(\mathbf{X}^{(n)})$$

$$= -n^{-1/2}\sum_{t=1}^{n}\phi_f(Z_t(\theta))\text{grad}_\theta(Z_t(\theta)).$$

(3.2)

The function ϕ_f in the traditional terminology is known as the score function associated with f. The innovation density f, however, at best, is specified up to a scale factor. Let f_1 be the standard version of $f : f(x) = \sigma^{-1}f_1(x/\sigma)$. Then, it is easily checked that $\phi_f(x) = \sigma^{-1}\phi_{f_1}(x/\sigma)$, so that

$$\sigma^2 \mathscr{I}_f = \mathscr{I}_{f_1}.$$

(3.3)

Denote by g a density satisfying Eq. (2.4), and such that

$$\mathscr{I}_g(f) \overset{\text{def}}{=} \int_{-\infty}^{\infty}\phi_{f_1}^2(z)g_1(z)\,dz - \left(\int_{-\infty}^{\infty}\phi_{f_1}(z)g_1(z)\,dz\right)^2 < \infty.$$

(3.4)

Then,

$$\Delta_{f;\theta}^{(n)} = \frac{n^{-1/2}}{\sigma}\sum_{t=1}^{n}\phi_{f_1}\left(\frac{Z_t(\theta)}{\sigma}\right)\begin{pmatrix}X_{t-1}\\ \vdots \\ X_{t-p}\end{pmatrix}$$

$$\overset{\mathscr{L}}{\to} \mathscr{N}(\mathbf{0}, \mathscr{I}_g(f)\Gamma_{\mathscr{G}}(\theta)),$$

(3.5)

under $H_g^{(n)}(\theta)$ as $n \to \infty$.

A consistent (under $H_g^{(n)}(\theta)$) empirical version of $\mathscr{I}_g(f)$ is

$$\hat{\mathscr{I}}_f^{(n)} \overset{\text{def}}{=} n^{-1}\sum_{t=1}^{n}\phi_{f_1}^2\left(\frac{Z_t(\theta)}{\hat{\sigma}_n}\right) - \left(n^{-1}\sum_{t=1}^{n}\phi_{f_1}\left(\frac{Z_t(\theta)}{\hat{\sigma}_n}\right)\right)^2.$$

(3.6)

Note however that

$$\Pi_{(\hat{\mathscr{I}}_f^{(n)}\Gamma_{\mathscr{G}}(\theta))^{1/2}\Omega} = \Pi_{(\mathscr{I}_f^{(n)}\Gamma_{\mathscr{G}}(\theta))^{1/2}\Omega} = \Pi_{\Gamma_{\mathscr{G}}^{1/2}(\theta)\Omega}.$$

For Gaussian densities f, $\mathscr{I}_g(f) = \int x^2 g_1(x)\, dx \equiv 1$, irrespective of g, so that an empirical version of $\mathscr{I}_g(\mathscr{G})$ is superfluous in the definition of the Gaussian Lagrange multiplier test statistic in Eq. (2.6); but the estimated $\hat{\mathscr{I}}_f^{(n)}$ is required in the corresponding non-Gaussian statistic in Eq. (3.7). Also observe that

$$\mathscr{I}_f(f) \overset{\text{def}}{=} \int_{-\infty}^{\infty} \phi_{f_1}^2(z) f_1(z)\, dz - 0 = \mathscr{I}_{f_1}$$

is the Fisher information quantity associated with μ in the location family $\{f_1(x - \mu) | \mu \in \mathbb{R}\}$.

Examples of score functions are

(i) f standard normal: $\phi_{f_1}(z) = z$; $\mathscr{I}_{f_1} = 1$;
(ii) f standard logistic: $\phi_{f_1}(z) = \pi(2F_1(z) - 1)/\sqrt{3}$; $\mathscr{I}_{f_1} = \pi^2/9$;
(iii) f standard double exponential: $\phi_{f_1}(z) = \sqrt{2}\,\text{sign}(z)$; $\mathscr{I}_{f_1} = 2$;

F_1 stands for the distribution function associated with f_1. Note however that the double exponential density does not meet the required smoothness assumptions, since it is not differentiable at the origin. As we shall see in Sec. 4, this differentiability assumption is too strong and can be weakened (into an assumption of mean-square differentiability of $f^{1/2}$) in order to accommodate the double exponential case.

Denote by $\hat{\theta}^{(n)}$ the constrained maximum likelihood (obtained from maximizing Eq. (1.7) under the condition $\theta - \theta_0 \in \mathscr{M}(\Omega)$): the Lagrange multiplier test of $H_f^{(n)}(\theta_0; \Omega)$, at asymptotic level α, then consists in rejecting the null hypothesis whenever $Q_f^{(n)}(\hat{\theta}^{(n)})$ exceeds the $(1 - \alpha)$-quantile $\chi^2_{p-r;1-\alpha}$ of a chi-square distribution with $p - r$ degrees of freedom, where

$$Q_f^{(n)}(\theta) \overset{\text{def}}{=} \Delta_{f:\theta}^{(n)\,'} [(\hat{\mathscr{I}}_f^{(n)}\Gamma_{\mathscr{G}}(\theta))^{-1} - \Omega(\Omega'(\hat{\mathscr{I}}_f^{(n)}\Gamma_{\mathscr{G}}(\theta))\Omega)^{-1}\Omega']\Delta_{f:\theta}^{(n)}$$

$$= (\hat{\mathscr{I}}_f^{(n)})^{-1}\Delta_{f:\theta}^{(n)\,'}\Gamma_{\mathscr{G}}^{-1/2}(\theta)[\mathbf{I}_{p\times p} - \Pi_{\Gamma_{\mathscr{G}}^{1/2}(\theta)\Omega}]\Gamma_{\mathscr{G}}^{-1/2}(\theta)\Delta_{f:\theta}^{(n)}. \qquad (3.7)$$

Here again, the Lagrange multiplier test has been designed as a parametric procedure. But, if we denote by $\mathscr{F}_f$ the class of densities g satisfying Eqns (2.4) and (3.4), the same test can be considered for the semi-parametric null hypothesis $H^{(n)}(\theta_0; \Omega)$. Remark that, for Gaussian f, Eq. (3.4) follows from the assumption in Eq. (2.4).

3.2   A Non-Gaussian Correlogram Concept

A correlogram-based version of Eq. (3.7) is also possible, suggesting the
following generalized residual correlogram concept. Let

$$
r_{f:u}^{(n)}(\boldsymbol{\theta}) \overset{\text{def}}{=} (n-u)^{-1} \sum_{t=u+1}^{n} \frac{\phi_f(Z_t(\boldsymbol{\theta}))}{\hat{\sigma}_n^{-1}(\hat{\mathscr{I}}_f^{(n)})^{1/2}} \frac{Z_{t-u}(\boldsymbol{\theta})}{\hat{\sigma}_n}
$$

$$
= (n-u)^{-1}(\hat{\mathscr{I}}_f^{(n)})^{-1/2} \sum_{t=u+1}^{n} \phi_{f_1}\left(\frac{Z_t(\boldsymbol{\theta})}{\hat{\sigma}_n}\right) \frac{Z_{t-u}(\boldsymbol{\theta})}{\hat{\sigma}_n}. \tag{3.8}
$$

Parallel to Eq. (2.8), it is easily seen that, under $H_f^{(n)}(\boldsymbol{\theta})$,

$$
\Delta_{f:\boldsymbol{\theta}}^{(n)} = n^{1/2} \begin{pmatrix} \displaystyle\sum_{u=1}^{n-1} g_{u-1}(\boldsymbol{\theta}) r_{f:u}^{(n)}(\boldsymbol{\theta}) \\ \vdots \\ \displaystyle\sum_{u=p}^{n-1} g_{u-p}(\boldsymbol{\theta}) r_{f:u}^{(n)}(\boldsymbol{\theta}) \end{pmatrix} + o_P(1). \tag{3.9}
$$

The generalized-correlogram-based form of the non-Gaussian Lagrange
multiplier test statistic in Eq. (3.7) readily follows.

For Gaussian f, $\phi_{f_1}(z) = z$, $\hat{\mathscr{I}}_f^{(n)}$ can be replaced with 1, and $r_{f:u}^{(n)}(\boldsymbol{\theta})$ coincides with the traditional residual autocorrelation $r_u^{(n)}(\boldsymbol{\theta})$. Furthermore,
$r_{f:u}^{(n)}(\boldsymbol{\theta})$, under $H_f^{(n)}(\boldsymbol{\theta})$ enjoys the same intuitive interpretation, and the
same inferential properties as its traditional counterpart—with increased
efficiency, though (see Sec. 4). For instance, for any k, the joint distribution,
under $H_g^{(n)}(\boldsymbol{\theta})$, $g \in \mathscr{F}_f$, of any k-tuple

$$
((n-1)^{1/2} r_{f:i_1}^{(n)}(\boldsymbol{\theta}), \ldots, (n-k)^{1/2} r_{f:i_k}^{(n)}(\boldsymbol{\theta}))',
$$

just as that of

$$
((n-1)^{1/2} r_{i_1}^{(n)}(\boldsymbol{\theta}), \ldots, (n-k)^{1/2} r_{i_k}^{(n)}(\boldsymbol{\theta}))',
$$

(where $i_1, \ldots, i_k$ are all distinct), is asymptotically multinormal, with mean $\mathbf{0}$
and unit covariance matrix. This allows, for instance, for non-Gaussian
portmanteau statistics, and for a correlogram-based interpretation of Eq.
(3.7).

Another interesting feature of the generalized autocorrelations in Eq.
(3.8) is that the present and the past in general do not play symmetric
roles in the definition. More precisely, only the Gaussian autocorrelations
(f Gaussian in Eq. (3.8)) are time-reversible, since the Gaussian density is
the unique solution of the differential equation $d \log f(x)/dx = -x$. This

asymmetry in the definition is perfectly consistent with the fact that only the Gaussian linear processes are time-reversible, Weiss (1975).

The terminology f-autocorrelation coefficient thus seems perfectly appropriate for $r_{f:u}^{(n)}(\theta)$, whereas the traditional correlograms clearly appear as a strongly Gaussian-flavored tool, much in the same spirit as arithmetic means, Student or Fisher statistics, least square regression, etc.

4. THE LE CAM APPROACH

4.1 Local Asymptotic Normality (LAN)

Local Asymptotic Normality (LAN) is the simplest and most general structure for standard asymptotic inference. When LAN holds, most traditional estimation and hypothesis testing problems are virtually solved (from an asymptotic point of view). Only a very basic knowledge of LAN and its consequences is required here; the reader who is not familiar with the concept is referred to Le Cam and Yang (1990).

Though LAN in the sequel will be treated as a parametric property— LAN holds, for AR models, with respect to the parametrization $\theta \in \Theta$, at fixed $f \in \mathcal{F}$—it also constitutes, as a basic tool in adaptive estimation, the key to the semi-parametric analysis of a variety of statistical models: see, e.g., Bickel, Klaassen, Ritov, and Wellner (1993) or Drost, Klaassen, and Werker (1997).

In the particular context of AR models, LAN has a long history, since autoregressive models provide one of the simplest cases, beyond the traditional location models with independent observations, where nontrivial applications of Le Cam's asymptotic theory of statistical experiments are possible. Among the many existing results, let us mention Davies (1973) and Dzhaparidze (1986) for Gaussian ARMA processes; Roussas (1979) for AR(1), Akritas and Johnson (1982) for general AR, and Swensen (1985) for general AR processes with a linear trend; Kreiss (1987a, 1990b) for ARMA and AR(∞) processes; Garel (1989) for the MA case; Hallin and Puri (1994) for a rank-based approach of the univariate ARMA case with a linear trend; Garel and Hallin (1995) for the multivariate ARMA case. A review of more general results in the ARMA and ARIMA contexts also can be found in Jeganathan (1995).

In this section, we will concentrate on the parametric testing problem— i.e., on null hypotheses of the form $H_f^{(n)}(\theta_0; \Omega)$. Semi-parametric problems will be considered in Sec. 4.3, as well as in Sec. 5–7. Though the presentation is original, the treatment is essentially equivalent to the one developed by Kreiss (1987a). The basic ingredients are

(I1) A LAN property, with central sequence $\Delta_{f;\theta}^{(n)}$ (the notational similarity with Eq. (3.2) is not just accidental),

(I2) A local asymptotic linearity result (with respect to θ) for the central sequence $\Delta_{f;\theta}^{(n)}$, $\theta \in \Theta$, and

(I3) A root-n consistent, locally discrete (this concept will be discussed more precisely below), constrained (i.e., satisfying $\hat{\theta}^{(n)} - \theta_0 \in \mathcal{M}(\Omega)$) estimate $\hat{\theta}^{(n)}$ of θ.

Once these three ingredients are available, the Le Cam theory of locally asymptotically normal experiments allows for constructing tests which are locally asymptotically optimal (maximin or most stringent), at asymptotic level α, for (sequences of) linear null hypotheses of the form $H_f^{(n)}(\theta_0; \Omega)$, against alternatives of the form $H_f^{(n)}(\theta_0; \Omega)$, $\theta - \theta_0 \notin \mathcal{M}(\Omega)$.

Some analytical assumptions of course are needed if ingredients (I1)–(I3) are to be available. Assume that f is such that

(A1) $f(x) > 0, x \in \mathbb{R}$; $\int_{-\infty}^{\infty} x f(x)\, dx = 0$; $\int_{-\infty}^{\infty} x^2 f(x)\, dx = \sigma^2 < \infty$;

(A2) f is absolutely continuous on finite intervals, i.e., there exists $\dot{f}$ such that, for all $\infty < a < b < \infty$, $f(b) - f(a) = \int_a^b \dot{f}(x)\, dx$;

(A3) Letting $\phi_f \overset{\text{def}}{=} -\dot{f}/f$, the generalized Fisher information $\int_{-\infty}^{\infty} \phi_f^2(x) f(x)\, dx \overset{\text{def}}{=} \mathscr{I}_f = \sigma^{-2} \mathscr{I}_{f_1}$ is finite.

Finally, if the influence of starting values on residual autocorrelations is to be asymptotically negligible, an additional assumption is convenient—though not required for the LAN result below:

(A4) The score function ϕ_f is piecewise Lipschitz, i.e., there exist a finite partition of $\mathbb{R}$ into nonoverlapping intervals $J_1, \ldots, J_k$ and a constant A_f such that

$$|\phi_f(x) - \phi_f(y)| \leq A_f |x - y| \qquad \forall x, y \in J_i, \ \forall i = 1, \ldots, k.$$

Note that assumption (A2) is considerably less restrictive than the differentiability assumption required in Sec. 3 for the non-Gaussian Lagrange multiplier methods. The (non-differentiable) double exponential density, for instance, satisfies (A1)–(A3) (with the score function $\phi(x) = \text{sign}(x)$), though it does not qualify for Lagrange multiplier tests.

The following LAN property then holds (whenever the notation $\theta + n^{-1/2}\tau^{(n)}$ is used, it is tacitly assumed that $(\tau^{(n)})'\tau^{(n)}$ is uniformly bounded, and that n is large enough for $\theta + n^{-1/2}\tau^{(n)} \in \Theta$ to hold).

PROPOSITION 4.1   Let f satisfy (A1)–(A3). Then, for all $\theta \in \Theta$,

$$\Lambda_{f:\boldsymbol{\theta}+n^{-1/2}\boldsymbol{\tau}^{(n)}/\boldsymbol{\theta}}^{(n)}(\mathbf{X}^{(n)}) \overset{\text{def}}{=} \log \frac{L_{f:\boldsymbol{\theta}+n^{-1/2}\boldsymbol{\tau}^{(n)}}^{(n)}(\mathbf{X}^{(n)})}{L_{f:\boldsymbol{\theta}}^{(n)}(\mathbf{X}^{(n)})}$$

$$= \boldsymbol{\tau}^{(n)\prime}\Delta_{f:\boldsymbol{\theta}}^{(n)}(\mathbf{X}^{(n)}) - \tfrac{1}{2}\boldsymbol{\tau}^{(n)\prime}\Gamma_{f:\boldsymbol{\theta}}\boldsymbol{\tau}^{(n)} + o_P(1) \qquad (4.1)$$

and

$$\Delta_{f:\boldsymbol{\theta}}^{(n)}(\mathbf{X}^{(n)}) \overset{\mathscr{L}}{\to} \mathscr{N}(\mathbf{0};\Gamma_{f:\boldsymbol{\theta}}) \qquad (4.2)$$

under $H_f^{(n)}(\boldsymbol{\theta})$, as $n \to \infty$, where

$$\Delta_{f:\boldsymbol{\theta}}^{(n)}(\mathbf{X}^{(n)}) \overset{\text{def}}{=} \frac{n^{-1/2}}{\sigma} \sum_{t=1}^{n} \phi_{f_1}\left(\frac{Z_t(\boldsymbol{\theta})}{\sigma}\right) \begin{pmatrix} X_{t-1} \\ \vdots \\ X_{t-p} \end{pmatrix}$$

$$= (n\mathscr{I}_{f_1})^{1/2} \begin{pmatrix} \sum_{u=1}^{n-1} g_{u-1}(\boldsymbol{\theta}) r_{f:u}^{(n)}(\boldsymbol{\theta}) \\ \vdots \\ \sum_{u=p}^{n-1} g_{u-p}(\boldsymbol{\theta}) r_{f:u}^{(n)}(\boldsymbol{\theta}) \end{pmatrix} + o_P(1), \qquad (4.4)$$

and

$$\Gamma_{f:\boldsymbol{\theta}} \overset{\text{def}}{=} \mathscr{I}_{f_1} \Gamma_{\mathscr{G}}(\boldsymbol{\theta}); \qquad (4.5)$$

$\Gamma_{\mathscr{G}}(\boldsymbol{\theta})$ denotes the Gaussian information matrix, as given in Eq. (2.3), so that $\Gamma_{f:}$ is a continuous function of $\boldsymbol{\theta} \in \Theta$.

Proposition 4.1 essentially takes care of (I1). As for the asymptotic linearity property required in (I2), it is provided by the following proposition; (I1) and (I2) often jointly are referred to as ULAN—Uniform Local Asymptotic Normality.

PROPOSITION 4.2   Under the same assumptions as in Proposition 4.1,

$$\Delta_{f:\boldsymbol{\theta}+n^{-1/2}\boldsymbol{\tau}^{(n)}}^{(n)}(\mathbf{X}^{(n)}) - \Delta_{f:\boldsymbol{\theta}}^{(n)}(\mathbf{X}^{(n)}) = \Gamma_{f:\boldsymbol{\theta}}\boldsymbol{\tau}^{(n)} + o_P(1) \qquad (4.6)$$

under $H_f^{(n)}(\boldsymbol{\theta})$, as $n \to \infty$.

As for (I3), let $\hat{\boldsymbol{\theta}}^{(n)}$ be such that

(B1) $\hat{\boldsymbol{\theta}}^{(n)} - \boldsymbol{\theta}_0 \in \mathscr{M}(\Omega)$ for all n, and (root-n consistency), under $H_f^{(n)}(\boldsymbol{\theta}_0;\Omega)$, as $n \to \infty$, $n^{1/2}(\hat{\boldsymbol{\theta}}^{(n)} - \boldsymbol{\theta}) = O_P(1)$;

assume moreover that

(B2) $\hat{\theta}^{(n)}$ is locally asymptotically discrete, i.e., $\forall \theta \in \theta_0 + \mathcal{M}(\Omega)$, $\forall c > 0$, there exists $K = K(\theta; c)$ such that the number of possible values of $\hat{\theta}^{(n)}$ in balls of the form

$$\{t \in \mathbb{R}^p : \| n^{1/2}(t - \theta) \| \leq c\} \tag{4.7}$$

is bounded by K, uniformly as $n \to \infty$.

These conditions imposed on $\hat{\theta}^{(n)}$ are rather mild, and all traditional estimates of θ in AR(p) models with finite innovation variance are root-n convergent: least squares and Yule–Walker estimators, exact or approximate (pseudo-)maximum likelihood estimators, M-estimators, R-estimators, ... all qualify for (B1); (B2) at first sight looks less classical, but has little practical implications.

4.2   Weak Convergence of Statistical Experiments

Local asymptotic normality implies, for all $\theta \in \Theta$, the weak convergence of the sequence of local experiments (localized at θ)

$$\mathscr{E}^{(n)} \overset{\text{def}}{=} \{\mathbb{R}^n, \mathscr{B}^n, \{P^{(n)}_{f;\theta+n^{-1/2}\tau}|\tau \in \mathbb{R}^p\}\} \tag{4.8}$$

to the p-dimensional Gaussian shift (the p-dimensional Gaussian location model)

$$\mathscr{E} \overset{\text{def}}{=} \{\mathbb{R}^p, \mathscr{B}^p, \{\mathcal{N}(\Gamma_{f;\theta}\tau; \Gamma_{f;\theta})|\tau \in \mathbb{R}^p\}\}. \tag{4.9}$$

This concept of convergence is based on a pseudo-distance between the sets of all possible risk functions (from $\mathbb{R}^p$ to $\mathbb{R}^+$) that can be implemented, under the experiments under consideration, for bounded loss functions. Roughly speaking, this implies, in the hypothesis testing context, that all power functions that are implementable from the statistical model $\mathscr{E}^{(n)}$ converge (as $n \to \infty$), pointwise in τ but uniformly with respect to the set of all possible testing procedures, to the power functions associated with the limit Gaussian location model $\mathscr{E}$. Conversely, all risk functions associated with $\mathscr{E}$ can be obtained as limits of sequences of risk functions associated with $\mathscr{E}^{(n)}$. For a more rigorous definition of this concept of weak convergence, the reader is referred to Le Cam (1986), or to Le Cam and Yang (1990).

Denoting by Δ the observation in Eq. (4.9), it follows that, if any test $\phi^*(\Delta)$ enjoys some exact optimality property in the limit Gaussian experiment $\mathscr{E}$, then the sequence $\phi^*(\Delta_{f;\theta}^{(n)})$ inherits, under local and asymptotic form, the same optimality results in the local sequence of experiments $\mathscr{E}^{(n)}$.

4.3   Maximin and Most Stringent Tests in The Gaussian Shift Experiment

Let θ be fixed, and consider the Gaussian shift model in Eq. (4.9), with mean $\Gamma_{f;\theta}\tau$ and covariance matrix $\Gamma_{f;\theta}$. Denote by Δ the observation in this model. Testing problems for this experiment are well-known problems, for which optimal solutions in general are easily obtained (see, e.g., Le Cam (1986), Chap. 11).

Consider, for instance, the problem of testing $\mu \overset{\text{def}}{=} \Gamma_{f;\theta}\tau = 0$ in $\mathscr{E}$. More precisely, define the null hypothesis $\mathscr{H} : \tau = 0$ and the alternative

$$\mathscr{H}_c : \mu'\Gamma_{f;\theta}^{-1}\mu = \tau'\Gamma_{f;\theta}\tau > c, \tag{4.10}$$

where $c > 0$ is an arbitrary positive constant. Recall that a test ϕ^* is maximin, at probability level α, for $\mathscr{H}$ against $\mathscr{H}_c$, iff (the notation $\mathrm{E}_P[\ ]$, $P \in \mathscr{H}$, and $P \in \mathscr{H}_c$ are used in an obvious fashion)

$$\begin{cases} \mathrm{E}_P[\phi^*] \leq \alpha & P \in \mathscr{H} \\ \mathrm{E}_P[\phi^*] = \sup\left\{ \inf_{P\in\mathscr{H}_c} \mathrm{E}_P[\phi] : \mathrm{E}_P[\phi] \leq \alpha \quad \forall P \in \mathscr{H} \right\} & P \in \mathscr{H}_c. \end{cases} \tag{4.11}$$

It follows from a classical result by Hunt and Stein (Lehmann (1986), Sec. 9.5) that the test rejecting $\mathscr{H}$ whenever

$$\Delta'\Gamma_{f;\theta}^{-1}\Delta > \chi_{p;1-\alpha}^2, \tag{4.12}$$

where $\chi_{p;1-\alpha}^2$ is the $(1 - \alpha)$-quantile of a chi-square distribution with p degrees of freedom, is maximin against $\mathscr{H}_c$ in the class of tests having probability level α under $\mathscr{H}$. In the univariate case, Eq. (4.12) also yields the uniformly most powerful unbiased test of $\mathscr{H}$ at probability level α.

The choice of c has no influence on the form of Eq. (4.12), despite its role in the definition of $\mathscr{H}_c$. The concept of stringency conveniently allows for avoiding the problem of bounding away, by means of an appropriate ellipsoid associated with c, the alternative $\mathscr{H}$ from the null hypothesis $\mathscr{H}$. The most stringent α-level test for $\mathscr{H}$ against $\mathscr{H} : \Gamma_{f;\theta}\tau \neq 0$ indeed coincides with the maximin test in Eq. (4.12), irrespective of c. For the sake of convenience, in the sequel, we therefore restrict ourselves to the concept of stringency.

Denote by $\mathscr{H}$ and $\mathscr{H}$, respectively, a null hypothesis and an alternative associated with the Gaussian shift model in Eq. (4.9). Let ϕ_0 belong to some class $\mathscr{C}$ of tests (e.g., the class of all tests having level α over $\mathscr{H}$). The regret $r_{\phi_0}(\tau)$ of ϕ_0 within $\mathscr{C}$, at $\tau \in \mathscr{H}$, is defined as the maximal loss of power resulting from using ϕ_0 rather than any other test in $\mathscr{C}$ (at $\tau \in \mathscr{H}$):

$$r_{\phi_0}(\tau) = \sup\{E_\tau(\phi) | \phi \in \mathscr{C}\} - E_\tau(\phi_0).$$

A test $\phi^* \in \mathscr{C}$ is called most stringent, within $\mathscr{C}$ and against $\mathscr{K}$, if its maximal (over $\mathscr{K}$) regret is minimal (over $\mathscr{C}$):

$$\sup_{\tau \in \mathscr{K}} r_{\phi^*}(\tau) \le \sup_{\tau \in \mathscr{K}} r_\phi(\tau) \qquad \forall \phi \in \mathscr{C}.$$

This concept of stringency goes back to Wald (1943). In the problem of testing the null hypothesis $\mathscr{H}$ that $\mu \stackrel{\text{def}}{=} \Gamma_{f;\theta}\tau$ in Eq. (4.9) satisfies a set of linear restrictions of the form $\tau \in \mathscr{M}(\Omega)$, i.e., $\mu \in \mathscr{M}(\Gamma_{f;\theta}\Omega)$, the most stringent test (against $\mathscr{K} : \mu \notin \mathscr{M}(\Gamma_{f;\theta}\Omega)$) consists in rejecting $\mathscr{H}$ whenever we have (same notation as above)

$$\Delta'[\Gamma_{f;\theta}^{-1} - \Omega(\Omega'\Gamma_{f;\theta}\Omega)^{-1}\Omega']\Delta$$

$$= \Delta'\Gamma_{f;\theta}^{-1/2}[\mathbf{I}_{p\times p} - \Gamma_{f;\theta}^{1/2}\Omega(\Omega'\Gamma_{f;\theta}\Omega)^{-1}\Omega'\Gamma_{f;\theta}^{1/2}]\Gamma_{f;\tau}^{-1/2}\Delta$$

$$> \chi^2_{p-r:1-\alpha}. \qquad (4.13)$$

4.4   Locally Asymptotically Optimal Tests

Combining the results of the previous two sections then yields the result that the sequence of tests rejecting $H_f^{(n)}(\theta_0; \Omega)$ whenever we have

$$Q_f^{(n)}(\theta) \stackrel{\text{def}}{=} (\Delta_{f;\theta}^{(n)})'\Gamma_{f;\theta}^{-1/2}[\mathbf{I}_{p\times p} - \Gamma_{f;\theta}^{1/2}\Omega(\Omega'\Gamma_{f;\theta}\Omega)^{-1}\Omega'\Gamma_{f;\theta}^{1/2}]\Gamma_{f;\theta}^{-1/2}\Delta_{f;\theta}^{(n)}$$

$$> \chi^2_{p-r:1-\alpha} \qquad (4.14)$$

is locally and asymptotically most stringent against

$$\bigcup \{H_f^{(n)}(\theta) : \theta \notin \theta_0 + \mathscr{M}(\Omega)\}$$

within the class of test having asymptotic level α—at arbitrary but fixed $\theta \in \theta_0 + \mathscr{M}(\Omega)$.

Since θ, even under $H_f^{(n)}(\theta_0; \Omega)$, remains unspecified, this latter result at first sight has little practical interest (except for the special case where Ω has rank 0, so that $\theta_0 + \mathscr{M}(\Omega) = \{\theta_0\}$). Note however that the test statistic, in Eq. (4.14), is nothing else but the Euclidean modulus of

$$[\mathbf{I}_{p\times p} - \Pi_{\Gamma_{f;\theta}^{1/2}\Omega}]\Gamma_{f;\theta}^{-1/2}\Delta_{f;\theta}^{(n)},$$

where $\Pi_{\Gamma_{f;\theta}^{1/2}\Omega}$ denotes the projection matrix onto $\mathscr{M}(\Gamma_{f;\theta}^{1/2}\Omega)$. The local discreteness property (B2) of $\hat{\theta}^{(n)}$ allows for applying the asymptotic linearity property (I3) of $\Delta_{f;\theta}^{(n)}$ with $\tau^{(n)}$ replaced with $n^{1/2}(\hat{\theta}^{(n)} - \theta^{(n)})$, which (Assumption (B1)) is $O_P(1)$ under $H_f^{(n)}(\theta_0; \Omega)$. Hence,

$$\Delta_{f;\hat{\theta}^{(n)}}^{(n)} - \Delta_{f;\theta}^{(n)} = n^{1/2}\Gamma_{f;\theta}(\hat{\theta}^{(n)} - \theta) + o_P(1), \tag{4.15}$$

still under $H_f^{(n)}(\theta_0;\Omega)$, as $n \to \infty$. Now, in view of Assumption (B1), $\Gamma_{f;\theta}(\hat{\theta}^{(n)} - \theta) \in \mathcal{M}(\Gamma_{f;\theta}\Omega)$; accordingly,

$$[\mathbf{I}_{p\times p} - \Pi_{\Gamma_{f;\theta}^{1/2}\Omega}]\Gamma_{f;\theta}^{-1/2}(\Delta_{f;\hat{\theta}^{(n)}}^{(n)} - \Delta_{f;\theta}^{(n)})$$

$$= n^{1/2}[\mathbf{I}_{p\times p} - \Pi_{\Gamma_{f;\theta}^{1/2}\Omega}]\Gamma_{f;\theta}^{1/2}(\hat{\theta}^{(n)} - \theta) + o_P(1) = o_P(1). \tag{4.16}$$

Moreover, Γ being a continuous function, $\Gamma_{f;\hat{\theta}^{(n)}}$ converges to $\Gamma_{f;\theta}$ as $n \to \infty$. This and Eq. (4.16) imply that substituting $\hat{\theta}^{(n)}$ for θ has no influence, locally and asymptotically (type 1 errors, local powers, optimality, . . .) on the most stringent test statistic in Eq. (4.14). Summing up: the test rejecting $H_f^{(n)}(\theta_0;\Omega)$ whenever $Q_f^{(n)}(\hat{\theta}^{(n)}) > \chi_{p-r;1-\alpha}^2$ is locally asymptotically most stringent at asymptotic probability level α.

4.5 Semi-Parametric Extensions and Local Powers

The optimal tests just described again have a strong parametric flavor. Throughout, the innovation density f has been fixed (up to a scale parameter), and local asymptotic optimality properties only hold at f. Whereas such a limitation of the optimality properties is inherent to non-adaptive methods (see Sec. 6 for an overview of adaptivity), the validity of the tests which are locally asymptotically optimal under f in general extends to a much broader set of innovation densities $g \neq f$.

Define $\mathscr{F}_f$ as the class of all densities g satisfying conditions (2.4) and (3.4), and such that

$$\Delta_{f;\theta+n^{-1/2}\tau}^{(n)}(\mathbf{X}^{(n)}) - \Delta_{f;\theta}^{(n)}(\mathbf{X}^{(n)}) = C_{f;g}\Gamma_{g;\theta}\tau^{(n)} + o_P(1), \tag{4.17}$$

for some constant $C_{f;g}$, under $H_g^{(n)}(\theta)$, as $n \to \infty$. Clearly, the test in Eq. (4.14) remains valid (asymptotic level α) under the semi-parametric null hypothesis $\bigcup_{g\in\mathscr{F}_f} H_g^{(n)}(\theta)$.

Local powers also can be explicitly derived under LAN, see Hallin and Puri (1994) for details.

PROPOSITION 4.3 Let f and $g(g \in \mathscr{F}_f)$ both satisfy assumptions (A1)–(A3). Then, under $H_g^{(n)}(\theta + n^{-1/2}\tau)$, with $\theta - \theta_0 \in \mathcal{M}(\Omega)$ and $\tau \notin \mathcal{M}(\Omega)$, the test statistic in Eq. (4.14) is asymptotically noncentral chi-square, with non-centrality parameter

$$[\mathscr{I}_g(f;g)/(\mathscr{I}_g(f))^{1/2}]^2 \mathcal{Q}(\theta;\tau), \tag{4.18}$$

Table 1. Asymptotic Relative Efficiencies (AREs) $[\mathscr{I}_g(f;g)/ (\mathscr{I}_g(f))^{1/2}]^2$, with Respect to Gaussian Lagrange Multipliers, of the Gaussian and Some Non-Gaussian Optimal Parametric Tests (Associated with Normal, Logistic and Double Exponential Densities f), Under Cosine, Normal, Logistic and Double Exponential Densities g, Respectively.

Scores	Cosine g	Normal g	Logistic g	Double exponential g
Normal f	1.000	1.000	1.000	1.000
Logistic f	1.008	0.941	1.098	1.385
Double exponential f	0.935	0.637	0.823	2.000

where $\mathscr{Q}(\theta; \tau)$ is a positive definite quadratic form depending on θ and τ, but not on f and g, and

$$\mathscr{I}_g(f;g) \stackrel{\text{def}}{=} \int_{-\infty}^{\infty} \phi_{f_1}(z)\phi_{g_1}(z)g_1(z)\,dz$$

$$= \int_0^1 \phi_{f_1}[G_1^{-1}(u)]\phi_{g_1}[G_1^{-1}(u)]\,du. \qquad (4.19)$$

Note that, in the case of a Gaussian density f,

$$\mathscr{I}_g(\mathscr{G};g) = \int_{-\infty}^{\infty} z\phi_{g_1}(z)g_1(z)\,dz = 1 = \mathscr{I}_g(\mathscr{G}) = \mathscr{I}_{\mathscr{G}},$$

and does not depend on g. Irrespective of the actual innovation density, the Gaussian methods thus achieve the same asymptotic performances as in the Gaussian case, for which they are optimal. It follows that Gaussian AR models are the least favourable ones: the Fisher information quantity $\mathscr{I}_{f_1}$ indeed is always larger than one, except at the normal density, where it equals one. Gaussian tests in this respect are maximin.

Moreover, the quadratic form $\mathscr{Q}(\theta; \tau)$ in Eq. (4.18) is nothing else but the noncentrality parameter of the asymptotic chi-square distribution of the traditional Gaussian Lagrange multiplier test statistic under local alternatives of the form $H_g^{(n)}(\theta + n^{-1/2}\tau)$. It follows that the asymptotic relative efficiency (ARE), under innovation density g, of (4.14) with respect to Gaussian Lagrange multipliers is $[\mathscr{I}_g(f;g)/(\mathscr{I}_g(f))^{1/2}]^2$. Some numerical values are provided in Table 1.

The cosine density reported in Table 1 (as well as in Tables 2 and 3) has standardized form

$$g_1(z) \overset{\text{def}}{=} \frac{\sqrt{\pi^2 - 8}}{4} \cos\left(\frac{\sqrt{\pi^2 - 8}}{2} z\right) I\left[-\frac{\pi}{\sqrt{\pi^2 - 8}} \leq z \leq \frac{\pi}{\sqrt{\pi^2 - 8}}\right], \quad (4.20)$$

with score function

$$\phi_{g_1}(z) \overset{\text{def}}{=} \frac{\sqrt{\pi^2 - 8}}{2} \tan\left(\frac{\sqrt{\pi^2 - 8}}{2} z\right) I\left[-\frac{\pi}{\sqrt{\pi^2 - 8}} \leq z \leq \frac{\pi}{\sqrt{\pi^2 - 8}}\right].$$

This score function is not square-integrable, and the cosine density does not satisfy assumption (A3) (the Fisher information is infinite). The results of Proposition 4.3 thus do not hold under this cosine density, and the corresponding figures in Table 1 are not, strictly speaking, asymptotic relative efficiencies. However, the cosine density can be obtained as the limit of a sequence of densities satisfying the assumptions required in Proposition 4.3, so that the corresponding sequence of AREs converges to the figures reported in Table 1. These figures accordingly can be interpreted as (limits of) AREs obtained under light-tailed innovation densities.

The cosine density in Eq. (4.20) has been chosen because of its extremal property under the Spearman correlograms defined in Eq. (5.12): indeed it is shown that it achieves (still, as a limiting case) the infimum of the ARE (with respect to Gaussian Lagrange multipliers), when Spearman correlograms are substituted for the traditional ones—see Eq. (5.27) and Sec. 5.5.

Some of the integrals computed here, as well as in Tables 2 and 3, are numerically unstable, and the third decimal figures, in some places, are not fully reliable.

4.6 Conclusions

The locally asymptotically optimal tests described in this section provide a generalization of the classical Lagrange multiplier testing method, and Le Cam's Local Asymptotic Normality constitutes the appropriate framework for an asymptotic approach to the problem. Compared with the Lagrange multiplier method, the tests derived in this section indeed

(i) require less restrictive assumptions on the underlying innovation densities, and

(ii) require less restrictive assumptions on the estimate $\hat{\theta}^{(n)}$ of θ under the null. Moreover,

(iii) they are much more flexible, since central sequences are only defined up to $o_P(1)$ (under the innovation density f for which optimality is achieved) quantities.

In addition, the LAN approach

(iv) also allows for well identified asymptotic optimality properties, and
(v)   provides explicit information about local powers and asymptotic rela-
      tive efficiencies.

Among all these interesting features, the most appealing one may well be
(iii), and this flexibility of central sequences will be exploited further in the
next sections, where rank-based (Sec. 5.4), adaptive (Sec. 6.2), and adaptive
rank-based (Sec. 6.5) central sequences will be constructed.

5. RANK-BASED TESTS

5.1 Testing for White Noise

Semi-parametric null hypotheses of the form $H^{(n)}(\theta_0)$, under which the
autoregression parameter θ is fixed but the innovation density remains
unspecified (within some class $\mathscr{F}$ of densities), thus playing the role of a
nuisance parameter, actually reduce to a null hypothesis of white noise.
Indeed, $X_1, \ldots, X_n$ satisfy Eq. (1.1) with $\theta = \theta_0$ and unspecified innovation
density $g \in \mathscr{F}$ iff the residuals

$$\mathbf{Z}^{(n)}(\theta_0) \overset{\text{def}}{=} (Z_1(\theta_0), \ldots, Z_n(\theta_0)) \tag{5.1}$$

defined in Eq. (1.6) are independent and identically distributed, with unspecified density $g \in \mathscr{F}$, i.e., iff they constitute a finite realization of some
(independent) white noise process. Testing $H^{(n)}(\theta_0)$ from $X_1, \ldots, X_n$ thus
reduces to testing that $\mathbf{Z}^{(n)}(\theta_0)$ is white noise.

Denote by $\mathscr{H}$ this hypothesis of white noise; $\mathscr{H}$ does not involve any
structural parameter anymore, but the nuisance (the density g) is still present, and somehow should be taken care of. Two main attitudes are possible
in this respect:

(i)   either the nuisance is estimated, in some adequate, consistent way: this
      is the solution adopted in adaptive methods (see Sec. 6);
(ii) or the nuisance is eliminated by means of some appropriate statistical
      principle. Classical statistics mainly offer two such principles: the
      unbiasedness principle, and the invariance principle. As we shall see,
      unbiasedness leads to permutation tests, whereas invariance yields the
      rank tests.

5.2 Unbiasedness and Permutation Tests

It is a well-known fact in the theory of hypothesis testing (see, e.g., Lehmann
(1986), Chap. 4) that, in the presence of nuisance parameters, the unbiasedness argument, via similarity and Neyman structure, suggests conditioning

with respect to some sufficient and (boundedly) complete statistic. This argument is classically used in the derivation of optimal unbiased tests in exponential families.

Sufficiency and completeness here are to be understood in the submodel that constitutes the common boundary $\bar{\mathscr{H}} \cap \bar{\mathscr{K}}$ (with respect to the coarsest topology ensuring the continuity of expectations of bounded functions) between the null $\mathscr{H}$ and the alternative $\mathscr{K}$. In the particular case of the null hypothesis of white noise, $\bar{\mathscr{H}} \cap \bar{\mathscr{K}}$ and $\mathscr{H}$ coincide. The conditioning statistic allowing for Neyman structure thus should be sufficient and boundedly complete for $\mathscr{H}$ itself.

The order statistic is sufficient and complete for the hypothesis $\mathscr{H}$ of white noise (see, e.g., Lehmann (1986), p. 143). Hence, in the problem of testing $H^{(n)}(\boldsymbol{\theta}_0)$, only those tests $\phi = \psi(\mathbf{Z}^{(n)}(\boldsymbol{\theta}_0))$ which satisfy the α-structure condition

$$E[\phi | \mathbf{Z}^{(n)}_{(\cdot)}(\boldsymbol{\theta}_0)] = \alpha \qquad \text{a.s. under } H^{(n)}(\boldsymbol{\theta}_0), \tag{5.2}$$

where

$$\mathbf{Z}^{(n)}_{(\cdot)}(\boldsymbol{\theta}_0) \stackrel{\text{def}}{=} (Z^{(n)}_{(1)}(\boldsymbol{\theta}_0), \ldots, Z^{(n)}_{(n)}(\boldsymbol{\theta}_0))$$

is the order statistic computed from $\mathbf{Z}^{(n)}(\boldsymbol{\theta}_0)$, qualify for unbiasedness (α-structure actually is necessary, but not sufficient). Now, denoting by $\boldsymbol{\Pi} \stackrel{\text{def}}{=} \{\pi = (\pi(1), \ldots, \pi(n))\}$ the set of all permutations of $(1, \ldots, n)$, condition (5.2) takes the form

$$\frac{1}{n!} \sum_{\pi \in \boldsymbol{\Pi}} \phi(Z^{(n)}_{\pi(1)}(\boldsymbol{\theta}_0), \ldots, Z^{(n)}_{\pi(n)}(\boldsymbol{\theta}_0)) = \alpha \qquad \text{a.s. under } H^{(n)}(\boldsymbol{\theta}_0); \tag{5.3}$$

a test ϕ satisfying (5.3) is called a permutation test. If unbiasedness is to be achieved—a requirement which, in a parametric context, is usually regarded as reasonable—the only tests to be taken into account (in the problem of testing $H^{(n)}(\boldsymbol{\theta}_0)$) thus are the permutation tests.

Note that permutation tests are entirely distribution-free. And, it can be shown (Hoeffding (1952)) that, under very general conditions, the permutation tests based on a sequence of test statistics $T^{(n)}$ (i.e., on the conditional distribution of $T^{(n)}$) are asymptotically as powerful as the sequence based on $T^{(n)}$'s unconditional distributions (which, in a semi-parametric context, typically are not known).

Consider, as an illustration, the problem of testing $\theta = 0$ in the AR(1) model

$$X_t - \theta X_{t-1} = \varepsilon_t, \qquad t = 0, \pm 1, \pm 2, \ldots \tag{5.4}$$

$(H^{(n)}(0)$ in our notation). Here the residuals under the null coincide with the observations. The usual parametric test is based on the approximate standard normal distribution of the first-order autocorrelation coefficient

$$r_1^{(n)} \overset{\text{def}}{=} \frac{\sum_{t=1}^{n} X_t X_{t-1}}{\sum_{t=1}^{n} X_t^2}. \tag{5.5}$$

This test, by the way, is also the locally asymptotically optimal one in the case of Gaussian innovations ($\theta = 0$ against $\theta \neq 0$ in Eq. (5.4), with Gaussian ε_t's). Unfortunately, in our semi-parametric problem (unspecified innovation density), this test is biased, since it does not satisfy condition (5.3). This bias can be removed if the test statistic is considered under its permutational distribution. Exact permutational critical values are easily obtained, either by enumerating the $n!$ possible permutations of $(X_1, \ldots, X_n)$, or by sampling them.

In case the innovation density f can be assumed to be symmetric with respect to the origin, the order statistic of the absolute values $|Z_t(\boldsymbol{\theta}_0)|$ is sufficient and complete, and the signs also enter the definition of permutation tests: see Dufour and Hallin (1991 and 1993) for further detail.

Of course, the domain of application of exact permutation tests so far is restricted to null hypotheses of the form $H^{(n)}(\boldsymbol{\theta}_0)$. This restriction somehow will be abandoned in Sec. 6, where adaptive rank tests will be considered for more general null hypotheses.

5.3 Invariance and Ranks

The second argument used in classical parametric hypothesis testing in the presence of nuisance parameters is invariance (Lehmann (1986), Chap. 6). This argument requires that the null hypothesis under study be generated by some group of transformations $\mathscr{G}, \circ$ acting on the observations.

Denote by $\mathscr{G}^{(n)}, \circ \overset{\text{def}}{=} \{\mathbf{h}^{(n)}\}, \circ$ the group of transformations $\mathbf{h}^{(n)} : \mathbb{R}^n \to \mathbb{R}^n$ defined by

$$\mathbf{h}^{(n)}\mathbf{x} = (h(x_1), \ldots, h(x_n)), \quad \mathbf{x} = (x_1, \ldots, x_n) \in \mathbb{R}^n,$$

where $h : \mathbb{R} \to \mathbb{R}$ is continuous, monotone increasing, and such that $\lim_{x \to \pm\infty} h(x) = \pm\infty : \mathscr{G}^{(n)}, \circ$ is known as the group of order-preserving transformations acting on $\mathbb{R}^n$. It is easily seen that this group generates the hypothesis of independent white noise with unspecified density $f \in \mathscr{F}$, where $\mathscr{F}$ contains all densities (with respect to the Lebesgue measure) which are positive over $\mathbb{R}$ (this latter limitation is not essential, and could easily be omitted).

The invariance argument recommends that only those tests which are invariant under the action of $\mathscr{G}^{(n)}, \circ$ should be considered. And, a statistic is invariant iff it is measurable with respect to some maximal invariant—any statistic generating the same σ-algebra as the orbits of $\mathcal{G}^{(n)}, \circ$ in $\mathbb{R}^n$. In the case of the hypothesis of white noise, a maximal invariant is the vector of ranks; accordingly, the invariant tests of the null hypothesis of white noise are the rank tests.

In the problem of testing $H^{(n)}(\boldsymbol{\theta}_0)$, invariance thus leads to rank tests, of the form $\phi(\mathbf{R}^{(n)}(\boldsymbol{\theta}_0))$, where $\mathbf{R}^{(n)}(\boldsymbol{\theta}_0) \stackrel{\text{def}}{=} (R_1^{(n)}(\boldsymbol{\theta}_0), \dots, R_n^{(n)}(\boldsymbol{\theta}_0))$, and $R_t^{(n)}(\boldsymbol{\theta}_0)$ denotes the rank of $Z_t^{(n)}(\boldsymbol{\theta}_0)$ among the residuals $Z_1^{(n)}(\boldsymbol{\theta}_0), \dots, Z_n^{(n)}(\boldsymbol{\theta}_0)$. Being invariant, such tests also are distribution-free, hence similar (under $H^{(n)}(\boldsymbol{\theta}_0)$): thus, they are a particular case of permutation tests.

The invariance argument also can be justified in the following, nontechnical way. Let $h : \mathbb{R} \to \mathbb{R}$ denote an arbitrary continuous, monotonic increasing function. A n-tuple of residuals $Z_1, \dots, Z_n$ is white noise iff the transformed n-tuple $h(Z_1), \dots, h(Z_n)$ is. Moreover, an observed value $z_1, \dots, z_n$, of $Z_1, \dots, Z_n$, is "equally close to", or "equally far from" being white noise as $h(z_1), \dots, h(z_n)$. Hence, if $Z_1, \dots, Z_n$, allows for rejecting the hypothesis of white noise, so should $h(Z_1), \dots, h(Z_n)$: reasonable tests should be such that $\phi(Z_1, \dots, Z_n) = \phi(h(Z_1), \dots, h(Z_n))$ for any h, i.e., they should be invariant under $\mathcal{G}^{(n)}, \circ$. And, the class of invariant tests is the class of rank tests.

5.4 Rank-Based Central Sequences and Residual Correlograms

If rank-based tests are to be considered, and if (local asymptotic) optimality nevertheless is to be achieved, in a LAN context, one clearly should try to construct rank-based versions $\underset{\sim}{\Delta}_{f;\boldsymbol{\theta}}^{(n)}$ of the central sequences $\Delta_{f;\boldsymbol{\theta}}^{(n)}$, then apply the general results of Sec. 4. More specifically, ingredients (I1)–(I3) (for testing semi-parametric linear hypotheses of the form $H^{(n)}[\boldsymbol{\theta}_0; \mathcal{M}(\Omega)]$) of Sec. 4 here should be replaced with

(J1)   sequences of $\mathbf{R}^{(n)}(\boldsymbol{\theta})$-measurable random vectors $\underset{\sim}{\Delta}_{f;\boldsymbol{\theta}}^{(n)}$ such that

$$\underset{\sim}{\Delta}_{f;\boldsymbol{\theta}}^{(n)} - \Delta_{f;\boldsymbol{\theta}}^{(n)} = o_P(1) \qquad \text{under } H_f^{(n)}(\boldsymbol{\theta}), \text{ as } n \to \infty, \tag{5.6}$$

(J2)   a local asymptotic linearity result (with respect to $\boldsymbol{\theta}$) for $\underset{\sim}{\Delta}_{f;\boldsymbol{\theta}}^{(n)}$, under $H_g^{(n)}(\boldsymbol{\theta})$ (not just under $H_f^{(n)}(\boldsymbol{\theta})$), and

(J3)   the same constrained, root-n consistent, locally discrete sequence of estimates $\hat{\boldsymbol{\theta}}^{(n)}$ of $\boldsymbol{\theta} \in \boldsymbol{\theta}_0 + \mathcal{M}(\Omega)$ as in (I3).

The end of this subsection will be devoted to (J1); (J3) coincides with (I3), which has been discussed before, whereas (J2) is related to the alignment problem, to be treated in Sec. 5.5.

The basic idea for building a rank-based central sequence consists in replacing, in the correlogram-based definition (3.9) of $\Delta_{f;\theta}^{(n)}$, the parametric generalized residual autocorrelation coefficients $r_{f;u}^{(n)}(\theta)$ with the rank-based ones (the ranks $R_t^{(n)} = R_t^{(n)}(\theta)$ are those of the residuals $Z_t^{(n)}(\theta)$):

$$\underline{r}_{f;u}^{(n)}(\theta) \stackrel{\text{def}}{=} \left\{ (n-u)^{-1} \sum_{t=u+1}^{n} \phi_{f_1} \circ F_1^{-1}\left(\frac{R_t^{(n)}}{n+1}\right) F_1^{-1}\left(\frac{R_{t-u}^{(n)}}{n+1}\right) - m_f^{(n)} \right\} \bigg/ s_{f;u}^{(n)}, \tag{5.7}$$

with $F_1(x) \stackrel{\text{def}}{=} \int_{-\infty}^{x} f_1(z)\, dz$; $m_f^{(n)}$ and $s_{f;u}^{(n)}$ are exact centering and scaling constants, so that $(n-u)^{1/2} \underline{r}_{f;u}^{(n)}(\theta)$ is exactly standardized under $H^{(n)}(\theta)$. Asymptotic standardizing constants also can be substituted for the exact ones, since

$$m_f^{(n)} = O(n^{-1/2}) \quad \text{and} \quad (s_{f;u}^{(n)})^2 = \mathscr{I}_{f_1} + O(n^{-(1/2)-2\delta}) \tag{5.8}$$

as $n \to \infty$, under appropriate smoothness assumptions on the scores (including a definition of $\delta \in (0, \frac{1}{4})$); see Hallin and Mélard (1988) for explicit values, and Hallin and Rifi (1994) for the rates of convergence in Eq. (5.8).

Particularizing the density f yields a variety of rank-based correlogram concepts:

(i)   van der Waerden autocorrelations (associated with a Gaussian innovation density f)

$$\begin{aligned}
\underline{r}_{vdW;u}^{(n)} = \Bigg[(n-u)^{-1} \sum_{t=u+1}^{n} \Phi^{-1}\left(\frac{R_t^{(n)}}{n+1}\right) \\
\times \Phi^{-1}\left(\frac{R_{t-u}^{(n)}}{n+1}\right) - m_{vdW}^{(n)} \Bigg] (s_{vdW}^{(n)})^{-1},
\end{aligned} \tag{5.9}$$

where Φ, as usual, stands for the standard normal distribution function,

(ii)   Wilcoxon autocorrelations (associated with a logistic innovation density f)

$$\begin{aligned}
\underline{r}_{\mathscr{W};u}^{(n)} = \Bigg[(n-u)^{-1} \sum_{t=u+1}^{n} \left(\frac{R_t^{(n)}}{n+1} - \frac{1}{2}\right) \\
\times \log\left(\frac{R_{t-u}^{(n)}}{n+1-R_{t-u}^{(n)}}\right) - m_{\mathscr{W}}^{(n)} \Bigg] (s_{\mathscr{W}}^{(n)})^{-1},
\end{aligned} \tag{5.10}$$

(iii) Laplace autocorrelations (associated with a double-exponential innovation density f)

$$
\underline{r}_{\mathscr{L}:u}^{(n)} = \left\{ (n-u)^{-1} \sum_{t=u+1}^{n} \operatorname{sign}\left(\frac{R_t^{(n)}}{n+1} - \frac{1}{2} \right) \right.
$$
$$
\times \left[\log\left(2\frac{R_{t-u}^{(n)}}{n+1} \right) I\left[\frac{R_{t-u}^{(n)}}{n+1} \leq \frac{1}{2} \right] \right.
$$
$$
\left. \left. - \log\left(2 - 2\frac{R_{t-u}^{(n)}}{n+1} \right) I\left[\frac{R_{t-u}^{(n)}}{n+1} > \frac{1}{2} \right] \right] - m_{\mathscr{L}}^{(n)} \right\} (s_{\mathscr{L}}^{(n)})^{-1}. \quad (5.11)
$$

Testing procedures based on van der Waerden, Wilcoxon, and Laplace autocorrelations achieve optimality under normal, logistic, and double exponential innovation densities, respectively. However, other score functions also can be adopted because of their simplicity, even though the resulting tests cannot be optimal at any density:

(iv) Spearman-Wald-Wolfowitz autocorrelations (Wald and Wolfowitz (1943))

$$
\underline{r}_{\mathscr{S}:u}^{(n)} = \left[(n-u)^{-1} \sum_{t=u+1}^{n} R_t^{(n)} R_{t-u}^{(n)} - \frac{1}{12}(n+1)(3n+2) \right] (s_{\mathscr{S}}^{(n)})^{-1},
$$
$$
\quad (5.12)
$$

where $(s_{\mathscr{S}}^{(n)})^2 = (n^4/144) + O(n^3)$ as $n \to \infty$; $\underline{r}_{\mathscr{S}:u}^{(n)}$ clearly is a serial version of the classical Spearman correlation coefficient, and

(v) sign autocorrelations (equivalently, generalized runs: Dufour (1981); Dufour, Hallin and Mizera (1997); Boldin and Stute (1997))

$$
\underline{r}_{\text{signs}:u}^{(n)} = \left[(n-u)^{-1} \sum_{t=u+1}^{n} \operatorname{sign}\left(\frac{R_t^{(n)}}{n+1} - \frac{1}{2} \right) \right.
$$
$$
\left. \times \operatorname{sign}\left(\frac{R_{t-u}^{(n)}}{n+1} - \frac{1}{2} \right) + \frac{1}{n-1} \right] (s_{\text{signs}}^{(n)})^{-1}, \quad (5.13)
$$

where $s_{\text{signs}}^{(n)} \to 1$ as $n \to \infty$; $\underline{r}_{\text{signs}:u}^{(n)}$ actually is the autocorrelation of the series of signs obtained from centering the residuals about their empirical median.

The following result describes the asymptotic behavior of rank-based autocorrelations under white noise. In addition to assumptions (A1)–(A3), we also will assume that

(A5) f is strongly unimodal, i.e., ϕ_f is monotone increasing.

PROPOSITION 5.1   Let the densities f and g both satisfy assumptions (A1)–(A3); assume moreover that f satisfies (A5). Then, under $H_g^{(n)}(\boldsymbol{\theta})$, as $n \to \infty$,

$$\underline{r}_{f;u}^{(n)}(\boldsymbol{\theta}) = \left\{ (n-u)^{-1} \sum_{t=u+1}^{n} \phi_{f_1} \circ F_1^{-1}(G(\varepsilon_t)) F_1^{-1}(G(\varepsilon_{t-u})) \right\} \Big/ \mathscr{I}_{f_1}^{1/2} + o_P(n^{-1/2}),$$

$$(5.14)$$

with $G(x) \stackrel{\text{def}}{=} \int_{-\infty}^{x} g(z)\,dz$.

COROLLARY 5.1   Under the same assumptions as in Proposition 5.1, any k-tuple $(i_1, \ldots, i_k$ all distinct)

$$((n - i_1)^{1/2} \underline{r}_{f;i_1}^{(n)}(\boldsymbol{\theta}), \ldots, (n - i_k)^{1/2} \underline{r}_{f;i_k}^{(n)}(\boldsymbol{\theta}))'$$

is asymptotically $\mathscr{N}(\mathbf{0}, \mathbf{I}_{p \times p})$ under $H_g^{(n)}(\boldsymbol{\theta})$, as $n \to \infty$.

It follows from this corollary that rank-based autocorrelation coefficients under white noise exhibit exactly the same asymptotic behavior, and thus admit exactly the same intuitive interpretation in correlogram inspection techniques, as traditional correlograms. Their advantage over the latter is that they are entirely distribution-free, and exactly standardized. Moreover, they can be expected to be more robust, (see Hallin and Mélard (1988); Garel and Hallin (1997)). And, as we shall see (Table 2), they can be more efficient—even, uniformly more efficient—than the usual "parametric" ones.

A proof of the asymptotic representation result Eq. (5.14), in the spirit of Hájek's classical projection theorem for nonserial rank statistics, can be found in Hallin and Vermandele (1996). The same result could be derived from the more general representation theorem for serial rank statistics proven in Hallin, Ingenbleek, and Puri (1985), where however the slightly stronger condition is required that ε_t and $\phi_f(\varepsilon_t)$ possess finite moments of order $2 + \delta$ (instead of 2) for some strictly positive δ.

Letting $g = f$ in Eq. (5.14) yields the result that $\underline{r}_{f;u}^{(n)}(\boldsymbol{\theta})$, under $H_f^{(n)}(\boldsymbol{\theta})$, is asymptotically equivalent to its parametric counterpart $r_{f;u}^{(n)}(\boldsymbol{\theta})$. Substituting $\underline{r}_{f;u}^{(n)}(\boldsymbol{\theta})$ for $r_{f;u}^{(n)}(\boldsymbol{\theta})$ in Eq. (3.9) accordingly yields the desired result (J1) that $\underline{\Delta}_{f;\theta}^{(n)}$ is a (rank-based) central sequence.

This asymptotic reconstruction of central sequences, hence of log-likelihoods, along with the invariance arguments of Sec. 5.3, provides a theoretical justification of rank-based techniques. Another, more heuristic justification is that, when $Z_1, \ldots, Z_n$, are iid, with distribution function G, then, denoting by $R_1^{(n)}, \ldots, R_n^{(n)}$ their ranks, $R_t^{(n)}/(n+1) \approx G(Z_t)$, where $G(Z_1), \ldots, G(Z_n)$ are iid, uniformly over $[0, 1]$. Hence, considering, for instance, the van der Waerden scores,

$$\left(\Phi^{-1}\left(\frac{R_1^{(n)}}{n+1} \right), \ldots, \Phi^{-1}\left(\frac{R_n^{(n)}}{n+1} \right) \right) \approx (\Phi^{-1}[G(Z_1)], \ldots, \Phi^{-1}[G(Z_n)]),$$

a standard normal iid n-tuple: replacing an observed n-tuple with the corresponding n-tuple of van der Waerden scores thus can be considered as a way of Gaussianizing possibly non-Gaussian observations.

Note however that the asymptotic equivalence in Eq. (5.14) between the rank-based central sequence $\underset{\sim}{\Delta}_{f;\boldsymbol{\theta}}^{(n)}$ and its parametric counterpart $\Delta_{f;\boldsymbol{\theta}}^{(n)}$ only holds under density f, and is no longer true under $g \neq f$. The asymptotic behavior of $\underset{\sim}{\Delta}_{f;\boldsymbol{\theta}}^{(n)}$, as well as that of the rank-based autocorrelations $\underset{\sim}{r}_{f;u}^{(n)}(\boldsymbol{\theta})$ thus very much depends on the true underlying density, g, say.

This "semi-adaptive" behavior is an extremely important feature of rank-based statistics and rank-based tests, which, contrary to their parametric competitors, automatically take into account some of the available information about the underlying density g. As we shall see, this semi-adaptivity feature can be exploited to produce test statistics which perform equally well, under density f, as their parametric counterparts, but, under $g \neq f$, perform uniformly better.

5.5 Aligned Ranks

Coming back to the problem of testing $H^{(n)}(\boldsymbol{\theta}_0; \Omega)$, and assuming ((J3) $\equiv$ (I3)) that an adequate estimator $\hat{\boldsymbol{\theta}}^{(n)}$ of $\boldsymbol{\theta}$ is available (assumptions (B1)–(B2)), the general results of Sec. 4 still apply, since $\underset{\sim}{\Delta}_{f;\boldsymbol{\theta}}^{(n)}$ is a central sequence. Therefore, provided (I2) holds, the sequence of tests rejecting the parametric hypothesis $H_f^{(n)}(\boldsymbol{\theta}_0; \Omega)$ whenever

$$\underset{\sim}{Q}_f^{(n)}(\hat{\boldsymbol{\theta}}^{(n)}) \overset{\text{def}}{=} (\underset{\sim}{\Delta}_{f;\hat{\boldsymbol{\theta}}^{(n)}}^{(n)})' \Gamma_{f;\hat{\boldsymbol{\theta}}^{(n)}}^{-1/2}$$

$$\times [\mathbf{I}_{p \times p} - \Gamma_{f;\hat{\boldsymbol{\theta}}^{(n)}}^{1/2} \Omega (\Omega' \Gamma_{f;\hat{\boldsymbol{\theta}}^{(n)}} \Omega)^{-1} \Omega' \Gamma_{f;\hat{\boldsymbol{\theta}}^{(n)}}^{1/2}] \Gamma_{f;\hat{\boldsymbol{\theta}}^{(n)}}^{-1/2} \underset{\sim}{\Delta}_{f;\hat{\boldsymbol{\theta}}^{(n)}}^{(n)}$$

$$> \chi_{p-r;1-\alpha}^2 \tag{5.15}$$

is locally and asymptotically most stringent against

$$\bigcup \{ H_f^{(n)}(\boldsymbol{\theta}) : \boldsymbol{\theta} \notin \boldsymbol{\theta}_0 + \mathcal{M}(\Omega) \} \tag{5.16}$$

within the class of tests having asymptotic level α—at any fixed $\boldsymbol{\theta} \in \boldsymbol{\theta}_0 + \mathcal{M}(\Omega)$.

In Eq. (5.15), the ranks $R_t^{(n)}(\boldsymbol{\theta})$ of the exact residuals $Z_t(\boldsymbol{\theta})$ are replaced by aligned ranks, i.e., the ranks $R_t^{(n)}(\hat{\boldsymbol{\theta}}^{(n)})$ of the estimated residuals $Z_t(\hat{\boldsymbol{\theta}}^{(n)})$. The asymptotic linearity result of Proposition 4.2 guarantees that this substitution has no influence, asymptotically, on the test statistic Eq. (5.15), i.e., that $Q_f^{(n)}(\hat{\boldsymbol{\theta}}^{(n)}) - Q_f^{(n)}(\boldsymbol{\theta}) = o_P(1)$, under density f. But it does not say anything about asymptotic linearity under $g \neq f$. The following result has been proved in Hallin and Puri (1994), and fills this gap.

PROPOSITION 5.2   Assume that f and g both satisfy (A1)–(A4); assume that f moreover satisfies (A5). For all $\boldsymbol{\tau} = (\tau_1, \ldots, \tau_p) \in \mathbb{R}^p$, consider the sequence $a_u(\boldsymbol{\tau}; \boldsymbol{\theta})$ characterized by

$$\sum_{u=1}^{\infty} a_u(\boldsymbol{\tau}; \boldsymbol{\theta}) z^u \overset{\text{def}}{=} \frac{\sum_{i=1}^{p} \tau_i z^i}{1 - \sum_{i=1}^{p} \theta_i z^i} = \left(\sum_{i=1}^{p} \tau_i z^i\right)\left(\sum_{u=0}^{\infty} g_u(\boldsymbol{\theta}) z^u\right), \qquad |z| < 1.$$

(5.17)

Then,

$$n^{1/2}[\underline{r}_{f:u}^{(n)}(\hat{\boldsymbol{\theta}}^{(n)}) - \underline{r}_{f:u}^{(n)}(\boldsymbol{\theta})]$$

$$= \varrho(f;g)\mathscr{J}(f;g)(\mathscr{I}_{f_1})^{-1/2} a_u(n^{1/2}(\hat{\boldsymbol{\theta}}^{(n)} - \boldsymbol{\theta}); \boldsymbol{\theta}) + o_P(1), \quad (5.18)$$

and

$$\underline{\Delta}_{f:\hat{\boldsymbol{\theta}}^{(n)}}^{(n)} - \underline{\Delta}_{f:\boldsymbol{\theta}}^{(n)} = \varrho(f;g)\mathscr{J}(f;g)\Gamma_{\mathscr{G}}(\boldsymbol{\theta}) n^{1/2}(\hat{\boldsymbol{\theta}}^{(n)} - \boldsymbol{\theta}) + o_P(1), \quad (5.19)$$

under $H_g^{(n)}(\boldsymbol{\theta})$, as $n \to \infty$, where

$$\varrho(f;g) \overset{\text{def}}{=} \int_0^1 F_1^{-1}(u) G_1^{-1}(u)\, du \quad (5.20)$$

and

$$\mathscr{J}(f;g) \overset{\text{def}}{=} \int_0^1 \phi_{f_1}(F_1^{-1}(u)) \phi_{g_1}(G_1^{-1}(u))\, du; \quad (5.21)$$

$\Gamma_{\mathscr{G}}(\boldsymbol{\theta})$ and $\mathscr{I}_f$ are defined in Eqns (2.5) and (3.1), respectively.

Since estimated residuals are no longer iid, aligned rank statistics cannot be expected to be exactly invariant. Nevertheless, one might hope for some kind of asymptotic invariance—namely, asymptotic equivalence with a genuinely invariant statistic, measurable with respect to exact residual ranks. It is an unfortunate consequence of Proposition 5.2 that neither $\underline{r}_{f:u}^{(n)}(\hat{\boldsymbol{\theta}}^{(n)})$ nor $\underline{\Delta}_{f:\hat{\boldsymbol{\theta}}^{(n)}}^{(n)}$ are asymptotically invariant in this sense, since the right-hand sides of Eqns (5.18) and (5.19) both depend on the unspecified underlying density g. The invariance argument accordingly loses most of its relevance. Moreover, and for the same reason, neither $\underline{r}_{f:u}^{(n)}(\hat{\boldsymbol{\theta}}^{(n)})$ nor $\underline{\Delta}_{f:\hat{\boldsymbol{\theta}}^{(n)}}^{(n)}$ are asymptotically distribution-free, which apparently irremediably precludes using them for testing purposes in a semi-parametric context. But, a closer look at Eq. (5.19) reveals that, irrespectively of g, $\Delta_{f;\hat{\theta}^{(n)}}^{(n)} - \Delta_{f;\theta}^{(n)}$ asymptotically belongs to the same linear subspace of $\mathbb{R}^p$ spanned by the columns of $\Gamma_{\mathscr{G}}(\theta)\Omega$ as in the parametric case, see Eqns (4.15) and (4.16). Hence, $Q_f^{(n)}(\hat{\theta}^{(n)}) - Q_f^{(n)}(\theta) = o_P(1)$ under $H^{(n)}(\theta_0; \Omega)$: although $\Delta_{f;\hat{\theta}^{(n)}}^{(n)}$ fails to be either asymptotically invariant or asymptotically distribution-free, the quadratic form $Q_f^{(n)}(\hat{\theta}^{(n)})$ is.

PROPOSITION 5.3 Under the assumptions of Proposition 5.2, the sequence of tests rejecting the semi-parametric hypothesis $H^{(n)}(\theta_0; \Omega)$ whenever $Q_f^{(n)}(\hat{\theta}^{(n)}) > \chi_{p-r;1-\alpha}^2$

(i) is asymptotically invariant, hence asymptotically distribution-free, and has asymptotic size α;
(ii) is locally asymptotically most stringent against the alternative (5.16);
(iii) under $H_g^{(n)}(\theta + n^{-1/2}\tau)$, with $\theta - \theta_0 \in \mathcal{M}(\Omega)$ and $\tau \notin \mathcal{M}(\Omega)$, $Q_f^{(n)}(\hat{\theta}^{(n)})$ is asymptotically noncentral chi-square, with noncentrality parameter

$$\left[\frac{\underline{\sigma}(f;g)\mathscr{I}(f;g)}{\mathscr{I}_{f_1}^{1/2}} \right]^2 \mathscr{Q}(\theta;\tau), \tag{5.22}$$

where $\mathscr{Q}(\theta;\tau)$ is the same positive definite quadratic form as in Eq. (4.18), depending on θ and τ, but not on f and g,

$$\underline{\sigma}(f;g) \stackrel{\text{def}}{=} \int_0^1 F_1^{-1}(u)G_1^{-1}(u)\,du \tag{5.23}$$

and

$$\mathscr{I}(f;g) \stackrel{\text{def}}{=} \int_0^1 \phi_{f_1}(F_1^{-1}(u))\phi_{g_1}(G_1^{-1}(u))\,du \tag{5.24}$$

(the notation ϕ_{g_1}, G_1 is used in an obvious fashion).

It follows that the asymptotic relative efficiency, under innovation density g, of the rank test just described with respect to the Gaussian Lagrange multiplier procedure of Sec. 2 is precisely the quantity

$$\left[\underline{\sigma}(f;g)\mathscr{I}(f;g)/\mathscr{I}_{f_1}^{1/2} \right]^2 \tag{5.25}$$

appearing in the noncentrality parameter Eq. (5.22). Some ARE values are listed in Table 2.

Table 2. Asymptotic Relative Efficiencies (AREs) $[\varrho(f;g)\mathcal{J}(f;g)]^2/\mathcal{I}_{f_i}$ of Rank Tests Based on Various Score Functions (van der Waerden, Wilcoxon, ...; see Eqns (5.9)–(5.13)) with Respect to Gaussian Lagrange Multipliers, Under Cosine, Normal, Logistic, and Double Exponential Densities, Respectively.

Scores (f)	Cosine g	Normal g	Logistic g	Double exponential g
van der Waerden	∞	1.000	1.039	1.226
Wilcoxon	0.405	0.947	1.098	1.483
Laplace	0.198	0.613	0.813	2.000
Spearman	0.856	0.912	1.000	1.226
Signs	0.152	0.405	0.481	1.000

Inspection of Table 2 reveals that rank tests can be substantially more powerful (in the local, Pitman sense) than their classical Gaussian counterparts. The asymptotic relative efficiency of rank tests of the Laplace type with respect to classical procedures is as high as 2, which means that they consume about 50 per cent less observations to achieve the same asymptotic result! The van der Waerden tests yield AREs strictly larger than one under logistic and double exponential densities. It can be shown, Hallin (1994), that this is a general result, extending to the time-series context the traditional Chernoff–Savage (1958) property of normal scores:

$$\inf_g \left[\varrho(\mathcal{G};g)\mathcal{J}(\mathcal{G};g)/\mathcal{I}_{\mathcal{G}}^{1/2}\right]^2 = \inf_g \left[\varrho(\mathcal{G};g)\mathcal{J}(\mathcal{G};g)\right]^2 = 1, \qquad (5.26)$$

where the $\inf_g$ is attained at Gaussian g only. The very strong implication of this generalized Chernoff–Savage result is that van der Waerden tests are always strictly more powerful, in the Pitman sense, than the traditional ones—except, of course, under Gaussian densities, where both are asymptotically equivalent.

Though performing slightly worse, under Gaussian densities, than the van der Waerden ones, the Spearman scores have an excellent overall performance. The celebrated Hodges–Lehmann 0.864 lower bound for the asymptotic relative efficiency of Wilcoxon tests with respect to the Student ones in the nonserial context, Hodges and Lehmann (1956), can be generalized, Hallin and Tribel (1997), yielding, for the Spearman tests,

$$\inf_{g}\left[\underline{\sigma}(\mathscr{S};g).\mathscr{I}(\mathscr{S};g)/\mathscr{I}_{\mathscr{S}}^{1/2}\right]^{2} = 144\inf_{g}\left[\underline{\sigma}(\mathscr{S};g)\mathscr{I}(\mathscr{S};g)\right]^{2}$$

$$= \left(\frac{3\pi^{2}}{32}\right)^{2} \approx 0.856, \tag{5.27}$$

where $\underline{\sigma}(\mathscr{S};g) \overset{\text{def}}{=} \int_{0}^{1} uG_{1}^{-1}(u)\,du$, $\mathscr{I}(\mathscr{S};g) \overset{\text{def}}{=} \int_{0}^{1} u\phi_{g_{1}}(G_{1}^{-1}(u))\,du$ and $\mathscr{I}_{\mathscr{S}} \overset{\text{def}}{=} 1/144$. The cost, in terms of AREs, of substituting Spearman correlograms for the classical ones thus is never more than 14 per cent. The least favorable density, achieving (as a limit: see the comments after Table 1) this lower bound is the cosine density given in Eq. (4.20).

When the innovation density can be assumed to be symmetric, ranks lose their maximal invariance property to the benefit of signed ranks, i.e., the vector of the ranks associated with the absolute values of the residuals, along with the vector of their signs. Signed rank procedures analogue to those described here can be found in Hallin and Puri (1991 and 1994).

5.6   Kreiss' Ranked Residual Tests

Kreiss (1987a), in a point estimation context, proposes a mixed method, where both the ranks of the residuals and the observations themselves are taken into account. Adapted to the testing context, and with the notation used here, this method would consist in considering central sequences of the form

$$\underline{\Delta}_{f;\theta}^{(n)} \overset{\text{def}}{=} \frac{n^{-1/2}}{\sigma}\sum_{t=1}^{n}\phi_{f_{1}}\left[F_{1}^{-1}\left(\frac{R_{t}^{(n)}(\theta)}{n+1}\right)\right]\begin{pmatrix}X_{t-1}\\ \vdots \\ X_{t-p}\end{pmatrix} \tag{5.28}$$

or residual autocorrelations

$$\underline{r}_{f;u}^{(n)}(\theta) \overset{\text{def}}{=} \left\{(n-u)^{-1}\sum_{t=u+1}^{n}\phi_{f_{1}}\circ F_{1}^{-1}\left(\frac{R_{t}^{(n)}}{n+1}\right)\frac{Z_{t}^{(n)}(\theta)}{\hat{\sigma}_{n}}\right\}(\mathscr{I}_{f_{1}})^{-1/2}. \tag{5.29}$$

The same central sequences (hence, implicitly, the same residual autocorrelations) also are considered by Koul and Saleh (1993) in their construction of R-estimates for AR models (with the additional restriction, which rules out, for instance, the van der Waerden scores, that $\phi_{f_{1}}$ is a uniformly bounded function).

Denote by $Q_{f}^{(n)}(\hat{\theta}^{(n)})$ the quadratic statistic obtained from substituting $\underline{\Delta}_{f;\theta}^{(n)}$ for $\Delta_{f;\theta}^{(n)}$ in Eq. (5.15). It can be shown that, under assumptions similar

Table 3. Asymptotic Relative Efficiencies (AREs) $\mathscr{I}^2(f;g)\mathscr{I}_{f_1}^{-1}$ of Ranked Residual Tests Based on Various Score Functions (van der Waerden, Wilcoxon, Laplace) with Respect to Gaussian Lagrange Multipliers, Under Normal, Logistic and Double Exponential Densities, Respectively.

Scores (f)	Normal g	Logistic g	Double exponential g
van der Waerden	1.000	1.049	1.280
Wilcoxon	0.955	1.098	1.500
Laplace	0.637	0.823	2.000

to those of Proposition 5.3, the resulting tests, based on the asymptotic χ_{p-r}^2 distribution of $\underline{Q}_f^{(n)}(\hat{\boldsymbol{\theta}}^{(n)})$, yield AREs of the form

$$\left[\mathscr{I}(f;g)/\mathscr{I}_{f_1}^{1/2}\right]^2. \tag{5.30}$$

Statistics based on Kreiss' ranked residuals are somewhat hybrid, since they involve both the residuals and their ranks. They do not satisfy the invariance principle (the group invariance principle described in Sec. 5.3); neither do they follow from any other sound decision-theoretic argument. They are likely to be less robust than the rank statistics proposed in the previous sections, since the possible leverage effect of the observations in Eq. (5.28) is not taken care of as it is in $\Delta_{f;\theta}^{(n)}$. However, one important advantage of $\Delta_{f;\theta}^{(n)}$ over $\Delta_{f;\theta}^{(n)}$ is that the asymptotic relative efficiency (under innovation density g) of the corresponding test with respect to its counterpart based on Eq. (5.15) is $(\underline{\sigma}(f;g))^{-2} \geq 1$. Some numerical values of the AREs of ranked residual tests with respect to Gaussian Lagrange multipliers are given in Table 3.

5.7 Autoregression Rank Scores

5.7.1 Regression and autoregression rank scores

Autoregression rank scores offer an interesting alternative to the alignment device. The concept of regression rank scores was first introduced by Gutenbrunner and Jurečková (1992) in the context of linear models with independent observations. Under the terminology of autoregression rank scores, the same concept has been extended to the context of autoregression processes by Koul and Saleh (1995). The assumptions made by these authors unfortunately rule out unbounded score functions such as the van der Waerden ones—a restriction which is removed by Hallin and Jurečková (1997), where locally optimal tests (based on autoregression rank scores) are explicitly derived for linear constraints on $\boldsymbol{\theta}$. The corresponding problem for linear models with independent observations is treated in Gutenbrunner, Jurečková, Koenker, and Portnoy (1993); see also Koenker (1996), and Jurečková (1984, 1997), where we refer to for further bibliographical information.

Recall that $\mathbf{X}^{(n)} \stackrel{\text{def}}{=} (X_1, \ldots, X_n)'$; define $\mathbf{W}_t \stackrel{\text{def}}{=} (X_t, \ldots, X_{t-p+1})'$ and $\mathbf{W}^{(n)} \stackrel{\text{def}}{=} (\mathbf{W}_0, \ldots, \mathbf{W}_{n-1})'$, of dimensions $p \times 1$ and $n \times p$, respectively. Consider again the problem of testing the hypothesis $H^{(n)}(\boldsymbol{\theta}_0; \Omega)$ under which $\boldsymbol{\theta} - \boldsymbol{\theta}_0 \in \mathcal{M}(\Omega)$. The autoregression rank scores for this problem are defined as the components $\hat{a}_t^{(n)}(\alpha) = \hat{a}_{\boldsymbol{\theta}_0;\Omega;t}^{(n)}(\alpha), t = 1, \ldots, n$ of the optimal solution $\hat{\mathbf{a}}_{\boldsymbol{\theta}_0;\Omega}^{(n)}(\alpha)$ of the linear program (parametrized by $\alpha \in [0,1]$)

$$\begin{cases} (\mathbf{X}^{(n)} - \mathbf{W}^{(n)}\boldsymbol{\theta}_0)'\mathbf{a} := \max \\ \Omega'(\mathbf{W}^{(n)})'(\mathbf{a} - (1-\alpha)\mathbf{1}_n) = \mathbf{0} \\ \mathbf{a} \in [0,1]^n, \end{cases} \quad (5.31)$$

where $\mathbf{1}_n \stackrel{\text{def}}{=} (1, \ldots, 1)'$. Efficient computer programs for solving the parametric linear program (5.31) can be found in Koenker and D'Orey (1987 and 1994).

The intuitive interpretation of $\hat{a}_t^{(n)}(\alpha)$ is as follows. Consider another linear program (parametrized by $\alpha \in [0,1]$):

$$\begin{cases} \alpha \mathbf{1}_n'\boldsymbol{\mu}^+ + (1-\alpha)\mathbf{1}_n'\boldsymbol{\mu}^- := \min \\ \mathbf{X}^{(n)} - \mathbf{1}_n\rho_0 - \mathbf{W}^{(n)}(\boldsymbol{\theta}_0 + \Omega\boldsymbol{\rho}_1) = \boldsymbol{\mu}^+ - \boldsymbol{\mu}^- \\ (\rho_0, \boldsymbol{\rho}_1')' \in \mathbb{R}^{r+1}, \boldsymbol{\mu}^- \in \mathbb{R}_+^n, \boldsymbol{\mu}^+ \in \mathbb{R}_+^n, \alpha \in (0,1). \end{cases} \quad (5.32)$$

This program is the dual program of (5.31). Denote by $(\hat{\rho}_0, \hat{\boldsymbol{\rho}}_1, \hat{\boldsymbol{\mu}}^-, \hat{\boldsymbol{\mu}}^+)$ the corresponding solution: $\hat{\boldsymbol{\rho}}_1 = \hat{\boldsymbol{\rho}}_{\boldsymbol{\theta}_0;\Omega}^{(n)}(\alpha)$ is called the autoregression quantile of order α of the observed series $(X_{-p+1}, \ldots, X_0, \mathbf{X}^{(n)})$, under the constraint $\boldsymbol{\theta} - \boldsymbol{\theta}_0 \in \mathcal{M}(\Omega)$. This terminology has been introduced by Koenker and Bassett (1978) in the context of traditional multiple regression. It is justified by the fact that, in a $(p+1)$-dimensional plot of the observed n-tuple $(X_t, \mathbf{W}_{t-1}')' \in \mathbb{R}^{p+1}, t = 1, \ldots, n$, about αn observations lie below the autoregression hyperplane $x = \hat{\rho}_0 + \mathbf{w}'(\boldsymbol{\theta}_0 + \Omega\hat{\boldsymbol{\rho}}_1)$, and $(1-\alpha)n$ above. Just as the quantile of order α, in the one-dimensional case, divides a n-tuple into two parts, of respective sizes αn and $(1-\alpha)n$ (when no ties are present and αn is an integer), the autoregression quantile of order α divides an observed $(p+1)$-dimensional scatterplot into two parts of (approximate) sizes αn and $(1-\alpha)n$, respectively. Moreover, the autoregression quantile of order α is unaffected if the observations are moved up or down in the scatterplot, provided that they remain on the same side of the corresponding autoregression hyperplane.

Now, for all $t, \hat{a}_t^{(n)}(\alpha)$ is a piecewise linear function of $\alpha \in (0,1)$; and it follows from duality that

$$a_t^{(n)}(\alpha) = \begin{cases} 1 & \text{if} \quad X_t < \hat{\rho}_0 + \mathbf{Y}'_{t-1}[\boldsymbol{\theta}_0 + \Omega\hat{\boldsymbol{\rho}}_{\boldsymbol{\theta}_0;\Omega}^{(n)}(\alpha)] \\ 0 & \text{if} \quad X_t > \hat{\rho}_0 + \mathbf{Y}'_{t-1}[\boldsymbol{\theta}_0 + \Omega\hat{\boldsymbol{\rho}}_{\boldsymbol{\theta}_0;\Omega}^{(n)}(\alpha)] \end{cases}. \tag{5.33}$$

Another basic property of (auto)regression rank scores is their (auto)regression invariance: it is easily seen indeed that the solution of the linear program in Eq. (5.31) is not affected if $\mathbf{X}^{(n)}$ in the linear criterion is modified by adding a quantity of the form $\mathbf{W}^{(n)}\Omega\mathbf{r}$, $\mathbf{r} \in \mathbb{R}^r$. As a consequence, if $\boldsymbol{\theta} - \boldsymbol{\theta}_0 \in \mathcal{M}(\Omega)$, the vector of regression rank scores $\hat{\mathbf{a}}_{\boldsymbol{\theta}_0;\Omega}^{(n)}(\alpha)$ is also the solution of

$$\begin{cases} (\boldsymbol{\varepsilon}^{(n)})'\mathbf{a} := \max \\ \Omega'(\mathbf{W}^{(n)})'(\mathbf{a} - (1-\alpha)\mathbf{1}_n) = \mathbf{0} \\ \mathbf{a} \in [0,1]^n, \end{cases} \tag{5.34}$$

where $\boldsymbol{\varepsilon}^{(n)}$ is the unobservable vector of errors. The nuisance (the exact, unspecified value of $\boldsymbol{\theta}$ under null hypotheses of the form $H^{(n)}(\boldsymbol{\theta}_0; \Omega)$) thus has been, quite conveniently, eliminated rather than estimated—just as location parameters are eliminated, in the one-dimensional case, when ordinary ranks are considered. Therefore, under $H^{(n)}(\boldsymbol{\theta}_0; \Omega)$, for each value of t, the autoregression score $\hat{a}_t^{(n)}(\alpha)$, as a function of α, tells us whether ε_t belongs to the group of αn smallest errors (lying below the autoregression hyperplane of order α) or not, providing an indication on the rank $R_t^{(n)}$ of ε_t. Actually, for all values of α such that $X_t \neq \hat{\rho}_0 + \mathbf{Y}'_{t-1}(\boldsymbol{\theta}_0 + \Omega)\hat{\boldsymbol{\rho}}_{\boldsymbol{\theta}_0;\Omega}(\alpha)$, $\hat{a}_t^{(n)}(\alpha)$ coincides with the so-called Hájek rank score process (Hájek (1965); see Hájek and Šidák (1967), Sec. V.3.5)

$$a_{*t}^{(n)}(\alpha) = \begin{cases} 1 & \text{if} \quad 0 \leq \alpha \leq (R_t^{(n)} - 1)/n \\ R_t^{(n)} - n\alpha & \text{if} \quad (R_t^{(n)} - 1)/n < \alpha \leq R_t^{(n)}/n \\ 0 & \text{if} \quad R_t^{(n)}/n < \alpha \leq 1 \end{cases}. \tag{5.35}$$

Autoregression rank scores can be used instead of aligned ranks in the construction of locally asymptotically optimal tests of $H^{(n)}(\boldsymbol{\theta}_0; \Omega)$. Letting $J : (0,1) \to \mathbb{R}$ be a square-integrable score-generating function, with $J(u) = -J(1-u)$, hence $\int_0^1 J(u)\,du = 0$ and $\int_0^1 J^2(u)\,du \stackrel{\text{def}}{=} \sigma_J^2 < \infty$, define the scores

$$\hat{j}_t^{(n)} \stackrel{\text{def}}{=} \int_0^1 J(u)\,d\hat{a}_{\boldsymbol{\theta}_0;\Omega;t}^{(n)}(u). \tag{5.36}$$

Choose $J(u) = \phi_{f_1}(F_1^{-1}(u))$, and denote by $\underline{Q}_J^{*(n)}$ the test statistic resulting from substituting the scores $\hat{j}_t^{(n)}$ for the aligned rank scores

$$\phi_{f_1}\left[F_1^{-1}\left(\frac{R_t^{(n)}(\hat{\boldsymbol{\theta}}^{(n)})}{n+1}\right)\right]$$

in Kreiss' ranked residual quadratic test statistic $Q_f^{(n)}(\hat{\boldsymbol{\theta}}^{(n)})$; note that the resulting statistic does not involve any estimate $(\hat{\boldsymbol{\theta}}^{(n)})$ of $\boldsymbol{\theta}$ anymore.

It can be shown (Koul and Saleh (1995) for bounded J; Hallin and Jurečková (1997) for unbounded J) that, under innovation densities satisfying some tail conditions,

$$\underline{Q}_f^{*(n)} - \underline{Q}_f^{(n)}(\hat{\boldsymbol{\theta}}^{(n)}) = o_P(1),$$

under $H^{(n)}(\boldsymbol{\theta}_0; \Omega)$, as $n \to \infty$ ($\hat{\boldsymbol{\theta}}^{(n)}$ is assumed to satisfy assumptions (B1) and (B2)). The autoregression rank score tests based on the asymptotic χ^2_{p-r} distribution of $\underline{Q}_f^{*(n)}$ thus inherit all the optimality properties of the corresponding ranked residual tests (Sec. 5.6), and the ARE values of Table 3 still apply. However, due to its dependence on $\mathbf{W}^{(n)}$, $\underline{Q}_f^{*(n)}$ also inherits the less pleasant lack of asymptotic invariance, with respect to order-preserving transformations performed on the residuals, of $Q_f^{(n)}(\hat{\boldsymbol{\theta}}^{(n)})$. The construction of asymptotically invariant autoregression rank score test statistics, asymptotically equivalent to $Q_f^{(n)}(\hat{\boldsymbol{\theta}}^{(n)})$, is the subject of ongoing research.

Summing up, autoregression rank score tests provide an attractive alternative to the aligned rank methods when nuisances are present under the null hypothesis. Indeed, the aligned rank method consists in replacing ranks with ranks (namely, replacing the ranks $\mathbf{R}^{(n)}$ of the unobservable residuals with those $\hat{\mathbf{R}}^{(n)} = \mathbf{R}^{(n)}(\hat{\boldsymbol{\theta}}^{(n)})$ of estimated residuals):

(i)   the rank score nature of the aligned rank scores $\{J[\hat{R}_1^{(n)}/(n+1)], \ldots, J[\hat{R}_n^{(n)}/(n+1)]\}$—i.e., being a permutation of $\{J[1/(n+1)], \ldots, J[n/(n+1)]\}$—is thus preserved; on the other hand,

(ii) the aligned rank scores $J(\hat{R}_t^{(n)}/(n+1))$ do not coincide with the genuine ones $J(R_t^{(n)}/(n+1))$, computed from the ranks of the unobservable errors; moreover,

(iii) they rely on a preliminary root-n consistent estimation of $\boldsymbol{\theta}$, the result of which is likely to have a substantial impact (for finite n) on the resulting test.

Quite on the contrary, the autoregression rank score approach

(i)   renounces the idea of replacing ranks with ranks: $(\hat{J}_1^{(n)}, \ldots, \hat{J}_n^{(n)})$ is no longer a permutation of $\{J[1/(n+1)], \ldots, J[n/(n+1)]\}$; but,

(ii) the autoregression rank scores $\hat{J}_t^{(n)}$ do coincide with their genuine counterparts, computed from the unobservable errors; moreover,

(iii)   since the nuisance parameter has no influence upon $\hat{J}_t^{(n)}$, no preliminary estimation of θ is required.

5.7.2   Autoregression rank score test of $\mathrm{AR}(p-1)$ against $\mathrm{AR}(p)$ dependence

As an illustration, let us consider the particular case of testing $\mathrm{AR}(p-1)$ against $\mathrm{AR}(p)$ dependence, the parametric version of which was treated in Sec. 2.2. The appropriate Ω matrix here is given in Eq. (1.5), and $\theta_0 = 0$. The special form of Ω will allow for a very simple and intuitively interpretable one-dimensional autoregression rank score test statistic

$$S_{J;n} = n^{-1/2} \sum_{t=1}^{n} X_{t-p} \hat{J}_{n;t}, \tag{5.37}$$

with the scores $\hat{J}_t^{(n)}$ defined in Eq. (5.36) from the autoregression rank scores $\hat{\mathbf{a}}^{(n)}$ resulting from the linear program

$$\begin{cases} (\mathbf{X}^{(n)})' \mathbf{a} := \max \\ \sum_{t=1}^{n} a_t = n(1 - \alpha) \\ \sum_{t=1}^{n} X_{t-i} a_t = (1 - \alpha) \sum_{t=1}^{n} X_{t-i}, \quad i = 1, \ldots, p-1 \\ \mathbf{a} \in [0, 1]^n, \ 0 \le \alpha \le 1. \end{cases} \tag{5.38}$$

Writing $\mathbf{W}_*^{(n)} \stackrel{\mathrm{def}}{=} (\mathbf{1}_n \vdots \mathbf{W}^{(n)}) \stackrel{\mathrm{def}}{=} (\mathbf{W}_I^{(n)} \vdots \mathbf{W}_{II}^{(n)})$, with $\mathbf{W}_I^{(n)}$ of dimension $n \times (p-1)$ and $\mathbf{W}_{II}^{(n)} \stackrel{\mathrm{def}}{=} (X_{-p+1}, \ldots, X_{n-p})'$ of dimension $n \times 1$, let

$$T_J^{(n)} \stackrel{\mathrm{def}}{=} n^{1/2} \sigma_J^{-1} D_{\mathbf{W}}^{-1} S_J^{(n)}, \tag{5.39}$$

where

$$D_{\mathbf{W}}^2 \stackrel{\mathrm{def}}{=} (\mathbf{W}_{II}^{(n)})' [\mathbf{I}_{n \times n} - \mathbf{W}_I^{(n)} ((\mathbf{W}_I^{(n)})' \mathbf{W}_I^{(n)})^{-1} (\mathbf{W}_I^{(n)})'] \mathbf{W}_{II}^{(n)}.$$

The following result then is proved (under mild conditions on J and f which we do not specify here) in Hallin and Jurečková (1997).

PROPOSITION 5.4   The test rejecting the null hypothesis of $\mathrm{AR}(p-1)$ dependence whenever $|T_J^{(n)}| > z_{\alpha/2} = \Phi^{-1}(1 - \alpha/2)$ (Φ, as usual, denotes the standard normal distribution function)

(i)   has asymptotic level α, and
(ii)   has the same local asymptotic power as the corresponding ranked residual test described in Sec. 5.6 (see Table 3 for AREs).

The most striking fact about the test statistic in Eq. (5.39) is its extreme simplicity: compare, for instance, $(T_J^{(n)})^2$ and the parametric quadratic form of Eq. (2.9). In addition to the benefits (asymptotic invariance, improved power, increased robustness, . . .) of rank-based procedures, and in addition to its specific methodological advantages (no estimation of the nuisance, autoregression invariance, . . .), indeed, the autoregression rank score method also automatically performs the projection required in all other test statistics.

The tests described in Proposition 5.4 have been applied in Hallin, Jurečková, Kalvová, Picek, and Zahaf (1997) as a tool in the analysis of meteorological time series. Another application of autoregression rank score methods to time series can be found in Hallin, Jurečková, Picek, and Zahaf (1997), where the problem of testing for mutual independence of two AR series is treated.

6. ADAPTIVE TESTS

6.1 Adaptivity

Throughout this section, we explicitly consider the AR(p) model in Eq. (1.1) from a semi-parametric point of view, i.e., we suppose that the innovation density f is unknown (up to the regularity conditions outlined in (A1)–(A3)). The testing procedures considered so far (Sec. 4.5 and 5) are semi-parametric as far as their validity is concerned. But, they remain parametric (even the rank-based ones) from the point of view of optimality, in the sense that their optimality properties only hold for the innovation density they were constructed for. Recall indeed that only optimality—not validity—is affected if the innovation density f used in the construction of these tests does not coincide with the true density that governs the distribution of the innovations.

The fully semi-parametric approach to the problem consists in constructing inference procedures that are optimal in the semi-parametric model $\{P_{f;\theta}^{(n)} | \theta \in \Theta; f \in \mathscr{F}\}$. In general, enlarging the model induces an efficiency loss in the sense that, e.g., optimal tests in the enlarged model have less local power than the optimal tests in the model where the innovation density is correctly specified. On the other hand, the semi-parametric approach has the obvious advantage that there is less "risk" that the model is misspecified. Together with many other important models, the AR(p) model enjoys a very desirable property. For the AR(p) model as specified in Eq. (1.1), optimal inference procedures exist that are (from a local and asymptotic point of view) as efficient as the optimal procedures in the parametric model with correctly specified innovation density. This phenomenon is called adaptivity.

In adaptive cases, one is able to construct semi-parametric inference procedures that behave asymptotically as well as the optimal parametric procedures. Extending a parametric model into a semi-parametric one in this respect can be achieved at no (asymptotic) cost. One can also turn around this argument. If, in a semi-parametric model, additional information about the true innovation density were available, one would not be able to construct (asymptotically) "better" inference procedures. From an asymptotic point of view, the semi-parametric and parametric models thus are equally difficult.

The best known model under which adaptivity occurs is undoubtedly the so-called symmetric location model. Stone (1975) constructs an adaptive estimator for the location parameter under conditions close to (A1)–(A3) on the innovations density. The regression model has been studied by Bickel (1982), under both symmetric and non-symmetric error distributions. Kreiss (1987a) shows essentially that adaptive estimation is possible in ARMA models with symmetric innovations, whereas Kreiss (1987b) shows how to treat the general AR case without symmetry. A general overview of the analysis of (adaptive and non-adaptive) semi-parametric procedures in iid models is given in Bickel, Klaassen, Ritov, and Wellner (1993). Also in the econometrics literature, semi-parametric models have recently received a lot of attention, see, e.g., Steigerwald (1992). A step towards unifying approach for time series and iid models has been undertaken in Drost, Klaassen, and Werker (1997).

In the next subsection, we consider the construction of such adaptive procedures from an intuitive point of view, that is, we will see how one essentially can construct central sequences that do not depend on the unknown innovation density. Some of the technical issues arising here are briefly discussed in Sec. 6.3. In that section, we will also see when adaptivity holds in slightly more general models.

6.2 Intuitive Construction of Adaptive Tests

The Local Asymptotic Normality property of the parametric AR(p) model, as described in Sec. 4, allows for the construction of (locally and asymptotically) optimal parametric inference procedures. These are essentially based on the central sequence $\Delta_{f;\theta}^{(n)}(\mathbf{X}^{(n)})$, which, for the AR($p$) model, is given in Eq. (4.3). This version of the central sequence (recall that central sequences are only defined up to $o_P(1)$ terms) depends on the innovation density f and, hence, cannot be evaluated if f is unknown. Semi-parametric analysis removes the dependence of the central sequence on f, by replacing f with an estimate $\hat{f}^{(n)}$. Although the idea is simple enough, two questions arise. First, what properties should $\hat{f}^{(n)}$ possess in order that $\Delta_{\hat{f}^{(n)}:\theta}^{(n)}(\mathbf{X}^{(n)}) - \Delta_{f:\theta}^{(n)}(\mathbf{X}^{(n)}) = o_P(1)$? Secondly do such estimators of f exist? We will see that for the AR(p) model this latter question can be answered affirmatively.

The construction of adaptive tests is then completely straightforward: simply replace all occurrences of the central sequence in the procedures by the adaptive central sequence (i.e., any version of the central sequence not depending on f). Note that, besides the central sequence, also the Fisher information matrices depend on f, through $\mathscr{I}_f$. We will also see how, for the AR(p) model, a consistent estimator of $\mathscr{I}_f$ can be constructed.

Recall from Eq. (4.3) that a central sequence for the AR(p) model is given by

$$\Delta_{f:\theta}^{(n)}(\mathbf{X}^{(n)}) = n^{-1/2} \sum_{t=1}^{n} \phi_f(Z_t(\theta))\mathbf{W}_{t-1}, \tag{6.1}$$

where $\mathbf{W}_{t-1} = (X_{t-1}, \ldots, X_{t-p})'$. Note that $\mathrm{E}[\mathbf{W}_t] = 0$. Replacing f in this version of the central sequence by $\hat{f}^{(n)}$, we obtain another version of the central sequence if and only if

$$n^{-1/2} \sum_{t=1}^{n} [\phi_f(Z_t(\theta)) - \phi_{\hat{f}^{(n)}}(Z_t(\theta))]\mathbf{W}_{t-1} = o_P(1) \tag{6.2}$$

under $H_f^{(n)}(\theta)$, as $n \to \infty$. The goal thus consists in finding an estimator $\hat{f}^{(n)}$ such that Eq. (6.2) is satisfied. Note that the left-hand side in Eq. (6.2) would have a martingale difference structure if $\mathrm{E}[\phi_{\hat{f}^{(n)}}(Z_t(\theta))]$ were equal to zero. In virtually all constructions of adaptive inference procedures, it is this problem that complicates matters. In the AR(p) case, however, there is a simple solution that we discuss now.

For the moment, let us make the foolish but convenient assumption that the estimator $\hat{f}^{(n)}$ is independent of the observed series. This, of course, is a highly unrealistic assumption, but we will see in the next section how a trick, called sample splitting, essentially allows to obtain exactly such independence.

The following proposition is proven in Drost, Klaassen, and Werker (1997).

PROPOSITION 6.1 Assume that the assumptions of Proposition 5.2 are satisfied, and let $\hat{f}^{(n)}$ be an estimator of f independent of all observations. Suppose that $\hat{f}^{(n)}$ is consistent in the sense that, if ε has density f,

$$\mathrm{E}[|\phi_{\hat{f}^{(n)}}(\varepsilon) - \phi_f(\varepsilon)|^2|\hat{f}^{(n)}] = o_P(1), \tag{6.3}$$

as $n \to \infty$. Moreover, define $\bar{\mathbf{W}}_n = n^{-1} \sum_{t=1}^{n} \mathbf{W}_{t-1}$. Then, for all $\boldsymbol{\theta} \in \Theta$,

$$n^{-1/2} \sum_{t=1}^{n} \phi_{\hat{f}^{(n)}}(Z_t(\boldsymbol{\theta}))[\mathbf{W}_{t-1} - \bar{\mathbf{W}}_n] = n^{-1/2} \sum_{t=1}^{n} \phi_f(Z_t(\boldsymbol{\theta}))\mathbf{W}_{t-1} + o_P(1),$$

(6.4)

and

$$\hat{\mathscr{I}}^{(n)} \overset{\text{def}}{=} n^{-1} \sum_{t=1}^{n} \phi_{\hat{f}^{(n)}}^2(Z_t(\boldsymbol{\theta})) = \mathscr{I}_f + o_P(1).$$

(6.5)

Proposition 6.1 essentially shows how to construct an adaptive central sequence for the AR(p) model and a consistent estimator for the Fisher information for location $\mathscr{I}_f$. The key condition is the consistency requirement in Eq. (6.3). Under the conditions (A1)–(A3) on the density f, a nonparametric estimator $\hat{f}^{(n)}$ satisfying Eq. (6.3) can be constructed. A possible example is given in Proposition 7.8.1 of Bickel, Klaassen, Ritov, and Wellner (1993). This $\hat{f}^{(n)}$ consists of a kernel density estimator for f with appropriate trimming and scaling. In general, of course, many other density estimation methods are available, which in principle also can be considered, as long as condition (6.3) is fulfilled.

Plugging $\Delta_{\hat{f}^{(n)};\boldsymbol{\theta}}^{(n)}$ and $\hat{\mathscr{I}}^{(n)}$ into Eq. (4.4) yields a quadratic function $Q^{(n)}(\boldsymbol{\theta})$ which does not depend on f and is such that $Q^{(n)}(\hat{\boldsymbol{\theta}}^{(n)})$ is asymptotically equivalent, under any semi-parametric null hypothesis of the form $H^{(n)}(\boldsymbol{\theta}_0;\Omega)$ as well as under local alternatives, to $Q_f^{(n)}(\hat{\boldsymbol{\theta}}^{(n)})$, irrespective of the underlying density f (satisfying the required regularity assumptions). We thus have the following result.

PROPOSITION 6.2   Under the assumptions of Proposition 6.1, the sequence of tests rejecting the semi-parametric hypothesis $H^{(n)}(\boldsymbol{\theta}_0;\Omega)$ whenever $Q^{(n)}(\hat{\boldsymbol{\theta}}^{(n)})$ exceeds $\chi_{p-r;1-\alpha}^2$

(i)   has asymptotic size α;
(ii) is locally asymptotically most stringent against

$$\bigcup \{H^{(n)}(\boldsymbol{\theta}) : \boldsymbol{\theta} \notin \boldsymbol{\theta}_0 + \mathscr{M}(\Omega)\}.$$

(6.6)

Moreover,

(iii) $Q^{(n)}(\hat{\boldsymbol{\theta}}^{(n)})$, under $H_g^{(n)}(\boldsymbol{\theta} + n^{-1/2}\boldsymbol{\tau})$, with $\boldsymbol{\theta} - \boldsymbol{\theta}_0 \in \mathscr{M}(\Omega)$ and $\boldsymbol{\tau} \notin \mathscr{M}(\Omega)$, is asymptotically noncentral chi-square, with noncentrality parameter $\mathscr{I}_{g_1}\mathscr{Q}(\boldsymbol{\theta};\boldsymbol{\tau})$, where $\mathscr{Q}(\boldsymbol{\theta};\boldsymbol{\tau})$ still denotes the same function of $\boldsymbol{\theta}$ and $\boldsymbol{\tau}$ as in Proposition 4.3.

It follows that the asymptotic relative efficiency, under innovation density g, of the adaptive test just described with respect to the Gaussian Lagrange multiplier procedure of Sec. 2 is precisely the Fisher information quantity $\mathscr{I}_{g_1}$: 1.000 under Gaussian g, 1.098 under logistic g, and 2.000 under double-exponential g. Note that each of these figures is maximal in its column, in all previous ARE tables.

6.3 Sample Splitting and Other Technicalities

Although Proposition 6.1 gives the correct intuition about how to construct adaptive tests, some technical problems are still to be solved.

As mentioned above, the requirement, in Proposition 6.1, that the estimator $\hat{f}^{(n)}$ be independent of the observations (of the residuals $Z_t(\theta)$) is extremely unsatisfactory. However, the proof relies on a martingale central limit theorem in which this independence plays a crucial role. This independence can be obtained by means of a trick called sample splitting.

Consider the residuals $Z_t(\theta), t = 1, \ldots, n$. Split this n-tuple into two groups. Estimate f from the first half of the sample, and use this estimated density $\hat{f}_1^{(n)}$ in order to compute the central sequence for the second half. To be specific (assuming that n is even), we use

$$n^{-1/2} \sum_{t=(n/2)+1}^{n} \phi_{\hat{f}_1^{(n)}}(Z_t(\theta)) \mathbf{W}_{t-1}. \tag{6.7}$$

Simultaneously, also estimate f from the second half of the residuals $[Z_t(\theta), t = (n/2) + 1, \ldots, n]$, and use this estimated density $\hat{f}_2^{(n)}$ to compute the central sequence for the first part of the observations, i.e.,

$$n^{-1/2} \sum_{t=1}^{n/2} \phi_{\hat{f}_2^{(n)}}(Z_t(\theta)) \mathbf{W}_{t-1}. \tag{6.8}$$

Along the lines of Proposition 6.1, it can then be shown that the sum of Eqns (6.7) and (6.8) forms a central sequence for the model under consideration. The independence needed in Proposition 6.1 follows from the independence among the residuals $Z_t(\theta)$ under $P_{f;\theta}^{(n)}$. The construction of optimal inference procedures then proceeds exactly as in Sec. 4.

Although this sample splitting device allows for the rigorous construction of an adaptive central sequence, it is somewhat artificial. One usually can dispense with it, at the cost, however, of extra regularity conditions, such as symmetry; in practice, sample splitting is seldom considered. Another way of getting rid of it without any additional regularity assumption on f is the use of adaptive rank tests: see Sec. 6.5.

Another remark concerns the $\bar{\mathbf{W}}_n$ term appearing in Eq. (6.4) as a centering quantity. The presence of this term may look surprising, as it is known that $E(\mathbf{W}_{t-1}) = \mathbf{0}$, so that, for the AR($p$) model, $\bar{\mathbf{W}}_n = o_P(1)$. However, it can be shown that the condition $E(\mathbf{W}_{t-1}) = \mathbf{0}$ in general is not sufficient for dropping this centering with respect to $\bar{\mathbf{W}}_n$. Moreover, in the case of the $AR(p)$ model, including $\bar{\mathbf{W}}_n$, simplifies proofs. Again, just as with sample splitting, including $\bar{\mathbf{W}}_n$, even when required by the theory, has only very marginal effects in practice .

Finally, we would like to emphasize once more that adaptivity is a property of the model under study. The $AR(p)$ model discussed here fortunately enjoys this property. However, adding a simple drift parameter is sufficient to destroy it—unless additional assumptions are made on innovation densities. Consider indeed the simple example of an iid sample with finite variance and unspecified expectation. It is well known that, under Gaussian densities, the optimal estimator for this expectation is the sample mean. Obviously, the sample mean is not adaptive. Under known non-Gaussian distributions, the true maximum likelihood estimator is asymptotically more efficient. This lack of adaptivity carries over to the $AR(p)$ model with a constant term, i.e., to model in Eq. (1.1) where the innovations have unknown expectation μ. In that case, it can be shown that θ still can be estimated adaptively, whereas (in analogy with the iid location model) μ cannot: no fully adaptive method thus exists for (θ, μ). One still may try to find optimal semi-parametric procedures, but they will be less efficient than the optimal parametric procedures under correct specification of the model. Since the focus here is on adaptive testing, we will not go further into this.

If, however, the assumption can be made that the innovation density is symmetric about μ, then adaptive inference again is possible for the full parameter (θ, μ). The score function for a symmetric density indeed is an odd function of the innovations. If the corresponding estimate also is odd, it is easily seen that the expectation of the estimated score does equal zero (recall that the expectation of an odd function of a symmetric random variable is zero when it exists). This means that the central limit theorem argument with estimated scores goes through, and that adaptive inference is possible. It also means that it is crucial that the density be estimated by a symmetric function: then, Proposition 6.1 still holds.

6.4 Adaptive Generalized Residual Correlograms

In Sec. 3.2, a non-Gaussian correlogram concept was introduced. We saw how non-Gaussian Lagrange multiplier tests can be rewritten and reinterpreted using this generalized correlogram concept. The main point is that the traditional correlogram is a highly Gaussian idea. If one knows to be dealing with another innovation density f, the correlogram defined in Eq.

(3.8) seems more natural. From a semi-parametric point of view, this is of course still unsatisfactory, since the correct density is usually unknown. However, the ideas put forward in the previous section about adaptivity, suggest the construction of adaptive correlograms. Such correlograms, from a local and asymptotic point of view, are uniformly sufficient, and allow for building adaptive tests. Recall that the generalized correlogram concept is given, for a density f, by

$$r_{f:u}^{(n)}(\boldsymbol{\theta}) = (n-u)^{-1}(\hat{\mathcal{F}}_{f_1}^{(n)})^{-1/2} \sum_{t=u+1}^{n} \phi_{f_1}\left(\frac{Z_t(\boldsymbol{\theta})}{\hat{\sigma}_n}\right) \frac{Z_{t-u}(\boldsymbol{\theta})}{\hat{\sigma}_n}. \tag{6.9}$$

Again, a martingale difference argument could be used to obtain the limit distribution of $r_{f:u}^{(n)}(\boldsymbol{\theta})$ under $H_f^{(n)}(\boldsymbol{\theta})$. More precisely, along the same lines as in Proposition 6.1, one can show the following.

PROPOSITION 6.3 Under the assumptions of Proposition 5.2, the existence of an estimator $\hat{f}^{(n)}$ for f_1 independent of the residuals $Z_t(\boldsymbol{\theta})$ satisfying

$$E[|\phi_{\hat{f}^{(n)}}(\varepsilon) - \phi_{f_1}(\varepsilon)|^2 | \hat{f}^{(n)}] = o_P(1), \tag{6.10}$$

implies that

$$r_{\hat{f}^{(n)}:u}^{(n)}(\boldsymbol{\theta}) = r_{f:u}^{(n)}(\boldsymbol{\theta}) + o_P(n^{-1/2}), \quad n \to \infty. \tag{6.11}$$

Of course, the density estimator $\hat{f}^{(n)}$ mentioned above should be adequately scaled (so that $\hat{f}^{(n)}$ estimates f_1). Secondly, the independence assumption needed for this proposition can again be obtained using the sample splitting device. Again, one may show that (see Eq. (3.9)), under $H_f^{(n)}(\boldsymbol{\theta})$,

$$\Delta_{f:\boldsymbol{\theta}}^{(n)} = n^{1/2} \begin{pmatrix} \sum_{u=1}^{n-1} g_{u-1}(\boldsymbol{\theta}) r_{\hat{f}^{(n)}:u}^{(n)}(\boldsymbol{\theta}) \\ \vdots \\ \sum_{u=p}^{n-1} g_{u-p}(\boldsymbol{\theta}) r_{\hat{f}^{(n)}:u}^{(n)}(\boldsymbol{\theta}) \end{pmatrix} + o_P(n^{-1/2}), \tag{6.12}$$

which provides again a correlogram-based interpretation of the adaptive test statistics.

6.5 Adaptive Rank Tests

The previous sections discussed two inference principles. In Sec. 5.4 we have seen how to construct central sequences which are measurable with respect to the ranks of the residuals calculated at θ (or, at $\hat{\theta}^{(n)}$). Section 6.1 discussed the construction of central sequences not depending on the unknown innovation density f. As we have seen, both approaches have some clear advantages. A natural question would be to ask whether both principles can be combined, that is: can we construct a central sequence based on an estimated innovation density $\hat{f}^{(n)}$ but (apart from $\hat{f}^{(n)}$) involving only the ranks of the residuals?

The answer, given in Hallin and Werker (1998), is positive. This combination of the two principles turns out to have some highly desirable properties, which we discuss now.

A convenient way to consider adaptive rank-based tests is to consider an adaptive version of the rank-based residual correlations in Eq. (5.7). The adaptive rank-based correlograms then can be used to form the test statistics using Eq. (3.9). The obvious idea is again to replace f_1 (and the corresponding distribution function F_1) in Eq. (5.7) with an estimator $\hat{f}^{(n)}$. In order that this substitution have no influence on the asymptotic behavior of the generalized correlograms, we need again some form of consistency of the density estimator. Hallin and Werker (1998) prove the following result.

PROPOSITION 6.4   Let the assumptions of Proposition 5.2 be satisfied. Suppose that the estimator $\hat{f}^{(n)}$ of f_1 is based on the order statistics of the residuals $Z_t(\theta)$ and is consistent in the sense that

$$
E_{f_1}\left\{\left[\left[\phi_{f_1} \circ F_1^{-1}\left(\frac{R_t^{(n)}}{n+1}\right) F_1^{-1}\left(\frac{R_{t-u}^{(n)}}{n+1}\right)\right.\right.\right.
$$
$$
\left.\left.\left.-\phi_{\hat{f}^{(n)}} \circ (\hat{F}^{(n)})^{-1}\left(\frac{R_t^{(n)}}{n+1}\right)(\hat{F}^{(n)})^{-1}\left(\frac{R_{t-1}^{(n)}}{n+1}\right)\right]^2 \middle| \hat{f}^{(n)}\right\} = o_P(1),
$$

$$\tag{6.13}$$

where $\hat{F}^{(n)}$ denotes the distribution function associated with $\hat{f}^{(n)}$. Then, under $H_f^{(n)}(\theta)$,

$$
r_{\hat{f}^{(n)};u}^{(n)}(\theta) = r_{f;u}^{(n)}(\theta) + o_P(n^{-1/2}).
\tag{6.14}
$$

Let us conclude this section with some remarks about the resulting adaptive rank-based inference procedures. First of all, conditionally on the estimated density $\hat{f}^{(n)}$, the procedures based on $\underline{r}_{f^{(n)}:u}^{(n)}(\boldsymbol{\theta})$ are genuinely rank-based, hence inherit all the properties of rank-based methods: exact (conditional) means and variances can be computed, the statistics are (conditionally) distribution-free, etc. Adaptive tests thus can be considered as (aligned) permutation tests. Secondly, in case the model is such that adaptive estimation is possible, an adaptive testing procedure is again constructed, so that, asymptotically and locally, there is no loss of efficiency with respect to the case of a specified density f. Thirdly, it turns out that sample splitting becomes also formally unnecessary. To see this, suppose that the estimated density of the innovations is based on the order statistics of the innovations (which form a sufficient statistic for this density). As ranks and order statistics are independent, we automatically obtain the independence needed in Proposition 6.1. Of course, the independence between ranks and order statistics can be viewed as another sample splitting technique: instead of splitting the residuals into two groups, the whole vector of residuals is split into an order statistic and a vector of ranks.

Finally, in case the model under study is such that adaptivity is not possible, the adaptive rank-based procedure just described automatically constructs efficient semi-parametric inference procedures. Intuitively, this can be explained as follows. The order statistics of the innovations contain all the information available about the innovation density f. In order to construct efficient testing procedures for the finite dimensional parameter of interest $\boldsymbol{\theta}$, we must make sure that local alternatives with respect to f (at fixed $\boldsymbol{\theta}$) be not mistaken for local alternatives with respect to $\boldsymbol{\theta}$ (at fixed f). As the ranks are independent of the order statistics, it seems intuitive that they contain all the information about $\boldsymbol{\theta}$ that cannot be attributed to the unknown f. In that sense, the rank-based procedures of Sec. 5.4 are semi-parametrically efficient. However, these procedures still contain the unknown density f that we again will replace by an estimate $\hat{f}^{(n)}$. It can be shown that under quite general conditions, this replacement in the rank-based sequence asymptotically has no effect. For further details, we refer to Hallin and Werker (1998).

7. CONCLUSIONS

Mainly, three distinct categories of testing procedures have been described: the parametric ones (Sec. 2–4), the rank-based ones (Sec. 5), and the adaptive ones (Sec. 6).

Although they usually remain valid under a broad range of densities—and hence also could be considered as semi-parametric methods—the parametric tests fail to take any advantage of the information, available in the data, about the actual underlying innovation density. The asymptotic performance of the parametric Gaussian methods—the traditional correlogram-based ones by far the most frequently considered in everyday practice—does not depend at all on the underlying density. The Gaussian tests thus are maximin among all the parametric methods described in Sec. 3 and 4, which also emphasizes the all too often ignored fact that Gaussian linear processes, from the point of view of statistical inference, are the worst ones (yielding smallest Fisher information)—contrary to the widespread opinion that the Gaussian case is most favorable. If however rank-based and adaptive tests are also taken into consideration, the parametric Gaussian tests, despite their maximin property, cease to be admissible (they are outperformed by the adaptive tests, of course, but also by the van der Waerden tests).

Adaptive tests, on the other hand, fully exploit the available information about the underlying innovation density. Being asymptotically as efficient as if the actual density were known, they clearly cannot be outperformed by any other, non adaptive method.

Rank-based tests are somewhat intermediate in this respect. While also taking into account, implicitly, some information about innovation densities (see the comments at the end of Sec. 5.4 about the semi-adaptivity of rank tests), they only use part of it. The van der Waerden tests share the maximin property of the parametric Gaussian ones, but uniformly dominate them (see Eq. (5.26)). Several variations of rank-based methods are described, of which the autoregression rank score one probably is the most general and most attractive.

Figure 1 summarizes the efficiency comparisons between the various methods described in this paper. From an asymptotic point of view, adaptive tests undisputably are best—under any of the proposed forms. Adaptive rank tests maybe are the most reliable ones, since they combine the optimality features of adaptivity with the reliability and robustness aspects of rank statistics: except indeed for the substitution of an estimate $\hat{\theta}^{(n)}$ for the "true" θ, adaptive rank tests are permutation tests, hence have exact level, are similar, and unbiased. Moreover, they avoid some unpleasant technicalities, such as sample splitting. Note that they could be combined with the autoregression rank score approach, thus also avoiding nuisance estimation.

Estimating densities, on the other hand, typically requires a very large number of observations, and involves some arbitrary choices such as (if kernel estimation is considered) a kernel or a bandwidth. In case the number of observations is not large enough, one might prefer the nonadaptive rank tests, either based on aligned ranks or on autoregression rank scores, under, for instance, their van der Waerden version. A few existing simulations, (Hallin and Mélard (1988); Garel and Hallin (1997)) indicate that, even in

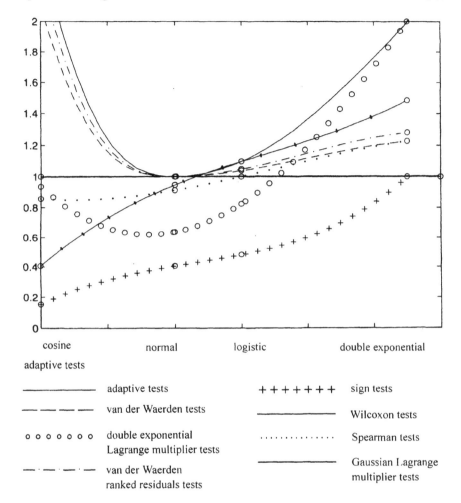

Figure 1. Some of the AREs (with respect to Gaussian Lagrange multipliers) reported in Tables 1, 2, and 3 (AREs are plotted against underlying innovation densities).

the case of very short series (as short as $n = 25$), rank-based methods exhibit substantially better performances than their classical correlogram-based counterparts, while being, at the same time, considerably more robust against the presence of possible outliers, atypical startup behavior, and so on.

Summing up, everyday time series practice, based on traditional correlograms, can be improved upon easily, often quite substantially so, by refining the concept of autocorrelation (cf. non Gaussian, rank-based, adaptive correlograms, . . .). Quite remarkably, efficiency and robustness here are not conflicting objectives, as both can benefit from the suggested modification—which moreover can be achieved without perturbing practitioners' intuition and habits, related with correlogram inspection.

Acknowledgment

The authors express their deep gratitude to Abdelghafour Ayadi, Faouzi El Bantli, and Toufik Zahaf, for their numerical derivation of ARE figures, and their computing assistance. The comments by several referees helped in clarifying the presentation. Some of them were communicated at a late stage, and could not be fully incorporated into the final version; all of them are gratefully acknowledged.

REFERENCES

Akritas, M. G. and R. A. Johnson (1982). Efficiencies of tests and estimators for p-order autoregressive processes when the error is non normal. *Ann. Inst. Statis. Math. 34*: 79–589.

Bickel, P. J. (1982). On adaptive estimation. *Ann. Statis. 10*: 647–671.

Bickel, P. J., C. A. J. Klaassen, Y. Ritov, and J. A. Wellner (1993). *Efficient and Adaptive Estimation for Semiparametric Models*. Johns Hopkins University Press, Baltimore.

Billingsley, P. (1961). *Statistical Inference for Markov Processes*. University of Chicago Press, Chicago.

Brockwell, P. J. and R. A. Davis (1991). *Time Series: Theory and Methods*, 2nd edition. Springer-Verlag, New York.

Boldin, M. V. and W. Stute (1997). On nonparametric sign tests in ARMA models. Preprint, University of Giessen.

Chernoff, H. and Savage, I. R. (1958). Asymptotic normality and efficiency of certain nonparametric tests. *Ann. Math. Stat. 29*: 972–994.

Davies, R. B. (1973). Asymptotic inference in stationary Gaussian series. *Adv. Appl. Prob. 5*: 469–497.

Drost, F. C., C. A. J. Klaassen, and B. J. M. Werker (1997). Adaptive estimation in time series models. *Ann. Statist. 25*: 786–818.

Dufour, J.-M. (1981). Rank tests for serial dependence, *J. Time Ser. Anal.* 2: 117–128.

Dufour, J.-M. and M. Hallin (1991). Nonuniform bounds for nonparametric *t*-tests. *Econometric Theory* 7: 253–256.

Dufour, J.-M. and M. Hallin (1993). Improved Eaton bounds for linear combinations of bounded random variables, with statistical applications. *J. Am. Statist. Assoc.* 88: 1026–1033.

Dufour, J.-M., M. Hallin, and I. Mizera (1997). Generalized runs test for heteroscedastic time series. *J. Nonparametric Statist.* 9: 39–86.

Dzhaparidze, K. O. (1986). *Parameter Estimation and Hypothesis Testing in Spectral Analysis of Stationary Time Series.* Springer-Verlag, New York.

Garel, B. (1989). The asymptotic distribution of the likelihood ratio for *MA* processes with a regression trend. *Statistics & Decisions* 7: 167–184.

Garel, B. and M. Hallin (1995). Local asymptotic normality of multivariate ARMA processes with a linear trend. *Ann. Inst. Statist. Math.* 47: 551–579.

Garel, B. and M. Hallin (1997). Rank-based AR order identification. Preprint, Institut de Statistique, ULB, Brussels.

Godfrey, L.G. (1979). Testing the adequacy of a time series model. *Biometrika* 66: 67–72.

Gutenbrunner, C. and J. Jurečková (1992). Regression rank scores and regression quantiles. *Ann. Statist.* 20: 305–330.

Gutenbrunner, C., J. Jurečková, R. Koenker and S. Portnoy (1993). Tests of linear hypotheses based on regression rank scores. *J. Nonparametric Statist.* 2: 307–331.

Hájek, J. (1965). Extension of the Kolmogorov–Smirnov test to regression alternatives. *Bernoulli-Bayes-Laplace* (L. Le Cam and J. Neyman, eds), Springer-Verlag, New York, 45–60.

Hájek, J. and Z. Šidák (1967). *Theory of Rank Tests.* Academic Press, New York.

Hallin, M. (1994). On the Pitman-nonadmissibility of correlogram-based methods. *J. Time Ser. Anal.* 15: 607–612.

Hallin, M., J.-Fr. Ingenbleek, and M. L. Puri (1985). Linear serial rank tests for randomness against ARMA alternatives. *Ann. Statist.* 13: 1156–1181.

Hallin, M. and J. Jurečková (1997). Optimal tests for autoregressive models based on autoregression rank scores. Preprint, Institut de Statistique, ULB, Brussels.

Hallin, M., J. Jurečková, J. Kalvová, J. Picek, and T. Zahaf (1997). Nonparametric tests in AR models, with applications to climatic data. *Environmetrics 8*: 651–660.

Hallin, M., J. Jurečková, J. Picek, and T. Zahaf (1997). Nonparametric tests of independence of two autoregressive time series based on autoregression rank scores. *J. Statist. Plan. and Inf.*, to appear.

Hallin, M. and G. Mélard (1988). Rank-based tests for randomness against first-order serial dependence. *J. Am. Statist. Assoc. 83*: 1117–1129.

Hallin, M. and M. L. Puri (1988). Optimal rank-based procedures for time-series analysis: testing an ARMA model against other ARMA models. *Ann. Statist. 16*: 402–432.

Hallin, M. and M. L. Puri (1991). Time-series analysis via rank-order theory: signed-rank tests for ARMA models. *J. Mult. Anal. 39*: 1–29.

Hallin, M. and M. L. Puri (1992). Rank tests for time series analysis: a survey. In *New Directions In Time Series Analysis* (D. Brillinger, E. Parzen and M. Rosenblatt, eds), Springer-Verlag, New York, 111–154.

Hallin, M. and M. L. Puri (1994). Aligned rank tests for linear models with autocorrelated error terms. *J. Mult. Anal. 50*: 175–237.

Hallin, M. and Kh. Rifi (1994). Comportement asymptotique de la moyenne et de la variance d'une statistique de rangs sérielle simple. In *Hommage à Simone Huyberechts, Cahiers du C. E. R. O. 36*: 189–201.

Hallin, M. and O. Tribel (1997). The efficiency of some nonparametric competitors to correlogram-based methods. In *Festschrift for T. S. Ferguson* (F. T. Bruss and L. Le Cam, eds), IMS Lecture Notes—Monograph Series, to appear.

Hallin, M. and C. Vermandele (1996). A simple proof of asymptotic normality for simple serial rank statistics. In *Research Developments in Probability and Statistics, Festschrift in Honor of M. L. Puri* (E. Brunner and M. Denker, eds), V. S. P., Utrecht, 163–191.

Hallin, M. and B. J. M. Werker (1998). Efficient rank-based semi-parametric inference. Preprint, Institut de Statistique, ULB, Brussels.

Hodges, J. L. and E. L. Lehmann (1956). The efficiency of some nonparametric competitors of the *t*-test. *Ann. Math. Statist. 58*: 324–335.

Hoeffding, W. (1952). The large-sample power of tests based on permutations of observations. *Ann. Math. Statist. 23*: 169–192.

Hosking, J. R. M. (1980). Lagrange multiplier tests of time series model. *J. Roy. Stat. Soc. Ser. B 42*: 170–181.

Jeganathan, P. (1995). Some aspects of asymptotic theory with applications to time series models. *Econometric Theory 11*: 818–887.

Jeganathan, P. (1997). On asymptotic inference in linear cointegrated time series systems. *Econometric Theory 13*: 692–746.

Jurečková, J. (1984). Regression quantiles and trimmed least squares estimator under a general design. *Kybernetika 20*: 345–357.

Jurečková, J. (1997). Regression rank scores against heavy-tailed alternatives. *Bernoulli*, to appear.

Koenker, R. (1996). Rank tests for linear models. In *Handbook of Statistics Vol. 15* (C. R. Rao and G. S. Maddala, eds). Elsevier, New York.

Koenker, R. and G. Bassett (1978). Regression quantiles. *Econometrica 46*: 33–50.

Koenker, R. and V. D'Orey (1987). Computing regression quantiles. *Appl. Statist. 36*: 383–393.

Koenker, R. and V. D'Orey (1994). A remark on algorithm AS 229: Computing dual regression quantiles and regression rank scores. *Appl. Statist. 43*: 410–414.

Koul, H. L. and A. K. Md. E. Saleh (1993). R-estimation of the parameter of autoregressive AR(p) model. *Ann. Statist. 21*: 464–472.

Koul, H. L. and A. K. Md. E. Saleh (1995). Autoregression quantiles and related rank scores processes. *Ann. Statist. 23*: 670–689.

Koul, H. L. and A. Schick (1997). Efficient estimation in nonlinear autoregressive time-series models. *Bernoulli 3*: 247–277.

Kreiss, J. P. (1987a). On adaptive estimation in stationary ARMA processes. *Ann. Statist. 15*: 112–133.

Kreiss, J. P. (1987b). On adaptive estimation in autoregressive models when there are nuisance functions. *Statistics & Decisions 5*: 59–76.

Kreiss, J. P. (1990a). Testing linear hypotheses in autoregression. *Ann. Statist. 18*: 1470–1482.

Kreiss, J. P. (1990b). Local asymptotic normality for autoregression with infinite order. *J. Stat. Plan. and Inf. 26*: 185–219.

Le Cam, L. (1986). *Asymptotic Methods in Statistical Decision Theory.* Springer-Verlag, New York.

Le Cam, L. and G. L. Yang (1990). *Asymptotics in Statistics.* Springer-Verlag, New York.

Lehmann, E. L. (1986). *Testing Statistical Hypotheses*, 2nd edition. J. Wiley, New York.

Pham, D. T. (1987). Exact maximum likelihood and Lagrange multiplier test statistic for ARMA models. *J. Time Ser. Anal. 8*: 61–78.

Pötscher, B. M. (1983). Order estimation in ARMA models by Lagrangian multiplier tests. *Ann. Statist. 11*: 872–885.

Pötscher, B. M. (1985). The behaviour of the Lagrange multiplier test in testing the orders of an ARMA model. *Metrika 32*: 129–150.

Rao, C. R. (1947). Large sample tests of statistical hypotheses concerning several parameters, with applications. *Proc. Cambridge Philosophical Soc. 44*: 50–57.

Roussas, G. G. (1979). Asymptotic distribution of the log-likelihood function for stochastic processes. *Zeitschrift für Wahrscheinlichkeitstheorie und verwandte Gebiete 47*: 31–46.

Swensen, A. R. (1985). The asymptotic distribution of the likelihood ratio for autoregressive time series with a regression trend. *J. Mult. Anal. 16*: 54–70.

Steigerwald, D. G. (1992). Adaptive estimation in time series regression models. *J. Econometrics 54*: 251–175.

Stone, C. J. (1975). Adaptive maximum likelihood estimators of a location parameter. *Ann. Statist. 3*: 267–284.

Wald A. (1943). Tests of statistical hypotheses concerning several parameters when the number of observations is large. *Trans. Amer. Math. Soc. 54*: 426–482.

Wald, A. and J. Wolfowitz (1943). An exact test for randomness in the nonparametric case based on serial correlation. *Ann. Math. Statist. 14*: 378–388.

Weiss, G. (1975). Time-reversibility of linear stochastic processes. *J. Appl. Probab. 12*: 831–836.

10

Statistical Analysis Based on Functionals of Nonparametric Spectral Density Estimators

MASANOBU TANIGUCHI   Osaka University, Toyonaka, Japan

1. INTRODUCTION

It seems that the origin of semiparametric estimation stems from Hannan (1963). He considered the problem of estimating the coefficient of a time series regression model with stationary dependent residuals. An estimator generalizing the ordinary least squares estimator was proposed by inserting nonparametric estimators for the regression and residual spectra. Under regularity conditions this estimator was shown to be a consistent estimator with "parametric consistent order" despite substitution of nonparametric spectral estimators. For independent identically distributed observations Beran (1977) proposed a new estimator $\hat{\theta}$ of θ by minimizing the Hellinger distance between a specified parametric probability density model f_θ and a suitable nonparametric probability density estimator. He showed that $\hat{\theta}$ is asymptotically efficient under the specified regular parametric family of densities and is minimax robust in a small Hellinger metric neighborhood of the given family. Although it is known that the maximum likelihood estimator of θ has full asymptotic efficiency among regular estimators, it should be noted that maximum likelihood estimators do not, in general, possess the property of the minimax robustness. The point of the results by Hannan and Beran is that their estimators are asymptotically written as the integral functionals of nonparametric estimators, which have "parametric consistent order" ($\sqrt{n}$-consistency).

For a stationary process with spectral density $f(\lambda)$, Taniguchi (1987) discussed fitting a parametric spectral model $f_\theta(\lambda)$ by a criterion $D(f_\theta, f)$ which measures the nearness of f_θ to f. Then a new estimator $\hat{\theta}$ of θ was proposed by the value minimizing $D(f_\theta, \hat{f}_n)$ with respect to θ, where $\hat{f}_n$ is a suitable nonparametric spectral estimator of f based on n observed stretch. Under appropriate conditions it was shown that the main order term of $\sqrt{n}(\hat{\theta} - \theta)$ can be written as $F = \sqrt{n} \int_{-\pi}^{\pi} \psi(\lambda)\{\hat{f}_n(\lambda) - f(\lambda)\} d\lambda$ where $\psi(\lambda)$ is an integrable function. Although nonparametric spectral estimators deviate from $f(\lambda)$ by a larger probability order than $n^{-1/2}$, Taniguchi (1987) showed that the integral functionals obey the $\sqrt{n}$-consistent asymptotics, and that $\hat{\theta}$ is asymptotically efficient if $f = f_\theta$. Thus we can see that the integral functional F is the key quantity. Hence, the purpose of this paper is to develop various statistical analyses based on integral functionals of spectral densities.

In Sec. 2, for a vector-valued linear process with spectral density matrix $\mathbf{f}(\lambda)$, the central limit theorem for $L = \sqrt{n} \int_{-\pi}^{\pi} [K\{\hat{\mathbf{f}}_n(\lambda)\} - K\{\mathbf{f}(\lambda)\}] d\lambda$ is given, where $K\{\cdot\}$ is a smooth function and $\hat{\mathbf{f}}_n(\lambda)$ is a suitable nonparametric spectral density matrix estimator. This result is fundamental and important because the main stream of this paper is based on the integral functional L. In Sec. 3, we review semiparametric estimations by Hannan (1963), Beran (1977), Taniguchi (1987) and Robinson (1988), which motivate our theme. Also some merits of semiparametric estimation are explained.

In Sec. 4 we consider the testing problem $H: \int_{-\pi}^{\pi} K\{\mathbf{f}(\lambda)\} d\lambda = c$ against $A: \int_{-\pi}^{\pi} K\{\mathbf{f}(\lambda)\} d\lambda = c$, where c is a given constant. For this, we propose a test statistic T_n based on $\int_{-\pi}^{\pi} K\{\hat{\mathbf{f}}_n(\lambda)\} d\lambda$, and define an efficacy of T_n for a sequence of contiguous spectral density matrices. The asymptotics of T_n are described by that of the integral functional L. Because the functional of spectral density matrix $\int_{-\pi}^{\pi} K\{\mathbf{f}(\lambda)\} d\lambda$ can represent various important indices in time series, the test setting is unexpectedly wide. The results are applied to the testing problems for Burg's entropy, spectral moment, interpolation error, two sample, measure of linear dependence and principal component analysis in time series, etc. The approach here has the following advantages:

(i)   It is designed for essentially nonparametric hypotheses. So we do not need any parametric assumptions on the spectral density matrix.

(ii) T_n is based on the nonparametric spectral estimator $\hat{\mathbf{f}}_n(\lambda)$. So we do
       not need iterative methods to calculate it.
(iii) Since T_n is based on the integral of $\hat{\mathbf{f}}_n(\lambda)$ we can develop the $\sqrt{n}$-
       consistent asymptotic theory (although $\hat{\mathbf{f}}_n(\lambda)$ is not $\sqrt{n}$-consistent).

In Sec. 5, we discuss the problems of classifying a stationary vector pro-
cess $\{\mathbf{X}(t)\}$ into one of two categories described by two hypotheses $\pi_1 : \mathbf{f}(\lambda)$,
$\pi_2 : \mathbf{g}(\lambda)$, where $\mathbf{f}(\lambda)$ and $\mathbf{g}(\lambda)$ are spectral density matrices. When the process concerned is a scalar-valued Gaussian stationary process, Capon
(1965), Liggett (1971) and Shumway and Unger (1974) discussed this problem. Here, we introduce an approximated Gaussian likelihood ratio classification statistic $I(\mathbf{f} : \mathbf{g})$ although the process concerned is not Gaussian.
Then it is shown that the misclassification probabilities by $I(\mathbf{f} : \mathbf{g})$ tend to
zero as $n \to \infty$. Also a goodness of $I(\mathbf{f} : \mathbf{g})$ is evaluated when $\mathbf{g}(\lambda)$ is contiguous to $\mathbf{f}(\lambda)$. Second, a disparity measure $D_H(\mathbf{f} : \mathbf{g})$ between $\mathbf{f}(\lambda)$ and $\mathbf{g}(\lambda)$
is introduced. $D_H(\mathbf{f} : \mathbf{g})$ is written as an integral functional of $\mathbf{f}$ and $\mathbf{g}$. Let
$\hat{\mathbf{f}}_n(\lambda)$ be a nonparametric spectral density estimator constructed by a sample
which should be classified into π_1 or π_2. Then it is shown that the misclassi-
fication probabilities by $D_H = D_H(\hat{\mathbf{f}}_n : \mathbf{g}) - D_H(\hat{\mathbf{f}}_n : \mathbf{f})$ tend to zero as
$n \to \infty$. Since D_H is an integral functional of $\hat{\mathbf{f}}_n$, the asymptotics are
based on L. When $\mathbf{g}(\lambda)$ is contiguous to $\mathbf{f}(\lambda)$, the goodness of D_H is equivalent to that of $I(\mathbf{f} : \mathbf{g})$ under appropriate conditions. A merit of D_H is discussed when the spectral densities have a sharp peak. As an actual
application we mention discriminating between earthquakes and mining
explosions by use of $I(\mathbf{f} : \mathbf{g})$ and D_H.

In Sec. 6 we briefly review a few papers, e.g., Robinson (1995), Linton
(1995) which deal with higher order asymptotics of semiparametric estimators. Although they are for independent observations they suggest the possibility of a higher order asymptotic theory for semiparametric estimators of
time series models.

2.   ASYMPTOTIC THEORY FOR FUNCTIONALS OF SMOOTHED PERIODOGRAMS

In this Section we review asymptotic theory for integral functionals of periodograms. Throughout this paper, we denote the set of all integers by $\mathbf{Z}$,
and denote Kronecker's delta by $\delta(t, s)$.

Let $\{\mathbf{X}(t) : t \in \mathbf{Z}\}$ be an m-dimensional linear process generated by

$$\mathbf{X}(t) = \sum_{j=0}^{\infty} G(j)\mathbf{e}(t-j), \qquad t \in \mathbf{Z}, \tag{2.1}$$

where $\mathbf{X}(t) = (X_1(t), \ldots, X_m(t))'$ and $\mathbf{e}(t) = (e_1(t), \ldots, e_r(t))'$ such that $E\{\mathbf{e}(t)\} = \mathbf{0}$ and $E\{\mathbf{e}(t)\mathbf{e}(s)'\} = \delta(t, s)\Omega$ with $\Omega = \{\Omega_{ab}\}$ a nonsingular $r \times r$ matrix, $G(j) = \{G_{ab}(j); a = 1, \ldots, m, b = 1, \ldots, r\}$'s are $m \times r$ matrices satisfying $\sum_{j=0}^{\infty} trG(j)\Omega G(j)' < \infty$. Then the process $\{\mathbf{X}(t); t \in \mathbf{Z}\}$ is a second–order stationary process with spectral density matrix

$$\mathbf{f}(\lambda) = \frac{1}{2\pi} k(\lambda)\Omega k(\lambda)^*, \qquad -\pi < \lambda \leq \pi, \tag{2.2}$$

where $k(\lambda) = \sum_{j=0}^{\infty} G(j)e^{i\lambda j}$. For a partial realization $\{\mathbf{X}(1), \ldots, \mathbf{X}(n)\}$, the periodogram matrix is defined as

$$\mathbf{I}_n(\lambda) = \frac{1}{2\pi n} \left\{ \sum_{t=1}^{n} \mathbf{X}(t)e^{i\lambda t} \right\} \left\{ \sum_{t=1}^{n} \mathbf{X}(t)e^{i\lambda t} \right\}^*.$$

Denote the (a, b) component of $\mathbf{f}(\lambda)$ and $\mathbf{I}_n$ by $f_{ab}(\lambda)$ and $I_n^{(a,b)}(\lambda)$, respectively. Since Gaussianity is not assumed for the process we need the following assumptions.

(A.1) $\{\mathbf{e}(t): t \in \mathbf{Z}\}$ is fourth-order stationary.

(A.2) The joint fourth-order cumulants $c_{abcd}^e(t_1, t_2, t_3)$ of $e_a(t)$, $e_b(t + t_1)$, $e_c(t + t_2)$, $e_d(t + t_3)$ satisfy

$$\sum_{t_1, t_2, t_3 = -\infty}^{\infty} |c_{abcd}^e(t_1, t_2, t_3)| < \infty, \qquad a, b, c, d = 1, \ldots, r.$$

Then $\{\mathbf{e}(t)\}$ has the fourth-order cumulant spectral density

$$f_{abcd}^e(\lambda) = \left(\frac{1}{2\pi} \right)^3 \sum_{t_1, t_2, t_3 = -\infty}^{\infty} c_{abcd}^e(t_1, t_2, t_3)e^{-i(\lambda_1 t_1 + \lambda_2 t_2 + \lambda_3 t_3)}.$$

Similarly we can define $c_{abcd}^x(t_1, t_2, t_3)$ and $f_{abcd}^x(\lambda_1, \lambda_2, \lambda_3)$, respectively, the fourth-order cumulant and spectral density of the process $\{\mathbf{X}(t)\}$. Denoting by $\mathcal{F}(t)$ the σ–field generated by $\{\mathbf{e}(s): s \leq t\}$, we now impose further assumptions on $\{G(j)\}$ and $\{\mathbf{e}(t)\}$.

(A.3) There exist $C_1 < \infty$ and $C_2 > 0$ such that

$$\sum_{j=0}^{\infty} (1 + j^2)\|G(j)\| \leq C_1,$$

and

$$\left| \det\left\{ \sum_{j=0}^{\infty} G(j)z^j \right\} \right| \geq C_2, \qquad \text{for all} \quad |z| \leq 1,$$

where $\|G(j)\|$ is the norm of the matrix $G(j)$.

(A.4)   For each $a, b, = 1, \ldots, r$ and $\ell \in \mathbf{Z}$,

$$\text{Var}[E\{e_a(t)e_b(t + \ell)|\mathcal{F}(t - \tau)\} - \delta(\ell, 0)\Omega_{ab}] = O(\tau^{-2-\varepsilon}), \quad \varepsilon > 0,$$

uniformly in t.

(A.5)

$$E|E\{e_a(t_1)e_b(t_2)e_c(t_3)e_d(t_4)|\mathcal{F}(t_1 - \tau)\}$$

$$- E\{e_a(t_1)e_b(t_2)e_c(t_3)e_d(t_4)\}| = O(\tau^{-1-\eta}),$$

uniformly in t_1, where $t_1 \leq t_2 \leq t_3 \leq t_4$ and $\eta > 0$.

(A.6)   For any $\eta > 0$ and any integer $L \geq 0$, there exists $\xi_\eta > 0$ such that, for every $a, b = 1, \ldots, r$,

$$E[T_{ab}(n, s)^2 \chi\{T_{ab}(n, s) > \xi_\eta\}] < \eta,$$

uniformly in n and s, where $\chi\{\cdot\}$ is the indicator function and

$$T_{ab}(n, s) = \left[\sum_{r=0}^{L} \left\{ \sum_{t=1}^{n} (e_a(t + s)e_b(t + s + r) - \delta(r, 0)\Omega_{ab})/\sqrt{n} \right\}^2 \right]^{1/2}.$$

Hosoya and Taniguchi (1982) introduced the assumptions (A.4)–(A.6) to prove the central limit theorem for an integral functional of a periodogram. In this topic "higher order martingale difference conditions" were usually imposed, i.e.,

$$E\{\mathbf{e}(t)|\mathcal{F}(t - 1)\} = \mathbf{0} \quad \text{a.s.}, \qquad E\{\mathbf{e}(t)\mathbf{e}'(t)|\mathcal{F}(t - 1)\} = \Omega \quad \text{a.s.},$$
$$E\{e_a(t)e_b(t)e_c(t)|\mathcal{F}(t - 1)\} = \rho_{abc} \quad \text{a.s., etc.}$$

e.g. Dunsmuir (1979). Here it may be noted that the ordinary linear process generated by independent and identically distributed (iid) $\mathbf{e}(t)$ with finite fourth-order moment satisfies (A.4)–(A.6). The following is a very fundamental result.

THEOREM 2.1   Hosoya–Taniguchi (1982). Assume that $\{\mathbf{X}(t)\}$ defined in Eq. (2.1) satisfies the conditions (A.1)–(A.6). Let $\phi_j(\lambda)$, $j = 1, \ldots, q$, be $m \times m$ matrix–valued continuous functions on $[-\pi, \pi]$ such that $\phi_j(\lambda) = \phi_j(\lambda)^*$. Then

(1) $p - \lim\limits_{n \to \infty} \int_{-\pi}^{\pi} \text{tr}\{\mathbf{I}_n(\lambda)\phi_j(\lambda)\} \, d\lambda = \int_{-\pi}^{\pi} \text{tr}\{\mathbf{f}(\lambda)\phi_j(\lambda)\} \, d\lambda,$

(2)   the quantites $\sqrt{n} \int_{-\pi}^{\pi} tr[\{\mathbf{I}_n(\lambda) - \mathbf{f}(\lambda)\}\phi_j(\lambda)] \, d\lambda, \quad j = 1, \ldots, q,$

have, asymptotically, a normal distribution with zero mean vector and co-variance matrix $\mathbf{V}$ whose (j, ℓ) element is

$$4\pi \int_{-\pi}^{\pi} \text{tr}\{\mathbf{f}(\lambda)\phi_j(\lambda)\mathbf{f}(\lambda)\phi_\ell(\lambda)\} \, d\lambda + 2\pi$$

$$\times \sum_{a,b,c,d=1}^{m} \iint_{-\pi}^{\pi} \phi_{ab}^{(j)}(\lambda_1)\phi_{cd}^{(\ell)}(\lambda_2) f_{abcd}^{x}(-\lambda_1, \lambda_2, -\lambda_2) \, d\lambda_1 \, d\lambda_2,$$

where $\phi_{ab}^{(j)}(\lambda)$ is the (a, b) element of $\phi_j(\lambda)$.

Assuming that all cumulant spectra, of all orders, exist for $\{\mathbf{X}(t)\}$, Brillinger (1975) proved the essentially same result as Theorem 2.1. For a data tapered periodogram Dahlhaus (1988) proved the weak convergence result for empirical spectral processes indexed by classes of functions.

We next consider the problem of nonparametric spectral estimation. We estimate $\mathbf{f}(\lambda)$ by weighted averages of the periodogram matrix $\mathbf{I}_n(\lambda)$ with a spectral window $W_n(\lambda)$ as weight, i.e.,

$$\hat{\mathbf{f}}_n(\lambda) = \int_{-\pi}^{\pi} W_n(\lambda - \mu)\mathbf{I}_n(\mu) \, d\mu. \tag{2.2}$$

The following condition is imposed on $W_n(\cdot)$.

(A.7)   (i) $W_n(\lambda)$ can be expanded as

$$W_n(\lambda) = \frac{1}{2\pi} \sum_{\ell=-M}^{M} w\left(\frac{\ell}{M}\right) \exp(-i\ell\lambda).$$

   (ii) $w(x)$ is a continuous, even function with $w(0) = 1$, and satisfies

$$\begin{cases} |w(x)| \leq 1, \\ \displaystyle\int_{-\infty}^{\infty} w(x)^2 \, dx < \infty, \\ \displaystyle\lim_{x \to 0} \frac{1 - w(x)}{|x|^2} = \kappa_2 < \infty. \end{cases}$$

   (iii) $M = M(n)$ satisfies

$$n^{1/4}/M + M/n^{1/2} \to 0 \qquad \text{as} \quad n \to \infty.$$

Concrete examples of $W_n(\cdot)$ satisfying (A.7) are found in Hannan (1970), Brillinger (1975) and Robinson (1983). Then, it is known that

$$E[\|\hat{\mathbf{f}}_n(\lambda) - \mathbf{f}(\lambda)\|^2] = O\left(\frac{M}{n}\right) + O(M^{-4}), \tag{2.3}$$

uniformly in λ, e.g., Hannan (1970). Hence, $\hat{\mathbf{f}}_n(\lambda) - \mathbf{f}(\lambda) = O_p\{(M/n)^{1/2}\}$, so $\hat{\mathbf{f}}_n(\lambda)$ is a $\sqrt{n/M}$-consistent estimator of $\mathbf{f}(\lambda)$. This is a disadvantage of nonparametric spectral estimators because parametric estimators usually have $\sqrt{n}$-consistency. But we will show in Sec. 3 that this disadvantage disappears if we consider appropriate integral functionals of $\hat{\mathbf{f}}_n(\lambda)$.

Let D be an open subset of $\mathbf{C}^{m^2}$. We set down the following assumption.

(A.8) A mapping $K : D \to \mathbb{R}$ is holomorphic.

The integral functional of $\mathbf{f}(\lambda)$

$$\int_{-\pi}^{\pi} K\{\mathbf{f}(\lambda)\}\, d\lambda \tag{2.4}$$

represents a wide variety of important time series indices. The examples will be given in Sec. 4. Suppose that we are interested in estimation of $\int_{-\pi}^{\pi} K\{\mathbf{f}(\lambda)\}\, d\lambda$. Then we propose $\int_{-\pi}^{\pi} K\{\hat{\mathbf{f}}_n(\lambda)\}\, d\lambda$ as an estimator of it. From Taniguchi, Puri and Kondo (1996) it can be shown that

$$\sqrt{n}\left[\int_{-\pi}^{\pi} K\{\hat{\mathbf{f}}_n(\lambda)\}\, d\lambda - \int_{-\pi}^{\pi} K\{\mathbf{f}(\lambda)\}\, d\lambda\right]$$
$$= \sqrt{n}\int_{-\pi}^{\pi} [\mathrm{tr}\{\mathbf{I}_n(\lambda) - \mathbf{f}(\lambda)\}K^{(1)}\{\mathbf{f}(\lambda)\}]\, d\lambda + o_P(1). \tag{2.5}$$

where $K^{(1)}\{\mathbf{f}(\lambda)\}$ is the first derivative of $K\{\mathbf{f}(\lambda)\}$ at $\mathbf{f}(\lambda)$ (see Magnus and Neudecker (1988)). Henceforth we denote the (a, b) element of $K^{(1)}\{\mathbf{f}(\lambda)\}$ by $K_{ab}^{(1)}\{\mathbf{f}(\lambda)\}$. In view of Eq. (2.5) and Theorem 2.1 we obtain

THEOREM 2.2   Taniguchi, Puri and Kondo (1996). Assume that the conditions (A.1)–(A.8) hold. Then

$$S_n = \sqrt{n}\left[\int_{-\pi}^{\pi} K\{\hat{\mathbf{f}}_n(\lambda)\}\, d\lambda - \int_{-\pi}^{\pi} K\{\mathbf{f}(\lambda)\}\, d\lambda\right] \tag{2.6}$$

has asymptotically a normal distribution with mean zero and variance $\nu_1(\mathbf{f}) + \nu_2(f_4^x)$, where

$$\nu_1(\mathbf{f}) = 4\pi \int_{-\pi}^{\pi} \mathrm{tr}[\mathbf{f}(\lambda)K^{(1)}\{\mathbf{f}(\lambda)\}]^2\, d\lambda \tag{2.7}$$

and

$$\nu_2(f_4^x) = 2\pi \sum_{a,b,c,d=1}^{m} \iint_{-\pi}^{\pi} K_{ab}^{(1)}\{\mathbf{f}(\lambda)\}K_{cd}^{(1)}\{\mathbf{f}(\lambda)\}f_{abcd}^x(-\lambda_1, \lambda_2, -\lambda_2)\, d\lambda_1 d\lambda_2. \tag{2.8}$$

Here it should be noted that $\sqrt{n}$-consistency holds in Theorem 2.2 despite Eq. (2.3) ($\sqrt{n/M}$-consistency of $\hat{\mathbf{f}}_n(\lambda)$). This is due to the fact that integration of $\hat{\mathbf{f}}_n$ recovers $\sqrt{n}$-consistency.

3. SEMIPARAMETRIC ESTIMATION

As far as I know it seems that the origin of semiparametric estimation stems from Hannan (1963). He considered the problem of estimating the matrix of the regression coefficient, **B**, in the relation

$$\mathbf{Y}(t) = \mathbf{B}\mathbf{Z}(t) + \mathbf{X}(t), \tag{3.1}$$

where $\mathbf{Y}(t)$ is a vector of m components $y_j(t)$, $\mathbf{Z}(t)$ is a vector of q components $z_k(t)$, **B** is a $(m \times q)$ matrix of real numbers b_{jk}, and $\mathbf{X}(t)$ is a vector of m components $X_j(t)$. The following conditions are assumed.

(B.1) $\{\mathbf{X}(t)\}$ is a linear process of the form defined by Eq. (2.1) whose coefficients satisfy

$$\sum_{j=0}^{\infty} \|G(j)\| < \infty, \tag{3.2}$$

where $\|G(j)\|$ is the norm of the matrix $G(j)$.
  Let

$$a_{jk}^n(h) = \begin{cases} \displaystyle\sum_{t=1}^{n-h} z_j(t+h)z_k(t), & h = 0, 1, \ldots, \\ \displaystyle\sum_{t=1-h}^{n} z_j(t+h)z_k(t), & h = 0, -1, \ldots. \end{cases} \tag{3.3}$$

(B.2) (Grenander's conditions)
  (i) $a_{jj}^n(0) \to \infty$ as $n \to \infty$, $j = 1, \ldots, q$.
  (ii) $\displaystyle\lim_{n\to\infty} z_j^2(n+1)/a_{jj}^n(0) = 0$, $j = 1, \ldots, q$.
  (iii) The limit

$$\lim_{n\to\infty} \frac{a_{jk}^n(h)}{\sqrt{a_{jj}^n(0)a_{kk}^n(0)}} = \rho_{jk}(h),$$

  exists for every $j, k = 1, \ldots, q$ and $h = 0, \pm 1, \ldots$. Let $\mathbf{R}(h) = \{\rho_{jk}(h); j, k = 1, \ldots, q\}$.
  (iv) $\mathbf{R}(0)$ is nonsingular.

Then there exists a matrix–valued function $\mathbf{M}(\lambda)$ whose increments are Hermitian nonnegative such that

$$\mathbf{R}(h) = \int_{-\pi}^{\pi} e^{ih\lambda} d\mathbf{M}(\lambda). \tag{3.4}$$

We denote by $\hat{\mathbf{B}}$ the least squares estimator (LSE) of $\mathbf{B}$ based on $\{\mathbf{Y}(1), \ldots, \mathbf{Y}(n)\}$ and $\{\mathbf{Z}(1), \ldots, \mathbf{Z}(n)\}$. Let $\hat{\mathbf{X}}(t) = \mathbf{Y}(t) - \hat{\mathbf{B}}\mathbf{Z}(t)$ and $\mathbf{D}_n = \mathrm{diag}\{\sqrt{a_{11}''(0)}, \ldots, \sqrt{a_{qq}''(0)}\}$. We define nonparametric spectral estimators $\hat{\mathbf{f}}_{YY}(\lambda)$, $\hat{\mathbf{f}}_{ZZ}(\lambda)$ and $\tilde{\mathbf{f}}_{XX}(\lambda)$ based on $\{\mathbf{Y}(1), \ldots, \mathbf{Y}(n)\}$, $\{\mathbf{Z}(1), \ldots, \mathbf{Z}(n)\}$ and $\{\hat{\mathbf{X}}(1), \ldots, \hat{\mathbf{X}}(n)\}$, respectively, in the manner of Eq. (2.2), where the spectral window function satisfies (A.7) with the sample lag M. Similarly we define a cross spectral estimator $\hat{\mathbf{f}}_{YZ}(\lambda)$ based on $\{\mathbf{Y}(1), \ldots, \mathbf{Y}(n)\}$ and $\{\mathbf{Z}(1), \ldots, \mathbf{Z}(n)\}$. Let us arrange the elements of the matrix $\mathbf{B}$ in a column vector $\boldsymbol{\beta}$ with b_{jk} in row $q(j-1) + k$. As an estimator of $\boldsymbol{\beta}$, Hannan (1963) uses

$$\hat{\mathbf{b}} = \left\{ \frac{1}{2M} \sum_{k=-M+1}^{M-1} \left[\tilde{\mathbf{f}}_{XX}\left(\frac{\pi k}{M}\right)^{-1} \otimes \hat{\mathbf{f}}_{ZZ}\left(-\frac{\pi k}{M}\right) \right] \right\}^{-1}$$

$$\times \left\{ \frac{1}{2M} \sum_{k=-M+1}^{M-1} \left[\tilde{\mathbf{f}}_{XX}\left(\frac{\pi k}{M}\right) \otimes \mathbf{I}_q \right]^{-1} \hat{\mathbf{f}}_{YZ}\left(\frac{\pi k}{M}\right) \right\}, \tag{3.5}$$

where $\mathbf{A} \otimes \mathbf{B}$ is the Kronecker product of the matrices $\mathbf{A}$ and $\mathbf{B}$, and $\mathbf{I}_q$ is the q dimensional identity matrix.

THEOREM 3.1 Hannan (1963). Under (B.1) and (B.2), $(\mathbf{I}_m \otimes \mathbf{D}_n)(\hat{\mathbf{b}} - \boldsymbol{\beta})$ is asymptotically normal with mean vector null and covariance matrix

$$\left[\frac{1}{2\pi} \int_{-\pi}^{\pi} \mathbf{f}_{XX}^{-1}(\lambda) \otimes d\mathbf{M}'(\lambda) \right]^{-1},$$

where $\mathbf{f}_{XX}(\lambda)$ is the spectral density matrix of $\{\mathbf{X}(t)\}$.

Although $\hat{\mathbf{b}}$ uses nonparametric estimators, it recovers the "parametric consistent" order.

For independent identically distributed (iid) observations Beran (1977) discusses a semiparametric estimation procedure. Suppose $X_1, \ldots, X_n$ are iid random variables with the probability density function $f(x)$. A specified parametric family $\{f_\theta : \theta \in \Theta\}$ is postulated. We recognize that lack of information, data contamination, and other factors beyond our control make it certain that the model is not strictly correct. How are we to estimate θ in order to investigate the fit of the model to the data?

A good estimator of θ would have two essential properties: it would be efficient if the postulated model for the data were in fact true and its distribution would not be greatly perturbed if the assumed model were only approximately true. It has long been known that the maximum likelihood estimator of θ has full asymptotic efficiency among regular estimators. However, it has been recognized that maximum likelihood estimators do not possess the property of stability under small perturbations in the underlying model. Beran (1977) proposed a new semiparametric estimator $\hat{\theta}_n$ of θ by the requirement that

$$\|f_{\hat{\theta}_n}^{1/2} - \hat{f}_n^{1/2}\| = \min_{\theta \in \Theta} \|f_\theta^{1/2} - \hat{f}_n^{1/2}\|, \tag{3.6}$$

where $\| \cdot \|$ denotes the L_2-norm and $\hat{f}_n$ is a suitable nonparametric estimator of the density f. He showed that $\hat{\theta}_n$ is asymptotically efficient under a specified regular parametric family of densities and is minimax robust in a small Hellinger metric neighborhood of the given family.

For a linear process Taniguchi (1987) studied semiparametric estimation as follows. Let $\{X(t); t \in \mathbf{Z}\}$ be a linear process

$$X(t) = \sum_{j=0}^{\infty} G(j)e(t - j), \qquad t \in \mathbf{Z}, \tag{3.7}$$

which is supposed to satisfy (A.1)–(A.6) with $Ee(t)^2 = \sigma^2$ and $m = 1$. We introduce $\mathcal{F}$, the space of spectral densities defined by

$$\mathcal{F} = \left\{ f; f(\lambda) = \frac{\sigma^2}{2\pi} \left| \sum_{j=0}^{\infty} G(j)e^{-ij\lambda} \right|^2 \right\}.$$

We propose to fit some parametric family $\mathcal{P} = \{f_\theta; f_\theta \in \mathcal{F}, \theta \in \Theta \subset \mathbf{R}^p\}$ of spectral densities to the true spectral density $f(\lambda)$ by minimizing a criterion. Initially, we make the following assumption.

(B.3) $K(x)$ is a three times continuously differentiable function on $(0, \infty)$, and has a unique minimum at $x = 1$.

We then define the criterion which measures the nearness of f_θ to f by

$$D(f_\theta, f) = \int_{-\pi}^{\pi} K\{f_\theta(\lambda)/f(\lambda)\} \, d\lambda. \tag{3.8}$$

The following are examples of $K(\cdot)$:

(i) $K(x) = \log x + x^{-1}$
(ii) $K(x) = -\log x + x$,
(iii) $K(x) = (\log x)^2$,
(iv) $K(x) = x \log x - x$,
(v) $K(x) = (x^\alpha - 1)^2$, $0 < \alpha < \infty$.

To estimate θ, since $f(\lambda)$ is unknown, we estimate $f(\lambda)$ by a nonparametric estimator $\hat{f}(\lambda)$ defined by Eq. (2.2) satisfying (A.7) with $m = 1$. Therefore, a semiparametric estimator $\hat{\theta}_n$ of θ is defined by the requirement that

$$D(f_{\hat{\theta}_n}, \hat{f}_n) = \min_{\theta \in \Theta} D(f_\theta, \hat{f}_n). \tag{3.9}$$

Assuming that f_θ is smooth with respect to θ, it can be shown that

$$\sqrt{n}(\hat{\theta}_n - \underline{\theta}) = \sqrt{n} \int_{-\pi}^{\pi} \psi(\lambda)\{\hat{f}_n(\lambda) - f(\lambda)\}\, d\lambda + o_P(1), \tag{3.10}$$

where $\psi(\lambda)$ is written in terms of f_θ and f, and $\underline{\theta}$ is defined by the relation $D(f_{\underline{\theta}}, f) = \min_{\theta \in \Theta} D(f_\theta, f)$. Recalling Theorem 2.2 we can prove the central limit theorem for $\sqrt{n}(\hat{\theta}_n - \underline{\theta})$; especially, if $f = f_\theta$ (i.e., $\underline{\theta} = \theta$), then $\sqrt{n}(\hat{\theta}_n - \theta)$ converges in distribution to $\mathcal{N}(0, F(\theta)^{-1})$, where

$$F(\theta) = \frac{1}{4\pi} \int_{-\pi}^{\pi} \frac{\partial}{\partial \theta} \log f_\theta(\lambda) \frac{\partial}{\partial \theta'} \log f_\theta(\lambda)\, d\lambda,$$

which is called the Fisher information matrix in time series analysis, see Taniguchi (1987). This result means that we can construct infinitely many (Gaussian) asymptotically efficient estimators. Here we say that an estimator $\hat{\theta}$ is asymptotically efficient if $\sqrt{n}(\hat{\theta} - \theta)$ has the asymptotic distribution $\mathcal{N}(0, F(\theta)^{-1})$.

Except for autoregressive models, (quasi) maximum likelihood estimation requires iterative computational procedures. On the contrary, the estimator $\hat{\theta}_n$ above has a merit, i.e., if we choose appropriate $K(\cdot)$, the estimator $\hat{\theta}_n$ gives an explicit, non-iterative and efficient estimator for various spectral parametrization.

Suppose that the spectral density $f_\theta(\lambda)$ is parametrized as

$$f_\theta(\lambda) = S\{A_\theta(\lambda)\}, \tag{3.10}$$

where $A_\theta(\lambda) = \sum_j \theta_j \exp(ij\lambda)$ and $S\{\cdot\}$ is a bijective continuously three times differentiable function. To give non-iterative estimators, the following relation should be imposed:

$$K\left[\frac{S\{A_\theta(\lambda)\}}{f(\lambda)}\right] = c_1(\lambda)A_\theta(\lambda)^2 + c_2(\lambda)A_\theta(\lambda) + c_3(\lambda) + c_4 \log S\{A_\theta(\lambda)\},$$

$$\tag{3.11}$$

where $c_1(\lambda)$, $c_2(\lambda)$ and $c_3(\lambda)$ are functions which are independent of θ, and c_4 is a constant which is independent of θ and λ. For a spectral density $f_\theta(\lambda)$ given in Eq. (3.10), choose the function $K(\cdot)$ so that Eq. (3.11) and (B.3) hold. Then, checking the above procedure we get the following results. Let $\hat{\theta}_n^{(\#)}$ be the estimator which minimizes the criterion

$$\int_{-\pi}^{\pi} K_{\#}[f_\theta(\lambda)/\hat{f}_n(\lambda)]\, d\lambda$$

with respect to θ. We use a $p \times p$-matrix $R_{\#} = [R_{\#}(j, \ell)]$, and a $p \times 1$-vector $r_{\#} = [r_{\#}(\ell)]$.

(i) In the case $f_\theta(\lambda) = \sigma^2/2\pi |\sum_{j=0}^{p} \theta_j e^{-ij\lambda}|^2$, where $\theta_0 = 1$ and $\sum_{j=0}^{p} \theta_j z^j \neq 0$ for $|z| \leq 1$, choose $K_{MA}(x) = -\log x + x$, then the non-iterative estimator is given by $\hat{\theta}_n^{(MA)} = R_{MA}^{-1} \cdot r_{MA}$, where

$$R_{MA}(j - \ell) = \int_{-\pi}^{\pi} \hat{f}_n(\lambda)^{-1} \cos(j - \ell)\lambda \, d\lambda$$

$$r_{MA}(\ell) = \int_{-\pi}^{\pi} \hat{f}_n(\lambda)^{-1} \cos \ell\lambda \, d\lambda .$$

(ii) In the case $f_\theta(\lambda) = \sigma^2 \exp[\sum_{j=0}^{p} \theta_j \cos j\lambda]$, $\theta_0 = 1$, (cf. Bloomfield (1973)), choose $K_E(x) = (\log x)^2$, then the non-iterative estimator is given by $\hat{\theta}_n^{(E)} = R_E^{-1} \cdot r_E$, where

$$R_E(j - \ell) = \pi\delta(j, \ell),$$

$$r_E(\ell) = \int_{-\pi}^{\pi} \cos \ell\lambda \cdot \log \hat{f}_n(\lambda) \, d\lambda.$$

(iii) Let $\Psi(x)$ be a three times continuously differentiable bijective function on $(0, \infty)$ and $\Psi(1) = 1$. We assume that there exists a function $h(\cdot)$ such that $\Psi(xy) = h(y)\Psi(x)$, for all $x, y \in (0, \infty)$. As a special case we can take $\Psi(x) = x^{1/\beta}$, $(\beta > 0)$. In the case $f_\theta(\lambda) = \Psi^{-1}[\sum_{j=0}^{p} \theta_j \cos j\lambda]$, where $\theta_0 = 1$ and $\theta_1, \ldots, \theta_p$ satisfy $\sum_{j=0}^{p} \theta_j \cos j\lambda > 0$ for all $\lambda \in [-\pi, \pi]$, choose $K_\Psi(x) = [\Psi(x) - 1]^2$, then the non-iterative estimator is given by $\hat{\theta}_n^{(\Psi)} = R_\Psi^{-1} \cdot r_\Psi$, where

$$R_\Psi(j - \ell) = \int_{-\pi}^{\pi} h[\hat{f}_n(\lambda)^{-1}]^2 \cos j\lambda \cos \ell\lambda \, d\lambda,$$

$$r_\Psi(\ell) = \int_{-\pi}^{\pi} h[\hat{f}_n(\lambda)^{-1}] \cos \ell\lambda \, d\lambda.$$

If $f = f_\theta$, and if the process $\{X(t)\}$ is Gaussian, all the above estimators are asymptotically efficient.

As the other semiparametric approach we briefly mention Robinson (1988), which considered a semiparametric form, such as the regression function

$$E(Y|X, Z) = \beta'X + \theta(Z) \quad \text{a.s.,} \tag{3.12}$$

where (X, Y, Z) is an $\mathbf{R}^p \times \mathbf{R} \times \mathbf{R}^q$ –valued observable random variable, β is an $\mathbf{R}^p$–valued unknown parameter, and θ is an unknown real function. Based on the independent observations $(X_1, Y_1, Z_1), \dots, (X_n, Y_n, Z_n)$, he proposed an estimator $\hat{\beta}$ generalizing the ordinary least squares estimator of β by inserting nonparametric regression estimators in the nonlinear orthogonal projection on Z. Under regularity conditions $\hat{\beta}$ was shown to be $\sqrt{n}$-consistent for β and asymptotically normal, and a consistent estimator of its limiting covariance matrix was given.

4. APPLICATIONS TO TESTING THEORY

The ordinary nonparametric approach for iid random variables has developed in various directions. For example Hallin et al. (1985) introduced a class of linear serial rank statistics for the problem of testing white noise against alternatives of autoregressive moving average (ARMA) serial dependence. They derived the asymptotic distributions of the proposed test statistics under the null as well as alternative hypotheses, and gave an explicit formulation of the asymptotically most powerful test under a sequence of contiguous ARMA alternatives. Dzhaparidze (1986) considered a class $\mathcal{F}$ of goodness-of-fit tests for testing a simple hypothesis about the form of the spectral density. He investigated the asymptotic properties of the test $T \in \mathcal{F}$ under a sequence of nonparametric contiguous alternatives.

Let $\{X_t : t = 0, \pm 1, \dots\}$ be a Gaussian stationary process with mean zero and spectral density $f(\lambda)$. Initially, we make the following assumption.

(C.1) (i) There exists a positive number δ such that $f(\lambda) \geq \delta$, $-\pi \leq \lambda \leq \pi$.

     (ii)

$$\sum_{t=1}^{\infty} t|R(t)| < \infty,$$

where

$$R(t) = E\{X_s X_{s+t}\}$$

We consider the testing problem

$$H : \int_{-\pi}^{\pi} K\{f(\lambda)\} \, d\lambda = c$$

against   (4.1)

$$A : \int_{-\pi}^{\pi} K\{f(\lambda)\} \, d\lambda \neq c,$$

where c is a given constant, and $K(x)$ satisfies the following condition.

(C.2) $K(x)$ is a continuously twice–differentiable function on $(0, \infty)$.

Denote by $I_n(\lambda)$ the periodogram constructed from a partial realization $\{X_1, \ldots, X_n\}$, namely

$$I_n(\lambda) = \frac{1}{2\pi n} \left| \sum_{t=1}^{n} X_t \exp(it\lambda) \right|^2, \quad -\pi \leq \lambda \leq \pi.$$

To estimate $f(\lambda)$ we use

$$\hat{f}_n(\lambda) = \int_{-\pi}^{\pi} W_n(\lambda - \mu) I_n(\mu) d\mu, \tag{4.2}$$

where $W_n(\lambda)$ satisfies (A.7).

THEOREM 4.1   Taniguchi–Kondo (1993). Assume that (C.1), (C.2) and (A.7) hold. Then, under H,

$$T_n = \frac{\sqrt{n}[\int_{-\pi}^{\pi} K\{\hat{f}_n(\lambda)\} \, d\lambda - c]}{[4\pi \int_{-\pi}^{\pi} K'\{\hat{f}_n(\lambda)\}^2 \hat{f}_n(\lambda)^2 \, d\lambda]^{1/2}} \tag{4.3}$$

has asymptotically the standard normal distribution $\mathcal{N}(0, 1)$.

In view of Theorem 4.1 we can propose the test of H given by the critical region

$$[|T_n| > t_\alpha], \tag{4.4}$$

where t_α is defined by

$$\int_{t_\alpha}^{\infty} (2\pi)^{-1/2} \exp\left(-\frac{x^2}{2}\right) dx = \frac{\alpha}{2}.$$

Next we evaluate the local power of test in Eq. (4.4) under a sequence of contiguous spectral densities. Henceforth we denote the probability density of $\mathbf{X}_n = (X_1, \ldots, X_n)$ by $p_h^n(\cdot)$ if the process $\{X_t\}$ is assumed to have a spectral density $h(\lambda)$. Let $a(\lambda)$ be a square integrable function on $[-\pi, \pi]$. Putting $g_n(\lambda) = f(\lambda)\{1 + a(\lambda)/\sqrt{n}\}$, we define

$$\Lambda(f, g_n) = \log\left\{\frac{p_{g_n}^n(\mathbf{X}_n)}{p_f^n(\mathbf{X}_n)}\right\}. \tag{4.5}$$

Then we have,

LEMMA 4.1   Dzhaparidze (1986). Suppose that (C.1) holds and that

$$\sum_{t=1}^{\infty} t|\rho(t)|^2 < \infty, \tag{4.6}$$

where

$$\rho(t) = \frac{1}{2\pi}\int_{-\pi}^{\pi}\frac{a(\lambda)}{f(\lambda)}\exp(i\lambda t)\,d\lambda.$$

Then

(i) p_f^n and $p_{g_n}^n$ are contiguous.
(ii)

$$\Lambda(f, g_n) - \frac{\sqrt{n}}{4\pi}\int_{-\pi}^{\pi}\frac{I_n(\lambda) - f(\lambda)}{f(\lambda)}\,a(\lambda)\,d\lambda + \frac{1}{8\pi}\int_{-\pi}^{\pi}a^2(\lambda)\,d\lambda \to 0 \quad \text{in } p_f^n. \tag{4.7}$$

This lemma, together with LeCam's third lemma, e.g., Hájek and Šidák (1967) yields the following theorem.

THEOREM 4.2   Taniguchi–Kondo (1993). Assume that (C.1), (C.2), (A.7) and (4.6) hold. Then T_n defined in Theorem 4.1 is under $p_{g_n}^n$ asymptotically normal $\mathcal{N}\{\mu(f, a), 1\}$, where

$$\mu(f, a) = \int_{-\pi}^{\pi}K'\{f(\lambda)\}a(\lambda)f(\lambda)\,d\lambda\bigg/\left\{4\pi\int_{-\pi}^{\pi}[K'\{f(\lambda)\}]^2 f(\lambda)^2\,d\lambda\right\}^{1/2}.$$

Therefore the local power is given by

$$\lim_{n\to\infty}P_{g_n}[|T_n| \ge t_\alpha] = \int_{|x|\ge t_\alpha}\frac{1}{\sqrt{2\pi}}\exp[-\tfrac{1}{2}\{x - \mu(f, a)\}^2]\,dx. \tag{4.8}$$

This theorem enables us to evaluate the local power of T_n for a sequence of "nonparametric alternatives". Let $E_{g_n}(\cdot)$ and $V_f(\cdot)$ denote the expectation under $p_{g_n}^n$ and the variance under p_f^n respectively. It is natural to define an efficacy of T_n by

$$\text{eff}(T_n) = \lim_{n \to \infty} \frac{\{E_{g_n}(T_n)\}^2}{V_f(T_n)}, \tag{4.9}$$

in line with the usual definition for a sequence of "parametric alternatives", see e.g., Randles and Wolfe (1976). Then we see that

$$\text{eff}(T_n) = \mu(f, a)^2. \tag{4.10}$$

For another test T_n^* we can define an asymptotic relative efficiency (ARE) of T_n relative to T_n^* by

$$\text{ARE}(T_n, T_n^*) = \frac{\text{eff}(T_n)}{\text{eff}(T_n^*)}. \tag{4.11}$$

If we take the parametric alternative $p_{f_{\theta + \varepsilon/\sqrt{n}}}^n$ (i.e., $g_n = f_{\theta + \varepsilon/\sqrt{n}}$), this asymptotic relative efficiency is equivalent to the ordinary efficiency, e.g., Hannan (1955). The test setting in Eq. (4.1) is unexpectedly wide and can be applied to many problems in time series. Therefore, we next give some examples given by Taniguchi and Kondo (1993).

If we put $K(x) = \log x$, the testing problem becomes

$$H : \int_{-\pi}^{\pi} \log f(\lambda)\, d\lambda = c$$

against   (4.12)

$$A : \int_{-\pi}^{\pi} \log f(\lambda)\, d\lambda \neq c.$$

It is known that the quantity $\int_{-\pi}^{\pi} \log f(\lambda)\, d\lambda$ is important and is called Burg's entropy. This is a key parameter of $\{X_t\}$ which is specified by the variance σ^2 of the prediction error for which the one–step–ahead best linear predictor of X_t, i.e.,

$$\sigma^2 = \exp\left\{ \frac{1}{2\pi} \int_{-\pi}^{\pi} \log 2\pi f(\lambda)\, d\lambda \right\}$$

e.g., Hannan (1970). Then the test statistic for Eq. (4.12) becomes

$$T_n = \frac{\sqrt{n}\,[\int_{-\pi}^{\pi} \log \hat{f}_n(\lambda)\, d\lambda - c]}{2\sqrt{2\pi}}. \tag{4.13}$$

Under the sequence of spectral densities g_n, we can evaluate the local power of T_n. In fact, from Theorem 4.2 we have

$$\lim_{n \to \infty} P_{g_n}[|T_n| \geq t_\alpha] = \int_{|x| \geq t_\alpha} dN\left[\frac{1}{2\sqrt{2\pi}} \int_{-\pi}^{\pi} a(\lambda)\, d\lambda,\ 1 \right]. \tag{4.14}$$

The efficacy of T_n is given by

$$\text{eff}(T_n) = \frac{1}{8\pi^2} \left\{ \int_{-\pi}^{\pi} a(\lambda) \, d\lambda \right\}^2. \tag{4.15}$$

For the testing problem in Eq. (4.12) we can construct another test. Let

$$T_n^* = \sqrt{n} \left\{ \int_{-\pi}^{\pi} \log \delta I_n(\lambda) \, d\lambda - c \right\} \frac{\sqrt{3}}{2\pi^2},$$

where $\delta = \exp \gamma$ ($\gamma \approx 0.57721$, Euler's constant). Under H, we can write

$$T_n^* = \sqrt{n} \left\{ \int_{-\pi}^{\pi} \log \delta I_n(\lambda) \, d\lambda - \int_{-\pi}^{\pi} \log f(\lambda) \, d\lambda \right\} \frac{\sqrt{3}}{2\pi^2}.$$

In view of the estimation for $\int_{-\pi}^{\pi} \log f(\lambda) \, d\lambda$, it is shown that T_n^* is asymptotically normal under H with zero mean and variance 1, see Bloomfield (1973), Hannan and Nicholls (1977) and Taniguchi (1980). Then the efficacy of T_n^* is

$$\text{eff}(T_n^*) = \frac{3}{4\pi^4} \left\{ \int_{-\pi}^{\pi} a(\lambda) \, d\lambda \right\}^2.$$

Therefore,

$$\text{ARE}(T_n, T_n^*) = \frac{\text{eff}(T_n)}{\text{eff}(T_n^*)} = \frac{\pi^2}{6} > 1.$$

which implies that T_n is better than T_n^*.

Hallin et al. (1985) investigated various asymptotic properties of a class of linear serial rank statistics for the problem of testing white noise against ARMA alternatives. They gave an explicit formulation of the asymptotically locally most powerful test under a sequence of contiguous ARMA alternatives.

Here we apply the results to testing for independence. The merit of the approach here is that we can deal with "essentially nonparametric alternatives". Let $\{X_t\}$ be the Gaussian process satisfying (C.1). We consider the testing problem

$H: X_t$; $t = 0, \pm 1, \ldots$, are mutually independent with a given variance $R(0) = R$, say against

$A : X_t$; $t = 0, \pm 1, \ldots$, are not mutually independent.

The hypothesis H is written in terms of the spectral density $f(\lambda)$ and the autocovariance functions $R(j)$ as follows.

$$H \Leftrightarrow R(\pm 1) = R(\pm 2) = \cdots = 0,$$

$$\Leftrightarrow \frac{1}{2\pi} \sum_{j \neq 0} R(j)^2 = 0, \tag{4.16}$$

$$\Leftrightarrow \int_{-\pi}^{\pi} f(\lambda)^2 \, d\lambda - \frac{1}{2\pi} R^2 = 0.$$

The last equivalence Eq. (4.16) follows from the relation

$$\int_{-\pi}^{\pi} f(\lambda)^2 \, d\lambda = \frac{1}{2\pi} \sum_{j=-\infty}^{\infty} R(j)^2.$$

Hence the testing for independence is written as

$$H : \int_{-\pi}^{\pi} f(\lambda)^2 \, d\lambda = \frac{1}{2\pi} R^2$$

against

$$A : \int_{-\pi}^{\pi} f(\lambda)^2 \, d\lambda \neq \frac{1}{2\pi} R^2.$$

By virtue of Theorem 4.1 we can propose the following test statistic

$$T_n = \frac{\sqrt{n} \{ \int_{-\pi}^{\pi} \hat{f}_n(\lambda)^2 \, d\lambda - (1/2\pi) R^2 \}}{\{ 16\pi \int_{-\pi}^{\pi} \hat{f}_n(\lambda)^4 \, d\lambda \}^{1/2}}, \tag{4.17}$$

which has asymptotically the standard normal $\mathcal{N}(0, 1)$. Also, from Theorem 4.2, T_n is asymptotically normal under $p_{g_n}^n$ with mean $\nu_{T_n} = \int_{-\pi}^{\pi} a(\lambda) \, d\lambda / (2\sqrt{2\pi})$ and variance 1. Thus the efficacy is

$$\mathrm{eff}(T_n) = \frac{1}{8\pi^2} \left\{ \int_{-\pi}^{\pi} a(\lambda) \, d\lambda \right\}^2. \tag{4.18}$$

Next we consider another test statistic. Let us put

$$\hat{R}(1) = n^{-1} \sum_{t=1}^{n-1} X_t X_{t-1},$$

$$U = \sqrt{n} \, \hat{R}(1).$$

Then

$$U = \sqrt{n} \int_{-\pi}^{\pi} \exp(-i\lambda) \{ I_n(\lambda) - f(\lambda) \} \, d\lambda$$

under H. From Theorem 2.1 it follows that U is asymptotically normal with zero mean and variance

$$2\pi \int_{-\pi}^{\pi} (1 + \cos 2\lambda) f(\lambda)^2 \, d\lambda \tag{4.19}$$

under H, which is equal to R^2. Let

$$DW = \frac{\sqrt{n}\hat{R}(1)}{R},$$ (4.20)

which is essentially equivalent to the Durbin–Watson statistic for our problem. It is not difficult to show that DW is asymptotically normal with zero
mean and variance 1 under H, and that under $p_{g_n}^n$, DW is asymptotically
normal with mean

$$\frac{1}{2\pi}\int_{-\pi}^{\pi}\cos\lambda\, a(\lambda)\, d\lambda$$

and variance 1. Hence

$$\text{eff(DW)} = \frac{1}{4\pi^2}\left\{\int_{-\pi}^{\pi}\cos\lambda\, a(\lambda)\, d\lambda\right\}^2.$$ (4.21)

From Eqns (4.18) and (4.21), T_n is asymptotically more powerful than DW
if

$$\int_{-\pi}^{\pi}\frac{1}{\sqrt{2}}a(\lambda)\, d\lambda \geq \left|\int_{-\pi}^{\pi}\cos\lambda\, a(\lambda)\, d\lambda\right|,$$ (4.22)

and vice versa.

Consider the following sequence of alternative spectral densities

$$g_n(\lambda) = \frac{1}{2\pi}\left\{1 + \frac{1}{\sqrt{n}}a(\lambda)\right\},$$ (4.23)

where $a(\lambda) = \sigma^2\exp(\theta\cos\lambda)$, $|\theta| \leq 2$ (exponential type alternative). In this
case it is shown that the inequality in Eq. (4.22) always holds. Therefore the
test T_n is always asymptotically more powerful than DW under the sequence
of alternatives in Eq. (4.23). We evaluate the simulated power of tests under
Eq. (4.23). For $n = 1024$, $M = 32$, $\sigma = 8$ and $\theta = 1.0$ (0.25) 2.0, we calculated the tests T_n and DW, and repeated this procedure 100 times. Table 1
gives the simulated powers of T_n and DW with level $\alpha = 0.05$. This simulation result confirms the theoretical results. As a related work, Hong (1996)
discussed tests based on a quadratic norm, the Hellinger metric and the
Kullback–Leibler distance, for serial correlation of unknown form.

The results above are extended to the case of the two sample problem,
which is mainly due to Kondo and Taniguchi (1993). Let $\{X_t;\ t = 0, \pm1, \ldots\}$
and $\{Y_t;\ t = 0, \pm1, \ldots\}$ be mutually independent Gaussian stationary process with $E(X_t) = 0$ and $E(Y_t) = 0$ and spectral densities $f(\lambda)$ and $g(\lambda)$,
respectively. We assume that $\{X_t\}$ and $\{Y_t\}$ satisfy (C.1). Suppose that

Table 1. Simulated Powers with Level $\alpha = 0.05$

θ	Power of T_n	Power of DW
1.00	0.61	0.34
1.25	0.70	0.44
1.50	0.81	0.63
1.75	0.89	0.75
2.00	0.95	0.88

$K(x)$ is a continuously three times differentiable function on $(0, \infty)$ and $K'(x) \neq 0$ a.e.. The testing problem considered is

$$H : \int_{-\pi}^{\pi} K\{f(\lambda)\} \, d\lambda = \int_{-\pi}^{\pi} K\{g(\lambda)\} \, d\lambda,$$

against

$$A : \int_{-\pi}^{\pi} K\{f(\lambda)\} \, d\lambda \neq \int_{-\pi}^{\pi} K\{g(\lambda)\} \, d\lambda.$$

This setting of the test can be also applied to many problems in time series. Since the testing problem is written nonparametrically, we construct a nonparametric test statistic.

Denote by $I_n^X(\lambda)$ and $I_n^Y(\lambda)$ the periodograms constructed from partial realizations $\{X_1, \ldots, X_n\}$ and $\{Y_1, \ldots, Y_n\}$; namely

$$I_n^X(\lambda) = \frac{1}{2\pi n} \left| \sum_{t=1}^{n} X_t \, e^{it\lambda} \right|^2, \qquad \lambda \in [-\pi, \pi],$$

and

$$I_n^Y(\lambda) = \frac{1}{2\pi n} \left| \sum_{t=1}^{n} Y_t \, e^{it\lambda} \right|^2, \qquad \lambda \in [-\pi, \pi].$$

To estimate $f(\lambda)$ and $g(\lambda)$ we use

$$\hat{f}_n(\lambda) = \int_{-\pi}^{\pi} W_n(\lambda - \mu) I_n^X(\mu) \, d\mu,$$

$$\hat{g}_n(\lambda) = \int_{-\pi}^{\pi} W_n(\lambda - \mu) I_n^Y(\mu) \, d\mu,$$

where $W_n(\cdot)$ satisfies (A.7). As in Theorem 4.1, we can show that, under H,

$$T_n = \frac{\sqrt{n}[\int_{-\pi}^{\pi} K\{\hat{f}_n(\lambda)\}\, d\lambda - \int_{-\pi}^{\pi} K\{\hat{g}_n(\lambda)\}\, d\lambda]}{\sqrt{4\pi \int_{-\pi}^{\pi}[K'\{\hat{f}_n(\lambda)\}]^2 \hat{f}_n(\lambda)^2\, d\lambda + 4\pi \int_{-\pi}^{\pi}[K'\{\hat{g}_n(\lambda)\}]^2 \hat{g}_n(\lambda)^2\, d\lambda}}$$

(4.25)

has, asymptotically, the standard normal distribution $\mathcal{N}(0,1)$. Thus we can propose the test of H given by the critical region

$$[|T_n| > t_\alpha],$$

(4.26)

where t_α is given in Eq. (4.4).

Next we evaluate the asymptotic power of the test in Eq. (4.25). Denote by $p''_{(h_1, h_2)}(\cdot)$ the probability density function of $(\mathbf{X}_n, \mathbf{Y}_n) = ((X_1, \ldots, X_n), (Y_1, \ldots, Y_n))$ if the processes $\{X_t\}$ and $\{Y_t\}$ are assumed to have spectral densities $h_1(\lambda)$ and $h_2(\lambda)$, respectively. Let $a(\lambda)$ and $b(\lambda)$ be square integrable functions on $[-\pi, \pi]$ satisfying that

$$\sum_{t=1}^{\infty} t|\rho_f(t)|^2 < \infty, \qquad \sum_{t=1}^{\infty} t|\rho_g(t)|^2 < \infty,$$

(4.27)

where

$$\rho_f(t) = \frac{1}{2\pi} \int_{-\pi}^{\pi} \frac{a(\lambda)}{f(\lambda)} e^{i\lambda t}\, d\lambda, \qquad \rho_g(t) = \frac{1}{2\pi} \int_{-\pi}^{\pi} \frac{b(\lambda)}{g(\lambda)} e^{i\lambda t}\, d\lambda.$$

Putting $f_n(\lambda) = f(\lambda)\{1 + a(\lambda)/\sqrt{n}\}$ and $g_n(\lambda) = g(\lambda)\{1 + b(\lambda)/\sqrt{n}\}$, in the same way as Theorem 4.2, we can prove that T_n defined in Eq. (4.25) is under $p''_{(f_n, g_n)}$ asymptotically normal $\mathcal{N}\{\mu(f, g, a, b), 1\}$ where

$$\mu(f, g, a, b) = \int_{-\pi}^{\pi} \nu(f, g)^{-1/2} K'\{f(\lambda)\}a(\lambda)f(\lambda)\, d\lambda$$

$$- \int_{-\pi}^{\pi} \nu(f, g)^{-1/2} K'\{g(\lambda)\}b(\lambda)g(\lambda)\, d\lambda,$$

with

$$\nu(f, g) = 4\pi \int_{-\pi}^{\pi} [K'\{f(\lambda)\}]^2 f(\lambda)^2\, d\lambda + 4\pi \int_{-\pi}^{\pi} [K'\{g(\lambda)\}]^2 g(\lambda)^2\, d\lambda.$$

Therefore the asymptotic power is given by

$$\lim_{n \to \infty} P_{(f_n, g_n)}[|T_n| \geq t] = \int_{|x| \geq t} \frac{1}{\sqrt{2\pi}} \exp{-\frac{1}{2}\{x - \mu(f, g, a, b)\}^2}\, dx.$$

(4.28)

Let $E_{(f_n, g_n)}(\cdot)$ and $V_{(f, g)}(\cdot)$ denote the expectation under $p''_{(f_n, g_n)}$ and the variance under $p''_{(f, g)}$, respectively. An efficacy of T_n is defined by

$$\text{eff}(T_n) = \lim_{n \to \infty} \frac{\{E_{(f_n,g_n)}(T_n)\}^2}{V_{(f,g)}(T_n)}, \tag{4.29}$$

(c.f., see Eq. (4.9)). Then it is shown that

$$\text{eff}(T_n) = \mu(f,g,a,b)^2. \tag{4.30}$$

For another test T_n^* we define an asymptotic relative efficiency (ARE) of T_n relative to T_n^* by

$$\text{ARE}(T_n, T_n^*) = \frac{\text{eff}(T_n)}{\text{eff}(T_n^*)}. \tag{4.31}$$

EXAMPLE 4.1 (Testing for spectral moments). Let $K(x) = x^{\beta}$ ($\beta > 0$), then the testing problem becomes

$$H : \int_{-\pi}^{\pi} f(\lambda)^{\beta} d\lambda = \int_{-\pi}^{\pi} g(\lambda)^{\beta} d\lambda,$$

against $\tag{4.32}$

$$A : \int_{-\pi}^{\pi} f(\lambda)^{\beta} d\lambda \neq \int_{-\pi}^{\pi} g(\lambda)^{\beta} d\lambda.$$

The test statistic for Eq. (4.32) is

$$T_n = \frac{\sqrt{n}\,[\int_{-\pi}^{\pi} \hat{f}_n(\lambda)^{\beta} d\lambda - \int_{-\pi}^{\pi} \hat{g}_n(\lambda)^{\beta} d\lambda]}{\sqrt{4\pi \int_{-\pi}^{\pi} \beta^2 \hat{f}_n(\lambda)^{2\beta} d\lambda + 4\pi \int_{-\pi}^{\pi} \beta^2 \hat{g}_n(\lambda)^{2\beta} d\lambda}}. \tag{4.33}$$

The efficacy of T_n is given by

$$\text{eff}(T_n) = \frac{\{\int_{-\pi}^{\pi} f(\lambda)^{\beta} a(\lambda) d\lambda - \int_{-\pi}^{\pi} g(\lambda)^{\beta} b(\lambda) d\lambda\}^2}{4\pi\{\int_{-\pi}^{\pi} f(\lambda)^{2\beta} d\lambda + \int_{-\pi}^{\pi} g(\lambda)^{2\beta} d\lambda\}}. \tag{4.34}$$

We can construct another test:

$$T_n^* = \frac{\sqrt{n}\left[\int_{-\pi}^{\pi} \frac{I_n^X(\lambda)^{\beta}}{\Gamma(\beta+1)} d\lambda - \int_{-\pi}^{\pi} \frac{I_n^Y(\lambda)^{\beta}}{\Gamma(\beta+1)} d\lambda\right]}{\sqrt{4\pi\Psi(\beta)\left\{\int_{-\pi}^{\pi} \frac{I_n^X(\lambda)^{2\beta}}{\Gamma(2\beta+1)} d\lambda + \int_{-\pi}^{\pi} \frac{I_n^Y(\lambda)^{2\beta}}{\Gamma(2\beta+1)} d\lambda\right\}}}, \tag{4.35}$$

where $\Psi(\beta) = \{\Gamma(2\beta+1)/\Gamma(\beta+1)^2 - 1\}$. In view of estimation for $\int_{-\pi}^{\pi} f(\lambda)^{\beta} d\lambda - \int_{-\pi}^{\pi} g(\lambda)^{\beta} d\lambda$, it is shown that T_n^* is, under H, asymptotically normal with mean 0 and variance 1, see Taniguchi (1980). Then the efficacy of T_n^* is

$$\text{eff}(T_n^*) = \frac{\beta^2 \{\int_{-\pi}^{\pi} f(\lambda)^3 a(\lambda)\, d\lambda - \int_{-\pi}^{\pi} g(\lambda)^3 b(\lambda)\, d\lambda\}^2}{4\pi\Psi(\beta)\{\int_{-\pi}^{\pi} f(\lambda)^{2.3}\, d\lambda + \int_{-\pi}^{\pi} g(\lambda)^{2.3}\, d\lambda\}}.$$

Therefore,

$$\text{ARE}(T_n, T_n^*) = \frac{\Psi(\beta)}{\beta^2} > 1, \qquad (\beta > 1).$$

EXAMPLE 4.2 (Testing for interpolation error). Let $K(x) = x^{-1}$, then the testing problem becomes

$$H: \int_{-\pi}^{\pi} f(\lambda)^{-1}\, d\lambda = \int_{-\pi}^{\pi} g(\lambda)^{-1}\, d\lambda,$$

against   (4.36)

$$A: \int_{-\pi}^{\pi} f(\lambda)^{-1}\, d\lambda \neq \int_{-\pi}^{\pi} g(\lambda)^{-1}\, d\lambda.$$

The quantity $[1/2\pi \int_{-\pi}^{\pi} \{2\pi f(\lambda)\}^{-1}\, d\lambda]^{-1}$ is known to be the interpolation error of X_0 by the best linear interpolation based on X_t, $t = \pm 1, \pm 2, \ldots$, see Hannan (1970, p. 165). The test statistic for Eq. (4.36) is

$$T_n = \frac{\sqrt{n}\, [\int_{-\pi}^{\pi} \hat{f}_n(\lambda)^{-1}\, d\lambda - \int_{-\pi}^{\pi} \hat{g}_n(\lambda)^{-1}\, d\lambda]}{\sqrt{4\pi\{\int_{-\pi}^{\pi} \hat{f}_n(\lambda)^{-2}\, d\lambda + \int_{-\pi}^{\pi} \hat{g}_n(\lambda)^{-2}\, d\lambda\}}}.$$ (4.37)

The efficacy of T_n is given by

$$\text{eff}(T_n) = \frac{\{\int_{-\pi}^{\pi} f(\lambda)^{-1} a(\lambda)\, d\lambda - \int_{-\pi}^{\pi} g(\lambda)^{-1} b(\lambda)\, d\lambda\}^2}{4\pi\{\int_{-\pi}^{\pi} f(\lambda)^{-2}\, d\lambda + \int_{-\pi}^{\pi} g(\lambda)^{-2}\, d\lambda\}}.$$ (4.38)

If we assume the parametric spectral densities $f(\lambda) = f_\theta(\lambda)$ and $g(\lambda) = g_\mu(\lambda)$, we can construct another test. Let $\hat{\theta}$ and $\hat{\mu}$ be the quasi–maximum likelihood estimators of θ and μ, respectively, see Hosoya and Taniguchi (1982). Then we can propose

$$T_n^* = \frac{\sqrt{n}[\int_{-\pi}^{\pi} f_{\hat{\theta}}(\lambda)^{-1}\, d\lambda - \int_{-\pi}^{\pi} g_{\hat{\mu}}(\lambda)^{-1}\, d\lambda]}{\sqrt{\nu(f_{\hat{\theta}}, g_{\hat{\mu}})}},$$ (4.39)

where

$$\nu(f_\theta, g_\mu) = 4\pi \mathbf{f}' M_f^{-1} \mathbf{f} + 4\pi \mathbf{g}' M_g^{-1} \mathbf{g},$$

$$\mathbf{f}' = \int_{-\pi}^{\pi} \frac{\partial}{\partial \theta'} f_\theta(\lambda)^{-1} \, d\lambda,$$

$$\mathbf{g}' = \int_{-\pi}^{\pi} \frac{\partial}{\partial \mu'} g_\mu(\lambda)^{-1} \, d\lambda,$$

$$M_f = \int_{-\pi}^{\pi} \frac{\partial}{\partial \theta} \log f_\theta(\lambda) \frac{\partial}{\partial \theta'} \log f_\theta(\lambda) \, d\lambda,$$

$$M_g = \int_{-\pi}^{\pi} \frac{\partial}{\partial \mu} \log g_\mu(\lambda) \frac{\partial}{\partial \mu'} \log g_\mu(\lambda) \, d\lambda.$$

It is shown that T_n^* is under H asymptotically normal with mean 0 and variance 1, and that the efficacy of T_n^* is

$$\mathrm{eff}(T_n^*) = 16\pi^2 \{ \mathbf{f}' M_f^{-1} \int_{-\pi}^{\pi} (\partial/\partial\theta) f_\theta(\lambda)^{-1} a(\lambda) \, d\lambda$$

$$- \mathbf{g}' M_g^{-1} \int_{-\pi}^{\pi} (\partial/\partial\mu) g_\mu(\lambda)^{-1} b(\lambda) \, d\lambda \}^2 \sqrt{(f_0, g_\mu)}^{-1}.$$

Concerning Example 4.1, we give a numerical study. Let

$$f_n(\lambda) = \frac{1}{2\pi} \exp(0.5\lambda), \tag{4.40}$$

and

$$g_n(\lambda) = \frac{1}{2\pi} \exp(0.5\lambda) \left\{ 1 + \frac{0.5}{2\pi\sqrt{n}} \exp(\theta \cos \lambda) \right\}. \tag{4.41}$$

Under the sequences of spectral densities in Eqs (4.40) and (4.41), for $n = 1024$, $M = 32$ and $\beta = 2$, we calculated the tests T_n and T_n^* defined by Eqns (4.33) and (4.35), respectively, and repeated this procedure 1000 times. Table 2 gives the simulated powers of T_n and T_n^* with level $\alpha = 0.05$. This result agrees with the theoretical result.

Kondo (1994) discussed the k-sample problem in time series. For $i = 1, 2, \ldots, k$, let $\{X_{i,t}; \, t = 0, \pm 1, \ldots\}$ be mutually independent Gaussian stationary processes with $E(X_{i,t}) = 0$ and spectral densities $f_i(\lambda)$, respectively. He considered the testing problem

$$H : \int_{-\pi}^{\pi} K\{f_1(\lambda)\} \, d\lambda = \cdots = \int_{-\pi}^{\pi} K\{f_k(\lambda)\} \, d\lambda = c$$

against

$$\tag{4.42}$$

A : At least, one of the equalities in the above is violated,

Table 2.   Simulated Powers with Level $\alpha = 0.05$

θ	Power of T_n	Power of T_n^*
1.5	0.35	0.31
2.0	0.54	0.48
2.5	0.84	0.78

where c is an unknown parameter. Then a test statistic for Eq. (4.42) was proposed, and the local power of it was evaluated, Kondo (1994).

Next we develop our discussion to the case when the process concerned is multivariate and non-Gaussian. Let $\{\mathbf{X}(t); t \in \mathbf{Z}\}$ be an m-dimensional non-Gaussian linear process generated by Eq. (2.1) satisfying (A.1)–(A.6) with spectral density matrix $\mathbf{f}(\lambda)$. Suppose that $K\{\cdot\}$ is defined on $\{\mathbf{f}(\lambda); \lambda \in [-\pi, \pi]\}$ and satisfies (A.8). We consider the testing problem

$$H : \int_{-\pi}^{\pi} K\{\mathbf{f}(\lambda)\}\, d\lambda = c,$$

against   (4.43)

$$A : \int_{-\pi}^{\pi} K\{\mathbf{f}(\lambda)\}\, d\lambda \neq c,$$

where c is a given constant. Recalling Theorem 2.2 we make a test statistic for Eq. (4.43) based on S_n given by Eq. (2.6). For this we need the following assumptions.

(C.3) (i)   The cumulants

$$c_{a_1 \ldots a_k}^X(t_1, \ldots, t_{k-1}) \equiv \mathrm{cum}\{X_{a_1}(0), X_{a_2}(t_1), \ldots, X_{a_k}(t_{k-1})\}$$

exist for all $a_1, \ldots, a_k = 1, \ldots, m$, and $k = 1, 2, \ldots, 8$.

(ii) For each $j = 1, 2, \ldots, k - 1$ and any k–tuple $a_1, a_2, \ldots, a_k$, we have

$$\sum_{t_1, \ldots, t_{k-1} = -\infty}^{\infty} |t_j c_{a_1 \ldots a_k}^X(t_1, \ldots, t_{k-1})| < \infty,$$

$$k = 2, 3, \ldots, 8.$$

(C.4) (i) $H(x)$ is a real-valued function, even, of bounded variation, and such that

$$\int_{-\pi}^{\pi} H(x)\, dx = 1,$$

and $H(x) = 0$ for $x \in [-\pi, \pi]$.

(ii) $\{B_n\}$ is a sequence which satisfies $B_n \to 0$, $B_n^2 n \to \infty$ as $n \to \infty$.

Define $d_r(\lambda) = \sum_{t=1}^{n} X_r(t)e^{it\lambda}$, for $r = 1, \ldots, m$, and $H_n(x) = B_n^{-1}H(B_n^{-1}x)$, and let $f_{ab}(\lambda)$ be the (a, b) element of $\hat{\mathbf{f}}_n(\lambda)$ given in Eq. (2.2). Recalling Theorem 2.2, we obtain,

THEOREM 4.3 Taniguchi, Puri and Kondo (1996). Assume that (A.1)–(A.8), (C.3) and (C.4) hold. Then

(i) $\nu_2(f_4^X)$ defined by (2.8) is consistently estimated by

$$\begin{aligned}
\hat{\nu}_2 = 2\pi \sum_{a,b,c,d=1}^{m} \left(\frac{2\pi}{n}\right)^3 \sum_{j_1} \sum_{j_2} \sum_{j_3} H_n\left\{\frac{2\pi(j_2 + j_3)}{n}\right\} \\
\times K_{ab}^{(1)}\left\{\hat{\mathbf{f}}_n(\lambda)\right\} K_{cd}^{(1)}\left\{\hat{\mathbf{f}}_n(\lambda)\right\} \\
\times \frac{1}{8\pi^3 n} d_a\left\{\frac{2\pi(-j_1 + j_2 + j_3)}{n}\right\} d_b\left(\frac{2\pi j_1}{n}\right) d_c\left(\frac{-2\pi j_2}{n}\right) d_d\left(\frac{-2\pi j_3}{n}\right) \\
- \int_{-\pi}^{\pi}\left[K_{ab}^{(1)}\left\{\hat{\mathbf{f}}_n(\lambda)\right\} K_{cd}^{(1)}\left\{\hat{\mathbf{f}}_n(-\lambda)\right\}\hat{f}_{ac}(-\lambda)\hat{f}_{bd}(\lambda)\right. \\
\left. + K_{ab}^{(1)}\left\{\hat{\mathbf{f}}_n(\lambda)\right\} K_{cd}^{(1)}\left\{\hat{\mathbf{f}}_n(-\lambda)\right\}\hat{f}_{ad}(-\lambda)\hat{f}_{bc}(\lambda)\right] d\lambda \\
- \frac{H(0)}{B_n}\left(\frac{2\pi}{n}\right)^2 \sum_{j_1} \sum_{j_2} K_{ab}^{(1)}\left\{\hat{\mathbf{f}}_n\left(\frac{2\pi j_1}{n}\right)\right\} K_{cd}^{(1)}\left\{\hat{\mathbf{f}}_n\left(\frac{2\pi j_2}{n}\right)\right\} \\
\times \left(\frac{1}{2\pi n}\right)^2 d_a\left(\frac{-2\pi j_1}{n}\right) d_b\left(\frac{2\pi j_1}{n}\right) d_c\left(\frac{-2\pi j_2}{n}\right) d_d\left(\frac{2\pi j_2}{2}\right), \quad (4.44)
\end{aligned}$$

where sums with respect to $j_1, \ldots$ are from $[-n/2] + 1$ to $[n/2]$.

(ii) Under H,

$$T_n = \frac{\sqrt{n}\left[\int_{-\pi}^{\pi} K\{\hat{\mathbf{f}}_n(\lambda)\}\, d\lambda - c\right]}{\sqrt{\nu_1(\hat{\mathbf{f}}_n) + \hat{\nu}_2}}, \qquad (4.45)$$

has asymptotically the standard normal distribution $\mathcal{N}(0, 1)$.

From Theorem 4.3 we can propose the test of H given by the critical region

$$[|T_n| > t_\alpha], \qquad (4.46)$$

where t_α is given in Eq. (4.4).

Next we introduce a measure of goodness of our test. Let $\mathbf{a}(\lambda)$ be an $m \times m$ matrix whose entries $a_{k\ell}(\lambda)$ $(k, \ell = 1, \ldots, m)$ are square integrable functions on $[-\pi, \pi]$. We assume that $\mathbf{a}(\lambda)$ is positive definite for each $\lambda \in [-\pi, \pi]$. For a sequence of alternative spectral density matrices

$$\mathbf{g}_n(\lambda) = \mathbf{f}(\lambda) + \frac{1}{\sqrt{n}} \, \mathbf{a}(\lambda),$$

we can define an efficacy of T_n in the same way as Eq. (4.9). Then it is shown that

$$\text{eff}(T_n) = \frac{\int_{-\pi}^{\pi} \text{tr}[K^{(1)}\{\mathbf{f}(\lambda)\}\mathbf{a}(\lambda)] \, d\lambda}{\sqrt{\nu_1(\mathbf{f}) + \nu_2(f_4^X)}}. \tag{4.47}$$

Since the functional of the spectral density matrix $\int_{-\pi}^{\pi} K\{\mathbf{f}(\lambda)\} \, d\lambda$ represents so many important indices in times series, the test setting in Eq. (4.43) is very wide. We give a few examples. Suppose that the process $\{\mathbf{X}(t) : t \in \mathbf{Z}\}$ has the form

$$\mathbf{X}(t) = \begin{bmatrix} \mathbf{x}(t) \\ \mathbf{y}(t) \end{bmatrix}$$

with $\mathbf{x}(t)$, q vector-valued, and $\mathbf{y}(t)$, r vector-valued; $q + r = m$ and has the spectral density matrix

$$\mathbf{f}(\lambda) = \begin{bmatrix} f_{xx}(\lambda) & f_{xy}(\lambda) \\ f_{yx}(\lambda) & f_{yy}(\lambda) \end{bmatrix}. \tag{4.48}$$

Denote by $H\{\cdot\}$ the linear closed manifold generated by $\{\cdot\}$ and denote by $\text{proj}[\mathbf{x}(t)|H\{\cdot\}]$ the projection of $\mathbf{x}(t)$ on $H\{\cdot\}$. We consider the residual process

$$\mathbf{u}_1(t) = \mathbf{x}(t) - \text{proj}[\mathbf{x}(t)|H\{\mathbf{x}(t-1), \mathbf{x}(t-2), \ldots\}],$$

$$\mathbf{v}_1(t) = \mathbf{y}(t) - \text{proj}[\mathbf{y}(t)|H\{\mathbf{y}(t-1), \mathbf{y}(t-2), \ldots\}],$$

$$\mathbf{u}_2(t) = \mathbf{x}(t) - \text{proj}[\mathbf{x}(t)|H\{\mathbf{x}(t-1), \mathbf{x}(t-2), \ldots; \mathbf{y}(t-1), \mathbf{y}(t-2), \ldots\}],$$

$$\mathbf{v}_2(t) = \mathbf{y}(t) - \text{proj}[\mathbf{y}(t)|H\{\mathbf{x}(t-1), \mathbf{x}(t-2), \ldots; \mathbf{y}(t-1)\mathbf{y}(t-2), \ldots\}],$$

and

$$\mathbf{u}_3(t) = \mathbf{x}(t) - \text{proj}[\mathbf{x}(t)|H\{\mathbf{x}(t-1), \mathbf{x}(t-2), \ldots; \mathbf{y}(t), \mathbf{y}(t-1), \ldots\}]$$

The measure of linear feedback from $Y = \{\mathbf{y}(t)\}$ to $X = \{\mathbf{x}(t)\}$ is defined by

$$F_{Y \to X} = \log[\det\{\text{Var}(\mathbf{u}_1(t))\} / \det\{\text{Var}(\mathbf{u}_2(t))\}]. \tag{4.49}$$

Symmetrically,

$$F_{X \to Y} = \log[\det\{\mathrm{Var}(\mathbf{v}_1(t))\}/\det\{\mathrm{Var}(\mathbf{v}_2(t))\}] \tag{4.50}$$

The measure of instantaneous linear feedback

$$F_{X \cdot Y} = \log[\det\{\mathrm{Var}(\mathbf{u}_2(t))\}/\det\{\mathrm{Var}(\mathbf{u}_3(t))\}]$$

has motivation similar to that of the above two measures. The following

$$F_{X.Y} = \log[\det\{\mathrm{Var}(\mathbf{u}_1(t))\}\det\{\mathrm{Var}(\mathbf{v}_1(t))\}/\det\{\mathrm{Var}(\mathbf{e}(t))\}]$$

is called the measure of linear dependence. Then it is known that

$$F_{X.Y} = F_{Y \to X} + F_{X \to Y} + F_{X \cdot Y} \tag{4.51}$$

and

$$F_{X.Y} = \int_{-\pi}^{\pi} K\{\mathbf{f}(\lambda)\} \, d\lambda, \tag{4.52}$$

where

$$K\{\mathbf{f}(\lambda)\} = -\frac{1}{2\pi}\log[\det\{I_q - f_{xy}(\lambda)f_{yy}(\lambda)^{-1}f_{yx}(\lambda)f_{xx}(\lambda)^{-1}\}],$$

see Geweke (1982). Since $F_{Y \to X}$, $F_{X \to Y}$, and $F_{X \cdot Y}$ are important econometric measures which represent "strength of causality", the testing problem

$$H : F_{X.Y} = c$$

against   (4.53)

$$A : F_{X.Y} \neq c$$

is important. This is exactly an example of our testing problem. We give a numerical result for Eq. (4.53). Let $\mathbf{X}(t) = (\mathbf{x}(t)', \mathbf{y}(t)')'$ be a two-dimensional linear process generated by

$$\mathbf{X}(t) = \mathbf{e}(t) + \begin{pmatrix} 1 & \varepsilon \\ \varepsilon & 1 \end{pmatrix} \mathbf{e}(t-1), \tag{4.54}$$

where the $\mathbf{e}(t)$'s are iid $\mathcal{N}(\mathbf{0}, I_2)$. Generating $\mathbf{X}(1), \ldots, \mathbf{X}(1024)$, for $\varepsilon = 1.15$, 1.20, 1.25, we calculated T_n for Eq. (4.53) and iterated this procedure 500 times. The results are summarized in Table 3. They seem to agree with the theoretical results.

Next, we turn to an investigation of another interrelation analysis. Let $\mathbf{X}(t) = \{\mathbf{u}(t)', \mathbf{v}(t)', \mathbf{w}(t)'\}'$ be an m-dimensional process satisfying all the assumptions in Theorem 4.3, where $\mathbf{u}(t)$, $\mathbf{v}(t)$ and $\mathbf{w}(t)$ are q, r and s component processes, respectively. Correspondingly we have the partition

Table 3 Simulated Behavior of T_n

ε	Mean of T_n	Variance of T_n	Frquency of $T_n < -1.64$	Frequency of $T_n > 1.64$
1.15	0.077	1.073	0.058	0.058
1.20	0.067	1.006	0.042	0.056
1.25	0.076	1.000	0.050	0.052

$$\mathbf{f}(\lambda) = \begin{bmatrix} f_{uu}(\lambda) & f_{uv}(\lambda) & f_{uw}(\lambda) \\ f_{vu}(\lambda) & f_{vv}(\lambda) & f_{vw}(\lambda) \\ f_{wu}(\lambda) & f_{wv}(\lambda) & f_{ww}(\lambda) \end{bmatrix}$$

and the spectral representation

$$\mathbf{X}(t) = \int_{-\pi}^{\pi} e^{-it\lambda}\, d\xi(\lambda),$$

with $d\xi(\lambda) = (d\xi_u(\lambda)', d\xi_v(\lambda)', d\xi_w(\lambda)')'$. Hannan (1970) considered a test for association for $\mathbf{u} = \{\mathbf{u}(t)\}$ with $\mathbf{v} = \{\mathbf{v}(t)\}$ (at frequency λ) after allowing for any effects of $\mathbf{w} = \{\mathbf{w}(t)\}$. The hypothesis is given by

$$H_\lambda : f_{uv}(\lambda) - f_{uw}(\lambda)f_{ww}(\lambda)^{-1}f_{wv}(\lambda) = 0, \tag{4.55}$$

which means that $d\xi_u(\lambda) - f_{uw}(\lambda)f_{ww}(\lambda)^{-1}d\xi_w(\lambda)$ is incoherent with $d\xi_v(\lambda) - f_{vw}(\lambda)f_{ww}(\lambda)^{-1} d\xi_w(\lambda)$ and all of the apparent association between $\mathbf{u}$ and $\mathbf{v}$ is truly due only to their common association with $\mathbf{w}$. For a given λ, Hannan (1970) developed the testing theory for Eq. (4.55) based on the asymptotic normality of the finite Fourier transformations of $\{\mathbf{X}(t)\}$ in a neighborhood of λ. In view of our testing problem we can consider a test for association for $\mathbf{u}$ with $\mathbf{v}$ at "all the frequency $\lambda \in [-\pi, \pi]$" after allowing for any effects of $\mathbf{w}$. The hypothesis is written as

$$H : f_{uv}(\lambda) - f_{uw}(\lambda)f_{ww}(\lambda)^{-1}f_{wv}(\lambda) = 0 \quad \text{for all } \lambda \in [-\pi, \pi],$$

which is equivalent to

$$H : \int_{-\pi}^{\pi} K\{\mathbf{f}(\lambda)\}\, d\lambda = 0, \tag{4.56}$$

where

$$K\{\mathbf{f}(\lambda)\} = \text{tr}[\{f_{uv}(\lambda) - f_{uw}(\lambda)f_{ww}(\lambda)^{-1}f_{wv}(\lambda)\} \\ \times \{f_{vu}(\lambda) - f_{vw}(\lambda)f_{ww}(\lambda)^{-1}f_{wu}(\lambda)\}].$$

Therefore we can apply all the results in Theorem 4.3 to this testing problem.

It is well known that the eigenvalues play a fundamental role in multivariate problems. We next investigate the principal components analysis of vector time series, which is related to the eigenvalues of the spectral density matrix. Suppose that $\{\mathbf{X}(t)\}$ is an m–dimensional process satisfying all the assumptions in Theorem 4.3, and that the spectral density matrix $\mathbf{f}(\lambda)$ has the simple eigenvalues $\mu_1(\lambda) > \mu_2(\lambda) > \cdots > \mu_m(\lambda)$. Then the variance of the error series by the q ($q < m$) principal components is given by

$$\int_{-\pi}^{\pi} \left\{ \sum_{j=q+1}^{m} \mu_j(\lambda) \right\} d\lambda, \tag{4.57}$$

see Brillinger (1975). If we are interested in the degree of the above measure we can set down the testing problem

$$H : \int_{-\pi}^{\pi} K\{\mathbf{f}(\lambda)\} \, d\lambda = c,$$

against   (4.58)

$$A : \int_{-\pi}^{\pi} K\{\mathbf{f}(\lambda)\} \, d\lambda \neq c,$$

where $K\{\mathbf{f}(\lambda)\} = \sum_{j=q+1}^{m} \mu_j(\lambda)$. From Magnus and Neudecker (1988) the derivative of $K\{\cdot\}$ is given by

$$K^{(1)}\{\mathbf{f}(\lambda)\} = \sum_{i=q+1}^{m} \left[\prod_{\substack{j=1 \\ j \neq i}}^{m} \left\{ \frac{\mu_j(\lambda) I_m - \mathbf{f}(\lambda)}{\mu_j(\lambda) - \mu_i(\lambda)} \right\} \right].$$

Thus we may construct the test T_n of Theorem 4.3 for Eq. (4.58).

5. DISCRIMINANT ANALYSIS FOR TIME SERIES

To begin with, suppose that $\{X_t; t = 0, \pm 1, \ldots\}$ is a scalar-valued stationary process with zero mean and spectral density $f(\lambda)$. Let $\mathbf{X} = (X_1, X_2, \ldots, X_n)'$ be a partial realization of the process, and $\sum_f = \{r(t - t'); t, t' = 1, \ldots, n\}$ be the covariance matrix of $\mathbf{X}$, where

$$r(t - t') = E(X_t X_{t'}) = \int_{-\pi}^{\pi} f(\lambda) e^{i\lambda(t-t')} \, d\lambda.$$

Let Π_1 and Π_2 be two populations with probability density function $p_1(\mathbf{X})$ and $p_2(\mathbf{X})$, respectively. We assign the observed stretch $\mathbf{X}$ to population Π_1

of $\mathbf{X}$ falls into region R_1; otherwise we assign it to Π_2, where R_1 and R_2 are exclusive and exhaustive regions in $\mathbf{R}^n$. Let the probability of misclassifying the observation from Π_i into population Π_j be

$$P(j|i) = \int_{R_i} p_i(\mathbf{X}) \, d\mathbf{X}.$$

We choose R_1 and R_2 to minimize $P(2|1) + P(1|2)$. It is well known that the classification regions defined by

$$R_k = \left[\mathbf{X} : n^{-1} \log\left\{ \frac{p_k(\mathbf{X})}{p_j(\mathbf{X})} \right\} > 0, j \neq k \right], \quad (k = 1, 2),$$

give the optimal classification, see Anderson (1984). Therefore we need to investigate the log-likelihood ratio $\mathrm{LR} = \log\{p_k(\mathbf{X})/p_j(\mathbf{X})\}$. However, if n is large, LR is intractable because it includes the inverse and determinant of the $n \times n$ matrix $\sum_f$.

Series of work for the discriminant analysis of time series have been done by several authors. Capon (1965) set up a method called an asymptotic simultaneous diagonalization procedure (ASDP) for discriminant analysis of Gaussian stationary time series, and gave an approximation for LR in terms of the finite Fourier transform (FFT) of $\mathbf{X}$. In multiple alternative classification problems for Gaussian stationary time series, Liggett (1971) showed that the posterior probabilities of the hypotheses based on spectral estimators are asymptotically equivalent to those based on the original data. For the problems of discriminating two Gaussian processes by linear filtering, Shumway and Unger (1974) gave certain spectral approximations of Kullback–Leibler information and J–divergence. They introduced linear discriminant filters maximizing these approximations and applied them to seismic records. Shumway (1982) gave an extensive review of various discriminant problems in time series.

First, we give a motivation of our classification statistic. Suppose that $\{X_t\}$ is a Gaussian process whose autocovariances and spectral density satisfy

$$\sum_{|t| \geq n} |t| \, |r(t)| \leq C n^{-\alpha}, \tag{5.1}$$

for a finite positive constant C, and $0 < \alpha < 1$, and $f(\lambda) \geq \delta$ on $[-\pi, \pi]$, for $\delta > 0$, respectively. Then, by Lemmas 1 and 2 of Liggett (1971), we can prove

$$n^{-1} \log\{p_f(\mathbf{X})\} = -\log 2\pi - \frac{1}{4\pi} \int_{-\pi}^{\pi} \left[\log f(\lambda) + \frac{I_n(\lambda)}{f(\lambda)} \right] d\lambda + o_p(1), \tag{5.2}$$

where $p_f(\mathbf{X})$ is the likelihood function based on $\mathbf{X}$ under the assumption that the spectral density of $\{X_t\}$ is $f(\lambda)$, and $I_n(\lambda)$ is the periodogram of $\mathbf{X}$. Therefore we adopt

$$I(f:g) = \frac{1}{4\pi} \int_{-\pi}^{\pi} \left[\log \frac{g(\lambda)}{f(\lambda)} + I_n(\lambda) \left\{ \frac{1}{g(\lambda)} - \frac{1}{f(\lambda)} \right\} \right] d\lambda \qquad (5.3)$$

as an approximation for $n^{-1}LR$, where $f(\lambda)$ and $g(\lambda)$ are the spectral densities associated with Π_1 and Π_2, respectively.

Now we turn to discuss the problems of classifying a stationary vector process $\{\mathbf{X}(t) = (X_1(t), \ldots, X_m(t))'\}$ into one of two categories described by two hypotheses:

$$\Pi_1 : \mathbf{f}(\lambda) \qquad \Pi_2 : \mathbf{g}(\lambda) \qquad (5.4)$$

where $\mathbf{f}(\lambda)$ and $\mathbf{g}(\lambda)$ are $m \times m$ spectral density matrices. For a partial realization $\{\mathbf{X}(1), \ldots, \mathbf{X}(n)\}$ of $\{\mathbf{X}(t)\}$, the periodogram matrix is defined as

$$\mathbf{I}_n(\lambda) = \frac{1}{2\pi n} \left\{ \sum_{t=1}^{n} \mathbf{X}(t) e^{i\lambda t} \right\} \left\{ \sum_{t=1}^{n} \mathbf{X}(t) e^{i\lambda t} \right\}^*. \qquad (5.5)$$

A natural extension of Eq. (5.3) to the case of vector process is given by

$$I(\mathbf{f}:\mathbf{g}) = \frac{1}{4\pi} \int_{-\pi}^{\pi} \left(\log \frac{|\mathbf{g}(\lambda)|}{|\mathbf{f}(\lambda)|} + tr \left[\mathbf{I}_n(\lambda) \{ \mathbf{g}(\lambda)^{-1} - \mathbf{f}(\lambda)^{-1} \} \right] \right) d\lambda. \qquad (5.6)$$

We use $I(\mathbf{f}:\mathbf{g})$ as a classification statistic for the problem in Eq. (5.4). That is, if $I(\mathbf{f}:\mathbf{g}) > 0$ we choose category Π_1. Otherwise we choose category Π_2. Initially, we make the following assumptions.

(D.1)   There exists $\eta > 0$ such that the minimum eigenvalues of $\mathbf{f}(\lambda)$ and $\mathbf{g}(\lambda)$ are greater than η for all $\lambda \in [-\pi, \pi]$.

(D.2) $\{\mathbf{X}(t); t \in \mathbf{Z}\}$ is generated by

$$\mathbf{X}(t) = \sum_{j=0}^{\infty} G(j) \mathbf{e}(t - j), \quad t \in \mathbf{Z}, \qquad (5.7)$$

which satisfies (A.1)–(A.6) in Sec. 2. Furthermore, the fourth-order cumulants of $\{\mathbf{e}(t)\}$ satisfy

$$\operatorname{cum}\{e_a(t), e_b(t + t_1), e_c(t + t_2), e_d(t + t_3)\}$$

$$= \begin{cases} \kappa_{abcd} & \text{for } t_1 = t_2 = t_3 = 0 \\ 0 & \text{otherwise.} \end{cases}$$

From the assumptions above, we can see that $\{\mathbf{X}(t)\}$ is second-order stationary and that the spectral density matrix under Π_1 is written as

$$\mathbf{f}(\lambda) = \frac{1}{2\pi} k_f(\lambda)\Omega\, k_f(\lambda)^*, \quad \lambda \in [-\pi, \pi], \tag{5.8}$$

where $k_f(\lambda) = \sum_{j=0}^{\infty} G_f(j)e^{i\lambda j}$, and $G_f(j)$'s are the coefficients $\{G(j)\}$ of the linear process in Eq. (5.7) under Π_1. For the problem in Eq. (5.4) the misclassification probabilities of $I(\mathbf{f}:\mathbf{g})$ are defined as

$$P_I(2|1) = P\{I(\mathbf{f}:\mathbf{g}) \leq 0|\Pi_1\}, \quad P_I(1|2) = P\{I(\mathbf{f}:\mathbf{g}) > 0|\Pi_2\}. \tag{5.9}$$

THEOREM 5.1 Zhang and Taniguchi (1994). Under the assumptions (D.1) and (D.2), if $\mathbf{f}(\lambda) \not\equiv \mathbf{g}(\lambda)$ on a set of positive Lebesgue measures, then the misclassification probabilities tend to zero as $n \to \infty$. That is,

$$\lim_{n\to\infty} P_I(2|1) = 0, \quad \lim_{n\to\infty} P_I(1|2) = 0. \tag{5.10}$$

Remark. Theorem 5.1 implies that $I(\mathbf{f}:\mathbf{g})$ is a consistent classification criterion even if the process is not Gaussian.

To evaluate the goodness of $I(\mathbf{f}:\mathbf{g})$ we consider the case when $\mathbf{g}(\lambda)$ is contiguous to $\mathbf{f}(\lambda)$. Then non–Gaussian robustness of $I(\mathbf{f}:\mathbf{g})$ will be discussed. $I(\mathbf{f}:\mathbf{g})$ is said to be non-Gaussian robust if the asymptotic behavior of the misclassification probabilities $P_I(1|2)$ and $P_I(2|1)$ is independent of the fourth-order cumulants κ_{abcd} of the process.

Now, let the spectral densities associated with Π_1 and Π_2 be

$$\Pi_1 : \mathbf{f}(\lambda) = \mathbf{f}(\lambda|\theta) \qquad \Pi_2 : \mathbf{g}(\lambda) = \mathbf{f}(\lambda|\theta + \mathbf{h}/\sqrt{n}) \tag{5.11}$$

where $\theta \in \Theta \subset \mathbf{R}^q$ and $\mathbf{h} = (h_1, \ldots, h_q)'$. The following condition is assumed.

(D.3) $\mathbf{f}(\lambda|\theta)$ is continuously three times differentiable with respect to $\theta \in \Theta$. There exists a finite constant M, such that

$$\sum_{i=1}^{q}\sum_{j=1}^{q}\sum_{k=1}^{q} \left\| \frac{\partial^3}{\partial\theta_i\partial\theta_j\partial\theta_k} \mathbf{f}(\lambda|\theta) \right\| \leq M, \quad \text{for all } \lambda \in [-\pi, \pi].$$

Henceforth the parameter θ is said to be innovation-free if the variance matrix Ω of $\{\mathbf{e}(t)\}$ is independent of θ. We set down

$$F(\theta) = \frac{1}{4\pi}\int_{-\pi}^{\pi} \mathrm{tr}\left\{ \sum_{i=1}^{q} h_i \frac{\partial}{\partial\theta_i} \mathbf{f}(\lambda|\theta) \cdot \mathbf{f}(\lambda|\theta)^{-1} \right\}^2 d\lambda,$$

$$D(\theta) = \frac{1}{2\pi}\sum_{i=1}^{q} h_i \int_{-\pi}^{\pi} \left[k_f^*(\lambda)\left\{ \mathbf{f}(\lambda|\theta)^{-1} \frac{\partial}{\partial\theta_i} \mathbf{f}(\lambda|\theta) \cdot \mathbf{f}(\lambda|\theta)^{-1} \right\} k_f(\lambda) \right] d\lambda.$$

The misclassification probabilities can be evaluated as follows.

THEOREM 5.2   Zhang and Taniguchi (1994)). Under Eq. (5.11), assume (D.1)–(D.3).

(i)   If we use $I(\mathbf{f}:\mathbf{g})$ as a classification criterion, then

$$\lim_{n\to\infty} P_I(2|1)$$

$$= \lim_{n\to\infty} P_I(1|2)$$

$$= \Phi\left[\frac{-F(\theta)/2}{\{F(\theta) + (16\pi^2)^{-1}\sum_{a,b,c,d=1}^{m}\kappa_{abcd}[D(\theta)]_{ab}[D(\theta)]_{cd}\}^{1/2}}\right],$$

(5.12)

where $\Phi(\cdot)$ is the probability distribution function of the standard normal distribution, and $[D(\theta)]_{ab}$ is the (a, b) component of $D(\theta)$.

(ii)   If θ is innovation–free, then $I(\mathbf{f}:\mathbf{g})$ is non–Gaussian robust (i.e., the right-hand side of Eq. (5.12) is equal to $\Phi[-\frac{1}{2}\{F(\theta)\}^{1/2}]$).

Zhang and Taniguchi (1994) gave numerical studies confirming (ii) in Theorem 5.2.

In the iid case it has been recognized that the log-likelihood ratio is not stable under small perturbations in the underlying model Huber (1972). This motivates us to seek a new discriminant statistic. Suppose that $\{X(t) : t = 0, \pm 1, \ldots\}$ is a stationary process with zero mean and spectral densities $f(\lambda)$ and $g(\lambda)$ under hypotheses Π_1 and Π_2, respectively. Let $\mathbf{X}_n = \{X(1), \ldots, X(n)\}'$ be a stretch of the series $\{X(t)\}$, and let $p_f(\cdot)$ and $p_g(\cdot)$ be the probability densities of $\mathbf{X}_n$ under Π_1 and Π_2, respectively. For $\alpha \in (0, 1)$, the quantity

$$\int \{p_f(\mathbf{X}_n)/p_g(\mathbf{X}_n)\}^{\alpha} p_g(\mathbf{X}_n)\, d\mathbf{X}_n = \exp\{H(p_f, p_g)\}$$

is called the α–entropy of $p_f(\cdot)$ with respect to $p_g(\cdot)$. When $\{X(t)\}$ is a Gaussian process, Albrecht (1983) showed

$$\lim_{n\to\infty} -\frac{1}{n} H(p_f, p_g) = e_\alpha(f, g),$$

(5.13)

where

$$e_\alpha(f, g) = \frac{1}{4\pi}\int_{-\pi}^{\pi}\left[\log\left\{(1-\alpha) + \alpha\frac{g(\lambda)}{f(\lambda)}\right\} - \alpha\log\frac{g(\lambda)}{f(\lambda)}\right]d\lambda.$$

Using $e_\alpha(f,g)$ and without assuming Gaussianity of stationary process $\{X(t)\}$, we propose a new discriminant criterion

$$B_\alpha = e_\alpha(\hat{f}_n, g) - e_\alpha(\hat{f}_n, f),$$ (5.14)

where $\hat{f}_n(\lambda)$ is a nonparametric spectral estimator based on $\mathbf{X}_n$. Although B_α is no longer interpreted as a function of α-entropy based criterion, as we state below, it has the goodness of discriminant criterion.

First, we state some fundamental properties of $e_\alpha(f,g)$. For $y_\alpha(x) = \log\{(1-\alpha) + \alpha x\} - \alpha \log x$, it follows from the arithmetic-geometric-mean inequality that $y_\alpha(x) \geq 0$, and $y_\alpha(x)$ is equal to zero at $x = 1$. Therefore

$$e_\alpha(f,g) = \frac{1}{4\pi} \int_{-\pi}^{\pi} y_\alpha \left\{ \frac{g(\lambda)}{f(\lambda)} \right\} d\lambda \geq 0,$$ (5.15)

where the equality holds if $f(\lambda) \equiv g(\lambda)$ a.e.. Let

$$m(f,g) = \frac{1}{4\pi} \int_{-\pi}^{\pi} \left[\log \frac{g(\lambda)}{f(\lambda)} + \frac{f(\lambda)}{g(\lambda)} - 1 \right] d\lambda,$$

and let $I(f:g)$ be the quantity defined in Eq. (5.3). Then differentiation of $e_\alpha(f,g)$ with respect to α yields

$$\lim_{\alpha \to 1} \frac{\partial}{\partial \alpha} e_\alpha(f,g) = -\lim_{n \to \infty} E_f[I(f:g)] = -m(f,g),$$
$$\lim_{\alpha \to 0} \frac{\partial}{\partial \alpha} e_\alpha(f,g) = -\lim_{n \to \infty} E_g[I(f:g)] = m(f,g),$$ (5.16)

where E_f and E_g denote the expectations with respect to $p_f(\cdot)$ and $p_g(\cdot)$, respectively. For $\alpha \in (-1, 1)$, the α-divergence is defined by

$$D_\alpha(f,g) = E_f \left[\phi_\alpha \left\{ \frac{p_g(\mathbf{X}_n)}{p_f(\mathbf{X}_n)} \right\} \right],$$ (5.17)

with $\phi_\alpha(u) = (4/\{1 - \alpha^2\})\{1 - u^{(1+\alpha)/2}\}$, see e.g., Amari (1985). Then it is easily shown that $D_\alpha(f,g)$ and $e_{\alpha'}(f,g)$ with $\alpha' = (1 + \alpha)/2$ are equivalent.

To develop the asymptotic theory for B_α, we set down the following assumptions.

(D.4) $\{X(t)\}$ is generated by the linear process given in (D.2) with $m = 1$.

(D.5) $\hat{f}_n(\lambda)$ is defined by Eq. (2.2) with the window function satisfying (A.7).

Now, consider the problems of classifying the process $\{X(t)\}$ into one of two categories described by two spectral hypotheses

$$\Pi_1 : f(\lambda) \qquad \Pi_2 : g(\lambda).$$ (5.18)

As we said we use B_α as a classification statistic for the problem in Eq. (5.18). That is, if $B_\alpha > 0$ we choose category Π_1. Otherwise, we choose category Π_2. This classification rule has the following intuitive appeal. The relation $B_\alpha > 0$ implies $e_\alpha(\hat{f}_n, g) > e_\alpha(\hat{f}_n, f)$. Because $e_\alpha(f, g)$ represents a kind of distance between f and g, the rule implies $\hat{f}_n$ is nearer to f than g. The associated misclassification probabilities of B_α:

$$P_B(2|1) \equiv P[B_\alpha \le 0|\Pi_1], \qquad P_B(1|2) \equiv P[B_\alpha > 0|\Pi_2].$$

are evaluated as follows.

THEOREM 5.3 Zhang and Taniguchi (1995). Under the assumptions (D.4) and (D.5), if $f(\lambda) \not\equiv g(\lambda)$ on a set of positive Lebesgue measures, then

$$\lim_{n\to\infty} P_B(2|1) = \lim_{n\to\infty} P_B(1|2) = 0. \tag{5.19}$$

Theorem 5.3 shows that B_α is a consistent classification statistic. However, we can not evaluate the degree of goodness of B_α under the situation of $f(\lambda) \not\equiv g(\lambda)$. Thus we consider the case when $f(\lambda)$ is contiguous to $g(\lambda)$. Let the spectral densities associated with Π_1 and Π_2 be

$$\Pi_1 : f(\lambda) = f(\lambda|\theta), \qquad \Pi_2 : g(\lambda) = f(\lambda|\theta + \mathbf{h}/\sqrt{n}), \tag{5.20}$$

where $f(\lambda|\theta)$ is continuously three times differentiable with respect to $\theta \in \Theta \subset \mathbf{R}^q$ and $\mathbf{h} = (h_1, \ldots, h_q)'$, and all the derivatives are bounded on $[-\pi, \pi]$.

THEOREM 5.4 Zhang and Taniguchi (1995). Under Eq. (5.20) and the assumptions (D.4) and (D.5),

$$\lim_{n\to\infty} P_B(2|1) = \lim_{n\to\infty} P_I(2|1) = \lim_{n\to\infty} P_B(1|2) = \lim_{n\to\infty} P_I(1|2),$$

for all $\alpha \in (0, 1)$. That is, the asymptotic misclassification probabilities based on B_α are exactly the same as those computed from $I(f:g)$ for all $\alpha \in (0, 1)$ when $f(\lambda)$ is contiguous to $g(\lambda)$.

Next we consider a case where the spectral density of $\{X(t)\}$ is contaminated by a sharp peak. In this case it will be shown that B_α is robust with respect to the peak, but $I(f:g)$ is not so.

THEOREM 5.5 Zhang and Taniguchi (1995): Define

$$\tilde{f}(\lambda) = \begin{cases} f(\lambda) & \text{on } A = [-\pi, \pi] - A_\varepsilon. \\ f(\lambda)/\varepsilon^r & \text{on } A_\varepsilon, \end{cases}$$

where $A_\varepsilon = [\lambda_0, \lambda_0 + \varepsilon]$ is an interval in $[-\pi, \pi]$ for sufficiently small $\varepsilon > 0$ and $r > 1$. Suppose that $f(\lambda) \neq g(\lambda)$ on a set of positive Lebesgue measure. Then, under (D.4),

$$\lim_{\varepsilon \to 0} |B_\alpha(\tilde{f}, f, g) - e_\alpha(f, g)| = 0, \quad \text{for } \alpha \in (0, 1),$$

$$\lim_{\varepsilon \to 0} |I(\tilde{f}, f : g) - m(f, g)| = \infty,$$

(5.21)

where

$$B_\alpha(\tilde{f}, f, g) = e_\alpha(\tilde{f}, g) - e_\alpha(\tilde{f}, f),$$

and

$$I(\tilde{f}, f : g) = \frac{1}{4\pi} \int_{-\pi}^{\pi} \left[\log \frac{g(\lambda)}{f(\lambda)} + \tilde{f}(\lambda) \left\{ \frac{1}{g(\lambda)} - \frac{1}{f(\lambda)} \right\} \right] d\lambda.$$

Theorem 5.5 shows that B_α is insensitive to a peak in the spectral density, while $I(f : g)$ is sensitive. Thus B_α is better than $I(f : g)$ if the spectral density of $\{X(t)\}$ is contaminated by a sharp peak. Simulation studies confirming this were given in Zhang and Taniguchi (1995).

For the discriminant problem in Eq. (5.4), Kakizawa (1996) introduced the following generalized disparity measure between spectral density matrices $\mathbf{f}(\lambda)$ and $\mathbf{g}(\lambda)$;

$$D_H(\mathbf{f} : \mathbf{g}) = \frac{1}{4\pi} \int_{-\pi}^{\pi} H\{\mathbf{f}(\lambda) \mathbf{g}(\lambda)^{-1}\} \, d\lambda,$$

(5.22)

where $H(\mathbf{Z})$ has a unique minimum zero at $\mathbf{Z} = I_m$ ($m \times m$ identity matrix). Setting

$$H_1(\mathbf{Z}) = -\log|\mathbf{Z}| + tr(\mathbf{Z}) - m,$$

$$H_2(\mathbf{Z}) = \tfrac{1}{2} tr(\mathbf{Z} - I_m)^2,$$

$$H_3(\mathbf{Z}) = H_3(\mathbf{Z}; \alpha) = \frac{1}{\alpha(1 - \alpha)}$$

$$\times \{\log|(1 - \alpha)I_m + \alpha\mathbf{Z}| - \alpha \log|\mathbf{Z}|\}, \quad \alpha \in (0, 1),$$

the concrete form of $H(\mathbf{Z})$ is defined by

$$H(\mathbf{Z}) = \sum_{l \neq 0} \sum_{j=1}^{3} p_{j,l} \frac{H_j(\mathbf{Z}^l)}{l^2},$$

(5.23)

with $p_{j,l} \geq 0$ for $j = 1, 2, 3$; $l = \pm 1, \pm 2, \ldots$ and $\sum_l \sum_{j=1}^{3} p_{j,l} = 1$. Thus $D_H(\mathbf{f} : \mathbf{g})$ includes many other criteria (e.g., $e_\alpha(\cdot, \cdot)$) as special cases. It is shown that $D_H(\mathbf{f} : \mathbf{g}) \geq 0$, and that $D_H(\mathbf{f} : \mathbf{g}) = 0$ iff $\mathbf{f}(\lambda) = \mathbf{g}(\lambda)$ a.e. Let

$$D_H = D_H(\hat{\mathbf{f}}_n : \mathbf{g}) - D_H(\hat{\mathbf{f}}_n : \mathbf{f}),$$

(5.24)

where $\hat{\mathbf{f}}_n = \hat{\mathbf{f}}_n(\lambda)$ is defined by Eq. (2.2). For the problem in Eq. (5.4) we assign the partial realization $\{\mathbf{X}(1), \ldots, \mathbf{X}(n)\}$ to Π_1 or Π_2 according as $D_H > 0$ or $D_H \leq 0$. The asymptotic misclassification probabilities

$$P_{D_H}(2|1) = P[D_H \leq 0|\Pi_1],$$

and

$$P_{D_H}(1|2) = P[D_H > 0|\Pi_2].$$

are evaluated as follows.

THEOREM   5.6   Kakizawa (1996). Assume (D.1), (D.2) and (A.7).

(i)   If $\mathbf{f}(\lambda) \not\equiv \mathbf{g}(\lambda)$ on a set of positive Lebesgue measures, then

$$\lim_{n \to \infty} P_{D_H}(2|1) = \lim_{n \to \infty} P_{D_H}(1|2) = 0. \tag{5.25}$$

(ii)   If we assume the contiguous condition in Eq.(5.11) and (D.3), and that θ is innovation-free, then

$$\lim_{n \to \infty} P_{D_H}(2|1) = \lim_{n \to \infty} P_{D_H}(1|2)$$
$$= \Phi[-\tfrac{1}{2}F(\theta)^{1/2}]. \tag{5.26}$$

In the above discussion we assumed that the spectral density matrices $\mathbf{f}(\lambda)$ and $\mathbf{g}(\lambda)$ associated with Π_1 and Π_2 are known. However they are never known in practice, and must be inferred from samples. If the training samples of Π_1 and Π_2 are available, we can make estimators $\bar{\mathbf{f}}(\lambda)$ and $\bar{\mathbf{g}}(\lambda)$ for $\mathbf{f}(\lambda)$ and $\mathbf{g}(\lambda)$, respectively, from them. Then the plug-in version of D_H is given by

$$\hat{D}_H = D_H(\hat{\mathbf{f}}_n : \bar{\mathbf{g}}) - D_H(\hat{\mathbf{f}}_n : \bar{\mathbf{f}}). \tag{5.27}$$

We choose Π_1 if $\hat{D}_H > 0$, and choose Π_2 if $\hat{D}_H \leq 0$. The asymptotics of $\hat{D}_H$ were illuminated in Kakizawa (1996).

Using the J-divergence and quasi-distance defined as

$$J(\mathbf{f} : \mathbf{g}) = I(\mathbf{f} : \mathbf{g}) + I(\mathbf{g} : \mathbf{f})$$

and

$$JB_\alpha(\mathbf{f} : \mathbf{g}) = D_{H_3}(\mathbf{f} : \mathbf{g}) + D_{H_3}(\mathbf{g} : \mathbf{f}),$$

respectively, Kakizawa, Shumway and Taniguchi (1998) discussed discriminating between earthquakes and mining explosions. Here each record is composed of two phases, the P group of first arrivals and the S group of latter arrivals. Thus we regard each record as a bivariate (P, S) time series.

They reported that both criteria $J(\mathbf{f}:\mathbf{g})$ and $JB_\alpha(\mathbf{f}:\mathbf{g})$ work well, especially $JB_\alpha(\mathbf{f}:\mathbf{g})$ does quite well.

6. HIGHER ORDER ASYMPTOTIC THEORY FOR SEMIPARAMETRIC ESTIMATION

There are a few papers dealing with higher order asymptotics of semiparametric estimators. Although they are for independent observations they suggest the possibility of a higher order asymptotic theory for semiparametric estimators of time series models.

First, we proceed in line with Robinson (1995). Let $(\mathbf{X}_t', Y_t)$, $t = 1, \ldots, n$, represent independent observations on an $\mathbf{R}^{d+1}$-valued variate $(\mathbf{X}', Y)$. Let $K : \mathbf{R}^d \to \mathbf{R}^1$ be a differentiable kernel function satisfying $\int_{\mathbf{R}^d} K(\mathbf{u})\, d\mathbf{u} = 1$, and having column vector of partial derivatives $K'(\mathbf{u}) = (\partial/\partial\mathbf{u})K(\mathbf{u})$. For $h > 0$, consider the statistic

$$U = \binom{n}{2}^{-1} h^{-d-1} \sum_{\substack{t,s=1 \\ t \neq s}}^{n} Y_t K'\left[\frac{\mathbf{X}_t - \mathbf{X}_s}{h}\right]. \tag{6.1}$$

We assume that $\mathbf{X}$ has the probability density function $f(\cdot)$, and set down

$g(\mathbf{X}) = E(Y|\mathbf{X})$,

$e(\mathbf{X}) = f(\mathbf{X})g(\mathbf{X})$,

$\mu_t = Y_t f'(Y_t) - e'(\mathbf{X}_t)$,

$\mu = -E\{g'(\mathbf{X})f(\mathbf{X})\}$

and

$$\sum = 4E(\mu_1 - \mu)(\mu_1 - \mu)'.$$

Some semiparametric models have the index form $g(\mathbf{X}) = G(\boldsymbol{\beta}'\mathbf{X})$, where $\boldsymbol{\beta}$ is an unknown $\mathbf{R}^d$-valued parameter and G is a differentiable nonparametric function. Then $\mu = -\boldsymbol{\beta}\, E\{G'(\boldsymbol{\beta}'\mathbf{X})f(\mathbf{X})\}$, which is called the average derivative, see Hardle and Stoker (1989). Since U is shown to be a $\sqrt{n}$-consistent estimator of μ in the below, it thus estimates $\boldsymbol{\beta}$ up to scale. The following assumptions are imposed.

(E.1) $E|Y|^3 < \infty$.

(E.2) $\sum$ is finite and positive definite.

(E.3)   The underlying measure of $(\mathbf{X}', Y)$ can be written as $\mu_{\mathbf{X}} \times \mu_Y$, where $\mu_{\mathbf{X}}$, μ_Y are Lebesgue measures on $\mathbf{R}^d$, $\mathbf{R}^1$; $(\mathbf{X}_t', Y_t)$, $t = 1, 2, \ldots$, are independent observations of $(\mathbf{X}', Y)$.

(E.4) f is $(L+1)$–times differentiable, and f and its first $(L+1)$ partial derivatives are bounded, for $2L > d + 2$.

(E.5) g is $(M+1)$–times differentiable, and the first $(M+1)$ partial derivatives of e are bounded, for $M \geq 1$.

(E.6) The conditional second moment function $q(x) = E(Y^2 | \mathbf{X} = x)$ is differentiable, and qf, $q'f$ and qf' are bounded.

(E.7) f, gf, and qf vanish on the boundaries of their convex (possibly infinite) supports.

(E.8) $K(\mathbf{u})$, $\mathbf{u} = (u_1, \ldots, u_d)'$, is differentiable,

$$\int_{\mathbf{R}^d} \{(1 + \|\mathbf{u}\|^L)|K(\mathbf{u})| + \|K'(\mathbf{u})\|\} \, d\mathbf{u} + \sup_{\mathbf{u} \in \mathbf{R}^d} \|K'(\mathbf{u})\| < \infty,$$

and, for the same L as in (E.4)

$$\int_{\mathbf{R}^d} u_1^{\ell_1} \cdots u_d^{\ell_d} K(\mathbf{u}) \, d\mathbf{u} = \begin{cases} 1, & \text{if } \ell_1 + \cdots + \ell_d = 0, \\ 0, & \text{if } \ell_1 + \cdots + \ell_d < L, \\ \neq 0, & \text{if } \ell_1 + \cdots + \ell_d = L. \end{cases} \tag{6.2}$$

(E.9) $(\log n)^5/nh^{d+2} = O(1)$, $nh^{2L} \to 0$, as $n \to \infty$.

For $\lambda \in \mathbf{R}^d$ we denote the distribution function of $\sqrt{n}\,\lambda'(U - \mu)$ by $F_\lambda(\cdot)$.

THEOREM 6.1   Robinson (1995). Assume that (E.1)–(E.9) hold. Then

(i)

$$\sup_{\lambda : \lambda' \Sigma \lambda} \sup_{z \in \mathbf{R}^1} |F_\lambda(z) - \Phi(z)| = O(n^{-1/2} + n^{-1}h^{-d-2} + \sqrt{n}\,h^L + h^M). \tag{6.3}$$

(ii) If there exists $C \in (1, \infty)$ such that, for all sufficiently large n, h satisfies

$$(Cn^{(1/2d+4)})^{-1} \leq h \leq Cn^{-1/L}, \quad M \geq L/2,$$

the bound in Eq. (6.3) is replaced by $O(n^{-1/2})$.

The result (ii) implies that the rate of convergence of finite–sample distribution to the normal limit distribution is equal to that of standard parametric statistics, and that the order $O(n^{-1/2})$ corresponds to the second-order asymptotics (for time series situation, see Taniguchi (1991)). It may be noted that Hardle and Stoker (1989) proposed another type of estimator for μ, and derived its "first-order" asymptotic distribution.

Linton (1995) discussed semiparametric estimation for a partially linear regression model

$$Y_1 = \beta' Y_2 + \theta(Z) + \varepsilon, \tag{6.4}$$

where $\theta(\cdot)$ is an unknown scalar function and ε is a zero mean error orthogonal to both Y_2 and $\theta(\cdot)$. For a semiparametric estimator $\hat{\beta}$ of β, he showed that a standardized version of $\hat{\beta}$ and its second–order stochastic expansion have the same distribution to order $n^{-2\mu}$ ($\mu < \frac{1}{2}$). This result also motivates the higher order asymptotic theory for semiparametric estimators as Taniguchi (1991) developed for parametric estimators.

Acknowledgments

The author is grateful to Professor Madan L. Puri, Indiana University, for his kind support. Thanks are also due to Ms. Ginny Jones for her excellent typing of the manuscript of this chapter.

REFERENCES

Albrecht, V. (1983). On the convergence rate of probability of error in Bayesian discrimination between two Gaussian processes. *Asymptotic Statistics 2, Proceedings of the Third Prague Symposium on Asymptotic Statistics*,165–175.

Amari, S. I. (1985). Differential-Geometrical Methods in Statistics. *Lecture Notes in Statistics, 28*, Springer-Verlag.

Anderson, T. W. (1984). *An Introduction to Multivariate Statistical Analysis*. New York: Wiley.

Beran, R. (1977). Minimum Hellinger distance estimates for parametric models. *Ann. Statist. 5*: 445–463.

Bloomfield, P. (1973). An exponential model for the spectrum of a scalar time series. *Biometrika 60*: 217–226.

Brillinger, D. R. (1975). *Time Series: Data Analysis and Theory*. New York: Holt, Rinehart and Winston.

Capon, J. (1965). An asymptotic simultaneous diagonalization procedure for pattern recognition. *Inf. Control 8*: 284–291.

Dahlhaus, R. (1988). Empirical spectral processes and their applications to time series analysis. *Stochastic Processes and Their Appl. 30*: 69–83.

Dunsmuir, W. (1979). A central limit theorem for parameter estimation in stationary vector time series and its application to models for a signal observed with noise. *Ann. Statist.* 7: 490–506.

Dzhaparidze, K. (1986). *Parameter Estimation and Hypothesis Testing in Spectral Analysis of Stationary Time Series.* New York: Wiley.

Geweke, J. (1982). Measurement of linear dependence and feedback between multiple time series. *J. Amer. Statist. Assoc.* 77: 304-324.

Hájek, J. and Šidák, Z. (1976). *Theory of Rank Tests.* New York: Academic Press.

Hallin, M., Ingenbleek, J.-F. and Puri, M. L. (1985). Linear serial rank tests for randomness against ARMA alternatives. *Ann. Statist.* 13: 1156–1181.

Hannan, E. J. (1955). Exact tests for serial correlation. *Biometrika 42*: 133–142.

Hannan, E. J. (1963). Regression for time series. In: *Time Series Analysis,* Ed. by M. Rosenblatt. New York: Wiley, 17–37.

Hannan, E. J. (1970). *Multiple Time Series.* New York: Wiley.

Hannan, E. J. and Nicholls, D. F. (1977). The estimation of the prediction error variance. *J. Amer. Statist. Assoc.* 72: 834–840.

Hardle, W. and Stoker, T. M. (1989). Investigating smooth multiple regression by the method of average derivatives. *J. Amer. Statist. Assoc.* 84: 986–995.

Hong, Y. (1996). Consistent testing for serial correlation of unknown form. *Econometrica, 64*: 837–864.

Hosoya, Y. and Taniguchi, M. (1982). A central limit theorem for stationary processes and the parameter estimation of linear processes. *Ann. Statist.* 21: 1115–1117.

Huber, P. J. (1972). Robust statistics: a review. *Ann. Math. Statist. 43*: 1042–1067.

Kakizawa, Y. (1996). Discriminant analysis for non-Gaussian vector stationary processes. *J. Nonparametric Statist.* 7: 187–203.

Kakizawa, Y., Shumway, R. H. and Taniguchi, M. (1998). Discrimination and clustering for multivariate time series. *J. Amer. Statist. Assoc.* 93: 328–340.

Kondo, M. and Taniguchi, M. (1993). Two sample problem in time series analysis. *Stat. Sci. and Data Anal.* Eds.: Matsusita et al. VSP, Utrecht, 165–174.

Kondo, M. (1994). *k* sample problem in time series analysis. *Math. Japon.* *39*: 537–544.

Liggett, W. S. (1971). On the asymptotic optimality of spectral analysis for testing hypotheses about time series. *Ann. Math. Statist.* *42*: 1348–1358.

Linton, O. (1995). Second-order approximation in the partially linear regression model. *Econometrica 63*: 1079–1112.

Magnus, J. R. and Neudecker, H. (1988). *Matrix Differential Calculus with Applications in Statistics and Econometrics.* New York: Wiley.

Randles, R. H. and Wolfe, D. A. (1979). *Introduction to the Theory of Nonparametric Statistics.* New York: Wiley.

Robinson, P. M. (1983). Review of various approaches to power spectrum estimation. *Handbook of Statistics* Vol. 3, Eds.: Brillinger and Krishnaiah. Amsterdam: North-Holland, 343–368.

Robinson, P. M. (1988). Root-*N*-consistent semiparametric regression. *Econometrica 56*: 931–954.

Robinson, P. M. (1995). The normal approximation for semiparametric averaged derivatives. *Econometrica 63*: 667–680.

Shumway, R. H. and Unger, A. N. (1974). Linear discriminant functions for stationary time series. *J. Amer. Statist. Assoc. 69*: 948–956.

Shumway, R. H. (1982). Discriminant analysis for time series. In: *Handbook of Statistics*, Vol. 2. Eds.: Krishnaiah and Kanal. Amsterdam: North-Holland, 1–46.

Taniguchi, M. (1980). On estimation of the integrals of certain functions of spectral density. *J. Appl. Prob. 17*: 73–83.

Taniguchi, M. (1987). Minimum contrast estimation for spectral densities of stationary processes. *J. Roy. Statist. Soc. Ser. B 49*: 315–325.

Taniguchi, M. (1991). *Higher Order Asymptotic Theory for Time Series Analysis. Lecture Notes in Statistics. Vol. 68.* Heidelberg: Springer-Verlag.

Taniguchi, M. and Kondo, M. (1993). Non-parametric approach in time series analysis. *J. Time Ser. Anal. 14*: 397–408.

Taniguchi, M., Puri, M. L. and Kondo, M. (1996). Nonparametric approach for non-Gaussian vector stationary processes. *J. Multivariate Anal.*, *56*: 259–283.

Zhang, G. and Taniguchi, M. (1994). Discriminant analysis for stationary vector time series. *J. Time Ser. Anal.* *15*: 117–126.

Zhang, G. and Taniguchi, M. (1995). Nonparametric approach for discriminant analysis in time series. *J. Nonparametric Statist.* *5*: 91–101.

11

Efficient Estimation in a Semiparametric Additive Regression Model with ARMA Errors

ANTON SCHICK   Binghamton University, Binghamton, New York

1. INTRODUCTION

This chapter characterizes and constructs efficient estimates of the finite dimensional parameters in semiparametric additive regression models with stationary causal autoregressive moving average errors. Such models are useful in economics and other fields.

In recent years semiparametric models have become a popular choice in statistical modeling. They now play a major role in many areas of statistics for they are more realistic and flexible than parametric models. An elegant asymptotic theory has emerged over the last two decades aiding in the understanding of these models. This theory allows for an easy characterization of efficient estimates of the parameter of interest in semiparametric models if the efficiency criterion used is either the concept of a least dispersed regular estimator or the notion of a locally asymptotically minimax estimator. These criteria are based on the convolution theorems and on the lower bounds on the local asymptotic risk, respectively, in locally asymptotically normal (LAN) and locally asymptotically mixed normal (LAMN) families. See Begun et al. (1983), Pfanzagl and Wefelmeyer (1982) and the recent monograph by Bickel et al. (1993) for an overview of the underlying theory in models with independent (and identically distributed) observations, and Schick (1988) for a treatment in the framework of the more general LAMN models. See also Jeganathan (1995) who gives a treatment of related topics with an emphasis on time series models and Choi, Hall and Schick (1996) who present a theory of testing in semiparametric models.

At the present there exists a general abstract theory for the construction of efficient estimates of the parameter of interest in a semiparametric model in the case of independent and identically distributed (iid) data. Schick (1986) showed how to construct efficient estimates in this case. His construction requires the availability of a preliminary $\sqrt{n}$-consistent estimator of the parameter of interest and an appropriate estimator of the efficient score function. Klaassen (1987) showed that these conditions are also necessary. The construction given in Schick (1986) is not quite satisfactory from a practical point of view. It relies on a sample splitting trick and uses two estimates of the efficient score function each based on only half the sample. By imposing stronger conditions on the estimates of the efficient score function, Schick (1987) overcame this difficulty and was able to construct an efficient estimate that uses *one* estimate of the efficient score function based on the *entire* sample. This approach was successfully applied to semiparametric regression models in Schick (1993).

In summary, the work of Klaassen (1987) and Schick (1986, 1987) reduces the problem of constructing efficient estimates to that of constructing preliminary $\sqrt{n}$-consistent estimates of the parameter of interest and of appropriate estimates of the efficient score function. It relies exclusively on the independence assumption and thus extends to the case of independent but not identically distributed observations.

Not much is known in general about the construction of efficient estimates when one observes dependent data. An interesting special case, however, has been treated by Kreiss (1987a, b) who constructed efficient estimates of the parameters in ARMA and AR models. His work has been recently generalized by Drost et al. (1997), Jeganathan (1995) and Koul and Schick (1997) to nonlinear autoregressive models, and by Schick (1996a, b) to regression models with autoregressive errors. These authors exploit the independence structure of the underlying innovations for the time series models considered. Relying on discretized preliminary estimates and contiguity arguments, they essentially reduce their problems to the iid case and then are able to employ the arguments studied for this case.

In this article we shall present how this theory can be applied to construct efficient estimates in a semiparametric regression model with a stationary causal autoregressive moving average (ARMA) error structure. To keep our approach concrete we shall only focus on a specific semiparametric additive regression model. This model was introduced by Engle et al. (1986) with time series errors and has potential applications in many fields.

The semiparametric additive regression model is defined by the structural relation

$$Y_j = \beta^T U_j + \gamma(V_j) + X_j, \qquad j = 1, 2, \ldots,$$

where β is an unknown vector in $\mathbb{R}^m$, γ is an unknown Lipschitz-continuous function from $[0, 1]$ to $\mathbb{R}$, $(U_1, V_1), (U_2, V_2), \ldots$ are independent $\mathbb{R}^m \times [0, 1]$-valued random vectors with common distribution G and are independent of the errors $X_1, X_2, \ldots$ which have zero means and finite variances.

This model has received considerable attention in the case when $X_1, X_2, \ldots$ are independent and identically distributed. Early work sought to estimate the regression function $(u, v) \rightarrow \beta^T u + \gamma(v)$: Wahba (1984) and Green et al. (1985). More recent work dealt with the estimation of β at a parametric rate. Chen (1988), Chen and Shiau (1991), Heckman (1986, 1988), Robinson (1988) and Speckman (1988) constructed $\sqrt{n}$-consistent estimates of β under the nonsingularity of the matrix

$$\Lambda = E[(U_1 - w(V_1))(U_1 - w(V_1))^T],$$

which guarantees the identifiability of the parameter β, and under various conditions on w and γ, where $w(V_1)$ is the conditional expectation of U_1 given V_1. More precisely, these authors constructed estimates which satisfy

$$\hat{\beta}_n = \beta + \frac{1}{n} \sum_{j=1}^{n} \Lambda^{-1} U_j^*(Y_j - \beta^T U_j - \gamma(V_j)) + o_p(n^{-1/2}),$$

with

$$U_j^* = U_j - w(V_j), \qquad j = 1, 2, \ldots.$$

As pointed out by Chen (1988) such estimates are efficient (in the sense of being least dispersed regular estimates) if the error density f is normal. Cuzick (1992a) constructed efficient estimates $\hat{\beta}_n$ of β if f is known and has finite Fisher information. Such estimates are characterized by

$$\hat{\beta}_n = \beta + \frac{1}{n} \sum_{j=1}^{n} (J\Lambda)^{-1} U_j^* \ell[Y_j - U_j^T \beta - \gamma(V_j)] + o_p(n^{-1/2}), \tag{1.1}$$

where J is the Fisher information and ℓ the score function for location, i.e.,

$$J = \int \ell^2(x) f(x)\, dx \qquad \text{and} \qquad \ell = -\frac{f'}{f}\, \mathbf{1}_{\{f > 0\}}.$$

Cuzick (1992b) and Schick (1993) constructed efficient estimates $\hat{\beta}_n$ of β when the error distribution is unknown. Such estimates are still characterized by Eq. (1.1). Thus not knowing f does not result in a loss of efficiency.

More recently, Schick (1996a) considered this model with autoregressive errors of order 1. In this form, the model was introduced by Engle et al. (1986) to study the effect of weather on electricity demand. Schick (1996a) characterized efficient estimates of β and α, the parameter of the autoregressive error process, in the presence of the unknown innovation density f and the unknown regression function γ and then constructed an efficient estimate of β. This amounted to the construction of an estimate $\hat{\beta}_n$ satisfying

$$\hat{\beta}_n = \beta + [(1 + \alpha^2)J\Lambda]^{-1} \frac{1}{n} \sum_{j=2}^{n} (U_j^* - \alpha U_{j-1}^*)\ell(X_j - \alpha X_{j-1}) + o_p(n^{-1/2}).$$

(1.2)

Note that Eq. (1.2) reduces to Eq. (1.1) if $\alpha = 0$. Thus not knowing that $\alpha = 0$ does not result in a loss of efficiency. He also showed that an efficient estimate $\hat{\alpha}_n$ of the autoregression parameter α must satisfy

$$\hat{\alpha}_n = \alpha + \frac{\sum_{j=2}^{n} X_j \ell(X_j - \alpha X_{j-1})}{\sum_{j=2}^{n} X_j^2 J} + o_p(n^{-1/2}).$$

(1.3)

Since this is also the characterization when the regression parameters and the error density f are known, this shows that adaptive estimation of α should be possible. In other words, there should be no loss of efficiency (at least asymptotically) in estimating α due to the fact that these parameters are unknown. Schick (1996b) verified that the characterization in Eq. (1.3) is also valid for other regression models and showed how to construct such an estimate of α for these models if appropriate estimates of the regression function are available. He demonstrated that such estimates are available under mild assumptions on the regression function and the underlying design distribution. Thus he verified that adaptive estimation of α is possible under mild additional assumptions. It was already shown by Kreiss (1987b) that the autoregression parameter can be estimated adaptively if the regression function is known.

In this article we shall consider the above semiparametric additive regression model with stationary and causal ARMA errors and address the question of efficient estimation of the ARMA parameter ψ and the regression parameter β in the presence of the nuisance parameters γ, f and G, where f is the density of the innovations of the ARMA process. To keep the notation simple and explicit we shall only treat the ARMA(1, 1) model and assume that the dimension of the regression parameter is 1. Thus we assume from now on that $X_1, X_2, \ldots, X_n$ are from the ARMA process $\{X_t : t \in Z\}$ defined by

$$X_t - \rho X_{t-1} = \eta_t - \mu\eta_{t-1}, \qquad t \in \mathbb{Z}, \tag{1.4}$$

where ρ and μ are two distinct elements of the interval $(-1, 1)$ and $\{\eta_t : t \in Z\}$ is a collection of independent random variables with common density f. This density is assumed to have mean zero, finite variance σ^2 and finite Fisher information for location.

The parameter $\psi = (\rho, \mu)^T$ can be estimated adaptively based on $X_1, \ldots, X_n$. This problem has just recently been solved under no additional assumptions in Koul and Schick (1997) by constructing an estimate $\langle \hat\psi_n \rangle$ which satisfies

$$\hat\psi_n = \psi + [JW]^{-1} \frac{1}{n} \sum_{j=1}^{n} A_j \ell(\eta_j) + o_p(n^{-1/2}) \tag{1.5}$$

where A_j is the gradient of the map $(a, b)^T \mapsto \sum_{i=1}^{j-1}(a - b)b^{i-1}X_{j-i}$ at ψ:

$$A_j = \sum_{i=1}^{j-1} \begin{pmatrix} \mu^{i-1} \\ \rho(i-1)\mu^{i-2} - i\mu^{i-1} \end{pmatrix} X_{j-i} \tag{1.6}$$

and W is the nonsingular matrix

$$W = \begin{bmatrix} \dfrac{\sigma^2}{1 - \rho^2} & -\dfrac{\sigma^2}{1 - \rho\mu} \\ -\dfrac{\sigma^2}{1 - \rho\mu} & \dfrac{\sigma^2}{1 - \mu^2} \end{bmatrix}. \tag{1.7}$$

Previously, Kreiss (1987a) constructed such an estimate for symmetric innovation densities if initial "innovations" are available. Drost et al. (1997) were able to do so without the symmetry assumption by utilizing a sample splitting technique, but their construction still uses initial "innovations". In contrast, the construction in Koul and Schick (1997) does not utilize sample splitting techniques nor does it require initial "innovations". Simulations carried out in their paper indicate that in moderate sample sizes estimates which avoid sample splitting perform better than those based on sample splitting.

Since we do not know the regression parameters β and γ, we cannot observe the errors $X_1, \ldots, X_n$. Thus it is not clear whether estimates satisfying Eq. (1.5) can be constructed based on the observations $Y_1, \ldots, Y_n$. The obvious approach is as follows: Construct estimates $\hat\beta_n$ and $\hat\gamma_n$ of the regression parameters β and γ and use $Y_j - \hat\beta_n U_j - \hat\gamma_n(V_j)$ in place of the unobservable error variables $X_j = Y_j - \beta U_j - \gamma(V_j)$. We shall show in this chapter that this approach works provided appropriate estimates of γ are chosen. This shows that ρ and μ can be estimated adaptively even if the regression parameters are unknown.

We shall also construct efficient estimates of the regression parameter β. It will be shown that this task requires the construction of estimates $\hat{\beta}_n$ that satisfy

$$\hat{\beta}_n = \beta + \left(\frac{1 - 2\mu\rho + \rho^2}{1 - \mu^2} J\Lambda\right)^{-1} \frac{1}{n} \sum_{j=1}^{n} B_j^* \ell(\eta_j) + o_p(n^{-1/2}) \qquad (1.8)$$

with

$$B_j^* = U_j^* + (\mu - \rho)U_{j-1}^* + (\mu - \rho)\mu U_{j-2}^* + \cdots + (\mu - \rho)\mu^{j-2} U_1^*.$$

Our asymptotic considerations are based on the local asymptotic normality (LAN) of our model. In Sec. 2 we shall establish LAN in both the parameter of interest and the nuisance parameter, with uniformity in the former. This semiparametric version is needed to characterize efficient estimates. The uniformity is helpful in the construction of efficient estimates. Various LAN results have been proved in the literature for related models. See, for example, Akritas and Johnson (1980), Swensen (1985), Kreiss (1987a), Drost et al. (1997), Jeganathan (1995), Koul and Schick (1997). Among these only the last three are concerned with uniformity and only the last one deals with parametrizations of the error density. Our LAN results are deduced from those in Koul and Schick (1997).

In Sec. 3 we carry out the efficiency considerations and derive the characterizations in Eqns (1.5) and (1.8) of efficient estimates for the parameter $\theta = (\psi^T, \beta)^T = (\rho, \mu, \beta)^T$. As efficiency criterion we use the notion of a locally asymptotically minimax (LAM) estimator for semiparametric models rather than the notion of a least dispersed regular estimator which is frequently used in semiparametric models.

In Sec. 4 we describe an efficient estimate of θ. The construction follows closely the one given in Schick (1993) in regression models with independent and identically distributed errors. It was already shown in Schick (1996a, b) that these ideas carry over to regression models with autoregressive errors. Thus it should be no surprise that this approach also works in the present more general context. The construction requires the availability of $\sqrt{n}$-consistent preliminary estimates of θ to obtain estimates of the regression function γ and the underlying innovations. The estimates of the innovations are used to estimate the score function ℓ and the Fisher information J. The efficiency of our estimate is proved in Sec. 8.

Section 5 provides a preliminary $\sqrt{n}$-consistent estimate of the regression parameter. The estimate is analogous to the one introduced by Schick (1996c). Sections 6 and 7 provide $\sqrt{n}$-consistent estimates of the ARMA parameters ρ and μ, respectively. We start with well-known $\sqrt{n}$-consistent estimates for ARMA processes and replace the observations X_j by the residuals $\hat{X}_j$. To establish the required $\sqrt{n}$-consistency for this approach we rely on ideas of Schick (1994) who treated such a problem with an AR(1) process. See also Truong (1991) and Truong and Stone (1994) for related work.

We shall carry out this program under the following assumptions on the error density f and the covariate distribution G.

ASSUMPTION 1.1   The density f satisfies the moment conditions

$$\int x f(x)\, dx = 0 \qquad \text{and} \qquad \sigma^2 = \int x^2 f(x)\, dx < \infty$$

and has finite Fisher information for location.

ASSUMPTION 1.2   The distribution Q of the random variable V_1 has a Lebesgue density that is bounded and bounded away from 0 on $[0,1]$, the random variable U_1 is bounded, and $\Lambda = E[(U_1 - w(V_1))^2]$ is positive.

The above are assumed throughout without further reference. The distribution associated with the density f is denoted by F. The normal distribution with mean vector ν and dispersion matrix Σ is denoted by $\mathcal{N}(\nu, \Sigma)$. The symbol $\mathfrak{L}(X|P)$ stands for the distribution of a random vector X under a probability measure P. We write $\xi_n \xrightarrow{P_n} c$ to indicate that the sequence $\langle \xi_n \rangle$ of random vectors convergences to the vector c in probability under the sequence $\langle P_n \rangle$ of probability measures, i.e., $P_n(\|\xi_n - c\| > \epsilon) \to 0$ for all positive ϵ. If $\{a_n\}$ and $\{b_n\}$ are sequences of positive numbers, then $a_n \sim b_n$ means that $a_n/b_n + b_n/a_n$ is bounded.

2. LOCAL ASYMPTOTIC NORMALITY

The parameter of interest in our model is $\theta = (\psi^T, \beta)^T = (\rho, \mu, \beta)^T$ and the nuisance parameter is $\phi = (\gamma, f, G)$. The parameter set for the parameter of interest is $\Theta = \{(a,b,c) \in \mathbb{R}^3 : -1 < a < 1, -1 < b < 1, a \neq b\}$, while the parameter set for the nuisance parameter is $\Phi = \Gamma \times \mathfrak{F} \times \mathfrak{G}$, where Γ is the set of all Lipschitz continuous functions from $\mathbb{R}$ to $\mathbb{R}$, $\mathfrak{F}$ is the set of all Lebesgue densities with zero means, finite variances and finite Fisher informations for location, and $\mathfrak{G}$ is the set of all distributions with the same properties as G. We write $P_{\vartheta, \varphi}$ for the measure with parameter $(\vartheta, \varphi) \in \Theta \times \Phi$ and abbreviate $P_{\vartheta, \phi}$ by P_ϑ. We write $E_{\vartheta, \varphi}$ and E_ϑ for the corresponding expectations.

Our goal is to establish LAN for regular parametric subproblems. For this purpose, let $a \mapsto \gamma_a$ be a map from $(-1, 1)$ into Γ such that $\gamma_0 = \gamma$, $b \mapsto f_b$ be a map from $(-1, 1)$ into $\mathfrak{F}$ such that $f_0 = f$, and $c \mapsto G_c$ be a map from $(-1, 1)$ into $\mathfrak{G}$ such that $G_0 = G$ and G_c has a density g_c with respect to G. Then we call the function π which maps (a, b, c) to (γ_a, f_b, G_c) a *path* and denote its domain by Δ.

We shall now restrict attention to the (finite-dimensional) submodel

$$\mathfrak{P}_\pi = \{P_{\vartheta, \pi(\delta)} : \vartheta \in \Theta, \delta \in \Delta\}$$

and assume that the following three conditions hold. Recall that Q denotes the distribution of V_1. Let λ denote the Lebesgue measure.

(R1) There exists a measurable function ξ from $[0, 1]$ to $\mathbb{R}$ such that $0 < \int \xi^2 \, dQ < \infty$ and

$$\int (\gamma_a - \gamma_0 - a\xi)^2 \, dQ = o(a^2).$$

(R2) The map $b \mapsto \int x^2 f_b(x) \, d\lambda(x)$ is continuous at 0 and there exists a measurable function ζ from $\mathbb{R}$ to $\mathbb{R}$ such that $0 < \int \zeta^2 f \, d\lambda < \infty$ and

$$\int (f_b^{1/2} - f_0^{1/2} - \tfrac{1}{2} b\zeta f_0^{1/2})^2 \, d\lambda = o(b^2).$$

(R3) There exists a measurable function χ from $\mathbb{R} \times [0, 1]$ to $\mathbb{R}$ such that $0 < \int \chi^2 \, dG < \infty$ and

$$\int (g_c^{1/2} - 1 - \tfrac{1}{2} c\chi)^2 \, dG = o(c^2).$$

REMARK 2.1 We should point out that the map ζ appearing in (R2) must satisfy

$$\int \zeta(x) f(x) \, d\lambda(x) = 0 \qquad \text{and} \qquad \int x\zeta(x) f(x) \, d\lambda(x) = 0. \tag{2.1}$$

Indeed, these two integrals are the derivatives of the maps $b \mapsto \int f_b(x) \, d\lambda(x)$ and $b \mapsto \int x f_b(x) \, d\lambda(x)$ at 0 and thus have to be zero as these maps are constant. Similarly, the map χ appearing in (R3) must satisfy $\int \chi \, dG = 0$.

We assume from now on that we can also observe the random variables X_0 and η_0. This will allow us to get an explicit formula for the likelihood. The idea of including X_0 and η_0 to define the likelihood goes back to Kreiss (1987a).

Note that

$$\eta_j = X_j - (\rho - \mu) \sum_{i=1}^{j-1} \mu^{i-1} X_{j-i} - \mu^{j-1} (\rho X_0 - \mu \eta_0), \qquad j = 1, 2, \ldots.$$

Replacing X_j by $Y_j - \beta U_j - \gamma(V_j)$ and setting $Y_0 = X_0$ and $Y_{-1} = \eta_0$, we see that

$$\eta_j = \varepsilon_j(\theta, \gamma) = Y_j - R_j(\theta, \gamma) - \mu^{j-1}(\rho Y_0 - \mu Y_0), \qquad j = 1, 2, \ldots,$$

where

$$R_j(\theta, \gamma) = \beta U_j + \gamma(V_j)$$

$$+ (\rho - \mu) \sum_{i=1}^{j-1} \mu^{i-1}(Y_{j-i} - \beta U_{j-i} - \gamma(V_{j-i})), \qquad j = 1, 2, \ldots.$$

Let $p_{v,f}$ denote the joint density of X_0 and η_0. Then the likelihood in the submodel $\mathfrak{P}_\pi$ is given by

$$L_n(\vartheta, \delta) = p_{(\vartheta_1, \vartheta_2), f_h}(Y_{-1}, Y_0)$$

$$\times \prod_{j=1}^{n} f_h(\varepsilon_j(\vartheta, \gamma_a)) g_c(U_j, V_j), \qquad \vartheta \in \Theta, \delta = (a, b, c)^T \in \Delta.$$

By a *local* sequence for θ we mean a sequence $\langle \theta_n \rangle$ in Θ such that $n^{1/2}(\theta_n - \theta)$ is bounded. We shall now prove the following LAN result.

THEOREM 2.2   There exist 6-dimensional random vectors $Z_n(\vartheta)$ and a nonsingular 6×6 matrix Σ such that

$$\log \frac{L_n(\theta_n + n^{-1/2} t_n, n^{-1/2} u_n)}{L_n(\theta_n, 0)} - v_n^T Z_n(\theta_n) + \tfrac{1}{2} v_n^T \Sigma v_n \xrightarrow{P_{\theta_n}} 0 \tag{2.2}$$

and

$$\mathfrak{L}(Z_n(\theta_n) | P_{\theta_n}) \Rightarrow \mathcal{N}(0, \Sigma) \tag{2.3}$$

for every local sequence $\langle \theta_n \rangle$ for θ and every bounded sequence $\langle v_n \rangle = \langle (t_n^T, u_n^T)^T \rangle$ in $\mathbb{R}^6$.

Proof.   It follows from (R3) that for each bounded sequence $\langle c_n \rangle$ in $\mathbb{R}$

$$\sum_{j=1}^{n} \log g_{n^{-1/2} c_n}(U_j, V_j) - \frac{1}{\sqrt{n}} \sum_{j=1}^{n} c_n \chi(U_j, V_j) + \frac{1}{2} c_n^2 \int \chi^2 \, dG \xrightarrow{P_\theta} 0. \tag{2.4}$$

It was shown in Koul and Schick (1997) that (R2) implies

$$\iint |p_{v_n, f_{h_n}}(x, y) - p_{v, f}(x, y)| \, dx dy \to 0 \quad \text{as } \psi_n \to \psi \text{ and } b_n \to 0. \tag{2.5}$$

It is easily checked that $\sum_{j=1}^{n}(\mu_n^{j-1}(\rho_n X_0 - \mu_n \eta_0) - \mu^{j-1}(\rho X_0 - \mu \eta_0))^2 \overset{P_\theta}{\to} 0$. Assume now that there are a $\nu \in \mathbb{R}^4$, a 4×4 positive definite matrix M and 4-dimensional random vectors $\dot{R}_j(\vartheta)$ such that

$$\sum_{j=1}^{n} |R_j(\vartheta_n, \gamma_{a_n}) - R_j(\theta, \gamma) - \binom{\vartheta_n - \theta}{a_n} \dot{R}_j(\theta)|^2 \overset{P_\theta}{\to} 0, \tag{2.6}$$

$$\max_{1 \le j \le n} n^{-1/2} \|\dot{R}_j(\theta)\| \overset{P_\theta}{\to} 0, \tag{2.7}$$

$$\frac{1}{n} \sum_{j=1}^{n} \dot{R}_j(\theta) \overset{P_\theta}{\to} \nu, \tag{2.8}$$

$$\frac{1}{n} \sum_{j=1}^{n} \dot{R}_j(\theta) \dot{R}_j^T(\theta) \overset{P_\theta}{\to} M, \tag{2.9}$$

$$\frac{1}{n} \sum_{j=1}^{n} \|\dot{R}_j(\vartheta_n) - \dot{R}_j(\theta)\|^2 \overset{P_\theta}{\to} 0. \tag{2.10}$$

for all local sequences $\langle \vartheta_n \rangle$ for θ and every sequence $\langle a_n \rangle$ in $\mathbb{R}$ such that $n^{1/2} a_n$ is bounded. Using Eqns (2.4)–(2.10) and the arguments in Remark 2.6 and the Proof of Theorem 2.4 of Koul and Schick (1997), we can then conclude the desired result with

$$Z_n(\vartheta) = \frac{1}{\sqrt{n}} \sum_{j=1}^{n} \begin{pmatrix} \dot{R}_j(\vartheta) \ell(\varepsilon_j(\vartheta, \gamma)) \\ \zeta(\varepsilon_j(\vartheta, \gamma)) \\ \chi(U_j, V_j) \end{pmatrix}$$

and

$$\Sigma = \begin{bmatrix} JM & \nu \int \ell \zeta \, dF & 0 \\ \int \ell \zeta \, dF \nu^T & \int \zeta^2 \, dF & 0 \\ 0 & 0 & \int \chi^2 \, dG \end{bmatrix}.$$

We are left to verify Eqns (2.6)–(2.10). This can be done with

$$\dot{R}_j(\theta) = \begin{pmatrix} A_j(\theta, \gamma) \\ B_j(\theta, U) \\ B_j(\theta, \xi(V)) \end{pmatrix}$$

where

$$A_j(\theta, \gamma) = \sum_{k=1}^{j-1} \binom{\mu^{k-1}}{\rho(k-1)\mu^{k-2} - k\mu^{k-1}} [Y_{j-k} - \beta U_{j-k} - \gamma(V_{j-k})],$$

and, for a G-square integrable function h,

$$B_j(\theta, h(U, V)) = h(U_j, V_j) + (\mu - \rho) \sum_{i=1}^{j-1} \mu^{i-1} h(U_{j-i}, V_{j-i}).$$

We shall not provide the details, but only identify the vector ν and the matrix M. Some of the details can be found in Koul and Schick (1997). Since, for each G-square integrable function h, the series

$$B(h(U, V)) = h(U_1, V_1) + (\mu - \rho) \sum_{i=1}^{\infty} \mu^{i-1} h(U_{1+i}, V_{1+i})$$

converges almost surely and in L_2, we obtain

$$\frac{1}{n} \sum_{j=1}^{n} B_j(\theta, h(U, V)) \xrightarrow{P_\theta} E[B(h(U, V))],$$

$$\frac{1}{n} \sum_{j=1}^{n} B_j(\theta, h(U, V)) B_j(\theta, k(U, V)) \xrightarrow{P_\theta} E[B(h(U, V)) B(k(U, V))],$$

$$\frac{1}{n} \sum_{j=1}^{n} B_j(\theta, h(U, V)) A_j(\theta, \gamma) \xrightarrow{P_\theta} 0,$$

whenever h and k are G-square integrable. Here we have written E for E_θ. In view of this and the results of Koul and Schick (1997) we find that $\nu = (0, 0, E[B(U)], E[B(\xi(V))])^T$ and M is block-diagonal:

$$M = \begin{bmatrix} W & 0 \\ 0 & M_{22} \end{bmatrix} \quad \text{with} \quad M_{22} = \begin{bmatrix} E[(B(U))^2] & E[B(U)B(\xi(V))] \\ E[B(\xi(V))B(U)] & E[(B(\xi(V)))^2] \end{bmatrix}$$

and W as in Eq. (1.7). Recall that $w(V_1) = E(U_1|V_1)$. It is easy to check that

$$E[B(U - w(V))B(d(V))] = 0 \tag{2.11}$$

for all Q-square integrable functions d. Since $B(U) = B(U - w(V)) + B(w(V))$, we get

$$M_{22} = \begin{bmatrix} E[(B(U - w(V)))^2] + E[(B(w(V)))^2] & E[B(w(V))B(\xi(V))] \\ E[B(\xi(V))B(w(V))] & E[(B(\xi(V)))^2] \end{bmatrix}.$$

As

$$E[(B(U - w(V)))^2] = \left(1 + \frac{(\mu - \rho)^2}{1 - \mu^2}\right) E\{[U_1 - w(V_1)]^2\} > 0,$$

we see that M_{22} is positive definite. Thus M is positive definite. $\qquad\square$

REMARK 2.3   We should point out that the matrix Σ is block-diagonal

$$\Sigma = \begin{bmatrix} JW & 0 & 0 \\ 0 & D & 0 \\ 0 & 0 & \int \chi^2 \, dG \end{bmatrix} \tag{2.12}$$

with

$$D = \begin{bmatrix} JE[(B(U))^2] & JE[B(U)B(\xi(V))] & E[B(U)] \int \ell\zeta \, dG \\ JE[B(U)B(\xi(V))] & JE[(B(\xi(V)))^2] & E[B(\xi(V))] \int \ell\zeta \, dF \\ E[B(U)] \int \ell\zeta \, dF & E[B(\xi(V))] \int \ell\zeta \, dF & \int \zeta^2 \, dF \end{bmatrix}.$$

Note that D is the dispersion matrix of the random vector

$$\begin{pmatrix} B(U)\ell(\eta_0) \\ B(\xi(V))\ell(\eta_0) \\ \zeta(\eta_0) \end{pmatrix}. \tag{2.13}$$

This will be exploited in the next section when we search for a least favorable path.

REMARK 2.4   Partition $Z_n(\vartheta)$ and Σ as

$$Z_n(\vartheta) = \begin{pmatrix} Z_n^{(1)}(\vartheta) \\ Z_n^{(2)}(\vartheta) \end{pmatrix} \quad \text{and} \quad \Sigma = \begin{bmatrix} \Sigma_{11} & \Sigma_{12} \\ \Sigma_{21} & \Sigma_{22} \end{bmatrix},$$

where $Z_n^{(1)}(\vartheta)$ is 3-dimensional and Σ_{11} is a 3×3 matrix. Then

$$I(\pi) = \Sigma_{11} - \Sigma_{12}\Sigma_{22}^{-1}\Sigma_{21}$$

is called the information matrix for estimating θ in the submodel $\mathfrak{P}_\pi$ and

$$S_n(\theta; \pi) = Z_n^{(1)}(\theta) - \Sigma_{12}\Sigma_{22}^{-1}Z_n^{(2)}(\theta)$$

the score for estimating θ in the submodel $\mathfrak{P}_\pi$. It follows from Theorem 2.2, that the score is regular in the sense that

$$S_n(\theta_n; \pi) - S_n(\theta; \pi) + I(\pi)n^{1/2}(\theta_n - \theta) \xrightarrow{P_\theta} 0 \tag{2.14}$$

for every local sequence $\langle \theta_n \rangle$ for θ.

REMARK 2.5   Theorem 2.2 implies that the sequences of distributions $\langle F_{n.\theta_n} \rangle$ and $\langle F_{n.\theta} \rangle$ are mutually contiguous for each local sequence $\langle \theta_n \rangle$ for θ, where

$$F_{n.\vartheta} = \mathfrak{L}((Y_1, Y_0, \ldots, Y_n, U_1, V_1, \ldots, U_n, V_n)|P_\vartheta).$$

3. EFFICIENCY CONSIDERATIONS

Call a path π with satisfies (R1)-(R3) a regular path. In this case call ξ its (R1)-characteristic, ζ its (R2)-characteristic, χ its (R3)-characteristic, and (ξ, ζ, χ) its characteristics. One can show that for each measurable function ξ with $0 < \int \xi^2 \, dQ < \infty$, each measurable ζ satisfying $0 < \int \zeta^2 \, dF < \infty$ and (2.1), and each measurable function χ with $0 < \int \chi^2 \, dG < \infty$ and $\int \chi \, dG = 0$, there exists a regular path with characteristics (ξ, ζ, χ).

Let Π denote the collection of all regular paths. Our first goal is to show that there exists a least favorable path in Π. A path π_* is called least favorable if it is regular and if it has the smallest information matrix among all regular paths in the sense that $I(\pi) - I(\pi_*)$ is nonnegative definite for every regular π.

Now fix a regular path π. It follows from the representation in Eq. (2.12) that

$$
I(\pi) = \begin{bmatrix} JW & 0 \\ 0 & D_{11} - D_{12}D_{22}^{-1}D_{21} \end{bmatrix}
$$

with $D_{11} = E[(B(U)\ell(\eta_0))^2]$ and D_{12}, D_{22} D_{21} the other blocks of the corresponding partition of D. Since D is the dispersion matrix of the random vector in Eq. (2.13), we can interpret $D_{12}D_{22}^{-1}D_{21}$ as the second moment of the projection (in L_2) of $B(U)\ell(\eta_0)$ onto the linear span of $B(\xi(V))\ell(\eta_0)$ and $\zeta(\eta_0)$. Since $B(U)\ell(\eta_0) = B(U - w(V))\ell(\eta_0) + B(w(V))\ell(\eta_0)$ and $B(U - w(V))\ell(\eta_0)$ is orthogonal to $B(\xi(V))\ell(\eta_0)$ and $\zeta(\eta_0)$ in view of Eq. (2.11), $D_{12}D_{22}^{-1}D_{21}$ is also the second moment of the projection of $B(w(V))\ell(\eta_0)$ onto the linear span of $B(\xi(V))\ell(\eta_0)$ and $\zeta(\eta_0)$ and is thus bounded by the second moment of $B(w(V))\ell(\eta_0)$. This bound is sharp for the choice $\xi = w$. Since $B(U - w(V))\ell(\eta_0)$ and $B(w(V))\ell(\eta_0)$ are also orthogonal by Eq. (2.11), we see that

$$
D_{11} - D_{12}D_{22}^{-1}D_{21} \leq E[(B(U - w(V))\ell(\eta_0))^2] = \frac{1 - 2\rho\mu + \rho^2}{1 - \mu^2} \Lambda J
$$

with equality if π has (R1)-characteristic $\xi = w$. This shows that $I(\pi)$ is minimized by any regular path π_* whose (R1)-characteristic is w with minimal value $I_* = I(\pi_*) = JH_*$, where

$$
H_* = H(\rho, \mu, \sigma^2, \Lambda) = \begin{bmatrix} \dfrac{\sigma^2}{1 - \rho^2} & -\dfrac{\sigma^2}{1 - \rho\mu} & 0 \\[3mm] -\dfrac{\sigma^2}{1 - \rho\mu} & \dfrac{\sigma^2}{1 - \mu^2} & 0 \\[3mm] 0 & 0 & \dfrac{1 - 2\rho\mu - \rho^2}{1 - \mu^2} \Lambda \end{bmatrix}.
$$

The matrix I_* is called the efficient information (for estimating θ). The above shows that a regular path is least favorable if its (R1)-characteristic is w. The corresponding score for estimating θ is

$$S_n^*(\theta) = S_n(\theta; \pi_*) = \frac{1}{\sqrt{n}} \sum_{j=1}^n \binom{A_j(\theta, \gamma)}{B_j(\theta, U - w(V))} \ell(\varepsilon_j(\theta, \gamma)).$$

By a loss function we mean a Borel measurable function L from $\mathbb{R}^3$ into $[0, \infty)$ such that $L(x) = L(-x)$ for all $x \in \mathbb{R}^m$ and the set $\{L \leq u\}$ is convex for each $u \geq 0$. By an estimate of θ we mean a sequence $\langle \hat{\theta}_n \rangle$ of 3-dimensional random vectors with $\hat{\theta}_n$ a measurable function of $(Y_{-1}, Y_0, \ldots, Y_n, U_1, V_1, \ldots U_n, V_n)$.

THEOREM 3.1 Let $\langle \hat{\theta}_n \rangle$ be an estimate of θ. Then

$$\sup_{\pi \in \Pi} \lim_{C \to \infty} \liminf_{n \to \infty} \sup_{\|\vartheta - \theta\| + \|\delta\| \leq C/\sqrt{n}} E_{\vartheta, \pi(\delta)} L(\sqrt{n}\,(\hat{\theta}_n - \vartheta)) \geq \int L \, d\mathcal{N}(0, I_*^{-1})$$

$$(3.1)$$

for every loss function L. Moreover, if $\langle \hat{\theta}_n \rangle$ satisfies

$$\sqrt{n}\,(\hat{\theta}_n - \theta) - I_*^{-1} S_n^*(\theta) \xrightarrow{P_\theta} 0, \tag{3.2}$$

then

$$\mathfrak{L}(\sqrt{n}\,(\hat{\theta}_n - \theta_n)|P_{\theta_n, \pi(u_n/\sqrt{n})}) \Rightarrow \mathcal{N}(0, I_*^{-1}) \tag{3.3}$$

for every local sequence $\langle \theta_n \rangle$, every $\pi \in \Pi$ and every bounded sequence u_n in $\mathbb{R}^3$. The latter implies

$$\lim_{C \to \infty} \limsup_{n \to \infty} \sup_{\|\vartheta - \theta\| + \|\delta\| \leq C/\sqrt{n}} E_{\vartheta, \pi(\delta)} L(\sqrt{n}\,(\hat{\theta}_n - \vartheta)) \leq \int L \, d\mathcal{N}(0, I_*^{-1}) \quad (3.4)$$

for every bounded loss function L and every regular path π.

Proof. The above can be deduced from the results in Schick (1988). Alternatively and more directly, we can proceed as follows. It follows from Theorem 6 in Fabian and Hannan (1982) that

$$\lim_{C \to \infty} \liminf_{n \to \infty} \sup_{\|\vartheta - \theta\| + \|\delta\| \leq C/\sqrt{n}} E_{\vartheta, \pi, (\delta)} L(\sqrt{n}\,(\hat{\theta}_n - \vartheta)) \geq \int L \, d\mathcal{N}(0, I_*^{-1})$$

for every loss function L. This immediately implies the lower bound Eq. (3.1).

To verify Eq. (3.3) fix a regular path π. It follows from Eq. (3.2) that

$$\mathcal{L}\left(\left(\begin{array}{c}\sqrt{n}\,(\hat{\theta}_n - \theta)\\ Z_n(\theta)\end{array}\right)\Big|P_\theta\right) \Rightarrow \mathcal{N}\left(0, \begin{bmatrix}I_*^{-1} & C\\ C^T & \Sigma\end{bmatrix}\right),$$

where $C = [I \ 0]$ with I the 3×3 identity matrix. The desired result now follows from Theorem 2.2 and an application of Le Cam's Third Lemma, Le Cam (1960, Theorem 2.1) or Hájek and Šidák (1967). Since loss functions are almost surely continuous with respect to $\mathcal{N}(0, I_*^{-1})$ as shown in Fabian and Hannan (1982, p. 467), Eq (3.3) implies 3.4. $\quad\square$

DEFINITION 3.2 In view of the above result an estimate $\langle\hat{\theta}_n\rangle$ of θ that satisfies Eq. (3.2) will be called Π-efficient or simply efficient.

REMARK 3.3 If we write $\hat{\theta}_n = (\hat{\rho}_n, \hat{\mu}, \hat{\beta}_n)^T$ and set $\hat{\psi}_n = (\hat{\rho}_n, \hat{\mu}_n)^T$, then Eq. (3.2) is equivalent to the statements in Eqns (1.5) and (1.8).

4. CONSTRUCTION OF EFFICIENT ESTIMATES

An important technical tool in the construction of efficient estimates in semi-parametric models is the use of discretized $\sqrt{n}$-consistent estimates of the finite dimensional parameter. Such estimates can be treated as nonstochastic local sequences in proofs and combined with contiguity arguments lead to considerable simplifications in the proofs. See e.g. Bickel (1982) and Schick (1986, 1987) for this approach. To implement this approach we assume that we have at our disposal discretized $\sqrt{n}$-consistent estimates $\langle\tilde{\theta}_n\rangle = \langle(\tilde{\rho}_n, \tilde{\mu}_n, \tilde{\beta}_n)\rangle$ of θ. Possible choices will be discussed in the next sections.

Recall a sequence of random vectors $\{\xi_n\}$ is discretized if for every $\epsilon > 0$ there exists an integer q and a sequence $\langle C_n \rangle$ of events such that $P_\theta(C_n) \geq 1 - \epsilon$ and the image $\{\xi_n(\omega): \omega \in C_n\}$ of C_n under ξ_n has at most q elements for all n. A possible way to discretize a d-dimensional $\sqrt{n}$-consistent estimate is to round the estimate to the closest point on the grid $\{(a_1, \ldots, a_d)/\sqrt{n}: a_1, \ldots, a_d = 0, 1, -1, 2, -2, \ldots\}$. More sophisticated methods are described in Fabian and Hannan (1982). We should stress that the following results do not depend on the way the estimates are discretized. We also believe that discretization can be avoided at the expense of more complicated proofs.

To simplify our treatment we shall only give a construction under the additional assumption that w is Hölder-continuous of order α greater than $1/2$. Recall, the function w is Hölder-continuous of order α if for some constant c_H

$$|w(x) - w(y)| \leq c_H |x - y|^\alpha, \qquad 0 \leq x, y \leq 1.$$

For a positive integer K, let $\mathfrak{G}_K$ denote the set of all continuous functions from $[0, 1]$ which are piecewise linear on the subintervals $[(i - 1)/K, i/K]$. This is a vector space of dimension $K + 1$. Its elements are called linear splines (for the equidistant knots $0, 1/K, 2/K, \ldots, 1$). A basis of $\mathfrak{G}_K$ is given by the B-splines $s_0, \ldots, s_K$, where for $i = 0, \ldots, K$, s_i is the map defined by $s_i(x) = \max\{0, (1 - |Kx - i|)\}$, $0 \le x \le 1$. We shall need the fact that

$$\inf_{s \in \mathfrak{G}_K} \sup_{0 \le v \le 1} |h(v) - s(v)| = O(K^{-\alpha}), \tag{4.1}$$

for every Hölder-continuous function h of order α. See De Boor (1978) or Schumaker (1981) for these and other facts about splines.

Let $\langle K_n \rangle$ be a sequence of positive integers. We shall estimate the functions w by a $\mathfrak{G}_{K_n}$-valued estimate $\hat{w}_n$ which satisfies

$$\sum_{j=1}^{n} |U_j - \hat{w}_n(V_j)|^2 = \inf_{s \in \mathfrak{G}_{K_n}} \sum_{j=1}^{n} |U_j - s(V_j)|^2$$

and the regression function γ by a $\mathfrak{G}_{K_n}$-valued estimate $\hat{\gamma}_n$ which satisfies

$$\sum_{j=1}^{n} |Y_j - \tilde{\beta}_n U_j - \hat{\gamma}_n(V_j)|^2 = \inf_{s \in \mathfrak{G}_{K_n}} \sum_{j=1}^{n} |Y_j - \tilde{\beta}_n U_j - s(V_j)|^2.$$

While the random functions $\hat{w}_n$ and $\hat{\gamma}_n$ are not uniquely defined, the quantities $\hat{w}_n(V_j)$ and $\hat{\gamma}_n(V_j)$ are uniquely determined by

$$\begin{pmatrix} \hat{w}_n(V_1) \\ \vdots \\ \hat{w}_n(V_n) \end{pmatrix} = Q_n \begin{pmatrix} U_1 \\ \vdots \\ U_n \end{pmatrix} \quad \text{and} \quad \begin{pmatrix} \hat{\gamma}_n(V_1) \\ \vdots \\ \hat{\gamma}_n(V_n) \end{pmatrix} = Q_n \begin{pmatrix} Y_1 - \tilde{\beta}_n U_1 \\ \vdots \\ Y_n - \tilde{\beta}_n U_n \end{pmatrix},$$

where

$$Q_n = D_n (D_n^T D_n)^- D_n^T$$

is the projection matrix associated with the $n \times (K_n + 1)$ random design matrix

$$D_n = \begin{bmatrix} s_0(V_1) & \cdots & s_{K_n}(V_1) \\ \vdots & \ddots & \vdots \\ s_0(V_n) & \cdots & s_{K_n}(V_n) \end{bmatrix}.$$

We shall need the fact that

$$\max_{1 \le i, j \le n} |Q_{n.i.j}| = \max_{1 \le i \le n} Q_{n.i.i} = \max_{1 \le i \le n} \sum_{j=1}^{n} Q_{n.i.j}^2 = O_{P_\theta}(n^{-2/3}) \tag{4.2}$$

if $K_n \sim n^{1/3}$. This rate was proved by Stone (1985).

Let $\langle a_n \rangle$ and $\langle b_n \rangle$ be sequences of positive numbers tending to zero and $\langle d_n \rangle$ a sequence of positive integers tending to infinity. Let $N_n = n - d_n - 1$. Based on the estimated innovations

$$\hat{\eta}_{n.j} = Y_j - R_j(\tilde{\theta}_n, \hat{\gamma}_n), \qquad j = d_n, \ldots, n,$$

we construct the kernel density estimate $\hat{f}_n$ with kernel k and window length a_n by

$$\hat{f}_n(x) = \frac{1}{N_n a_n} \sum_{j=d_n}^{n} k\left(\frac{x - \hat{\eta}_{n.j}}{a_n}\right), \qquad x \in \mathbb{R}.$$

We do not use the first few estimated innovations as they may not be good approximations of the actual innovations. Estimate the score function ℓ by

$$\hat{\ell}_n(x) = -\frac{\hat{f}_n'(x)}{b_n + \hat{f}_n(x)}, \qquad x \in \mathbb{R},$$

and the efficient information I_* by $\hat{I}_{*.n} = \hat{J}_n H(\tilde{\rho}_n, \tilde{\mu}_n, \hat{\sigma}_n^2, \hat{\Lambda}_n)$ where

$$\hat{J}_n = \frac{1}{N_N} \sum_{j=d_n}^{n} \hat{\ell}_n^2(\hat{\eta}_{n.j}), \qquad \hat{\sigma}_n^2 = \frac{1}{N_N} \sum_{j=d_n}^{n} \hat{\eta}_{n.j}^2$$

and

$$\hat{\Lambda}_n = \frac{1}{n} \sum_{j=1}^{n} (U_j - \hat{w}_n(V_j))^2.$$

Finally construct the estimate

$$\hat{\theta}_n = \tilde{\theta}_n + \frac{1}{N_n} \sum_{j=d_n}^{n} \hat{I}_{*.n}^{-1} \begin{pmatrix} \hat{A}_{n.j} - \hat{A}_{n.\cdot} \\ \hat{B}_{n.j} - \hat{B}_{n.\cdot} \end{pmatrix} \hat{\ell}_n(\hat{\eta}_{n.j}), \tag{4.3}$$

where $\hat{A}_{n.j} = A_j(\tilde{\theta}_n, \hat{\gamma}_n)$, $\hat{B}_{n.j} = B_j(\tilde{\theta}_n, U - \hat{w}_n(V))$,

$$\hat{A}_{n.\cdot} = \frac{1}{N_n} \sum_{j=d_n}^{n} \hat{A}_{n.j} \qquad \text{and} \qquad \hat{B}_{n.\cdot} = \frac{1}{N_n} \sum_{j=d_n}^{n} \hat{B}_{n.j}.$$

THEOREM 4.1 The estimate $\langle \hat{\theta}_n \rangle$ defined in Eq. (4.3) satisfies Eq. (3.2) under the following conditions.

(1) The sequences $\langle a_n \rangle$, $\langle b_n \rangle$, $\langle d_n \rangle$ and $\langle K_n \rangle$ satisfy $a_n \to 0$, $b_n \to 0$, $b_n \geq a_n^3$, $n^{-1/6} a_n^{-2} \to 0$, $K_n \sim n^{1/3}$ and $d_n \sim \log^2(n)$.
(2) The Lebesgue density k is positive, symmetric, three times continuously differentiable, has finite variance $\int x^2 k(x)\, dx$ and satisfies

$$|k^{(i)}(x)| \leq c_0 k(x), \qquad x \in \mathbb{R},\ i = 1, 2, 3,$$

for some positive constant c_0.

(3) $\langle \bar{\theta}_n \rangle$ is a discretized $\sqrt{n}$-consistent estimate of θ.
(4) The map w is Hölder-continuous of order greater than $1/2$.

A proof of this theorem is given in Sec. 8. Choices of $\sqrt{n}$-consistent estimates of θ are provided in the next sections. Note that our construction only utilizes the data $(Y_1, U_1, V_1, \ldots, Y_n, U_n, V_n)$ and not the unobservable X_0 and η_0.

REMARK 4.2   The assumption that w is Hölder-continuous of order greater than $1/2$ will be only used to conclude Eq. (5.2) below. This conclusion can also be reached under weaker assumptions provided we replace the sequence $\langle K_n \rangle$ by a faster growing sequence $\langle K_n' \rangle$ of knots. For example, if one takes $K_n' \sim n^{2/3}$, one only needs Hölder-continuity of order $\alpha > 1/4$ to conclude Eq. (5.2).

REMARK 4.3   Since the above procedure requires nonparametric estimates of the curves γ and ℓ, it might require rather large sample sizes to work. If one is willing to impose additional smoothness assumptions on γ such as Lipschitz continuous higher order derivatives, one can use nonparametric estimates of γ with a faster rate of convergence. Possible candidates are higher order least squares spline estimates. The use of such faster estimates should result in a better performance of the proposed estimate in moderate sample sizes.

REMARK 4.4   The construction of the proposed estimator requires the selection of proper tuning constants a_n, b_n, K_n. Of course, the conclusions of the above theorem remain valid if these constants are replaced by *discretized* stochastic versions. For example, one can use bootstrap techniques to select these quantities from properly chosen grids subject to minimizing say the bootstrapped variance of the proposed estimate.

5.   A PRELIMINARY ESTIMATE OF β

In this section we shall describe a $\sqrt{n}$-consistent estimate of β. We shall do so without imposing any smoothness assumptions on w. Let $U_{n.j}^* = U_j - \hat{w}_n(V_j)$ and set

$$\bar{\beta}_n = \frac{\sum_{j=1}^n U_{n.j}^* Y_j}{\sum_{j=1}^n U_{n.j}^* U_j}.$$

This is the estimate studied in Schick (1996c) with independent errors.

Suppose now that $K_n/n \to 0$ and $K_n/\sqrt{n} \to \infty$. Then $\langle \bar{\beta}_n \rangle$ is a $\sqrt{n}$-consistent estimate of β. We demonstrate this by showing that

$$\mathfrak{L}(n^{1/2}(\bar{\beta}_n - \beta)|P_\theta) \Rightarrow \mathcal{N}\left(0, \frac{\sigma^2(1 - 2\rho\mu + \mu^2)}{(1 - \rho^2)\Lambda}\right).$$

Our proof relies on the fact that

$$\frac{1}{n} \sum_{j=1}^n |\hat{w}_n(V_j) - w(V_j)|^2 \xrightarrow{P_\theta} 0, \tag{5.1}$$

which was shown in Schick (1996c). Under the assumptions of Theorem 4.1 one can even show that

$$\frac{1}{n} \sum_{j=1}^n |\hat{w}_n(V_j) - w(V_j)|^2 = o_{P_\theta}(n^{-1/3}), \tag{5.2}$$

which will be used in the proof of this theorem, but is not needed here. From Eq. (5.1) one derives

$$\frac{1}{n} \sum_{j=1}^n U_{n.j}^* U_j \xrightarrow{P_\theta} \Lambda \tag{5.3}$$

and

$$\frac{1}{n} \sum_{j=1}^n (U_{n.j}^* - U_j^*)X_j \xrightarrow{P_\theta} 0. \tag{5.4}$$

For the latter calculate the conditional second moment given $U_1, V_1, \ldots, U_n, V_n$ and use the fact that the largest eigenvalue λ_n of the dispersion matrix of $(X_1, \ldots, X_n)^T$ under P_θ is bounded by

$$\max_{1 \le i \le n} \sum_{j=1}^n |E_\theta(X_i X_j)| \le 2 \sum_{j=1}^n |E_\theta(X_1 X_j)|$$

$$\le 2\sigma^2 \left(1 + \frac{|\rho - \mu|}{1 - |\rho|} + \frac{(\rho - \mu)^2}{(1 - \rho^2)(1 - |\rho|)}\right).$$

By the definition of $\hat{w}_n$,

$$\sum_{j=1}^n U_{n.j}^* s(V_j) = 0 \qquad \text{for every } s \in \mathfrak{G}_{K_n}. \tag{5.5}$$

By the Lipschitz continuity of γ, there exist $\gamma_n \in \mathfrak{G}_{K_n}$ such that

$$\sup_{0 \le v \le 1} |\gamma(v) - \gamma_n(v)| = O(K_n^{-1}). \tag{5.6}$$

It follows from Eqns (5.5), (5.6) and $n^{1/2}/K_n \to 0$ that

$$\frac{1}{\sqrt{n}} \sum_{j=1}^{n} U_{n.j}^* \gamma(V_j) = \frac{1}{\sqrt{n}} \sum_{j=1}^{n} U_{n.j}^* (\gamma(V_j) - \gamma_n(V_j)) \xrightarrow{P_\theta} 0.$$

Combining the above shows that

$$\frac{1}{\sqrt{n}} \sum_{j=1}^{n} U_{n.j}^* Y_j - \frac{1}{\sqrt{n}} \sum_{j=1}^{n} U_{n.j}^* U_j \beta - \frac{1}{\sqrt{n}} \sum_{j=1}^{n} U_j^* X_j \xrightarrow{P_\theta} 0.$$

The desired result now follows from this (5.3), (5.4) and the fact that

$$\mathfrak{L}\left(\frac{1}{\sqrt{n}} \sum_{j=1}^{n} U_j^* X_j | P_\theta \right) \Rightarrow N\left(0, \sigma^2 \Lambda \frac{1 - 2\rho\mu + \mu^2}{1 - \rho^2} \right).$$

To see the latter note that conditionally given $X_1, \ldots, X_n$ the random variables $U_1^*, \ldots, U_n^*$ are iid with mean vector 0 and covariance Λ and that

$$\frac{1}{n} \sum_{j=1}^{n} X_j^2 \xrightarrow{P_\theta} E_\theta[X_1^2] = \sigma^2 \frac{1 - 2\rho\mu + \mu^2}{1 - \rho^2}.$$

6.   A PRELIMINARY ESTIMATE   OF ρ

In this section we shall describe a preliminary estimate of ρ. If β and γ were known, we could use the estimate

$$\rho_n^* = \frac{\sum_{j=3}^{n} X_j X_{j-2}}{\sum_{j=3}^{n} X_{j-1} X_{j-2}} = \rho + \frac{\sum_{j=3}^{n} (X_j - \rho X_{j-1}) X_{j-2}}{\sum_{j=3}^{n} X_{j-1} X_{j-2}}.$$

This estimate is $\sqrt{n}$-consistent as

$$\frac{1}{n} \sum_{j=3}^{n} X_{j-1} X_{j-2} \xrightarrow{P_\theta} E_\theta(X_2 X_1) = \sigma^2 \frac{(\rho - \mu)(1 - \rho\mu)}{1 - \rho^2} \neq 0$$

and

$$E_\theta \left(\frac{1}{\sqrt{n}} \sum_{j=3}^{n} (X_j - \rho X_{j-1}) X_{j-2} \right)^2 = O(1).$$

We shall now mimic this estimate.

Let $\langle \tilde{\beta}_n \rangle$ denote a $\sqrt{n}$-consistent estimate of β and $\langle \hat{\gamma}_n \rangle$ the estimate of γ described in Sec. 4 with $K_n \sim n^{1/3}$. The estimate $\tilde{\beta}_n$ does not have to be discretized. Set

$$\hat{X}_{n,j} = Y_j - \tilde{\beta}_n U_j - \hat{\gamma}_n(V_j), \qquad j = 1, 2, \ldots, n.$$

Then we can estimate ρ by

$$\bar{\rho}_n = \frac{\sum_{j=3}^{n} \hat{X}_{n,j} \hat{X}_{n,j-2}}{\sum_{j=3}^{n} \hat{X}_{n,j-1} \hat{X}_{n,j-2}} = \rho + \frac{\sum_{j=3}^{n} (\hat{X}_{n,j} - \rho \hat{X}_{n,j-1}) \hat{X}_{n,j-2}}{\sum_{j=3}^{n} \hat{X}_{n,j-1} \hat{X}_{n,j-2}}.$$

Let us now show that this estimate is $\sqrt{n}$-consistent.

In view of the $\sqrt{n}$-consistency of $\langle \rho_n^* \rangle$, it suffices to show that

$$\frac{1}{\sqrt{n}} \sum_{j=3}^{n} \hat{X}_{n,j} \hat{X}_{n,j-2} - \frac{1}{\sqrt{n}} \sum_{j=3}^{n} X_j X_{j-2} \xrightarrow{P_\theta} 0, \tag{6.1}$$

$$\frac{1}{\sqrt{n}} \sum_{j=3}^{n} \hat{X}_{n,j-1} \hat{X}_{n,j-2} - \frac{1}{\sqrt{n}} \sum_{j=3}^{n} X_{j-1} X_{j-2} \xrightarrow{P_\theta} 0. \tag{6.2}$$

Since the proofs of Eqns (6.1) and (6.2) are similar, we shall only give the proof of Eq. (6.1).

The left-hand-side of Eq. (6.1) can be expressed as the sum $T_{n,1} + T_{n,2} + T_{n,3}$, where

$$T_{n,1} := \frac{1}{\sqrt{n}} \sum_{j=3}^{n} \Delta_{n,j} X_{j-2}, \quad T_{n,2} := \frac{1}{\sqrt{n}} \sum_{j=3}^{n} X_j \Delta_{n,j-2}, \quad T_{n,3} := \frac{1}{\sqrt{n}} \sum_{j=3}^{n} \Delta_{n,j} \Delta_{n,j-2},$$

with $\Delta_{n,i} = \hat{X}_{n,i} - X_i$ for $i = 1, \ldots, n$. Let $Q_{n,i,j}$ denote the (i,j)-entry of Q_n. Easy calculations show that

$$\Delta_{n,i} = -U_{n,i}(\tilde{\beta}_n - \beta) + \gamma_{n,i} - \sum_{j=1}^{n} Q_{n,i,j} X_j, \qquad i = 1, \ldots, n,$$

with $U_{n,i} = U_i - \sum_{j=1}^{n} Q_{n,i,j} U_j$ and $\gamma_{n,i} = \gamma(V_i) - \sum_{j=1}^{n} Q_{n,i,j} \gamma(V_j)$. The Lipschitz continuity of γ and Eq. (4.1) yield

$$\frac{1}{n} \sum_{j=1}^{n} \gamma_{n,j}^2 = \inf_{s \in \mathbb{G}_{K_n}} \frac{1}{n} \sum_{j=1}^{n} (\gamma(V_j) - s(V_j))^2 = O(K_n^{-2}). \tag{6.3}$$

Since Q_n is a projection matrix, so is $I - Q_n$, and this gives

$$\frac{1}{n} \sum_{j=1}^{n} U_{n,j}^2 \le \frac{1}{n} \sum_{j=1}^{n} U_j^2 = O_{P_\theta}(1). \tag{6.4}$$

The argument used in the proof of Eq. (5.4) yields

$$E_\theta\left(\frac{1}{n}\sum_{i=1}^{n}\left(\sum_{j=1}^{n}H_{n.i.j}X_j\right)^2\bigg| U_1, V_1, \ldots, U_n, V_n\right) = O_{P_\theta}\left(\frac{1}{n}\sum_{j=1}^{n}\sum_{i=1}^{n}H_{n.i.j}^2\right)$$

(6.5)

and

$$E_\theta\left(\left(\frac{1}{\sqrt{n}}\sum_{j=1}^{n}H_{n.j}X_j\right)^2\bigg| U_1, V_1, \ldots, U_n, V_n\right) = O_{P_{\theta_n}}\left(\frac{1}{n}\sum_{j=1}^{n}H_{n.j}^2\right) \qquad (6.6)$$

whenever $H_{n.j}$ and $H_{n.j.i}$ are random variables based on the covariates $(U_1, V_1, \ldots, U_n, V_n)$ only. Since Q_n is a projection matrix of rank at most $K_n + 1$, we find that

$$\frac{1}{n}\sum_{i=1}^{n}\sum_{j=1}^{n}Q_{n.i.j}^2 = \frac{1}{n}\sum_{i=1}^{n}Q_{n.i.i} = \frac{1}{n}\mathrm{trace}(Q_n) \le \frac{K_n+1}{n} = O(n^{-2/3}). \qquad (6.7)$$

It follows from Eqns (6.3)–(6.5) and (6.7) that

$$\frac{1}{n}\sum_{j=1}^{n}\Delta_{n.j}^2 \le (\bar{\beta}_n - \beta)^2\frac{3}{n}\sum_{j=1}^{n}U_{n.j}^2 + \frac{3}{n}\sum_{j=1}^{n}\gamma_{n.j}^2 + \frac{3}{n}\sum_{i=1}^{n}\left(\sum_{j=1}^{n}Q_{n.i.j}X_j\right)^2$$
$$= O_{P_\theta}(n^{-2/3}).$$

This yields $T_{n.3} = o_{P_\theta}(1)$ in view of the Cauchy-Schwarz inequality. Next, Eqns (6.3), (6.4) and (6.6) yield

$$\frac{1}{\sqrt{n}}\sum_{j=3}^{n}U_{n.j}X_{j-2} = O_{P_\theta}(1) \qquad \text{and} \qquad \frac{1}{\sqrt{n}}\sum_{j=3}^{n}\gamma_{n.j}X_{j-2} = O_{P_\theta}(n^{-1/3}).$$

Use the fact that Q_n is idempotent and symmetric, the Cauchy-Schwarz inequality, Eqns (6.5) and (6.7) to conclude that

$$\left|\frac{1}{n}\sum_{j=3}^{n}\sum_{k=1}^{n}Q_{n.j.k}X_kX_{j-2}\right|^2 \le \frac{1}{n}\sum_{i=1}^{n}\left(\sum_{j=3}^{n}Q_{n.i.j}X_{j-2}\right)^2\frac{1}{n}\sum_{i=1}^{n}\left(\sum_{k=1}^{n}Q_{n.i.k}X_k\right)^2$$
$$= O_{P_{\theta_n}}\left(\left(\frac{1}{n}\sum_{i=1}^{n}\sum_{j=1}^{n}Q_{n.i.j}^2\right)^2\right) = O_{P_\theta}(n^{-4/3}).$$

Combining the above yields $T_{n.1} = o_{P_\theta}(1)$. Similarly, one verifies $T_{n.2} = o_{P_\theta}(1)$. This establishes Eq. (6.1) and completes the proof.

7. A PRELIMINARY ESTIMATE OF μ

In this section we shall describe a preliminary estimate of the parameter μ. Let $\langle \tilde{\beta}_n \rangle$ and $\langle \tilde{\rho}_n \rangle$ denote $\sqrt{n}$-consistent estimates of β and ρ and let $\langle \hat{\gamma}_n \rangle$ be the estimate defined in Sec. 4 with $K_n \sim n^{1/3}$. We do not require that $\tilde{\beta}_n$ and $\tilde{\rho}_n$ are discretized. With $\hat{X}_{n,j} = Y_j - \tilde{\beta}_n U_j - \hat{\gamma}(V_j)$ as in the previous section, set

$$\hat{\alpha}_n = \frac{\sum_{j=3}^n (\hat{X}_{n,j} - \tilde{\rho}_n \hat{X}_{n,j-1})(\hat{X}_{n,j-1} - \tilde{\rho}_n \hat{X}_{n,j-2})}{\sum_{j=3}^n (\hat{X}_{n,j} - \tilde{\rho}_n \hat{X}_{n,j-1})^2}.$$

As estimate $\bar{\mu}_n$ we take the solution to the equation

$$\frac{-\bar{\mu}_n}{1 + \bar{\mu}_n^2} = \hat{\alpha}_n$$

closest to 0. Then $\langle \bar{\mu}_n \rangle$ is $\sqrt{n}$-consistent.

As in the previous section, we can prove that

$$\frac{1}{\sqrt{n}} \sum_{j=3}^n \hat{X}_{n,j}^2 - \frac{1}{\sqrt{n}} \sum_{j=3}^n X_j^2 = o_{P_\theta}(1).$$

This, Eqns (6.1), (6.2) and the $\sqrt{n}$-consistency of $\tilde{\rho}_n$ can now be used to show that

$$\hat{\alpha}_n = \alpha_n + O_{P_\theta}(n^{-1/2}), \tag{7.1}$$

where

$$\alpha_n = \frac{(1/n) \sum_{j=3}^n (X_j - \rho X_{j-1})(X_{j-1} - \rho X_{j-2})}{(1/n) \sum_{j=3}^n (X_j - \rho X_{j-1})^2} \tag{7.2}$$

is the usual estimate of the autocorrelation coefficient of lag 1 based on the MA(1) process $\{X_t - \rho X_{t-1} : t = 1, \ldots, n\}$. Note that the numerator on the right-hand-side of Eq. (7.2) can be written as

$$\frac{1}{n} \sum_{j=3}^n (\eta_j - \mu\eta_{j-1})(\eta_{j-1} - \mu\eta_{j-2}) = -\mu \frac{1}{n} \sum_{j=3}^n \eta_{j-1}^2 + O_{P_\theta}(n^{-1/2}),$$

while the denominator can be expressed as

$$\frac{1}{n} \sum_{j=3}^n (\eta_j - \mu\eta_{j-1})^2 = (1 + \mu^2) \frac{1}{n} \sum_{j=3}^n \eta_j^2 + O_{P_\theta}(n^{-1/2}).$$

This shows that

$$\alpha_n = \frac{-\mu}{1 + \mu^2} + O_{P_\theta}(n^{-1/2}). \tag{7.3}$$

The $\sqrt{n}$-consistency of $\langle \bar{\mu}_n \rangle$ follows now from Eqns (7.1), (7.3) and the δ-method.

8. PROOF OF THEOREM 4.1

Let $\theta_n = \langle (\rho_n, \mu_n, \beta_n) \rangle$ be a local sequence for θ. Set $\eta_{n,0} = \rho_n Y_0 - \mu_n Y_{-1}$,

$$X_{n,j} = Y_j - \beta_n U_j - \gamma(V_j)$$

and

$$\eta_{n,j} = Y_j - R_j(\theta_n, \gamma) - \mu_n^{j-1} \eta_{n,0}, \qquad j = 1, \ldots, n.$$

Under the measure P_{θ_n}, $(U_1, V_1), \ldots, (U_n, V_n), \eta_{n,0}, \eta_{n,1}, \ldots, \eta_{n,n}$ are independent, the random variables $X_{n,1}, \ldots, X_{n,n}$ are from an ARMA(1,1) process with parameter (ρ_n, μ_n) and innovations $\eta_{n,1}, \ldots, \eta_{n,n}$. Moreover, we have the representations

$$X_{n,i} = \sum_{r=1}^{i} \psi_{n,i-r} \eta_{n,r} + \rho_n^{i-1} \eta_{n,0}, \qquad i = 1, \ldots, n, \tag{8.1}$$

$$\eta_{n,i} = \sum_{r=1}^{i} \phi_{n,i-r} X_{n,r} - \mu^{i-1} \eta_{n,0}, \qquad i = 1, \ldots, n, \tag{8.2}$$

where $\psi_{n,0} = \phi_{n,0} = 1$,

$$\psi_{n,r} = (\rho_n - \mu_n)\rho_n^{r-1} \quad \text{and} \quad \phi_{n,r} = (\mu_n - \rho_n)\mu_n^{r-1}, \qquad r = 1, \ldots, n-1.$$

Next, let

$$\cdot \tilde{\eta}_{n,i} = Y_j - R_i(\theta_n, \tilde{\gamma}_n) = \sum_{r=1}^{i} \phi_{n,i-r}(Y_r - \beta_n U_r - \tilde{\gamma}_n(V_r)), \qquad i = 1, \ldots, n,$$

where $\tilde{\gamma}_n$ is defined as $\hat{\gamma}_n$ but with $\tilde{\beta}_n$ replaced by β_n so that

$$\tilde{\gamma}_n(V_i) = \sum_{j=1}^{n} Q_{n,i,j}(Y_j - \beta_n U_j) = \sum_{j=1}^{n} Q_{n,i,j} X_{n,j} + \sum_{j=1}^{n} Q_{n,i,j} \gamma(V_j).$$

Using this and the identities in Eqns (8.1) and (8.2), we find that

$$r_{n,i} := \tilde{\eta}_{n,i} - \eta_{n,i} = c_{n,i} \eta_{n,0} + \sum_{r=1}^{i} \phi_{n,i-r} \gamma_{n,r} - \sum_{j=1}^{n} M_{n,i,j} \eta_{n,j},$$

with $\gamma_{n,i} = \gamma(V_i) - \sum_{j=1}^{n} Q_{n,i,j} \gamma(V_j)$ as in Sec. 6,

$$c_{n,i} = \mu_n^{i-1} - \sum_{r=1}^{i} \phi_{n,i-r} \sum_{b=1}^{n} Q_{n,r,b} \rho_n^{b-1}$$

and

$$M_{n,i,j} = \sum_{r=1}^{i} \phi_{n,i-r} \sum_{b=j}^{n} Q_{n,r,b} \psi_{n,b-j}.$$

Of course, $M_{n,i,j}$ is just the (i,j)-entry of the matrix $\Phi_n Q_n \Psi_n$, where Φ_n and Ψ_n are the lower triangular matrices with entries $\Phi_{n,i,j} = \phi_{n,i-j}$ and $\Psi_{n,i,j} = \psi_{n,i-j}$ for $i \geq j$. The operator norm $\|\Phi_n\|_0 = \sup\{\|\Phi_n x\| : \|x\| = 1\}$ of the matrix Φ_n is bounded by $\sum_{r=0}^{n-1} |\phi_{n,k}|$ so that $\|\Phi_n\|_0 = O(1)$. Similarly, $\|\Psi_n\|_0 = O(1)$. From this we obtain that $\operatorname{trace}(\Phi_n Q_n \Psi_n \Psi_n^T Q_n \Phi_n) \leq \|\Psi_n\|_0^2 \operatorname{trace}(\Phi_n Q_n^2 \Phi_n^T) \leq \|\Psi_n\|_0^2 \operatorname{trace}(Q_n \Phi_n^T \Phi_n Q_n) \leq \|\Psi_n\|_0^2 \|\Phi_n\|_0^2 \operatorname{trace}(Q_n)$ and thus

$$\sum_{i=1}^{n} \sum_{j=1}^{n} M_{n,i,j}^2 = O_p(n^{1/3}).$$

Furthermore, since $|M_{n,i,j}| \leq \sum_{r=0}^{n-1} |\phi_{n,r}| \sum_{r=0}^{n-1} |\psi_{n,r}| \max_{i,j} |Q_{n,i,j}|$, we obtain from Eq. (4.2) that

$$\sum_{i=1}^{n} M_{n,i,i}^2 \leq n \max_{1 \leq i,j \leq n} |M_{n,i,j}| = O_p(n^{1/3}).$$

Next, use $\|\Phi_n\|_0 = O(1)$, Eq. (4.2), $\sum_{i=0}^{n-1} |\phi_{n,i}| + |\rho_n^{i-1}| = O(1)$, Eq. (6.3) and the asymptotic behaviour of d_n to conclude that

$$\frac{1}{N_n} \sum_{j=d_n}^{n} c_{n,i}^2 = O_p(n^{-2/3}).$$

Combining the above shows that with $\hat{r}_{n,i,j} = \hat{r}_{n,i} - M_{n,i,j} \eta_{n,j}$

$$\sum_{j=d_n}^{n} E_{\theta_n}((r_{n,j} - \hat{r}_{n,j,j})^2 | \mathbf{Z}_n) = O_{P_{\theta_n}}(n^{-1/3}), \tag{8.3}$$

$$\frac{1}{N_n} \sum_{j=d_n}^{n} E_{\theta_n}(\hat{r}_{n,j}^2 | \mathbf{Z}_n) = O_{P_{\theta_n}}(n^{-2/3}), \tag{8.4}$$

$$\frac{1}{N_n} \sum_{j=d_n}^{n} \sum_{j=1}^{n} E_{\theta_n}((\hat{r}_{n,i} - \hat{r}_{n,i,j})^2 | \mathbf{Z}_n) = O_{P_{\theta_n}}(n^{-2/3}), \tag{8.5}$$

where $\mathbf{Z}_n = (U_1, V_1, \ldots, U_n, V_n \eta_{n,0})$. This is the analogue of Condition R in Schick (1993) with weights $w_{n,j} = 0$ for $j < d_n$ and 1 otherwise, with his r_n taken to be 0 and his Y_j replaced by our $\eta_{n,j}$. Arguing as in the proof of his Theorem 5.3 we can now show that

$$\frac{1}{N_n} \sum_{j=d_n}^{n} (\tilde{\ell}_n(\tilde{\eta}_{n,j}) - \ell(\eta_{n,j}))^2 = o_{P_{\theta_n}}(1), \tag{8.6}$$

with $\tilde{\ell}_n$ defined as in Sec. 5 but with $\tilde{\eta}_{n,j}$ in place of $\hat{\eta}_{n,j} = Y_j - R_j(\bar{\theta}_n, \hat{\gamma}_n)$. Arguing as in this proof also yields

$$\frac{1}{N_n} \sum_{j=d_n}^{n} (\hat{s}_{n,j} - \hat{s}_{n,.}) \tilde{\ell}_n(\tilde{\eta}_{n,j}) - s_{n,j}\ell(\eta_{n,j}) = O_{P_{\theta_n}}(n^{-1/2})$$

whenever $\hat{s}_{n,j}$ and $s_{n,j}$ are random variables based on $(\eta_{n,1}, \ldots, \eta_{n,n}, \mathbf{Z}_n)$ and $\mathbf{Z}_n$, respectively, which satisfy $E_{\theta_n}[s_{nj}] = 0$,

$$\frac{1}{N_n} \sum_{j=d_n}^{n} (\hat{s}_{n,j} - \hat{s}_{n,.}) \hat{r}_{n,j} = o_{P_{\theta_n}}(n^{-1/2}) \tag{T}$$

and the following conditions for some random variables $\tilde{s}_{n,j}$ and $\tilde{s}_{n,j,i}$ that do not depend on $\eta_{n,j}$ and $(\eta_{n,j}, \eta_{n,i})$, respectively, and some positive C:

$$\max_{d_n \leq j \leq n} |\tilde{s}_{n,j}| \leq C, \tag{S0}$$

$$\frac{1}{N_n} \sum_{j=d_n}^{n} E_{\theta_n}(|\tilde{s}_{n,j} - s_{n,j}|^2 |\mathbf{Z}_n) = o_{P_{\theta_n}}(1), \tag{S1}$$

$$\sum_{j=d_n}^{n} |\hat{s}_{n,j} - \tilde{s}_{n,j}|^2 = o_{P_{\theta_n}}(1), \tag{S2}$$

$$\frac{1}{N_n} \sum_{j=d_n}^{n} \sum_{i=1}^{n} E_{\theta_n}(|\tilde{s}_{n,j} - \tilde{s}_{n,j,i}|^2 |\mathbf{Z}_n) = o_{P_{\theta_n}}(a_n^2). \tag{S3}$$

For the choices $\hat{s}_{n,j} = \hat{B}_{n,j} = B_j(\theta_n, U - \hat{w}_n(V))$ and $s_{n,j} = B_{n,j} = B_j(\theta_n, U - w_n(V))$, Eqns (S0)–(S3) are easily verified. Let us now verify Eq. (T). In view of Eqns (5.2), (8.4) and $\|\Phi_n\|_0 = O(1)$, one finds

$$\frac{1}{N_n} \sum_{j=d_n}^{n} \hat{s}_{n,.} \hat{r}_{n,j} = o_{P_{\theta_n}}(n^{-1/2})$$

and

$$\frac{1}{N_n} \sum_{j=d_n}^{n} \sum_{i=1}^{j} \phi_{n,j-i}(w(V_i) - \hat{w}_n(V_i)) \hat{r}_{n,j} = o_{P_{\theta_n}}(n^{-1/2}).$$

Thus Eq. (T) follows if we show that

$$T_n = \frac{1}{N_n} \sum_{j=d_n}^{n} \sum_{i=1}^{j} U_i^* \phi_{n,j-i} \hat{r}_{n,j} = o_{P_{\theta_n}}(n^{-1/2}).$$

But this follows as

$$E_{\theta_n}(T_n^2|\eta_n, \ldots, \eta_{n,n}, V_1, \ldots, V_n) = \frac{1}{N_n^2} \sum_{i=1}^n \Lambda \left(\sum_{j=d_n \vee i}^n \phi_{n,j-i} \hat{r}_{n,j} \right)^2$$

$$= O_{P_{\theta_n}}(n^{-5/3}).$$

The above show that

$$\frac{1}{N_n} \sum_{j=d_n}^n (\tilde{B}_{n,j} - \tilde{B}_{n,\cdot}) \tilde{\ell}_n(\tilde{\eta}_{n,j}) - B_{n,j} \ell(\eta_{n,j}) = o_{P_{\theta_n}}(n^{-1/2}). \tag{8.7}$$

Now let $\hat{s}_{n,j} = \sum_{i=1}^{j=1} \alpha_{n,j-i}(\gamma(V_i) - \tilde{\gamma}_n(V_i))$ and $s_{n,i} = 0$, where the numbers $\alpha_{n,i}$ are such that $\sum_{i=1}^n |\alpha_{n,i}| = O(1)$. Then one can again verify Eq. (T) and Eqns (S0)–(S3). Taking $\alpha_{n,i} = \mu_n^{-1}$ and then $\alpha_{n,i} = \rho_n(i-1)\mu_n^{i-2} - i\mu_n^{i-1}$, one gets

$$\frac{1}{N_n} \sum_{j=d_n}^n (\tilde{A}_{n,j} - A_{n,j} - \tilde{A}_{n,\cdot} + A_{n,\cdot}) \tilde{\ell}_n(\hat{\eta}_{n,j}) = O_{P_{\theta_n}}(n^{-1/2}) \tag{8.8}$$

with $\tilde{A}_{n,j} = A_j(\theta_n, \tilde{\gamma}_n)$ and $A_{n,j} = A_j(\theta_n, \gamma)$.

Finally, let us show that

$$\frac{1}{N_n} \sum_{j=d_n}^n (A_{n,j} - A_{n,\cdot}) \tilde{\ell}_n(\tilde{\eta}_{n,j}) - A_{n,j} \ell(\eta_{n,j}) = o_{P_{\theta_n}}(n^{-1/2}). \tag{8.9}$$

In view of Eq. (8.6) and the fact that $A_{n,\cdot} = O_{P_{\theta_n}}(n^{-1/2})$, one needs only show that

$$\frac{1}{N_n} \sum_{j=d_n}^n A_{n,j}(\tilde{\ell}_n(\tilde{\eta}_{n,j}) - \ell(\eta_{n,j})) = o_{P_{\theta_n}}(n^{-1/2}).$$

Since each coordinate of $A_{n,j}$ can be written as $\sum_{k=0}^{j=1} \eta_{n,k} \alpha_{n,j-k}$ where $\sum_{k=1}^n |\alpha_{n,k}| = O(1)$, it suffices to prove

$$\frac{1}{n} \sum_{i=1}^{n-1} \eta_{n,i} L_{n,i} = o_{P_{\theta_n}}(n^{-1/2}), \tag{8.10}$$

where

$$L_{n,i} = \sum_{j=(i+1) \vee d_n}^n \alpha_{n,j-i}(\tilde{\ell}_n(\tilde{\eta}_{n,j}) - \ell(\eta_{n,j})).$$

Using the definition of $L_{n,j}$ and Lemmas 10.1, 10.2 and 10.5 of Schick (1993) one finds that there are random variables $\tilde{L}_{n,j}$ and $\tilde{L}_{n,j,i}$ that are not based on $\eta_{n,j}$ and $(\eta_{n,j}, \eta_{n,i})$, respectively, such that

$$\sum_{j=1}^{n-1} E_{\theta_n}(|L_{n,j} - \tilde{L}_{n,j})|^2|\mathbf{Z}_n) \xrightarrow{P_{\theta_n}} 0,$$

$$\frac{1}{n}\sum_{j=1}^{n-1} E_{\theta_n}(\tilde{L}_{n,j}^2|\mathbf{Z}_n) \xrightarrow{P_{\theta_n}} 0,$$

$$\frac{1}{n}\sum_{i \neq j} E_{\theta_n}(|\tilde{L}_{n,j} - \tilde{L}_{n,j,i})|^2|\mathbf{Z}_n) \xrightarrow{P_{\theta_n}} 0.$$

The desired result in Eq. (8.10) now follows from these and an argument similar to the one given in Lemma 10.3 of Schick (1993) with $\hat{h}_{n,j}(y) = yL_{n,j}$, $\tilde{h}_{n,j}(y) = y\tilde{L}_{n,j}$ and $r_n = 0$.

Combining the above shows that

$$\frac{1}{\sqrt{n_n}}\sum_{j=d_n}^{n} \begin{pmatrix} \tilde{A}_j - \tilde{A}_{n,\cdot} \\ \tilde{B}_{n,j} - \tilde{B}_{n,\cdot} \end{pmatrix} \tilde{\ell}_n(\tilde{\eta}_{n,j}) - S_n^*(\theta_n) \xrightarrow{P_{\theta_n}} 0.$$

It follows from Eqns (8.6), (8.4) and the results in Sec. 5 that

$$\tilde{J}_n = \frac{1}{N_n}\sum_{j=d_n}^{n} \tilde{\ell}_n^2(\tilde{\eta}_{n,j}) \xrightarrow{P_{\theta_n}} 0, \qquad \tilde{\sigma}_n^2 = \frac{1}{N_n}\sum_{j=d_n}^{n} \tilde{\eta}_{n,j}^2 \xrightarrow{P_{\theta_n}} \sigma^2, \qquad \hat{\Lambda}_n \xrightarrow{P_{\theta_n}} \Lambda.$$

This shows that

$$\tilde{I}_{*,n} = \tilde{J}_n H(\rho_n, \mu_n, \tilde{\sigma}_n^2, \hat{\Lambda}_n) \xrightarrow{P_{\theta_n}} I_*.$$

It follows from Eq. (2.14) applied with a least favorable path that

$$S_n^*(\theta_n) - S_n(\theta) - I_*(\theta_n - \theta) \xrightarrow{P_{\theta_n}} 0.$$

Now use the contiguity of $\langle F_{n,\theta_n} \rangle$ and $\langle F_{n,\theta} \rangle$ (see Remark 2.5) to conclude from the above that the estimate

$$\theta_n + \frac{1}{N_n}\sum_{j=d_n}^{n} \tilde{I}_{*,n}^{-1} \begin{pmatrix} \tilde{A}_{n,j} - \tilde{A}_{n,\cdot} \\ \tilde{B}_{n,j} - \tilde{B}_{n,\cdot} \end{pmatrix} \tilde{\ell}_n(\tilde{\eta}_{n,j})$$

satisfies Eq. (3.2). This gives the desired result if the preliminary estimate is replaced by a local sequence. Thus it holds for every preliminary discretized and $\sqrt{n}$-consistent estimate $\langle \tilde{\theta}_n \rangle$.

Acknowledgement

I would like to thank my colleague Hira Koul for helpful comments on an earlier version of this article.

REFERENCES

Akritas, M. G. and Johnson, R. A. (1980). Efficiencies of tests and estimators of p-order autoregressive processes when the error distribution is nonnormal. *Ann. Inst. Statist. Math., 34*: 579-589.

Begun, J., Huang, W-M. and Wellner, J. (1983): Information and asymptotic efficiency in parametric-nonparametric models. *Ann. Statist., 11*: 432–452.

Bickel, P. J. (1982): On adaptive estimation. *Ann. Statist., 10*: 647–671.

Bickel, P. J., Klaassen, C. A. J, Ritov, Y. and Wellner, J. (1993): Efficient and Adaptive Estimation for Semiparametric Models. Baltimore, Johns Hopkins University Press.

Chen, H. (1988). Convergence rates for parametric components in a partly linear model. *Ann. Statist., 16*: 136–146.

Chen, H. and Shiau, H. J. (1991): A two-stage spline smoothing method for partially linear models. *J. Statist. Plann. Inference, 27*: 187–201.

Choi, S., Hall, W. J. and Schick, A. (1996). Asymptotically uniformly most powerful tests in parametric and semiparametric models. *Ann. Statist., 24*: 841–862.

Cuzick, J. (1992a): Semiparametric additive regression. *J. R. Statist. Soc. Ser. B., 54*: 831–843.

Cuzick, J. (1992b): Efficient estimates in semiparametric additive regression models with unknown error distribution. *Ann. Statist., 20*: 1129–1136.

De Boor, C. (1978): A practical Guide to Splines. New York, Springer.

Drost. F. C, Klaassen, C. A. J. and Werker, B. J. M. (1997): Adaptive estimation in time-series models. *Ann. Statist., 25*: 786–817.

Engle, R. F., Granger, C. W. J., Rice, J. and Weiss, A. (1986): Semiparametric estimates of the relation between weather and electricity sales. *J. Amer. Statist. Assoc., 81*: 310–320.

Fabian, V. and Hannan, J. (1982): On estimation and adaptive estimation for locally asymptotically normal families. *Z. Wahr. verw. Gebiete, 59*: 459–478.

Green, P., Jennison, C. and Seheult, A. (1985): Analysis of field experiments by least squares smoothing. *J. R. Statist. Soc. Ser. B., 47*: 299–315.

Hájek, J. and Šidák, Z. (1967): Theory of Rank Tests. New York, Academic Press.

Heckman, N. E. (1986): Spline smoothing in partly linear models. *J. R. Statist. Soc. Ser. B.*, *48*: 244–248.

Heckman, N. E. (1988): Minimax estimates in a semiparametric model. *J. Amer. Statist. Assoc.*, *83*: 1090–1096.

Jeganathan, P. (1995). Some aspects of asymptotic theory with applications to time series models. *Econometric Theory, 11*: 818–887.

Koul, H. L. and Schick, A. (1997). Efficient estimation in nonlinear autoregressive time series models. *Bernoulli, 3*: 247–277.

Klaassen, C. A. J. (1987). Consistent estimation of the influence function of locally asymptotically linear estimators. *Ann. Statist.*, *15*: 1548–1562

Kreiss, J. P. (1987a): On adaptive estimation in stationary ARMA processes. *Ann. Statist.*, *15*: 112–133.

Kreiss, J. P. (1987b): On adaptive estimation in autoregressive models when there are nuisance functions. *Statist. & Decisions, 15*: 112–133, 1987.

Le Cam, L. (1960): Locally asymptotically normal families of distributions. Univ. Calif. Publ. Statist., *3*: 37–98.

Pfanzagl, J. and Wefelmeyer, W. (1982). Contributions to a general asymptotic statistical theory. *Lecture Notes in Statistics, 13*: Springer-Verlag.

Robinson, P. M. (1988): Root-n-consistent parametric regression. *Econometrica, 56*: 931–954.

Schick, A. (1986): On asymptotically efficient estimation in semi-parametric models. *Ann. Statist.*, *14*: 1139–1151.

Schick, A. (1987): A note on the construction of asymptotically linear estimators. *J. Statist. Plann. Inference, 16*: 89–105. Correction *22*: 269–270, 1989.

Schick, A. (1988): On estimation in LAMN families when there are nuisance parameters present, *Sankhya, Ser. A, 50*: 249–268.

Schick, A. (1993): On efficient estimation in regression models. *Ann. Statist., 21*: 1486–1521. Correction and Addendum *23*: 1862–1863.

Schick, A. (1994): Estimation of the autocorrelation coefficient in the presence of aregression trend. *Statist. Probab. Letters, 21*: 371–380.

Schick, A. (1996a): Efficient estimation in a semiparametric additive regression model with autoregressive errors. *Stoch. Proc. Appl., 61*: 339–361.

Schick, A. (1996b): An adaptive estimator of the autocorrelation coefficient in regression models with autoregressive errors. To appear in *J. Time Ser. Anal.*

Schick, A. (1996c): Root-*n* consistent estimation in partly linear regression models. *Statist. Probab. Letters, 28*: 353–358.

Schumaker, L. L. (1981): Spline Functions: Basic Theory. New York, Wiley.

Speckman, P. (1988). Kernel smoothing in partial linear models. *J. R. Statist. Soc. Ser. B., 50*: 413–436.

Stone, C. J. (1985). Additive regression and other nonparametric models. *Ann. Statist., 3*: 267–284.

Swensen, A. R. (1985). The asymptotic distribution of the likelihood ratio for autoregressive time series with a regression trend. *J. Multivariate Anal., 16*: 54–70.

Truong, Y. (1991): Nonparametric curve estimation with time series errors. *J. Statist. Plann. Inference, 28*: 167–183.

Truong, Y. and Stone, C. J. (1994): Semi-parametric time series regression. *J. Time Ser. Anal., 15*: 405–428.

Wahba, G. (1984): Cross validated spline methods for the estimation of multivariate functions from data on functionals, In: H. A. David and H. T. David, eds., Statistics: An Appraisal, Proceedings 50th Anniversary Conference Iowa State Statistical Laboratory, pp. 205–235.

12
Efficient Estimation in Markov Chain Models: An Introduction

WOLFGANG WEFELMEYER   University of Siegen, Siegen, Germany

1. INTRODUCTION

We outline the theory of efficient estimation for semiparametric Markov chain models, and illustrate in a number of simple cases how the theory can be used to determine lower bounds for the asymptotic variance of estimators and to construct efficient estimators. In particular, we consider estimation of stationary distributions of Markov chains, of autoregression parameters and innovation distributions in AR- and ARCH-models and more general time series, and of parameters in quasi-likelihood models.

The basic example of a time series is the autoregressive process

$$X_i = \alpha X_{i-1} + \varepsilon_i,$$

where the innovations ε_i are independent and identically distributed with mean zero and finite variance. For $|\alpha| < 1$ this is an ergodic Markov chain. We may want to estimate the autoregression parameter α, the distribution of the ε_i, or the stationary distribution of the X_i. The classical estimator for α is the least squares estimator $\hat{\alpha} = \sum X_{i-1} X_i / \sum X_{i-1}^2$. The distribution function of the innovations can be estimated by the empirical distribution function $(1/n) \sum 1(\hat{\varepsilon}_i \leq r)$ based on the estimated innovations $\hat{\varepsilon}_i = X_i - \hat{\alpha} X_{i-1}$. The obvious estimator for the stationary distribution function is the empirical distribution function $(1/n) \sum 1(X_i \leq r)$.

None of these three estimators is efficient. How far are they from being efficient? How can one find better estimators? In the following Sections we outline how one determines lower bounds for the asymptotic variance of estimators in general semiparametric Markov chain models. Then we char-

427

acterize efficient estimators and indicate how one uses the characterization to construct such estimators in specific cases.

Our approach requires the model to be 'locally asymptotically normal'. In Sec. 2 we introduce this concept for general Markov chain models with possibly infinite-dimensional parameter and illustrate it with a few simple examples. In Sec. 3 we consider the problem of estimating a one-dimensional function of the parameter and determine an optimal estimator within a simple class of estimators: the 'asymptotically linear' and 'regular' ones. Section 4 considers martingale estimating equations and indicates when they lead to asymptotically linear and regular estimators. Section 5 shows that the optimal asymptotically linear and regular estimator is already 'efficient', i.e., optimal among all regular estimators. The presentation is rigorous in Sections 2, 3 and 5, and heuristic in the others.

In Sec. 6 to 9 we treat examples: nonparametric models in Sec. 6, autoregressive models in Sec. 7, quasi-likelihood models in Sec. 8. Section 9 briefly describes some other types of Markov chain models that are amenable to our approach. To keep the exposition simple, we restrict attention to first-order Markov chains and estimation of one-dimensional functions of the parameter. The extension to higher-order Markov chains and higher-dimensional functions is straightforward. Extensions to infinite-dimensional functions and to other types of processes are also possible.

A basic reference for convergence of Markov chains is Meyn and Tweedie (1994). The results on efficient estimation in Sec. 2, 3 and 5 generalize those for independent and identically distributed observations, for which we refer to Bickel et al. (1993).

2. LOCAL ASYMPTOTIC NORMALITY FOR MARKOV CHAINS

In this Section we give conditions under which a general Markov chain model with possibly infinite-dimensional parameter is locally asymptotically normal. This means that if we fix a parameter and rescale the parametrization approximately, the model is approximated by a Gaussian shift model as the sample size increases. The setting will be rather abstract in order to cover all applications in the following Sections. Before we describe the general setting, we begin with a simple case, the full nonparametric model, in which the transition distribution itself plays the role of a parameter.

Let $X_0, \ldots, X_n$ be realizations of a homogeneous Markov chain on some measurable state space $(E, \mathcal{E})$. For notational simplicity we will always assume that the chain starts at a fixed value $X_0 = x_0$, but we continue to write X_0.

Let us first look at the full nonparametric model, in which no structural assumptions are made on the transition distribution. We 'parametrize' the model by the transition distribution and view the 'parameter space', the family of all transition distributions, as a manifold.

We fix a transition distribution $Q(x, dy)$ and assume that under it the chain is ergodic with invariant distribution $\pi(dx)$. We write $\pi \otimes Q(dx, dy) = \pi(dx)Q(x, dy)$ and $Q(x, f) = \int Q(x, dy)f(x, y)$ and introduce the Hilbert space

$$H = \{h \in L_2(\pi \otimes Q) : Q(x, h) = 0 \quad \text{for } x \in E\}.$$

The manifold of transition distributions is smooth with tangent space H in the following sense. For $h \in H$ bounded,

$$Q_{nh}(x, dy) = Q(x, dy)(1 + n^{-1/2}h(x, y)) \tag{2.1}$$

defines a transition distribution with derivative h. We will find it convenient to consider sequences indexed by $n^{-1/2}$.

If h is not bounded, we must modify the definition in Eq. (2.1) to obtain a true conditional distribution. We will see later, however, that it suffices to restrict attention to bounded h because they are dense in H.

REMARK. To interpret H as a tangent space, we identify a transition distribution $R(x, dy)$ with $(dR/dQ)(x, dy)^{1/2} - 1$. This function is automatically in $L_2(\pi \otimes Q)$ and is 0 for $R = Q$. The identification implies in particular that we ignore all transition distributions $R(x, dy)$ which are not absolutely continuous with respect to $Q(x, dy)$. We will see later that our 'tangent space' is still big enough for our purposes. $\square$

Let P denote the joint law of $X_0, \ldots, X_n$ if the chain is generated by Q, and P_{nh} the corresponding law if Q_{nh} is true. By ergodicity,

$$\frac{1}{n}\sum_{i=1}^{n} h(X_{i-1}, X_i)^2 \to \pi \otimes Qh^2. \tag{2.2}$$

From this result and a Taylor expansion we obtain a stochastic approximation of the log-likelihood,

$$\log \frac{dP_{nh}}{dP} = \sum_{i=1}^{n} \log \frac{dQ_{nh}}{dQ}(X_{i-1}, X_i)$$

$$= n^{-1/2} \sum_{i=1}^{n} h(X_{i-1}, X_i) - \tfrac{1}{2}\pi \otimes Qh^2 + o_P(1). \tag{2.3}$$

By a martingale central limit theorem,

$$n^{-1/2} \sum_{i=1}^{n} h(X_{i-1}, X_i) \Rightarrow (\pi \otimes Qh^2)^{1/2} N \qquad \text{under } P. \tag{2.4}$$

Here N denotes a random variable with standard normal distribution. Relations in Eqns (2.3) and (2.4) constitute a nonparametric version of local asymptotic normality. The reason for choosing the rate $n^{-1/2}$ in Eq. (2.1) is now apparent: with this choice, the likelihood ratio has a nondegenerate limit.

Let us now consider submodels of the full nonparametric model. Such submodels may be described in different ways, three of which will appear in the following Sections:

1. There is a restriction on the transition distribution, say the chain is reversible (Example 5 in Sec. 6), or $Q(x, dy)$ is symmetric about x.
2. There is a parametric family of restrictions on the transition distribution, say on the conditional mean,

$$\int Q(x, dy)y = \alpha x,$$

   and we may want to estimate that parameter. See Sec. 8, and also Example 10 in Section 9.
3. The transition distribution is parametrized (with a possibly infinite-dimensional parameter), as in the AR(1)-model,

$$Q(x, dy) = p(y - \alpha x)\, dy,$$

   with α unknown and p a mean 0 density which is possibly also unknown. See Example 3 below and Sec. 7, and also Example 11 in Sec. 9.

For the first two types of model we will continue using the transition distribution as a parameter. This can also be done for the third type, but is usually more convenient, and certainly more common, to parametrize by the obvious parameters.

Let Θ be a possibly infinite-dimensional set, the parameter space, and Q_ϑ, $\vartheta \in \Theta$, a family of transition distributions on the state space $(E, \mathcal{E})$. We view Θ as a manifold, fix $\vartheta \in \Theta$ so that $Q = Q_\vartheta$ is ergodic with invariant distribution $\pi = \pi_\vartheta$, and assume that Θ is smooth in the following sense. There is a linear space K, the tangent space, and a linear map $D : K \to H$, and for each $k \in K$ there is a sequence ϑ_{nk} such that $Q_{nk} = Q_{\vartheta_{nk}}$ is Hellinger differentiable with derivative Dk,

$$\int Q(x, dy) \left(\left(\frac{dQ_{nk}}{dQ}(x, y) \right)^{1/2} - 1 - \tfrac{1}{2} n^{-1/2} (Dk)(x, y) \right)^2 \leq n^{-1} r_n(x), \qquad (2.5)$$

where r_n decreases to 0 pointwise and is π-integrable for large n. This version of Hellinger differentiability is due to Höpfner et al. (1990).

REMARK. Our description of the tangent space omits one important feature: The space K should contain all directions k from which we can approximate ϑ within Θ. We have already mentioned in the previous Remark that this is not always possible. In the applications we simply try to make K as large as we can. In Sec. 3 we will determine a lower bound for the asymptotic variance of estimators of functions $t(\vartheta)$. The bound depends on K. It is the larger the larger K. We will know that K was chosen large enough whenever we can construct an estimator which attains the bound.

□

The operator induces an inner product on K,

$$\langle k, k' \rangle = \pi \otimes Q(Dk \cdot Dk'), \qquad (2.6)$$

and a corresponding norm $\|k\| = \langle k, k \rangle^{1/2}$. Write P_{nk} for the joint law of $X_0, \ldots, X_n$ if Q_{nk} is true. A Taylor expansion now gives local asymptotic normality of the form

$$\log \frac{dP_{nk}}{dP} = n^{-1/2} \sum_{i=1}^{n} (Dk)(X_{i-1}, X_i) - \tfrac{1}{2} \|k\|^2 + o_P(1). \qquad (2.7)$$

As in (2.4),

$$n^{-1/2} \sum_{i=1}^{n} (Dk)(X_{i-1}, X_i) \Rightarrow \|k\| N \qquad \text{under } P. \qquad (2.8)$$

Local asymptotic normality for Markov chains is basically due to Roussas (1965). The nonparametric version in Eqns (2.3) and (2.4) is in Penev (1991). The parametric version may be obtained by modifying Höpfner (1993a) who treats Markov step processes; see also Höpfner (1993b).

The nonparametric model may be considered as a special case of the (infinite-dimensional) parametric model:

EXAMPLE 1. (*Full nonparametric model.*) If the transition distribution Q is completely unknown, parametrize Q by itself, set $K = H$ and define, for bounded $h \in H$, the perturbed transition distribution Q_{nh} through Eq. (2.1). Then Hellinger differentiability in Eq. (2.5) holds with $Dh = h$, and the inner product in Eq. (2.6) is

$$\langle h, h' \rangle = \pi \otimes Q(hh'),$$

the natural inner product on $L_2(\pi \otimes Q)$. $\square$

At the other end of the spectrum are the models with a one-dimensional parameter.

EXAMPLE 2 (*One-dimensional parameter.*) If $\Theta \subset \mathbf{R}$, set $K = \mathbf{R}$ and $\vartheta_{nr} = \vartheta + n^{-1/2}r$. Write ℓ' for the Hellinger derivative of Q_τ at $\tau = \vartheta$,

$$\int Q(x, dy) \left(\left(\frac{dQ_\tau}{dQ}(x,y) \right)^{1/2} - 1 - \tfrac{1}{2}(\tau - \vartheta)\ell'(x,y) \right)^2 \leq (\tau - \vartheta)^2 r_\tau(x)$$

(2.9)

with r_τ decreasing to 0 pointwise for $\tau \to \vartheta$ and π-integrable for τ close to ϑ. Then $Dr = r\ell'$, and the inner product in Eq. (2.6) is

$$\langle r, r' \rangle = rr'I$$

with $I = \pi \otimes Q\ell'$ the Fisher information. The inner product is not the natural one. We have local asymptotic normality of the form

$$\log \frac{dP_{nr}}{dP} = rn^{-1/2} \sum_{i=1}^{n} \ell'(X_{i-1}, X_i) - \tfrac{1}{2}r^2 I + o_P(1).$$

(2.10)

The multivariate version of this example is straightforward. $\square$

Example 3 describes a whole class of models, in particular the following autoregressive model.

EXAMPLE 3 (*AR(1) with normal innovations.*) Let $X_i = \vartheta X_{i-1} + \varepsilon_i$ with ε_i iid and standard normal. The X_i form a Markov chain with transition distribution

$$Q(x, dy) = \varphi(y - \vartheta x) \, dx,$$

where φ is the standard normal density. If $\vartheta < 1$, then the chain is ergodic, and Hellinger differentiability holds with $\ell'(x,y) = x(y - \vartheta x)$. Hence $(Dr)(x,y) = rx(y - \vartheta x)$, the Fisher information is

$$I = \pi \otimes Q\ell'^2 = \int \pi(dx)x^2 \int \varphi(y - \vartheta x) \, dy(y - \vartheta x)^2 = \mathrm{E}\, X^2,$$

and the inner product in Eq. (2.6) is

$$\langle r, r' \rangle = rr' \mathrm{E}\, X^2.$$

We have local asymptotic normality of the form

$$\log \frac{dP_{nr}}{dP} = rn^{-1/2} \sum_{i=1}^{n} X_{i-1}(X_i - \vartheta X_{i-1}) - \tfrac{1}{2} r^2 \mathrm{E}\, X^2 + o_P(1). \quad \square$$

3. ASYMPTOTICALLY LINEAR AND REGULAR ESTIMATORS

As in the previous Section, we consider a general Markov chain model with possibly infinite-dimensional parameter. We now want to estimate a one-dimensional function of the parameter. We introduce a class of estimators, the "asymptotically linear" estimators, whose properties are particularly easy to study. Our aim is to find an optimal estimator in this class. We show that this problem has a meaningful answer if we further restrict attention to "regular" estimators. The restriction to asymptotically linear estimators is justified by the characterization in Eq. (5.2) of efficient estimators.

Let Q_ϑ, $\vartheta \in \Theta$, be a family of transition distributions as in Sec. 2. In addition, we consider a function $t: \Theta \to \mathbf{R}$ which we want to estimate. Fix $\vartheta \in \Theta$. We call an estimator T_n asymptotically linear for t at ϑ with influence function f if $f \in H$ and

$$n^{1/2}(T_n - t(\vartheta)) = n^{-1/2} \sum_{i=1}^{n} f(X_{i-1}, X_i) + o_P(1). \tag{3.1}$$

From the martingale central limit theorem used in Eq. (2.4),

$$n^{1/2}(T_n - t(\vartheta)) \Rightarrow (\pi \otimes Q f^2)^{1/2} N \qquad \text{under } P.$$

Hence the asymptotic variance of an asymptotically linear estimator with influence function f is $\pi \otimes Q f^2$

What is a good asymptotically linear estimator? Without further restricting the class of estimators, this question has no meaningful answer: the asymptotic variance can be made 0 at ϑ by taking $f = 0$. To exclude estimators which have small variance for certain ϑ at the expense of other parameters, we will restrict attention to 'regular' estimators, in the sense that their distribution converges continuously in the parameter to a limit distribution.

For a proper definition, pick a tangent space K and sequences ϑ_{nk} as in Sec. 2. An estimator T_n is called regular for t at ϑ with limit L if

$$n^{1/2}(T_n - t(\vartheta_{nk})) \Rightarrow L \qquad \text{under } P_{nk} \text{ for } k \in K. \tag{3.2}$$

If the function t is smooth, regularity implies a restriction on the influence function of an asymptotically linear estimator. To describe what we mean by smooth, we recall that $Q_{nk} = Q_{\vartheta_{nk}}$ has derivative Dk in the sense of Eq. (2.5). The function t is called differentiable at ϑ with gradient h' if $h' \in H$ and

$$n^{1/2}(t(\vartheta_{nk}) - t(\vartheta)) \to \pi \otimes Q(Dk \cdot h') \qquad \text{for } k \in K. \tag{3.3}$$

The canonical gradient is the projection Dk' of an arbitrary gradient h' into DK. For the canonical gradient, Eq. (3.3) reads

$$n^{1/2}(t(\vartheta_{nk}) - t(\vartheta)) \to \langle k, k' \rangle \qquad \text{for } k \in K. \tag{3.4}$$

Conversely, any gradient h' has the property that $h' - Dk'$ is orthogonal to DK,

$$\pi \otimes Q(Dk \cdot (h' - Dk')) = 0 \qquad \text{for } k \in K. \tag{3.5}$$

In particular, the canonical gradient is the shortest gradient,

$$\|k'\| = \pi \otimes Q(Dk')^2 \leq \pi \otimes Q(h')^2. \tag{3.6}$$

REMARK   The canonical gradient is not always easy to calculate. Sometimes it is easier to find another gradient first (which may, in turn, be canonical in a larger model). One can then try to project that gradient into the tangent space.

Another approach is possible when the tangent space K comes equipped with some inner product $\langle k, k' \rangle_0$, say, as is usually the case. It may then be possible to find the gradient k_0' of t with respect to this inner product,

$$n^{1/2}(t(\vartheta_{nk}) - t(\vartheta)) \to \langle k, k_0' \rangle_0 \qquad \text{for } k \in K.$$

Comparing with Eqn (3.3) and (3.5), we see that the canonical gradient Dk' is now determined by

$$\langle k, k_0' \rangle_0 = \langle Dk, Dk' \rangle \qquad \text{for } k \in K.$$

If D has an adjoint $D^*: H \to K$, we have

$$\langle Dk, Dk' \rangle = \langle k, D^*Dk' \rangle_0.$$

Hence, if D^*D has an inverse, the canonical gradient is Dk' with

$$k' = (D^*D)^{-1}k_0'.$$

It suffices to find an operator C such that

$$\langle Dk, Dk' \rangle = \langle k, Ck' \rangle_0.$$

Then the canonical gradient is $k' = C^{-1}k_0'$. It may happen that C^{-1} is difficult to determine but that C can be written as a perturbation of the identity operator, say $C = I - B$. If B is not too large, C^{-1} may then be written as the von Neumann series $C^{-1} = \sum_{j=0}^{\infty} B^j$. For an application of this approach see Greenwood et al. (1997). $\square$

We can now characterize the regular among the asymptotically linear estimators:

If T_n is asymptotically linear for t at ϑ with influence function f, then T_n is regular if and only if f is a gradient for t at ϑ, and then the limit of T_n is $L = (\pi \otimes Qf^2)^{1/2} N$.

The proof is simple and instructive: By the martingale central limit theorem,

$$n^{-1/2} \sum Dk(X_{i-1}, X_i) \quad \text{and} \quad n^{-1/2} \sum f(X_{i-1}, X_i)$$

are jointly asymptotically normal under P. The means are 0 and the covariance is $\pi \otimes Q(Dk \cdot f)$. Because of asymptotic linearity in Eq. (3.1) and local asymptotic normality in Eqns (2.3), (2.4), LeCam's third lemma (see, e.g., Bickel et al. 1993, p. 503, Lemma 3) implies

$$n^{1/2}(T_n - t(\vartheta)) \Rightarrow (\pi \otimes Qf^2)^{1/2} N + \pi \otimes Q(Dk \cdot f) \qquad \text{under } P_{nk}.$$

With differentiability Eq. (3.13) of t,

$$n^{1/2}(T_n - t(\vartheta_{nk})) \Rightarrow (\pi \otimes Qf^2)^{1/2} N + \pi \otimes Q(Dk \cdot (f - Dk')) \quad \text{under } P_{nk}.$$

Hence T_n is regular for t at ϑ if and only if

$$\pi \otimes Q(Dk \cdot (f - Dk')) = 0 \qquad \text{for } k \in K.$$

By the characterization in Eq. (3.5), the last relation says that f is a gradient for t at ϑ. $\square$

By the characterization of regular estimators, we now have a meaningful answer to the question of which influence function is optimal. Call an influence function regular if the corresponding estimator is regular. By Eq. (3.6):

The optimal regular influence function is the canonical gradient Dk', and the corresponding estimator has asymptotic variance $\|k'\|^2$.

REMARK The canonical gradient is defined as the projection of an arbitrary gradient into the tangent space. When the tangent space is not closed, it suffices to take the projection into its closure. In particular, it suffices to check regularity of an estimator for a dense subset of K. We have already seen in the full nonparametric model that this is convenient: In Eq. (2.1) it suffices to construct Q_{nh} for bounded h. □

EXAMPLE 2 (continued) (*One-dimensional parameter.*) Let Q_ϑ, $\vartheta \in \Theta$, be a one-dimensional parametric model for the transition distributions. We want to estimate the parameter ϑ. By Eq. (3.6) and Example 2 in Sec. 2, the canonical gradient is of the form $r'\ell'$, and r' solves

$$n^{1/2}(\vartheta_{nr} - \vartheta) = r \stackrel{!}{=} \langle r, r' \rangle = rr'I;$$

hence $r' = I^{-1}$. Here $I = \pi \otimes Q\ell'^2$ is the Fisher information. □

EXAMPLE 3 (continued) (*AR(1) with normal innovations.*) Let $X_i = \vartheta X_{i-1} + \varepsilon_i$ with ε_i iid and standard normal, and $\vartheta < 1$. We want to estimate the parameter ϑ. The canonical gradient is $(Dr')(x,y) = r'x(y - \vartheta x)$ with r' determined by Eq. (3.4),

$$n^{1/2}(\vartheta_{nr} - \vartheta) = r \stackrel{!}{=} \langle r, r' \rangle = rr'\mathrm{E}\,X^2 \qquad \text{for } r \in \mathbf{R}.$$

We obtain $r' = (\mathrm{E}\,X^2)^{-1}$ and

$$(Dr')(x,y) = (\mathrm{E}\,X^2)^{-1}x(y - \vartheta x).$$

The optimal regular influence function equals the gradient. Hence the least squares estimator

$$T_n = \frac{\sum X_{i-1} X_i}{\sum X_{i-1}^2} \tag{3.7}$$

is an optimal regular and asymptotically linear estimator. This follows immediately by ergodicity,

$$n^{1/2}(T_n - t(\vartheta)) = \frac{n^{-1/2} \sum X_{i-1}(X_i - \vartheta X_{i-1})}{(1/n) \sum X_{i-1}^2}$$

$$= (\mathrm{E}\,X^2)^{-1} n^{-1/2} \sum X_{i-1}(X_i - \vartheta X_{i-1}) + o_P(1). \tag{3.8}$$

Of course, we expect that the least squares estimator is even efficient in this model because it is the maximum likelihood estimator. See Example 2 (continued) in Section 4. □

4. MARTINGALE ESTIMATING EQUATIONS

In the setting of the previous section, we consider again estimation of a real-valued function $t(\vartheta)$ of a possibly infinite-dimensional parameter. We indicate how asymptotically linear and regular estimators can be obtained as solutions of estimating equations. We refer to Andrews (1994) for a study of estimating equations in general semiparametric time series models. Estimating equations usually cannot lead to efficient estimators unless t is one-one. However, in certain models, efficient estimators may be obtained from "adaptive" versions of estimating equations; see Sec. 8.

Let Q_ϑ, $\vartheta \in \Theta$, be a family of transition distributions and $t: \Theta \to \mathbf{R}$ a function which we want to estimate.

A martingale estimating equation is based on a function $e: \mathbf{R} \times E \times E \to \mathbf{R}$ such that $e_{t(\vartheta)} \in H$ for $\vartheta \in \Theta$. Then $\sum e_{t(\vartheta)}(X_{i-1}, X_i)$ is a martingale, and an estimator T_n of $t(\vartheta)$ is obtained as a solution $t = T_n$ of the equation

$$\sum_{i=1}^{n} e_t(X_{i-1}, X_i) = 0. \tag{4.1}$$

From the point of view of the theory of estimating functions, martingale estimating equations have been studied in very general settings. We refer to Godambe (1985), Godambe and Heyde (1987), Godambe and Thompson (1989), and to the book Godambe (1991). Under appropriate conditions on the function e, a Taylor expansion gives

$$0 = n^{-1/2} \sum_{i=1}^{n} e_{T_n}(X_{i-1}, X_i)$$

$$= n^{-1/2} \sum_{i=1}^{n} e_{t(\vartheta)}(X_{i-1}, X_i) + n^{1/2}(T_n - t(\vartheta)) \frac{1}{n} \sum_{i=1}^{n} e'_{t(\vartheta)}(X_{i-1}, X_i) + o_P(1) \tag{4.2}$$

By ergodicity,

$$\frac{1}{n} \sum_{i=1}^{n} e'_{t(\vartheta)}(X_{i-1}, X_i) \to \pi \otimes Q e'_{t(\vartheta)} \quad (P).$$

We obtain

$$n^{1/2}(T_n - t(\vartheta)) = -(\pi \otimes Q e'_{t(\vartheta)})^{-1} n^{-1/2} \sum_{i=1}^{n} e_{t(\vartheta)}(X_{i-1}, X_i) + o_P(1).$$

Hence T_n is asymptotically linear for t at ϑ with influence function

$$f(x, y) = -(\pi \otimes Q e'_{t(\vartheta)})^{-1} e_{t(\vartheta)}(x, y). \tag{4.3}$$

In particular, the asymptotic variance of T_n is

$$(\pi \otimes Q e'_{t(\vartheta)})^{-2} \pi \otimes Q e^2_{t(\vartheta)}. \tag{4.4}$$

The influence function in Eq. (4.3) is not automatically regular or, equivalently, a gradient. Note, however, that so far we have used the condition $e_{t(\vartheta)} \in H$ only for a fixed parameter ϑ. We indicate now that under appropriate differentiability conditions, the assumption

$$Q_{nk}(x, e_{t(\vartheta_{nk})}) = 0 \qquad \text{for } k \in K \tag{4.5}$$

implies that the influence function is regular. Since t is differentiable in Eq. (3.4) with canonical gradient Dk', we obtain

$$e_{t(\vartheta_{nk})} \approx e_{t(\vartheta)} + n^{-1/2} \langle k, k' \rangle e'_{t(\vartheta)}.$$

Since Q_{nk} is differentiable in Eq. (2.1) with derivative Dk, relation in Eq. (4.5) implies

$$\begin{aligned}
0 &= Q_{nk}(x, e_{t(\vartheta_{nk})}) \\
&\approx Q(x, e_{t(\vartheta)}) + n^{-1/2}(Q(x, Dk \cdot e_{t(\vartheta)}) + \langle k, k' \rangle Q(x, e'_{t(\vartheta)})).
\end{aligned}$$

Taking expectations with respect to π,

$$\pi \otimes Q(Dk \cdot e_{t(\vartheta)}) = -\langle k, k' \rangle \pi \otimes Qe'_{t(\vartheta)} \qquad \text{for } k \in K. \tag{4.6}$$

The characterization in Eq. (3.5) now implies that the influence function in Eq. (4.3) is a gradient.

Using Eq. (4.6) again, now with $k = k'$, we can write the asymptotic variance in Eq. (4.4) of the estimator as

$$\frac{\|k'\|^2 \pi \otimes Qe^2_{t(\vartheta)}}{\pi \otimes Q(Dk \cdot e_{t(\vartheta)})^2}. \tag{4.7}$$

REMARK   By the Schwarz inequality or by Eq. (3.6), the asymptotic variance in Eq. (4.7) is minimal if e is proportional to the canonical gradient Dk', and the minimal asymptotic variance is $\|k'\|^2$. Of course, such a function e cannot, in general, be used in an estimating equation because it will depend on ϑ not only through $t(\vartheta)$. $\square$

EXAMPLE 2 (continued)   (*One-dimensional parameter.*) Let $Q_\vartheta, \vartheta \in \Theta$, be a one-dimensional parametric model for the transition distributions. We want to estimate the parameter ϑ. By Example 2 in Sec. 2, the tangent space is $r\ell'$, $r \in \mathbf{R}$. By Example 2 (continued) in Sec. 3, the canonical gradient of ϑ is $I^{-1}\ell'_\vartheta$, with $I = \pi \otimes Q\ell'^2_\vartheta$ the Fisher information. The maximum likelihood estimator T_n solves the martingale estimating equation

$$\sum \ell'_\vartheta(X_{i-1}, X_i) = 0.$$

By Eq. (4.3), T_n is asymptotically linear with influence function $-(\pi \otimes Q\ell_\vartheta'')^{-1}\ell_\vartheta'$. Differentiating with respect to ϑ under the integral,

$$0 = (Q_\vartheta(x, \ell_\vartheta'))' = Q_\vartheta(x, \ell_\vartheta'') + Q_\vartheta(x, \ell_\vartheta'^2).$$

Hence $-\pi \otimes Q\ell_\vartheta'' = I$, and the influence function is seen to equal the canonical gradient, so that the maximum likelihood estimator is efficient. (We note that $-\pi \otimes Q\ell_\vartheta'' = I$ is also obtained from Eq. (4.6).) Efficiency of the maximum likelihood estimator has been proved for many specific models. We refer in particular to Hwang and Basawa (1994) who assume that $Q_\vartheta(x, dy)$ is an exponential family for each x. $\square$

5. A CHARACTERIZATION OF EFFICIENT ESTIMATORS

We have seen in Sec. 3 that the best regular and asymptotically linear estimator has an influence function which is equal to the canonical gradient. In this Section we show that this estimator is already optimal among all regular estimators.

This is a consequence of Hàjek's (1970) convolution theorem. In our context, it reads as follows. (For a proof see, e.g., Bickel et al. 1993, p. 63, Theorem 2A.)

Let Q_ϑ, $\vartheta \in \Theta$, be Hellinger differentiable in Eq. (2.5), let $t : \Theta \to \mathbf{R}$ be differentiable in Eq. (3.3) with canonical gradient Dk', and let T_n be regular in Eq. (3.2) with limit L. Then

$$\left(n^{-1/2} \sum_{i=1}^{n} (Dk')(X_{i-1}, X_i), \; n^{1/2}(T_n - t(\vartheta)) - n^{-1/2} \sum_{i=1}^{n} (Dk')(X_{i-1}, X_i) \right)$$

$$\Rightarrow (\|k'\|N, M) \qquad \text{under } P, \tag{5.1}$$

with N standard normal and M independent of N. In particular,

$$L = \|k'\|N + M \qquad \text{in distribution.}$$

By Anderson (1955), L is more spread out than $\|k'\|N$,

$$P(|L| \le c) \le P(|\|k'\|N| \le c) \qquad \text{for } c > 0.$$

This justifies calling T_n efficient if

$$L = \|k'\|N \qquad \text{in distribution.}$$

It follows from Eq. (5.1) that a regular and efficient estimator is asymptotically linear with influence function equal to the canonical gradient. We have seen in Sec. 3 that the converse is also true:

An estimator is regular and efficient if and only if it is asymptotically linear with influence function equal to the canonical gradient,

$$n^{1/2}(T_n - t(\vartheta)) = n^{-1/2} \sum_{i=1}^{n} (Dk')(X_{i-1}, X_i) + o_P(1). \tag{5.2}$$

REMARK   In the full nonparametric model (see Example 1 in Sec. 2) we have $K = H$, and the gradient h' is unique by Eq. (3.5). In particular, there is only one regular influence function, namely h', and the optimality result for regular influence functions is rather uninteresting. In the iid case, this is sometimes used as an argument against results on regular estimators in nonparametric models. However, the class of regular estimators is much larger than the class of regular and asymptotically linear estimators. One might object that reasonable estimators should be asymptotically linear. It should however be noted that the concept of asymptotic linearity can be considerably extended. This is probably not so obvious in the iid case, but in our definition in Eq. (3.1) for Markov chains it suggests itself to allow functions f with more than two arguments. This point was made in Wefelmeyer (1991).   □

6. NONPARAMETRIC MODELS AND MARTINGALE APPROXIMATIONS

Let $X_0, \ldots, X_n$ be observations from an ergodic Markov chain with unknown transition distribution $Q(x, dy)$ and invariant distribution $\pi(dx)$. We are interested in estimating π or, more specifically, the expectation πf of some square-integrable function $f(x)$.

By ergodicity, a consistent estimator is the empirical estimator

$$E_n f = \frac{1}{n} \sum_{i=1}^{n} f(X_i).$$

More generally, the expectation $\pi \otimes Qf$ of $f \in L_2(\pi \otimes Q)$ is estimated by the empirical estimator

$$E_n f = \frac{1}{n} \sum_{i=1}^{n} f(X_{i-1}, X_i).$$

We expect $E_n f$ to be efficient, but it is not even clear that it is asymptotically linear in the sense of Eq. (3.1): We can write

$$n^{1/2}(E_n f - \pi \otimes Qf) = n^{-1/2} \sum_{i=1}^{n} (f(X_{i-1}, X_i) - \pi \otimes Qf),$$

but $f - \pi \otimes Qf$ is not in H. Note, however, that we can expand $\sum f(X_{i-1}, X_i)$ into a series of martingales, the first step being

$$\sum_{i=1}^{n} f(X_{i-1}, X_i) = \sum_{i=1}^{n}(f(X_{i-1}, X_i) - Q(X_{i-1}, f))$$

$$+ \sum_{i=1}^{n} Q(X_i, f) + Q(X_n, f) - Q(X_0, f).$$

Under an appropriate geometric ergodicity assumption, if we continue adding and subtracting higher order conditional expectations, we obtain the martingale approximation

$$n^{1/2}(E_n f - \pi \otimes Qf) = n^{-1/2} \sum_{i=1}^{n} (Af)(X_{i-1}, X_i) + o_P(1)$$

with

$$(Af)(x, y) = f(x, y) - Q(x, f) + \sum_{j=1}^{\infty}(Q^j(y, f) - Q^{j+1}(x, f)).$$

Note that $Af \in H$. Hence $E_n f$ has influence function Af. The expansion is due to Gordin (1969); see also Gordin and Lifšic (1978), Maigret (1978), Dürr and Goldstein (1986) and Meyn and Tweedie (1994, Sec. 17.4).

To prove that the empirical estimator $E_n f$ is efficient, it remains to show that its influence function Af is the canonical gradient of $\pi \otimes Qf$. As in the first part of Sec. 2, we parametrize the model by the distribution function, take H as tangent space and pick sequences Q_{nh} with derivative h in the sense of Eq. (2.1). We view $\pi \otimes Qf$ as a function t of Q. By Eq. (3.4) and Example 1 in Sec. 2, the canonical gradient h' is defined by

$$n^{1/2}(t(Q_{nh}) - t(Q))$$

$$= n^{1/2}(\pi_{nh} \otimes Q_{nh} f - \pi \otimes Qf) \to \pi \otimes Q(hh') \qquad \text{for } h \in H.$$

By a perturbation expansion of π_{nh} due to Kartashov (1985a), (1985b), the left side converges to $\pi \otimes Q(hAf)$, and we obtain $h' = Af$ as expected. The above proof is due to Greenwood and Wefelmeyer (1995). Other proofs, for $f(x)$ with one argument, are in Penev (1991) and Bickel (1993). When f has more than two arguments, the corresponding empirical estimator $E_n f$ is *not* efficient; it is, however, efficient in the appropriate larger model of higher order Markov chains.

EXAMPLE 4   (*Countable state space.*) Suppose the state space E is countable. Then the transition distribution $Q(x, dy)$ is determined by the matrix of transition probabilities q_{xy}, and the stationary distribution $\pi(dx)$ by a vector of probabilities p_x. The above result, applied for $f = \delta_x$ and $f = \delta_{(x,y)}$, shows that

$$N_n^x = \frac{1}{n} \#\{i : X_i = x\} \qquad \text{is efficient for } p_x$$

and

$$N_n^{xy} = \frac{1}{n} \#\{i : X_{i-1} = x, X_i = y\} \qquad \text{is efficient for } p_x q_{xy}.$$

Hence N_n^{xy}/N_n^x is efficient for q_{xy}. Similar results for Markov step processes and semi-Markov processes are in Greenwood and Wefelmeyer (1994) and (1996a). $\square$

EXAMPLE 5 (*Reversible chains.*) Suppose we know that the chain is reversible,

$$\pi(dx)Q(x, dy) = \pi(dy)Q(y, dx).$$

Can we do better than the empirical estimator? It suggests itself to use a symmetrized version,

$$\frac{1}{2n} \sum_{i=1}^{n} (f(X_{i-1}, X_i) + f(X_i, X_{i-1})).$$

Greenwood and Wefelmeyer (1996b) show that this estimator is efficient. Although the result seems fairly obvious, the proof is not: It is not simple to translate the condition

$$\pi_{nh}(dx)Q_{nh}(x, dy) = \pi_{nh}(dy)Q_{nh}(y, dx)$$

into a restriction on h. $\square$

REMARK   Consider the empirical measure

$$M_n = \frac{1}{n} \sum_{i=1}^{n} \varepsilon_{(X_{i-1}, X_i)}, \tag{6.1}$$

with ε_a denoting the one-point probability measure in a. The empirical estimator $E_n f$ is the expectation $M_n f$ of f under the empirical measure. Since $E_n f$ is efficient for $\pi \otimes Q f$ for all (square-integrable) f, we expect that sufficiently smooth functions $s(M_n)$ are efficient for $s(\pi \otimes Q)$. The following Example 6 illustrates this point. $\square$

EXAMPLE 6 (*Misspecified parametric models.*) Suppose we have a one-dimensional parametric model Q_α, $\alpha \in A \subset \mathbf{R}$, for the transition distribution. Let ℓ'_α denote the Hellinger derivative of Q_α in the sense of Eq. (2.9). Let $q_\alpha(x, y)$ be a density of $Q_\alpha(x, dy)$. The maximum likelihood estimator maximizes

$$\frac{1}{n} \sum_{i=1}^{n} \log q_\alpha(X_{i-1}, X_i) = E_n \log q_\alpha.$$

Suppose now that the model is misspecified, and that the true transition distribution is Q. Let $s(\pi \otimes Q)$ denote the parameter α which maximizes the Kullback-Leibler information $\pi \otimes Q \log q_\alpha$. Then the maximum likelihood estimator is $s(M_n)$, where M_n is the empirical measure in Eq. (6.1). By the preceding Remark, we expect $s(M_n)$ to be efficient. It is, however, not clear how to check that s is sufficiently smooth. A proof avoiding this problem is due to Beran (1977) in the iid case, and to Greenwood and Wefelmeyer (1997) in the Markov chain setting. Related results for stationary Gaussian time series are in Dahlhaus and Wefelmeyer (1996). The asymptotic distribution of the maximum likelihood estimator under misspecification was derived by Huber (1967) in the iid case, by Ogata (1980) for Markov chains, and by Hosoya (1989) for general stationary linear processes. □

7. AUTOREGRESSION

Consider the linear first-order autoregressive model

$$X_i = \alpha X_{i-1} + \varepsilon_i,$$

where the ε_i are iid with a density p which has mean 0 and finite variance, and where $|\alpha| < 1$. This is an ergodic Markov chain with transition distribution

$$Q(x, dy) = p(y - \alpha x) \, dy.$$

In Example 3 we have treated the case that p is known (and standard normal). Now we treat p as an unknown nuisance parameter.

To prove local asymptotic normality, fix α and write π for the invariant distribution of $Q = Q_\alpha$. The parameter of the model is $\vartheta = (\alpha, p)$. Perturb α and p separately: $\alpha_{na} = \alpha + n^{-1/2} a$ and, for a bounded function $b(x)$,

$$p_{nb}(x) = p(x)(1 + n^{-1/2} b(x)).$$

Since p_{nb} must be a probability density, we must have $E\, b(\varepsilon) = 0$. Since p_{nb} must have mean 0, we must have $E\, \varepsilon b(\varepsilon) = 0$. We obtain the tangent space $K = \mathbf{R} \times B$ with

$$B = \{b \in L_2(p) : E\, b(\varepsilon) = 0 \quad \text{and} \quad E\, \varepsilon b(\varepsilon) = 0\}.$$

Write $\vartheta_{nk} = (\alpha_{na}, p_{nb})$. With $\ell' = p'/p$, the corresponding transition distribution $Q_{nk} = Q_{\vartheta_{nk}}$ is

$$Q_{nk}(x, dy) = p_{nb}(y - \alpha_{na}x)\, dy$$

$$\approx Q(x, dy)(1 + n^{-1/2}(-ax\ell'(y - \alpha x) + b(y - \alpha x))).$$

Hence the derivative of Q_{nk} is

$$(Dk)(x, y) = -ax\ell'(y - \alpha x) + b(y - \alpha x).$$

Since $E\varepsilon = 0$, the two functions on the right are orthogonal, and the inner product in Eq. (2.6) induced on K is

$$\langle k, k' \rangle = aa'\, E\, X^2 E\, \ell'(\varepsilon)^2 + E\, b(\varepsilon)b'(\varepsilon), \tag{7.1}$$

where the expectation of X is taken with respect to the stationary distribution. We obtain local asymptotic normality

$$\log \frac{dP_{nk}}{dP} = -an^{-1/2} \sum_{i=1}^{n} X_{i-1}\ell'(X_i - \alpha X_{i-1}) + n^{-1/2} \sum_{i=1}^{n} b(X_i - \alpha X_{i-1})$$

$$- \frac{1}{2}a^2 E\, X^2 E\, \ell'(\varepsilon)^2 - \frac{1}{2}E\, b(\varepsilon)^2 + o_P(1).$$

For a proof see Huang (1986) or Kreiss (1987b). For known innovation distribution see Akahira (1976) and Akritas and Johnson (1982).

We consider first the problem of estimating the autoregression parameter α. The classical estimator for α is the least squares estimator

$$\hat{\alpha}_n = \frac{\sum X_{i-1}, X_i}{X_{i-1}^2}.$$

As in Eq. (3.8) we see that $\hat{\alpha}_n$ is asymptotically linear with influence function

$$f(x, y) = (E\, X^2)^{-1}x(y - \alpha x).$$

Hence its asymptotic variance is $(E\, X^2)^{-1}E\, \varepsilon^2$.

To calculate a variance bound, we view the parameter α of interest as a function t of the parameter $\vartheta = (\alpha, p)$ of the model. By (3.4), the canonical gradient Dk', with $k' = (a', b')$, solves

$$n^{1/2}(t(\vartheta_{nk}) - t(\vartheta)) = n^{1/2}(\alpha_{na} - \alpha) = a$$

$$\overset{!}{=} \langle k, k' \rangle = aa'E\, X^2 E\, \ell'(\varepsilon)^2 + E\, b(\varepsilon)b'(\varepsilon) \qquad \text{for } a \in \mathbf{R},\ b \in B.$$

The solution is $a' = (E\, X^2 E\, \ell'(\varepsilon)^2)^{-1}$, $b' = 0$. Hence by Eq. (5.2) a regular and efficient estimator T_n for α is characterized by

$$n^{1/2}(T_n - \alpha) = (E\, X^2 E\, \ell'(\varepsilon)^2)^{-1}n^{-1/2} \sum_{i=1}^{n} X_{i-1}\ell'(X_i - \alpha X_{i-1}) + o_P(1).$$

Such an estimator is constructed in Kreiss (1987b). For an efficient estimator under the stronger model assumption that p is symmetric see Kreiss (1987a). The asymptotic variance of the efficient estimator is $(\mathrm{E}\,X^2\mathrm{E}\,\ell'(\varepsilon)^2)^{-1}$. Hence the relative efficiency of the least squares estimator is $(\mathrm{E}\,\varepsilon^2\mathrm{E}\,\ell'(\varepsilon)^2)^{-1}$. This equals the relative efficiency of the empirical estimator in the i.i.d. location model generated by the density p.

Consider now the problem of estimating the distribution of the innovations ε_i. To be specific, we estimate the expectation $\mathrm{E}f(\varepsilon)$ of some square-integrable function f. We do not observe the ε_i and cannot use the empirical estimator $(1/n)\sum_{i=1}^n f(\varepsilon_i)$. With $\hat{\alpha}_n = \sum X_{i-1}X_i/\sum X_{i-1}^2$ denoting the least squares estimator, we can replace the ε_i by estimated innovations $\hat{\varepsilon}_{ni} = X_i - \hat{\alpha}_n X_{i-1}$ and use the corresponding empirical estimator

$$E_n f = \frac{1}{n}\sum_{i=1}^n f(\hat{\varepsilon}_{ni}).$$

For $f(x) = 1(x \le r)$ this is the empirical distribution function. It is well studied. Functional central limit theorems are obtained by Boldin (1982), (1983) for autoregressive processes, by Boldin (1989) for moving average processes and by Kreiss (1991) for general linear processes. Explosive autoregressive processes, with $|\alpha| > 1$, are treated in Koul and Leventhal (1989). See also the monograph by Koul (1992). It can be shown that the asymptotic variance $\mathrm{E}f(\varepsilon)^2$ of the estimator is not changed if we replace ε_i by $\hat{\varepsilon}_{ni}$. Nevertheless, $E_n f$ is not efficient. The reason is that we have not used the information that the innovations have mean 0. To see how we might improve the empirical estimator, we calculate the canonical gradient of $\mathrm{E}f(\varepsilon)$, viewed as a function t of (α, p). By Eq. (3.4), the canonical gradient Dk', with $k' = (a', b')$, solves

$$n^{1/2}(t(\vartheta_{nk}) - t(\vartheta)) = n^{1/2}\left(\int p_{nb}(x)\,dxf(x) - \int p(x)\,dxf(x)\right) = \mathrm{E}\,b(\varepsilon)f(\varepsilon)$$

$$\overset{!}{=} \langle k, k'\rangle = aa'\mathrm{E}\,X^2\mathrm{E}\,\ell'(\varepsilon)^2 + \mathrm{E}\,b(\varepsilon)b'(\varepsilon) \qquad \text{for } a \in \mathbf{R},\; b \in B.$$

We obtain $a' = 0$ and must solve

$$\mathrm{E}\,b(\varepsilon)f(\varepsilon) = \mathrm{E}\,b(\varepsilon)b'(\varepsilon) \qquad \text{for } b \in B.$$

Using both restrictions $\mathrm{E}\,b(\varepsilon) = 0$ and $\mathrm{E}\,\varepsilon b(\varepsilon) = 0$, one checks that $b'(x) = f(x) - \mathrm{E}f(\varepsilon) - cx$ with $c = \mathrm{E}\,\varepsilon f(\varepsilon)/\mathrm{E}\,\varepsilon^2$. Hence the canonical gradient is

$$(Dk')(x, y) = f(y - \alpha x) - \mathrm{E}f(\varepsilon) - c(y - \alpha x).$$

It is easy to see that an efficient estimator for $\mathrm{E}f(\varepsilon)$, with influence function equal to the canonical gradient, is

$$E_n f - \hat{c}_n \frac{1}{n} \sum_{i=1}^{n} \hat{\varepsilon}_{ni} \qquad \text{with } \hat{c}_n = \frac{\sum \hat{\varepsilon}_{ni} f(\hat{\varepsilon}_{ni})}{\sum \hat{\varepsilon}_{ni}^2}.$$

For a proof see Wefelmeyer (1994). The tangent space $K = \mathbf{R} \times B$ is also calculated in Huang (1986) and Kreiss (1987b). These authors forget the restriction $\mathrm{E}\,\varepsilon b(\varepsilon) = 0$. One would conclude from their result that the usual empirical estimator is efficient.

REMARK   The tangent space is $K = \mathbf{R} \times B$, and by Eq. (7.1) the spaces $\mathbf{R}$ and B are orthogonal. As a consequence, α has a canonical gradient Dk' with k' of the form $(a', 0)$, while k' is of the form $(0, b')$ for $\mathrm{E}f(\varepsilon)$, a function of the second parameter, p. This means that, asymptotically, each of the two parameters can be estimated just as well knowing as not knowing the other parameter. Such a model is called adaptive, and efficient estimators for these parameters are also called adaptive.

Suppose we have proved, for each fixed p, local asymptotic normality of the one-dimensional model with transition distribution $p(y - \alpha x)\,dy$. Suppose we have constructed an estimator not depending on p which is efficient in all of these one-dimensional models. Then the parameter α is adaptive, and the estimator is efficient in the model with both α and p unknown. It is not necessary to prove local asymptotic normality of this model. Kreiss (1987a) uses this approach for ARMA-models, and Gassiat (1993) for non-causal AR-models. $\square$

EXAMPLE 7   (*ARCH.*) The first-order ARCH-model is

$$X_i = \sigma(1 + \alpha X_{i-1}^2)^{1/2} \varepsilon_i,$$

where the ε_i are iid with a density p which has mean 0 and variance 1. This is a Markov chain with transition distribution

$$Q(x, dy) = \frac{1}{\sigma(1 + \alpha x^2)^{1/2}} p\left(\frac{y}{\sigma(1 + \alpha x^2)^{1/2}}\right) dy.$$

It can be treated in a similar way as the AR(1)-model above. A review of ARCH-models is Bollerslev et al. (1992). Efficient estimators in this model and generalized ARCH-models are constructed in Engle and González-Rivera (1991), Linton (1993) and Drost and Klaassen (1997) under increasingly weaker assumptions. $\square$

EXAMPLE 8   (*Nonlinear autoregression.*) A generalization of AR- and ARCH-models are nonlinear and heteroscedastic autoregressive models

$$X_i = m_\alpha(X_{i-1}) + v_\alpha(X_{i-1})^{1/2}\varepsilon_i,$$

where the ε_i are iid with a density p which has mean 0 and variance 1 as in Example 7. Efficient estimation in nonlinear autoregressive models is studied by Hwang and Basawa (1993), Drost et al. (1994), (1997), Jeganathan (1995) and Koul and Schick (1997). □

8. MODELING CONDITIONAL MOMENTS

Let $X_0, \ldots, X_n$ be observations from a real-valued ergodic Markov chain with transition distribution $Q(x, dy)$ and invariant distribution $\pi(dx)$. Suppose we have a parametric model for the conditional mean, or autoregression function, but that the transition distribution is unspecified otherwise. A simple such specification is

$$\int Q(x, dy)y = \alpha x. \tag{8.1}$$

(The linear autoregressive model of Sec. 7 is a submodel, with a transition distribution of the form $Q(x, dy) = p(y - \alpha x)\, dy$, where p is a mean 0 density.)

Such a model suggests a special class of martingale estimating equations. Note that the model can be described by saying that the innovations $\varepsilon_i = X_i - \alpha X_{i-1}$ are martingale increments. In particular, "stochastic integrals" of the form

$$\sum_{i=1}^{n} w_\alpha(X_{i-1})(X_i - \alpha X_{i-1}) \tag{8.2}$$

are martingales. We obtain martingale estimating equations of the form

$$\sum e_\alpha(X_{i-1}, X_i) = 0$$

as in Sec. 4, now with $t(\alpha) = \alpha$ and $e_\alpha(x, y) = w_\alpha(x)(y - \alpha x)$. As before, we write T_n for the corresponding estimator. The choice $w_\alpha(x) = x$ leads to the least squares estimator, $T_n = \sum X_{i-1}X_i/X_{i-1}^2$. In general, to solve the estimating equation, we expand it as in Eq. (4.2),

$$0 = n^{-1/2} \sum_{i=1}^{n} e_{T_n}(X_{i-1}, X_i)$$

$$= n^{-1/2} \sum_{i=1}^{n} e_\alpha(X_{i-1}, X_i) + n^{1/2}(T_n - \alpha)\frac{1}{n}\sum_{i=1}^{n} e'_\alpha(X_{i-1}, X_i) + o_P(1).$$

Here $e'_\alpha(x, y) = -xw_\alpha(x) + w'_\alpha(x)(y - \alpha x)$, and we observe a special feature of the estimating function: The terms $w_\alpha'(X_{i-1})(X_i - \alpha X_{i-1})$ are martingale increments; hence

$$\frac{1}{n}\sum_{i=1}^{n} w_\alpha'(X_{i-1})(X_i - \alpha X_{i-1}) = o_P(1)$$

and the influence function of T_n does not contain the derivative of w_α:

$$n^{1/2}(T_n - \alpha) = (\mathrm{E}\, Xw_\alpha(X))^{-1} n^{-1/2} \sum_{i=1}^{n} w_\alpha(X_{i-1})(X_i - \alpha X_{i-1}) + o_P(1).$$

$$(8.3)$$

The asymptotic variance of T_n is

$$(\mathrm{E}\, Xw_\alpha(X))^{-2} \pi(vw_\alpha^2),$$

$$(8.4)$$

where v denotes the conditional variance

$$v(x) = \int Q(x, dy)(y - \alpha x)^2.$$

Are there better estimators than the least squares estimator? By the Schwarz inequality, the variance in Eq. (8.4) is minimized by $w_\alpha(x) = v(x)^{-1} x$, but this function cannot be used in an estimating Eq. (8.2) because $v(x)$ is unknown. However, as shown in Wefelmeyer (1997a), the conditional variance $v(x)$ can be replaced by an estimator $\hat{v}_n(x)$ without changing the asymptotic variance of the estimator which solves the corresponding estimating equation. In this sense, the estimator is "adaptive". The solution of this estimating equation is the weighted least squares estimator

$$T_n = \frac{\sum \hat{v}_n(X_{i-1})^{-1} X_{i-1} X_i}{\sum \hat{v}_n(X_{i-1})^{-1} X_{i-1}^2}.$$

By Eq. (8.4), its asymptotic variance is $(\mathrm{E}\, v(X)^{-1} X^2)^{-1}$.

Are there better estimators than the optimal weighted least squares estimator? To decide this, we calculate the canonical gradient of α. As in the first part of Sec. 2, we parametrize the model by the transition distribution, fix Q and consider perturbations in Eq. (2.1)

$$Q_{nh}(x, dy) = Q(x, dy)(1 + n^{-1/2}h(x, y))$$

with $h \in H$. The transition distribution Q_{nh} must fulfill the model assumption Eq. (8.1), with a possibly perturbed $\alpha_{nh} = \alpha + n^{-1/2}a$,

$$\int Q_{nh}(x, dy)y = \alpha_{nh}x,$$

i.e.,

$$\int Q(x, dy)y + n^{-1/2} \int Q(x, dy)h(x, y)y = \alpha x + n^{-1/2}ax.$$

It follows that the tangent space is the union of the affine spaces

$$H_a = \{h \in H : \int Q(x, dy)h(x, y)y = ax\}, \qquad a \in \mathbf{R}.$$

By Eq. (3.4), the canonical gradient h' of α, viewed as a function t of Q, is in one of these affine spaces and solves

$$n^{1/2}(t(Q_{nh}) - t(Q)) = n^{1/2}(\alpha_{nh} - \alpha) = a$$
$$\overset{!}{=} \langle h, h' \rangle = \pi \otimes Q(hh') \qquad \text{for } a \in \mathbf{R}, \ h \in H_a. \quad (8.5)$$

Our candidate is the influence function of the optimal weighted least squares estimator,

$$g(x, y) = (E v(X)^{-1} X^2)^{-1} v(x)^{-1} x(y - \alpha x).$$

This function is clearly in H, and since

$$\int Q(x, dy)v(x)^{-1}x(y - \alpha x)y = v(x)^{-1}x \int Q(x, dy)(y - \alpha x)^2 = x,$$

we also have $g \in H_a$ for $a = (E v(X)^{-1} X^2)^{-1}$. It remains to check whether $h' = g$ solves Eq. (8.5). But

$$\pi \otimes Q(hg) = (E v(X)^{-1} X^2)^{-1} \int \pi(dx)v(x)^{-1}x \int Q(x, dy)h(x, y)(y - \alpha x) = a.$$

Hence the optimal weighted least squares estimator is efficient.

REMARK We have just checked that the influence function of the optimal weighted least squares estimator is a gradient. By the characterization in Sec. 3, this would be unnecessary if we knew that the estimator is regular. Indeed, every estimating equation based on a martingale in Eq. (8.2) gives a regular estimator, provided w_α is so smooth that the estimator is asymptotically linear of the form in Eq. (8.3). In other words:
    Influence functions of the form $f(x, y) = (E X w_\alpha(X))^{-1} w_\alpha(x)(y - \alpha x)$ are regular.

By Eq. (3.5), we must check that

$$\pi \otimes Q(hf) = \pi \otimes Q(hg) \qquad \text{for } a \in \mathbf{R}, \ h \in H_a.$$

We calculate both sides for $h \in H_a$, ignoring constants:

$$\int \pi(dx)w_\alpha(x) \int Q(x, dy)h(x, y)(y - \alpha x) = aE X w_\alpha(X)$$

and

$$\int \pi(dx)v(x)^{-1} \int Q(x, dy)h(x, y)(y - \alpha x) = a\mathrm{E}\, v(X)^{-1} X^2.$$

Taking into account the constants in f and g, the check is finished. $\quad\square$

REMARK If the innovations $\varepsilon_i = X_i - \alpha X_{i-1}$ happen to be independent, the model reduces to the AR(1)-model of Sec. 7, and the conditional variance $v(x)$ reduces to

$$v(x) = \int p(y - \alpha x)(y - \alpha x)^2 = \mathrm{E}\,\varepsilon^2$$

and does not depend on x. Then the optimal weighted least squares estimator is asymptotically equivalent to the ordinary least squares estimator. $\quad\square$

EXAMPLE 9 (*Quasi-likelihood models.*) A quasi-likelihood model for Markov chains is described by parametric models for both the conditional mean and the conditional variance,

$$\int Q(x, dy)y = m_\alpha(x),$$

$$\int Q(x, dy)(y - m_\alpha(x))^2 = v_\alpha(x),$$

with a common parameter α. For examples see Zeger and Qaqish (1988) and Huhtala (1992). A submodel is the nonlinear and heteroscedastic autoregression model of Example 8 in Sec. 7, with transition distribution

$$Q(x, dy) = \frac{1}{v_\alpha(x)^{1/2}} p\left(\frac{y - m_\alpha(x)}{v_\alpha(x)^{1/2}}\right) dy.$$

The approach of the present section works for quasi-likelihood models; see Wefelmeyer (1996). Wefelmeyer (1997b) treats a generalization in which m_α and v_α also depend on covariates. $\quad\square$

9. OTHER TYPES OF MARKOV CHAIN MODELS

In this section we briefly describe further models which also fall under our Markov chain setting.

EXAMPLE 10 (*Discretely observed diffusions.*) Suppose we observe an ergodic diffusion process

$$dX_t = b_\alpha(X_t) + \sigma_\alpha(X_t)\, dW_t$$

at discrete time points $0, \Delta, 2\Delta, \ldots$. The observations form a Markov chain $X_0, \ldots, X_n$. The model is in principle parametrized by α. However, while the stationary distribution of X_i is the same as for the diffusion process, the transition distribution may be difficult to calculate. Recent references on estimators for α are Bibby and Sørensen (1995), Pedersen (1995), Kessler and Sorensen (1995) and Bibby and Sørensen (1996). In the context of the present review, we mention the approach of Kessler, Schick and Wefelmeyer (1998): Consider the transition distribution as unknown. We have a parametric family of restrictions on the transition distributions, namely a parametric model for the stationary distribution of X_i. A simple estimator for α is the estimator that would be the maximum likelihood estimator if the observations were independent, but this estimator turns out to be inefficient. □

EXAMPLE 11 (*Markov chain Monte Carlo.*) Suppose we want to calculate the expectation πf of a function f on a product space $E = E_1 \times \cdots \times E_k$. This may be difficult directly or numerically, or even by ordinary Monte Carlo integration. If we can simulate the conditional distributions $p_j(x_{-j}, dx_j)$ of the jth component of π given the other components x_{-j}, we may generate a k-dimensional Markov chain based on updates of a single component by

$$Q_j(x, dy) = p_j(x_{-j}, dy_j)\varepsilon_{x_{-j}}(dy_{-j}).$$

An introduction to such Markov chain Monte Carlo procedures is the monograph Gilks et al. (1996). The Gibbs sampler with deterministic sweep uses the transition distribution $Q = Q_1 \cdots Q_k$, the one with random sweep uses $Q = (1/k) \sum Q_j$. If we denote the simulations from the corresponding Markov chain by $X^0, \ldots, X^n$, the expectation πf can be estimated by the empirical estimator $(1/n) \sum_{i=1}^{n} f(X^i)$. Its asymptotic variance is studied, e.g., by Peskun (1973), Frigessi, Hwang and Younes (1992), Green and Han (1992), Liu, Wong and Kong (1994) and (1995) and Clifford and Nicholls (1995). Does the empirical estimator make effective use of the simulations? Greenwood et al. (1997) view π as the unknown parameter and calculate variance bounds for estimators of πf. It turns out that it is best not to use the usual empirical estimator but to use the empirical estimator for deterministic sweep which considers the update of a single component as a new observation:

$$\frac{1}{k}\sum_{j=1}^{k}\frac{1}{n}\sum_{i=1}^{n} f(X_1^i, \ldots, X_j^i, X_{j+1}^{i-1}, \ldots, X_k^{i-1}).$$

This estimator is close to efficient. □

EXAMPLE 12   (*Random coefficient autoregression.*) We observe $X_0, \ldots,$ X_n with

$$X_i = (\alpha + Z_i)X_{i-1} + \varepsilon_i,$$

where the ε_i are iid with a density p which has mean 0 and finite variance, and the Z_i are iid with mean 0 and distribution function G such that $\alpha + \operatorname{var} G < 1$. The X_i are a Markov chain with transition distribution

$$Q(x, dy) = \int p(y - (\alpha + z)x)\, dG(z)\, dy.$$

The model is studied in the monographs by Nicholls and Quinn (1982) and Tong (1990). Efficient estimators for α are constructed in Koul and Schick (1996). Weighted least squares estimators and local asymptotic normality for generalized random coefficient autoregressive models are studied in Hwang and Basawa (1996), (1997).   □

EXAMPLE 13   (*Regression with autoregressive errors.*) We observe (U_i, Y_i), $i = 1, \ldots, n$, where the Y_i follow a linear regression model

$$Y_i = \beta U_i + X_i,$$

the covariates U_i are iid, with density g, and the errors X_i are AR(1),

$$X_i = \alpha X_{i-1} + \varepsilon_i,$$

where the ε_i are iid with a density p which has mean 0 and finite variance, and where $|\alpha| < 1$. Writing

$$Y_i = \beta U_i + \alpha(Y_{i-1} - \beta U_{i-1}) + \varepsilon_i,$$

we see that the observations (U_i, Y_i) are a Markov chain with transition distribution

$$Q(u_{i-1}, y_{i-1}, du_i, dy_i) = g(u_i)p(y_i - \beta u_i - \alpha(y_{i-1} - \beta u_{i-1}))\, du_i\, dy_i$$

and four parameters, α, β, p, g. Local asymptotic normality for deterministic U_i is proved in Swensen (1985), for moving average X_i in Garel (1989). For the additive regression model $Y_i = \beta U_i + \gamma(V_i) + X_i$ of Engle et al. (1986), efficient estimators for α and β, respectively, are constructed in Schick (1993) and (1996). Quite different efficient estimators are obtained for long range stationary Gaussian errors X_i: see Dahlhaus (1995).   □

Acknowledgment

I thank the referee for a number of suggestions which improved the presentation.

REFERENCES

Akahira, M. (1976). On the asymptotic efficiency of estimators in an autoregressive process. *Ann. Inst. Statist. Math. 28*: 35–48.

Akritas, M. G. and Johnson, R. A. (1982). Efficiencies of tests and estimators for *p*-order autoregressive processes when the error distribution is nonnormal. *Ann. Inst. Statist. Math. 34*: 579–589.

Anderson, T. W. (1955). The integral of a symmetric unimodal function over a symmetric convex set and some probability inequalities. *Proc. Amer. Math. Soc. 6*: 170–176.

Andrews, D. W. K. (1994). Asymptotics for semiparametric econometric models via stochastic equicontinuity. *Econometrica 62*: 43–72.

Beran, R. (1977). Minimum Hellinger distance estimates for parametric models. *Ann. Statist. 5*: 445–463.

Bibby, B. M. and Sørensen, M. (1995). Martingale estimating functions for discretely observed diffusion processes. *Bernoulli 1*: 17–39.

Bibby, N. and Sørensen, M. (1996). On estimation for discretely observed diffusions: a review. Theory of Stochastic Processes *2*: 49–56.

Bickel, P. J. (1993). Estimation in semiparametric models. In: Multivariate Analysis: Future Directions (C. R. Rao, ed.) 55–73, North-Holland, Amsterdam.

Bickel, P. J., Klaassen, C. A. J., Ritov, Y. and Wellner, J. A. (1993). Efficient and Adaptive Estimation for Semiparametric Models. Johns Hopkins University Press, Baltimore.

Boldin, M. V. (1982). Estimation of the distribution of noise in an autoregressive scheme. *Theory Probab. Appl. 27*: 866–871.

Boldin, M. V. (1983). Testing hypotheses in autoregressive schemes by the Kolmogorov and ω^2 tests. *Soviet Math. Dokl. 28*: 550–553.

Boldin, M. V. (1989). On testing hypotheses in the sliding average scheme by the Kolmogorov-Smirnov and ω^2 tests. *Theory Probab. Appl. 34*: 758–746.

Bollerslev, T., Chou, R. Y. and Kroner, K. (1992). ARCH modeling in finance. A review of the theory and empirical evidence. *J. Econometrics 52*: 5–59.

Clifford, P. and Nicholls, G. (1995). A Metropolis sampler for polygonal image reconstruction. Technical report, Department of Statistics, Oxford University.

Dahlhaus, R. (1995). Efficient location and regression estimation for long range dependent regression models. *Ann. Statist. 23*: 1029–1047.

Dahlhaus, R. and Wefelmeyer, W. (1996). Asymptotically optimal estimation in misspecified time series models. *Ann. Statist. 24*: 952–974.

Drost, F. C. and Klaassen, C. A. J. (1997). Efficient estimation in semiparametric GARCH models. CentER discussion paper 9638, Tilburg University. *J. Econometrics 81*: 193–221.

Drost, F. C., Klaassen, C. A. J. and Werker, B. J. M. (1994). Adaptiveness in time series models. In: Asymptotic Statistics (P. Mandl and M. Hušková, eds.), 467–474. Physica-Verlag, Heidelberg.

Drost, F. C., Klaassen, C. A. J. and Werker, B. J. M. (1997). Adaptive estimation in time-series models. *Ann. Statist. 25*: 786–817.

Dürr, D. and Goldstein, S. (1986). Remarks on the central limit theorem for weakly dependent random variables. In: Stochastic Processes — Mathematics and Physics (S. Albeverio, P. Blanchard and L. Streit, eds.), 104–118, Lecture Notes in Mathematics 1158, Springer, Berlin.

Engle, R. F. and González-Rivera (1991). Semiparametric ARCH models. *J. Business Economic Statist. 9*: 345–359.

Engle, R. F., Granger, C. W. J., Rice, J. and Weiss, A. (1986). Semiparametric estimates of the relation between weather and electricity sales. *J. Amer. Statist. Assoc. 81*: 310–320.

Frigessi, A., Hwang, C.-R. and Younes, L. (1992). Optimal spectral structure of reversible stochastic matrices, Monte Carlo methods and the simulation of Markov random fields. *Ann. Appl. Probab. 2*: 610–628.

Garel, B. B. (1989). The asymptotic distribution of the likelihood ratio for M. A. processes with a regression trend. *Statist. Decisions 7*: 167–184.

Gassiat, E. (1993). Adaptive estimation in noncausal stationary AR processes. *Ann. Statist. 21*: 2022–2042.

Gilks, W. R., Richardson, S. and Spiegelhalter, D. J., eds. (1996). Markov Chain Monte Carlo in Practice. Chapman & Hall, London.

Godambe, V. P. (1985). The foundations of finite sample estimation in stochastic processes. *Biometrika 72*: 419–428.

Godambe, V. P., ed. (1991). Estimating Functions. Oxford University Press.

Godambe, V. P. and Heyde, C. C. (1987). Quasi-likelihood and optimal estimation. *Int. Stat. Rev. 55*: 231–244.

Godambe, V. P. and Thompson, M. E. (1989). An extension of quasi-likelihood estimation. *J. Stat. Plan. Inf. 22*: 137–152.

Gordin, M. I. (1969). The central limit theorem for stationary processes. *Soviet Math. Dokl. 10*: 1174–1176.

Gordin, M. I. and Lifšic, B. A. (1978). The central limit theorem for stationary Markov processes. *Soviet Math. Dokl. 19*: 392–394.

Green, P. J. and Han, X.-l. (1992). Metropolis methods, Gaussian proposals and antithetic variables. In: Stochastic Models, Statistical Methods, and Algorithms in Image Analysis (P. Barone, A. Frigessi and M. Piccioni, eds.), 142–164, Lecture Notes in Statistics 74, Springer-Verlag, Berlin.

Greenwood, P. E., McKeague, I. W. and Wefelmeyer, W. (1997). Information bounds for Gibbs samplers. *To appear in: Ann. Statist.*

Greenwood, P. E. and Wefelmeyer, W. (1994). Nonparametric estimators for Markov step processes. *Stochastic Process. Appl. 52*: 1–16.

Greenwood, P. E. and Wefelmeyer, W. (1995). Efficiency of empirical estimators for Markov chains. *Ann. Statist. 23*: 132–143.

Greenwood, P. E. and Wefelmeyer, W. (1996a). Empirical estimators for semi-Markov processes. *Math. Meth. Statist. 5*: 299–315.

Greenwood, P. E. and Wefelmeyer, W. (1996b). Reversible Markov chains and optimality of symmetrized empirical estimators. *To appear in: Bernoulli*

Greenwood, P. E. and Wefelmeyer, W. (1997). Maximum likelihood estimator and Kullback–Leibler information in misspecified Markov chain models. *Teor. Veroyatnost. i Primenen. 42:* 169–178.

Hájek, J. (1970). A characterization of limiting distributions of regular estimates. *Z. Wahrsch. Verw. Gebiete 14*: 323–330.

Höpfner, R. (1993a). On statistics of Markov step processes: representation of log-likelihood ratio processes in filtered local models. *Probab. Theory Related Fields 94*: 375–398.

Höpfner, R. (1993b). Asymptotic inference for Markov step processes: observation up to a random time. *Stochastic Process. Appl. 48*: 295–310.

Höpfner, R., Jacod, J. and Ladelli, L. (1990). Local asymptotic normality and mixed normality for Markov statistical models. *Probab. Theory Related Fields 86*: 105–129.

Hosoya, Y. (1989). The bracketing condition for limit theorems on stationary linear processes. *Ann. Statist. 17*: 401–418.

Huang, W.-M. (1986). A characterization of limiting distributions of estimators in an autoregressive process. *Ann. Inst. Statist. Math. 38*: 137–144.

Huber, P. J. (1967). The behavior of maximum likelihood estimates under nonstandard conditions. Proc. Fifth Berkeley Symp. Math. Statist. Probab. 1, 221–233.

Huhtala, K. (1992). A quasi-likelihood Markov model for the hardenability of steel. In: Proceedings of the Sixth European Conference on Mathematics in Industry (F. Hodnett, ed.), 191–194. Teubner, Stuttgart.

Hwang, S. Y. and Basawa, I. V. (1993). Asymptotic optimal inference for a class of nonlinear time series models. *Stochastic Process. Appl. 46*: 91–113.

Hwang, S. Y. and Basawa, I. V. (1994). Large sample inference for conditional exponential families with applications to nonlinear time series. *J. Stat. Plan. Inf. 38*: 141–158.

Hwang, S. Y. and Basawa, I. V. (1996). Parameter estimation for generalized random coefficient autoregressive processes. To appear in: Proceedings of the Franco-Belgian Conference on Nonlinear Time Series.

Hwang, S. Y. and Basawa, I. V. (1997). The local asymptotic normality of a class of generalized random coefficient autoregressive processes. *Statist. Probab. Lett. 34*: 165–170.

Jeganathan, P. (1995). Some aspects of asymptotic theory with applications to time series models. *Econometric Theory 11*: 818–887.

Kartashov, N. V. (1985a). Criteria for uniform ergodicity and strong stability of Markov chains with a common phase space. *Theory Probab. Math. Statist. 30*: 71–89.

Kartashov, N. V. (1985b). Inequalities in theorems of ergodicity and stability for Markov chains with common phase space. I. *Theory Probab. Appl. 30*: 247–259.

Kessler, M., Schick, A. and Wefelmeyer, W. (1998). The information in the marginal law of a Markov chain. Unpublished manuscript.

Kessler, M. and Sørensen, M. (1995). Estimating equations based on eigenfunctions for a discretely observed diffusion process. Research Report 332, Department of Theoretical Statistics, University of Aarhus, Denmark. To appear in: *Bernoulli.*

Koul H. L. (1992). Weighted Empiricals and Linear Models. IMS Lecture Notes—Monograph Series 21, Hayward, California.

Koul, H. L. and Leventhal, S. (1989). Weak convergence of the residual empirical process in explosive autoregression. *Ann. Statist. 17*, 1784–1784.

Koul, H. L. and Schick, A. (1996). Adaptive estimation in a random coefficient autoregressive model. *Ann. Statist. 24*: 1025–1052.

Koul, H. L. and Schick, A. (1997). Efficient estimation in nonlinear autoregressive time series models. *Bernoulli 3*: 247–277.

Kreiss, J.-P. (1987a). On adaptive estimation in stationary ARMA processes. *Ann. Statist. 15*: 112–133.

Kreiss, J.-P. (1987b). On adaptive estimation in autoregressive models when there are nuisance functions. *Statist. Decisions 5*: 59–76.

Kreiss, J.-P. (1991). Estimation of the distribution function of noise in stationary processes. *Metrika 38*: 285–297.

Linton, O. (1993). Adaptive estimation in ARCH models. *Econometric Theory 9*: 539–569.

Liu, J. S., Wong, W. H. and Kong, A. (1994). Covariance structure of the Gibbs sampler with applications to the comparisons of estimators and augmentation schemes. *Biometrika 81*: 27–40.

Liu, J. S., Wong, W. H. and Kong, A. (1995). Covariance structure and convergence rate of the Gibbs sampler with various scans. *J. Roy. Statist. Soc. Ser. B 57*: 157–169.

Maigret, N. (1978). Thérème de limite centrale fonctionnel pour une chaîne de Markov récurrente au sens de Harris et positive. *Ann. Inst. H. Poincaré Probab. Statist. 14*: 425–440.

Meyn, S. P. and Tweedie, R. L. (1994). Markov Chains and Stochastic Stability. Second Printing. Springer, London.

Nicholls, D. F. and Quinn, B. G. (1982). Random Coefficient Autoregressive Models: An Introduction. Lecture Notes in Statistics 11. Springer, New York.

Ogata, Y. (1980). Maximum likelihood estimates of incorrect Markov models for time series and the derivation of AIC. *J. Appl. Probab. 17*: 59–72.

Pedersen, A. R. (1995). Consistency and asymptotic normality of an approximate maximum likelihood estimator for discretely observed diffusion processes. *Bernoulli 1*: 257–279.

Penev, S. (1991). Efficient estimation of the stationary distribution for exponentially ergodic Markov chains. *J. Stat. Plan. Inf. 27*: 105–123.

Peskun, P. H. (1973). Optimum Monte Carlo sampling using Markov chains. *Biometrika 60*: 607–612.

Roussas, G. G. (1965). Asymptotic inference in Markov processes. *Ann. Math. Statist. 36*: 987–992.

Schick, A. (1993). An adaptive estimator of the autocorrelation coefficient in a semiparametric regression model with autoregressive errors. Unpublished manuscript.

Schick, A. (1996). Efficient estimation in a semiparametric additive regression model with autoregressive errors. *Stochastic Process. Appl. 61*: 339–361.

Swensen, A. R. (1985). The asymptotic distribution of the likelihood ratio for autoregressive time series with a regression trend. *J. Multivariate Anal. 16*: 54–70.

Tong, H. (1990). Nonlinear Time Series: A Dynamical Approach. Oxford Statistical Science Series 6. Oxford University Press.

Wefelmeyer, W. (1991). A generalization of asymptotically linear estimators. Statist. *Probab. Lett. 11*: 195–199.

Wefelmeyer, W. (1994). An efficient estimator for the expectation of a bounded function under the residual distribution of an autoregressive process. *Ann. Inst. Statist. Math. 46*: 309–315.

Wefelmeyer, W. (1996). Quasi-likelihood models and optimal inference. *Ann. Statist. 24*: 405–422.

Wefelmeyer, W. (1997a). Adaptive estimators for parameters of the auto-regression function of a Markov chain. *J. Stat. Plan. Inf. 58*: 389–398.

Wefelmeyer, W. (1997b). Quasi-likelihood regression models for Markov chains. In: Selected Proceedings of the Symposium on Estimating Functions (I. V. Basawa, V. P. Godambe and R. L. Taylor, eds.) 149–173, IMS Lecture Notes—Monograph Series 32, IMS, Hayward.

Zeger, S. L. and Qaqish, B. (1988). Markov regression models for time series: a quasi-likelihood approach. *Biometrics 44*: 1019–1031.

13

Nonparametric Functional Estimation: An Overview

B. L. S. PRAKASA RAO   Indian Statistical Institute, New Delhi, India

1. INTRODUCTION

There are three basic problems in the subject of Statistics:

1. Devising scientific and valid statistical techniques for data collection.
2. Developing methods for statistical inference for the analysis of data so collected and
3. Improving the statistical models for future use based on these experiences.

Apart from its traditional applications of the subject of statistics in social sciences, statistics per se is no longer just agricultural statistics, bio-statistics, industrial statistics or medical statistics but its techniques has penetrated

An abridged version of this paper was presented as the Sectional Presidential address of the Statistics Section of the 83rd Session, Indian Science Congress Association held at Patiala, India (cf. Prakasa Rao (1996a)).

461

more recently into subjects such as image processing, pattern recognition, signal processing, remote sensing etc., in the area of electrical engineering and communication sciences (cf. Prakasa Rao (1995a)). The classical statistical methods were parametric. The assumption was that the sample of observations were from a population with a known parametric family of distributions. The problem was either to estimate the unknown parameters or to devise tests and confidence regions for the unknown parameters based on the sample. The assumption that the model is parametric is rather strong because the assumed model need not be the "actual" one if there is one and the statistical methods developed for a particular probability model could lead to erroneous conclusions when applied to a slightly perturbed model. These problems brought about the trend to develop nonparametric statistical methods for data analysis. One of the fundamental problems is the estimation of the distribution function or the density function, when it exists, of the population from which the sample is obtained. Histograms which are a type of density estimators have been in use for many years and have been used both by the statisticians and other research workers in the social, physical and biological sciences as tools for drawing inferences from data. Density estimators are potentially of use in all the three stages of statistical treatment of data, namely, exploratory, confirmatory and presentational (cf. Silverman (1986)). At the exploratory stage, density estimators give an indication of the multimodality, skewness or dispersion of the data. For confirmatory purposes, they can be used for instance in nonparametric discriminant analysis for classification purposes. For data presentation, density estimators are the best information—preserving transformation given the data.

After my book on "Nonparametric Functional Estimation" (Academic Press, Orlando) was published in 1983, the subject has grown enormously over the years and books by Silverman (1986), Devroye and Gyorfi (1985), Devroye (1987), Thompson and Tapia (1990), Härdle (1990), Scott (1994) and others have been published later in the area of nonparametric density and regression estimation and application of the results to subjects such as time series, econometrics, image processing and pattern recognition are discussed extensively in the literature.

I will restrict attention to the work some of my colleagues and I have done during the past few years and some recent work by others which is of importance from my view in the area of nonparametric functional estimation. The overview is necessarily personal and not comprehensive.

2. NONPARAMETRIC FUNCTIONAL ESTIMATION

2.1 Density Estimation

Given a sample of observations $X_1, \ldots, X_n$ independent and identically distributed with distribution function $F(x)$ and density function $f(x)$, it is known that there does not exist an estimator $\hat{f}_n(x)$ depending on $X_1, \ldots, X_n$ such that

$$E[\hat{f}_n(x)] = f(x)$$

for all x under some reasonable conditions. Several methods of density estimation are known in the literature. For our purposes, we shall briefly describe two such methods which have been widely used. For complete details, see Prakasa Rao (1983).

(a) The Method of kernels

Since f is the derivative of F almost everywhere, it is natural to expect that $[F(x + h) - F(x - h)]/2h$ to be close to $f(x)$ for sufficiently small h. Let $F_n(x)$ be the empirical distribution function based on the observed sample $(X_1, \ldots, X_n)$. One can consider

$$f_n^*(x) = \frac{F_n(x + h_n) - F_n(x - h_n)}{2h_n}$$

as an estimator of $f(x)$. This estimator can also be written in the form

$$f_n^*(x) = \frac{1}{nh_n} \sum_{i=1}^{n} K_0\left(\frac{x - X_i}{h_n}\right) \tag{2.1}$$

where

$$K_0(x) = \begin{cases} \frac{1}{2} & \text{if } x \in [-1, 1) \\ 0 & \text{otherwise} \end{cases}.$$

In general, let $K(\cdot)$ be a suitable density function and define $f_n(x)$ by

$$f_n(x) = \frac{1}{nh_n} \sum_{i=1}^{n} K\left(\frac{x - X_i}{h_n}\right) \tag{2.2}$$

where $h_n \downarrow 0$ as $n \to \infty$. The function $K(\cdot)$ is called the kernel and the sequence $\{h_n\}$ is called the window or bandwith of the estimator. It is known that the asymptotic properties of the estimator $f_n(x)$ depend crucially on the bandwidth $\{h_n\}$ and the choice of kernel $K(\cdot)$ is relatively less important (cf. Prakasa Rao (1983)).

(b) The Method of orthogonal series

Suppose $f \in L^2(R)$. Let $\{e_i, u \geq 1\}$ be an orthonormal basis for $L^2(R)$. Then f can be represented in the form

$$f = \sum_{i=1}^{\infty} a_i e_i(x). \tag{2.3}$$

An orthogonal series estimator f_n of f is defined by

$$\tilde{f}_n = \sum_{i=1}^{q(n)} \hat{a}_{in} e_i(x) \tag{2.4}$$

where

$$\hat{a}_{in} = \frac{1}{n} \sum_{i=1}^{n} e_i(X_i) \tag{2.5}$$

and $q(n) \to \infty$ as $n \to \infty$. It is easy to see that $E(\hat{a}_{in}) = a_i$.

It can be shown that the mean integrated squared error (MISE)

$$J_n^2 = E\left[\int_{-\infty}^{\infty} [\tilde{f}_n(x) - f(x)]^2 \, dx \right] \to 0 \tag{2.6}$$

iff

$$\lim_{n \to \infty} \int_{-\infty}^{\infty} \frac{1}{n} \left[\sum_{i=1}^{q(n)} e_i^2(x) \right] f(x) \, dx = 0. \tag{2.7}$$

For further details and properties of these estimators, see Prakasa Rao (1983).

Extension of properties of kernel type and orthogonal series type of density estimators for the multidimensional case are discussed in Prakasa Rao (1983).

2.2 Estimation of Regression Function

Suppose (X_i, Y_i), $i \geq 1$ are iid bivariate random vectors distributed as (X, Y). Further suppose that $E(Y|X = x) = m(x)$ exists. $m(\cdot)$ is called the regression function of Y on X. We will not go into the importance of the study of regression function in statistical inference. The problem is to estimate $m(\cdot)$ when the joint density of (X, Y), assumed to exist, is unknown. Under some regularity conditions, one can write

$$m(x) = \frac{\int_{-\infty}^{\infty} y f_{X,Y}(x, y) \, dy}{f_X(x)}$$

where $f_{X,Y}(x,y)$ and $f_X(x)$ are the joint density of (X, Y) and marginal density of X respectively. It is clear that one can get an estimator for $m(x)$ by choosing suitable nonparametric estimators for $f_{XY}(x,y)$ and $f_X(x)$. Motivated by this consideration, Nadaraja (1964) and Watson (1964) suggested the nonparametric regression estimator

$$m_n(x) = \frac{\sum_{i=1}^{n} Y_i K\left(\frac{x - X_i}{h_n}\right)}{\sum_{i=1}^{n} K\left(\frac{x - X_i}{h_n}\right)}$$

where $K(\cdot)$ is a suitable kernel and $\{h_n\}$ is a bandwidth sequence. In general, an estimator for a function of the form

$$E[g(Y)|\mathbf{X} = \mathbf{x}],$$

when $(\mathbf{X}, Y)$ is a $(p + 1)$-dimensional random vector, can be chosen to be

$$\frac{\sum_{i=1}^{n} g(Y_i) K\left(\frac{\mathbf{x} - \mathbf{X}_i}{h_n}\right)}{\sum_{i=1}^{n} K\left(\frac{\mathbf{x} - \mathbf{X}_i}{h_n}\right)}.$$

This estimator is not recursive and computationally cumbersome. If additional data is obtained on line, the estimator has to be computed all over again. However, one can consider recursive estimator for $E[g(Y)|\mathbf{X}= \mathbf{x}]$ of the form

$$\frac{\sum_{i=1}^{n} \frac{1}{h_i^p} g(Y_i) K\left(\frac{\mathbf{x} - \mathbf{X}_i}{h_i}\right)}{\sum_{i=1}^{n} \frac{1}{h_i^p} K\left(\frac{\mathbf{x} - \mathbf{X}_i}{h_i}\right)}.$$

The numerator and the denominator of this estimator can be computed recussively as and when the data is received on line as in the case of satellite data. For instance, speed in computation is essential for quick analysis of the satellite data, which is on-line, for useful decision making. Recussive methods are also of great importance when the computational costs and algorithmic aspects of computation and other issues are to be taken into account. For additional discussion of recursive estimation, see Prakasa Rao (1983) and Prakasa Rao (1979b). The problem of estimation of the regression function can be further generalized as follows.

One can consider estimation of functionals of the type

$$m^g(t) = E[g(Y_1, \ldots, Y_k)|X_1 = t_1, \ldots, X_p = t_p], \mathbf{t} \in R^p$$

where $g: R^k \to R$ such that $E|g(Y_1, \ldots, Y_k)| < \infty$. Stute (1991) considered this conditional problem and proposed the following conditional U-statistic as an estimator of $m^g(t)$:

$$Y_n^g(\mathbf{t}) = U_n(g, \mathbf{t}) / U_n(1, \mathbf{t}), \quad \mathbf{t} \in R^p$$

where

$$U_n(g, \mathbf{t}) = \frac{1}{(n)_p} \frac{1}{h_n^p} \sum_{\beta(n,p)} g(Y_{\beta(1)}, \ldots, Y_{\beta(p)}) \prod_{j=1}^p K\left(\frac{t_j - X_{\beta(j)}}{h_n}\right), \quad n \geq p$$

and $U_n(1, \mathbf{t})$ is obtained by taking $g \equiv 1$ in $U_n(g, \mathbf{t})$. Here

$$\beta(n,p) = \{(\beta(1), \ldots, \beta(p)) : 1 \leq \beta(i) \leq n, \ \beta(i) \neq \beta(j) \ \forall \ i \neq j\},$$

$$(n)_p = card(\beta(n,p)) = \frac{n!}{(n-p)!}.$$

Stute (1991) established strong consistency and asymptotic normality of $U_n^g(t)$ as an estimator of $m^g(t)$ under some conditions. Arushaka Sen (1994) obtained uniform strong consistency over compact subsets of R^p. Note that if

$$g(Y_1, Y_2) = \frac{(Y_1 - Y_2)^2}{2},$$

then

$$m_1^g(t, t) = E[g(Y_1, Y_2)]X_1 = t, \ X_2 = t) = Var(Y_1|X_1 = t)$$

and if

$$g(Y_1, Y_2, Y_3) = I(Y_1 - Y_2 - Y_3 > 0),$$

then

$$m_2^g(t, t, t) = P(Y_1 > Y_2 + Y_3|X_1 = t, X_2 = t, X_3 = t).$$

The functional $m_1^g(\cdot, \cdot)$ is the conditional variance of Y given X and the functional $m_2^g(\cdot, \cdot)$ can be estimated by $U_n(g, t)$ which is a conditional analogue of Hollander-Proschan test statistic (cf. Hollander and Proschan (1972), Bochynak (1987)). Prakasa Rao and Arusharka Sen (1995a) have obtained general description of the limit distributions of such conditional U-statistics in terms of multiple Wiener integrals.

3. NONPARAMETRIC DENSITY ESTIMATION FOR POSITIVE VALUED RANDOM VARIABLES

If a random variable is positive valued, it is clear that its density function whenever it exists will have its support contained in $(0, \infty)$. In such a case, it would be reasonable to expect that any estimator of such a density should also have its support in $(0, \infty)$. If an estimator of kernel type as described in Sec. 2.1 is used, then the estimator $f_n(x)$ might take positive values even for $x \in (-\infty, 0]$ which is not desirable. Silverman (1986) suggests some adaptations of existing methods to avoid this problem. One possible way to ensure that $f_n(x)$ is zero for negative x is to calculate the estimator for positive x ignoring the boundary condition and then to set $f_n(x)$ to be zero for negative x. A drawback of this estimator is that the estimator so obtained may not be a probability density as it might not integrate to unity. Another possible approach is to transform the data, say for example, by the logarithmic transformation, estimate the density from the transformed data set and then transform the density estimator so obtained to that corresponding to the original data set. In a recent paper, Loader (1996) introduced a local likelihood density estimation method which involves smoothing by local polynomials. This method models the logarithm of the density and is useful when estimating the tails of densities. We now present an alternate way for the estimation of the density of a positive random variable due to Bagai and Prakasa Rao (1995a).

Our proposal is to start with a kernel $K(\cdot)$ with support on $(0, \infty)$ satisfying the condition

$$\int_0^\infty x^2 K(x)\, dx < \infty \tag{3.1}$$

and define a kernel type estimator for $f(x)$ given by

$$f_n(x) = \frac{1}{nh_n} \sum_{i=1}^n K\left(\frac{x - X_i}{h_n}\right), \quad x \ge 0. \tag{3.2}$$

Assume that f is twice continuously differentiable on $(0, \infty)$. Then it can be shown that

$$E[f_n(x) - f(x)]^2 \simeq (nh_n)^{-1} f(x) \int_0^\infty K^2(y)\, dy + h_n^2 (f'(x))^2 \gamma_1^2 \tag{3.3}$$

where $f'(\cdot)$ is the first derivation of $f(\cdot)$ and

$$\gamma_1 = \int_0^\infty x K(x)\, dx. \tag{3.4}$$

The integrated mean square error (IMSE) is given by

$$U_n^2 = \int_0^\infty E[f_n(x) - f(x)]^2 \, dx$$

$$\simeq (nh_n)^{-1} \int_0^\infty K^2(y) \, dy + h_n^2 \gamma_1^2 \int_0^\infty (f'(y))^2 \, dy$$

It is easy to see that U_n^2 is minimum when

$$h_n \simeq \left(\frac{\beta_0}{\gamma_1^2 \int_0^\infty (f'(y))^2 \, dy} \right)^{1/3} n^{-1/3} \tag{3.5}$$

where

$$\beta_0 = \int_0^\infty K^2(y) \, dy. \tag{3.6}$$

It follows by standard arguments using the calculus of variations that the optimal kernel minimizing IMSE is given by

$$K(x) = \begin{cases} \dfrac{2}{3a} - \dfrac{2x}{9a^2}, & 0 \le x \le 3a \\ 0 & \text{otherwise} \end{cases} \tag{3.7}$$

among the class of all probability densities $K(\cdot)$ satisfying

$$\int_0^\infty K(x) \, dx = 1, \quad \int_0^\infty x K(x) \, dx = a, \tag{3.8}$$

where a is fixed positive number.

Based on simulation results, we recommend the choice of exponential kernel with mean a for the estimation of positive densities. The behavior of the kernel estimator with the exponential kernel is as good as that with the optimal kernel given by Eq. (3.7) subject to Eq. (3.8).

4. NONPARAMETRIC ESTIMATION FOR DISCRETE TIME STOCHASTIC PROCESSES

Apart from the intrinsic mathematical interest in generalization of the results in Sec. 2 to the case of estimation of finite dimensional densities for stochastic process framework, these results are of great interest and are of practical use due to the recent upsurge in using nonparametric techniques in image processing, pattern recognition and time series analysis. We will come back to this aspect later in this review. In the chapter on "Estimation for Stochastic Processes" in Prakasa Rao (1983), we have given results on estimation of marginal densities for discrete and continuous time stationary Markov processes and stationary ϕ-mixing processes. The problem of nonparametric estimation of the drift coefficient for diffusion processes was also discussed. We now present some recent work in the area.

4.1 Associated Sequences

Let $\{X_n, n \geq 1\}$ be a strictly stationary sequence of associated random variables with the one-dimensional marginal density f, distribution function F, survival function $\bar{F} = 1 - F$ and the failure rate function $r_F = f/\bar{F}$. The random variables $\mathbf{X} = \{X_1, \ldots, X_n\}$ are said to be associated if for every pair of functions $h(\mathbf{x})$ and $g(\mathbf{x})$ from R^n to R which are nondecreasing componentwise

$$\text{Cov}(h(\mathbf{X}), g(\mathbf{X})) \geq 0 \qquad (4.1)$$

whenever it is finite. An infinite family is said to be associated if every finite subfamily is associated.

Probabilistic properties of associated sequences have been discussed by Newman (1980, 1984) and Birkel (1988a, b) among others. In reliability studies, the random variables which are generally the life times of components may not be independent but associated. For example, if the failure times of a system follow the multivariate exponential distribution (cf. Marshall and Olkin (1967)), then the component life times are associated. Order statistics corresponding to a finite set of random variables are associated. If independent components of a system are subject to the same stress, then their life times are associated. Thus there is a need to study the problem of estimation of survival function, density function etc. for such sequences. This was done by Bagai and Prakasa Rao (1991, 1995b). As far as we are aware, the only other publications dealing with this inferential problem are by Roussas (1991, 1993). Recently Yu (1993) studied the weak convergence for empirical processes of associated sequences. It is known that the covariances structure of associated random variables controls the nature of their approximate independence. For instance if two random variables are associated and uncorrelated, then they are independent. Some probabilistic properties of associated sequences useful for nonparametric inference were derived in Newman and Wright (1981), Prakasa Rao (1993) and Dewan and Prakasa Rao (1997, 1997a). Strong law of large numbers and central limit theorem for U-statistics of associated random variables were proved in Dewan and Prakasa Rao (1997b, 1997c). Recent results on associated sequences and related inference problems are surveyed in Prakasa Rao and Dewan (1998).

Nonparametric density estimation

Suppose $\{X_n, n \geq 1\}$ is a strictly stationary associated sequence and the problem is estimation of the one-dimensional marginal density of X_1 (say) f. Bagai and Prakasa Rao (1995b) proposed a kernel type estimator

$$f_n(x) = \frac{1}{nh_n} \sum_{i=1}^{n} K\left(\frac{x - X_i}{h_n}\right). \tag{4.2}$$

THEOREM 4.1.1   Bagai and Prakasa Rao (1995b): Assume that the following conditions hold:

(A_0) Support of f is a closed interval $I = [a, b]$ in R and f is Lipschitzian;
$(A_1) K(\cdot)$ is a bounded density function and is of bounded variation on R satisfying

(i) $\lim_{|u| \to \infty} u|K(u) = 0$

(ii) $\int_{-\infty}^{\infty} u^2 K(u)\, dy < \infty;$ (4.3)

(B) for all ℓ and $r \geq 0$,

$$\sum_{j: |\ell - j| \geq r} \mathrm{Cov}(X_j, X_\ell) \leq u(r) = O(e^{-\alpha r}) \tag{4.4}$$

for some $\alpha > 0$. Then

$$\sup_{x \in I} |f_n(x) - f(x)| \to 0 \text{ as } n \to \infty \tag{4.5}$$

provided

$$h_n^{-4} = O(n^\gamma) \text{ for some } \gamma > 0. \tag{4.6}$$

A general method of density estimation for associated random variables is studied in Dewan and Prakasa Rao (1998). Sufficient conditions for the exponential rate for uniform consistancy of density estimator are discussed here.

Nonparametric estimation of failure rate

An obvious estimator for the failure rate

$$r(X) = \frac{f(x)}{\bar{F}(x)} \qquad \text{of } X_1 \text{ is} \qquad r_n(x) = \frac{f_n(x)}{\bar{F}_n(x)}$$

where $f_n(x)$ is a kernel type density estimator of $f(x)$ as discussed above and $\bar{F}_n(x)$ is the empirical survival function. Asymptotic properties of $r_n(x)$ can be derived from those of $f_n(x)$ and $\bar{F}_n(x)$. For details, see Bagai and Prakasa Rao (1995b).

4.2 Exchangeable sequences

Suppose $\{X_n, n \geq 1\}$ forms an exchangeable (interchangeable) sequence of random variables in the sense that the joint distribution of any finite subset of k of these random variables depends only on k and not on the particular subset for every $k \geq 1$.

Kingman (1978, 1982) gives applications of the notion of exchangeability in population genetics in the study of evolution of large populations. For a comprehensive discussion of earlier work on exchangeability, see Koch and Spizzichino (1982).

It is known that if $\{X_n, n \geq 1\}$ is exchangeable, then there exists a random variable V such that conditional on V, the random variables $\{X_n, n \geq 1\}$ are independent and identically distributed (cf. Chow and Teicher (1978)).

Suppose a mixing experiment is conducted in two stages. Let V be a random variable observed at the first stage with density $g_V(\cdot)$ and given $V = v$, suppose that a sample of iid observations $X_1, \ldots, X_n$ are obtained with density $f_{X|V}(x|v)$. The marginal density of X is

$$f_X(x) = \int_k f_{X|V}(x|v) g_V(v)\, dv. \tag{4.7}$$

Observe that $X_1, X_2, \ldots, X_n$ is an exchangeable sequence unconditionally. Another possible interpretation of the model is that V can be though of as a nuisance parameter with prior density $g_V(\cdot)$ and $X_1, X_2, \ldots, X_n$ are iid random variables with density $f_{X|Y}(x|v)$ given $V = v$.

The problem is to estimate $f_X(\cdot)$ based on the sample $X_1, \ldots, X_n$. Following the earlier discussions, one can choose

$$\hat{f}_n(x) = \frac{1}{nh_n} \sum_{i=1}^{n} K\left(\frac{x - X_i}{h_n}\right) \tag{4.8}$$

as an estimator for $f(x)$ where $\{h_n\}$ is a suitable bandwidth and $K(\cdot)$ is an appropriate kernel. Under some regularity conditions on $\{h_n\}, K(\cdot)$ and $f_{X|V}(x|v)$, it can be shown that

(i) $\int_{-\infty}^{\infty} |\hat{f}_n(x) - f(x)|\, dx \to 0$ a.s as $n \to \infty$ and

(ii) $E[\hat{f}_n(x)] \to f(x)$ at all continuity points x of $f_X(\cdot)$. However $(nh_n)^{1/2}[\hat{f}_n(x) - f_{X|V}(x|V)]$ converges asymptotically to a mixture of normal densities unlike in the iid case. In other words, the kernel type density estimator $f_n(x)$ is not asymptotically normal. The model discussed here is a type of non-ergodic model in the nonparametric framework. For further details, see Prakasa Rao (1990a).

4.3   Markov Processes

The problem of estimation of the marginal densities for stationary Markov processes was extensively discussed in the literature. See Prakasa Rao (1983) for some details. The basic assumption used in all the earlier discussion is that the Markov process satisfies the Doeblin's condition (cf. Doob (1953)) which in turn implies geometric ergodicity for the process. In their recent work, Athreya and Atuncar (1994) have dispensed with this condition. For instance, the autoregressive process $\{X_n\}$ where

$$X_n + 1 = \rho X_n + \varepsilon_{n+1} \tag{4.9}$$

with $|\rho| < 1$ and $\{\varepsilon_n, n \geq 1\}$ are iid, uniform $(-1, 1)$ does not satisfy the Doeblin's condition. They have studied nonparametric estimation of marginal density for such Markov processes which are Harris recurrent (cf. Athreya and Ney (1978)).

It can be seen that the usual kernel type density estimator is not recursive. Acquisition of additional observations necessitates computation of the estimator all over again. In order to avoid this problem, recursive kernel type density estimators were studied in the case of iid observations. For a detailed survey, see Prakasa Rao (1983). These type of estimators are amenable for analysis in the dependent case, have been studied for Markov processes and have been found applicable in the recent literature on nonparametric inference for time series analysis (cf. Prakasa Rao (1996b)).

DEFINITION 4.3.1   Let $\{X_n, n \geq 1\}$ be a stationary Markov process and define the transition operator H_n by

$$(H_n g)(x) = E[g(X_{n+1})|X_1 = x]$$

where g is any bounded measurable function on the real line. Define

$$|H_n|_2 = \sup_{\{g : E[g(X_1)]=0\}} \{E^{1/2}(H_n g)^2 / E^{1/2}(g^2)\} \tag{4.10}$$

(cf. Prakasa Rao (1983), p. 322). The transition operator is said to satisfy $G_2(m, \alpha)$-condition if there exists a positive integer m such that

$$|H_m|_2 \leq \alpha \tag{4.11}$$

where $0 < \alpha < 1$.

The following result gives equivalence relation on the universal consistency of a recursive density estimator of kernel type for f.

THEOREM 4.3.1 Mishra and Prakasa Rao (1995). Suppose the process $\{X_n\}$ is a stationary Markov process satisfying the condition $G_2(m, \alpha)$ with one-dimensional marginal density f assuming that it exists. Further suppose that the kernel $K(\cdot)$ is a bounded density function satisfying

$$\int_0^\infty \gamma(u)\, du < \infty \quad \text{where} \quad \gamma(u) = \sup_{|x| \ge u} K(x), \qquad u \ge 0 \tag{4.12}$$

and the sequence $\{h_n\}$ satisfies the conditions

$$h_n \downarrow 0 \text{ and } \sum_{i=1}^n h_i \simeq n^r \quad \text{where} \quad \frac{3}{4} < r < 1 \tag{4.13}$$

as $n \to \infty$. Define

$$f_n(x) = \frac{1}{\sum_{i=1}^n h_i} \sum_{i=1}^n K\left(\frac{x - X_i}{h_i}\right). \tag{4.14}$$

Then the following conditions are equivalent;

(i) $f_n(x) \to f(x)$ almost surely, almost all x, all f;
(ii) $f_n(x) \to f(x)$ in probability, almost all x, some f;
(iii) $\lim_{n\to\infty} \{\sum_{i=1}^n h_i I(h_i > \varepsilon)/\sum_{i=1}^n h_i\} = 0$ for all $\varepsilon > 0$;
(iv) $\int_{-\infty}^\infty |f_n(x) - f(x)|\, dx \to 0$ almost surely, all f;
(iv) $\int_{-\infty}^\infty |f_n(x) - f(x)|\, dx \to 0$ in probability, some f.

4.4 Time Series

A time series is a set of observations recorded over a period of time. Examples of such series include monthly rain fall data or daily river flow data or decennial population census of a country recorded over a long period of time, etc. Traditionally, analysis of such time series data has been carried out by selecting suitable parametric models such as autoregressive or moving average processes and then drawing inference from such series.

Suppose the series $\{Z_t, 0 \le t < \infty\}$ is modeled by a discrete time real-valued stationary process. If the process is a gaussian process, then it is determined by the mean and covariance function which in turn can be estimated efficiently under some weak dependence conditions. It has been the practice to use the methods developed for the gaussian case for other models simply because of lack of suitable methods in the latter case.

Robinson (1983) gives several reasons why a nonparametric approach to time series analysis is of interest. For instance, nonparametric methods can be used to estimate the finite dimensional densities and these can be used for detecting gaussianity or non-gaussianity of the process. Nonparametric approach can also be used for prediction, extending the nonparametric methods of estimation of the regression function, of some processes satisfying weak dependence conditions such as mixing.

Suppose $\{Z_i, i \geq 1\}$ is a real-valued stationary second order process. The problem of interest is to predict Z_{n+1} given $Z_1, \ldots, Z_n$. Since the process is a second order process, the best predictor (with respect to the quadratic loss function) is the conditional expectation $E(Z_{n+1}|Z_1, \ldots, Z_n)$ which is nothing but the regression function of Z_{n+1} given $Z_1, \ldots, Z_n$. If the process is also Markov of order k (say), then the best predictor has the representation $R(Z_{n-k+1}, \ldots, Z_n) = E(Z_{n+1}|Z_{n-k+1}, \ldots, Z_n)$. Since the probability structure of the process $\{Z_n\}$ is unknown, one has to estimate the best predictor given $Z_{n-k+1}, \ldots, Z_n$. Hence, a natural predictor of Z_{n+1} given $Z_1, \ldots, Z_n$ is an "estimator" of $R(Z_{n-k+1}, \ldots, Z_n)$. The regression function $R(Z_{n-k+1}, \ldots, Z_n)$ can be estimated by $R_n(Z_{n-k+1}, \ldots, Z_n)$ where R_n is a nonparametric regression estimator using the techniques described in Sec. 2.2. Note that Z_{n+1} replaces $g(Y)$ and the vector $(Z_{n-k+1}, \ldots, Z_n)$ replaces $\mathbf{X}$ in that discussion.

Properties of such estimtors and other related results are reviewed in Prakasa Rao (1996b). For an application to time series, see Yakowitz (1985).

4.5 Galton–Watson Type Process

Let $\{X_n, n \geq 0\}$ be a non-negative valued Markov process such that $X_0 = 1$ and

$$E[e^{-tX_{n+1}}|X_n] = e^{h(t)X_n}, \qquad n \geq 0, t \geq 0 \tag{4.15}$$

where $h(t)$ is the cumulant generating function of X_1 with an infinitely divisible off-spring distribution. Such a process is called a Galton–Watson type process. If X_1 is a non-negative integer-valued random variable, then the process $\{X_n\}$ is an example of the classical Galton–Watson branching process. Since the basic relation defining the process $\{X_n, n \geq 0\}$ is through the cumulant generating function $h(\cdot)$ of X_1, it is of interest to estimate $h(\cdot)$ based on the observations $X_1, \ldots, X_n$. Note that $\{X_n, n \geq 0\}$ is a non-stationary Markov process.

Kallenberg (1979) studied such processes naming them as branching processes with continuous state space generalizing the earlier work of Jirina

(1958). The motivation for the study of such processes is that the size of the population at any stage can be measured by its total weight or volume for example, instead of just by the number of individuals present at that stage as in the context of a classical branching process.

In analogy with the nonparametric estimation of the regression function as discussed in Sec. 2.2 (cf. Prakasa Rao (1983)), define $\hat{h}_n(t)$ as an estimate of $h(t)$ by the relations

$$\hat{a}_n(x) \equiv \hat{q}_n(x)e^{-\hat{h}_n(t)x} = \frac{1}{nb_n}\sum_{i=1}^{n} K\left(\frac{x-X_i}{b_n}\right)e^{-tX_{i+1}}\, d_i, \qquad (4.16)$$

and

$$\hat{q}_n(x) = \frac{1}{nb_n}\sum_{i=1}^{n} K\left(\frac{x-X_i}{b_n}\right)d_i \qquad (4.17)$$

where $K(\cdot)$ is a suitable kernel, $0 < b_n \to 0$ as $n \to \infty$ and $\{d_i\}$ is sequence of real numbers related to the densities f_i of X_i satisfing

$$\frac{\sum_{j=1}^{n} d_j f_j(y)}{n} \to \psi(y) \quad \text{as} \quad n \to \infty, \qquad y \in R \qquad (4.18)$$

for some function $\psi(\cdot)$. Note that x is arbitrary and the above definition of $\hat{h}_n(t)$ gives a family of estimators for $h(t)$. It can be shown that

$$-\frac{1}{x}\log\frac{\hat{a}_n(x)}{\hat{q}_n(x)} \xrightarrow{p} h(t) \quad \text{as} \quad n \to \infty \qquad (4.19)$$

for all x at which ψ is continuous and $\psi(x) \neq 0$ under some regularity conditions. For details, see Prakasa Rao (1992a). The results obtained are not satisfactory and the problem of obtaining suitable estimators and study of their asymptotic properties remains open.

4.6 Processes With Long-Range Dependence

As was mentioned at the beginning of this section, results of nonparametric density estimation in the classical iid case have been extended to processes which are short range dependent such as mixing procersses provided the mixing coefficient tends to zero at a sufficiently fast rate. Detailed discussion of such results is given in Prakasa Rao (1983) and more recently in Gyorfi et al. (1989). Rosenblatt (1991) indicated that the problem is difficult for processes which are long-range dependent. An example of such a process which is long-range dependent is $X_k = Y_k^2 - 1, -\infty < k < \infty$ where $\{Y_k\}$ is a gaussian process with $EY_k = 0$ and covariance $E(Y_\ell Y_{\ell+k}) = (1+k^2)^{-\gamma}, \gamma > 0, -\infty < \ell, k < \infty$ (cf. Rosenblatt (1961)). This process is not strong mixing and in fact $n^{-(1+2\gamma)}\sum_{k=1}^{n}X_k$ has a non-gaussian limiting distribution (cf. Rosenblatt (1991)).

Consider the stationary process $Y_k = G(X_k)$, $-\infty < k < \infty$ where $\{X_k\}$ is a stationary gaussian process with mean zero and covariance $E(X_k X_{k+\ell}) = g|\ell|^{-\alpha}$ with $0 < \alpha < 1$. It is known that the process $\{X_k\}$ has asymptotic correlation zero with mixing coefficient, $\rho(s) = o(|s|^{-\beta})$, $\beta < \alpha - 1$ (cf. Ibragimov and Rozanov (1978)). If $\alpha > 1$, then the process is mixing and hence short-range dependent and one can estimate the marginal densities by methods given in Prakasa Rao (1983). However if $0 < \alpha < 1$, then the process $\{X_k\}$ is long-range dependent and so is $\{Y_k\}$. Suppose we want to estimate the density f of Y_k by the kernel method. Let

$$f_n(y) = \frac{1}{nh_n}\sum_{j=1}^{n}K\left(\frac{y-Y_j}{h_n}\right).$$

Suppose $n^{1-\alpha}h_n \to 0$ as $n \to \infty$ and $G^{-1}(y) \neq 0$. Then Rosenblatt (1991) proved that

$$n^{\alpha/2}[f_n(y) - Ef_n(y)] \xrightarrow{L} N(0,\sigma_y^2) \quad \text{as} \quad n \to \infty \tag{4.20}$$

where

$$\sigma_y = G^{-1}(y)\exp\left(-\frac{G^{-1}(y)^2}{2}\right)\frac{1}{\sqrt{2\pi}}|G'(G^{-1}(y))|^{-1}.$$

It would be interesting to study the problems of density estimation and estimation of regression function when the observations form a long-range dependent sequence. For recent work, see Cheng and Robinson (1991) Csorgo and Mielniczuk (1995a,b).

5. NONPARAMETRIC ESTIMATION FOR CONTINUOUS TIME STOCHASTIC PROCESSES

Nonparametric estimation for continuous time stationary processes has been the subject of investigation for the last several years. Banon (1977) and Prakasa Rao (1979a) studied the problem for continuous time stationary Markov processes. Results for continuous time stationary processes which are of mixing type were obtained by Delecroix (1980) among others. For a comprehensive discussion, see Prakasa Rao (1983). One of the basic assumptions used in all the papers cited is that the process is completely observable over any given period. In practice, it is impossible to observe the path continuously for various reasons such as lack of precision of measuring instruments or unavailability of observations at all time points as in the medical studies etc. In other words the problem of the estimation of density from sampled data of the process is of much importance in practical problems. Sampling instants for observing the process may be of several types. They might be regularly (equally) spaced or irregularly spaced. Sampling may be done at fixed times or at random times. Random times of observation may depend on the process under observation or may be independent of the process.

Most of the results on estimation for continuous time processes assume some type of weak dependence or equivalently asymptotic independence of some type. It is not clear whether conditions of mixing type for asymptotic independence for a process are inherited by the process obtained by sampling from the original process at random times. Some aspects of this problem were investigated in Prakasa Rao (1990b).

We now briefly discuss these results.

Let $(\Omega, \mathscr{F}, P)$ be a probability space. Let $\{\mathscr{F}_t, t \geq 0\}$ be an increasing flow of sub σ-algebra of $\mathscr{F}$ and $\{\zeta_t, t \geq 0\}$ be a decreasing flow of sub σ-algebras of $\mathscr{F}$. For any real-valued non-negative random variable τ defined on $(\Omega, \mathscr{F}, P)$, let $\mathscr{F}_\tau$ be the σ-algebra generated by sets $A \in \mathscr{F}$ such that $A \cap [\tau \leq t] \in \mathscr{F}_t$ for every $t \geq 0$ and ζ_τ be the σ-algebra generated by sets $B \in \mathscr{F}$ such that $B \cap [\tau \geq s] \in \zeta_s$, $s \geq 0$.

DEFINITION 5.1 Let $\{\tau_n, n \geq 1\}$ and $\{S_n, n \geq 1\}$ be increasing sequences of non-negative random variables defined on $(\Omega, \mathscr{F}, P)$. The increasing flow $\{\mathscr{F}_t\}$ is said to be ϕ-mixing strongly with the decreasing flow $\{\zeta_s\}$ with respect to the sequences $\{\tau_n\}$ and $\{S_n\}$ if for every $A \in \mathscr{F}_{\tau_n}$, $n \geq 1$ and $B \in \zeta_{S_m}$, $m \geq 1$.

(i) $|P(A \cap B) - P(A)P(B)| \leq E\{\phi(|\tau_n - S_m|)\}P(A)$

and

(ii) $E\{\phi(|\tau_n - S_m|)\} \to 0$ as $|\tau_n - S_m| \overset{p}{\to} \infty$ as $m \to \infty$.

If the condition (i) holds for every pair $\{\tau_n\}$ and $\{S_m\}$ satisfying (ii), then the increasing flow $\{\mathscr{F}_t\}$ is said to be ϕ-mixing strongly with the decreasing flow $\{\zeta_s\}$.

Examples of families of sub σ-algebras which are ϕ-mixing strongly are discussed in Prakasa Rao (1990b). An example which is of interest in the present context is the following. We first extend the notion of ϕ-mixing for stochastic processes.

DEFINITION 5.2 Suppose $\{X_t, t \geq 0\}$ is a progressively measurable stochastic process defined on a probability space $(\Omega, \mathscr{F}, P)$. Let $\{\tau_n, n \geq 1\}$ be an increasing sequence of non-negative random variables defined on

$(\Omega, \mathcal{F}, P)$ such that $|\tau_n - \tau_{n+m}|p \xrightarrow{p} \infty$ as $m \to \infty$ for every $n \geq 1$ Let $\mathcal{F}_t^X$ be the σ-algebra generated by $X_u, 0 \leq u \leq t$ and ζ_s^X be the σ-algebra generated by $X_v, v \geq s$. Define $\mathcal{F}_\tau^X$ and ζ_S^X as before for non-negative random variable τ and S defined on $(\Omega, \mathcal{F}, P)$. If $\{\mathcal{F}_{\tau_n}\}$ is ϕ-mixing strongly with the flow $\{\zeta_{S_{m,n}}^X\}$ with respect to $\{\tau_n\}$ and $\{S_{m,n}\}$ where $S_{m,n} = \tau_{n+m}$ for every fixed $n \geq 1$, then $\{X_t\}$ is said to be ϕ-mixing strongly with respect to $\{\tau_n\}$.

THEOREM 5.1   Suppose $\{X_t, t \geq 0\}$ is a stationary ϕ-mixing progressively measurable stochastic process defined on a probability space $(\Omega, \mathcal{F}, P)$. Let $\{\tau_n\}$ be an increasing sequence of non-negative random variables defined on $(\Omega, \mathcal{F}, P)$ and independent of $\{X_t, t \geq 0\}$. Further suppose that $|\tau_n - \tau_{n+m}| \xrightarrow{p} \infty$ as $m \to \infty$ for every $n \geq 1$. Then $\{X_t, t \geq 0\}$ is ϕ-mixing strongly with respect to $\{\tau_n\}$.

For the proof of Theorem 5.1 and related results, see Prakasa Rao (1990b).

Ait-Sahalia (1996) considered the problem of nonparametric estimation of the drift and the volatility function from a discrete data sampled from the interest rates process $\{r_t\}$ which is modeled as a diffussion process satisfying the stochastic differential equation.

$$dr_t = \mu(r_t)\, dt + (r_t)\, dW_t, \qquad t \geq 0.$$

In the course of the discussion, it is assumed that the observed data sequence $\{r_k, 1 \leq k \leq n\}$ is a strictly stationary β-mixing sequence satisfying $k^\delta \beta_k \to 0$ from some fixed $\delta > 1$ as $k \to \infty$. In order that the diserelity observed process $\{r_k \geq 1\}$ satisfy this condition, it is sufficient that the stochastic process $\{r_t, t \geq 0\}$ verifies the stronger continuous-time mixing condition as given in Prakasa Rao (1990b). Ait-Sahalia (1996) discussed an alternate condition ensuring the mixing property for the discrete sequence $\{r_k, k \geq 1\}$.

Another important aspect there should be checked is whether the "information" obtained from the sampled data retains all the "information" contained in the process. For instance if $\{X_n\}$ is any process, then all the "information" about $\{X_n\}$ is not recoverable in general from a subsequence $\{X_{k_n}\}$. For instance if $k_n = nk$ for some fixed $k > 1$, it may be impossible to estimate the joint distribution of (X_1, X_2) from observation on $\{X_{k_n}\}$. Problem of the nature was investigated for the discrete processes by Blum and Rosenblatt (1964) and for continuous time processes by Blum and Boyles (1981).

A third aspect is the choice of optimum sampling design so that there is no loss of information as discussed earlier. We will not discuss this issue here. Some work in this direction is due to Masry (1983).

We shall now briefly dicuss some work in Prakasa Rao (1990c).

Suppose a stationary process $\mathbf{X} = \{X_t, t \geq 0\}$ is observed at times $\{\tau_i, i \geq 0\}$ with $\tau_0 = 0$. The problem is to estimate the one-dimensional (say) marginal density of X_t (assuming that it exists) from $\{X_{\tau_i}, 0 \leq i \leq n\}$.

DEFINITION 5.3 A family of functions $\{\delta_t(x), t \geq 0\}$ is said to be a family of delta-type if

(i) $\int_{-\infty}^{\infty} |\delta_t(x)| \, dx < A < \infty$, $\quad t \geq 0$,
(ii) $\int_{-\infty}^{\infty} \delta_t(x) \, dx = 1$, $\quad t \geq 0$,
(iii) $\delta_t(x) \to 0$ as $t \to \infty$ uniformly for $|x| > \lambda$, $\lambda > 0$, and
(iv) $\int_{|t| \geq \lambda} |\delta_t(x)| \, dx \to 0$ as $t \to \infty$ for all $\lambda > 0$.

An example of a family of delta type is

$$\delta_t(x) = \frac{1}{h_t} K\left(\frac{x}{h_t}\right), \qquad t \geq 0$$

where $K(\cdot)$ is a bounded probability density such that $|xK(x)| \to 0$ as $|x| \to \infty$, $h_t > 0$, $h_t \to 0$ as $t \to \infty$. Density estimation for continuous time Markov process via delta families is discussed in Prakasa Rao (1979a). It is assumed that the family of functions $\{\delta_t(x), t \geq 0\}$ is such that the process $\{\delta_t(x - X_t), 0 \leq t \leq T\}$ is $\mathscr{F}_t$-predictable for every x and every $T \geq 0$. Let

$$\hat{f}_T(x) = \frac{1}{T} \sum_i^{*} \delta_T(x - X_{\tau_i})$$

where Σ^* is the sum over i with jump times τ_i in $[0, T]$. Properties of $\hat{f}_T(x)$ as an estimator of $f(x)$ are discussed in Prakasa Rao (1990c). The interesting point is that $\hat{f}_T(x)$ can be written in the form

$$\hat{f}_T(x) = \frac{1}{T} \int_0^T \delta_T(x - X_s) \, dN_s,$$

and one can bring in stochastic calculus and martingale techniques to study its properties. Let

$$f_T^*(x) = \frac{1}{T} \int_0^T \delta_T(x - X_s) \lambda_s \, ds.$$

Then

$$J_T(x) = \hat{f}_T(x) - f_T^*(x) = \frac{1}{T} \int_0^T \delta_T(x - X_s) \, dM_s,$$

where $M_s = N_s - \int_0^s \lambda_t \, dt$, $s \geq 0$ is a zero-mean martingale.

Extensive discussion on the behaviour of $\{J_T(x), \ T \geq 0\}$ and the properties of the estimator $\hat{f}_T(x)$ are given in Prakasa Rao (1990c). A comprehensive review of statistical inference for sampled data for stochastic processes both from the parametric and nonparametric points of view is given in Prakasa Rao (1988).

6. NONPARAMETRIC ESTIMATION IN ECONOMETRICS

In an extensive review, Ullah (1987) discussed the need for the use of nonparametric methods for estimation of econometric functionals. Our discussion here is partly based on Ullah (1987). Since the introduction of econometrics as a subject, the estimation and testing of econometric models is done under many strong parametric assumptions. In the econometric model

$$Y = m(\mathbf{X}) + \varepsilon, \tag{6.1}$$

Y is the response (dependent) variable, $\mathbf{X}$ is a vector of regressions, ε is the disturbance (noice, error) and $m(\mathbf{X}) = E(Y|\mathbf{X})$ is an unspecified regression function. The main assumption of parametric econometrics is that the form of $m(\mathbf{x})$ (usually linear) is known. It is also assumed that the conditional variance of Y given $\mathbf{X}$ and the autocovariance of ε are also specified. Finally, it is assumed that the joint density of $(Y, \mathbf{X})$ is multivariate normal. Obviously all these assumptions are made to circumvent the problem of lack of knowledge of the functional form of $m(\mathbf{x})$. It is clear that any misspecification in this regard may lead to erroneous conclusions.

Apart from the regression function $m(\mathbf{x})$, there are other function such as the response function (regression coefficient) corresponding to x_j viz

$$\beta_j(\mathbf{x}) = \frac{\partial m(\mathbf{x})}{\partial x_j},$$

or the average derivative (fixed response coefficient) $p_j = E[\beta_j(\mathbf{X})]$ which are of interest in econometric studies. Economic theory imposes conditions such as concavity or convexity on $m(\mathbf{x})$. For instance, the demand function is assumed to be concave or the profit function convex in price. If the form of $m(\mathbf{x})$ is known, this can be verified possibly from the second partial derivatives. The problem is whether or not the unknown $m(\mathbf{x})$ satisfies these curvature conditions. This may be verified by estimating its second-order derivatives. There are several other such functions in econometric studies which need to be estimated form the data. A common feature of these functions is that they depend on the unknown joint density. If the joint density, marginal density and conditional densities of $(Y, \mathbf{X})$ can be estimated from the data, then one can get an idea of the forms of these functionals.

Our purpose here is not to go into a full discussion on this topic. We illustrate the results by looking at one particular aspect namely, estimation of the average derivative.

Estimation of average derivative

Let $(\mathbf{X}, Y) = (X_1, \ldots, X_p, Y)$ denote a random vector where Y is the response. Let

$$m(\mathbf{X}) = E(Y|\mathbf{X}).$$

Then the vector of "average derivatives" is given by

$$\delta = E(m')$$

where $m' = (\partial m/\partial x_1, \ldots, \partial m/\partial x_p)$ is the vector of partial derivatives and expectation is taken with respect to the marginal distributions of $\mathbf{X}$.

Suppose $m(\mathbf{x}) = g(\mathbf{x}'\beta)$. Here $\mathbf{x}'$ denotes the transpore of the vector $\mathbf{x}$. Then

$$\delta = E\left[\frac{dg}{d(\mathbf{X}'\beta)}\right]\beta = \gamma\beta$$

where γ is a scalar. Hence δ is proportional to the coefficient vector β when $\gamma \neq 0$. Härdle and Stoker (1989) give several examples of stochastic models such as the classical linear model $Y = \alpha + \mathbf{X}'\beta + \varepsilon$ or the model which is linear upto transformation such as

$$\phi(Y) = \psi(\mathbf{X}'\beta) + \varepsilon$$

where $\psi(\cdot)$ is not a constant and $\phi(\cdot)$ is invertible and where the above concept of average derivative can be interpreted as "coefficients" or index of changes in $\mathbf{X}$ and Y. Härdle and Stoker (1989) suggested a kernel method for estimation of average derivatives.

Let $f(\mathbf{x})$ denote the marginal density of $\mathbf{X}$ and $f' \equiv \partial f/\partial \mathbf{x}$ be the vector of partial derivative of f with respect to $\mathbf{x}$. Let

$$\ell(\mathbf{x}) \equiv \frac{-\partial \log f}{\partial \mathbf{x}} = \frac{-f'}{f}.$$

Assuming that $f(\mathbf{x}) \to 0$ as $|\mathbf{x}| \to \infty$, integration by parts, gives the relation,

$$\delta = E(m') = E[\ell(\mathbf{X})Y]. \tag{6.2}$$

An estimate for $\ell(\mathbf{x})$ is

$$\ell_n(\mathbf{x}) = \frac{f_n'(\mathbf{x})}{f_n(\mathbf{x})}$$

where

$$f_n(\mathbf{x}) = \frac{1}{nh_n} \sum_{i=1}^{n} K\left(\frac{\mathbf{x} - \mathbf{X}_i}{h_n}\right) \tag{6.3}$$

is a kernel type estimator for $f(\mathbf{x})$ and $f_n'(\mathbf{x}) \equiv (\partial f_n(\mathbf{x})/\partial \mathbf{x})$. Relation in Eq. (6.2) suggests the estimator

$$\delta_n^* = \frac{1}{n} \sum_{i=1}^{n} \ell_n(\mathbf{X}_i) Y_i \tag{6.4}$$

as an estimator for δ defined by Eq. (6.2). Since $f_n(\mathbf{x})$ may take possibly zero values, Härdle and Stroker (1989) modify the estimator to

$$\delta_n = \frac{1}{n} \sum_{i=1}^{n} \ell_n(\mathbf{X}_i) Y_i I(f_n(\mathbf{X}_i) > b_n) \tag{6.5}$$

where $0 < b_n \to 0$ as $n \to \infty$. They prove the following theorem under some technical conditions on $m(\cdot)$, $f(\cdot)$ and the following condition on the kernel $K(\cdot)$: The kernel $K(\cdot)$ has support $\{u| \; |u| \le 1\}$, is symmetric, has p moments and $K(u) = 0$ for $|u| = 1$. Further $K(u)$ is of order k i.e. $\int_0^T \cdots \int_0^T K(u) \, du = 1$, $\int_0^T \cdots \int_0^T u_1^{\ell_1} \cdots u_p^{\ell_p} K(u) \, du = 0$ for $\ell_1 + \cdots + \ell_p < k$ and $\ne 0$ for $\ell_1, \ldots + \ell_p = k$.

THEOREM 6.1 If

(i) $h_n \to 0, b_n \to 0$ such that $b_n^{-1} h_n \to 0$
(ii) for some $\varepsilon > 0, b_n^4 n^{1-\varepsilon} h_n^{2p+2} \to \infty$ and
(iii) $nh_n^{2k-2} \to 0$, then

$$\sqrt{n}(\hat{\delta}_n - \delta) \xrightarrow{\mathscr{L}} N(0, \Sigma) \tag{6.6}$$

where Σ is the covariance matrix of $R(Y, \mathbf{X})$ with

$$R(y, \mathbf{x}) = m'(\mathbf{x}) + (y - m(\mathbf{x}))\ell(\mathbf{x}). \tag{6.7}$$

Estimation of density-weighted average derivative

Powell, Stock and Stoker (1989) considered the problem of estimation coefficients of index models through the estimation of density-weighted average derivative of a regression function. The density weighted average derivative is defined by the relation

$$\gamma = E\left[f(\mathbf{X}) \frac{\partial m}{\partial \mathbf{X}}\right] \tag{6.8}$$

where f is the density of $\mathbf{X}$ and $m(\mathbf{X}) = E[Y|\mathbf{X}]$. It is easy to see, using integration by parts, that

$$\gamma = -2E\left[Y\frac{\partial f}{\partial \mathbf{X}}\right] \tag{6.9}$$

under some smoothness conditions on $f(\mathbf{x})$ and $m(\mathbf{x})$ and existence of appropriate second moments. Powell et al. (1989) suggest

$$\hat{\gamma}_n = -\frac{2}{n}\sum_{i=1}^{n}\frac{\partial \hat{f}_i}{\partial \mathbf{X}}\bigg|_{\mathbf{X}=\mathbf{X}_i} Y_i \tag{6.10}$$

as an estimator of γ where

$$\hat{f}_i(\mathbf{x}) = \frac{1}{n-1}\sum_{\substack{j=1\\j\neq i}}^{n}\frac{1}{h_n^p}K\left(\frac{\mathbf{x}-\mathbf{X}}{h_n}\right), \qquad 1 \leq i \leq n \tag{6.11}$$

are the kernel type density estimators of $f(\mathbf{x})$. Note that $\hat{f}_i(\mathbf{x})$ is independent of $\mathbf{X}_i$. It is assumed that $K(\cdot)$ is symmetric and differentiable. The estimator $\hat{\gamma}_n$ can be written in the form

$$\hat{\gamma}_n = \frac{2}{n(n-1)}\sum_{i=1}^{n}\sum_{\substack{j=1\\j\neq i}}^{n}\frac{1}{h_n^{p+1}}K'\left(\frac{\mathbf{X}_i-\mathbf{X}_j}{h_n}\right)Y_i \tag{6.12}$$

$$= -\binom{n}{2}^{-1}\sum_{i=1}^{n-1}\sum_{j=i+1}^{n}\frac{1}{h_n^{p+1}}K'\left(\frac{\mathbf{X}_i-\mathbf{X}_j}{h_n}\right)(Y_i - Y_j) \tag{6.13}$$

where $K'(\cdot) \equiv \partial K/\partial \mathbf{u}$ is the derivative of $K(\cdot)$ using the fact that $K'(\mathbf{u}) = -K'(-\mathbf{u})$ from the symmetry of $K(\cdot)$. The latter representation leads to a U-statistic structure for $\hat{\gamma}_n$. Powell et al. (1989) assume some technical smoothness conditions on m, f and that the kernel $K(\cdot)$ is assumed to be of the order k.

THEOREM 6.2   Powell et al. (1989) If $h_n \to 0$ as $n \to \infty$ such that $nh_n^{p+2} \to \infty$ and $nh_n^{2k} \to 0$, then

$$\sqrt{n}(\hat{\gamma}_n - \gamma) \xrightarrow{\mathscr{L}} N(0, \Sigma_\gamma) \tag{6.14}$$

where

$$\Sigma_\gamma = 4E[R(\mathbf{Z}_i)R(\mathbf{Z}_i)'] - 4\gamma\gamma', \tag{6.15}$$

$$R(\mathbf{Z}_i) = f(\mathbf{X}_i)\frac{\partial m(\mathbf{X}_i)}{\partial \mathbf{X}} - [Y_i - m(\mathbf{X}_i)]\frac{\partial f(\mathbf{X}_i)}{\partial \mathbf{X}}, \tag{6.16}$$

and

$$\mathbf{Z}_i = (\mathbf{X}_i, Y_i) \tag{6.17}$$

An alternate method for estimation of density-weighted average by the method of orthogonal series was given in Prakasa Rao (1995b) when $p = 1$.

Suppose the density f has an orthogonal series expansion

$$f(x) = \sum_{i=1}^{\infty} a_i e_i(x) \tag{6.18}$$

with respect to an orthonormal basis $\{e_i(x)\}$. Define

$$\tilde{\gamma}_n = -\frac{2}{n} \sum_{i=1}^{n} Y_i, \frac{\partial \hat{f}_n}{\partial X}\bigg|_{x=X_i} \tag{6.19}$$

where

$$\hat{f}_{ni}(x) = \sum_{\ell=1}^{q(n)} \hat{a}_{\ell n}^{(i)} e_\ell(x), \tag{6.20}$$

$$\hat{a}_{\ell n}^{(i)} = \frac{1}{n-1} \sum_{\substack{j=1 \\ j \neq i}}^{n} e_\ell(X_j) \tag{6.21}$$

and $q(n) \to \infty$ as $n \to \infty$. If $\{e_i\}$ are differentiable, then

$$\tilde{\gamma}_n = -\frac{2}{n} \sum_{i=1}^{n} Y_i \left[\sum_{\ell=1}^{q(n)} \sum_{\ell=1}^{q(n)} a_{\ell n}^{(i)} e_i'(X_i) \right]$$

$$= -\frac{2}{n} \sum_{\ell=1}^{q(n)} \sum_{i=1}^{n} \psi_\ell(X_i, Y_i) \eta_\ell(\mathbf{X}_n^{(i)}) \tag{6.22}$$

where

$$\psi_\ell(X_i, Y_i) = Y_i e_\ell'(X_i), \tag{6.23}$$

$$\eta_\ell(\mathbf{X}_n^{(i)}) = \hat{a}_{\ell n}^{(i)}, \tag{6.24}$$

$$\mathbf{X}_n^{(i)} = (X_1, \ldots, X_i - 1, X_{i+1}, \ldots, X_n). \tag{6.25}$$

Note that η_ℓ and ψ_ℓ are stochastically independent. Using this property and the above representations for $\tilde{\gamma}_n$, the following result can be proved. We assumed that $\{e_i\}$ and $\{e_i'\}$ are uniformly bounded and some additional regularity conditions hold.

THEOREM 6.3 Prakasa Rao (1995b): If $q(n) \to \infty$ such that $(q^2(n)/n) \to 0$ as $n \to \infty$, then

$$\tilde{\gamma}_n \xrightarrow{p} \gamma \quad \text{as} \quad \to \infty. \tag{6.26}$$

The problem of finding the limiting distribution of $\tilde{\gamma}_n$ is still open.

7. NONPARAMETRIC ESTIMATION FROM CENSORED DATA

Density estimation from censored data is of importance as this type of data arises in problems of reliability and survival analysis. In clinical trials, patients may enter for treatment at different times and then either die from the disease under investigation during study or leave the study before it is terminated (for other reasons such as death due to another competing disease or moving away). In industrial life testing, items may be removed from the study at different times for more extensive analysis. For such cases, the study of nonparametric estimation of the density function and hazard or failure rate of the life time variable based on the censored data is of great interest. Padgett and McNichols (1984) gives a review of the work. We briefly discuss some recent results.

Randomly censored data

Suppose $X_1, \ldots, X_n$ are iid observations which are at risk of being censored from the right by another sequence of iid random variables $Y_1, \ldots, Y_n$ with $\{X_i\}$ independent of $\{Y_i\}$. Suppose that

$$Z_i = \min(X_i, Y_i) \quad \text{and} \quad \delta_i = I[X_i \leq Y_i], \qquad 1 \leq i \leq n \qquad (7.1)$$

are observable. Our interest is in estimation of the density f of X_1 and the failure rate $r = f/(1 - F)$ of X_1 based on the observed sample (Z_i, δ_i), $1 \leq i \leq n$. Here F denotes the distribution function of X_1. The nonparametric maximum likelihood estimator of F is the Kaplan-Meier (1958) product limit estimator $\hat{F}_n$ defined by

$$1 - \hat{F}_n(x) = \prod_{i=1}^{n} \left\{ 1 - \frac{\delta_{in}}{n - i + 1} \right\}^{I\{Z_{in} \leq x\}}. \qquad (7.2)$$

Here $Z_{1n} \leq Z_{2n} \leq \cdots \leq Z_{nn}$ are the ordered statistics corresponding to $Z_1, \ldots, Z_n$ where ties within life times within censoring times are ordered arbitrarily and ties among life times and censored times are treated as if the life time preceeds the censored time. δ_{in} is the concommitant of the ith order statistic i.e., $\delta_{in} = \delta_j$ if $Z_{i,n} = Z_j$.

An estimator for the density f can be taken to be

$$f_n(t) = \frac{1}{h_n} \int_{-\infty}^{\infty} K\left(\frac{t - x}{h_n}\right) \hat{F}_n(dx) \qquad (7.3)$$

where $\{h_n\}$ is a bandwidth sequence tending to zero and $K(\cdot)$ is a smooth kernel. Properties of this estimator are investigated by Diehl and Stute (1988) and Mielniczuk (1986). In recent papers Stute (1995a), Stute and Wang (1993) studied the central limit theorem and strong law of large numbers under random censorship. They investigated the asymptotic behavior of

$$S_n = \int \phi \, d\hat{F}_n \tag{7.4}$$

where ϕ is any function such that $\int \phi^2 \, dF < \infty$. Stute (1995b) gives a survey of statistical analysis of Kaplan-Meier integrals defined above.

Bivariate randomly censored data

Let us now consider the situation where the data (X_i, Y_i), $i \geq 1$ is subject to a bivarite right random censoring. We assume that $X_i \geq 0$ and $Y_i \geq 0$, $i \geq 1$. Examples of bivarite censoring include times for two types of non-catastrophic failure in a complex system, Cuzick (1982), life times of relatives, response times in two successive courses of treatment for the same patient, Mielniczuk (1991). Other examples are described in Prakasa Rao (1992b). The data (X_i, Y_i), $i \geq 1$ is subject to right random censoring means that there exist two other iid non-negative sequences of random variables $\{X_i', i \geq 1\}$ and $\{Y_i', i \geq 1\}$ independent of each other and of $\{(X_i, Y_i), i \geq 1\}$ such that we can only observe

$$\tilde{X}_i = \min(X_i, X_i'), \qquad \delta_i = I(X_i \leq X_i'), \qquad 1 \leq i \leq n \tag{7.5}$$

and

$$\tilde{Y}_i = \min(Y_i, Y_i'), \qquad \eta_i = I(Y_i \leq Y_i'), \qquad 1 \leq i \leq n. \tag{7.6}$$

Let (X_i, Y_i) have joint density $f(x, y)$, marginal densities $f_1(x)$ and $f_2(y)$ and marginal distribution functions $F_1(x)$ and $F_2(y)$ respectively for $i \geq 1$. Let X_i' and Y_i' have distribution functions $G_1(x)$ and $G_2(y)$ respectively for $i \geq 1$.

Assume that $E[Y_1] < \infty$. The problem of interest is in estimating the regression function

$$m(t) = E(Y_1 | X_1 = t), \qquad t \geq 0 \tag{7.7}$$

from the censored data $(\tilde{X}_i, \tilde{Y}_{ij}; \delta_i, \eta_i)$, $i \geq 1$. Mielniczuk (1991) studies asymptotic properties of a kernel type estimation of the regression function $m(t)$.

Prakasa Rao and Arusharka Sen (1995b) proposed a new class of estimators for $m(t)$ following a different approach using the theory of counting processes and properties of semimartingales. Let

$$N_n(t) = \sum_{i=1}^{n} I(\tilde{X}_i \leq t, \ \delta_i = 1), \qquad t \geq 0. \tag{7.8}$$

It is known that this counting process $\{N_n(t),\ t \geq 0\}$ has the compensator

$$\Delta_n(t) = \int_0^t V_n(u)_1(u)\, du \tag{7.9}$$

in the sense that

$$\{\ell_n(t) = N_n(t) - \Delta_n(t), t \geq 0\} \tag{7.10}$$

is a zero mean martingale where $\alpha_1(\cdot) = f_1(\cdot)/[1 - F_1(\cdot)]$ is the hazard rate and

$$V_n(t) = \sum_{i=1}^n I(\tilde{X}_i \geq t), \qquad t \geq 0. \tag{7.11}$$

The process $\{N_n(t),\ t \geq 0\}$ has the intensity structure of the form $\{V_n(t)\alpha_1(t),\ t \geq 0\}$ which follows the multiplicative intensity model of Aalen (1978). Note that the above relation can be written in the form of the stochastic differential

$$\begin{aligned} dN_n(t) &= d\Lambda_n(t) + d\ell_n(t) \\ &= V_n(t)\alpha_1(t)\, dt + d\ell_n(t), \qquad t \geq 0 \end{aligned} \tag{7.12}$$

where $\{\ell_n(t),\ t \geq 0\}$ is a zero mean martingale. This relation suggests that $\alpha_1(t)$ can be estimated by $[1/V_n(t)]\, dN_n(t)$ provided $V_n(t) > 0$. Following this, Ramlau-Hansen (1983) suggested the following estimator $\hat{\alpha}_n(t)$ for $\alpha_1(t)$:

$$\hat{\alpha}_n(t) = \frac{1}{h_n} \int_0^T K\left(\frac{t-u}{h_n}\right) J_n(u)(V_n(u))^{-1}\, dN_n(u) \tag{7.13}$$

where $0 \leq t \leq T$, $J_n(u) = I(V_n(u) > 0)$ and it is assumed that $F_1(T) < 1$, $G_1(T) < 1$ and $G_1(\cdot)$ is continuous. It can be shown that the process

$$P_n(t) = \sum_{i=1}^n \eta_i \tilde{Y}_i(\bar{G}_2(\tilde{Y}_i))^{-1} I(\tilde{X}_i \leq t,\ \delta_i = 1), \qquad t \geq 0 \tag{7.14}$$

has the compensator

$$A_n(t) = \int_0^t V_n(u) m(u)\alpha_1(u)\, du, \qquad t \geq 0 \tag{7.15}$$

where $\bar{G}_2 = 1 - G_2$ (cf. Prakasa Rao and Arusharka Sen (1995b)). Hence the process $\{P_n(t)\}$ has also an "intensity" of multiplicative intensity type and we proposed the following estimator for $m(\cdot)$:

$$\hat{m}_n(t) = \frac{\hat{\alpha}_n^{(0)}(t)}{\hat{\alpha}_n(t)}, \qquad 0 \leq t \leq T \tag{7.16}$$

where

$$\hat{\alpha}_n^{(0)}(t) = \frac{1}{h_n} \int_0^T K\left(\frac{t-u}{h_n}\right) J_n(u)(V_n(u))^{-1} dP_n^{(0)}(u) \tag{7.17}$$

and $P_n^{(0)}(t)$ is obtained from $P_n(t)$ by replacing $\bar{G}_2(\tilde{Y}_i)$ by $\bar{G}_{2n}(\tilde{Y}_i)$ which is the Kaplan-Meier product limit estimator for $\bar{G}_2(\cdot)$. It is assumed that $T < \min(T_{G_1}, T_{G_2})$ where

$$T_H = \sup\{t: H(t) < 1\} \tag{7.18}$$

for any distribution function H. Properties such as uniform consistency on compact sets and limiting distribution of the estimator are investigated in Prakasa Rao and Arusharka Sen (1995b).

8. WAVELETS AND THEIR APPLICATIONS TO NONPARAMETRIC INFERENCE

8.1 Introduction

Wavelets system is an orthogonal system. The subject of wavelets had its origins a few years back in the area of seismic analysis (cf. Grossman and Morlet (1984)). In addition to the applications in seismic analysis, wavelet representations have been found to be useful in image processing, signal analysis and data compression.

In this Section, we describe the applications of wavelets for nonparametric functional estimation. Walter and Ghorai (1992) discuss advantages and disadvantages of density estimation with wavelets. They indicate that application of wavelet bases give better asymptotic properties of the estimators but, for small samples, they have little advantage over the kernel methods and do not give as close an approximation to the true density.

A wavelet system in an infinite collection of translated and scaled versions of functions ϕ and ψ called the scaling function and the primary wavelet function respectively. The function $\phi(x)$ is a solution of the difference equation

$$\phi(x) = \sum_{k=-\infty}^{\infty} c_k \phi(2x - x) \tag{8.1}$$

with normalization

$$\int_{-\infty}^{\infty} \phi(x) \, dx = 1 \tag{8.2}$$

and the function $\psi(x)$ is defined by

$$\psi(x) = \sum_{k=-\infty}^{\infty} (-1)^k c_{-k+1} \, \phi(2x - k). \tag{8.3}$$

The coefficients c_k are called the filter coefficients. Choice of $\{c_k\}$ determine the wavelet system. It is easy to see that the condition in Eq. (8.2) implies that

$$\sum_{k=-\infty}^{\infty} c_k = 2. \tag{8.4}$$

Define

$$\phi_{jk}(x) = 2^{j/2}\phi(2^j x - k), \qquad -\infty < j, k < \infty \tag{8.5}$$

and

$$\psi_{jk}(x) = 2^{j}/2\psi(2^j x - k), \qquad -\infty < j, k < \infty. \tag{8.6}$$

Under some regularity conditions on ϕ, a condition on the filter coefficients $\{c_k\}$, namely,

$$\sum_{k=-\infty}^{\infty} c_k c_{k+2\ell} = 2 \quad \text{if} \quad \ell = 0$$

$$= 0 \quad \text{if} \quad \ell \neq 0 \tag{8.7}$$

implies that $\{\psi_{jk}, -\infty < j, k < \infty\}$ is an orthonormal basis for $L^2(R)$ and $\{\phi_{jk}, -\infty < k < \infty\}$ is an orthonormal system in $L^2(R)$ for each j (cf. Daubechies (1990), Lemma 3.4, p. 958).

If $c_0 = 1 = c_1$, then the following $\phi(x)$ given by Eq. (8.1) is the function

$$\phi(x) = \begin{cases} 1 & \text{if} \quad 0 \le x \le 1 \\ 0 & \text{otherwise} \end{cases}$$

and the corresponding wavelet system is the Haar system. In this case, the scaling function and the primary wavelet have the interval $[0, 1]$ as their support and the resulting set of functions is an orthonormal basis for $L^2(R)$.

DEFINITION 8.1.1 A scaling function $\phi \in C^{(q)}$ is said to be q-regular for an integer $q \ge 1$, if for every non-negative integer $\ell \le q$ and for any integer k,

$$\phi^{(\ell)}(x)| \le C_k(1 + |x|)^{-}k, \qquad -\infty < x < \infty \tag{8.8}$$

for some $C_k \ge 0$ depending only on k where $\phi^{(\ell)}$ denotes the ℓ-th derivative of ϕ.

DEFINITION 8.1.2   A multiresolution analysis of $L^2(R)$ consists of an increasing sequence of closed subspaces $\{V_j\}$ of $L^2(R)$ such that

(i) $\displaystyle\bigcap_{j=-\infty}^{\infty} V_j = \{0\},$ $\hspace{4cm}$ (8.9)

(ii) $\displaystyle\overline{\bigcup_{j=-\infty}^{\infty} V_j} = L^2(R),$ $\hspace{4cm}$ (8.10)

(iii)   there is a scaling function $\phi \in V_0$ such that $\{\phi(x - k), -\infty < k < \infty\}$ is an orthonormal basis for V_{oj}; and for all $h \in L^2(R),$

(iv)   for all $\quad -\infty < k < \infty, h(x) \in V_o \Rightarrow h(x - k) \in V_0,$ $\hspace{1cm}$ (8.11)

(v) $h(x) \in V_j \Rightarrow h(2x) \in V_j + 1.$ $\hspace{4cm}$ (8.12)

Mallat (1989) has shown that given any multiresolution analysis, it is possible to derive a function ψ (primary wavelet) such that for any fixed $j, -\infty < j < \infty$, the family $\{\psi_{jk}, -\infty k < \infty)$ is an orthonormal basis of the orthogonal complement W_j of V_j in V_{j+1} so that $\{\psi_{jk}, -\infty < j, k < \infty\}$ is an orthonormal basis of $L^2(R)$. Conversely, given any compactly supported wavelvet system, it gives rise to a multiresolution analysis of $L^2(R)$ (cf. Daubechies (1990, Theorem 3.6)). When the scaling function ϕ is q-regular, the corresponding multiresolution analysis is said to be q-regular.

Let H_2^s denote the Sobolev space of functions $g(\cdot)$ in $L^2(R)$ whose first $s - 1$ derivatives are absolutely continuous and define the norm

$$\|g\|_{H_2^s} = \sum_{j=0}^{s}\left[\int_{-\infty}^{\infty} |g^{(j)}(t)|^2\, dt\right]^{1/2}.$$ $\hspace{1cm}$ (8.13)

THEOREM 8.1.1   Mallat (1989). Let the multiresolution analysis be q-regular. Then, for every $0 < s < q$, any function $g \in L^2(R)$ belongs to H_2^s if and only if

$$\sum_{\ell=-\infty}^{\infty} e_\ell^2 e^{2st} < \infty$$ $\hspace{1cm}$ (8.14)

where $e_i^2 = \|g - g_\ell\|_2^2$ and g_ℓ is the orthogonal projection of g on V_t.

REMARKS   The above introduction is based on Autoniadis et al. (1994). For detailed introduction to wavelets, see Chui (1992) and Daubechies (1992). For a brief survey, see Strang (1989).

8.2 NONPARAMETRIC ESTIMATION OF DENSITY FUNCTION AND ITS DERIVATIVES

Suppose $X_1, X_2, \ldots, X_n$ are iid random variables with density f. Suppose f is d-times differentiable. The problem is to estimate the dth derivative $f^{(d)}$ of f. We interpret $f^{(0)}$ as f in the following discussion and the problem of estimation of $f^{(0)} \equiv f$ is the problem of estimation of density f. Suppose that $f^{(d)} \in L_2(R)$. Assume that there exists $D_j \geq 0$ and $\beta_j \geq 0$, $0 \leq j \leq d$ such that

$$|f^{(j)}(x)| \leq \frac{D_j}{|x|^{\beta_j}} \quad \text{for} \quad |x| \geq 1 \tag{8.15}$$

with $\beta_0 > 4d + 1$.

Consider a multiresolution as discussed in Sec. 8.1. Let ϕ be the corresponding scaling function. Suppose the multiresolution is r-regular for some $r \geq d$. Then, by definition, $\phi \in C^{(r)}$, ϕ and its derivatives $\phi^{(j)}$ up to order r are rapidly decreasing i.e., for every integer $m \geq 1$, there exists a constant $A_m > 0$ such that

$$|\phi^{(j)}(x)| \leq \frac{A_m}{(1 + |x|)^m}, \qquad 0 \leq j \leq r. \tag{8.16}$$

Let

$$\phi_{\ell k}(x) = 2^{\ell/2} \phi(2^{\ell} x - k), \qquad -\infty < \ell, \, k < \infty. \tag{8.17}$$

Then

$$\phi_{\ell k}^{(j)}(x) = 2^{(\ell/2) + \ell j} \phi^{(j)}(2^{\ell} x - k), \qquad 0 \leq j \leq r \tag{8.18}$$

and

$$|\phi_{\ell k}^{(j)}(x)| \leq \frac{2^{(\ell/2) + \ell j}}{(1 + |2^{\ell} x - k|)^m}, \qquad 0 \leq j \leq r. \tag{8.19}$$

If $d \geq 1$, then for any $0 \leq j \leq d - 1$,

$$|\phi_{\ell k}^{(j)}(x) f^{(d-j-1)}(x)| \leq \frac{2^{(\ell/2) + \ell j}}{(1 + |2^{\ell} x - k|)^m} \frac{D_{d-j-1}}{|x|^{\beta_{d-j-1}}}.$$

which implies that

$$\lim_{|x| \to x} |\phi_{\ell.k(x)}^{(j)} f^{(d-j-1)}(x)| = 0, \qquad 0 \leq j \leq d - 1 \tag{8.20}$$

for any fixed ℓ and k. Let $f_{\ell d}$ be the orthogonal projection of $f^{(d)}$ on V_t. Note that

$$f_{\ell d}(x) = \sum_{j=-\infty}^{\infty} a_{\ell j}\phi_{\ell j}(x) \tag{8.21}$$

where

$$a_{\ell j} = \int_{-\infty}^{\infty} f^{(d)}(u)\phi_{\ell j}(u)\,du \tag{8.22}$$

$$= (-1)^d \int_{-\infty}^{\infty} f(u)\phi_{\ell j}^{(d)}(u)\,du \tag{8.23}$$

by Eq. (8.20) using integration by parts. For $d = 0$, the relation in Eq. (8.23) holds clearly. Hence

$$a_{\ell j} = (-1)^d E[\phi_{\ell,j}^{(d)}(X_1)]. \tag{8.24}$$

Let

$$\hat{a}_{\ell j} = \frac{(-1)^d}{n}\sum_{i=1}^{n} \phi_{\ell,j}^{(d)}(X_i) \tag{8.25}$$

be an estimator of $a_{\ell j}$ based on the iid sample X_i, $1 \le i \le n$. Estimate $f^{(d)}(x)$ by

$$\hat{f}_{nd}(x) = \sum_{j=-q_n}^{q_n} \hat{a}_{\ell_n,j}\phi_{\ell_n,j}(x) \tag{8.26}$$

where $\{\ell_n\}$ and $\{q_n\}$ are suitably chosen with $q_n \to \infty$ as $n \to \infty$.

Following the techniques of Masry (1994), one can develop exact orders for

$$E\|f^{(d)} - \hat{f}_{nd}\|_2^2$$

which is the integrated mean squared error of the estimator $\hat{f}_{nd}$ for $f^{(d)}$. Under the assumption that $f^{(d)} \in L_2(R) \cap H_2^s$ for some $0 < s < r$ and that f is of bounded variation on R, it can be shown that

$$\frac{nE\|f^{(d)} - \hat{f}_{nd}\|_2^2}{n^{(1+2d)/2s+1}} \to \int_{-\infty}^{\infty} [\phi^{(d)}(v)]^2\,dv \quad \text{as} \quad n \to \infty \tag{8.27}$$

for $\{\ell_n\}$ such that $2^{\ell n} = n^{1/(2s+1)}$ and $\{q_n\}$ such that

$$q_n = 2^{((2d-1)+\beta_0+2s)\ell_n/(2\beta_0-1)}\,\log n.$$

Note that

$$\frac{2^{\ell_n(\beta_0-4d-1)}}{q_n^{\beta_0-1}} \to 0 \quad \text{as} \quad n \to \infty \quad \text{and} \quad \ell_n \to \infty. \tag{8.28}$$

Detailed proof of this result is given in Prakasa Rao (1996c). For related work on this problem, see Antoniadis and Carmona (1990) and Kerkyacharian and Picard (1992). Masry (1997) discusses multivariate density estimation using the wavelets approach.

8.3 Nonparametric Estimation of Integral of Square of Density

The motivation for estimation of the functional

$$I(f) = \int_{-\infty}^{\infty} f^2(x)\, dx \tag{8.29}$$

where f is a probability density function is well-known. For instance, it occurs in the expression for the asymptotic efficiency of rank for problems in location-shift (cf. Prakasa Rao (1983)). It is also of interest in the study of the relationship between information bounds and convergence rates (cf. Ritov and Bickel (1990)). Kernel type estimation for the functional $I(f)$ has been studied by several authors (cf. Prakasa Rao (1983)) earlier and more recently by Hall and Marron (1987), Bickel and Ritov (1988), Hall and Wolff (1995) among others. In a recent paper, Birge and Massert (1995) studied estimation of functionals of the type

$$T(f) = \int \phi(f(x), f^{(1)}(x), \dots, f^{(k)}(x), x)\, dx$$

where ϕ is a smooth function of $k + 2$ variables, $f^{(i)}$ denotes the i-th derivative of $f, 1 \leq i \leq k$ and f belongs to a class of probability densities of smoothness s. In Prakasa Rao (1997a), we develop an estimator $\tilde{I}(f)$ for $I(f)$ by the method of wavelets and obtain a precise asymptotic expression for its mean integrated squared error. We briefly describe these results.

Suppose $X_1, X_2, \dots, X_n$ are iid random variables with density $f \in L_2(R)$. The problem of interest is the estimation of $I(f)$ given by Eq. (8.29).

Consider a multiresolution as discussed in Sec. 8.1. Let ϕ be the corresponding scaling function. Suppose that the multiresolution analysis given by ϕ is r-regular where $r \geq 1$ is a positive integer. Further suppose that $f \in H_2^s$ where $0 < s < r$ where H_2^s is the Sobolev space. We assume further that there exists $1 < \beta < \frac{7}{6}$ such that

$$f(x) \leq D|x|^{-\beta} \quad \text{for} \quad |x| \geq 1 \tag{8.30}$$

for some $D > 0$ and that f is of bounded variation on $(-\infty, \infty)$.

Given the sample $X_1, X_2, \dots, X_n$, let

$$A_{\ell,k} = \frac{1}{(n-1)} \sum_{i=1}^{n} \sum_{\substack{j=1 \\ i \neq j}}^{n} \phi_{\ell k}(X_i) \phi_{\ell k}(X_j) \tag{8.31}$$

where

$$\phi_{\ell k}(x) = 2^{\ell/2} \phi(2^\ell x - k), \qquad -\infty < \ell, k < \infty. \tag{8.32}$$

We estimate $I(f)$ by

$$\tilde{I}(f) = \sum_{k=-K}^{K} A_{\ell,k}. \tag{8.33}$$

It is easy to see that

$$\lim_{\ell \to \infty} \lim_{K \to \infty} E(\tilde{I}(f)) = I(f). \tag{8.34}$$

Let

$$\varepsilon_n^2 = E|\tilde{I}(f) - I(f)|^2. \tag{8.35}$$

The following result holds.

THEOREM 8.3.1   Let $\ell_n \to \infty$ and

$$K_n = 2^{2[(\beta+s-1/2)/(2\beta-1)]\ell_n} \log n. \tag{8.36}$$

Define $\tilde{I}(f)$ as an estimator for $I(f)$ with $K = K_n$ and $\ell = \ell_n$. Then

$$\frac{n(n-1)}{2^{2\ell_n}} \varepsilon_n^2 \to 1 \quad \text{as} \quad n \to \infty. \tag{8.37}$$

For proof, see Prakasa Rao (1997).

8.4 Nonparametric Estimation of Regression Function

Consider the nonparametric regression model

$$Y_i = m(X_i) + \varepsilon_i, \qquad 1 \leq i \leq n. \tag{8.38}$$

Two versions of this model are of interest.

1. The fixed design model : Here the X_i's are nonrandom design points (In this case X_i are denoted by t_i with $0 \leq t_1 \leq \cdots \leq t_n \leq 1$) and ε_i are iid errors with mean zero and variance σ^2.
2. The random design model: Here the (X_i, Y_i)'s are iid bivariate random vectors identically distributed as (X, Y) with $m(x) = E(Y|X = x)$ and $\varepsilon_i = Y_i - m(X_i)$.

The problem is to estimate $m(x)$. We have discussed the second problem using kernel estiamtors in Sec. 2.2. Recently Antoniadis et al. (1994) discussed an application of wavelet methods for curve estimation in the context of models described above. We discuss their results briefly.

Consider the projection operator E_j mapping $h \in L^2(R)$ to V_j as defined in Sec. 8.1. In other words

$$E_j(h)(x) = \int_R E_j(\cdot, y)h(y)\, dy$$

It can be seen that $E_j(x, y) = 2^j E_0(2^j x, 2^j y)$ and $E_0(x + k, y + k) = E_0(x, y)$ for any integer $k, -\infty < k < \infty$. Antoniadis et al. (1994) propose the estimator

$$\hat{m}(t) = \sum_{i=1}^{n} Y_i \int_{A_i} E_r(t, s)\, ds$$

where $\{A_i\}$ is a partition of the interval $[0, 1]$ with $t_i \in A_i$ for the fixed design model and for the random design model, they propose the estimator

$$\tilde{m}(t) = n^{-1} \sum_{i=1}^{n} Y_i E_r(t, X_i)/\tilde{f}(t)$$

where $\tilde{f}(t)$ is a wavelet estimator of the density of X given by

$$\tilde{f}(t) = n^{-1} \sum_{i=1}^{n} E_r(t, X_i).$$

Antoniadis and Carmona (1990) introduced density estimator of the form $\tilde{f}$. The estimate $\hat{r}$ and $\tilde{r}$ can be calculated by observing that

$$E_r(t, s) = \sum_{k=-\infty}^{\infty} \phi(2^r t - k)\phi(2^r s - k)$$

where ϕ is the scaling function as discussed in Sec. 8.1. Antoniadis et al. (1994) discuss asymptotic properties of a modified versions of $\hat{m}$ and $\tilde{m}$ following techniques for delta-sequence type estimator in the regression case studied by Isogai (1990) and some techniques of Gasser and Muller (1979) for kernel-type estimators. Method of delta sequence for density estimators was discussed in Prakasa Rao (1978), Walter and Blum (1979) and Prakasa Rao (1979a,b).

8.5 Nonparametric Estimation for Stochastic Processes

Leblanc (1995) discusses general theory of wavelet density estimation for
continuous time processes with application to inference for diffusion processes. Masry (1994) discussed related results for discrete time mixing processes. We briefly discuss results from Leblanc (1995).

Let $\{X_t, t \geq 0\}$ be a stationary locally integrable real valued stochastic
process defined on a probability space $(\Omega, \mathscr{F}, P)$. Suppose the marginal density f of X_0 exists and is bounded and the process $\{X_t\}$ satisfies the following
conditions:

(i) For any $u > 0$, the joint density f_u of (X_0, X_u) exists and satisfies the
condition

$$\int_0^\infty |f_u(x, y) - f(x)f(y)| \, du \leq \gamma_q < \infty$$

for all u whenever $x, y \in [-q, q]$, and

(ii) the process is geometrically strong mixing that is, there exist constants
C and $\alpha > 1$ such that for any $\tau \geq 0$,

$$\alpha(\tau) = \sup_{A \in \mathscr{F}_0^t, \, B \in \mathscr{F}_{t+\tau}^\infty} |P(A \cap B) - P(A)P(B)| \leq C\alpha^{-\tau}$$

where $\mathscr{F}_t^m$ is the σ-algebra generated by $\{X_s, t \leq s \leq m\}$.

The problem is to estimate the marginal density f of X_t over an
interval $[-K, K]$ from the observed sample path $\{X_t, \leq t \leq T\}$. Since
$fI([-K, K]) \in L_2([-K, K])$, we have the wavelet expansion

$$fI([-K, K]) = \sum_{k \in K, 0} \alpha_{j_0, k} \phi_{j_0, k} + \sum_{j \geq j_0} \sum_{k \in K_{j_0}} \beta_{j, k} \psi_{j, k}$$

$$= P_{j_0} f + \sum_{j \geq j_0} D_j f$$

where

$$\phi_{j_0 k}(x) = 2^{j_0/2} \phi(2^{j_0} x - k)$$

and

$$\psi_{j, k}(x) = 2^{j/2} \psi(2^j x - k).$$

Here $\phi(\cdot)$ and $\psi(\cdot)$ are as defined in Sec. 8.1. By orthogonality of the wavelet
basis, it follows that

$$\alpha_{j_0, k} = \int f(x) \phi_{j_0, k}(dt) \quad \text{and} \quad \beta_{j, k} = \int f(x) \psi_{j, x}(dx).$$

Suppose the ϕ and $\psi \in C^{r+1}$ where r is an integer ≥ 1 and have compact support. Define

$$\hat{f}_T(x) = \sum_{k \in K_{j_0}} \hat{\alpha}_{k_0.k} \phi_{j_0.k}(x)$$

with

$$\hat{\alpha}_{j_0.k} = \frac{1}{T} \int_0^T \phi_{j_0.k}(X_s)\,ds.$$

Leblanc (1995) obtains upper bound for $E\|(f - \hat{f}_T)I(-K, K)\|_p^2$, for

$$f \in F_{s.p.a} = \{f \in B_{p.q}^s : \|f\|_{B_{p.q}^s} \leq M\}$$

for some $0 < s < r + 1$, $p \geq 1$, $q \geq 1$ where

$$\|f\|_{B_{p.s}^q} = \|P_0 f\|_p + \left(\sum_{j \geq 0} (\|D_j f\|_p 2^{js})^q \right)^{1/q}.$$

Diffusion processes are widely used for stochastic modelling (cf. Karlin and Taylor (1981)). Estimation of the limiting density or drift coefficient for diffusion process from a continuous sample path of the process has been discussed in Basawa and Prakasa Rao (1980) and Prakasa Rao (1985) among others.

The problem of estimation of the drift coefficient for diffusion process when a discrete sampling of the process is available has been investigated in Dorogoveev (1976), Prakasa Rao (1983a) and more recently by Genon-Catalot (1990) among others. The problem of the estimation of the diffusion coefficient for diffusion process is of interest in modeling financial data. In option pricing theory and in modeling interest rates, the time-dependent diffusion coefficient is unknown and is of major interest. Genon-Catalot et al. (1992) propose a nonparametric method for estimation of the diffusion coefficient by wavelet methods. A comprehensive discussion about all these results is given in Prakasa Rao (1999).

9. GENERALISATIONS AND EXTENSIONS

9.1 Nonparametric Density Estimation in the Multivariate Case

Extensions of different methods of density estimation to the multivariate case and study of their asymptotic properties is possible and has been done. For a comprehensive survey, see Prakasa Rao (1983). Silverman (1986) discusses application of the kernel method for the multivariate data. Estimation of the density in more than two dimensions is unlikely to be of use just for exploratory purpose. It might still be useful and easy to comprehend for a two dimensional data. However, if one is only interested in applications of the estimated density, then the problem of density estimation is useful in the case of higher dimensions. Due to the "curse of dimensionality", however computations involved to get the type of accuracy as in the one-dimensional case increase exponentially with dimension. Even with a high speed computer, inappropriate algorithm for calculating multivariate density estimates may lead to excessively large amount of calculation time. In spite of these difficulties, multivariate density estimation is useful and Scott (1994) gives a comprehensive discussion dealing with some aspects of applications of multivariate density estimation. He emphasises that density estimation is a powerful tool for analysis of even for trivariate and quadrivariate data.

9.2 Nonparametric Density Estimation for Directional Data

There are various situations where the observed data is in the form of directional cosines or in the form of vectors scaled by an unknown positive scalar so that only the direction is known. For example, a biologist might want to investigate the flight directions of birds, an ecologist might record the direction of wind or a geologist might measure the paleomagnectic directions etc. Problems of inference based on such data are of great interest. For a review, see Mardia (1972) and Rao (1984).

Suppose $X_1, \ldots, X_n$ are iid unit vectors with f as the density function of X_1 over the k-dimensional unit sphere Ω_k, $k > 1$. The problem of interest is the estimation for the density f. Bai et al. (1988) discuss kernel type of estimation for f. They propose the estimator

$$f_n(\mathbf{x}) = \frac{1}{nh_n^{k-1}} C(h_n) \sum_{i=1}^{n} K\left(\frac{1 - \mathbf{x}'\mathbf{X}_i}{h_n^2}\right), \qquad \mathbf{x} \in \Omega_k \qquad (9.1)$$

where $h_n > 0$, $K(u)$ is a non-negative function defined for $u \geq 0$ such that

$$0 < \int_0^\infty K(v) v^{(k-3)/2} \, dv < \infty, \qquad (9.2)$$

$C(h_n)$ is defined by

$$h_n^{k-1} (C(h_n))^{-1} = \int_{\Omega_k} K\left(\frac{(1 - \mathbf{x}'\mathbf{y})}{h_n^2}\right) dw(\mathbf{y}) \qquad (9.3)$$

and $w(\cdot)$ is the Haar measure on Ω_k. If $h_n \to 0$, then

$$(C(h_n))^{-1} \to \frac{(2\pi)^{(k-1)/2}}{\Gamma((k-1)/2)} \int_0^\infty K(v) v^{(k-3)/2} \, dv \tag{9.4}$$

following Watson (1983). For instance $K(\cdot)$ can be chosen to be

$$K(v) = e^{-v}, \qquad 0 \le v < \infty \tag{9.5}$$

or

$$K(v) = 1 \quad \text{for} \quad 0 \le v \le 1$$
$$= 0 \quad \text{otherwise} \tag{9.6}$$

THEOREM 9.2.1   Bai et al. (1988). Let $K(\cdot)$ and $\{h_n\}$ satisfy the following conditions:

(i) $K(\cdot)$ is bounded on R ;
(ii) $0 < \int_0^\infty K(v) v^{(k-3)/2} \, dv < \infty$;
(iii) $\lim_{v \to \infty} v^{(k-1)/2} K(v) = 0$ or f is bounded on Ω^k;
(iv) $\lim_{n \to \infty} h_n = 0$; and
(v) $\lim_{n \to \infty} (nh_n^{k-1} / \log n) = \infty$.

Then

$$\lim_{n \to \infty} f_n(\mathbf{x}) = f(\mathbf{x}) \text{ a.s.} \tag{9.7}$$

at all continuity points $\mathbf{x}$ of f.

Bai et al. (1988) have also studied the uniform consistency as well as the L_1-consistency of these estimators. The problem of finding the asymptotic distribution of the estimator f_n seems to be unsolved. For circular data, Watson (1983) proposed another method for estimation of density f when

$$f(\theta) = \frac{1}{2\pi} \sum_{m=-\infty}^{\infty} C_m e^{im\theta}, \qquad 0 \le \theta < 2\pi$$

by using an analogue of the method of orthogonal series. He observed that

$$E[e^{im\theta}] = C_m$$

and the problem reduces to estimation of the coefficients C_m.

9.3   Nonparametric Estimation When the Data are Curves

These are situations where the observed data are curves rather than single points or vectors. For instance, growth curves are a case in point where a seperate growth curve is observed for each individual. In missile tracking studies, there is a target track (curve) from the point where the missile is fired and the observed path is another curve. In other words, the sample path might be a realization of a stochastic process $\{X(t),\ 0 \le t \le T\}$ and the mean path may be denoted by (say) $\mu(t),\ 0 \le t \le T$. The problem is to estimate $\mu(\cdot)$ from an observation on the process $\{X(t),\ 0 \le t \le T\}$. This problem has been discussed in the context of gaussian processes (cf. Grenander (1951, 1981), Basawa and Prakasa Rao (1980)) and in the context of time series analysis and signal processing (cf. Parzen (1961)). Recent work in this area is due to Rice and Silverman (1991) and Ramsay and Dalzell (1991). Ramsay (1982) discusses several examples when the data are functions. Application of the theory of wavelets for this study have not yet been explored. Rice and Silverman (1991) discuss a method for smoothed principal component analysis for functional data. They follow the penalty function approach to the problem and the sample curves are modeled as independent realizations (trajectories) of a second-order stochastic process $\{X(t)\}$ that has mean $E(X(t)) = \mu(t)$. Observe that if $\mathrm{Cov}(X(s),\ X(t)) = (s, t)$ and

$$\gamma(s, t) = \sum \lambda_\nu \phi_\nu(s)\phi_\nu(t) \tag{9.8}$$

in the L^2-sense, then the process $X(t)$ can be represented in the form

$$X(t) = \mu(t) + \sum \xi_\nu \phi_n(t) \tag{9.9}$$

where $\{\xi_\nu\}$ are uncorrelated random variables with mean zero and variance $E\xi_\nu^2 = \lambda_\nu$. Assume that the mean function $\mu(\cdot)$ is sufficiently smooth. Let $\{X_i(t),\ 0 \le t \le T\},\ 1 \le i \le n$ be the observed sample curves. Let $\Phi(u) = \int_0^T (d^2u/dt^2)dt$. Rice and Silverman (1991) suggest the curve $\hat{\mu}(\cdot)$ that minimizes

$$\frac{1}{n}\sum_{i=1}^{n} \|X_i - \hat{\mu}\|^2 + \Phi(\hat{\mu}) \tag{9.10}$$

as an estimate of the mean function μ. For estimating the eigenfunction ϕ_ν, assume at first that mean of the process is zero. The leading eigen function can be estimated by maximizing

$$\sum (u, X_i)^2 - \lambda\Phi(u) \tag{9.11}$$

subject to $\|u\| = 1$. Higher order eigenfunctions are estimated in turn by performing similar calculations with the additional condition $(u_j, u_\ell) = 0$, $1 \le \ell \le j - 1$. Here $(u, v) = \int_0^T u(t)\bar{v}(t)\,dt$, $\|u\|^2 = (u, u)$ and α and λ are smoothing parameters which can be obtained by cross-validation techniques for instance.

Ramsay and Dalzell (1991) study the problem extending the concepts of linear modelling and the principal component analysis used in multivariate analysis to infinite dimensional spaces. Pezzulli and Silverman (1993) discuss some properties of the method for smoothed principal component analysis of functional data described above due to Rice and Silverman (1991).

An alternate modified approach to functional principal component analysis is studied in Silverman (1996).

REFERENCES

Aalen. O. (1978) Nonparametric inference for a family of counting processes, *Ann. Statist. 6*: 701–726.

Ait-Sahalia, Y. (1996) Nonparametric pricing of interest rate derivative securities, *Econometrica 64*: 527–560.

Antoniadis, A. and Carmona, R. (1990) Multiresolution analysis and wavelets for density estimation, Tech. Report, University of California, Irvine.

Antoniadis A., Gregoire, G., Mckeague, I. (1994) Wavelet methods for curve estimation, *J. Amer. Statist. Assoc. 89*: 1340–1353.

Arusharka Sen (1994) Uniform strong consistency rates for conditional U-statistics, *Sankhya Ser A 56*: 179–194.

Athreya, K. B. and Atuncar, G. S. (1994) Kernel estimation for real-valued Markov chains, Preprint, Iowa State University, Ames.

Athreya, K.B. and Ney, P. (1978) A new approach to the limit theory of recurrent Markov chains, *Ann. Probability 6* (1978), 788–797.

Bai, Z. D., Rao, C. R. Zhao, L. C. (1988) Kernel estimation of density function of directional data, *J. Multivariate Anal. 27*: 24–39.

Bagai, I. and Prakasa Rao, B.L.S. (1991) Estimation of the survival function for stationary associated processes, *Statist., Probab. Lett., 12*: 385–391.

Bagai, I. and Prakasa Rao, B.L.S. (1995a) Kernel-type density estimation for positive valued random variables, *Sankhya Ser. A, 57*: 56–67.

Bagai, I. and Prakasa Rao, B.L.S. (1995b) Kernel-type density and failure rate estimation for associated sequences, *Ann. Inst. Statist. Math. 47*: 253–266.

Banon, G. (1977) Nonparametric identification for diffusion processes. *SIAM J. Control Optim. 16*: 380–395.

Basawa, I. V. and Prakasa Rao, B. L. S. (1980) *Statistical Inference for Stochastic Processes,* Academic Press, London.

Bickel, P. and Ritov, Y. (1988) Estimating integrated squared density derivative: sharp best order of convergence estimates. *Sankhya* A *50*: 381–393.

Birge, L. and Massart, P. (1995) Estimation of integral functions of a density, *Ann. Statist. 23*: 11–29.

Birkel, T. (1988a) Moment bounds for associated sequences, *Ann. Probability 16*: 1184–1193.

Birkel, T. (1988b) On the convergence rate in the central limit theorem for associated processes, *Ann. Probability 16*: 1685–1693.

Blum, J.R. and Boyles, R.A. (1981) Random sampling from a continuous parameter stochastic processes, In *Analytical Methods in Probability Theory*. Lecture Notes in Mathematics No. 861, Springer, Berlin, 15–24.

Blum, J. R. and Rosenblatt, J. (1964) On random sampling from a stochastic process, *Ann. Math. Stat. 35*: 1713–1717.

Bochynek, J. (1987) Asymptotische Normalitat bedingter U-statistiken, *Ph.D. Dissertation Justus Liebing Universität*, Giessen.

Cheng, B. and Robinson, P. M. (1991) Density estimation in strongly dependent nonlinear time series, *Statistica Sinica 1*: 335–360.

Chow, Y. S. and Teicher, H. (1978) *Probability Theory*, Springer-Verlag, New York.

Chui, C.K. (1992) *An Introduction to Wavelets*, Academic Press, Boston.

Csorgo, S. and Mielniczuk, J. (1995a) Density estimation under long-range dependence, *Ann. Statistcs 23*: 990–999.

Csorgo, S. and Mielniczuk, J. (1995b) Nonparametric regression under long-range dependent normal errors, *Ann. Statistics, 23*: 1000–1014.

Cuzick, I. (1982) Rank tests for association with right-censored data, *Biometrika 69*: 351–364.

Daubechies, J. (1990) Orthonormal bases of compactly supported wavelets, *Communications on Pure and Applied Mathematics 49*: 906–996.

Daubechies, I. (1992) *Ten Lectures on Wavelets*, CBMF-NSF Regional conference series on Applied Math. SIAM, Philadelphia.

Delecroix, M. (1980) Sur ℓ'estimation des densites d'un processus stationnaire a temp continuous, *Publ. Inst. Statist. Univ. Paris 25*: 17–40.

Devroye, L. (1987) A *Course on Density Estimation*, Birkhauser, Boston.

Devroye, L. and Gyorfi, L. (1985) *Nonparametric Density Estimation: The L_1-view*, Wiley, New York.

Dewan, I. and Prakasa Rao, B. L. S. (1997) Remarks on the strong law of large numbers for a triangular array of associated random variables, *Metrika 45*: 225–234.

Dewan, I. and Prakasa Rao, B. L. S. (1997a) Remarks on the Berry–Esseen type bound for stationary associated random variables, *Gujarat Stat. Rev. 24*: 17–20.

Dewan, I. and Prakasa Rao, B. L. S. (1997b) The strong law of large numbers for U-statistic for associated random variables, Preprint, Indian Statistical Institute, New Delhi.

Dewan, I. and Prakasa Rao, B.L.S. (1997c) Central limit theorem for U-statistics of associated random variables, Preprint, Indian Statistical Institute, New Delhi.

Dewan, I. and Prakasa Rao, B.L.S. (1998) A general method of density estimator for associated random variables, *J. Nonparametric Statistics*. (To appear).

Diehl, S. and Stute, W. (1988) Kernel density and hazard function estimation in the presence of censoring, *J. Multivariate Anal. 25*: 299–310.

Doob, J. L. (1953) *Stochastic Processes*, Wiley, New York.

Dorogovcev, A. J. (1976) The consistency of an estimate of a parameter of a stochastic differential equation, *Theor. Probab. Math. Statist. 10*: 73–82.

Gasser, T. and Muller, H. (1979) Kernel estimtion of regression function, In *Lecture Notes in Mathematics* No. 757 Springer, Berlin; 23–68.

Genon-Catalot, V., (1990) Maximum contrast estimation for diffusion processes from discrete observations, *Statistics 21*: 99–116.

Genon-Catalot, V., Laredo, C. and Picard, D. (1992) Nonparametric estimation of the diffusion coefficient by wavelvet methods, *Scand. J. Statist. 19*: 317–335.

Grenander, U. (1951) Stochastic processes and statistical inference, *Ark Mat. 1*: 195–277.

Grenander, U. (1981) *Abstract Inference*, Wiley, New York.

Grossman, A. and Morlet, J. (1984) Decomposition of Hardy functions into square integrable wavelets of constant shape. *SIAM J. Math. Anal.* *15*: 723–736.

Gyorfi, L. Härdle, W., Sarda, P. and Vieu, P. (1989) *Nonparametric Curve Estimation from Time Series*, Lecture Notes in Statistics No. 60, Springer-Verlag, New York.

Hall, P. and Marron, J. S. (1987) Estimation of integrated squared density derivatives: *Statist. Prob. Lett. 6*: 109–115.

Hall, P. and Wolff, R. C. L. (1995) Estimation of integrals of powers of density derivatives, *Statist. Prob. Lett. 24*: 105–110.

Härdle, W. (1990) *Applied Nonparametric Regression*, Cambridge University Press, Cambridge, U.K.

Härdle W. and Stoker, T.M. (1989) Investigating smooth multiple regression by the method of average derivatives, *J. Amer. Statist. Assoc. 84*: 986–995.

Hollander, M. and Proschan, F. (1972) Testing whether new is better than used, *Ann. Math. Statist. 43*: 1136–1146.

Ibragimov, I.A. and Rozanov, Yu. A (1978) *Gaussian Random Processes*, Springer-Verlag, New York.

Isogai, E. (1990) Nonparametric estimation of a regression function by delta-sequences, *Ann. Inst. Statist. Math. 42*: 699–708.

Jirina, M. (1958) Stochastic branching processes with continuous state space, *Czechoslovak Math. J. 8 (83)*: 292–313.

Kallenberg, P. J. M. (1979) *Branching Processes with Continuous State Space*, Math. Centre Tracts No. 117, Mathematische Centrum, Amsterdam.

Kaplan, E. L. and Meier, P. (1958) Nonparametric estimation from incomplete observations, *J. Amer. Statist. Assoc. 53*: 457–482.

Karlin, S. and Taylor, H. (1981) *A Second Course in Stochastic Processes*, Academic Press, New York.

Kerkycharian G. and Picard D. (1992) Density estimation in Besov spaces, *Statist. Probab. Lett. 13*: 15–24.

Kingman, J. F. C. (1978) Uses of exchangeability, *Ann. Probability 6*: 183–197.

Kingman, J. F. C. (1982) Exchangeability and the evolution of large populations, In *Exchangeability in Probability and Statistics* (Ed. G. Koch and F. Spizzichino) North Holland, Amsterdam, 97–112.

Koch, G. and Spizzichino, F. (1982) *Exchangeability in Probability and Statistics*, North Holland, Amsterdam.

Leblanc, F. (1995) Wavelet density estimation of a continuous time process and application to diffusion process, *C. R. Acad. Sci. Paris,* Vol. 321, 345–350.

Loader, C. (1996) Local likelihood density estimation, *Ann. Statist. 24*: 1602–1618.

Malat, S. (1989) Multiresolution aproximations and wavelet orthonormal bases of $L^2(R)$. *Trans. Amer. Math. Soc. 315*: 69–87.

Mardia, K. V. (1972) *Statistics of Directional Data*, Academic Press, New York.

Marshall, A. W. and Olkin, I. (1967) A multivariate exponential distribution, *J. Amer. Statist. Assoc. 62*: 30–44.

Masry, E. (1983) Probability density estimation from sampled data, *IEEE Trans. Inform. Theory IT-29*: 696–709.

Masry, E. (1994) Probability density estimation from dependent observations using wavelets orthonormal bases, *Statist. Probab. Lett 21*: 181–194.

Masry, E. (1997) Multivariate probability density estimation by wavelet methods: Strong consistency and rates for stationary time series. *Stoch. Proc. and their Apprl. 67*: 177–193.

Mielniczuk, J. (1986) Some asymptotic properties of kernel estimators of a density function on the use of censored data, *Ann. Statistics 14*: 767–778.

Mielniczuk, J. (1991) Some asymptotic properties of nonparametric regression estimators in case of randomly censored data, *Statist. 22*: 85–93.

Mishra, M. N. and Prakasa Rao, B. L. S. (1995) On L_1-consistency of kernel-type density estimator for stationary Markov processes. *Comm. in Stat. Theory and Methods*, 24: 581–592.

Nadaraja, E. (1964) On regression estimators, *Theory Probab. Appl. 9*: 157–159.

Newman, C. M. (1980) Normal fluctuations and the FKG inequalities, *Comm. Math. Phys. 74*: 119–128.

Newman, C. M. (1984) Asymptotic independence and limit theorems for positively and negatively dependent random variables, In *Inequalities in Statistics and Probability* (Ed. Y. L. Tong), IMS, Hayward, California, 127–140.

Newman, C. M. and Wright, A. (1981) An invariance principle for certain dependent sequences, *Ann. Probability 9*: 671–677.

Padgett, W. J. and McNicholas, D. T. (1984) Nonparametric density estimation from censored data, *Comm. Statist. A Theory Methods 13*: 1581–1611.

Parzen, E. (1961) An approach to time series analysis, *Ann. Math. Statist 32*: 951–989.

Pezzulli, S. D. and Silverman, B. W. (1993) Some properties of smoothed principal components analysis for functional data. *Comput. Statist. Data Anal. 8*: 1–16.

Powell, U. L., Stock, J. H. and Stoker, T. M. (1989) Semiparametric estimation of index coefficients, *Econometrica 57*: 1403–1430.

Prakasa Rao, B. L. S. (1978) Density estimator for Markov processes via delta-sequences, *Ann. Inst. Statist. Math. 30*: 321–328.

Prakasa Rao, B. L. S. (1979a) Nonparametric estimation for continuous time Markov process via delta families, *Publ. Inst. Statist. Univ. Paris, 24*: 79–97.

Prakasa Rao, B. L. S. (1979b) Sequential nonparametric estimation of density via delta sequences, *Sankhya Ser. A 41*: 82–94.

Prakasa Rao, B. L. S. (1983) *Nonparametric Functional Estimation*, Academic Press, Orlando.

Prakasa Rao, B. L. S. (1983a) Asymptotic theory for nonlinear least squares estimator for diffusion processes. *Math. Operation forch. Stat. Ser. Statistics 12*: 195–210.

Prakasa Rao, B. L. S. (1985) Estimation of the drift for diffusion processes. *Statistics, 16*: 263–275.

Prakasa Rao, B. L. S. (1988) Statistical inference from sampled data for stochastic processes, *Contemporary Mathematics 80*: 249–284.

Prakasa Rao, B. L. S. (1990a) Nonparametric density estimation for exhcangeable sequences. *Proc. of R.C. Bose Symp. on Probability, Statis-*

tics and Design of Experiments (Ed. R. R. Bahadur), Wiley Eastern, New Delhi, 595–606.

Prakasa Rao, B. L. S. (1990b) On mixing for flows of σ-algebras, *Sankhya Ser. A 52*: 1–15.

Prakasa Rao, B. L. S. (1990c) Nonparametric density estimation for stochastic processes from sampled data, *Publ. Inst. Stat. Univ. Paris 35*: 51–83.

Prakasa Rao, B. L. S. (1992a) Nonparametric estimation for Galton–Watson type process, *Statist. Probab. Lett.*, *13*: 287–293.

Prakasa Rao, B. L. S. (1992b) *Identifiability in Stochastic Models*, Academic Press, Boston.

Prakasa Rao, B. L. S. (1993) Bernstein-type inequality for associated sequences, In *Statistics and Probability : A Raghuraj Bahadur Festschrift* (Ed. J. K. Ghosh et al.), Wiley Eastern, New Delhi, 499–509.

Prakasa Rao, B. L. S. (1995a) *Statistics and its Applications*, Indian National Science Academy, New Delhi.

Prakasa Rao, B. L. S. (1995b) Consistent estimation of density-weighted average derivative by orthogonal series method, *Statist. Probab. Lett* *22*: 205–212.

Prakasa Rao, B. L. S. (1996a) On some recent advances in nonparametric functional estimation, *Sectional Presidential Address*, 83rd Session of Indian Science Congress Association; pp. 1–42.

Prakasa Rao, B. L. S. (1996b) Nonparametric approach to time series analysis, In *Stochastic Processes and Statistical Inference* (Ed. B. L. S. Prakasa Rao and B. R. Bhat), New Age International, New Delhi, 73–89.

Prakasa Rao, B. L. S. (1996c) Nonparametric estimation of derivatives of density by the method of wavelets, *Bull. Inform. Cyb. 28*: 91–100.

Prakasa Rao, B. L. S. (1997) Estimation of integral of square of density by wavelets, *Publ. Inst. Statist. Univ. Paris, 41*: 29–48.

Prakasa Rao, B. L. S. (1999) *Statistical Inference for Diffusion Type Processes*. Arnold, London. (To appear).

Prakasa Rao, B. L. S. and Arusharka Sen (1995a) Limit distributions of conditional U-statistics, *Journal Theor. Probability 8*: 261–301.

Prakasa Rao, B. L. S. and Arusharka Sen (1995b) An application of the "martingale methods" to nonparametric regression based on censored data, *Statistics and Decisions 13*: 201–220.

Prakasa Rao, B. L. S. and Dewan, I. (1998) Associated sequences : applications to statistics and probability, In Handbook of Statistics "STOCHASTIC PROCESSES: THEORY AND METHODS", (Ed. C. R. Rao and D. N. Shanbag), Elsevier Science, Amsterdam. (To appear).

Ramlau-Hansen, H. (1983) Smoothing counting process intensities by means of kernel functions, *Ann. Statist. 11*: 453–466.

Ramsay, J. O. (1982) When the data are functions, *Psychometrika, 47*: 379–396.

Ramsay, J. O. and Delzell, C. J. (1991) Some tools for functional data analysis. *J. Roy. Statist. Soc. Ser. B 53*: 661–674.

Rao, J. S. (1984) Nonparametric methods in directional data analysis, In *Hand Book of Statistics, 4:* (Ed. P. R. Krishnaiah and P. K. Sen), 755–77, Elsevier, Amsterdam.

Rice, J. A. and Silverman, B. W. (1991) Estimating the mean and covariance structure nonparametrically when the data are curves, *J. Roy. Statist. Soc. Ser. B 53*: 233–243.

Ritov, Y. and Bickel, P. (1990) Achieving information bounds in non semiparametric models. *Ann. Statist. 18*: 925–938.

Robinson, P.M. (1983) Nonparametric estimators for time series, *J. Time Ser. Anal., 4*: 185–197.

Rosenblatt, M. (1961) Independence and dependence, *Proc. Fourth Berkeley Symp. Math. Statist. Prob.* 431–443.

Rosenblatt, M. (1991) *Stochastic Curve Estimation*, NSF-CBMS Conference Series in Probability and Statistics Vol. 2, Institute of Mathematics Statistics, Hayward, California.

Roussas, G. G. (1991) Kernel estimation under associatioon: strong uniform consistency, *Statist. Prob. Lett. 12*: 393–403.

Roussas, G. G. (1993) Curve estimation in random fields of associated processes, *Nonparametric Statistics, 2*: 215–224.

Scott, D. W. (1994) *Multivariate Density Estimation : Theory, Practice and Visualization*, Wiley, New York.

Silverman, B. W. (1986) *Density Estimation for Statistics and Data Analysis*, Chapman and Hall, London.

Silverman, B. W. (1996) Smoothed functional principal component analysis by choice of norm, *Ann. Statist.*, *24*: 1–24.

Strang, G. (1989) Wavelets and dilation equations : A brief introduction, *SIAM Review 31*: 614–627.

Stute, W. (1991) Conditional *U*-statistics, *Ann. Probability 19*: 812–825.

Stute, W. (1995a) The central limit theorem under random censorship, *Ann. Statistics*, *23*: 422–439.

Stute, W. (1995b) The statistical analysis of Kaplan-Meier integrals, In *Analysis of Censored Data*, IMS Lecture Notes, Vol. 27 (Ed. H. L. Koul and J. V. Deshpande), 231–254.

Stute, W. and Wang, J. L. (1993) The strong law under random censorship, *Ann. Statistics 21*: 1591–1607.

Thompson, J.R. and Tapia, R.A. (1990) *Nonparametric Function Estimation, Modelling and Simulation, SIAM*, Philadelphia.

Ullah, A. (1987) Nonparametric estimation of econometric functionals. *Canadian J. of Economics*.

Walter, G. and Blum, J.R. (1979) Probability density estimation using delta sequences, *Ann. Statist. 7*: 328–340.

Walter, G. and Ghorai, J. (1992) Advantages and disadvantages of density estimation with wavelets. In *Proceedings of the 24th Symp. on the Interface* (Ed. H. Joseph Newton) Interface FNA, VA *24*: 234–243.

Watson, G. S. (1964) Smooth regression analysis. *Sankhya Ser. A 26*: 359–372.

Watson, G. S. (1983) *Statistics on Spheres*, John Wiley & Sons, New York.

Yakowitz, S. (1985) Markov flow models and the flood warning problem. *Water Resources Research 21*: 81–88.

Yu, H. (1993) A Gilvenko-Cantelli lemma and weak convergence for empirical processes of associated sequences, *Probab. Thoery Relat. Fields 95*: 357–370.

14

Minimum Distance and Nonparametric Dispersion Functions

ÖMER ÖZTÜRK   Ohio State University, Marion, Ohio

THOMAS P. HETTMANSPERGER   Pennsylvania State University, University Park, Pennsylvania

JÜRG HÜSLER   University of Bern, Bern, Switzerland

1. INTRODUCTION

Three classes of criterion functions which measure the distance between two probability models are introduced. These criterion functions, in special cases, produce rank estimators, M-estimators and $L1$-estimators depending on the choice of the target model. It is shown that location and scale estimators for the one sample problem are asymptotically normal and have high efficiency. The location estimator produces the median functional for symmetric distributions and it is between the mean and median functionals for asymmetric distributions. The criterion functions are used to derive a goodness-of-fit test for the symmetric underlying probability model. The goodness-of-fit test has an asymptotically normal distribution. The results are extended to linear models and an example is provided.

In the context of minimum distance estimation, it is known that many robust estimators are asymptotically equivalent. The unweighted minimum Cramér–von Mises distance estimator, Hodges–Lehmann estimator, aligned and Wilcoxon rank estimators are all equivalent asymptotically when the target model is also the true model. To our knowledge, the exact relation among these estimators is not known. In this paper, we investigate three

511

different criterion functions. In some special cases, these criterion functions reduce to aligned and Wilcoxon rank estimators, $L1$ estimators and M-estimators.

Minimum distance functions were originally developed for goodness-of-fit purposes. The idea is to measure the distance between the data, summarized by the empirical distribution function, and the hypothesized probability model. Then depending on the magnitude of this distance measure, a decision is made either against or in favor of the hypothesized probability model. Well-known goodness of fit test statistics in this context are Kolmogorov–Smirnov, Anderson–Darling and Cramér–von Mises distance functions.

In the last two decades, it has been of interest to look at these goodness-of-fit tests for the purpose of robust estimation. Although minimum distance functions in goodness-of-fit tests are sensitive to small departures from the model, they have satisfactory robustness properties in estimation. Parr and Schucany (1980), in an extensive simulation study concluded that minimum distance estimators (MDE) perform very well among well-known robust estimators. Boos (1981) used a weighted Cramér–von Mises distance function to estimate the location parameter in a location family. Parr and De Wet (1981) generalized Boos' results to a non-location parameter. Wiens (1987) used Huber's minimax theory in order to find an optimal weight function in a weighted Cramér–von Mises distance function. Öztürk and Hettmansperger (1996) constructed a distance function that compares the tail probabilities of the underlying probability model. They showed that the minimizer of this distance function is robust and as efficient as the maximum likelihood estimator. Heathcote and Silvapulle (1981), Hettmansperger, Hueter and Hüsler (1994) used the unweighted Cramér–von Mises distance function for simultaneous estimation of location and scale parameters.

Several researchers combined the efficiency and robustness in the same estimator by using density based minimum distance functions; see Basu and Lindsay (1994), Lindsay (1994) and Donoho and Liu (1988a, 1988b). Although the robustness is achieved at the expense of efficiency in classical robustness theory, the minimum Hellinger distance function combines these two competing ideas in the same estimator.

Recently Öztürk and Hettmansperger (1997) used a class of minimum distance functions that contains weighted and unweighted Cramér–von Mises distance functions as special cases. They show that the weight function has a dramatic effect on the estimator. Robustness and efficiency depend on the choice of the criterion function.

The objective of this paper is to introduce three classes of criterion functions that can be used for estimation and goodness-of-fit purposes.

Section 2 discusses the motivation for the criterion functions. It is shown that choice of different target models produces M-estimators, rank estimators and $L1$-estimators. The estimation of location and scale parameters is discussed in Sec. 3. A goodness-of-fit test is constructed in Sec. 4. The results are extended to linear models in Sec. 5. An example is provided in Sec. 6. Detailed proofs of results are contained in a technical report, Öztürk, Hettmansperger and Hüsler (1997), and can be obtained from the first author.

2. MOTIVATION

Let $X_1, X_2, \ldots, X_n$ be iid observations from an absolutely continuous distribution F in the space of distribution functions. At the moment, we do not put any other constraint on the choice of F other than absolute continuity. For a given standardized location-scale model F_0 with mean zero and scale 1, we define a location and scale family as follows:

$$\mathscr{F} = \{F_0((t - \mu)/\sigma) : (\mu, \sigma) \in R \times R^+\}.$$

In the minimum distance methodology, we specify a target family $\mathscr{F}$ and minimize the difference between the data and target family. The success of the procedure depends on the type of criterion function and the choice of target family. Define

$$d_1(\theta, F_n, F_0) = \int [1 - F_n(t)]\left[1 - F_0\left(\frac{-t + \mu}{\sigma}\right)\right] dt$$

$$+ \int F_n(t)F_0\left(\frac{-t + \mu}{\sigma}\right) dt + C_1\sigma, \tag{1}$$

$$d_2(\theta, F_n, F_0) = \int [1 - F_n(t)]F_0\left(\frac{t - \mu}{\sigma}\right) dt$$

$$+ \int \left[1 - F_0\left(\frac{t - \mu}{\sigma}\right)\right]F_n(t) dt + C_2\sigma, \tag{2}$$

$$d_3(\theta, F_n, F_0) = d_1(\theta, F_n, F_0) + d_2(\theta, F_n, F_0) + C_3\sigma, \tag{3}$$

where F_n is the empirical distribution function (EDF) of $X_1, X_2, \ldots, X_n$ and F_0 is the assumed target model for the standardized random variable $e_1, e_2, \ldots, e_n$, where $e_i = (X_i - \mu)/\sigma$. Lemma 1 and the following discussion provides motivation for the choice of Eqns (1), (2) and (3). The choice of the target model F_0 depends on knowledge about the underlying true distribution. It may or may not be equal to F. On the other hand, it is believed to be close to the true model F. For example, if we believe that the true model is normal when it is actually the contaminated normal, then we use the standard normal distribution function as the target model. For possible choices of F_0; see Öztürk and Hettmansperger (1998). Coefficients C_1, C_2, C_3 are needed for correct centering of the scale estimating equations. The C_i's are chosen so that scale estimates are consistent at the target model.

Now we will look at the motivation behind these criterion functions. For example, in Eq. (2) if we replace F_n by F_0 which is centered at 0 with scale equal to 1, then $d_2(., ., .)$ is minimized by $\mu = 0$ and $\sigma = 1$. It is interesting to note that if we consider $(1/\sigma)/d_i(\theta, F_n, F_n^*)$ for $i = 1, 2, 3$ we will have rank based dispersion functions, where F_n^* is the empirical distribution function of $e_1, e_2, \ldots, e_n$.

LEMMA 1   Assume $C_i = 0$ for $i = 1, 2, 3$, then

$$\frac{1}{\sigma} d_1(\theta, F_n, F_n^*) = \frac{1}{n^2} \sum_{i=1}^{n} \sum_{j=1}^{n} |e_i + e_j|,$$

$$\frac{1}{\sigma} d_2(\theta, F_n, F_n^*) = \frac{4}{n^2} \sum_{i=1}^{n} e_i \left[R(e_i) - \frac{n+1}{2} \right],$$

$$\frac{1}{\sigma} d_3(\theta, F_n, F_n^*) = \frac{1}{\sigma} d_1(\theta, F_n, F_n^*) + \frac{1}{\sigma} d_2(\theta, F_n, F_n^*)$$

$$- \frac{2}{n^2} \sum_{i=1} |e_i| = \frac{4}{n^2} \sum |e_i| R(|e_i|)$$

where $R(e_i)$ is the rank of the observation e_i.

Lemma 1 indicates that our criterion functions are close to well-known dispersion functions in the nonparametric statistics literature, that is, if we choose the target model F_0 to be F_n^*, we exactly get dispersion functions used in Jaeckel (1972) and van Eeden (1972). The derivatives of these dispersion functions lead to Wilcoxon rank and signed rank statistics; see Hettmansperger and Aubuchon (1988). The dispersion function $(1/\sigma)/d_2(\theta, F_n, F_n^*)$ is invariant with respect to a location change. Therefore, it can not be used to estimate the location parameter. In order to avoid this deficiency our criterion functions in Eqns (1), (2) and (3) replace F_n^* with a target model F_0. This also provides another advantage in that the dispersion functions behave like goodness-of-fit test statistics. By introducing a target model, we change the dispersion functions in Lemma 1 from rank criterion functions to M-estimation criterion functions. In fact, a good choice of F_0 is the projection of F_n^* under an assumed model.

Consider Eqs (1), (2) and (3) evaluated at $t \equiv 0$, then $(1/\sigma)d_1(\boldsymbol{\theta}, F_n, F_n^*) = (1/n^2)[n^2 - 2na + 2a^2]$ and $(1/\sigma)d_2(\boldsymbol{\theta}, F_n, F_n^*) = (2/n^2)(1-a)a$, where, $a = nF_n^*(0)$. Let $\hat{\mu}_i$ be the estimator of μ based on the minimizer of $d_i(\boldsymbol{\theta}, F_n, F_n^*)$, evaluated at $t \equiv 0$, for $i = 1, 2, 3$. It is obvious from the above calculation that the estimator $\hat{\mu}_i$, for $i = 1, 2, 3$, is the median of $X_1, X_2, \ldots, X_n$.

We will look at the problem from another perspective. It is obvious that the choice of different target models leads to different types of estimators. We expect that $(1/\sigma)d_i(\boldsymbol{\theta}, F_n, F_0)$ should be close to $(1/\sigma)d_i(\boldsymbol{\theta}, F_n, F_n^*)$ if F_0 and F_n^* are close to each other. Assume $F_0 = F$ and $C_1 = C_2 = 0$, then integration in $d_i(\boldsymbol{\theta}, F_n, F)$ and $d_2(\boldsymbol{\theta}, F_n, F)$ yield the following:

$$\frac{1}{\sigma} d_2(\boldsymbol{\theta}, F_n, F) = \frac{1}{n} \sum_{i=1}^{n} [[2F(e_i) - 1]e_i - 2K(e_i) + E_F(X)], \tag{4}$$

$$\frac{1}{\sigma} d_1(\boldsymbol{\theta}, F_n, F) = \frac{1}{n} \sum_{i=1}^{n} [e_i[1 - 2F(-e_i)] - 2K(-e_i) + E_F(X)] \tag{5}$$

where $K(t) = \int_{-\infty}^{t} f(y)y \, dy$ and $E_F(X)$ is the expected value of X with respect to distribution F. Now we will show that the criterion functions $d_i(.,.,.)$ $i = 1, 2$ are the projections of the well-known norms in nonparametric statistics literature. Let $\tau = E|e_i + e_j|$,

$$U_n^{**} = \sum_{i=1}^{n} \sum_{j \neq i} [|e_i + e_j| - E|e_i + e_j|],$$

$$U_n^{*} = \sum_{i=1}^{n} \left[e_i \left[R(e_i) - \frac{n+1}{2} \right] - Ee[F(e) - 1/2] \right].$$

LEMMA 2   Assume F has a finite second moment, then V_{p1} and V_{p2} are the projections of $U_n^{**}/(n(n-1))$ and $U_n^{*}/(n(n-1))$, respectively, on the space of sum of independent random variables, where

$$V_{p1} = \frac{2}{\sigma} d_1(\boldsymbol{\theta}, F_n, F) - \tau,$$

$$V_{p2} = \frac{1}{2\sigma} d_2(\boldsymbol{\theta}, F_n, F) - 2Ee[1/2 - F(-e)].$$

The criterion functions $d_i(\boldsymbol{\theta}, F_n, F_0)$ for $i = 1, 2, 3$ generate a broad class of estimators. Some choices of the target model, such as F_n^*, lead to well-known rank estimates and some other choices of the target model lead to M estimates that we will discuss later. On the other hand, consideration of integrands evaluated at 0 leads to the $L1$ estimate. Thus, our criterion function provides a general class that combine all three types of estimates.

3. ONE SAMPLE PROBLEM

We assume the same set-up in the previous Section. Our concern, in this Section, is to estimate μ and σ simultaneously and explore the properties of these estimators and criterion functions. In Sec. 5, these results will be generalized to linear models. Define the functional $\mathbf{T}_i(F)$ as

$$\mathbf{T}_i(F) = (M_i(F), S_i(F))^T = \arg \min_{\mu \in R, \sigma \in R^+} d_i(\boldsymbol{\theta}, F, F_0) \; i = 1, 2, 3. \tag{6}$$

If F is replaced by the EDF F_n, then $\mathbf{T}_i(F_n)$ will be the estimator, $M_i(F_n)$ for the location, $S_i(F_n)$ for the scale. Equation (6) is not complete since it is not clear at this point if there exists a minimizer of the criterion function. For the uniqueness of the functional $\mathbf{T}_i(F)$ we assume that $F_0(x)$ and $F(x)$ have non-vanishing densities with continuous derivatives. Then, the Hessian matrix of $d_1(\boldsymbol{\theta}, F_n, F_0)$ is found to have determinant

$$|\mathbf{H}_1| = \frac{2}{n^2 \sigma^2} \sum_{i=1}^{n} \sum_{j=1}^{n} f_0(-e_i) f_0(-e_j)(e_i - e_j)^2,$$

which is strictly positive if there exist at least two distinct observations. Similarly the determinant of the Hessian matrix of $d_1(\boldsymbol{\theta}, F, F_0)$ is

$$|\mathbf{H}_1^*| = 2 \int \int (r - k)^2 f(\mu - r\sigma) f(\mu - k\sigma) f_0(r) f_0(k) \, dr \, dk.$$

This is also strictly positive. Consequently, criterion functions $d_1(\boldsymbol{\theta}, F_n, F_0)$ and $d_1(\boldsymbol{\theta}, F, F_0)$ are strictly convex; see Roberts and Varberg (1973, p.103). The same results also hold for the criterion functions $d_2(\boldsymbol{\theta}, F_n, F_0)$ and $d_2(\boldsymbol{\theta}, F, F_0)$. Then, the convexity of $d_3(\boldsymbol{\theta}, F_n, F_0)$ and $d_3(\boldsymbol{\theta}, F, F_0)$ follows from the fact that a sum of two convex functions is also convex. We can summarize these results in the following Lemma.

LEMMA 3   Let $X_1, \ldots, X_n$ be independent identically distributed observations from the distribution F. Assume F_0 and F have nonvanishing densities with continuous derivatives. Then $T_i(F_n)$ and $T_i(F)$ are unique.

Differentiation of the criterion functions leads to the estimating equations. For ease of presentation we define the following:

$$\left. \begin{aligned}
&\psi_1(z) = F_0(-z) - 1/2, \quad \psi_2(z) = 1/2 - F_0(z), \\
&\psi_3(z) = F_0(-z) - F_0(z), \quad \chi_1(z) = C_1 - K_0(-z), \\
&\chi_2(z) = C_2 - K_0(z), \quad \chi_3(z) = C_3 - K_0(z) - K_0(-z), \\
&K_0(t) = \int_{-\infty}^{t} f_0(y) y \, dy.
\end{aligned} \right\} \tag{7}$$

Now the functional $T_i(F)$ is equivalently defined as the solution of the following equations.

$$
\lambda_i(T_i(F)) = \left[\frac{\int \psi_i \left\{ \dfrac{t - M_i(F)}{S_i(F)} \right\} dF(t)}{\int \chi_i \left\{ \dfrac{t - M_i(F)}{S_i(F)} \right\} dF(t)} \right] = 0, \quad i = 1, 2, 3. \tag{8}
$$

In the defining equations above, $C_1 - C_3$ should be chosen so that estimating equations are correctly centered. This indicates that

$$
C_1 = \int K_0(-z) \, dF(z), \qquad C_2 = \int K_0(z) \, dF(z)
$$

and

$$
C_3 = \int K_0(z) \, dF(z) + \int K_0(-z) \, dF(z).
$$

It is evident that C_i, $i = 1, 2, 3$, depends on both F_0 and F and in practice F is unknown. On the other hand, we choose F_0 such that we believe $F_0 = F$ or at least close to F. Therefore, C_i's will be determined by replacing F with F_0. When $F \neq F_0$, this may introduce a small bias in the estimator.

Öztürk (1994) showed that the biases of the location functionals $M_i(F_n)$ are zero if both F and F_0 are symmetric at the same point. Under asymmetric ε-neighborhood contamination model, $F_a(t) = (1 - \varepsilon)F_0(t) + \varepsilon F_0(t - a)$, respective biases of $M_1(F_a)$, $M_2(F_a)$ and $M_3(F_a)$ are εa, εa and $2\varepsilon a$.

If the target model F_0 is symmetric then it is clear that all the estimating equations reduce to $\lambda_2(T_2(F))$ which is the same as the estimating equation of Heathcote and Silvapulle (1981) and Hettmansperger, Hueter and Hüsler (1994). Therefore, the criterion function $d_2(\theta, F_n, F_0)$ produces the same estimator as the unweighted Cramér–von Mises distance estimator.

In the context of defining a measure for the natural parameters of a population, one wants to define these measures for a family rather than a single population. The choice of suitable measures then depends strongly on the nature of this family. Bickel and Lehmann (1975a, 1975b, 1976) define such measures for the natural parameters location, scale, and kurtosis. Since our estimators also strongly depend on the target family, we will explore their properties within the context of the class of location functionals. We use the same definition for the location functional as in Bickel and Lehmann (1975b).

$M(F)$ is a location functional if the following conditions hold:

(i) S is stochastically larger than F implies $M(S) \geq M(F)$,
(ii) If X has distribution F then $M(aX + b) = aM(X) + b$, and
(iii) $M(-X) = -M(X)$.

THEOREM 1

(i) In Eq. (6), the functionals $M_i(F)$ are location functionals.
(ii) Suppose F and F_0 are in $\mathscr{F}$ and symmetric, then $M_i(F) = M_2(F)$, $i = 1, 2, 3$ and $M_2(F) = F^{-1}(1/2)$.

Since $M_2(F)$ is actually equal to the unweighted Cramér–von Mises distance function estimator, the result also holds for Boos's (1981) location estimator with his $w(x) \equiv 1$. The second part of the theorem might be anticipated since the median is the natural location parameter of a symmetric distribution. Therefore, any sensible estimator of location should produce the median functional. When the underlying distribution is asymmetric, the functional is defined implicitly through the estimating equations. The following results state the relationship between the mean, median, mode and $M_i(F)$.

THEOREM 2   Let m, μ be the median and mean functionals, respectively and F_0 be a symmetric unimodal distribution. Suppose F is an asymmetric unimodal distribution and $\int_{-\infty}^{0} f_0(t)[F(-t + \mu) + F(t + \mu) + 2F(\mu)]\,dt > 1$.

(i) If $m < \mu$ holds then $m < M_i(F) < \mu$,
(ii) If $\mu < m$ holds then $\mu < M_i(F) < m$ where $M_i(F)$ is defined in Eq. (6).

Theorem 2 puts our estimators between the mean and median functionals for asymmetric distributions.

Asymptotic results for $\mathbf{T}_i(F_n)$, $i = 1, 2, 3$, follow with proofs similar to those of Heathcote and Silvapulle (1981), by considering corresponding estimating equations. We now state the main results.

LEMMA 4   Assume that the assumptions of Lemma 3 hold and the true model F has a finite first moment. Then $\mathbf{T}_i(F_n)$ converges almost surely to $\mathbf{T}_i(F)$.

Finiteness of the first moment of F ensures that criterion function $d_i(\mathbf{\theta}, F_n, F_0)$ converges uniformly to $d_i(\mathbf{\theta}, F, F_0)$. Convexity and uniform convergence of the criterion functions lead to the convergence of the statistical functionals. For detailed proof, see Öztürk (1994).

THEOREM 3   Assume that the assumptions of Lemma 3 hold. Furthermore, assume that (i) $\int f'_0(z)z^2\,dF(z) < \infty$ and (ii) $\int f_0(z)z^2\,dF(z) < \infty$. Then the limiting distribution of $\sqrt{n}(\mathbf{T}_i(F_n) - \mathbf{T}_i(F))$ is normal with mean zero and covariance matrix $\mathbf{D}_i^{-1}(F)\mathbf{\Sigma}_i(F)\mathbf{D}_i^{-1}(F)$, where $Z_i(t) = [t - M_i(F)]/S_i(F)$ and

$$
\mathbf{D}_i(F) = \begin{bmatrix} E_F \dfrac{\partial \psi_i\{Z_i(t)\}}{\partial M_i(F)} & E_F \dfrac{\partial \psi_i\{Z_i(t)\}}{\partial S_i(F)} \\[2ex] E_F \dfrac{\partial \psi_i\{Z_i(t)\}}{\partial S_i(F)} & E_F \dfrac{\partial \chi_i\{Z_i(t)\}}{\partial S_i(F)} \end{bmatrix},
$$

$$
\mathbf{\Sigma}_i(F) = \begin{bmatrix} \mathrm{Var}_F(\psi_i\{Z_i(t)\}) & \mathrm{Cov}_F(\psi_i\{Z_i(t)\}, \chi_i\{Z_i(t)\}) \\[2ex] \mathrm{Cov}_F(\psi_i\{Z_i(t)\}, \chi_i\{Z_i(t)\}) & \mathrm{Var}_F(\chi_i\{Z_i(t)\}) \end{bmatrix},
$$

and $\psi_i(.;.)$ and $\chi_i(.;.)$ are defined in Eq. (7).

This result is proved for $\mathbf{T}_2(F_n)$ in Heathcote and Silvapulle (1981). For the general case, the proof follows from a usual Taylor series expansion of $\lambda_i(\mathbf{T}_i(F_n))$ around $\mathbf{T}_i(F)$. A detailed proof is given in Öztürk (1994).

4. GOODNESS OF FIT

In minimum distance estimation, the criterion function itself can be used to test hypotheses about the model as well as the parameters. Usually the asymptotic distribution of the distance function is not normal; see Boos (1981). It is expressed as the distribution of an infinite sum of weighted chi-square random variables and is not very practical. The criterion function that we use has asymptotically a normal distribution and it has a simple calculating form. Below we show that the test is consistent for a large class of possible alternatives. Consider $d_3(\mathbf{\theta}, F_n, F_0)$

$$
d_3(\mathbf{\theta}, F_n, F_0) = \frac{2\sigma}{n} \sum_{i=1}^{n} e_i[F_0(e_i) - F_0(-e_i)]
$$

$$
- \frac{2\sigma}{n} \sum_{i=1}^{n} [K_0(e_i) + K_0(-e_i)] + C_3\sigma.
$$

The above equation follows from Eqns (4) and (5). It is obvious that $d_3(\mathbf{\theta}, F_n, F_0)$ is an average of iid random variables and under mild regularity conditions almost surely converges to its mean, i.e,

$$
d_3(\mathbf{\theta}, F_n, F_0) \xrightarrow{\text{a.s}} \sigma \int 2z[F_0(z) - F_0(-z)]\, dF(z)
$$

$$
- 2\sigma \int [K_0(z) + K_0(-z)]\, dF(z) + C_3\sigma = A_F. \tag{9}
$$

The expectation in Eq. (9) depends on the underlying distribution F. If the target model F_0 is equal to the true model F, then we replace F by F_0. Now our concern is on the difference between $A_F - A_{F_0}$. In other words, how much reduction or increase results from not knowing the true model F.

In the next theorem, we will show that the difference $A_F - A_{F_0}$ will be smaller or larger depending on whether F is stochastically larger or smaller than F_0. Before we state the theorem, define the following: $\mathscr{F}_{F_0} = \{S : S \leq F_0 \text{ or } S \geq F_0 \text{ with strict inequality at some point}\}$, i.e, $\mathscr{F}_{F_0}$ is the set of all distribution functions stochastically smaller or larger than F_0.

THEOREM 4   Let $F \in \mathscr{F}_{F_0}$ and F_0 and F are symmetric. Then $A_F - A_{F_0} \neq 0$.

A detailed proof of this can be found in Öztürk (1994).

Theorem 4 provides a basis for a restricted goodness-of-fit test. By basing the test on the residuals after removing location and scale effects, we are able to detect differences from the target model due to stochastic ordering. This is not an omnibus test and we do not expect it to be sensitive to all possible alternative models. As we indicated earlier, we pick a model F_0 which we believe is close to the true model F. Under the symmetry condition on both F_0 and F, the criterion function $d_3(\boldsymbol{\theta}, F_n, F_0)$ becomes

$$d_3(\boldsymbol{\theta}, F_n, F_0) = \frac{4\sigma}{n} \sum_{i=1}^{n} [e_i[F_0(e_i) - 1/2] - K_0(e_i)] + C_3\sigma. \tag{10}$$

It is clear that $d_3(\boldsymbol{\theta}, F_n, F_0)$ will be close to A_{F_0} if $F_0 = F$. Therefore, we will reject the hypothesis that $F = F_0$ if $|d_3(\boldsymbol{\theta}, F_n, F_0) - A_{F_0}|$ is too big.

The following theorem indicates that the proposed test is consistent for stochastically ordered alternatives.

THEOREM 5   The goodness of fit test defined above is consistent for stochastically ordered alternatives.

The test statistic $d_3(\boldsymbol{\theta}, F_n, F)$ is not useful in its present form since the true value of the parameters are not known. A remedy for this problem is to replace $\boldsymbol{\theta}$ with the minimizer of the criterion function $d_3(\boldsymbol{\theta}, F_n, F)$. Let

$$Q(F) = \frac{4\sigma(F)}{\sqrt{n}} \sum_{i=1}^{n} [e_i\{F_0(e_i) - 1/2\} - K_0(e_i) - \tau] + \sqrt{n}C_3\{\sigma(F) - \sigma_0\}$$

$$\tag{11}$$

where $e_i = \{X_i - \mu(F)\}/\sigma(F)$ and τ is the expected value of $e\{F(e) - 1/2\} - K_0(e)$. It is obvious that $Q(F_n)$ is a test statistic. In order to find the asymptotic distribution of $Q(F_n)$, we need to find an asymptotically equivalent expression for $\sqrt{n}\{\sigma(F_n) - \sigma_0\}$. From the linearization of the scale estimating equation, we can write

$$\sqrt{n}(\sigma(F_n) - \sigma_0) = -\frac{\sigma(F)\sum_{i=1}^{n}\{C_3/4 - K_0(e_i)\}}{\sqrt{n}E_{F_0}f_0(t)t^2}. \tag{12}$$

By combining Eqns (11) and (12), for large n we have

$$Q(F) = \frac{4\sigma(F)}{\sqrt{n}}\sum_{i=1}^{n}$$

$$\times\left[e_i\{F_0(e_i) - 1/2\} - K_0(e_i) - \tau - \frac{C_3\sum_{i=1}^{n}\{C_3/4 - K_0(e_i)\}}{E_{F_0}f_0(t)t^2}\right]. \tag{13}$$

For the normal distribution, with appropriate standardization Eq. (13) reduces to

$$Q^* = \frac{\sum_{i=1}^{n}[4e_i\{\Phi(e_i) - 1/2\} + 12\phi(e_i) - A_\Phi^*]}{\sqrt{n}\sigma_\Phi^*},$$

where A_Φ^* and σ_Φ^* are the mean and standard deviation of $4e\{\Phi(e) - 0.5\} + 12\phi(e)$. Straightforward but tedious calculations yield that $A_\Phi^* = 8/\sqrt{\pi}$ and $\sigma_\Phi^* = 0.272$. Then our test statistic becomes

$$Q_n^* = \frac{\sum_{i=1}^{n}[4\hat{e}_i\{\Phi(\hat{e}_i) - 1/2\} + 12\phi(\hat{e}_i) - 8/\sqrt{\pi}]}{0.272\sqrt{n}},$$

where $\hat{e}_i = \{X_i - \mu(F_n)\}/\sigma(F_n)$. Our test rejects $H_0 : F = \Phi$ in favor of $H_1 : F \neq \Phi$ if $|Q_n^*| > Z_{\alpha/2}$, where $Z_{\alpha/2}$ is upper $\alpha/2$ percentile of the standard normal distribution.

For one sample inference, the proposed procedure can be used informally as follows: First calculate $\mu(F_n)$ and $\sigma(F_n)$ by minimizing the criterion function $d_3(\theta, F_n, F)$. Then compute Q_n^* by inserting $\mu(F_n)$ and $\sigma(F_n)$ in Q^*. If the test fails to reject $H_0 : F = \Phi$ then it suggests that the true model may be close to the normal model and there is statistical support for the appropriateness of the target model Φ. Hence, we use robust estimates $\mu(F_n)$ and $\sigma(F_n)$ which are appropriate in a neighborhood of the true model. If the test rejects $H_0 : F = \Phi$ then we should consider changing our target model. This suggests that we are not in a neighborhood of the true model.

In a simulation study, we investigate the behavior of the goodness-of-fit test statistic. We have calculated the power of the test for several distributions. These distributions include normal, Cauchy, logistic and t distribution

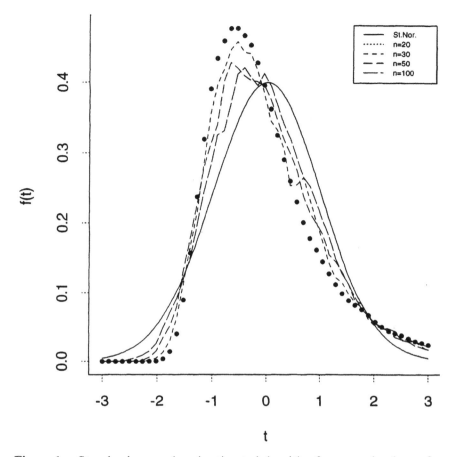

Figure 1.   Standard normal and estimated densities for several values of n, $n = 20, 30, 50, 100$.

with k degrees of freedom, $k = 3, 5$. Power is calculated for each distribution for sample sizes, 20, 30, 50, 100. For the null hypothesis, we have performed a density estimation and superimposed the standard normal density on the estimated densities, see Fig. 1.

Table 1 presents the simulation results. In the simulation study, the type I errors are slightly larger than the theoretical values of 0.05 and 0.01. Departures from the theoretical values are severe for the small sample sizes. For example, type I errors are 0.06 and 0.02 instead of 0.05 and 0.01, respectively when $n = 20$. This can be also observed from Fig. 1. Figure 1 indicates that

Table 1. Power of the Goodness-of Fit Test for Different Sample Sizes

Dist.	$n = 20$		$n = 30$		$n = 50$		$n = 100$	
	$\alpha = 0.05$	$\alpha = 0.01$	$\alpha = 0.05$	$\alpha = 0.01$	$\alpha = 0.05$	$\alpha = 0.01$	$\alpha = 0.05$	$\alpha = 0.01$
N	0.0624	0.0330	0.0532	0.0257	0.0519	0.0216	0.0527	0.0168
L	0.1771	0.1163	0.2111	0.1401	0.2732	0.1789	0.4107	0.2734
C	0.9162	0.8921	0.9810	0.9705	0.9988	0.9981	1.0000	1.0000
t_3	0.4462	0.3675	0.5765	0.4905	0.7460	0.6620	0.9331	0.8871
t_5	0.2642	0.1944	0.3356	0.2489	0.4540	0.3480	0.6662	0.5459

N: normal distribution; L: logistic distribution; C: Cauchy distribution; t_k: t distribution with k degrees of freedom

for small sample sizes the asymptotic distribution is slightly skewed to the right. As a result of this, the type I error becomes larger.

Table 1 indicates that the power of the test is very good for the Cauchy distribution. For $n = 20$, power is 0.9162 and it increases to 1 when n gets large. The goodness of fit test does not provide such high powers for the logistic and t distributions. This is not surprising since the logistic and t distributions are close to the normal distribution, especially when the degrees of freedom of t distribution is moderately large.

5. LINEAR MODELS

This Section extends the results of the one sample problem in the previous section to linear models. It is assumed that observations $Y_1, Y_2, \ldots, Y_n$ are generated by the following linear model:

$$Y_i = \mathbf{x}_i^T \boldsymbol{\beta} + \varepsilon_i \qquad i = 1, 2, \ldots, n, \tag{14}$$

where $Y_i - \mathbf{x}_i^T \boldsymbol{\beta}$ has distribution F with mean zero and scale σ. The model in Eq. (14) can equivalently can be written as

$$\mathbf{Y} = \mathbf{X}\boldsymbol{\beta} + \boldsymbol{\varepsilon}.$$

In this model we wish to estimate $\boldsymbol{\beta}^T = (\beta_0, \ldots, \beta_p)$ and σ jointly. As in Sec. 2, for the true distribution F, a known target model F_0 is chosen from the location and scale family. Criterion functions in the one sample problem can be extended easily to cover the linear model. In this section, we denote the parameter of interest $\theta^T = (\beta^T, \sigma)$. Although it is possible to consider all three criterion functions discussed in Sec. 2, we only consider $d_2(\theta, F_n, F_0)$ since the other two reduce to $d_2(\theta, F_n, F_0)$ for symmetric target models

$$
d_2(\theta, F_n, F_0) = \int [1 - F_n(t)] \left[F_0 \left(\frac{t - x_i^T \beta}{\sigma} \right) \right] dt
$$

$$
+ \int F_n(t) \left[1 - F_0 \left(\frac{t - x_i^T \beta}{\sigma} \right) \right] dt + \sigma C. \tag{15}
$$

Again the coefficient C is needed for correct centering of the scale estimating equation.

The estimator is defined as the minimizer of the criterion function in Eq. (15) as follows:

$$
\theta^T(F_n) = (\beta^T(F_n), \sigma(F_n)) = \arg \min_{\beta \in R^{p+1}, \sigma \in R^+} d_2(\theta, F_n, F_0)
$$

where F_n is EDF of $(Y_i - x_i^T \beta)/\sigma$. Before we consider the existence of the estimators, more notation is needed. Let $X^* = (X, e)$ where the residual vector is augmented with the design matrix X.

LEMMA 5   Assume the columns of X^* are linearly independent, $F_0(t)$ has non-vanishing density $f_0(t)$ and $f'_0(t)$ is continuous. Then $d_2(\theta, F_n, F_0)$ is strictly convex in θ.

Lemma 5 indicates that there exists a unique minimizer for the criterion function. It is convenient to define the estimators in terms of estimating equations.

$$
\frac{1}{n} \sum_i x_i^T \left\{ 1 - 2F_0 \left(\frac{y_i - x_i^T \beta_n}{\sigma_n} \right) \right\} = 0, \tag{16}
$$

$$
\frac{1}{n} \sum_i \left\{ C - 2K_0 \left(\frac{y_i - x_i^T \beta_n}{\sigma_n} \right) \right\} = 0. \tag{17}
$$

Equations (16) and (17) are M-estimating equations for the regression vector β and scale parameter σ with $\psi(t) = 1 - 2F_0(t)$ and $\chi(t) = C - 2K_0(t)$ where $K_0(t) = \int_{-\infty}^t f_0(y) y \, dy$ and $C = 2EK_0(e)$.

For the limiting distribution of the estimator, there exists a vast literature from the theory of M-estimation; see Huber (1981) and Hampel et. al. (1986). Assume $n^{-1} X^T X$ converges to a positive definite matrix Σ, then under some regularity conditions on F_0

$$n^{1/2}\{\beta(F_n) - \beta\} = \frac{\sigma}{2E f_0(\sigma^{-1}e)} \Sigma^{-1} n^{-1/2}$$

$$\times \sum_i \mathbf{x}_i \left\{ 1 - 2F_0\left(\frac{y_i - \mathbf{x}_i^T\beta}{\sigma}\right) \right\} + o_p(1), \tag{18}$$

$$n^{1/2}\{\sigma(F_n) - \sigma\} = \frac{\sigma}{2E f_0(\sigma^{-1}e)e^2} \Sigma^{-1} n^{-1/2}$$

$$\times \sum_i \left\{ C - 2K_0\left(\frac{y_i - \mathbf{x}_i^T\beta}{\sigma}\right) \right\} + o_p(1), \tag{19}$$

where $e_1, \ldots, e_n$ are iid F. From Eqns (18) and (19) the influence functions of $\beta(F_n)$ and $\sigma(F_n)$ can be constructed easily.

$$\mathbf{IF}(\beta, e_i, \mathbf{x}_i) = \frac{\sigma\left\{ 1 - 2F_0\left(\frac{y_i - \mathbf{x}_i^T\beta}{\sigma}\right) \right\}}{2E f_0(\sigma^{-1}e)} \Sigma^{-1}\mathbf{x}_i,$$

$$\mathbf{IF}(\sigma, e_i) = \frac{\sigma\left\{ C - 2K_0\left(\frac{y_i - \mathbf{x}_i^T\beta}{\sigma}\right) \right\}}{2E\{f_0(\sigma^{-1}e)e^2\}}.$$

The second factor in $\mathbf{IF}(\beta, e_i, \mathbf{x}_i)$ shows that the influence function is not bounded in factor space. The M-estimating equation can be easily modified to achieve bounded influence; see a general version of the estimating equation (16) in Öztürk (1994). Boundedness of the influence function of the scale depends on the tail behavior of the target model F_0. If F_0 has a finite expectation then the scale estimator has bounded influence function.

It follows from linear representations in Eqns (18) and (19) that $\sqrt{n}\{\beta(F_n) - \beta\}$ and $\sqrt{n}\{\sigma(F_n) - \sigma\}$ are asymptotically normally distributed with mean zero and covariance matrix $k^2\Sigma^{-1}$ and variance ω^2, respectively, where

$$k^2 = \frac{\sigma^2 E\{1 - 2F_0(e\sigma^{-1})\}^2}{\{2E f_0(e\sigma^{-1})\}^2},$$

$$\omega^2 = \frac{\sigma^2 E\{C - 2K_0(e\sigma^{-1})\}^2}{\{2E f_0(e\sigma^{-1})e^2\}^2}.$$

The efficiency of the estimators are determined by k^2 and ω^2 and they are the same as in the location-scale model. In a numerical comparison of the efficiency of location and scale estimators, Heathcote and Silvapulle (1981) and Hettmansperger, Hueter and Hüsler (1994) observed that the minimum distance estimator is very efficient for moderately heavy tailed distributions.

In order to construct confidence intervals and test the hypotheses of interest, the parameters k^2 and ω^2 must be estimated. For computational ease we recommend simple moment estimators

$$\hat{E}\{1 - 2F_0(e\sigma^{-1})\}^2 = \frac{1}{n} \sum_i \{1 - 2F_0(\hat{e}_i)\}^2,$$

$$\hat{E}f_0(e\sigma^{-1}) = \frac{1}{n} \sum_i f_0(\hat{e}_i),$$

$$\hat{E}\{C - 2K_0(e\sigma^{-1})\}^2 = \frac{1}{n} \sum_i \{C - K_0(\hat{e}_i)\}^2,$$

$$\hat{E}f_0(e\sigma^{-1})e^2 = \frac{1}{n} \sum_i f_0(\hat{e}_i)\hat{e}_i^2,$$

$$\hat{e}_i = \frac{y_i - \mathbf{x}_i^T \boldsymbol{\beta}_n}{\sigma_n}.$$

If the target model is normal $F_0(t) = \Phi(t)$, then $K(e) = -\phi(t)$ and k^2 and ω^2 simplify to

$$k^2 = \frac{\sigma^2 E\{1 - 2\Phi(\sigma^{-1}e)\}^2}{\{2E\phi(e\sigma^{-1})\}^2}$$

$$\omega^2 = \frac{\sigma^2 E\{-1/\sqrt{\pi} + 2\phi(e\sigma^{-1})\}^2}{\{2E\phi(e\sigma^{-1}e^2)\}^2},$$

where Φ and ϕ are the distribution and density functions of the standard normal distribution.

6. GOODNESS-OF-FIT IMPLEMENTATION

It is evident that the efficiency of the proposed estimator depends on the choice of the target model. If the target model is in a neighborhood of the true model then the estimator has high efficiency. The goodness-of-fit test proposed in Sec. 4 can be extended to linear models without any difficulty. In order to implement the procedure with a normal target distribution we recommend the following steps:

Step I: In this step, again first we calculate $\boldsymbol{\beta}_n$ and σ_n by minimizing $d_2(\boldsymbol{\beta}, F_n, \Phi)$. In order to calculate these, we solve

$$\sum_{i=1}^{n} \left\{ \Phi\left(\frac{y_i - \mathbf{x}_i^T\boldsymbol{\beta}}{\sigma}\right) - 1/2 \right\} \mathbf{x}_i = 0$$

$$\sum_{i=1}^{n} \left\{ \phi\left(\frac{y_i - \mathbf{x}_i^T\boldsymbol{\beta}}{\sigma}\right) \right\} = n/(2\pi^{1/2}).$$

These equations can be solved by using iterative procedures. For example, the kth iteration of $\boldsymbol{\beta}$ can easily be obtained by using iteratively reweighted least squares with weights

$$w_i(\boldsymbol{\beta}^{k-1}, \sigma) = \frac{\Phi((y_i - \mathbf{x}_i^T\boldsymbol{\beta}^{k-1})/\sigma) - 1/2}{(y_i - \mathbf{x}_i^T\boldsymbol{\beta}^{k-1})/\sigma}.$$

For the scale parameter, we recommend using Newton's method. It can be obtained by iterating the following equation.

$$\sigma^k = \frac{\sigma^{k-1} \sum_{i=1}^{n} \phi((y_i - \mathbf{x}_i^T\boldsymbol{\beta})/\sigma^{k-1})}{2\sum_{i=1}^{n} \phi((y_i - \mathbf{x}_i^T\boldsymbol{\beta})/\sigma^{k-1}) - n/(2\pi^{1/2})}.$$

Step II: We compute Q_n^* and proceed as in the one sample problem.

7. EXAMPLE

In order to illustrate the use of the procedure developed in this paper, we use the cross-country data on inflation and monetary growth. The data set consists of the inflation rate and monetary growth of 83 countries during the post World War II. The model is

$$P_i = \beta_0 + \beta_1 M_i + \varepsilon_i, \qquad i = 1, \ldots, 83,$$

where P_i the average annual growth rate of consumer prices and M_i is the growth rate of the stock of currency across 83 countries. The data set can be found in Barro (1995, p. 167).

One of the basic assumptions for our goodness-of-fit test is that the underlying distribution is symmetric. The same data set is analyzed by Fan and Gencay (1995) and they conclude that the residuals are from a symmetric nonnormal distribution.

Since the modal is a simple regression model, we look at visual inspection of the association between P_i and M_i. Figure 2 shows that there is strong positive association between inflation rate and growth rate.

Table 2 presents robust and least square estimates of the regression and scale parameters. The author's regression estimates are similar to the least

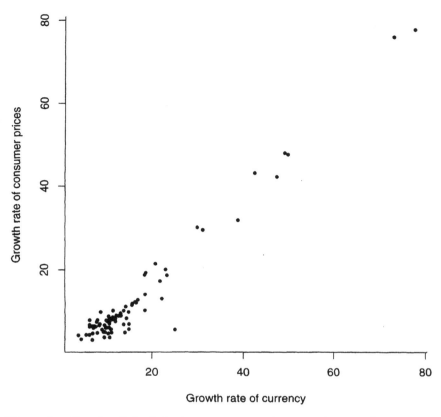

Figure 2. Graph of the inflation rate versus the growth rate of currency.

Table 2. Cross-Country Data Estimates

Parameters	LS	MDE
β_0	−3.936 (0.540)	−3.849 (0.451)
β_1	1.026 (0.026)	1.034 (0.022)
σ	3.145	2.578 (2.291)
Q_n^*		3.74

LS: least square estimates; MDE: minimum distance estimates; Q_n^* is the goodness-of-fit test statistic. Entries in the parentheses are the standard deviations of the estimates

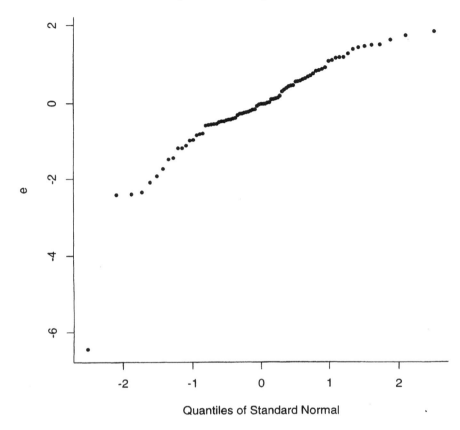

Figure 3.   QQ-plot of the regression residuals.

square estimates. On the other hand, our scale estimate is smaller than the least square estimate. Furthermore, our test statistic strongly rejects the normality assumption, again in agreement with Fan and Gencay (1995). The QQ-normal plot in Fig. 3 also suggests that the residuals are not normal.

REFERENCES

Barro, R. J. (1985). *Macroeconomics*. Wiley & Sons Inc., New York.

Basu, A., and Lindsay, B. G. (1994). Minimum disparity estimation for continuous models: Efficiency, distribution and robustness. *Ann. Inst. Statist. Math.*, *46*: 683–705.

Bickel, P. J., and Lehmann, E. L. (1975a). Descriptive statistics for non-parametric models, I. Introduction. *Ann. Statist.*, *3*: 1038–1044.

Bickel, P. J., and Lehmann, E. L. (1975b). Descriptive statistics for non-parametric models, II. Location. *Ann. Statist.*, *3*: 1045–1069.

Bickel, P. J., and Lehmann, E. L. (1976). Descriptive statistics for nonparametric models, III. Dispersion. *Ann. Statist.*, *4*: 1139–1158.

Boos, D. D. (1981). Minimum distance estimators for location and goodness of fit. *J. Amer. Statist. Assoc.*, *76*: 663–670.

Donoho, D. L., and Liu, R. C. (1988a). The "automatic" robustness of minimum distance functionals. *Ann. Statist.*, *16*: 552–586.

Donoho, D. L., and Liu, R. C. (1988b). Pathologies of some minimum distance estimators. *Ann. Statist.*, *16*: 587–608.

Fan, Y., and Gencay, R. (1995). A consistent nonparametric test of symmetry in linear regression models. *J. Amer. Statist.*, *90*: 551–557.

Hampel, F. R., Ronchetti, E. M., Rousseeuw, P. J., and Stahel, W. A. (1986). *Robust statistics: the approach based on influence functions*. John Wiley & Sons, New York.

Heathcote, C. R., and Silvapulle, M. J. (1981). Minimum mean squared estimation of location and scale parameters under misspecification of the model. *Biometrika.*, *68(2)*: 501–514.

Hettmansperger, T. P., and Aubuchon, J. C. (1988). Comments on "rank-based robust analysis of linear models. I. Exposition and review". *Statistical Science*, *3*: 262–263.

Hettmansperger, T. P., Hueter, I., and Hüsler, J. (1994). Minimum distance estimators. *J. Statist. Plan. and Inference.*, *41*: 291–302.

Huber, P. J. (1981). *Robust statistics*. John Wiley & Sons, New York.

Jaeckel, L. A. (1972). Estimating regression coefficients by minimizing the dispersion of the residuals. *Ann. Math. Statist.*, *43*: 1449–1458.

Lindsay, B. G. (1994). Efficiency versus robustness: The case for minimum Hellinger distance and related methods. *Ann. Statist.*, *22*: 1081–1114.

Öztürk, Ö. (1994). *Minimum distance estimation*. Ph.D. Thesis, The Pennsylvania State University.

Öztürk, Ö., and Hettmansperger, T. P. (1996). Almost fully efficient and robust simultaneous estimation of location and scale parameters: A minimum distance approach. *Statist. and Prob. Lett.*, *29*: 233–244.

Öztürk, Ö., and Hettmansperger, T. P. (1997). Generalized Cramer–von Mises Distance Estimators. *Biometrika*, *84*: 283–294.

Öztürk, Ö., and Hettmansperger, T. P. (1988). Simultaneous robust estimation of location and scale parameters: A minimum distance approach. To appear in *Canad. J. Statist.*, *26*: 217–229.

Öztürk, Ö., Hettmansperger, T. P., and Hüsler, J. (1997). Minimum distance and nonparametric dispersion functions. Tech. Rep. 621, The Ohio State University.

Parr, W. C., and De Wet, T. (1981). On Minimum Cramér–von Mises-norm parameter estimation. *Commun. Statist.-Theory. Meth.*, *A10(12)*: 1149–1166.

Parr, W. C., and Schucany, W. R. (1980). Minimum distance and robust estimation. *J. Amer. Statist. Assoc.*, *75(372)*: 616–624.

Roberts, A. W., and Varberg, D. E. (1973). *Convex functions*. Academic Press, New York.

van Eden, C. (1972). An analogue, for signed rank statistics, of Jureckova's asymptotic linearity theorem for rank statistics. *Ann. Math. Statist.*, *43*: 791–802.

Wiens, D. P. (1987). Robust weighted Cramér–von Mises estimators of location, with mini-max variance in ε-contamination neighborhoods. *Canad. J. Statist.*, *15*: 269–278.

15

Estimators of Changes

J. ANTOCH and M. HUŠKOVÁ   Charles University, Prague, Czech Republic

1. INTRODUCTION

This chapter concerns point estimators of the change point in various models. The least squares type estimators, M-estimators and R(rank)-estimators are introduced and studied. The confidence intervals for the change point are also constructed. Selected procedures are applied to two data sets (Nile data and Klementinum data) and their finite sample performance is checked in a simulation study.

Let $Y_{n1}, \ldots, Y_{nn}$ be independent observations having distribution function (cdf) F_n for $1 \leq i \leq m_n$ and cdf $G_n(\neq F_n)$ for $m_n < i \leq n$, where $m_n(<n)$ is unknown and called the change point. The cdf F_n and G_n may be known or unknown distributions, generally, they depend on n. Usually it is assumed that the differences $F_n - G_n(\neq 0)$ either do not depend on n or tend in a certain way to 0. The former case is usually called fixed changes while the latter one local changes.

We consider the problem of point and interval estimation of the change point m_n. Several classes of estimators of m_n are introduced and their limit distributions studied for various setups (parametric and nonparametric). The presented results are partially a survey of known results, but a number of new results is also shown. Theoretical results are illustrated on the analy-

Partially supported by grant GAČR 97/201/1163.

sis of two data sets (Nile data and Klementinum data) and on results of a simulation study.

This problem has attracted a number of researchers during last years. On one side this is a problem occurring in practice relatively often and on the other side it brings a number of interesting problems for theoretical statisticians. Parametric setup was considered by Hinkley (1970), Hinkley and Hinkley (1970), Cobb (1978), Worsley (1986), Yao Y.C. (1987) and Siegmund (1988). They dealt with the fixed changes only. Hinkley (1970) and Hinkley and Hinkley (1970) dealt with the maximum likelihood estimators when the observations are normally distributed. Assuming that the observations follow one-parameter exponential family distributions, Worsley (1986) derived both exact and approximate confidence regions for the change point based on the values accepted by a level α likelihood ratio tests. Cobb (1978) developed a method relying on a approximate ancillary statistics and applied it to the Nile data. Hinkley and Schechtman (1987) proposed various bootstrap schemes for the change point analysis. Siegmund (1988) discussed five types of confidence sets for the case when the distribution before and after the change are fully specified or have the distributions from an exponential family. Gombay and Horváth (1996) dealt with maximum likelihood estimators.

Nonparametric setups were studied by, e.g., Csörgő and Horváth (1988), Darkhovsky (1976), Carlstein (1988), Ritov (1990), Dümbgen (1991) and Brodsky and Darkhovsky (1993). Namely, they assumed that the cdf F_n and G_n satisfy some regularity conditions and are unknown otherwise. Carlstein (1988) proposed a strongly consistent nonparametric estimator for m_n based on the maximum (over k) of the distance between empirical measures based on $Y_1, \ldots, Y_k$ and $Y_{k+1}, \ldots, Y_n$. Ritov (1990) developed for local changes an asymptotically efficient estimator for m_n. Dümbgen (1991) obtained the limit distribution for a class of nonparametric estimators for m_n and derived some results on bootstrap confidence sets for m_n. He dealt with both fixed and local changes.

The procedures related to the U-statistics were studied by Ferger (1994a, b, 1995).

For more detailed reviews we refer to Csörgő and Horváth (1988) and Brodsky and Darkhovsky (1993).

In this chapter least squares type estimators of the change point m_n as well as of other parameters in location models are developed and studied in Sec. 2. Namely, we consider estimators based on maximum of the weighted partial sums of residuals and on moving sums of partial residuals. The limit distributions of these estimators are formulated for fixed changes, local changes and no changes (iid observations). It is quite interesting that the limit distributions for these situations differ substantially. These results immediately imply the rate of consistency.

In Sec. 3 the effect of dependence of observations on the limit distribution of these estimators for m_n is considered.

In Sec. 4 the estimators introduced in Sec. 2 are used to detect multiple changes.

In Sec. 5 M-type estimators of the change point m_n in one-parameter family of distributions are introduced as a generalization of the least squares estimators. Their limit properties are studied both for fixed and local changes. Remarks on the choice of the score function are made.

In Sec. 6 R-type estimators are developed along the lines of the least squares type and M-estimators. Quite general setup is considered.

The least squares type estimators are used to develop confidence intervals in Sec. 7. Three methods are presented, one based on the limit distributions of the (point) estimators and two relying on bootstrapping. These methods can be transferred to the case when M- or R-type estimators are used.

Section 8 is devoted to the applications to two data sets (Nile data and Klementinum data) and also to the results of a simulation study that checks the performance of the proposed methods for finite sample situation.

Finally, sketches of the proofs are in Sec. 9.

In the rest of the paper we will omit the index n whenever possible, i.e., we will write $Y_i, m, \ldots$, instead of $Y_{ni}, m_n, \ldots$, respectively.

Convention. We will typically work with a random variable V defined as a solution of the maximization problem

$$\max\{W(t),\ t \in T\},$$

where $\{W(t);\ t \in T\}$ is a random process or a sequence of random variables. If the solution is not uniquely determined, we always take the smallest solution and write shortly

$$V = V(W) = \arg\max\{W(t); t \in T\}.$$

2. CHANGES IN LOCATION MODEL

We assume that the observations follow the model

$$\begin{aligned} Y_i &= \mu + E_i, & i &= 1, \ldots, m, \\ &= \mu + \delta_n + E_i, & i &= m+1, \ldots, n, \end{aligned} \tag{2.1}$$

where $m(<n), \mu$ and δ_n are unknown parameters, $E_1, \ldots, E_n$ are independent identically distributed (iid) with zero mean and finite generally unknown variance $\sigma^2 > 0$. The least squares (LS) type estimators $\hat{m}_{LS}, \hat{\mu}_{LS}$ and $\hat{\delta}_{LS}$

of the parameters m, μ and δ are defined as solutions of the minimization problem

$$\min\left\{\sum_{i=1}^{k}\rho_{LS}(Y_i - \mu) + \sum_{i=k+1}^{n}\rho_{LS}(Y_i - \mu - \delta)\right\},$$

$$k \in \{1,\ldots,n\}, \qquad \mu \in R^1, \delta \in R^1, \tag{2.2}$$

where

$$\rho_{LS}(x) = x^2, \qquad x \in R^1. \tag{2.3}$$

It means that the parameters m, μ and δ are estimated in such a way that the sum of squares of residuals is minimal. In case of normally distributed errors E_i this estimate is also maximum likelihood estimators studied by Gombay and Horváth (1996).

Direct calculations give the explicit expressions for the estimators $\hat{\mu}_{LS}$ and $\hat{\delta}_{LS}$, while for $\hat{m}_{LS}$ only an implicit form can be obtained. Namely, we get

$$\hat{\mu}_{LS} = \bar{Y}_{\hat{m}_{LS}}, \tag{2.4}$$

$$\hat{\delta}_{LS} = \bar{Y}^o_{\hat{m}_{LS}} - \bar{Y}_{\hat{m}_{LS}} \tag{2.5}$$

and $\hat{m}_{LS}$ is a solution of the maximization problem

$$\max\left\{\sqrt{\frac{n}{k(n-k)}}\,|S_{k.LS}|\right\}, \qquad k \in \{1,\ldots,n-1\}, \tag{2.6}$$

where

$$S_{k.LS} = \sum_{i=1}^{k}(Y_i - \bar{Y}_n), \qquad k = 1,\ldots,n, \tag{2.7}$$

$$\bar{Y}_k = \frac{1}{k}\sum_{i=1}^{k}Y_i, \qquad k = 1,\ldots,n, \tag{2.8}$$

$$\bar{Y}^o_k = \frac{1}{n-k}\sum_{i=k+1}^{n}Y_i, \qquad k = 1,\ldots,n-1. \tag{2.9}$$

Sums $S_{k.LS}$, $k = 1,\ldots,n$, can be viewed as partial sums of residuals and $\sqrt{n/k(n-k)}$, $k = 1,\ldots,n-1$, as the weights.

According to Convention the estimator $\hat{m}_{LS}$ is defined as

$$\hat{m}_{LS} = \arg\max\left\{\sqrt{\frac{n}{k(n-k)}}\,|S_{k,LS}|; k = 1,\ldots,n-1\right\}. \tag{2.10}$$

Moreover, the estimator $\hat{m}_{LS}$ can be equivalently defined as

$$\arg\max\left\{\frac{k(n-k)}{n}\,(\bar{Y}_k - \bar{Y}_k^o)^2;\ k = 1,\ldots,n-1\right\}. \tag{2.11}$$

Next, we introduce several estimators based on the partial sums $S_{k.LS}$, $k = 1,\ldots,n$. Natural extension of $\hat{m}_{LS}$ is the estimator $\hat{m}_{LS}(\eta)$ defined as

$$\hat{m}_{LS}(\eta) = \arg\max\left\{\left(\frac{n}{k(n-k)}\right)^{\eta} |S_{k.LS}|;\ k = 1,\ldots,n-1\right\} \tag{2.12}$$

where $0 \le \eta \le 1/2$.

Another type of estimators of m is based on moving sums (MOSUM). These estimators will be denoted by $\hat{m}_{LS}(G)$ and defined as

$$\hat{m}_{LS}(G) = \arg\max\{|S_{k+G.LS} - 2S_{k.LS} + S_{k-G.LS}|;\ k = G+1,\ldots,n-G\}. \tag{2.13}$$

Notice that the estimator $\hat{m}_{LS}(G)$ is based on the second-order differences of $S_{k.LS}$'s.

To get an idea about the behavior of the estimators, we calculate the expectations of the partial sums of residuals and moving sums. We have

$$E\left(S_{k.LS}\left(\frac{n}{k(n-k)}\right)^{\eta}\right)$$

$$= -\delta_n k\,\frac{n-m}{n}\left(\frac{n}{k(n-k)}\right)^{\eta}, \qquad 1 \le k \le m,$$

$$= -\delta_n(n-k)\,\frac{m}{n}\left(\frac{n}{k(n-k)}\right)^{\eta}, \qquad m < k < n. \tag{2.14}$$

and

$$E(S_{k+G.LS} - 2S_{k,LS} + S_{k-G.LS})$$

$$= 0 \qquad\qquad\quad k = G+1,\ldots,m-G$$

$$= \delta_n(k+G-m) \qquad k = m-G+1,\ldots,m$$

$$= \delta_n(m+G-k) \qquad k = m+1,\ldots,m+G$$

$$= 0 \qquad\qquad\quad k = m+G+1,\ldots,n-G. \tag{2.15}$$

In both cases the maximum of the absolute values of the expectations is reached for $k = m$.

For the later convenience we put

$$
W_I(j, \mathbf{Z}_I) = \begin{cases} 0 & j = 0 \\ \displaystyle\sum_{i=j}^{0} Z_i & j = -1, -2, \ldots, \\ \displaystyle -\sum_{i=1}^{j} Z_i, & j = 1, 2, \ldots, \end{cases} \tag{2.16}
$$

and

$$
W_{II}(j, \mathbf{Z}_{II}) = \begin{cases} 0 & j = 0 \\ \displaystyle\sum_{i=j+1}^{0} (-Z_{i,1} + 2Z_{i,2} - Z_{i,3}), & j = -1, -2, \ldots, \\ \displaystyle\sum_{i=1}^{j} (Z_{i,1} - 2Z_{i,2} + Z_{i,3}), & j = 1, 2, \ldots. \end{cases} \tag{2.17}
$$

where $\mathbf{Z}_I = \{Z_j\}_{j=-\infty}^{\infty}$ and $\mathbf{Z}_{II} = \{Z_{j,1}, Z_{j,2}, Z_{j,3}\}_{j=-\infty}^{\infty}$ are sequences of random variables that will be specified on proper places.

Now, we will study limit properties (as $n \to \infty$) of the considered estimators. Namely, we formulate assertions on the limit behavior of these estimators in three situations:

(i) $\delta_n = \delta \neq 0$ is fixed (fixed change),
(ii) $\delta_n \neq 0$, $\delta_n \to 0$ as $n \to \infty$ (local change),
(iii) $\delta_n = 0$ (no change).

We start with case (i), i.e., fixed change.

THEOREM 2.1   (Fixed change). Let random variables $Y_1, \ldots, Y_n$ follow the model in Eq. (2.1) with $\delta_n = \delta \neq 0$ fixed and with $E_1, \ldots, E_n$ being iid random variables with zero mean, nonzero variance σ^2 and $E|E_i|^{2+\Delta} < \infty$ with some $\Delta > 0$, and assume

$$
m = [n\gamma], \qquad \gamma \in (0, 1), \tag{2.18}
$$

where $[a]$ denotes the integer part of a.

Then $\hat{m}_{LS}(\eta) - m$ converges in distribution to

$$
\arg\max\{\delta W_I(j, \mathbf{E}_I) - \delta^2 |j| g_{\eta,\gamma}(j); \; j = 0, \pm 1, \pm 2, \ldots\} \tag{2.19}
$$

as $n \to \infty$, where $\mathbf{E}_I = \{E_j\}_{j=-\infty}^{\infty}$ with E_i's being iid random variables distributed as E_i's in Eq. (2.1) and

$$
g_{\eta,\gamma}(s) = \begin{cases} (1 - \eta)(1 - \gamma) + \eta\gamma & s < 0 \\ (1 - \eta)\gamma + \eta(1 - \gamma) & s \geq 0. \end{cases} \tag{2.20}
$$

$\square$

THEOREM 2.2 (Fixed change). Let the assumptions of Theorem 2.1 be satisfied. Assume that, as $n \to \infty$,

$$G/n \le C \qquad \text{and} \qquad G^{-1}n^{2/(2+\Delta)}\log n \to 0 \qquad (2.21)$$

for some $C \in [0, 1/3)$.

Then, as $n \to \infty$, $\hat{m}_{LS}(G) - m$ converges in distribution to

$$\arg\max\{\delta W_{II}(j, \mathbf{E}_{II}) - \delta^2|j|; \ j = 0, \pm 1, \pm 2, \ldots\} \qquad (2.22)$$

where $\mathbf{E}_{II} = \{E_{j,1}, E_{j,2}, E_{j,3}\}_{j=-\infty}^{\infty}$ with $E_{j,v}$'s, $j = 0, \pm 1, \pm 2, \ldots$; $v = 1, 2, 3$, being iid random variables distributed as E_i's in Eq. (2.1). $\qquad \square$

Next we state the assertions on limit behavior under local changes, see (ii).

THEOREM 2.3 (Local change). Let random variables $Y_1, \ldots, Y_n$ follow the model in Eq. (2.1) with $E_1, \ldots, E_n$ being iid random variables with zero mean, nonzero variance σ^2 and $E|E_i|^{2+\Delta} < \infty$ for some $\Delta > 0$, and $m = [n\gamma]$ with some unknown $\gamma \in (0, 1)$. We assume $0 \le \eta < 1/2$, as $n \to \infty$,

$$\delta_n \to 0 \qquad \text{and} \qquad |\delta_n|\sqrt{n} \to \infty \qquad (2.23)$$

Then, as $n \to \infty$,

$$\frac{\delta_n^2}{\sigma^2}(\hat{m}_{LS}(\eta) - m) \xrightarrow{\mathcal{D}} V_{\eta,\gamma}, \qquad (2.24)$$

where

$$V_{\eta,\gamma} = \arg\max\{W(s) - |s|g_{\eta,\gamma}(s); \ s \in R^1\}, \qquad (2.25)$$

$g_{\eta,\gamma}(.)$ is defined by Eq. (2.20) and $\{W(s); s \in R^1\}$ is a two sided standard Wiener process, i.e.,

$$W(s) = \begin{cases} W_1(-s) & s < 0 \\ W_2(s) & s \ge 0, \end{cases} \qquad (2.26)$$

with $\{W_1(t); t \in [0, \infty)\}$ and $\{W_2(t); t \in [0, \infty)\}$ being independent standard Wiener processes.

The assertion in Eq. (2.24) remains true if $\eta = 1/2$ and, as $n \to \infty$,

$$\delta_n \to 0, \qquad |\delta_n|\sqrt{n}/\sqrt{\log\log n} \to \infty \qquad (2.27)$$

$\qquad \square$

THEOREM 2.4 (Local change). Let assumptions of Theorem 2.3 be satisfied and let, as $n \to \infty$,

$$G/n \to 0 \qquad \text{and} \qquad G^{-1}n^{2/(2+\Delta)}\log n \to 0. \qquad (2.28)$$

Then, as $n \to \infty$,

$$\frac{\delta_n^2}{6} \frac{\hat{m}_{LS}(G) - m}{\sigma^2} \xrightarrow{\mathcal{L}} V^o \tag{2.29}$$

where

$$V^o = \arg\max\{W(s) - |s|; \ s \in R^1\}, \tag{2.30}$$

with $\{W(t), t \in R^1\}$ being the two sided Wiener process defined in Eq. (2.26). $\qquad\square$

The limit behavior of the considered estimators of m is quite different in case of no change as shown in the next theorem.

THEOREM 2.5 (No change). Let $Y_1, \ldots, Y_n$ be iid random variables with zero mean, nonzero variance σ^2 and finite moment of order $2 + \Delta$ with some $\Delta > 0$. Then for any $\eta \in [0, 1/2]$, as $n \to \infty$,

$$\frac{\hat{m}_{LS}(\eta)}{n} \xrightarrow{\mathcal{L}} U(\eta)$$

where $U(\eta)$ is a random variable with distribution

$$P(U(1/2) = 0) = P(U(1/2) = 1) = 1/2,$$

and, if $\eta \in [0, 1/2)$, $U(\eta)$ has the distribution as

$$\arg\max\{|B(s)|(s(1-s))^{-\eta}; \ s \in R^1\}$$

with $\{B(t); t \in (0, 1)\}$ being a standard Brownian bridge.

If, moreover, Eq. (2.28) is satisfied, then, as $n \to \infty$,

$$\frac{\hat{m}_{LS}(\eta)}{n} \xrightarrow{\mathcal{L}} U^o$$

where U^o has the uniform distribution on $(0, 1)$. $\qquad\square$

Finally, we shall shortly discuss the problem of estimation of the parameters μ, δ_n and σ^2. There is a number of reasonable estimators of these parameters. The following theorems concerns the usual estimators of μ, δ_n and σ^2.

THEOREM 2.6 Let assumptions of Theorem 2.1 or Theorem 2.3 be satisfied. Let $\hat{m}$ be an estimator of m with the property, as $n \to \infty$,

$$\hat{m} - m = O_P(\delta_n^{-2}).$$

(i) Then $\hat{\mu}_n = \bar{Y}_{\hat{m}}$ and $\hat{\delta}_n = \bar{Y}_{\hat{m}}^o - \bar{Y}_{\hat{m}}$ are $\sqrt{n}$-consistent estimators of μ and δ_n, respectively. Moreover, $\sqrt{m}\,(\hat{\mu}_n - \mu)/\sigma$ and $\sqrt{n-m}\,(\hat{\delta}_n - \delta_n)/\sigma$ have asymptotically standardized normal distribution and are asymptotically independent.

(ii) The random variable

$$\hat{\sigma}_n^2 = \frac{1}{n}\left\{\sum_{i=1}^{\hat{m}}(Y_i - \bar{Y}_{\hat{m}})^2 + \sum_{i=\hat{m}+1}^{n}(Y_i - \bar{Y}_{\hat{m}}^o)^2\right\} \tag{2.31}$$

is n^{β}-consistent estimator of σ^2 for some $\beta > 0$. □

Remark 2.1 Notice that in case of fixed changes the limit distributions depend on the distribution F of the error terms E_i's, while in case of local changes and no changes do not depend.

Remark 2.2 The assertions of Theorem 2.1–2.4 remain true if the parameters δ_n and σ^2 are replaced by a suitable estimators. Namely, the parameter δ_n and σ^2 can be replaced by the estimators $\hat{\delta}_{LS}$ and $\hat{\sigma}_n^2$ defined in Eqns (2.5) and (2.31), respectively.

Remark 2.3 The explicit form of the distribution of $V_{\eta,\gamma}$, $\eta \in [0, 1/2]$, $\gamma \in (0,1)$, and V^o are known. They were derived independently by several authors, e.g., Bhattacharya and Brockwell (1976), Shepp (1979), Yao (1987) and Ferger (1994). Sthryn (1996) showed that random variable

$$V(c_1, c_2) = \arg\max\{W(s) - |s|(c_1 I\{s \le 0\} + c_2 I\{s > 0\};\ s \in R^1)\} \tag{2.32}$$

$c_1 > 0, c_2 > 0$, has the density

$$p(t; c_1, c_2) = \begin{cases} q(-t;\ c_1, c_2) & t \le 0 \\ q(t;\ c_2, c_1) & t > 0 \end{cases} \tag{2.33}$$

where for $t > 0$

$$q(t; c_1, c_2) = 2c_1(c_1 + 2c_2)\exp\{2c_2(c_1 + c_2)t\}\Phi(-(c_1 + 2c_2)\sqrt{t}) \\ - 2c_1^2\Phi(-c_1\sqrt{t}).$$

The corresponding distribution function can be calculated taking into account that for $t > 0$

$$
\begin{aligned}
\int_0^t q(x; c_1; c_2)\, dx = {}& c_1/(c_1 + c_2) + (2c_1/\sqrt{2\pi})\sqrt{t}\exp\{-c_1^2 t/2\} \\
& + [(c_1(c_1 + 2c_2))/(c_2(c_1 + c_2))] \\
& \times \exp\{2c_2(c_1 + c_2)t\}\Phi(-(c_1 + 2c_2)\sqrt{t}) \\
& - [2c_1^2 t + (c_1^2 + 2c_2^2 + 2c_1 c_2)/(c_2(c_1 + c_2))]\Phi(-c_1\sqrt{t}).
\end{aligned}
$$

This result can be used for construction of a confidence intervals for the change point m. Tables for $c_1 = c_2 = 1/2$ are in Horváth, Hušková and Serbinowska (1997).

3. DEPENDENT OBSERVATIONS

We investigate here limit properties of the estimators $\hat{m}_{LS}(\eta)$ and $\hat{m}_{LS}(G)$ (defined by Eqns (2.12) and (2.13), respectively) in the following time series model

$$
\begin{aligned}
Y_t &= \mu + E_t, & t &= 1, 2, \ldots, m \\
&= \mu + \delta_n + E_t, & t &= m + 1, \ldots, n
\end{aligned}
\tag{3.1}
$$

where μ, $\delta_n \neq 0$ and $m(<n)$ are unknown parameters and $\{E_t\}_{t=1}^{\infty}$ is a linear process satisfying

$$
E_t = \sum_{j=0}^{\infty} w_j \varepsilon_{t-j}, \qquad t = 1, 2, \ldots
\tag{3.2}
$$

where $\{\varepsilon_t\}_{t=-\infty}^{\infty}$ are iid random variables with $E\varepsilon_t = 0$, $\operatorname{var}\varepsilon_t = \sigma^2 > 0$, $E|\varepsilon_t|^{2+\triangle} < \infty$ for some $\triangle > 0$ and the weights $\{w_j\}_{j=0}^{\infty}$ satisfy

$$
\sum_{j=0}^{\infty} j|w_j| < \infty, \qquad \sum_{j=0}^{\infty} w_j \neq 0.
\tag{3.3}
$$

Next, we formulate two theorems that correspond to Theorems 2.1–2.4.

THEOREM 3.1   (Fixed change). Let Eqns (3.1)–(3.3) be satisfied with $\delta_n = \delta \neq 0$ fixed and let $m = [n\gamma]$, $\gamma \in (0, 1)$ for an unknown $\gamma \in (0, 1)$.

(i)   The assertion of Theorem 2.1 remains true with E_j's as in Eq. (3.2).

(ii)   If, moreover, Eq. (2.21) holds true, then the assertion of Theorem 2.2 remains true with E_j's as in Eq. (3.2).

$\square$

THEOREM 3.2 (Local change). Let Eqns (3.1)–(3.3) be satisfied with δ_n fulfilling Eq. (2.24) and let $m = [m\gamma]$ for an unknown $\gamma \in (0, 1)$. Then the assertion in Eq. (2.25) remains true if σ^2 is replaced by

$$\sigma_o^2 = \sigma^2 \left(\sum_{j=0}^{\infty} w_j \right)^2. \tag{3.4}$$

If, moreover, Eq. (2.28) holds true, then the assertion in Eq. (2.29) of Theorem 2.4 remains true if σ^2 is replaced by σ_o^2. $\qquad\square$

Remark 3.1 The assertions of Theorem 3.2 remain true if δ_n and σ_o^2 are replaced by proper estimators. Namely, δ_n can be estimated by $\hat{\delta}_{n,LS}$ defined by Eq. (2.5) and σ_o^2 by

$$\sigma_{o,n}^2(L) = \hat{R}(0) + 2 \sum_{k=1}^{L} \left(1 - \frac{k}{L} \right) \hat{R}(k), \qquad L < n \tag{3.5}$$

where

$$\hat{R}(k) = \frac{1}{n} \left\{ \sum_{t=1}^{\hat{m}-k} (Y_t - \bar{Y}_{\hat{m}})(Y_{t+k} - \bar{Y}_{\hat{m}}) + \sum_{t=\hat{m}+1}^{n-k} (Y_t - \bar{Y}_{\hat{m}}^o)(Y_{t+k} - \bar{Y}_{\hat{m}}^o) \right\},$$

$$k \geq 0.$$

Further discussions on choice of L together with other theoretical results as well as results of simulation study can be found in Antoch, Hušková and Prášková (1997).

Remark 3.2 Horváth and Shao (1996) and Horváth and Kokoszka (1997) treated the case when $\{e_t\}$ are Gaussian random variables with long range dependence. Bai (1994) treated the estimator $\hat{m}_{LS}(1/2)$. Brodsky and Dark-hovsky (1993) studied strong consistency of various modifications of the estimator $\hat{m}_{LS}(\eta)$, when $\{E_t\}_t$ forms a ψ-mixing sequence and the size of the change is fixed.

4. MULTIPLE CHANGES

Consider the model with multiple changes

$$Y_i = \mu_j + E_i, \qquad [n\gamma_{j-1}] < i \leq [n\gamma_j], \qquad j = 1, \ldots, q+1 \tag{4.1}$$

where $\mu_j \in R^1$, $\mu_j \neq \mu_{j+1}$, $j = 1, \ldots, q$, $0 = \gamma_0 < \gamma_1 < \ldots < \gamma_{q+1} = 1$ and $E_1, \ldots, E_n$ are iid random variables with zero mean, nonzero variance var $E_i = \sigma^2$ and $E|E_i|^{2+\Delta} < \infty$ for some $\Delta > 0$. Values $[n\gamma_j], j = 1, \ldots, q$, are change points. Their number q can be known or unknown, however an upper bound, say q_o, for q is supposed known.

The estimators $\hat{m}_{LS}(\eta)$, $\eta \in [0, 1/2]$, and $\hat{m}_{LS}(G)$ treated in previous sections can be used to estimate multiple changes. We restrict ourselves to the estimators $\hat{m}_{LS}(1/2)$ and $\hat{m}_{LS}(G)$.

By direct calculation we find that

$$ES_{k.LS} = \sum_{j=1}^{v} ([n\gamma_j] - [n\gamma_{j-1}])(\mu_j - \bar{\mu}) + (k - [n\gamma_v])(\mu_v - \bar{\mu})$$

$$\text{for} \quad [n\gamma_{v-1}] < k \leq [n\gamma_v], \quad v = 1, \ldots, q+1, \quad (4.2)$$

and

$$E(S_{k+G.LS} - 2S_{k.LS} + S_{k-G.LS})$$

$$= \begin{cases} 0 & |k - [n\gamma_j]| > G, \\ (\mu_j - \mu_{j-1}) \cdot ([n\gamma_j] - k + G) & [n\gamma_j] - G \leq k \leq [n\gamma_j], \\ (\mu_j - \mu_{j-1}) \cdot (k - [n\gamma_j] + G) & [n\gamma_j] \leq k \leq [n\gamma_j] + G, \end{cases} \quad (4.3)$$

for $j = 1, \ldots, q$.

The extremes in both expectations can be reached only for $k = [n\gamma_j]$, $j = 1, \ldots, q$. This leads to the following estimation procedures. At first find

$$\tilde{m}^{(1)} = \arg\max \left\{ \sqrt{\frac{n}{k(n-k)}} |S_{k.LS}|; k = 1, \ldots, n-1 \right\} \quad (4.4)$$

Then divide observations into two groups $Y_1, \ldots, Y_{\tilde{m}^{(1)}}$ and $Y_{\tilde{m}^{(1)}+1}, \ldots, Y_n$ and find the estimator in each group defined according to Eq. (4.4). The whole procedure is then repeated for each subset until a constant mean is obtained, i.e., after each step one should test whether there is a change in each group. We apply the test based on

$$\max_{k \in D} \left\{ \sqrt{\frac{n}{k(n-k)}} \frac{1}{\hat{\sigma}_n} \left| \sum_{i \in D} \left(Y_i - \frac{1}{\#D} \sum_{j \in D} Y_j \right) \right| \right\},$$

where D denotes the indexes corresponding to the particular group, $\#D$ its cardinality and $\hat{\sigma}_n^2$ is a suitable estimators of σ^2. Critical value can be determined from Theorem 9.2. and the level α_n has to be chosen such that, as $n \to \infty$, $\alpha_n \to 0$. Then this procedure estimates consistently all change points and also the number of changes. Vostrikova (1981) proposed estimators for the number of changes.

Motivated by Eq. (4.3) and Theorem 9.2 we propose the following estimation procedure. We find all pairs of indices v_j, w_j, $j = 1, \ldots, \hat{q}$, such that $w_j - v_j \geq G/2$, $j = 1, \ldots, \hat{q}$ and such that for $k = v_j, \ldots, w_j$

$$|S_{k+G.LS} - 2S_{k.LS} + S_{k-G.LS}| \leq D(n, G, \alpha_n)$$
$$|S_{v_j-1+G.LS} - 2S_{v_j-1.LS} + S_{v_j-1-G.LS}| \geq D(n, G, \alpha_n)$$
$$|S_{w_j+1+G.LS} - 2S_{w_j+1.LS} + S_{w_j+1-G.LS}| \geq D(n, G, \alpha_n)$$

where

$$D(n, G, \alpha_n) = \sigma_n \sqrt{2G}(2\log(n/G) + (1/2)\log\log(n/G)$$
$$- (1/2)\log(\pi/9) - \log\log(1 - \alpha_n)^{-1})(\sqrt{2\log(n/G)})^{-1}$$

and $\alpha_n \to 0$ as $n \to \infty$. This means that we find all segments where a certain critical value is exceeded. This critical value corresponds to the test no change versus there is a change based on

$$\max_{G<k<n_G} \{|S_{k+G.LS} - 2S_{k.LS}S_{k-G.LS}|\}$$

with asymptotic level α_n. The estimator of the number of the change points is $\hat{q}$.

Then $\hat{q}$ estimates of the number of change points and the index $k \in [v_j, w_j]$ for which the maximum over the set $\{v_j, \ldots, w_j\}$ is reached can serve as the estimator of one of the change points.

These two procedures, particularly, the second one, are quite simple and can certainly serve as a simple diagnostic tool. There is a problem to find a reasonable estimator of the variance σ^2. For that we need an upper bound, say q_0, of the number q of change points and then we can estimate σ^2 as follows:

$$\sigma_n^2 = \frac{1}{n - q_0 - 1} \min_{1 \leq k_1 < \ldots k_{q_0} \leq n} \left\{ \sum_{i=1}^{k_1} (Y_i - \bar{Y}_{k_1})^2 \right.$$
$$\left. + \sum_{i=k_1+1}^{k_2} (Y_i - \bar{Y}_{k_1,k_2})^2 + \cdots + \sum_{i=k_{q_0}+1}^{n} (Y_i - \bar{Y}_{k_{q_0},n})^2 \right\},$$

where

$$\bar{Y}_{k.s} = \frac{1}{s - k} \sum_{i=k+1}^{s} Y_i.$$

There are more sophisticated procedures for detection and identification of multiple changes, however, they need stronger assumptions, see Yao Y.-C. (1988).

Particular attention has been paid to the epidemic alternatives corresponding to the model in Eq. (4.1) with $q = 2$ and $\mu_1 = \mu_3 \neq \mu_2$. For more information see Bhattacharya and Brockwell (1976), Yao Q. (1993), Antoch and Hušková (1996) and Hušková (1995).

5. *M*-ESTIMATORS

We assume the model:

(M.1) $Y_1, \ldots, Y_n$ are independent random variables with distribution functions $F(.; \theta_o)$ for $1 \leq i \leq m$ and $F(.; \theta_o + \delta_n)$ for $m + 1 \leq i \leq n$, where $m(< n)$, θ_o and δ_n are unknown parameters, θ_o and $\theta_o + \delta_n$ belong to an open interval $\Theta \subseteq R^1$.

The *M*-type estimators can be introduced along the lines of Sec. 2. According to that we can develop three types of estimators corresponding to Eqns (2.2), (2.10) and (2.11). The definitions can be obtained by replacing the quadratic loss function ρ_{LS} by the loss function $\rho(x; \theta)$, $x \in R^1$, $\theta \in \Theta$, and the partial sums $S_{k,LS}, k = 1, \ldots, n$, by the partial sums of *M*-residuals

$$S_{k,M} = \sum_{i=1}^{k} \psi(Y_i; \hat{\theta}_{n,M}), \quad k = 1, \ldots, n, \tag{5.1}$$

where $\psi(x; \theta)$, $x \in R^1$, $\theta \in \Theta$, is the derivative of ρ w.r.t. the second argument, and $\hat{\theta}_{n,M}$ is the *M*-estimator generated by the score function ψ and defined as any solution of the equation

$$\sum_{i=1}^{n} \psi(Y_i; \theta) = 0 \tag{5.2}$$

or some asymptotically equivalent estimator.

In case of the squared loss function in Eqns (2.2), (2.10) and (2.11) are three equivalent definitions of the estimator $\hat{m}_{LS}$ of *m*, however, in case of a general loss function ρ this leads to three different (asymptotically equivalent) estimators.

We focus here on *M*-estimators corresponding to Eq. (2.12) (which is a generalization of Eq. (2.10)) and Eq. (2.13). They are defined as

$$\hat{m}_M(\eta) = \arg\max \left\{ \left(\frac{n}{k(n-k)} \right)^{\eta} |S_{k,M}|; \ k = 1, \ldots, n-1 \right\},$$

$$0 \leq \eta \leq 1/2, \tag{5.3}$$

and

$$\hat{m}_M(G) = \arg\max \{ |S_{k+G,M} - 2S_{k,M} + S_{k-G,M}|; \ k = G + 1, \ldots, n - G \}. \tag{5.4}$$

Clearly, these estimators reduce to the least squares type estimators for $F(x; \theta_o + I\{i > m\}\delta_n) = F(x - \theta_o - I\{i > m\}\delta_n)$, $x \in R^1, \theta \in \Theta$, $\psi(x; \theta) = x - \theta$, $x \in R^1, \theta \in \Theta$, and $\int x \, dF(x) = 0$. ($I\{A\}$ denotes the indicators of a set A.) Typical choices of the score function ψ are the same as in the problem of estimation of the location in the model in Eq. (1.1) with $\delta_n = 0$ (e.g, Huber's ψ, Andrew's ψ etc., for more information see Huber (1981)).

It can be anticipated that the M-estimators of the change point m have analogous limit properties to the least squares counterpart. The question is to find suitable assumptions.

Two main theorems of this Section concern the limit behavior of the estimators $\hat{m}_M(\eta)$ and $\hat{m}_M(G)$. Theorem 5.1 deals with fixed changes and is an analog of Theorems 2.1 and 2.2. Theorem 5.2 covers local changes and corresponds to Theorems 2.3 and 2.4. Versions of Theorem 2.5 and Theorem 2.6 are omitted for they are just simple modifications of these theorems.

We impose the following assumptions on cdf F and the score function ψ:

(M.2) $\int \psi(x; \theta) \, dF(x; \theta) = 0$, $\theta \in \Theta$.

(M.3)   There exists a constant $b_M = b_M(\psi, F) \neq 0$ depending on ψ and F such that, as $n \to \infty$,

$$\frac{n}{m(n - m)\delta_n} S_{m,M} \xrightarrow{\mathscr{P}} -b_M.$$

(M.4)   There are constants $c_{i,M}^2 = c_{i,M}^2(\psi, F) > 0$, $i = 1, 2$, depending on ψ and F such that, as $n \to \infty$,

$$\frac{1}{m} \sum_{i=1}^{m} (\psi(Y_i; \hat{\theta}_{n,M}) - \bar{\psi}_M)^2 \xrightarrow{\mathscr{P}} c_{1,M}^2$$

$$\frac{1}{n - m} \sum_{i=m+1}^{n} (\psi^2(Y_i; \hat{\theta}_{n,M}) - \bar{\psi}_M^0)^2 \xrightarrow{\mathscr{P}} c_{2,M}^2.$$

with

$$\bar{\psi}_M = \frac{1}{m} \sum_{i=1}^{m} \psi(Y_i; \hat{\theta}_{n,M}), \quad \bar{\psi}_M^0 = \frac{1}{n - m} \sum_{i=m+1}^{n} \psi(Y_i; \hat{\theta}_{n,M}).$$

(M.5)   There exists a constant $\Delta > 0$ such that, as $n \to \infty$,

$$\frac{1}{n} \sum_{i=1}^{n} \left| \psi(Y_i; \hat{\theta}_{n,M}) \right|^{2+\Delta} = O_p(1).$$

(M.6) There exist functions h_j, $j = 1, 2$, not depending on n such that, as $n \to \infty$,

$$(\psi(Y_1; \hat{\theta}_{n.M}) - \bar{\psi}_M) - h_1(Y_1) \xrightarrow{\mathscr{P}} 0$$

$$(\psi(Y_n; \hat{\theta}_{n.M}) - \bar{\psi}_M^0) - h_2(Y_n) \xrightarrow{\mathscr{P}} 0.$$

THEOREM 5.1 (Fixed change). Let assumptions (M.1)–(M.6) be satisfied with $\delta_n = \delta \neq 0$ fixed, and let $m = [n\gamma]$, $\gamma \in (0, 1)$.

Then $\hat{m}_M(\eta) - m$ converges in distribution to

$$\arg\max\{\delta b_M W_I(j, \mathbf{Z}_I(M)) - \delta^2 b_M^2 |j| g_{n.\gamma}(j); \; j = 0, \pm 1, \pm 2, \ldots\} \qquad (5.5)$$

where $g_{n.\gamma}(.)$ and $W_I(j, Z_I(M))$'s are defined by Eqns (2.20) and (2.16), respectively; the components $Z_i(M)$, $i = 0, \pm 1, \pm 2, \ldots$, of $\mathbf{Z}_I(M)$ are independent random variables having cdf as $h_1(Y_1)$ for $i \leq 0$ and as $h_2(Y_n)$ for $i > 0$.

If, moreover, Eq. (2.21) holds true, then $\hat{m}_M(G) - m$ converges in distribution to

$$\arg\max\{\delta b_M W_{II}(j, \mathbf{Z}_{II}(M)) - \delta^2 b_M^2 |j|; \; j = 0, \pm 1, \pm 2, \ldots\} \qquad (5.6)$$

where $W_{II}(j, \mathbf{Z}_{II}(M))$'s are defined by Eq. (2.17) and the components $Z_{i.j}(M)$, $i = 0, \pm 1, \pm 2, \ldots; j = 1, \ldots, 3$, of $\mathbf{Z}_{II}(M)$ are independent random variables with $Z_{i.1}$, $i = 0, \pm 1, \pm 2, \ldots$, $Z_{i.2}$, $i \leq 0$, having cdf as $h_1(Y_1)$ and with $Z_{i.3}$, $i = 0, \pm 1, \pm 2, \ldots$, $Z_{i.2}$, $i > 0$, having cdf as $h_2(Y_n)$. $\qquad \square$

Next theorem establishes the limit behavior of the M-estimators under local alternatives.

THEOREM 5.2 (Local change). Let assumptions (M.1)–(M.5) be satisfied with δ_n satisfying Eq. (2.24) and with $c_{1.M}^2 = c_{2.M}^2$ and let $m = [n\gamma]$, $\gamma \in (0, 1)$. Then, as $n \to \infty$,

$$\delta_n^2 \frac{b_M^2}{c_{1.M}^2} (\hat{m}_M(\eta) - m) \to^d V_{n.\gamma}, \qquad (5.7)$$

where $V_{n.\gamma}$ is described in Eq. (2.24).

If, moreover, Eq. (2.27) is fulfilled, then, as $n \to \infty$,

$$\delta_n^2 \frac{b_M^2}{c_{1.M}^2} (\hat{m}_M(G) - m) \to^d V^o, \qquad (5.8)$$

where V^o is defined in Eq. (2.30). $\qquad \square$

It can be easily seen that the asymptotic variance of the estimators in case of local changes (see Theorem 5.2) depends the ratio

$$\frac{b_M^2}{c_{1,M}^2} = \frac{b_M^2(\psi, F)}{c_{1,M}^2(\psi, F)},$$

i.e., depends both on the distribution F and the score function ψ.

The larger ratio the smaller is the asymptotic variance. Analogously as in the problem of estimation of location, the smallest asymptotic variance corresponds to

$$\psi(x; \theta) = \frac{\partial f(x; \theta)/\partial \theta}{f(x; \theta)}, \qquad x \in R^1, \qquad \theta \in \Theta,$$

where $f(.; \theta)$ is the density corresponding to F. If this density is unknown then an estimator of the optimal score function can be developed and this estimator can be used as the proper score function.

Assumptions (M.1)–(M.6) are satisfied for a number of situations. We will formulate the consequences of Theorems 5.1 and 5.2 to the location model.

Remark 5.1. Let assumption (M.1) be satisfied with $\delta_n = \delta \neq 0$ fixed and with $F(x; \theta) = F(x - \theta)$, $x \in R^1$, $\theta \in \Theta$. Suppose that $m = [n\gamma]$, $\gamma \in (0, 1)$ and that $\psi(x; \theta) = \psi(x - \theta)$, $x \in R^1$, $\theta \in \Theta$, where Θ is an open interval and ψ is a monotone function with the properties:

(i) the function $\lambda_{\psi, F}(t) = -\int \psi(x - t)\, dF(x)$, $t \in \Theta$, is strictly monotone and continuous, $\lambda_{\psi, F}(0) = 0$;

(ii) the function $\kappa_{\psi, F}(t) = \int \psi^2(x - t)\, dF(x)$, $t \in \Theta$, is positive and continuous;

(iii) there exists $\Delta > 0$ such that

$$\sup\left\{ \int |\psi(x - t)|^{2+\Delta}\, dF(x); \ t \in \Theta \right\} < \infty.$$

Then assumptions (M.2)–(M.6) are satisfied with

$$b_M = \frac{1}{\delta(1 - \gamma)} \lambda_{\psi, F}(d^o),$$

$$c_{1,M} = \kappa_{\psi, F}(d^o) - \lambda_{\psi, F}^2(d^o),$$

$$c_{2,M} = \kappa_{\psi, F}(d^o - \delta)\lambda_{\psi, F}^2(d^o - \delta),$$

$$h_1(Y_1) = \psi(Y_1 - \theta_o - d^o) - E\psi(Y_1 - \theta_o - d^o),$$

$$h_2(Y_n) = \psi(Y_n - \theta_o - d^o) - E\psi(Y_n - \theta_o - d^o),$$

where d^o is a solution of the equation

$$\gamma \lambda_{\psi,F}(t) + (1 - \gamma)\lambda_{\psi,F}(t - \delta) = 0.$$

Remark 5.2.   Let assumption (M.1) be satisfied with δ_n fulfilling Eq. (2.24) and with $F(x; \theta) = F(x - \theta)$, $x \in R^1$, $\theta \in \Theta$. Let the score function $\psi(x; \theta) = \psi(x - \theta)$, $x \in R^1$, $\theta \in \Theta$, where ψ is a monotone function such that

(iv)   function $\lambda_{\psi,F}(t)$ has continuous first derivative in a neighborhood of 0 and $\lambda_{\psi,F}(0) = 0$, $\lambda_{\psi,F}^{(1)}(0) \neq 0$, $\lambda_{\psi,F}^{(1)}$ denotes the first derivative of $\lambda_{\psi,F}$;

(v)   function $\kappa_{\psi,F}(t)$ is continuous in a neighborhood of $0, \kappa_{\psi,F}(0) > 0$;

(vi)   there exist $\Delta > 0$ and a neighborhood U_o of 0 such that:

$$\sup\left\{ \int |\psi(x - t)|^{2+\Delta}\, dF(x); \, t \in U_o \right\} < \infty,$$

where $\lambda_{\psi,F}(.)$ and $\lambda_{\psi,F}(.)$ are defined in Remark 5.1.

Then assumptions (M.2)–(M.5) are satisfied with

$$b_M = \lambda_{\psi,F}^{(1)}(0)$$

and

$$c_{1,M}^2 = c_{2,M}^2 = \kappa_{\psi,F}(0) = \int \psi^2(x)\, dF(x).$$

6.   RANK BASED ESTIMATORS

We assume the model:

(R.1)   Let $Y_1, \ldots, Y_n$ be independent random variables with continuous cdf F_n for $1 \leq i \leq m$ and with continuous cdf G_n for $m < i \leq n$.

Partial sums are here formed by the simple linear rank statistics

$$S_{k,R} = \sum_{i=1}^{k} (a_n(R_i) - \bar{a}_n), \quad k = 1, \ldots, n, \tag{6.1}$$

where $R_1, \ldots, R_n$ are the ranks corresponding to $Y_1, \ldots, Y_n; (a_n(1), \ldots, a_n(n))$ are scores and

$$\bar{a}_n = \frac{1}{n} \sum_{i=1}^{n} a_n(i).$$

The rank based estimators are introduced along the line of the least squares type estimators, particularly, we put

$$\hat{m}_R(\eta) = \arg\max\left\{\left(\frac{n}{k(n-k)}\right)^{\eta}|S_{k,R}|;\ k = 1,\ldots,n-1\right\} \qquad (6.2)$$

and

$$\hat{m}_R(G) = \arg\max\{|S_{k+G.R} - 2S_{k.R} - S_{k-G.R}|;\ k = G+1,\ldots,n-G\} \qquad (6.3)$$

These estimators have similar limit properties as the least squares type estimators and M-estimators.

We will work with the assumptions:

(R.2)   There is a sequence of constants $\{\delta_n\}$ and a constant $b_R \neq 0$ such that, as $n \to \infty$,

$$\frac{n}{m(n-m)\delta_n} S_{m.R} \xrightarrow{\mathscr{P}} -b_R.$$

(R.3)   There exist constants $c_{1.R}^2 > 0$ and $c_{2.R}^2 > 0$ such that, as $n \to \infty$,

$$\frac{1}{m}\sum_{i=1}^{m}(a_n(R_i) - \bar{a}_m)^2 \xrightarrow{\mathscr{P}} c_{1.R}^2$$

$$\frac{1}{n-m}\sum_{i=m+1}^{n}(a_n(R_i) - \bar{a}_m^0)^2 \xrightarrow{\mathscr{P}} c_{2.R}^2,$$

where

$$\bar{a}_m = \frac{1}{m}\sum_{i=1}^{m}a_n(R_i), \quad \bar{a}_m^0 = \frac{1}{n-m}\sum_{i=m+1}^{n}a_n(R_i).$$

(R.4)   There exists a constant $\Delta > 0$ such that, as $n \to \infty$,

$$\frac{1}{n}\sum_{i=1}^{n}|a_n(R_i)|^{2+\Delta} = O_p(1).$$

(R.5)   There exist functions k_1 and k_2 such that, as $n \to \infty$,

$$(a(R_1) - \bar{a}_m) - k_1(Y_1) \xrightarrow{\mathscr{P}} 0$$

$$(a(R_n) - \bar{a}_m^0) - k_2(Y_n) \xrightarrow{\mathscr{P}} 0.$$

Next, we state two theorems concerning the limit behavior of the estimators $\hat{m}_R(\eta)$ and $\hat{m}_R(G)$ both for fixed and local changes.

THEOREM 6.1   (Fixed change). Let assumptions (R.1)–(R.6) be satisfied with $\delta_n = \delta \neq 0$ fixed, and let $m = [n\gamma]$, $\gamma \in (0, 1)$.

Then $\hat{m}_R(\eta) - m$ has the limit distribution as

$$\arg\max\{\delta b_R W_I(j, \mathbf{Z}_I(R)) - \delta^2 b_R^2 |j| g_{\eta,\gamma}(j); \ j = 0, \pm 1, \pm 2, \ldots\} \qquad (5.5)$$

where $g_{\eta,\gamma}(.)$ and $W_I(j, \mathbf{Z}_I(R))$'s are defined by Eqns (2.20) and (2.16), respectively; the components $Z_i(R)$, $i = 0, \pm 1, \pm 2, \ldots$, of $\mathbf{Z}_I(R)$ are independent random variables having cdf as $k_1(Y_1)$ for $i \leq 0$ and having cdf as $k_2(Y_n)$ for $i > 0$.

If, moreover, Eq. (2.21) holds true, then $\hat{m}_R(G) - m$ has the limit distribution as

$$\arg\max\{\delta b_R W_{II}(j, \mathbf{Z}_{II}(R)) - \delta^2 b_R^2 |j|; \ j = 0, \pm 1, \pm 2, \ldots\} \qquad (5.6)$$

where $W_{II}(j, \mathbf{Z}_{II}(R))$'s are defined by Eq. (2.17) and the components $Z_{i,j}(R)$, $i = 0, \pm 1, \pm 2, \ldots; j = 1, \ldots, 3$, of $\mathbf{Z}_{II}(R)$ are independent random variables with $Z_{i,1}$, $i = 0, \pm 1, \pm 2, \ldots$, $Z_{i,2}$, $i \leq 0$, having cdf as $k_1(Y_1)$ and with $Z_{i,3}$, $i = 0, \pm 1, \pm 2, \ldots$, $Z_{i,2}$, $i > 0$, having cdf as $k_2(Y_n)$. $\qquad \Box$

THEOREM 6.2   (Local change). Let assumptions (R.1)–(R.4) be satisfied with δ_n satisfying Eq. (2.24) and let $m = [n\gamma]$, $\gamma \in (0, 1)$. Then the assertion of Theorem 5.2 remains true if the random variables with the index M are replaced be the random with the index R. $\qquad \Box$

Following two remarks deal with case when the scores $a_n(1), \ldots, a_n(n)$ are closed to a score function ϕ on $(0, 1)$:

Remark 6.1. (Fixed change). Let assumption (R.1) with $F_n = F$ and $G_n = G$, $F \neq G$, fixed be satisfied and let the scores fulfil:

(i)   the function $a_n(1 + [nu])$ has uniformly bounded variation on closed subintervals of $(0, 1)$;

(ii) there exists a measurable function $\phi(u), u \in (0, 1)$, such that, as $n \to \infty$,

$$\int_0^1 |a_n(1 + [un]) - \phi(u)|^{2+\Delta} \, du \to 0, \qquad \int_0^1 \phi^2(u) \, du = 0,$$

$$0 < \int_0^1 \phi^2(u) \, du, \qquad \int_0^1 |\phi(u)|^{2+\Delta} \, du < \infty$$

with some $\Delta > 0$.

Then assumptions (R.2)–(R.5) are satisfied with

$$b_R = -\frac{1}{\delta(1-\gamma)} \int \phi(H(x)) \, dF(x)$$

$$c_{1.R}^2 = \int \phi^2(H(x)) \, dF(x) - \left(\int \phi(H(x)) \, dF(x) \right)^2$$

$$c_{2.R}^2 = \int \phi^2(H(x)) \, dG(x) - \left(\int \phi(H(x)) \, dG(x) \right)^2.$$

Here $H(x) = \gamma F(x) + (1 - \gamma)G(x)$, $x \in R^1$.

Remark 6.2. (Local change). Let assumption (R.1) be satisfied with $F_n(x) = F(x)$, $x \in R^1$, and $G_n(x) = F(x - \delta_n)$, $x \in R^1$, where F has absolutely continuous density f and the finite Fisher information

$$0 < \int \frac{(f'(x))^2}{f(x)} \, dx < \infty.$$

Let assumptions (i) and (ii) of Remark 6.1 be satisfied.

Then assumptions (R.2)–(R.4) are satisfied with

$$b_R = - \int \phi(F(x)) f'(x) \, dx \quad \text{and} \quad c_{1.R}^2 = c_{2.R}^2 = \int_0^1 \phi^2(u) \, du.$$

Further discussions and results on limit behavior rank based estimators of the change points for local alternatives can be found in Gombay and Huškova (1996).

7. CONFIDENCE INTERVALS

Most of the results presented so far can be employed in construction of confidence intervals for the change point m. We focus on the estimators in the location model in Eq. (2.1) studied in Sec. 2, however, we attempt to keep in mind generalization to the estimators and models studied in Sec. 3–6 as well.

Three types of confidence intervals will be developed, one based on the limit distribution of the (point) estimators of m and two based on the bootstrap methods. All three methods are suitable for local changes while only the bootstrap constructions apply also to fixed changes.

Checking the assertions in Sec. 2 for the fixed changes (Theorems 2.1 and 2.2), the limit distribution of both types of estimators $\hat{m}_{LS}(\eta)$ and $\hat{m}_{LS}(G)$ depends on the distribution F of the error term E_i's. Under additional assumptions on the distribution F (F has to have some particular "nice form"), various approximations to the confidence interval of m were developed by several authors, e.g., by Hinkley (1970) and Siegmund (1986, 1988). Particularly, Siegmund (1988) discussed five types of confidence sets when the distribution functions before and after the change are completely specified. We develop here bootstrap approximations to the confidence intervals.

In case of the local changes the limit distributions of both types of estimators $\hat{m}_{LS}(\eta)$ and $\hat{m}_{LS}(G)$ do not depend on the distribution F. The explicit forms of the limit distributions are known, they are simple consequences of Eqns (2.26), (2.29) and (2.32). The assertions on the limit behavior of both types of estimators (Theorems 2.3 and 2.4) remain true if the unknown parameters are replaced by proper estimators. Then by Theorem 2.3, Remark 2.2 and Eq. (2.32) we get the $100(1 - \alpha)$ percent asymptotic confidence interval

$$\left(\hat{m}_{LS}(\eta) - \hat{v}_{\eta,\gamma}(\alpha/2)\, \frac{\hat{\sigma}_n^2}{\hat{\delta}_n^2},\ \hat{m}_{LS}(\eta) + \hat{v}_{\eta,\gamma}(1 - \alpha/2)\, \frac{\hat{\sigma}_n^2}{\hat{\delta}_n^2} \right), \tag{7.1}$$

$\eta \in [0, 1/2]$, where $\hat{m}_{LS}(\eta)$ and $\hat{\sigma}_n^2$ are defined in Eqns (2.12) and (2.31), respectively, and $\hat{v}_{\eta,\gamma}(\alpha)$ is the 100α percent quantile of the random variable $V_{\eta,\gamma}$, obtained from the distribution of $V_{\eta,\gamma}$ replacing the unknown parameter γ by its estimator $\hat{m}_{LS}(\eta)/n$. The density of $V_{\eta,\gamma}$ is given by Eq. (2.33) with

$$c_1 = (1 - \eta)(1 - \gamma) + \eta\gamma \quad \text{and} \quad c_2 = (1 - \eta)\gamma + \eta(1 - \gamma). \tag{7.2}$$

Analogously, by Theorem 2.4, Remark 2.2 and Eq. (2.32) we get the $100(1 - \alpha)$ percent asymptotic confidence interval

$$\left(\hat{m}_{LS}(G) - \hat{v}^o(\alpha/2)\, \frac{\hat{\sigma}_n^2}{\hat{\delta}_n^2},\ \hat{m}_{LS}(G) + \hat{v}^o(1 - \alpha/2)\, \frac{\hat{\sigma}_n^2}{\hat{\delta}_n^2} \right)$$

where $\hat{v}^o(1 - \alpha)$ is the $100(1 - \alpha)$ percent quantile of the random variable V^o. Its density is given by Eq. (2.33) with $c_1 = c_2 = 1$.

Now, we turn to the bootstrap approximations to the confidence intervals based on either of the estimators of m introduced in Sec. 2. Denote by $\hat{m}$ any of these estimators.

Bootstrap sampling scheme I

Take two independent samples $Y_1^*, \ldots, Y_{\hat{m}}^*$ and $Y_{\hat{m}+1}^*, \ldots, Y_n^*$ from the empirical cdf of $Y_1, \ldots, Y_{\hat{m}}$ and $Y_{\hat{m}+1}, \ldots, Y_n$, respectively.

Then the bootstrap estimator $\hat{m}^*$ corresponding to the estimator $\hat{m}$ is defined as $\hat{m}$ replacing Y_i by their bootstrap counterparts Y_i^*, $i = 1, \ldots, n$.

We can also modify the procedure taking the maximum only over the set $\{k; |k - \hat{m}| \le D_n\}$, with $\{D_n\}_n$ fulfilling, as $n \to \infty$,

$$D_n \le \min(m, n - m) \quad \text{and} \quad D_n \hat{\delta}_n^2 \xrightarrow{\mathscr{P}} \infty \tag{7.3}$$

where $\hat{\delta}_n$ is defined in Theorem 2.6.(i). If the amount of the change $\delta_n = \delta \ne 0$ is fixed, $\{D_n\}_n$ can be chosen to slowly tend to infinity arbitrarily, while in the case of local changes $\{D_n\}_n$ tends to infinity faster then $\hat{\delta}_n^{-2}$.

The bootstrap estimator $\hat{m}_{LS}^*(\eta)$ related to $\hat{m}_{LS}(\eta)$ is defined as

$$\hat{m}_{LS}^*(\eta) = \arg\max\left\{ \left(\frac{n}{k(n-k)}\right)^{\eta} \left| \sum_{i=1}^{k} (Y_i^* - \bar{Y}^*_n) \right|; |k - \hat{m}_{LS}(\eta)| \le D_n \right\}$$

$$\tag{7.4}$$

with D_n fulfilling (7.3). The bootstrap estimator $\hat{m}_{LS}^*(G)$ related to $\hat{m}_{LS}(G)$ is defined accordingly.

Bootstrap sampling scheme II

Define the estimated residuals

$$\tilde{E}_i = \begin{cases} Y_i - \bar{Y}_{\hat{m}}, & i = 1, \dots, \hat{m}, \\ Y_i - \bar{Y}_{\hat{m}}^o, & i = \hat{m} + 1, \dots, n, \end{cases}$$

and the centered residuals

$$\hat{E}_i = \tilde{E}_i - \frac{1}{n}\sum_{j=1}^{n} \tilde{E}_j, \quad i = 1, \dots, n,$$

where $\bar{Y}_{\hat{m}}$ and $\bar{Y}_{\hat{m}}^o$ are defined by Eqns (2.8) and (2.9), respectively, with $m = \hat{m}$.

Take $E_1^{**}, \dots, E_n^{**}$ iid from the empirical cdf of $\hat{E}_1, \dots, \hat{E}_n$ and consider the bootstrap observations

$$Y_i^{**} = \begin{cases} \hat{Y}_{\hat{m}} + E_i^{**}, & i = 1, \dots, \hat{m}, \\ \hat{Y}_{\hat{m}}^o + E_i^{**}, & i = \hat{m} + 1, \dots, n. \end{cases}$$

Then we proceed as in the bootstrap sampling scheme I.

We denote by P^* and $\hat{m}^*$ the probability corresponding the bootstrap sampling scheme I. given $Y_1, \dots, Y_n$, and corresponding estimators of m. Analogously, we denote by P^{**} and $\hat{m}^{**}$ the probability corresponding the bootstrap sampling scheme II. given $Y_1, \dots, Y_n$, and the corresponding bootstrap estimators, respectively.

In the following two theorems we examine the performance of $P^*(\hat{\delta}_n^2(\hat{m}_{LS}^*(\eta) - \hat{m}_{LS}(\eta)) \le x)$ and $P^{**}(\hat{\delta}_n^2(\hat{m}_{LS}^{**}(\eta) - \hat{m}_{LS}(\eta)) \le x)$ as estimators of $P(\delta_n^2(\hat{m}_{LS}(\eta) - m) \le x)$ both for fixed and local changes.

THEOREM 7.1 (Fixed changes)

(i) Let the assumptions of Theorem 2.1 be satisfied. Then for arbitrary $\eta \in [0, 1/2]$, as $n \to \infty$,

$$P^*(\hat{m}^*_{LS}(\eta) - \hat{m}_{LS}(\eta) \leq x) - P(\hat{m}_{LS}(\eta) - m \leq x) \xrightarrow{\mathscr{P}} 0, \qquad x \in R^1$$

(7.5)

with $D_n \to \infty$ and $D_n \leq \min(m, n - m)$. The assertion remains true if $*$ is replaced by $**$.

(ii) If assumptions of Theorem 2.2 are satisfied, then Eq. (7.5) remains true if $\hat{m}^*_{LS}(\eta)$ and $\hat{m}_{LS}(\eta)$ are replaced by $\hat{m}^*_{LS}(G)$ and $\hat{m}_{LS}(G)$, respectively. The assertion remains true if $*$ is replaced by $**$. ☐

THEOREM 7.2 (Local changes)

(i) Let the assumptions of Theorem 2.3 be satisfied. Then, as $n \to \infty$,

$$\sup_{x \in R^1} \{|P^*(\hat{\delta}^2_n(\hat{m}^*_{LS}(\eta) - \hat{m}_{LS}(\eta))/\hat{\sigma}^2_n \leq x)$$

$$- P(\delta^2_n(\hat{m}^*_{LS}(\eta) - m)/\sigma^2 \leq x)|\} \to 0 \quad \text{a.s.}$$

(7.6)

and

$$\sup_{x \in R^1} \{|P^*(\hat{\delta}^2_n(\hat{m}^*_{LS}(\eta) - \hat{m}_{LS}(\eta)) \leq x)$$

$$- P(\delta^2_n(\hat{m}_{LS}(\eta) - m) \leq x)|\} \to 0 \quad \text{a.s.}$$

(7.7)

where $\hat{\delta}_n$ and $\hat{\sigma}^2_n$ are described in Theorem 2.6 and $\{D_n\}_n$ fulfils Eq. (7.3).

(ii) If the assumptions of Theorem 2.4 are satisfied, then Eqns (7.6) and (7.7) remain true if $\hat{m}^*_{LS}(\eta)$ and $\hat{m}_{LS}(\eta)$ are replaced by $\hat{m}^*_{LS}(G)$ and $\hat{m}^*_{LS}(G)$, respectively.

The assertions of both parts remain true if $*$ is replaced by $**$. ☐

Remark 7.1. Theorem 2.3 and Theorem 7.2 (i) imply that for arbitrary $\eta \in [0, 1/2]$, as $n \to \infty$,

$$\sup_{x \in R^1} \{|P^*(\hat{\delta}^2_n(\hat{m}^*_{LS}(\eta) - \hat{m}_{LS}(\eta))/\hat{\sigma}^2_n \leq x) - P(V_{\eta,\gamma} \leq x)|\} \to 0 \quad \text{a.s.}$$

Remark 7.2. According to Theorem 7.1 and Theorem 7.2 the bootstrap works both for fixed and local changes. Both theorems provide a bootstrap approximation for the confidence interval for m.

In case of fixed changes, by Theorems 2.1 and 2.2 the limit cdf of $\hat{m}_{LS}(\eta) - m$ and $\hat{m}_{LS}(G) - m$ is not continuous and therefore the approximation to the quantiles does not exists for all $\alpha \in (0, 1)$. We usually work with nominal type of confidence level.

In case of local changes, Remarks 2.3 and 6.1 imply that the resulting limit cdf of the estimators is continuous and therefore both described bootstrap sampling schemes provide also bootstrap approximations to the $100(1 - \alpha)$ percent quantile $\hat{w}^*_{\eta,\gamma}(\alpha)$ of the distribution of $\hat{\delta}^2_n(\hat{m}_{LS}(\eta) - m)$ for arbitrary $\alpha \in (0, 1)$. This means that in case of local changes the bootstrap approximation to the $100(1 - \alpha)$ percent confidence interval has the form

$$(\hat{m}_{LS}(\eta) - \hat{w}^*_{\eta,\gamma}(\alpha/2)/\hat{\delta}^2_n, \hat{m}_{LS}(\eta) + \hat{w}^*_{\eta,\gamma}(1 - \alpha/2)/\hat{\delta}^2_n), \tag{7.8}$$

$\eta \in [0, 1/2]$. Analogous remark holds also for bootstrap sampling scheme II.

Remark 7.3. Bootstrap sampling scheme II for local changes was studied in detail in Antoch, Hušková and Veraverbeke (1995).

We shortly mention approximations to the confidence intervals based on the M-type estimators investigated in Sec. 5 for local changes. They can be incorporated along the above part of the Section. We focus on the confidence intervals based on $\hat{m}_M(\eta)$ and suppress the dependence on ψ whenever possible.

Namely, using the limit distribution (Theorem 5.2) for local changes we get

$$\left(\hat{m}_M(\eta) - \hat{v}_{\eta,\gamma}(\alpha/2) \frac{\hat{c}^2_{1.M}}{\hat{\delta}^2_n \hat{b}^2_M}, \hat{m}_{LS}(\eta) + \hat{v}_{\eta,\gamma}(1 - \alpha/2) \frac{\hat{c}^2_{1.M}}{\hat{\delta}^2_n \hat{b}^2_M} \right)$$

where $\hat{m}_M(\eta)$ is defined by Eq. (5.5), $\hat{v}_{\eta,\gamma}(\alpha)$ has the same meaning as in Eq. (7.1) and

$$\hat{c}^2_{1.M} = \frac{1}{n} \min \left\{ \sum_{i=1}^{k} \psi^2(Y_i; \hat{\theta}_{k.M}) + \sum_{i=k+1}^{n} \psi^2(Y_i; \hat{\theta}^o_{k.M}); k = 1, \ldots, n \right\}$$

$$\hat{\delta}^2_n \hat{b}^2_M = \left(\frac{n}{\hat{m}_M(\eta)(n - \hat{m}_M(\eta))} S_{m.M} \right)^2,$$

where $\hat{\theta}_{k.M}$ and $\hat{\theta}^o_{k.M}$ are M-estimators defined as solutions of

$$\sum_{i=1}^{k} \psi(Y_i; t_1) = 0 \qquad \text{and} \qquad \sum_{i=k+1}^{n} \psi(Y_i; t_2) = 0,$$

respectively.

To get the bootstrap approximation to the confidence interval we first describe the bootstrap sampling scheme. We define the bootstrap estimator

$\hat{m}_M^*(\eta)$ as $\hat{m}_{LS}^*(\eta)$ in both bootstrap sampling schemes with Y_i, $i = 1, \ldots, n$, replaced $\psi(Y_i; \hat{\theta}_{n.M})$. The bootstrap estimator of $\hat{m}_M^*(\eta)$ is defined accordingly.

Analogously to Theorem 7.1 we have for both bootstrap sampling schemes for M-type estimators:

THEOREM 7.3. (Local changes)

(i) Let assumptions of Theorem 5.2 be satisfied. Then, as $n \to \infty$,

$$\sup_{x \in R^1} \{ P^*(\hat{\delta}_n^2(\psi)(\hat{m}_M^*(\eta) - \hat{m}_M(\eta)) \le x)$$

$$- P(\delta_n^2(\hat{m}_M^*(\eta) - m) \le x) \} \xrightarrow{\mathscr{P}} 0, \tag{7.9}$$

where $\{D_n\}_n$ fulfils Eq. (7.3) and $\hat{\delta}_n(\psi) = \hat{\theta}_{\hat{m}_M(\eta)}^o - \hat{\theta}_{\hat{m}_M(\eta)}$ with $\hat{\theta}_{\hat{m}_M(\eta)}$ and $\hat{\theta}_{\hat{m}_M(\eta)}^o$ being M-estimators based on $Y_1, \ldots, Y_{\hat{m}_M(\eta)}$ and $Y_{\hat{m}_M(\eta)+1}, \ldots, Y_n$, respectively. $\qquad \square$

8. APPLICATIONS TO REAL DATA AND RESULTS OF SIMULATIONS

This Section is devoted to the applications to two data sets (Nile data and Klementinum data) and also to the results of a simulation study that checks the performance of the proposed methods for finite sample situation.

Nile data

This data set is probably one of the most frequently used in the given setup, for details see, e.g., Cobb (1978) or Hinkley and Schechtmann (1987). The data, i.e. the annual flows in the river Nile at Aswan during the years 1871–1970 are shown in Fig. 1.

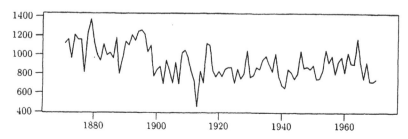

Figure 1. Nile data.

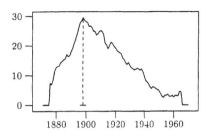

Figure 2. *LS* estimator, $\eta = 0$, $s = 1$.

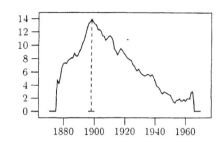

Figure 3. *LS* estimator, $\eta = 0.25$, $s = 1$.

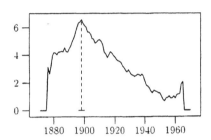

Figure 4. *LS* estimator, $\eta = 0.5$, $s = 1$.

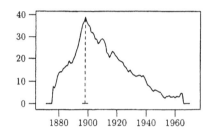

Figure 5. *LS* estimator, $\eta = 0$, $s = 2$.

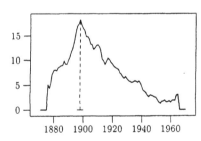

Figure 6. *LS* estimator, $\eta = 0.25$, $s = 2$.

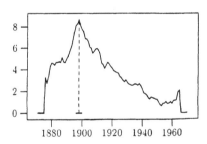

Figure 7. *LS* estimator, $\eta = 0.5$, $s = 2$.

In Figs 2–13 we illustrate:

The influence of the choice of η on the $m_{LS}(\eta)$; the values of η were fixed to 0, 0.25 and 0.5.

The influence of the proper normalization, namely,

$s = 1$ corresponds to the normalization by the standard sample deviation calculated from all the data;

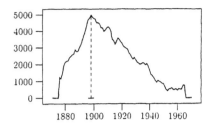

Figure 8. *LS* estimator, $\eta = 0$,
$s = 3$.

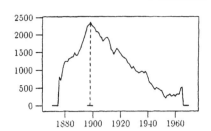

Figure 9. *LS* estimator, $\eta = 0.25$,
$s = 3$.

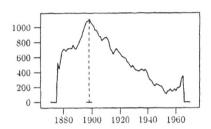

Figure 10. *LS* estimator, $\eta = 0.5$,
$s = 3$.

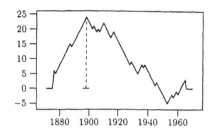

Figure 11. Robust estimator,
$\psi(x) = \text{sign}(x)$, $\eta = 0$, $s = 3$.

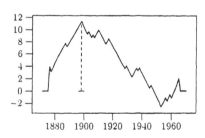

Figure 12. Robust estimator,
$\psi(x) = \text{sign}(x)$, $\eta = 0.25$, $s = 3$.

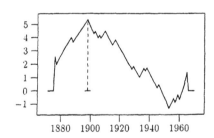

Figure 13. Robust estimator,
$\psi(x) = \text{sign}(x)$, $\eta = 0.5$, $s = 3$.

$s = 2$ corresponds to the normalization according to Eq. (2.31);
$s = 3$ corresponds to no normalization.
The difference between the classical least squares approach and the robust
one, i.e. the difference between the choice $\psi(x) = x$ and $\psi(x) = \text{sign}(x)$.

We can see that in all the cases the estimators detect the possible change
around the year 1898, which is in agreement with other authors. Concerning

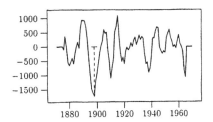

Figure 14. MOSUM, $\psi(x) = x$,
$G = 5, s = 3$.

Figure 15. MOSUM, $\psi(x) = x$,
$G = 10, s = 3$.

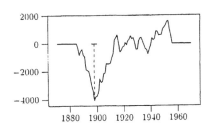

Figure 16. MOSUM, $\psi(x) = x$,
$G = 15, s = 3$.

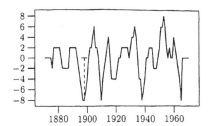

Figure 17. MOSUM,
$\psi(x) = \text{sign}(x), G = 5, s = 3$.

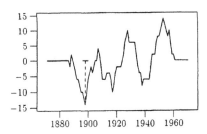

Figure 18. MOSUM,
$\psi(x) = \text{sign}(x), G = 10, s = 3$.

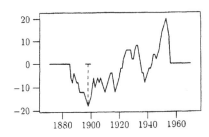

Figure 19. MOSUM,
$\psi(x) = \text{sign}(x), G = 15, s = 3$.

the normalization, it is evident that the use of $\hat{\sigma}_n^2$ calculated according to Eq. (2.31) seems to be (as almost always in the case of a change) preferable to the use of the classical standard sample deviation. As to the difference between the classical least squares and the robust approach, there is no big difference in this case.

Figs 14–21 demonstrate the behavior of the MOSUM type estimators.

Figure 20. MOSUM,
$\psi(x) = \text{sign}(x)$, $G = 20$, $s = 3$.

Figure 21. MOSUM,
$\psi(x) = \text{sign}(x)$, $G = 25$, $s = 3$.

They document the influence of two basic "parameters", i.e. the use of different ψ functions and the influence of enlarging the window width G. We can clearly see that the choice $\psi(x) = \text{sign}(x)$ typically requires larger value of G than the choice $\psi(x) = x$ to indicate the same conclusion. Generally, providing the value of G is large enough the change-point is (again) well located around 1898.

Finally, in Figs. 22–33 we demonstrate the use of the bootstrap for the establishing the confidence intervals for m.

Bootstrap I and II approximations to the probability distribution function of the estimator $\hat{m}_{LS}(\eta)$, $\eta \in \{0, 1/2\}$ and their difference, cf. Figs 22–24 and 28–30.

Bootstrap I and II approximations to the cumulative distribution function of the estimator $\hat{m}_{LS}(\eta)$, $\eta \in \{0, 1/2\}$ and their difference, cf. Figs 25–27 and 31–33.

We can see rather big difference between the choice $\eta = 0$ and $\eta = 1/2$, the later one being more realistic in fact.

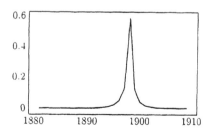

Figure 22. Bootstrap II approximations to pdf of $\hat{m}_{LS}$ given by Eq. (2.10).

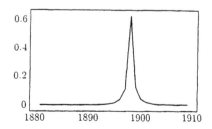

Figure 23. Bootstrap I approximations to pdf of $\hat{m}_{LS}$ given by Eq. (2.10).

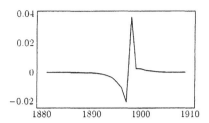

Figure 24. Difference between bootstrap II and I approximations to pdf of $\hat{m}_{LS}$.

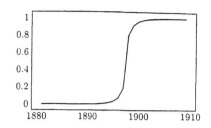

Figure 25. Bootstrap II approximations to cdf of $\hat{m}_{LS}$ given by Eq. (2.10).

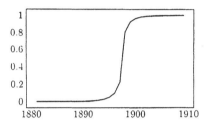

Figure 26. Bootstrap I approximations to cdf of $\hat{m}_{LS}$ given by Eq. (2.10).

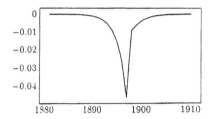

Figure 27. Difference between bootstrap II and I approximations to cdf of $\hat{m}_{LS}$.

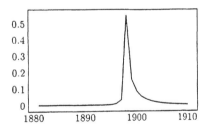

Figure 28. Bootstrap II approximations to pdf of $\hat{m}_{LS}(\eta)$ given by Eq. (2.12), $\eta = 0$.

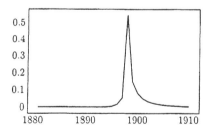

Figure 29. Bootstrap I approximations to pdf of $\hat{m}_{LS}(\eta)$ given by Eq. (2.12), $\eta = 0$.

Klementinum data

As the second example we consider the long temperature series measured in Klementinum (Prague). For the data, comprising annual temperature series measured during the years 1775–1992 see Fig. 34. These data sets were used many times in the literature and different authors applied different methods

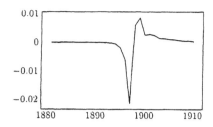

Figure 30. Difference between bootstrap II and I approximations to pdf of $\hat{m}_{LS}(0)$.

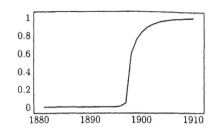

Figure 31. Bootstrap II approximations to cdf of $\hat{m}_{LS}(\eta)$ given by Eq. (2.12), $\eta = 0$.

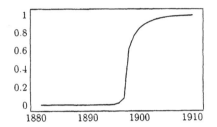

Figure 32. Bootstrap I approximations to cdf of $\hat{m}_{LS}(\eta)$ given by Eq. (2.12), $\eta = 0$.

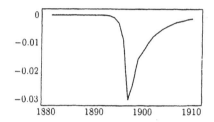

Figure 33. Difference between bootstrap II and approximations to cdf of $\hat{m}_{LS}(0)$.

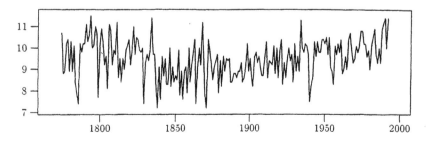

Figure 34. Klementinum data.

to them. For details see, e.g., Jarušková (1997). If we suppose that the first part of the series (years 1775–1900) are stationary, then the long-memory model, among others, can be applied. On the other hand, some statisticians prefer the model with changes occurring between the years 1835–1840 (Jarušková (1997)).

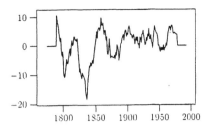

Figure 35. MOSUM, $\psi(x) = x$, $G = 15$, $s = 3$.

Figure 36. MOSUM, $\psi(x) = x$, $G = 25$, $s = 3$.

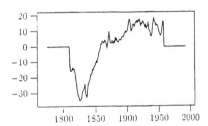

Figure 37. MOSUM, $\psi(x) = x$, $G = 35$, $s = 3$.

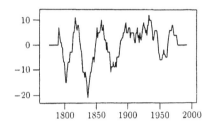

Figure 38. MOSUM, $\psi(x) = \text{sign}(x)$, $G = 15$, $s = 3$.

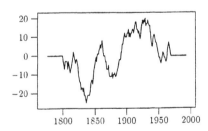

Figure 39. MOSUM, $\psi(x) = \text{sign}(x)$, $G = 25$, $s = 3$.

Figure 40. MOSUM, $\psi(x) = \text{sign}(x)$, $G = 35$, $s = 3$.

In Figs 35–40 we present the results demonstrating the behavior of the MOSUM type estimators. They document the influence of three basic "parameters", i.e. the use of different ψ functions, the influence of enlarging the window width G and the effect of the normalization (the encoding is exactly the same as in the case if Nile data). We can clearly see, that the choice $\psi(x) = \text{sign}(x)$ typically calls for larger value of G than the choice $\psi(x) = x$

(as also seen in the case of the analysis of Nile data). All applied estimators clearly detect the first change between the years 1835–1840. The second change in the twentieth century is not so well pronounced, nevertheless, a careful check of Fig. 34 shows slow gradual increase of the annual temperatures starting approximately from the first quarter of the twentieth century and is supported by the results on Fig. 40 (and partially also by Figs 39 and 37).

Simulation experiment

To see how the bootstrap approximations work, we prepared a small simulation experiment illustrating it. At first we fixed the values of the following constants, i. e.:

- number of the observations $n = 80$;
- location of the "true" change point $m = 20, 40, 60$;
- value of the location parameter $\mu = 0$;
- value of the size of the change amount in mean $\delta = 1$;
- value of the constant $\eta = 1/2$.

The errors followed standard normal distribution $N(0, 1)$.

For each fixed setup of constants n and m we run 5000 repetitions of the simulation experiment. In each repetition Y_i, $i = 1, \ldots, n$, have been simulated according to the model Eq. (2.1), estimator $\hat{m}_{(LS)} \equiv \hat{m}_{(LS)}(1/2)$ calculated and δ estimated by Eq. (2.5).

For both bootstrap scheme I and II we followed the Sec. 7 to generate corresponding 5000 bootstrapped samples. For each of these samples the change-point m has been estimated and stored for the further analysis. Figures 41–58 summarize for each of these situations:

Bootstrap I and II approximations to the probability distribution function of the estimator $\hat{m}_{LS}(1/2)$, cf. Figs 41, 42, 44, 45, 47 and 48;

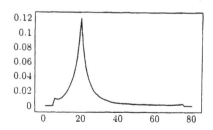
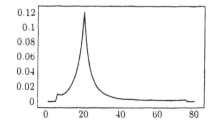

Figure 41. Bootstrap II approximations to pdf of $\hat{m}_{LS}$ given by Eq. (2.10), $n = 80$, $m = 20$.

Figure 42. Bootstrap I approximations to pdf of $\hat{m}_{LS}$ given by Eq. (2.10), $n = 80$, $m = 20$.

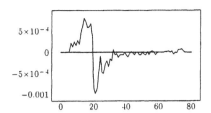

Figure 43. Difference between bootstrap II and I approximations to pdf of $\hat{m}_{LS}$, $n = 80$, $m = 20$.

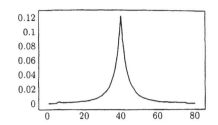

Figure 44. Bootstrap II approximations to pdf of $\hat{m}_{LS}$ given by Eq. (2.10), $n = 80$, $m = 40$.

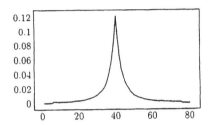

Figure 45. Bootstrap I approximations to pdf of $\hat{m}_{LS}$ given by Eq. (2.10), $n = 80$, $m = 40$.

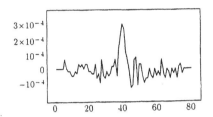

Figure 46. Difference between bootstrap II and I approximations to pdf of $\hat{m}_{LS}$, $n = 80$, $m = 40$.

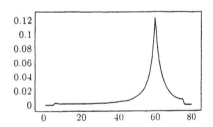

Figure 47. Bootstrap II approximations to pdf of $\hat{m}_{LS}$ given by Eq. (2.10), $n = 80$, $m = 60$.

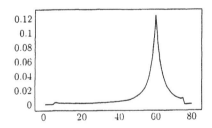

Figure 48. Bootstrap I approximations to pdf of $\hat{m}_{LS}$ given by Eq. (2.10), $n = 80$, $m = 60$.

Difference between the bootstrap I and II approximations to the probability distribution functions of the estimator $\hat{m}_{LS}(1/2)$, cf. Figs 43, 46 and 49; Bootstrap I and II approximations to the cumulative distribution function of the estimator $\hat{m}_{LS}(1/2)$, cf. Figs 50, 51, 53, 54, 56 and 57;

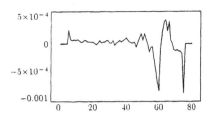

Figure 49. Difference between
bootstrap II and I approximations
to pdf of $\hat{m}_{LS}$, $n = 80$, $m = 60$.

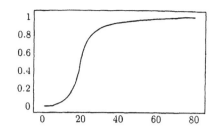

Figure 50. Bootstrap II
approximations to cdf of $\hat{m}_{LS}$ given
by Eq. (2.10), $n = 80$, $m = 20$.

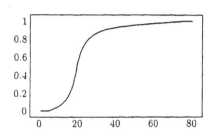

Figure 51. Bootstrap I
approximations to cdf of $\hat{m}_{LS}$ given
by Eq. (2.10), $n = 80$, $m = 20$.

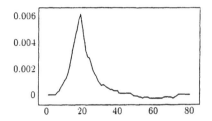

Figure 52. Difference between
bootstrap II and I approximations
to cdf of $\hat{m}_{LS}$, $n = 80$, $m = 20$.

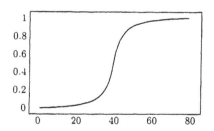

Figure 53. Bootstrap II
approximations to cdf of $\hat{m}_{LS}$ given
by Eq. (2.10), $n = 80$, $m = 40$.

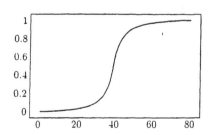

Figure 54. Bootstrap I
approximations to cdf of $\hat{m}_{LS}$ given
by Eq. (2.10), $n = 80$, $m = 40$.

Difference between the bootstrap I and II approximations to the cumulative
distribution functions of the estimator $\hat{m}_{LS}(1/2)$, cf. Figs 52, 55 and 58.

We can see that the difference between bootstrap schemes I and II
is practically negligible. The only slight difference is that for the case

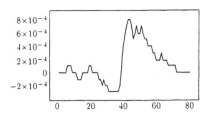

Figure 55. Difference between bootstrap II and I approximations to cdf of $\hat{m}_{LS}$, $n = 80$, $m = 40$.

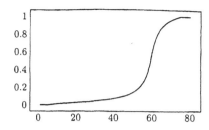

Figure 56. Bootstrap II approximations to cdf of $\hat{m}_{LS}$ given by Eq. (2.10), $n = 80$, $m = 60$.

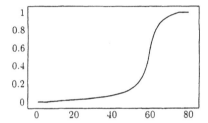

Figure 57. Bootstrap I approximations to cdf of $\hat{m}_{LS}$ given by Eq. (2.10), $n = 80$, $m = 60$.

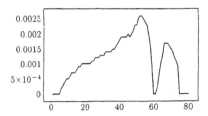

Figure 58. Difference between bootstrap II and I approximations to cdf of $\hat{m}_{LS}$, $n = 80$, $m = 60$.

$n = 80$ and $m = 20$ bootstrap scheme II results in slightly more detections of the change point below the real change point than the bootstrap scheme I, while for the case $n = 80$ and $m = 60$ the situation is just opposite.

Much more detailed results of the simulation study covering analogous situation can be found in Antoch, Hušková and Veraverbeke (1995).

9. PROOFS

Here we give sketches of the proofs of theorems trying to point out both main steps and ideas while omitting technical details.

First, we formulate two auxiliary assertions on behavior of functionals of partial sums $S_{k.LS}$, $k = 1, \ldots, n$, defined in Eq. (2.7).

THEOREM 9.1 Let $Y_1, \ldots, Y_n$ be iid random variables with zero mean, $0 < \operatorname{var} Y_i = \sigma^2$ and $E|Y_i|^{2+\Delta} < \infty$ with some $\Delta > 0$. Then, as $n \to \infty$,

$$\sup_{0 < t < 1} \left\{ \frac{|S_{[nt].LS}|}{\sqrt{n}\sigma(t(1-t))^\eta} \right\} \xrightarrow{\mathcal{L}} \sup_{0 < t < 1} \left\{ \frac{|B(t)|}{(t(1-t))^\eta} \right\}, \tag{9.1}$$

for $\eta \in [0, 1/2]$, where $\{B(t), t \in [0, 1]\}$ is a Brownian bridge.

Proof. The proof is, e.g., in Csörgő and Horváth (1993).

THEOREM 9.2 Let assumptions of Theorem 9.1 be satisfied.

(i) As $n \to \infty$, for any $y \in R^1$

$$P\left(\sqrt{2 \log\log n} \max_{1 \le k < n} \left\{ \sqrt{\frac{n}{k(n-k)}} \frac{1}{\sigma} |S_{k.LS}| \right\} \right.$$

$$\le y + 2\log\log n + \frac{1}{2}\log\log\log n - \frac{1}{2}\log\pi \Big)$$

$$\to \exp\{-2\exp\{-y\}\}. \tag{9.2}$$

(ii) If moreover (2.28) is satisfied, then, as $n \to \infty$, for any $y \in R^1$

$$P\left(\sqrt{2\log\frac{n}{G}} \max_{G < k < n-G} \right.$$

$$\times \left\{ \frac{1}{\sqrt{G}} \frac{1}{\sigma} |S_{k+G.LS} - 2S_{k.LS} + S_{k-G.LS}| \right\}$$

$$\le y + 2\log\frac{n}{G} + \frac{1}{2}\log\log\frac{n}{G} - \frac{1}{2}\log\frac{4\pi}{9} \Big)$$

$$\to \exp\{-2\exp\{-y\}\}. \tag{9.3}$$

Proof. Assertion (i) is the Erdős–Darling theorem, for the proof see, e.g., Shorack and Wellner (1986). The result in Eq. (9.3) is proved in Chen (1988).

Proof of Theorem 2.1. This is the case with the fixed change. For simplicity assume that $\eta = 1/2$. The estimator $\hat{m}_{LS}(1/2)$ can be equivalently defined as

$$\hat{m}_{LS}(1/2) = \arg\max \left\{ \frac{n}{k(n-k)} S_{k.LS}^2; k = 1, \ldots, n-1 \right\}. \tag{9.4}$$

We can write for $k \leq m$

$$\frac{n}{k(n-k)} \left(\sum_{i=1}^{k} (Y_i - \bar{Y}_n) \right)^2 = \frac{n}{k(n-k)} T_k^2 + 2 T_k Q_m \frac{n-m}{n-k}$$

$$+ (k-m) Q_m^2 \frac{n-m}{n-k} + Q_m^2 \frac{m(n-m)}{n}, \quad (9.5)$$

and for $m < k < n$

$$\frac{n}{k(n-k)} \left(\sum_{i=1}^{k} (Y_i - \bar{Y}_n) \right)^2 = \frac{n}{k(n-k)} T_k^{o2} + 2 T_k^o Q_m \frac{m}{k}$$

$$- (k-m) Q_m^2 \frac{m}{k} + Q_m^2 \frac{m(n-m)}{n} \quad (9.6)$$

where

$$Q_m = \frac{n}{m(n-m)} \sum_{j=1}^{m} (Y_j - \bar{Y}_n),$$

$$T_k = - \sum_{i=k+1}^{m} (Y_i - \bar{Y}_m), \quad k = 1, \ldots, m,$$

$$T_k^o = \sum_{i=m+1}^{k} (Y_i - \bar{Y}_m^o), \quad k = m+1, \ldots, n,$$

with $\bar{Y}_m$ and $\bar{Y}_m^o$ defined by Eqns (2.8) and (2.9), respectively. By the assumptions we have, as $n \to \infty$,

$$Q_m = -\delta + O_p(n^{-1/2}) \quad (9.7)$$

$$\max_{1 \leq k < m/2} \left\{ |T_k|^2 \frac{n}{k(n-k)} \right\} = O_p(\log \log n) \quad (9.8)$$

$$\max_{m/2 \leq k < m-D_n} \left\{ |T_k|^2 \frac{n}{k(n-k)} \right\} = O_p(1) \quad (9.9)$$

$$\max_{m-D_n \leq k \leq m} \left\{ |T_k|^2 \frac{n}{k(n-k)} \right\} = O_p\left(\frac{D_n}{n}\right) \quad (9.10)$$

$$\max_{(m+n)/2 \leq k < n} \left\{ |T_k^o|^2 \frac{n}{k(n-k)} \right\} = O_p(\log \log n) \quad (9.11)$$

$$\max_{m+D_n \leq k < (m+n)/2} \left\{ |T_k^o|^2 \frac{n}{k(n-k)} \right\} = O_p(1) \quad (9.12)$$

$$\max_{m < k < m+D_n} \left\{ |T_k^o|^2 \frac{n}{k(n-k)} \right\} = O_p\left(\frac{D_n}{n}\right) \quad (9.13)$$

for any $D_n \to \infty$ satisfying Eq. (7.3).

These relations together with standard arguments imply that, as $n \to \infty$,

$$\hat{m}_{LS}(\eta) - m = O_p(1). \tag{9.14}$$

Therefore, it suffices to deal with

$$\frac{n}{k(n-k)} \left(\sum_{i=1}^{k} (Y_i - \bar{Y}_n) \right)^2, \quad |k - m| \le D,$$

where D is an arbitrarily large but fixed positive number. We have, as $n \to \infty$,

$$\max_{|k-m|\le D} \left\{ \frac{n}{k(n-k)} (T_k^2 + T_k^{o2}) \right\} = o_p(1). \tag{9.15}$$

Moreover,

$$T_k = - \sum_{i=k+1}^{m} (E_i - \bar{E}_m)$$

and

$$T_k^o = \sum_{i=m+1}^{k} (E_i - \bar{E}_m^o).$$

Hence

$$\max_{|k-m|\le D} \left\{ \frac{n}{k(n-k)} \left(\sum_{i=1}^{k} (Y_i - \bar{Y}_n) \right)^2 \right.$$
$$\left. - \left(2\delta \left(\sum_{i=k+1}^{m} E_i I\{k \le m\} - \sum_{i=k+1}^{m} E_i I\{k > m\} \right) - |k - m|\delta^2 \right) \right\} = o_p(1).$$

This result together with Eqns (9.4), (9.14) and (9.15) imply the assertion of Theorem 2.1.   □

Proof of Theorem 2.2. The estimator $\hat{m}_{LS}(G)$ is equivalently defined as

$$\hat{m}_{LS}(G) = \arg\max \left\{ \frac{1}{2G} |S_{k+G,LS} - 2S_{k,LS} + S_{k-G,LS}|; \right.$$
$$\left. k = G + 1, ..., n - G \right\}. \tag{9.16}$$

By standard arguments we obtain

$$S_{m+G.LS} - 2S_{m.LS} + S_{m-G.LS}$$

$$= G\delta + O_p(\sqrt{G}) + \max_{G < k < n-G} \left\{ G^{-2} \left(\sum_{i=k+1}^{k+G} E_i - \sum_{i=k-G+1}^{k} E_i \right)^2 \right\}$$

$$= O_p(nG^{-2}) = o_p(1). \tag{9.17}$$

Since for $1 \leq k < m - G$ and $m + G < k \leq n$

$$S_{k+G.LS} - 2S_{k.LS} + S_{k-G.LS} = \sum_{i=k+1}^{k+G} E_i - \sum_{i=k-G+1}^{k} E_i,$$

using Eqns (9.16) and (9.17) we obtain, as $n \to \infty$,

$$P\left(\max_{G < k < n-G} \left\{ \frac{1}{2G} (S_{k+G.LS} - 2S_{k.LS} + S_{k-G.LS})^2 \right\} \right.$$

$$= \max_{-G \leq k-m \leq G} \left\{ \frac{1}{2G} (S_{k+G.LS} - 2S_{k.LS} + S_{k-G.LS})^2 \right\} \right) \to 0. \tag{9.18}$$

Denoting

$$\tilde{E}_i = \begin{cases} Y_i - \dfrac{1}{m} \displaystyle\sum_{j=1}^{m} Y_j, & i = 1, \ldots, m, \\[3ex] Y_i - \dfrac{1}{n-m} \displaystyle\sum_{j=m+1}^{n} Y_j, & i = m+1, \ldots, n, \end{cases}$$

after straightforward but tedious calculations we receive that for $|k - m| \leq G$

$$S_{k+G.LS} - 2S_{k.LS} + S_{k-G.LS} = \sum_{i=k+1}^{k+G} \tilde{E}_i - \sum_{i=k-G+1}^{k} \tilde{E}_i + GQ_m - |k - m|Q_m, \tag{9.19}$$

which in combination with Eq. (9.18) and the assumptions imply that, as $n \to \infty$,

$$\hat{m}_{LS}(G) - m = O_p(1). \tag{9.20}$$

Treating the right-hand-side of Eq. (9.19) similarly as in the proof of Theorem 2.1, we finish the proof.

Proof of Theorem 2.3. The proof follows the lines of the proof of Theorem 2.1, just one has to proceed more carefully since $\delta_n \to 0$. Details can be found in Antoch, Hušková, Veraverbeke (1995).

Proof of Theorem 2.4. The proof follows the lines of the proof of Theorem 2.2 with more careful treating the single decomposition terms in Eqns (9.4) and (9.5) taking into account the assumption $\delta_n \to \infty$. Details of the proof are in Antoch and Hušková (1994) and Hušková (1997 b).

Proof of Theorem 2.5. The assertion for $\eta \in [0, 1/2)$ is an immediate consequence of Theorem 9.1, while the assertion for $\eta = 1/2$ follows from Eq. (9.2) and the fact that, as $n \to \infty$,

$$\max_{n\epsilon \leq k \leq n(1-\epsilon)} \left\{ \sqrt{\frac{n}{k(n-k)}} |S_{k.LS}| \right\} = O_p(1) \quad \text{for any} \quad 1/2 > \epsilon > 0.$$

Last part of the theorem follows from the stationarity of the sequence of $\sum_{i=k+1}^{k+G} E_i - \sum_{i=k-G+1}^{k} E_i$, $k = G + 1, \ldots, n - G$, and the assumption that $G/n \to 0$.

Proof of Theorem 2.6. The proof is omitted since the assertions can be proved using simple standard arguments.

Proof of Theorem 3.1. The proof is very close to those of Theorem 2.1 and Theorem 2.2.

Proof of Theorem 3.2. See Antoch, Hušková and Práškova (1997).

Proofs of Theorems 5.1, 6.1, 7.1, 7.2 and 7.3 follow the lines of the proofs of the above theorems, however, deep properties of the rank statistics (Hájek (1961, 1974), Hušková (1997 a)) are utilized. Theorem 5.2 is proved in Hušková (1997 b) while Theorem 6.2 is shown in Gombay and Hušková (1998).

REFERENCES

Antoch J. and Hušková M.: Change-point problem, in Computational Aspects of Model Choice, ed. Antoch J., Physica-Verlag, Heidelberg, 1993, 11–38.

Antoch J. and Hušková M.: Procedures for detection of multiple changes in series of independent observations, in Proc. of 5th Prague Symp. on Asymptotic Statistics, eds. Hušková M. and Mandl P., Physica-Verlag, Heidelberg, 1994, 3–20.

Antoch J., Hušková M. and Veraverbeke N.: Change-point problem and bootstrap, *J. of Nonparametric Stat.* 5: 1995, 123–144.

Antoch J. and Hušková M: Tests and estimators for epidemic alternatives, Tatra Mountains Math. Publ. 7, 1996, 316–329.

Antoch J., Hušková M. and Prášková Z.: Effect of dependence on statistics for determination of change, *J. Stat. Plan. Infer.*, 60: 1997, 291–310.

Bai J.: Least squares estimation of a shift in linear processes, *J. Time Ser. Anal. 15*: 1994, 453–472.

Bhattacharya P. K. and Brockwell P. J.: The minimum of additive process with applications to signal estimation and storage theory, Z. Wahrsch. Verw. *Gebite 37*: 1976, 51–75.

Brodsky B. S. and Darkhovsky B. E.: Nonparametric Methods in Change-Point Problems, Kluwer Academic Publ., Dordrecht, 1993.

Carlstein E.: Nonparametric estimation of a change-point, *Ann. Statist. 16*: 1988, 188–197.

Chen X.: Inference in a simple change-point problem, *Scientia Sinica A 31*: 1988, 654–667.

Cobb G. W.: The problem of the Nil: Conditional solution to a change-point problem, *Biometrika 65*: 1978, 243–251.

Csörgő M. and Horváth L.: Nonparametric methods for the change-point problem, in Handbook of Statistics, vol. 7, eds. Krishnaiah P. R. and Rao C. R., North Holland, 1988, 403–425.

Csörgő M. and Horváth L.: Weighted Approximations in Probability and Statistics, J. Wiley, New York, 1993.

Csörgő M. and Horváth L.: Limit Theorems in Change-point Analysis, J. Wiley, New York, 1997.

Darkhovsky B. E.: A nonparametric method for a posteriori detection of the "disorder" time of a sequence of independent random variables, *Theory Probab. Appl. 21*: 1976, 178–183.

Dümbgen L.: The asymptotic behavior of some nonparametric change-point estimators, *Ann. Statist. 19*: 1991, 1471–1495.

Ferger D.: Change-point estimators in case of small disorders, *J. Stat. Plan. Infer. 40*: 1994, 33–49.

Ferger D.: On the rate of almost sure convergence of Dümbgen change-point estimators, *Statist. and Probab. Letters 25*: 1994a, 27–31.

Ferger D.: Asymptotic distribution theory of change-point estimators and confidence intervals based on bootstrap approximation, *Math. Methods of Statistics 3*: 1994b, 362–378.

Ferger D.: Nonparametric tests for nonstandard change-point problems, *Ann. Statist. 23*: 1995, 1848–1861.

Ferger D. and Stute W.: Convergence of change-point estimators, *Stoch. Proc. Appl. 42*: 1992, 345–351.

Gombay E., Horváth L.: Approximations for the time of change and the power function in change-point models. *J. Stat. Plan. Infer. 52*: 1996, 43–66.

Gombay E., Horváth L. and Hušková M.: Estimators and tests for change in variances, *Statistics and Decisions 14*: 1996, 145–159.

Gombay E. and Hušková M.: Rank based estimators of the change-point *J. Stat. Plan. Infer., 67*: 1998, 137–154.

Hájek J.: Some extensions of the Wald-Wolfowitz-Noether theorem, Ann. Math. Statist. 32, 1961, 506–523.

Hájek J.: Asymptotic sufficiency of the vector of ranks in the Bahadur sense, *Ann. Statist. 2*: 1974, 75–83.

Hinkley D. V.: Inference about the change-point in a sequence of random variables, *Biometrika 57*: 1970, 1–17.

Hinkley D. V. and Hinkley E. A.: Inference the change-point in a sequence of binomial variables, *Biometrika 57*: 1970, 477–488.

Hinkley D. and Schechtman E.: Conditional bootstrap methods in the mean-shift model, *Biometrika 74*: 1987, 85–93.

Horváth L. and Kokozska P.: The effect of long-range dependence on change-point estimators, to appear in *J. Stat. Plan. Infer.* 1998.

Horváth L. and Qi-Man Shao: Darling–Erdős type theorems for sums of Gaussian variables with long range memory, *Stoch. Proc. Appl. 63*: 1996, 117–137.

Huber P. J., Robust Statitics, J. Wiley, New York, 1981.

Hušková M.: Estimators for epidemic alternatives, Comment. Math. Univ. Carolin. 36, 1995, 279–291.

Hušková M.: Limit theorems for rank statistics, *Statist. and Probab. Letters 30*: 1997a, 45–55.

Hušková M.: Limit theorems for M-processes via rank statistics processes, in Advances in Combinatorial Methods with Applications to Probability and Statistics, ed. Balakrishnan N., 1997b, 527–533.

Jarušková D.: Some problems with application of change-point detection methods to environmental data, *Environmetrics*, 8, 1997, 469–483.

Ritov Y.: Asymptotic efficient estimation of the change-point with unknown distribution, *Ann. Statist. 18*: 1990, 1829–1839.

Shepp L. A.: The joint density of the maximum and its location for a Wiener process with drift, *J. Appl. Probab. 16*: 1979, 423–427.

Shorack G. R. and Wellner J. A.: Empirical Processes with Applications to Statistics, J. Wiley, New York, 1986.

Siegmund D.: Boundary crossing probabilities and statistical applications, *Ann. Statist. 14*: 1986, 361–404.

Siegmund D.: Confidence sets in change-point problems, *Int. Stat. Rev. 56*: 1988, 31–48.

Stryhn H.: The location of maximum of asymmetric two-sided Brownian motion with triangular drift, *Statist. and Probab. Letters 29*: 1996, 279–284.

Vostrikova L. Y.: Detection "disorder" in multidimensional random process, *Sov. Math. Dokl. 24*: 1981, 55–59.

Worsley K. J.: Confidence regions and tests for a change-point in a sequence of exponential family random variables, *Biometrika 73*: 1986, 91–104.

Yao Q.: Tests for change-points with epidemic alternatives, *Biometrika 80*: 1993, 179–191.

Yao Y.-C. Estimating the number of change-points via Schwarz' criterion, *Statist. and Probab. Letters 6*: 1988, 181–189.

Yao Y.-C. On asymptotic behavior of a class of nonparametric tests, *Statist. and Probab. Letters 19*: 1990, 173–177.

Yao Y.-C.: Approximating the distribution of maximum likelihood estimate of the change-point in a sequence of independent random variables, *Ann. Statist. 15*: 1987, 1321–1328.

16

On Inverse Estimation

ARNOUD C. M. VAN ROOIJ*   University of Nijmegen, Nijmegen, The Netherlands

FRITS H. RUYMGAART*   Texas Tech University, Lubbock, Texas

1. INTRODUCTION, MODEL, AND EXAMPLES

Inverse estimation concerns the recovery of an unknown input signal from blurred observations on the output, which is supposed to be a known transformation of that signal. The estimators considered in this chapter are based on a regularized inverse of the transformation involved, employing a Hilbert space set-up. Halmos's version of the spectral theorem turns out to be extremely useful to present a unified theory both with respect to the type of transformation involved and the type of regularization used. Results on the rate of convergence of the IMSE, optimality of this rate, data-driven selection of the regularization parameter, the Hájek–LeCam convolution theorem for linear functionals, and weak convergence are reviewed. Two important models considered are those of indirect density and indirect nonparametric regression estimation, and particular attention is paid to the special cases of compact and convolution operators. Relations with certain aspects of time series analysis are indicated, and some examples and open problems are discussed.

Ever since 1923, when Hadamard coined the term "ill-posed problem" for partial differential equations with solutions that do not depend continuously on the data, the topic has been one of great mathematical activity. Typically, ill-posed problems reduce to integral equations. Since in practice the data

*Partly supported by NSF grant DMS 95-04485.

are inevitably discrete and usually blurred by random disturbances, the integral equations to be solved are noisy and statistical considerations become pertinent. Statisticians may also refer to this area of noisy integral equations as inverse or ill-posed estimation, or indirect curve estimation. The last name correctly suggests that ordinary, direct, curve estimation is formally included as a special case. From a statistical perspective an impressive number of interesting contributions has been given by Wahba and her school, starting in the seventies. More recently a number of other prominent statisticians like e.g., Donoho, Efromovich, Fan, Groeneboom, Hall, Johnstone, Silverman, and Titterington have added impetus to the research in this field which is in rapid development.

On the one hand there is a great variety of challenging applications. O'Sullivan (1986) mentions areas like crystallography, geophysics, medical electrocardiograms, meteorology, microfluoroimagery, radio astronomy, reservoir engineering, and tomography. To this list one might add instances of physical systems like inverse heat conduction, Mair and Ruymgaart (1996), and imaging. Some of these examples will be considered more closely in this chapter. On the other hand there are many ramifications in other areas of mathematics such as operator theory, harmonic analysis, numerical analysis, and approximation theory. A recent survey of mathematical methodology is given by Bertero (1989); see also Kress (1989) for a recent monograph on integral equations.

A forerunner of inverse estimation is classical finite dimensional regression where the design matrix is often ill-conditioned although its inverse is still continuous. This led Hoerl and Kennard (1970a, b) to employ so-called ridge regression based on a stable Moore–Penrose type inverse of the design matrix. Ridge regression can be regarded as a Bayes procedure given a multivariate normal prior. The stabilization of the inverse of the design matrix introduces a bias term in the estimation procedure and risk analysis displays a trade-off between the variance and the bias component. If prior knowledge about the length of the unknown parameter is available, it is equivalent to a penalized least squares estimator and turns out to be linear minimax, Bunke (1975), Toutenburg (1982), Rao and Toutenburg (1995). All these aspects have analogues in the more general inverse estimation model where we will focus on the infinite dimensional case.

In the infinite dimensional case the inverse of the integral operator is usually not bounded, which is the source of the ill-posedness. This is why regularization of some sort is required. A widely employed regularization method is Tikhonov or penalized least squares regularization, Tikhonov and Arsenin (1977), Wahba (1977), Nychka and Cox (1989). Important further references are Titterington (1985), and, for application in image restoration, Hall and Titterington (1986), and Titterington (1991). The method is related to Grenander's (1981) method of sieves and has a Bayesian interpretation, Kimeldorf and Wahba (1971). For compact operators Johnstone and Silverman (1990, 1991) use the singular value decomposition combined with tapering of the reciprocals of the singular values for regularization. This method is extended by Donoho (1995) to classes of operators that need no longer be compact. This author employs a wavelet-vaguelette decomposition that yields sparse rather than strictly diagonal matrices. Data-driven thresholding yields nonlinear estimators with optimal properties over the entire Besov scale, including spaces of signals of bounded variation where these estimators perform better than the usual linear ones. Fourier methods naturally arise in all sorts of deconvolution and mixture problems. For deconvolution on the real line $\mathbb{R}$ we refer to, e.g., Carroll and Hall (1988), Stefanski and Carroll (1991), Devroye (1989), Zhang (1990), and Fan (1991). Deconvolution on the circle $\mathbb{T}$ is considered in Efromovich (1996), deconvolution on $\mathbb{Z}^k$ with application in image analysis in Hall (1990), and for general abstract deconvolution on locally compact abelian groups, see van Rooij and Ruymgaart (1995). Various other inversion methods are used, tailored to specific conditions. Thus, Silverman et al. (1990) employ a smoothed EM approach, and Masry and Rice (1990) expand the unknown signal in the system of Hermite polynomials along with differentiation for deconvolution when the kernel is Gaussian. Vardi and Lee (1993) propose a maximum likelihood procedure for certain inverse problems with positivity restriction. Groeneboom and Jongbloed (1995) employ isotonized estimators for Wicksell's unfolding problem in stereology.

Despite this wide variety of methods a great deal of unification can be obtained if one is willing to adopt a Hilbert space framework and linearity of the integral operator. In this set-up Halmos's (1963) special version of the spectral theorem, stating that any Hermitian operator is unitarily equivalent to a multiplication operator in the spectral domain, turns out to be extremely expedient. Unification can be obtained both regarding the type of bounded operator (compact or not) and the type of regularization (spectral cut-off or Moore–Penrose).

Throughout, $(\mathbb{D}, \mathscr{D}, \mu_{\mathbb{D}})$ and $(\mathbb{E}, \mathscr{E}, \mu_{\mathbb{E}})$ will denote σ-finite measure spaces and $L^2(\mu_{\mathbb{D}})$ and $L^2(\mu_{\mathbb{E}})$ the corresponding Hilbert function spaces. These Hilbert spaces are assumed to be over the real numbers and separable. The operator is supposed to be a bounded injective linear integral operator $K : L^2(\mu_{\mathbb{D}}) \to L^2(\mu_{\mathbb{E}})$. Without confusion integral operators and their kernels will be denoted by the same symbol. All kernels will be real-valued.

The equation to be solved is

$$p = K\theta, \qquad \theta \in \Theta \subset L^2(\mu_{\mathbb{D}}), \tag{1.1}$$

where K and Θ are known. The data consist of a random sample $X_1, \ldots, X_n$ of independent copies of a random variable X that assumes values in a σ-finite measure space $(\mathbb{X}, X, \mu_{\mathbb{X}})$. All random variables are defined on a sufficiently rich underlying probability space $(\Omega, \mathscr{W}, \mathbf{P})$, and X has a density f with respect to $\mu_{\mathbb{X}}$. The data are directly related to p in a sense that will be made clear. Therefore we sometimes call p "observable" and θ the "unobservable" parameter of actual interest. In the language of system theory we may also refer to θ as the unknown "input signal" and to p as the "output signal".

In general we will apply preconditioning and replace Eq. (1.1) with the equivalent equation

$$q := \Gamma p = \Gamma K \theta =: R\theta, \qquad \theta \in \Theta, \tag{1.2}$$

where $\Gamma : L^2(\mu_{\mathbb{E}}) \to L^2(\mu_{\mathbb{D}})$ is a bounded, injective linear integral operator. By preconditioning we intend to achieve two objectives. First we want R to be easier to deal with than K. For our purposes it will suffice to assume that

$$R : L^2(\mu_{\mathbb{D}}) \to L^2(\mu_{\mathbb{D}}) \text{ is strictly positive Hermitian.} \tag{1.3}$$

In the second place we want q to be easy to estimate. Therefore we assume that there exists a linear integral operator

$$Q : L^1(\mu_{\mathbb{X}}) \to L^2(\mu_{\mathbb{D}}), \tag{1.4}$$

such that the common density f of the sample elements is related to q according to

$$Qf = q. \tag{1.5}$$

It will, moreover, be assumed that

$$Q(\bullet, x) \in L^2(\mu_{\mathbb{D}}), \qquad x \in \mathbb{X}, \qquad \mathbf{E}\|Q(\bullet, X)\|^2 < \infty. \tag{1.6}$$

The problem is to recover θ from the data $X_1, \ldots, X_n$. In most models preconditioning can be performed in such a way that the present assumptions are fulfilled. This is in particular true for the two most important indirect curve estimation models.

Model A: indirect density estimation

In this case Θ is contained in the class of all probability densities with respect to $\mu_{\mathbb{D}}$ that are square integrable, and K is such that p is also a density. The $X_1, \ldots, X_n$ are a random sample from p, and the model can be summarized in the form

$$X \sim p = K\theta, \theta \in \Theta \subset L^2(\mu_{\mathbb{D}}), \qquad f = p. \tag{1.7}$$

Preconditioning with Γ entails that we simply have

$$Q = \Gamma, \tag{1.8}$$

and if one chooses Γ to be K^*, the adjoint of K, one obviously has that $R := K^*K$ is strictly positive Hermitian. A natural unbiased and $\sqrt{n}$-consistent estimator of q is

$$\hat{q} = \frac{1}{n} \sum_{k=1}^{n} \Gamma(\bullet, X_k), \tag{1.9}$$

and we have $\mathbb{X} = \mathbb{E}$ in this case.

Model B: indirect regression estimation

Now the data $X_1, \ldots, X_n$ are independent copies of $X := (Y, Z)$ satisfying

$$Y = (K\theta)(Z) + \varepsilon, \tag{1.10}$$

where Z is a random design variable with values in $\mathbb{E}$, and ε the real-valued error variable, independent of Z, with mean 0, variance $0 < \sigma^2 < \infty$, and density w with respect to Lebesgue measure λ on $\mathbb{R}$. In this model we usually have

$$\mu_{\mathbb{E}}(\mathbb{E}) < \infty, \tag{1.11}$$

so that we may and will choose Z such that it has a density h on $\mathbb{E}$ with respect to $\mu_{\mathbb{E}}$ which is bounded away from both 0 and ∞. In this model, moreover, we usually take $\Theta = L^2(\mu_{\mathbb{D}})$. In this situation, $\mathbb{X} = \mathbb{R} \times \mathbb{E}$ and X has density

$$f(y, z) = w(y - (K\theta)(z))h(z), \qquad y \in \mathbb{R}, z \in \mathbb{E}, \tag{1.12}$$

with respect to $\lambda \times \mu_{\mathbb{E}}$. This yields

$$p(z) = (K\theta)(z) = \int_{-\infty}^{\infty} \frac{y}{h(z)} f(y, z) \, dy, \qquad z \in \mathbb{E}, \tag{1.13}$$

and preconditioning with Γ gives

$$q(u) = (Qf)(u) = (\Gamma p)(u)$$
$$= \int_{\mathbb{E}} \int_{-\infty}^{\infty} \left\{ \frac{y}{h(z)} \Gamma(u, z) \right\} f(y, z) \, dy \, d\mu_{\mathbb{E}}(z), \qquad u \in \mathbb{D}. \tag{1.14}$$

Apparently, Q has kernel

$$Q(u, (y, z)) = \frac{y}{h(z)} \Gamma(u, z), \qquad u \in \mathbb{D}, (y, z) \in \mathbb{R} \times \mathbb{E}. \tag{1.15}$$

Choosing $\Gamma = K^*$ yields once again a strictly positive Hermitian R, and

$$\hat{q} = \frac{1}{n} \sum_{k=1}^{n} \frac{Y_k}{h(Z_k)} \Gamma(\bullet, Z_k), \qquad (1.16)$$

is a natural unbiased and $\sqrt{n}$-consistent estimator of q.

Parametric analogues of models A and B arise when Θ is a parametric family of functions

$$\Theta = \{\theta_t, t \in \mathcal{T}\}, \mathcal{T} \subset \mathbb{R}^k, k \in \mathbb{N}. \qquad (1.17)$$

Estimation of t might take place by first estimating the unknown function in Θ followed by estimating the Euclidean parameter via a minimum distance method.

Before proceeding with some concrete examples let us briefly comment on direct curve estimation models and their parametric counterparts which boil down to classical Euclidean parameter models. These models are formally obtained from those previously discussed by replacing K with the identity operator. By preconditioning with a genuine integral operator, direct curve estimation is reduced to inverse estimation. This approach to direct curve estimation is followed in Dey and Ruymgaart (1998). The resulting approach to estimation of Euclidean parameters may bear some resemblance to the minimum distance method in Beran (1977) or estimation methods based on the empirical characteristic function surveyed in Prakasa Rao (1987).

Let us now describe some examples and then conclude the present section with an overview of the contents of this chapter.

Examples

Can you see the weight of a cable? This question, a paraphrase of the title of Kac's famous (1966) paper will be considered in Sec. 4. There are many more examples from physical systems where the source term of a differential equation is to be recovered like, for instance, inverse heat conduction, Mair and Ruymgaart (1996). These examples fit within this framework but cannot be considered here.

Wicksell's unfolding problem, Nychka et al (1984), Nychka and Cox (1989), Johnstone and Silverman (1991) deals with recovering sizes of spherical particles from planar cuts and boils down to solving a noisy integral equation of Abel type. Here we might precondition with $(K^2)^*K$ rather than K^* as the action of the operator K^2 reduces to taking antiderivatives.

The problem of computerized tomography, Vardi et al. (1985), Johnstone and Silverman (1990) is to recover the density of an object from its line integrals and is in fact inversion of a noisy Radon transform.

In Sec. 5 we list a few examples of errors-in-variables or mixture models. These models often reduce to noisy deconvolution, Carroll and Hall (1988),

Stefanski and Carroll (1989), Devroye (1989), Zhang (1990), Fan (1991), and van Rooij and Ruymgaart (1995). A practical example is provided by the image degradation model

$$p_k = \sum_{j \in \mathbb{Z}^2} w_{k-j}\theta_j + \varepsilon_k, \qquad k \in \mathbb{Z}^2, \tag{1.18}$$

considered in Hall (1990), where w is the spread function, θ the actual image of interest, ε the error, and p the blurred picture.

A well-known problem in the theory of dynamical systems is the recovery of the input function θ from the output function p, where θ and p are functions on $(0, \infty)$ related through a differential equation. Application of the Laplace transform $\mathscr{L}$ reduces the differential equation to an integral equation, inversion of which contains multiplication with a polynomial. It turns out that this preconditioning with the Laplace transform does not increase the ill-posedness, although regularized inversion of this transform is part of the recovery procedure. Associating with θ the function

$$\theta^0(y) := y^{-1/2}\theta(y^{-1}), \qquad y \in (0, \infty), \tag{1.19}$$

we obtain

$$(\mathscr{L}\theta)(t) := \int_0^\infty e^{-tx}\theta(x)\,dx = \int_0^\infty e^{-t/y}\theta(y^{-1})y^{-2}\,dy \tag{1.20}$$

$$= t^{-1/2}\int_0^\infty \{e^{-(t\oplus y)}(t\ominus y)^{1/2}\}\theta^0(y)\,d\mu(y), \qquad 0 < t < \infty,$$

where μ denotes the Haar measure on the multiplicative group $(0, \infty)$ and $a \oplus b := ab$, $a \ominus b := ab^{-1}(a, b \in (0, \infty))$. Thus, the Laplace transform is related to convolution on $(0, \infty)$. This relation has been exploited in Chauveau et al. (1994) to construct a regularized inverse of $\mathscr{L}$, and applied to the dynamical system mentioned above in Dey et al. (1998). For related results on the Laplace transform see Gilliam et al. (1988) and Bertero (1989).

We briefly sketch the structure of the rest of this paper.

In Sec. 2 we will consider the general model as specified in Formulas (1.1)–(1.6). Since R^{-1} is usually unbounded, $\hat{q}$, the estimator of q, may not lie in its domain and if it does, $R^{-1}\hat{q}$ may not be close to $R^{-1}q = \theta$, not even if $\hat{q}$ were close to q. Therefore we propose the estimators

$$\hat{\theta}_\alpha := R_\alpha^{-1}\hat{q}, \qquad \alpha > 0, \tag{1.21}$$

where R_α^{-1} is a regularized inverse of R, constructed with the help of Halmos's (1963) result, and $\alpha > 0$ a regularization parameter not unlike the bandwidth in direct curve estimation. The estimators in Eq. (1.21) will be called regularized inverse type (RI)-estimators. Their integrated mean square error (IMSE) will be calculated and a lower bound to the minimax risk will be established. This lower bound displays a clear relation between the degree of ill-posedness, the size of the submodel, and the sample size. Asymptotic normality of suitably standardized $\langle \hat{\theta}_\alpha, \varphi \rangle$ for certain $\varphi \in L^2(\mu_\mathbb{D})$ will be discussed, and the Hájek–LeCam convolution theorem for estimators of the linear functional $\langle \theta, \varphi \rangle$ will be formulated. Since the rate of the asymptotic normality will usually depend on φ, there cannot in general be weak convergence in $L^2(\mu_\mathbb{D})$ of $\hat{\theta}_\alpha$ to some Gaussian process. We also describe a data-driven method, based on least-squares cross-validation, to select the regularization parameter α for given finite sample size.

The general model is not sufficiently specific to reach conclusions about convergence of the IMSE at the minimax rate or about asymptotic efficiency in the sense of the Hájek-LeCam theorem. If we specialize to models A and B and precondition with the adjoint operator, however, these asymptotic optimality results are within reach. This is considered in Sec. 3. In Sec. 4 and 5 we deal with compact and convolution operators, respectively, and return to some of the examples mentioned above.

This paper focuses on estimation of reasonably smooth parameters θ. The case of extremely smooth parameters is relatively simple, but the case of severely irregular parameters requires a considerable modification of the current methods. A short discussion of this kind of extension and a brief overview of some open problems form the contents of the final Sec. 6.

2. THE GENERAL MODEL

Our point of departure is the preconditioned Eq. (1.2), where the operator R is strictly positive Hermitian.

For any measure μ, by $L^2_\mathbb{C}(\mu)$ we denote the complex Hilbert space of all square-integrable complex-valued functions. By a slight extension of Halmos's version of the Spectral Theorem, see Halmos (1963), there exist a σ-finite measure space $(\mathbb{S}, \mathscr{S}, \mu_\mathbb{S})$, a unitary map U of $L^2(\mu_\mathbb{D})$ onto a closed $\mathbb{R}$-linear subspace $\mathbb{L}$ of $L^2_\mathbb{C}(\mu_\mathbb{S})$ and a function ρ in $L^\infty(\mu_\mathbb{S})$ satisfying $\rho > 0$ almost everywhere, such that

$$R = U^{-1} M_\rho U, \tag{2.1}$$

where the operator $M_\rho : L^2_\mathbb{C}(\mu_\mathbb{S}) \to L^2_\mathbb{C}(\mu_\mathbb{S})$ is pointwise multiplication by ρ. This M_ρ leaves $\mathbb{L}$ invariant. Moreover, if φ is any bounded Borel function $(0, \infty) \to (0, \infty)$ (so that $\varphi(\rho) \in L^\infty(\mu_\mathbb{S})$), then $M_{\varphi(\rho)}$ maps $\mathbb{L}$ into $\mathbb{L}$, and $U^{-1} M_{\varphi(\rho)} U$ is a positive Hermitian operator in $L^2(\mu_\mathbb{D})$. The space $\mathbb{L}$ is called the spectral domain, the measure $\mu_\mathbb{S}$, which is σ-finite, the spectral measure, and ρ the spectral density.

A regularized inverse of R is in fact a family of operators $\{R_\alpha^{-1}, \alpha > 0\}$ such that each $R_\alpha^{-1} : L^2(\mu_\mathbb{D}) \to L^2(\mu_\mathbb{D})$ is bounded and such that for each $f \in L^2(\mu_\mathbb{D})$:

$$\|R_\alpha^{-1} Rf - f\| \to 0, \quad \text{as } \alpha \downarrow 0, \tag{2.2}$$

where $\|\cdot\|$ is a generic notation for the norm in a Hilbert space. (To avoid confusion, in some instances we may label the norm with a characteristic of the space: in Formula (2.2) we might, e.g., write $\|\cdot\|_\mathbb{D}$.) For suitable families of functions $\varphi_\alpha : (0, \infty) \to [0, \infty)$, $\alpha > 0$, that in particular have the property $\varphi_\alpha(t) \to t^{-1}$, as $\alpha \downarrow 0$, for all $t > 0$, we obtain a regularized inverse by setting $R_\alpha^{-1} := \varphi_\alpha(R) := U^{-1} M_{\varphi_\alpha(\rho)} U$. Two important examples of such families of functions are

$$\varphi_{1,\alpha}(t) := \frac{1}{\alpha + t}, \qquad t > 0, \qquad \alpha > 0, \tag{2.3}$$

$$\varphi_{2,\alpha}(t) := \frac{1}{t} \mathbf{1}_{[\alpha,\infty)}(t), \qquad t > 0, \qquad \alpha > 0. \tag{2.4}$$

The first family yields the Moore–Penrose inverse of R. In fact, one can easily show that

$$\varphi_{1,\alpha}(R) = (\alpha I + R)^{-1}, \qquad \alpha > 0. \tag{2.5}$$

If R is obtained from preconditioning with $\Gamma = K^*$ the Moore–Penrose inverse yields an interesting interpretation as a penalized least-squares solution, since one can show that

$$(\alpha I + K^*K)^{-1} K^*g = \operatorname{argmin}_{f \in L^2(\mu_\mathbb{D})} \{\|Kf - g\|_\mathbb{E}^2 + \alpha\|f\|_\mathbb{D}^2\}, \qquad g \in \mathscr{R}_K, \tag{2.6}$$

where $\mathscr{R}_K$ is the range of K. The second family in Eq. (2.4) yields the so-called spectral cut-off type inverse of R. In this chapter we will be exclusively dealing with the second family and throughout employ the inverse

$$R_\alpha^{-1} := \varphi_{2,\alpha}(R) = U^{-1} M_{\{\mathbf{1}_{[\alpha,\infty)}(\rho)\}/\rho} U, \qquad \alpha > 0, \tag{2.7}$$

without further reference.

It will be assumed that $U : L^2(\mu_\mathbb{D}) \to L^2(\mu_\mathbb{S})$ is an integral operator with kernel $U(s, u) \in \mathbb{C}$, $(s, u) \in \mathbb{S} \times \mathbb{D}$, satisfying

$$\operatorname{ess\,sup}_{(s,u) \in \mathbb{S} \times \mathbb{D}} |U(s, u)| < \infty. \tag{2.8}$$

Since U is unitary we have $U^{-1} = U^*$, where the adjoint U^* has kernel

$$U^*(u, s) = \overline{U(s, u)}, \qquad (u, s) \in \mathbb{D} \times \mathbb{S}. \tag{2.9}$$

Returning now to the assumptions in Formulas (1.4)–(1.6) it is immediate that

$$\hat{q} := \frac{1}{n} \sum_{k=1}^{n} \hat{q}_k, \quad \hat{q}_k := Q(\bullet, X_k) \in L^2(\mu_\mathbb{D}), \tag{2.10}$$

is an unbiased and $\sqrt{n}$-consistent estimator of q, which is at the root of the construction of the estimators of θ.

DEFINITION 2.1   The RI-estimators of θ are given by

$$\hat{\theta}_\alpha := R_\alpha^{-1}\hat{q}, \quad \text{with } \theta_\alpha := \mathbf{E}\hat{\theta}_\alpha = R_\alpha^{-1}q, \quad \alpha > 0. \tag{2.11}$$

Since the kernels K and Γ are real-valued, the same holds true for the kernel of the strictly positive Hermitian operator $R := \Gamma K$. Although $\hat{q}$ is also real, the construction of R_α^{-1} in Eq. (2.7) does not immediately reveal that $\hat{\theta}_\alpha$ is real as well. To see that it is, we approximate $\varphi_{2,\alpha}$ in Eq. (2.4) by means of continuous functions $\varphi_n : [0, \|R\|] \to [0, 1/\alpha]$ chosen such that $\varphi_n(t) = 1/t$ for $\alpha \leq t \leq \|R\|$, and $\varphi_n(t) = 0$ for $0 \leq t \leq \alpha - n^{-1}$. Next we choose a real polynomial P_n such that $|P_n(t) - \varphi_n(t)| \leq 1/n$ for $0 \leq t \leq \|R\|$ and prove that $P_n(R) \to R_\alpha^{-1}$ pointwise on $L^2(\mu_\mathbb{D})$. Thus, we have the pleasant result that, when we start with a signal in a Hilbert function space of real-valued functions, the RI-estimator is also real-valued and in that space.

THEOREM 2.1   The RI-estimator $\hat{\theta}_\alpha$ is real-valued and, more precisely, $\hat{\theta}_\alpha \in L^2(\mu_\mathbb{D})$, $\alpha > 0$.

Let us note that

$$\mathbf{E}\hat{\theta}_\alpha =: \theta_\alpha = U^{-1}\mathbf{1}_{\{\rho \geq \alpha\}}U\theta =: I_\alpha\theta, \quad \alpha \geq 0, \tag{2.12}$$

so that it depends on the operator R only via $\{\rho \geq \alpha\}$. The operator I_α on the right is an approximation of the identity. Most calculations are conveniently carried out in the spectral domain. By way of a first example let us consider the IMSE. Because $\hat{q}$ is unbiased and U unitary we have

$$\mathbf{E}\|\hat{\theta}_\alpha - \theta\|_\mathbb{D}^2 = \mathbf{E}\|\hat{\theta}_\alpha - \theta_\alpha\|_\mathbb{D}^2 + \|\theta_\alpha - \theta\|_\mathbb{D}^2$$

$$= \mathbf{E}\|U\hat{\theta}_\alpha - U\theta_\alpha\|_\mathbb{S}^2 + \|U\theta_\alpha - U\theta\|_\mathbb{S}^2$$

$$= \mathbf{E}\|\rho^{-1}\mathbf{1}_{[\alpha,\infty)}(\rho)(U\hat{q} - Uq)\|_\mathbb{S}^2 + \|(1 - \mathbf{1}_{[\alpha,\infty)}(\rho))U\theta\|_\mathbb{S}^2.$$

By assumption, $Q(\bullet, X)$ is a random element of $L^2(\mu_\mathbb{D})$, and since U also has a kernel we find that $U(Q(\bullet, X)) = (UQ)(\bullet, X)$. We now have the following result.

THEOREM 2.2   The IMSE equals

$$\mathbf{E}\|\hat{\theta}_\alpha - \theta\|^2 = \frac{1}{n} \int_{\{\rho \geq \alpha\}} \rho(s)^{-2} \operatorname{Var}(UQ)(s, X) \, d\mu_{\mathbb{S}}(s)$$

$$+ \int_{\{\rho < \alpha\}} |(U\theta)(s)|^2 \, d\mu_{\mathbb{S}}(s), \qquad \alpha > 0. \tag{2.13}$$

At this point some comments are in order. Although Eqns (1.1) and (1.2) are equivalent, one must be careful: replacing q with the estimator $\hat{q}$ could be a source of undue extra ill-posedness. The present model is far too general for us to go into details, but we will see that in the indirect curve estimation models, preconditioned with $\Gamma = K^*$, the estimator $\hat{q}$ has such good properties that the extra ill-posedness is eliminated.

In order to find the smallest IMSE one needs to minimize the right-hand-side of Eq. (2.13) as a function of the regularization parameter α. This boils down to the balancing of the variance and the bias part, just as in direct curve estimation. But here too, the optimal value of α will depend on the unknown parameter θ to be estimated. The following pseudo-consistency result can be easily shown.

COROLLARY 2.3   For each $\theta \in \Theta$ there exist $\alpha_\theta(n) \downarrow 0$, as $n \to \infty$, such that

$$\mathbf{E}\|\hat{\theta}_{\alpha_\theta(n)} - \theta\|^2 \to 0, \qquad \text{as } n \to \infty. \tag{2.14}$$

To assess the performance of a curve estimator one usually restricts the model to a class of smooth functions, which allows one to control the bias term. One might consider hyperrectangles or hyperellipses in the spectral domain, see, e.g., Donoho et al. (1990). The type of submodel has a bearing on the convergence rate of the IMSE. Throughout this paper we will derive our subclass from a hyperrectangle according to

$$\Theta_\lambda := \Theta_{1,\lambda} \cap \Theta, \quad \Theta_{1,\lambda} := \{\theta \in L^2(\mu_{\mathbb{D}}) : |U\theta| \leq \lambda\}, \tag{2.15}$$

for some $\lambda \in L^2(\mu_{\mathbb{S}})$, $\lambda \geq 0$. Our first concern is the rate of convergence of the supremum of the IMSE over Θ_λ which usually follows easily from

$$\sup_{\theta \in \Theta_\lambda} \mathbf{E}\|\hat{\theta}_\alpha - \theta\|^2 \leq \frac{1}{n} \int_{\{\rho \geq \alpha\}} \rho^{-2} \operatorname{Var}(UQ)(\bullet, X) \, d\mu_{\mathbb{S}} + \int_{\{\rho < \alpha\}} \lambda^2 \, d\mu_{\mathbb{S}}. \tag{2.16}$$

The question that naturally emerges here is whether $\alpha = \alpha(n) > 0$ can be found such that the rate of convergence will be optimal.

Therefore, our next concern is to derive a lower bound to the minimax risk. The supremum of the IMSE will be taken over a subclass of type Eq. (2.15) constructed in the following manner. Let $e_1, e_2, \ldots$ be an orthonormal system in $L^2(\mu_{\mathbb{D}})$, not necessarily a basis, and $\lambda_1, \lambda_2, \ldots > 0$ numbers with $\sum_{m \geq 1} \lambda_m^2 < \infty$. Let us define

$$\Theta_{0,\lambda} := \left\{ \theta \in L^2(\mu_{\mathbb{D}}) : \theta = \sum_{m \geq 1} t_m e_m, |t_m| \leq \lambda_m \right\}$$

$$\subset \left\{ \theta \in L^2(\mu_{\mathbb{D}}) : |U\theta| \leq \sum_{m \geq 1} \lambda_m |\varphi_m| =: \lambda \right\} =: \Theta_{1,\lambda}, \qquad (2.17)$$

where the $\varphi_k := U e_k$ form an orthonormal system in $L^2(\mu_{\mathbb{S}})$. (See the preamble to Theorem 3.3.) For the lower bound we need to assume that

$$\Theta_{0,\lambda} \subset \Theta. \qquad (2.18)$$

The infimum of the IMSE will be taken over essentially all estimators, i.e. over the class of statistics

$$\mathcal{T} := \{ T : \mathbb{X}^n \to L^2(\mu_{\mathbb{D}}) : \mathbf{E} \| T(X_1, \ldots, X_n) \|^2 < \infty \text{ for all } \theta \in \Theta \}. \qquad (2.19)$$

A parameter $\theta \in \Theta_{0,\lambda}$ will be identified with the vector $t := (t_1, t_2, \ldots) \in I_1 \times I_2 \times \cdots =: I \subset l^2$, where $I_m := [-\lambda_m, \lambda_m]$. Let us write f_{t,t_m} to indicate the common density of the sample elements corresponding to $t \in I$ with all coordinates fixed except for t_m which varies in I_m. Hence

$$t_m \mapsto \sqrt{f_{t,t_m}}, \qquad t_m \in I_m, \qquad (2.20)$$

is a curve in $L^2(\mu_{\mathbb{X}})$, assumed to be strongly differentiable with respect to t_m at any $t \in I$ with partial derivative $\partial \sqrt{f_{t,t_m}}/\partial t_m \in L^2(\mu_{\mathbb{X}})$ satisfying

$$\sup_{t \in I} \left\| \frac{\partial}{\partial t_m} \sqrt{f_{t,t_m}} \right\|^2 =: \bar{\rho}_m < \infty, \quad m \in \mathbb{N}. \qquad (2.21)$$

This means that in each direction the Fisher information is uniformly bounded.

In this context the van Trees inequality, which is a global version of the Cramér–Rao inequality, turns out to be extremely useful. To prepare for this inequality let us choose a density

$$\pi \in C^{(1)}(\mathbb{R}) : \quad \pi(t) > 0 \text{ for } t \in (-1, 1), \pi(t) = 0 \text{ for } t \notin (-1, 1). \quad (2.22)$$

The density should have finite nonzero information:

$$0 < \mathcal{I} := \int_{-1}^{1} \left(\frac{\pi'(t)}{\pi(t)} \right)^2 \pi(t) \, dt < \infty. \qquad (2.23)$$

Adjusting the scale to the intervals I_k we obtain

$$\pi_m(t) := \frac{1}{\lambda_m} \pi\left(\frac{t}{\lambda_m}\right), t \in \mathbb{R}, \qquad \mathscr{I}_m := \left(\frac{1}{\lambda_m}\right)^2 \mathscr{I}, m \in \mathbb{N}. \tag{2.24}$$

For recent results regarding the van Trees inequality including application in nonparametrics and the choice of a sequence like in Eq. (2.24), see Gill and Levit (1995).

Let us write $T_m := \langle T, e_m \rangle$. It follows from Bessel's inequality that

$$\sup_{t \in I} \mathbf{E}\|T - t\|^2 \geq \sum_{m \geq 1} \int_{I_1} \int_{I_2} \cdots \{\mathbf{E}(T_m - t_m)^2\} \left\{\prod_{m \geq 1} \pi_m(t_m)\right\} dt_1\, dt_2 \ldots .$$

$$\tag{2.25}$$

Application of the van Trees inequality to the mth term yields

$$\int_{I_m} \mathbf{E}(T_m - t_m)^2 \pi_m(t_m) dt_m \geq \frac{\lambda_m^2}{n\lambda_m^2 \bar{\rho}_m + \mathscr{I}}. \tag{2.26}$$

Since this lower bound is independent of $t \in I$ and because Eq. (2.17) we have proved the following result, van Rooij and Ruymgaart (1996).

THEOREM 2.4   We have

$$\inf_{T \in \mathscr{T}} \sup_{\theta \in \Theta_\lambda} \mathbf{E}\|T - \theta\|^2 \geq \sum_{m \geq 1} \frac{\lambda_m^2}{n\lambda_m^2 \bar{\rho}_m + \mathscr{I}}. \tag{2.27}$$

The lower bound on the right in Formula (2.27) gives a clear insight into the impact of the size of the subclass, the sample size, and also the degree of ill-posedness. If the system $\{e_m\}$ is properly chosen the numbers $\bar{\rho}_m$ will turn out to be related to the spectral density of the operator K^*K. Such a proper choice is usually obtained if one ensures that the functions Ue_m have disjoint supports that are not "too large." In any specific case one might compare the upper bound in Formula (2.16) with the lower bound in Formula (2.27). Provided that we precondition with $\Gamma = K^*$ a general claim can be made that the minimax risk of RI-estimators converges to 0 at an optimal rate. We will return to this optimality in Sec. 3.

It has already been observed in the passage preceding Corollary 2.1 that minimization of the IMSE as a function of α, for given fixed sample size n, yields a parameter value that depends on the unknown θ. Therefore, this procedure is not suitable in practice. Here we will briefly review the least-squares cross-validation selection method developed in Dey et al. (1996), which is an extension of such methods by Bowman (1984) and

Rudemo (1982) for direct curve estimation. Let us write

$$\mathcal{M}_n(\alpha) := \int_{\{\rho \geq \alpha\}} \left(\frac{1}{\rho}\right)^2 \left\{ \frac{1}{n} \mathbf{E} \left|(UQ)(\bullet, X)\right|^2 - \frac{n+1}{n} \left|Uq\right|^2 \right\} d\mu_{\mathbb{S}}, \qquad \alpha > 0.$$

(2.28)

It is immediate from Eq. (2.13) that

$$\mathbf{E}\|\hat{\theta}_\alpha - \theta\|^2 = \mathcal{M}_n(\alpha) + \int |U\theta|^2 \, d\mu_{\mathbb{S}}, \qquad \alpha > 0.$$

(2.29)

One sees that the IMSE and the function $\mathcal{M}_n$ have the same minimizer. It will be convenient to introduce (cf. Eq. (2.10))

$$\hat{q}_{(k)} := \frac{1}{n-1} \sum_{j \neq k} \hat{q}_j, \qquad k = 1, \ldots n,$$

(2.30)

and observe that

$$\mathbf{E}|U\hat{q}_k|^2 = \mathbf{E}|(UQ)(\bullet, X)|^2,$$

(2.31)

by definition, and

$$\mathbf{E}(U\hat{q}_{(k)})(\overline{U\hat{q}_k}) = |Uq|^2,$$

(2.32)

exploiting the independence of $\hat{q}_{(k)}$ and $\hat{q}_k$. Therefore

$$\frac{1}{n} \sum_{k=1}^n |U\hat{q}_k|^2 \text{ and } \frac{1}{n} \sum_{k=1}^n (U\hat{q}_{(k)})(\overline{U\hat{q}_k}),$$

(2.33)

are unbiased estimators of the quantities in Eqns (2.31) and (2.32) respectively. This enables us to construct an estimator $\hat{\mathcal{M}}_n$, to be given below, of $\mathcal{M}_n$ and to use the minimizer $\hat{\alpha}(n)$ of $\hat{\mathcal{M}}_n$ as a data-driven substitute for the minimizer of $\mathcal{M}_n$. This method has been applied to a deconvolution problem, Dey et al. (1996) with rather good results. Although most of the properties of $\hat{\alpha}(n)$ such as those obtained in Hall (1983) and Fan and Marron (1992) for direct curve estimation are still a subject of research, we have, of course, the following.

THEOREM 2.5   The statistical function

$$\hat{\mathcal{M}}_n(\alpha) := \frac{1}{n} \sum_{k=1}^n \int_{\{\rho \geq \alpha\}} \rho^2 \left(\frac{1}{n} |U\hat{q}_k|^2 - \frac{n+1}{n} (U\hat{q}_{(k)})(\overline{U\hat{q}_{(k)}})\right) d\mu_{\mathbb{S}}, \qquad \alpha > 0,$$

(2.34)

is an unbiased estimator of $\mathcal{M}_n(\alpha)$, $\alpha > 0$.

Another way to look at asymptotic optimality is via the weak convergence of linear functionals. The celebrated Hájek–LeCam convolution theorem, see, e.g., Beran (1977), van der Vaart (1988) provides a criterion to verify optimality if asymptotic normality of the functional can be established. With a view towards the construction of confidence bands the limiting distribution of $\|\hat{\theta}_{\alpha(n)} - \theta\|^2$ might be a first step. This could be a hard problem (cf. Bickel and Rosenblatt (1973) for direct curve estimation). A fundamental issue behind all these questions is whether the process $\hat{\theta}_{\alpha(n)} - \theta$ is weakly convergent to a Gaussian process in the Hilbert space $L^2(\mu_{\mathbb{D}})$. We will see that in general this question must be answered in the negative.

First let us recall that in a Hilbert space a finite dimensional distribution (fidi) is a finite collection of inner products, Grenander (1981), and that taking an inner product is applying a smoothness operation due to which the speed of convergence usually will be increased. Although in classical, direct, curve estimation all fidis converge weakly at the $\sqrt{n}$-rate to a normal distribution, there is no weak convergence to a Gaussian process in the Hilbert space, Ruymgaart (1996). For the present indirect curve estimation models the fidis will not even converge weakly at the same rate if they converge at all, which excludes the possibility of weak convergence of the processes.

Let us recall from Theorem 2.1 that $\hat{\theta}_\alpha$ is real-valued. The fidis that we will consider here are vectors of linear functionals, determined by elements $g_j \in L^2(\mu_{\mathbb{D}})$, $j = 1, \ldots, m$, such that the technical condition

$$\int \left| \frac{Ug_j}{\rho} \left\{ (UQ)(\bullet, X) - Uq \right\} \right|^2 d\mu_{\mathbb{S}} < \infty, \qquad j = 1, \ldots, m, \tag{2.35}$$

is fulfilled. Under this condition, the random variables

$$T_n := \frac{1}{\sqrt{n}} \sum_{k=1}^{n} \left\{ \sum_{j=1}^{m} \lambda_j \int_{\mathbb{S}} \frac{\overline{Ug_j}}{\rho} (U\hat{q}_k - Uq) \, d\mu_{\mathbb{S}} \right\}, \tag{2.36}$$

with $\lambda := (\lambda_1, \ldots, \lambda_m)^* \in \mathbb{R}^m$, are well defined and the central limit theorem in its simplest form yields that

$$T_n \Rightarrow \mathcal{N}(0, \lambda^* \Delta \lambda), \qquad \text{as } n \to \infty, \tag{2.37}$$

where the $m \times m$ matrix Δ has elements

$$\Delta_{jl} := \int_{\mathbb{S}} \int_{\mathbb{S}} \frac{\overline{Ug_j(s)}}{\rho(s)} \operatorname{Cov}((UQ)(s, X), (UQ)(t, X)) \frac{Ug_l(t)}{\rho(t)} \, d\mu_{\mathbb{S}}(s) \, d\mu_{\mathbb{S}}(t). \tag{2.38}$$

Now choose any numbers $\alpha(n) \downarrow 0$, as $n \to \infty$. The condition in Formula (2.35) entails that we may apply dominated convergence to see that

$$\mathrm{Var}\left(T_n - \sqrt{n} \sum_{j=1}^{m} \lambda_j \langle \hat{\theta}_{\alpha(n)} - \theta_{\alpha(n)}, g_j \rangle \right) \to 0, \qquad \text{as} \quad n \to \infty. \tag{2.39}$$

Since these variables are centered at 0 they converge to 0 in probability, so that we have the following result.

THEOREM 2.6   Under assumption (2.35) it follows that

$$\sqrt{n}(\langle \hat{\theta}_{\alpha(n)} - \theta, g_1 \rangle, \dots, \langle \hat{\theta}_{\alpha(n)} - \theta, g_m \rangle) \Rightarrow \mathscr{N}(0, \Delta), \qquad \text{as} \quad n \to \infty, \tag{2.40}$$

provided that $\alpha(n) \downarrow 0$, as $n \to \infty$, at such a rate that

$$\sqrt{n}(\theta_{\alpha(n)} - \theta) \to 0, \qquad \text{as} \quad n \to \infty. \tag{2.41}$$

If the condition in Formula (2.35) is violated there can still be asymptotic normality and convergence rates different from $\sqrt{n}$ may show up. Asymptotic efficiency in the sense of Hájek–LeCam of a single linear functional $\langle \hat{\theta}_{\alpha(n)}, g \rangle$ as an estimator of $\langle \theta, g \rangle$ is another topic of interest. For such considerations, however, more specific assumptions have to be made like, in particular, that we precondition with $\Gamma = K^*$. In Sec. 3 we will return to these questions.

3.   INDIRECT CURVE ESTIMATION UNDER STANDARD PRECONDITIONING

In this section the generality will be reduced in two respects: in the first place we will precondition with K^*, which is in a certain sense the standard way to do, and in the second place we will restrict to the indirect density and regression estimation models. The vast majority of the models are of this type and Wicksell's unfolding problem is one of the rare examples where preconditioning with K^* does not yield a tractable operator.

In the present situation $R = \Gamma K = K^*K$ and the polar decomposition, Riesz and Sz-Nagy (1990) entails that there exists a partial isometry $V : L^2(\mu_\mathbb{D}) \to L^2(\mu_\mathbb{E})$ such that

$$K = V|K| = VU^{-1}M_{\sqrt{\rho}}U, \tag{3.1}$$

cf. Eq. (2.1). The operator V maps the range of $|K|$ isometrically onto the range of K.

Inspection of Eqns. (1.9) and (1.16) shows that in either curve estimation model there exist measurable

$$\eta : \mathbb{X} \to \mathbb{R}, \quad \zeta : \mathbb{X} \to \mathbb{E}, \tag{3.2}$$

such that

$$Q(u, X) = \eta(X)K^*(u, \zeta(X)), \qquad u \in \mathbb{D}. \tag{3.3}$$

In the case of indirect density estimation we have

$$\eta(x) = 1, \qquad \zeta(x) = x, \qquad x \in \mathbb{X} = \mathbb{E}, \tag{3.4}$$

and in the indirect regression estimation model

$$\eta(y, z) = \frac{y}{h(z)}, \qquad \zeta(y, z) = z, \qquad x := (y, z) \in \mathbb{R} \times \mathbb{E}. \tag{3.5}$$

Henceforth we will assume that the operator $UV^* : L^2(\mu_\mathbb{E}) \to L^2(\mu_\mathbb{S})$ is an integral operator with kernel satisfying

$$\text{ess sup}_{(v,s) \in \mathbb{E} \times \mathbb{S}} |(UV^*)(v, s)| < \infty. \tag{3.6}$$

In order to ensure that Formula (1.6) holds true we will require that

$$\text{ess sup}_{v \in \mathbb{E}} \int_{\mathbb{D}} K^2(v, u) \, d\mu_\mathbb{D}(u) < \infty, \tag{3.7}$$

i.e. that K satisfies a uniform Carleman condition. Let us define the generalized characteristic function of f by

$$\chi_f(s) := \int \eta(x)(UV^*)(s, \zeta(x)) f(x) \, d\mu_\mathbb{X}(x), \quad s \in \mathbb{S}, \tag{3.8}$$

and let us write

$$\hat{\chi}_X(s) := \eta(X)(UV^*)(s, \zeta(X)), \quad s \in \mathbb{S}. \tag{3.9}$$

Combination of the above easily leads to the following important property of Q.

THEOREM 3.1   Under the present assumptions both in model A and B the kernel satisfies

$$(UQ)(s, X) = \sqrt{\rho(s)}\hat{\chi}_X(s), \quad \text{Var}(UQ)(s, X) \le C\rho(s), \quad s \in \mathbb{S}, \tag{3.10}$$

for some number $0 < C < \infty$.

The expressions in Eq. (3.10) show that the potential extra ill-posedness due to the inversion of K^* in addition to that of K is compensated by the occurrence of the factors $\sqrt{\rho}$ and ρ respectively. The RI-estimators now assume the form

$$\hat{\theta}_\alpha = U^{-1} \frac{1}{\sqrt{\rho}} \mathbf{1}_{\{\rho \ge \alpha\}} \hat{\chi}, \qquad \text{where } \hat{\chi} := \frac{1}{n} \sum_{k=1}^{n} \hat{\chi}_{X_k}, \tag{3.11}$$

cf. Eqns (2.11), (3.9), and (3.10).

COROLLARY 3.2   For some number $0 < C < \infty$ we have

$$\sup_{\theta \in \Theta_\lambda} \mathbf{E}\|\hat{\theta}_\alpha - \theta\|^2 \leq \frac{C}{n} \int_{\{\rho \geq \alpha\}} \rho^{-1} d\mu_\mathbb{S} + \int_{\{\rho < \alpha\}} \lambda^2 d\mu_\mathbb{S}. \tag{3.12}$$

Because the operators K and R are now related in a simple manner, it is also possible to obtain a relation between the upper bound in Eq. (2.16) and the lower bound in Formula (2.27). At this still very general level some technical assumptions will be needed. Since we are more or less free to choose λ we will link it to the function ρ according to

$$\lambda = \Psi(\rho), \tag{3.13}$$

where $\Psi : [0, \infty) \to [0, \infty)$ is continuous and strictly increasing with $\Psi(0) = 0$. Usually one can obtain the functions λ one would like to consider in this manner. To construct a suitable orthonormal system $\{e_m\}$ in $L^2(\mu_\mathbb{D})$ we first construct an orthonormal system in $L^2(\mu_\mathbb{S})$ by choosing disjoint intervals I_m in the real line and defining

$$\varphi_m := \frac{1}{\sqrt{\mu_\mathbb{S}(S_m)}} \mathbf{1}_{S_m}, \quad S_m := \{\lambda \in I_m\}, \qquad m \in \mathbb{N}. \tag{3.14}$$

Then the system

$$e_m := U^{-1}\varphi_m, \quad m \in \mathbb{N}, \tag{3.15}$$

is orthonormal in $L^2(\mu_\mathbb{D})$. If the Fisher information in the mth direction, $\bar{\rho}_m$ (cf. Eq. (2.21)), is suitably related to the values of ρ on S_m it can be shown that

$$\inf_{T \in \mathcal{T}} \sup_{\theta \in \Theta_\lambda} \mathbf{E}\|T - \theta\|^2 \geq \int_\mathbb{S} \frac{\lambda^2}{A + nB\rho\lambda^2} \, d\mu_\mathbb{S}, \tag{3.16}$$

for numbers $A, B \in (0, \infty)$. See van Rooij and Ruymgaart (1995) for details. All such conditions are usually fulfilled and they yield the asymptotic optimality of the RI-estimators in the following sense.

THEOREM 3.3   Under the technical conditions sketched above, numbers $\alpha(n) > 0$ can be found such that the upper bound for the IMSE on the right in Formula (3.12) and the lower bound to the minimax risk on the right in Formula (3.16) converge to 0 at the same rate, as $n \to \infty$.

COROLLARY 3.4   The statistical function $\hat{\mathcal{M}}_n$ in Eq. (2.34) now assumes the form

$$\hat{\mathcal{M}}_n(\alpha) = \frac{1}{n} \sum_{k=1}^n \int_{\{\rho \geq \alpha\}} \frac{1}{\rho} \left\{ \frac{1}{n} |\hat{\chi}_{X_k}|^2 - \frac{n+1}{n} \hat{\chi}_{(k)}\bar{\hat{\chi}}_{X_k} \right\} d\mu_\mathbb{S}, \quad \alpha > 0, \tag{3.17}$$

where $\hat{\chi}_{(k)} = (n-1)^{-1} \sum_{j \neq k} \hat{\chi}_{X_j}$.

For weak convergence of linear functionals we introduce the linear submanifold

$$\mathscr{L} := \left\{ g \in L^2(\mu_{\mathbb{D}}) : \frac{Ug}{\sqrt{\rho}} \in L^2(\mu_{\mathbb{S}}) \right\}, \tag{3.18}$$

and define the covariance function

$$D(s,t) := \mathbf{E}\{\hat{\chi}_X(s) - \chi(s)\}\{\overline{\hat{\chi}_X(t) - \chi(t)}\}, \quad (s,t) \in \mathbb{S} \times \mathbb{S}, \tag{3.19}$$

where χ_f and $\hat{\chi}_X$ are defined in Eqns (3.8) and (3.9), respectively. Because of Eq. (3.10), the condition in Formula (2.35) is clearly satisfied for $g \in \mathscr{L}$ so that the following is immediate from Theorem 2.6.

COROLLARY 3.5   For $g \in \mathscr{L}$ we have

$$\sqrt{n}\langle \hat{\theta}_{\alpha(n)} - \theta, g \rangle \Rightarrow \mathcal{N}(0, \Delta^2(g)), \quad \text{as} \quad n \to \infty, \tag{3.20}$$

where $\Delta^2(g) := \int_{\mathbb{S}} \int_{\mathbb{S}} (Ug)(s)\sqrt{\rho}^{-1}(s) D(s,t)(Ug)(t)\sqrt{\rho}^{-1}(t)\, d\mu_{\mathbb{S}}(s)\, d\mu_{\mathbb{S}}(t)$, provided that $\alpha(n) \downarrow 0$, as $n \to \infty$, at such a rate that

$$\sqrt{n}(\theta_{\alpha(n)} - \theta) \to 0. \tag{3.21}$$

Asymptotic efficiency of $\langle \hat{\theta}_{\alpha(n)}, g \rangle$ as an estimator of $\langle \theta, g \rangle$ is within reach but it is again due to the great generality that some technical conditions are required that cannot be given here in full detail. First of all we shall now restrict ourselves to model A of indirect density estimation and in addition assume that

$$K = |K| = U^{-1} M_{\sqrt{\rho}} U. \tag{3.22}$$

These assumptions entail in particular that $\mathbb{X} = \mathbb{D} = \mathbb{E}$ and

$$\hat{\chi}_X = U(\bullet, X), \quad (UQ)(\bullet, X) = \sqrt{\rho} U(\bullet, X). \tag{3.23}$$

In this case $\{K\theta : \theta \in \Theta\}$ is a family of densities and we set

$$\mathscr{P} := \{\sqrt{K\theta} : \theta \in \Theta\} \subset L^2(\mu_{\mathbb{D}}). \tag{3.24}$$

For a p that is positive a.e. on $\mathbb{X}$, the tangent space to $\mathscr{P}$ at $\sqrt{p}$ will be denoted by $\mathbb{T}_{\sqrt{p}}(\mathscr{P})$. Under a technical condition (which is often fulfilled) it can be shown that

$$\mathbb{T}_{\sqrt{p}}(\mathscr{P}) = [\sqrt{p}]^{\perp}. \tag{3.25}$$

The functional that we are interested in can be written $T(\sqrt{p}) := \langle \theta, g \rangle = \langle K^{-1}p, g \rangle$, for some $g \in \mathscr{L}$. The derivative of this functional at $\sqrt{p}$

is the linear functional

$$\psi \mapsto \langle \psi, 2\sqrt{p}\, U^{-1}(\sqrt{\rho}^{-1} Ug)\rangle, \qquad \psi \in \mathbb{T}_{\sqrt{p}}(\mathscr{P}), \tag{3.26}$$

where the inner product is that of the ambient space $L^2(\mu_{\mathbb{D}})$.

The Hájek–LeCam convolution theorem yields that any regular sequence of estimators of the functional $T : \mathscr{P} \to \mathbb{R}$ has a limiting distribution of the form

$$\mathscr{N}(0, \|\sqrt{p}\, U^{-1}(\sqrt{\rho}^{-1} Ug) - \langle \sqrt{p}\, U^{-1}(\sqrt{\rho}^{-1} Ug), \sqrt{p}\rangle \sqrt{p}\|^2) * \mathscr{D}, \tag{3.27}$$

where $\mathscr{D}$ is a probability distribution on the real line, because of Eq. (3.25), cf. van der Vaart (1988). The norm and inner product are again those of $L^2(\mu_{\mathbb{D}})$. Straightforward calculation shows that the normal distribution in Formula (3.20) has variance satisfying

$$\Delta^2(g) = \|\sqrt{p}\, U^{-1}(\sqrt{\rho}^{-1} Ug) - \langle \sqrt{p}\, U^{-1}(\sqrt{\rho}^{-1} Ug), \sqrt{p}\rangle \sqrt{p}\|^2. \tag{3.28}$$

THEOREM 3.6   Let us focus on model A and assume Formula (3.21). Under suitable regularity conditions explained above it is possible to choose $\alpha(n) \downarrow 0$ in such a way that the statistics $\langle \hat{\theta}_{\alpha(n)}, g\rangle$ are efficient estimators of $\langle \theta, g\rangle$, $g \in \mathscr{L}$, in the sense that their asymptotic variance $\Delta^2(g)$ is optimal according to the Hájek–Le Cam convolution theorem.

4.   COMPACT OPERATORS; EXAMPLES

Throughout this section we will assume that K is compact, and precondition with $\Gamma = K^*$. The first results of the general theory will be illustrated for model B, assuming Formula (1.11), and the last results for model A. Since $R := K^*K$ will be also compact there exist an orthonormal basis $e_1, e_2, \ldots$ of eigenfunctions and corresponding eigenvalues $\rho_1 \geq \rho_2 \geq \ldots \downarrow 0$, each one repeated according to its finite multiplicity. The unitary operator U that reduces R to multiplication maps $L^2(\mu_{\mathbb{D}})$ onto $L^2(\mathbb{N}, \mathscr{P}(\mathbb{N}), \#) = \ell^2$ and has kernel

$$U(m, u) := \overline{e_m(u)}, \qquad (m, u) \in \mathbb{N} \times \mathbb{D}. \tag{4.1}$$

Note that $Ug = (\langle g, e_1\rangle, \langle g, e_2\rangle, \ldots), g \in L^2(\mu_{\mathbb{D}})$, and that

$$R = U^{-1} M_\rho U, \qquad \rho := (\rho_1, \rho_2, \ldots) \in \ell^\infty, \tag{4.2}$$

where $M_\rho \varphi = (\rho_1 \varphi_1, \rho_2 \varphi_2, \ldots), \varphi := (\varphi_1, \varphi_2, \ldots) \in \ell^2$.   Since $(UV^*)h = (\langle V^*h, e_1\rangle, \langle V^*h, e_2\rangle, \ldots) = (\langle h, Ve_1\rangle, \langle h, Ve_2\rangle, \ldots), h \in L^2(\mu_{\mathbb{E}})$ it follows that UV^* has kernel

$$(UV^*)(m, v) = \overline{(Ve_m)(v)}, \qquad (m, v) \in \mathbb{N} \times \mathbb{E}. \tag{4.3}$$

We will assume that all kernels satisfy the conditions formulated in the previous Sections.

From the above it is immediate that

$$\hat{\chi}_X(m) = \frac{Y}{h(Z)} \overline{(Ve_m)(Z)}, \qquad m \in \mathbb{N}, \tag{4.4}$$

cf. Eqns (3.5) and (3.9), which according to Eq. (3.11) entails that

$$\hat{\theta}_\alpha(u) = \sum_{m:\rho_m \geq \alpha} \frac{1}{\sqrt{\rho_m}} \hat{\chi}(m) e_m(u), \qquad u \in \mathbb{D}, \tag{4.5}$$

where $\hat{\chi}$ is the generalized empirical characteristic function.

It is now possible to show optimality of the rate of convergence of the IMSE under simple conditions. Recall that w is the density of the error ε in Eq. (1.10) and that we might take $\Theta = L^2(\mu_\mathbb{D})$. The joint density of (Y, Z) equals

$$f(y, z) = w(y - (K\theta)(z))h(z), \qquad (y, z) \in \mathbb{R} \times \mathbb{E}, \theta \in \Theta. \tag{4.6}$$

Letting

$$w_t(x) := w(x - t), \qquad x \in \mathbb{R}, t \in \mathbb{R}, \tag{4.7}$$

we will assume that there exists $v \in L^2(\mathbb{R})$ such that

$$\left\| \frac{1}{h} \left(\sqrt{w_{t+h}} - \sqrt{w_t} \right) - v \right\| \to 0, \qquad \text{as} \quad h \to 0. \tag{4.8}$$

Let us choose the submodel

$$\Theta_\lambda := \{ \theta \in L^2(\mu_\mathbb{D}) : |\langle \theta, e_m \rangle| \leq \lambda_m, \quad m \in \mathbb{N} \}, \tag{4.9}$$

for some $\lambda := (\lambda_1, \lambda_2, \ldots) \in \ell^2$ with $\lambda_m \geq 0$ for all m, where in this case $e_1, e_2, \ldots$ is the basis of eigenfunctions of R. Note that here $\Theta_\lambda = \Theta_{0,\lambda} = \Theta_{1,\lambda}$, see Eq. (2.17). It follows that

$$\left(\frac{\partial}{\partial t_m} \sqrt{f_{t,t_m}} \right)(y, z) = v(y - (K\theta)(z))\sqrt{\rho_m}(Ve_m)(z)h(z), \tag{4.10}$$

where we identify θ and $t := (\langle \theta, e_1 \rangle, \langle \theta, e_2 \rangle, \ldots)$ for $t \in \Theta_\lambda$. Hence the Fisher information in the mth direction (see Eq. (2.21)) satisfies

$$\bar{\rho}_m \leq \|v\|^2 \left\{ \operatorname*{ess\,sup}_{z \in \mathbb{E}} h(z) \right\} \left\{ \operatorname*{ess\,sup}_{m \in \mathbb{N}, z \in \mathbb{E}} |(Ve_m)(z)|^2 \right\} \leq C\rho_m, \tag{4.11}$$

for some $0 < C < \infty$, and hence is directly related to the mth eigenvalue of R. Thus a clear relation between the upper bound and the lower bound of the IMSE can be obtained which will be summarized in the optimality result below. For strictly positive numbers a_m, b_m we write $a_m \asymp b_m$, as $m \to \infty$, to mean that $0 < \inf_{m \in \mathbb{N}} a_m/b_m \leq \sup_{m \in \mathbb{N}} a_m/b_m < \infty$.

THEOREM 4.1 If $\rho_m \approx m^{-a}$, $\lambda_m \approx m^{-b}$, as $m \to \infty$, for some $a > 0$, $b > 1/2$, and provided that we choose $\alpha(n) \approx n^{-a/(a+2b)}$, as $n \to \infty$, it follows that

$$\sup_{\theta \in \Theta_\lambda} \mathbf{E}\|\hat{\theta}_{\alpha(n)} - \theta\|^2 = C\,(n^{-(2b-1)/(2b+a)}), \qquad \text{as} \quad n \to \infty. \tag{4.12}$$

This rate is optimal.

Again, when comparing this rate with rates in the literature one should realize that our submodel is a hyperrectangle. For data-driven parameter selection we simply follow Eq. (3.17) with $\hat{\chi}_X$ defined by Eq. (4.4). Regarding the weak convergence of linear functionals let us return to model A and observe that now

$$\hat{\chi}_X(m) = \overline{(Ve_m)(X)}, \quad m \in \mathbb{N}. \tag{4.13}$$

Because of Formula (1.11) it is not a priori excluded that a $p \in K\Theta$ may be found which is not only bounded away from ∞ but also from 0. When Θ is the family of all probability densities in $L^2(\mu_\mathbb{D})$, this simple condition on p entails that Eq. (3.23) is fulfilled.

THEOREM 4.2 Suppose there exists $p \in K\Theta$ which is bounded away from both 0 and ∞, and let $g \in \mathscr{L}$. Then $\alpha(n) \downarrow 0$ can be found such that the $\langle \hat{\theta}_{\alpha(n)}, g \rangle$ are asymptotically normal with optimal variance $\Delta^2(g)$ at this p.

EXAMPLE The shape of a cable suspended at its endpoints with coordinates $(0,0)$ and $(1,0)$ is given by the differential equation

$$-\frac{d^2 y}{du^2} = f(u), \qquad 0 \le u \le 1, \quad y(0) = y(1) = 0, \tag{4.14}$$

where apart from the sign the source term f represents the load per horizontal distance, Margenau and Murphy (1956). The problem is to recover the weight distribution, i.e., the source term from the data $X_1, \ldots, X_n$ that are independent copies of $X = (Y, Z)$, where

$$Y = y(Z) + \varepsilon. \tag{4.15}$$

Let us assume that Z has the uniform density on $[0, 1]$.

Using the Green's function we can rewrite the differential equation in the form

$$y(v) = \int_0^1 K(v, u) f(u)\, du =: (Kf)(v), \quad 0 \le v \le 1, \tag{4.16}$$

where K turns out to be Hermitian with kernel

$$K(v, u) := \begin{cases} v(1 - u), & 0 \le v \le u \\ u(1 - v), & u \le v \le 1 \end{cases}. \tag{4.17}$$

Preconditioning with $K^* = K$ yields

$$q = K^* y = K^2 f =: Rf, \quad f \in L^2(0, 1). \tag{4.18}$$

The operator R is compact with an orthonormal basis of eigenfunctions

$$e_m(u) := \sqrt{2} \sin m\pi u, \quad 0 \le u \le 1, \quad m \in \mathbb{N}, \tag{4.19}$$

and corresponding eigenvalues

$$\rho_m := \left(\frac{1}{\pi m} \right)^4, \quad m \in \mathbb{N}. \tag{4.20}$$

Clearly the RI-estimator equals ($V = I$ in this case)

$$\hat{f}_\alpha(u) = \sum_{m=1}^{M(\alpha)} (\pi m)^2 \left\{ \frac{1}{n} \sum_{k=1}^{n} Y_k \sin m\pi Z_k \right\} \sin m\pi u, \quad 0 \le u \le 1, \tag{4.21}$$

for suitable $M(\alpha) \in \mathbb{N}$.

5. CONVOLUTION OPERATORS; EXAMPLES

Let $\mathbb{G}$ denote a locally compact commutative topological group which is also supposed to be metrizable and σ-compact, Hewitt and Ross (1963). The addition will be denoted by $\oplus$, the subtraction by $\ominus$. Let $\mu_\mathbb{G}$ denote Haar measure on $\mathbb{G}$. Convolution of $w \in L^1(\mu_\mathbb{G})$ with $\theta \in L^2(\mu_\mathbb{G})$ yields a function $w * \theta \in L^2(\mu_\mathbb{G})$ defined by

$$(w * \theta)(u) := \int w(u \ominus v) \theta(v) \, d\mu_\mathbb{G}(v) =: (K\theta)(u), \quad u \in \mathbb{G}. \tag{5.1}$$

Since w is supposed to be real, the adjoint of K is easily seen to be convolution with $w^*(x) := w(\ominus x)$, $x \in \mathbb{G}$. We consider the equation

$$q = (w^* w) * \theta = K^* K \theta =: R\theta, \tag{5.2}$$

and will construct the RI-estimator of θ in model A.

By $L_\mathbb{C}^2(\mu_\mathbb{G})$ we indicate the space of complex-valued square integrable functions on $\mathbb{G}$, of which $L^2(\mu_\mathbb{G})$ is an $\mathbb{R}$-linear subspace. Similarly, we have the complex space $L_\mathbb{C}^1(\mu_\mathbb{G})$, containing $L^1(\mu_\mathbb{G})$.

A character of $\mathbb{G}$ is a continuous mapping $\gamma : \mathbb{G} \to \mathbb{T} := \{z \in \mathbb{C} : |z| = 1\}$ such that $\gamma(x \oplus y) = \gamma(x) \cdot \gamma(y)$ for $x, y \in \mathbb{G}$. Under pointwise multiplication, these characters form a locally compact commutative group $\tilde{\mathbb{G}}$, say, with Haar measure $\mu_{\tilde{\mathbb{G}}}$. In exceptional cases $\mathbb{G}$ and $\tilde{\mathbb{G}}$ are isomorphic; $\mathbb{G} = \mathbb{R}^d$ is such a case. The Fourier transform of an arbitrary $w \in L^1_{\mathbb{C}}(\mu_{\mathbb{G}})$ is the mapping $\tilde{w} : \tilde{\mathbb{G}} \to \mathbb{C}$ defined by

$$\tilde{w}(\gamma) := \int w(u)\gamma(u)\, d\mu_{\mathbb{G}}(u), \quad \gamma \in \tilde{\mathbb{G}}. \tag{5.3}$$

It can be shown that there exists a unique linear isometry $\mathscr{F}_{\mathbb{G}} : L^2_{\mathbb{C}}(\mu_{\mathbb{G}}) \to L^2_{\mathbb{C}}(\mu_{\tilde{\mathbb{G}}})$, called the Fourier–Plancherel transform, such that

$$\mathscr{F}_{\mathbb{G}}w = c_{\mathbb{G}}\tilde{w}, \quad w \in L^1_{\mathbb{C}}(\mu_{\mathbb{G}}) \cap L^2_{\mathbb{C}}(\mu_{\mathbb{G}}), \tag{5.4}$$

for a certain number $c_{\mathbb{G}}$.

For every $w \in L^1_{\mathbb{C}}(\mu_{\mathbb{G}})$ and $\theta \in L^2_{\mathbb{C}}(\mu_{\mathbb{G}})$ one can define $w * \theta \in L^2_{\mathbb{C}}(\mu_{\mathbb{G}})$ by Eq. (5.1); it turns out that

$$\mathscr{F}_{\mathbb{G}}(w * \theta) = \tilde{w} \cdot \mathscr{F}_{\mathbb{G}}(\theta), \quad w \in L^1_{\mathbb{C}}(\mu_{\mathbb{G}}), \quad \theta \in L^2_{\mathbb{C}}(\mu_{\mathbb{G}}). \tag{5.5}$$

Now let, again, $w \in L^1(\mu_{\mathbb{G}})$ and define the operators K and R by Eqns. (5.1) and (5.2). As usual, we assume K to be injective, which is the case if and only if $\tilde{w} \neq 0$ almost everywhere. It is not difficult to show that $\mathscr{F}_{\mathbb{G}}$ maps $L^2(\mu_{\mathbb{G}})$ onto

$$\mathbb{L} := \{\omega \in L^2_{\mathbb{C}}(\mu_{\mathbb{G}}) : \omega(\gamma^{-1}) = \overline{\omega(\gamma)} \quad \text{for all} \quad \gamma \in \tilde{\mathbb{G}}\}, \tag{5.6}$$

and that R reduces to a multiplication operator $\mathbb{L} \to \mathbb{L}$ according to

$$R\theta = \mathscr{F}_{\mathbb{G}}^{-1} M_{\tilde{r}} \mathscr{F}_{\mathbb{G}}\theta, \quad \theta \in L^2(\mu_{\mathbb{G}}), \text{ where } r := w^* * w. \tag{5.7}$$

$\mathscr{F}_{\mathbb{G}}$ acts on $L^1(\mu_{\mathbb{G}}) \cap L^2(\mu_{\mathbb{G}})$ as an integral operator with kernel

$$U(\gamma, u) = c_{\mathbb{G}}\gamma(u), \quad (\gamma, u) \in \tilde{\mathbb{G}} \times \mathbb{G}. \tag{5.8}$$

If $V : L^2(\mu_{\mathbb{G}}) \to L^2(\mu_{\mathbb{G}})$ is the partial isometry for which $K = V|K|$ it can easily be seen that $V = \mathscr{F}_{\mathbb{G}}^{-1} M_{\tilde{w}/|\tilde{w}|} \mathscr{F}_{\mathbb{G}}$, so that UV^* has kernel

$$(UV^*)(\gamma, u) = c_{\mathbb{G}} \frac{|\tilde{w}(\gamma)|}{\tilde{w}(\gamma)}\, \gamma(u), \quad (\gamma, u) \in \tilde{\mathbb{G}} \times \mathbb{G}, \tag{5.9}$$

when restricted to $L^1(\mu_{\mathbb{G}}) \cap L^2(\mu_{\mathbb{G}})$. Again we will assume that the kernels satisfy the conditions specified in the beginning of Sec. 3.

All ingredients for constructing the estimator are now available. In model A we have

$$\hat{\chi}_X(\gamma) := c_{\mathbb{G}} \frac{|\tilde{w}(\gamma)|}{\tilde{w}(\gamma)}\, \gamma(X), \quad \gamma \in \tilde{\mathbb{G}}, \tag{5.10}$$

which yields

$$\hat{\theta}_\alpha = \mathscr{F}_G^{-1} \frac{1}{\sqrt{\tilde{r}}} \mathbf{1}_{\{\tilde{r} \geq \alpha\}} \frac{|\tilde{w}|}{\tilde{w}} \hat{\chi} = \mathscr{F}_G^{-1} \frac{1}{\tilde{w}} \mathbf{1}_{\{\tilde{r} \geq \alpha\}} \hat{\chi}, \tag{5.11}$$

where $\hat{\chi}$ is the generalized empirical characteristic function.

Topological groups that occur in practice are $\mathbb{R}^d, (0,\infty)^d, \mathbb{Z}^d, \mathbb{T}^d$ with character groups $\mathbb{R}^d, \mathbb{R}^d, \mathbb{T}^d, \mathbb{Z}^d$ respectively $(d \in \mathbb{N})$. Identifying $\mathbb{T}$ with $[0, 2\pi)$, the Fourier–Plancherel transforms are given by

$$(\mathscr{F}_{\mathbb{R}^d} w)(s) = (2\pi)^{-d/2} \int_{\mathbb{R}^d} w(u) e^{i\langle s, u \rangle}\, du, \qquad s \in \mathbb{R}^d, \tag{5.12}$$

$$(\mathscr{F}_{(0,\infty)^d} w)(s) = (2\pi)^{-d/2} \int_{(0,\infty)^d} w(u) \exp\left(i \sum_{j=1}^d s_j \log u_j\right)\left(\prod_{j=1}^d \frac{1}{u_j}\right) du, \quad s \in \mathbb{R}^d, \tag{5.13}$$

$$(\mathscr{F}_{\mathbb{Z}^d} w)(s) = \sum_{n \in \mathbb{Z}^d} w(n) e^{i\langle s, n \rangle}, \quad s \in [0, 2\pi)^d, \tag{5.14}$$

$$(\mathscr{F}_{\mathbb{T}^d} w)(n) = (2\pi)^{-d} \int_{[0,2\pi)^d} w(u) e^{i\langle n, u \rangle}\, du, \quad n \in \mathbb{Z}^d, \tag{5.15}$$

for w in the proper domain.

EXAMPLES An immediate generalization of the univariate errors-in-variables model Carroll and Hall (1988), Zhang (1990), Fan (1991), van Rooij and Ruymgaart (1996) to higher dimension arises when we observe $X_1, \ldots, X_n$ that are independent copies of

$$X = Z + \varepsilon, \tag{5.16}$$

where Z and ε are stochastically independent random vectors in $\mathbb{R}^d (d \in \mathbb{N})$. The error variable ε has a known density $w \in L^1(\mathbb{R}^d) \cap L^2(\mathbb{R}^d)$, and Z has an unknown density $\theta \in L^1(\mathbb{R}^d) \cap L^2(\mathbb{R}^d)$. The density p of X is given by

$$p(x) = \int_{\mathbb{R}^d} w(x - y) \theta(y)\, dy, \quad x \in \mathbb{R}^d. \tag{5.17}$$

It has been observed in Springer (1979) that in many practical situations also products of positive random variables play a role. Extending to $(0,\infty)^d$ under coordinate-wise multiplication we now observe

$$X = Z \cdot \varepsilon, \tag{5.18}$$

which has density

$$p(x) = \int_{(0,\infty)^d} y^{-1} w(xy^{-1})\theta(y)dy$$

$$= \int_{(0,\infty)^d} w(xy^{-1})\theta(y)d\mu(y) = (w * \theta)(x), \quad x \in (0,\infty)^d, \tag{5.19}$$

where μ, defined by $d\mu(y) = y^{-1}dy$, denotes Haar measure on $(0,\infty)^d$.

In directional statistics, Mardia (1972), we observe random variables on the circle. Extending to the torus $\mathbb{T}^d$ with $\oplus$ denoting coordinatewise addition mod 2π the errors-in-variables model assumes the form

$$X = Z \oplus \varepsilon. \tag{5.20}$$

6. TIME SERIES AND SOME OPEN PROBLEMS

In this section we mention a few problems suitable for treatment with the methods we described above. The first is not an inverse problem, but presents another application of Halmos' version of the Spectral Theorem.

Representation of time series

Spectral theory is an important tool to construct regularized inverses of operators. Since spectral theory is also of great importance in time series analysis, it is not surprising that some of the present methodology could also be used in the latter area. The Halmos version of the spectral theorem in Eq. (2.1) may provide an alternative to the Herglotz–Bochner approach to representing time series. We will refrain from comparing these approaches and from specifying regularity conditions and satisfy ourselves with a brief sketch.

Consider a second-order stochastic process $\{X_t : t \in \mathbb{T}\}$, defined on $(\Omega, \mathcal{W}, \mathbf{P})$. Usually, $\mathbb{T}$ will be a space or, more specifically, a locally compact commutative group, and equipped with a σ-finite measure $\mu_{\mathbb{T}}$. Let the covariance function be given by

$$R(t,u) := \mathbf{E}X_t\bar{X}_u, \quad (t,u) \in \mathbb{T} \times \mathbb{T}. \tag{6.1}$$

Assume that this kernel defines a bounded integral operator $R : L^2(\mu_{\mathbb{T}}) \to L^2(\mu_{\mathbb{T}})$, so that it has representation Eq. (2.1). (If, more generally, R is selfadjoint, Eq. (2.1) remains true although now ρ no longer needs to be bounded, and it will be possible to generalize to this case.)

If the unitary operator U also has a kernel one obtains the representation

$$R(t,u) = \int_{\mathbb{S}} \overline{U(s,t)}U(s,u)\rho(s)\,d\mu_{\mathbb{S}}(s), \tag{6.2}$$

for the kernel. For the process this entails the general Karhunen representation

$$X_t = \int U(t,s) \sqrt{\rho(s)} \, dZ(s), \tag{6.3}$$

where Z is an orthonormal process.

It seems that the possibility of representing R in the form of Eq. (6.2) occurs in the literature as a condition for the Karhunen expansion (Grenander (1981, Theorem 2 of Sec. 2.3)). When a spectral density exists this condition is implied by Eq. (2.1). Priestley (1988) assumes representation of Eq. (2.1) in a somewhat more general form to arrive at oscillatory processes, and it might be worthwhile to consider whether some more insight could be gained from the connection with Eq. (2.1).

Prediction

For stationary processes the prediction problem leads to a Wiener–Hopf equation. For some remarks on a stable solution of this equation see Sec. 7.

Indirect regression with dependent errors

Let us formulate this problem in the continuous time setting and suppose we observe a stochastic process

$$Y(t) = (K\theta)(t) + N(t), \quad \text{or} \quad Y(t) = (K(\theta + N))(t), \quad a \le t \le b, \tag{6.4}$$

where N is a zero mean process with known covariance operator C, say. Like in ordinary finite dimensional regression with dependent errors, recovery of θ will involve regularized inversion of both K and C. Here progress has been made in the literature using Donoho's (1995) seminal paper: see Wang (1997).

Inverse filtering

Consider a linear process $\xi_k = \sum_{j=-\infty}^{\infty} a_{k-j} \varepsilon_j$, where the ε_j are iid innovations and suppose this process (the input) passes through a linear filter to yield the observable output process $X_k = \sum_{j=-\infty}^{\infty} b_{k-j} \xi_j$. Given a sample $X_1, \ldots, X_n$ from the output we wish to recover the input process. When we choose $a_0 = 1, a_m = 0$ otherwise, the problem reduces to recovering the innovations. Although the random vector $(\xi_j)_{j \in \mathbb{Z}}$ is not an element of ℓ^2, this problem is related to deconvolution.

7. SOME FURTHER TOPICS AND OPEN PROBLEMS

Prior information: smoothness

Let us consider a self-adjoint operator with dense domain $H \subset L^2(\mu_{\mathbb{D}})$ and spectral representation $L = U^{-1} M_\ell U$, for some measurable $\ell : \mathbb{S} \to [1, \infty)$. For arbitrary $\nu \geq 0$ we define the operator $L^\nu := U^{-1} M_{\ell^\nu} U$. The domain $H_\nu \subset L^2(\mu_{\mathbb{D}})$ of this operator, equipped with the inner product $\langle f, g \rangle_\nu = \langle L^\nu f, L^\nu g \rangle$, becomes a Hilbert space in its own right that will be denoted $\mathbb{H}_\nu$. The family of Hilbert spaces $\{\mathbb{H}_\nu, \nu \geq 0\}$ with $\mathbb{H}_0 = L^2(\mu_{\mathbb{D}})$ is called a Hilbert scale and may be used to conveniently accommodate prior information about the input signal θ. In most applications prior information is available regarding the degree of smoothness of θ and the Hilbert scale will be a family of Sobolev spaces.

If we know a priori that $\Theta \subset H_\nu$ we may prefer to consider Θ with the inner product of $\mathbb{H}_\nu$, in which case the restriction of K to H_ν will be considered as an operator on $\mathbb{H}_\nu$. This restriction, which is still bounded, will be denoted by K_ν. We will precondition with K_ν^* and write $R_\nu := K_\nu^* K_\nu : \mathbb{H}_\nu \to \mathbb{H}_\nu$. If we assume that $R := K^* K = U^{-1} M_\rho U$ with the same unitary operator U as used for L above it can be shown that

$$R_\nu = U_\nu^{-1} M_{\rho_\nu} U_\nu, \quad \rho_\nu := \ell^{-2\nu} \rho, \tag{7.1}$$

where $U_\nu := U L^\nu : \mathbb{H}_\nu \to L^2(\mu_{\mathbb{S}})$ is unitary. A regularized inverse is

$$R_{\nu,\alpha}^{-1} := U_\nu^{-1} M_{(1/\rho_\nu) \mathbf{1}_{\{\rho_\nu \geq \alpha\}}} U_\nu. \tag{7.2}$$

In the curve estimation models A and B we now have

$$(U_\nu Q)(s, X) = \eta(X) \ell^{-\nu}(s) \sqrt{\rho(s)} (UV^*)(s, \zeta(X)), \qquad s \in \mathbb{S}, \tag{7.3}$$

cf. Eq. (3.10), which yields the estimators

$$\hat{\theta}_{\nu,\alpha} := U_\nu^{-1} \frac{1}{\sqrt{\rho_\nu}} \mathbf{1}_{\{\rho_\nu \geq \alpha\}} \hat{\chi}, \tag{7.4}$$

where $\hat{\chi} = n^{-1} \sum_{k=1}^n \eta(X_k)(UV^*)(\bullet, \zeta(X_k))$ as in Eq. (3.11). For details and examples we refer to Mair and Ruymgaart (1996).

Prior information: irregularity and other metrics

It may, on the other hand, be a priori expected that the unknown input signal θ is irregular. Let us restrict this brief discussion to noisy deconvolution on the real line and assume that θ is in a suitably defined class of functions having discontinuities of the first kind. The bias term tells us much of the story. We have seen in Eq. (2.12) that the bias term depends on the operator R only via $\{\rho \geq \alpha\}$. In the case of deconvolution this set equals $\{\tilde{r} \geq \alpha\} = [-A, A]$ for some $0 < A := A(\alpha) < \infty$ if $\tilde{r}$ is strictly decreasing on $[0, \infty)$.

Therefore the expectation of the estimator reduces to

$$\mathscr{F}^{-1} \mathbb{1}_{[-A,A]} \mathscr{F} \theta =: \mathscr{F}_A^{-1} \mathscr{F} \theta =: I_A \theta, \qquad (7.5)$$

where the operator I_A is an approximation of the identity with kernel

$$I_A(x, y) = \frac{A}{\pi} \operatorname{sinc} A(x - y), \qquad x, y \in \mathbb{R}. \qquad (7.6)$$

This ordinary Fourier approximation is known to suffer from the Gibbs phenomenon near the discontinuity points of θ. To remedy this, one could introduce an averaging procedure which replaces $\mathscr{F}_A^{-1}$ with $\bar{\mathscr{F}}_A^{-1}$ and I_A with $\bar{I}_A$ where the last operator has kernel

$$\bar{I}_A(x, y) = \frac{A}{2\pi} \left\{ \operatorname{sinc} \frac{A}{2} (x - y) \right\}^2, \qquad x, y \in \mathbb{R}. \qquad (7.7)$$

This Fourier–Cesàro approximation is no longer corrupted by the Gibbs phenomenon.

Since integral norms are not sufficiently sensitive to detect the Gibbs phenomenon, in van Rooij and Ruymgaart (1997) the Hausdorff distance $d_{\mathscr{H}}$, say, between the closed extended graphs of two functions is used as a metric. Of course $d_{\mathscr{H}}(I_A \theta, \theta)$ does not converge at all, as $A \to \infty$, but it turns out that

$$d_{\mathscr{H}}(\bar{I}_A \theta, \theta) = \mathcal{O}\left(\sqrt{\frac{1}{A}} \right), \qquad \text{as } A \to \infty. \qquad (7.8)$$

In the statistical literature there is some controversy about which norm to use, Devroye and Györfi (1985). Recently Marron and Tsybakov (1995) advocated the use of metrics like $d_{\mathscr{H}}$ above. In Chandrawansa et al. (1996) the signal θ is recovered from convolution with w (cf. Eqns (5.2), (5.5), (5.9)) by means of

$$\hat{\hat{\theta}}_\alpha := \bar{\mathscr{F}}_A^{-1} \frac{1}{\tilde{w}} \hat{\chi}, \quad \hat{\chi}(t) = \frac{1}{n\sqrt{2\pi}} \sum_{k=1}^{n} e^{itX_k}, \qquad t \in \mathbb{R}, \qquad (7.9)$$

assuming model A. A speed of a.s. convergence of $d_{\mathscr{H}}(\hat{\hat{\theta}}_\alpha, \theta)$ is obtained under further specification of r. An open problem is to determine the optimal rate of this convergence.

Restriction of operators: the Wiener–Hopf equation

For a convenient description let $K = |K| : L^2(\mu_{\mathbb{D}}) \to L^2(\mu_{\mathbb{D}})$. In practice it often occurs that, due to experimental limitations, both input and output need to be restricted to a subset $\mathbb{D}_0$ of $\mathbb{D}$. For the model this means that we are dealing with

$$p = PKP\theta, \quad \theta \in L^2(\mu_\mathbb{D}), \tag{7.10}$$

where P is the orthogonal projection of $L^2(\mu_\mathbb{D})$ onto $L^2(\mu_{\mathbb{D}_0})$, and $\mu_{\mathbb{D}_0}$ the restriction of $\mu_\mathbb{D}$ to $\mathbb{D}_0$. The problem is now to recover $P\theta$ rather than θ from the data that refer to p.

A famous instance of this situation is the Wiener–Hopf equation

$$p(x) = \int_0^\infty r(x - y)\theta(y)\, dy, \quad 0 \le x < \infty, \quad \theta \in L^2(\mathbb{R}). \tag{7.11}$$

This problem plays a role in the prediction of time series. Priestley (1988), and the inverse problem of spectroscopy. Vapnik (1982). A standard reference is Gohberg and Fel'dman (1974).

When the Wiener–Hopf equation is noisy, regularization of the inverse becomes pertinent. Although there are still mathematical problems left, some progress is made by discretization. One chooses an orthonormal system

$$\varphi_{N,k} := \sqrt{N}\mathbf{1}_{[(2k-1)/2N,(2k+1)/2N]}, \quad k \in \mathbb{Z}. \tag{7.12}$$

This is a system of generalized eigenfunctions of the multiplication operator $M_{\tilde{r}}$ and the $e_{N,k} := \mathscr{F}_\mathbb{R}^{-1}\varphi_{N,k}$ form such a system for R with generalized eigenvalues $\tilde{r}(k/N)$. This means that

$$PRP \approx \sum_{k \in \mathbb{Z}} \tilde{r}\left(\frac{k}{N}\right) Pe_{N,k} \otimes \overline{Pe_{N,k}}. \tag{7.13}$$

The matrix $\Lambda_{m,k} := \langle Pe_{N,m}, Pe_{N,k}\rangle$ does not depend on N. It is not a diagonal matrix, however, but of Toeplitz form and reduces to multiplication with

$$\tilde{\lambda}(t) = \left(\frac{\sin \pi t}{\pi}\right)^2 \sum_{k=0}^\infty \frac{1}{(k+t)^2}, \quad 0 \le t < 1, \tag{7.14}$$

by application of $\mathscr{F}_\mathbb{Z}$, see Eq. (5.14). Hence, for Λ a regularized inverse can be specified and, of course, also for the diagonal part given by the $\tilde{r}(k/N)$. The problem is to combine these two regularized inverses and prove that a sufficiently large N exists such that a bounded approximation of $(PRP)^{-1}$ on $L^2(0,\infty)$ can be obtained. An alternative approach might be derived from Donoho (1995).

The general question of how the spectral properties of the unrestricted K could be used to gain insight into the restricted operator is an area where much research remains to be done.

Further open problems

Various open problems still exist in the following areas:

1. Indirect regression with dependent errors (the errors might, e.g., form an ARMA-process);
2. Indirect semiparametric models, e.g., the recovery of a point of discontinuity or any "change-point" in an indirectly observed regression function, see Wang (1995), and indirectly observed mixtures;
3. Construction of confidence intervals and tests of hypotheses including model tests, robustness, and prediction for indirect nonparametric regression;
4. Inverse estimation under order restrictions;
5. Models with nonlinear operators;
6. Deconvolution type models for homogeneous spaces;
7. Recovering the innovations when a sample from a linear process is given, and other inverse problems in time series analysis.

Besides these questions of general interest there are many specific models that require further exploration along the lines sketched in this chapter.

REFERENCES

Beran, R. (1977). Minimum Hellinger distance estimates for parametric models. *Ann. Statist. 5*: 445–463.

Bertero, M. (1989). Linear inverse and ill-posed problems. *Advances in Electronics and Electron Physics 75*: 1–120.

Bickel, P. J. and Rosenblatt, M. (1973). On some global measures of the deviations of density function estimates. *Ann. Statist. 1*: 1071–1095.

Bowman, A. W. (1984). An alternative method of cross-validation for the smoothing of density estimates. *Biometrika 71*: 353–360.

Bunke, O. (1975). Minimax linear, ridge and shrunken estimators for linear parameters. *Math. Operationsforschung Statistik 6*: 697–701.

Carroll, R. J. and Hall, P. (1988). Optimal rates of convergence for deconvolving a density. *J. Amer. Statist. Assoc. 83*: 1184–1186.

Chandrawansa, K., van Rooij, A. C. M. and Ruymgaart, F. H. (1996). Speed of convergence in the Hausdorff metric for estimators of irregular mixing densities. *J. Nonpar. Statist.,* to appear.

Chauveau, D. E., van Rooij, A. C. M. and Ruymgaart, F. H. (1994). Regularized inversion of noisy Laplace transforms. *Adv. Appl. Math.* *15*: 186–201.

Devroye, L. P. (1989). Consistent deconvolution in density estimation. *Canad. J. Statist.* *17*: 235–239.

Devroye, L. P. and Györfi, L. (1985). Nonparametric Density Estimation: the L_1-View. Wiley, New York.

Dey, A. K., Mair, B. A. and Ruymgaart, F. H. (1996). Cross-validation for parameter selection in inverse estimation. *Scand. J. Statist.* *23*: 609–620.

Dey, A. K., Martin, C. F. and Ruymgaart, F. H. (1998). Input recovery from noisy output data, using regularized inversion of the Laplace transform. *IEEE Trans. Inf. Th.*, *44:* 1125–1130.

Dey, A. K. and Ruymgaart, F. H. (1998). Direct curve estimation as an ill-posed inverse estimation problem. *Statist. Neerl.*, to appear.

Donoho, D. L. (1995). Nonlinear solutions of linear inverse problems by wavelet-vaguelette decomposition. *Appl. Comp. Harmon. Anal.* *2*: 101–126.

Donoho, D. L., Liu, R. C. and MacGibbon, B. (1990). Minimax risk over hyperrectangles and implications. *Ann. Statist.* *18*: 1416–1437.

Efromovich, S. (1996). Efficient nonparametric deconvolution. *Ann. Statist.*, to appear.

Fan, J. (1991). Global behavior of deconvolution kernel estimates. *Statist. Sin.* *1*: 541–551.

Fan, J. and Marron, J. S. (1992). Best possible constant for bandwidth selection. *Ann. Statist.* *20*: 2057–2070.

Gill, R. D. and Levit, B. Y. (1995). Applications of the van Trees inequality: a Bayesian Cramér–Rao bound. *Bernoulli 1*: 59–79.

Gilliam, D. S., Schulenberger, J. R. and Lund, J. L. (1988). Spectral representation of the Laplace and Stieltjes transforms. *Mat. Aplic. Comp.* *7*: 101–107.

Gohberg, I. C. and Fel'dman, I. A. (1974). Convolution Equations and Projection Methods for their Solution. Translations Math. Mon. #41, Amer. Math. Soc., Providence.

Grenander, U. (1981). Abstract Inference. Wiley, New York.

Groeneboom, P. and Jongbloed, G. (1995). Isotonic estimation and rates of convergence in Wicksell's problem. *Ann. Statist. 23*: 1518–1542.

Hall, P. (1983). Large sample optimality of least squares cross-validation in density estimation. *Ann. Statist. 11*: 1156–1174.

Hall, P. (1990). Optimal convergence rates in signal recovery. *Ann. Probab. 18*: 887–900.

Hall, P. and Titterington, D. M. (1986). On some smoothing techniques used in image restoration. *J. Roy. Statist. Soc. B, 48*: 330–343.

Halmos, P. R. (1963). What does the spectral theorem say? *Amer. Math. Monthly 70*: 241–247.

Hewitt, E. and Ross, K. A. (1963). Abstract Harmonic Analysis, I. Springer, New York.

Hoerl, A. and Kennard, R. W. (1970a). Ridge regression: biased estimation for nonorthogonal problems. *Technometrics 12*: 55–67.

Hoerl, A. and Kennard, R. W. (1970b). Ridge regression: applications to nonorthogonal problems. *Technometrics 12*: 69–82.

Johnstone, I. M. and Silverman, B. W. (1990). Speed of estimation in positron emission tomography and related inverse problems. *Ann. Statist. 18*: 251–280.

Johnstone, I. M. and Silverman, B. W. (1991). Discretization effects in statistical inverse problems. *J. Complex. 7*: 1–34.

Kac, M. (1966). Can you hear the shape of a drum? *Amer. Math. Monthly 73*: 1–23.

Kimeldorf, G. S. and Wahba, G. (1971). Some results on Tchebycheffian spline functions. *Math. Anal. Appl. 33*: 82–95.

Kress, R. (1989). Linear Integral Equations. Springer, New York.

Mair, B. A. and Ruymgaart, F. H. (1996). Statistical inverse estimation in Hilbert scales. *SIAM J. Appl. Math. 56*: 1424–1444.

Mardia, K. V. (1972). Statistics of Directional Data. Academic Press, New York.

Margenau, H. and Murphy, G. M. (1956). The Mathematics of Physics and Chemistry, 2nd edition. Van Nostrand, Princeton.

Marron, J. S. and Tsybakov, A. B. (1995). Visual error criteria for qualitative smoothing. *J. Amer. Statist. Assoc. 90*: 499–507.

Masry, E. and Rice, J. A. (1990). Gaussian deconvolution via differentiation. Tech. Rep., Depts. El. and Computer Eng. and Math., Univ. of Cal., San Diego.

Nychka, D. and Cox, D. D. (1989). Convergence rates for regularized solutions of integral equations from discrete noisy data. *Ann. Statist. 17*: 556–572.

Nychka, D., Wahba, G., Goldfarb, S. and Pugh, T. (1984). Cross-validated spline methods for the estimation of three-dimensional tumor size distributions from observations on two-dimensional cross sections. *J. Amer. Statist. Assoc. 79*: 832–846.

O'Sullivan, F. (1986). A statistical perspective on ill-posed inverse problems. *Statist. Sci. 4*: 502–527.

Prakasa Rao, B. L. S. (1987). Asymptotic Theory of Statistical Inference. Wiley, New York.

Priestley, M. B. (1988). Non-linear and Non-stationary Time Series Analysis. Academic Press, New York.

Rao, C. R. and Toutenburg, H. (1995). Linear Models: Least Squares and Alternatives. Springer, New York.

Riesz, F. and Sz.-Nagy, B. (1990). Functional Analysis. Dover, New York.

van Rooij, A. C. M. and Ruymgaart, F. H. (1995). Abstract inverse estimation with application to deconvolution on locally compact Abelian groups. Tech. Rep., Dept. Math., Texas Tech Univ.

van Rooij, A. C. M. and Ruymgaart, F. H. (1996). Asymptotic minimax rates for abstract linear estimators. *J. Statist. Pl. Inf. 53*: 389–402.

van Rooij, A. C. M. and Ruymgaart, F. H. (1997). Hausdorff distance and Fourier integrals. Tech. Rep. #9703, Dept. Math., Katholieke Universiteit Nijmegen.

Rudemo, M. (1982). Empirical choice of histograms and kernel density estimators. *Scand. J. Statist. 9*: 65–78.

Ruymgaart, F. H. (1998). A note on weak convergence of density estimators in Hilbert spaces. *Statistics 30*: 331–343.

Silverman, B. W., Jones, M. C., Wilson, J. D. and Nychka, D. W. (1990). A smoothed EM approach to indirect estimation problems, with particular reference to stereology and emission tomography (with discussion). *J. Roy. Statist. Soc. B 52*: 271–324.

Springer, M. D. (1979). The Algebra of Random Variables. Wiley, New York.

Stefanski, L. A. and Carroll, R. J. (1991). Deconvolution-based score tests in measurement error models. *Ann. Statist. 19*: 249–259.

Tikhonov, A. N. and Arsenin, V. Y. (1977). Solutions of Ill-Posed Problems. Wiley, New York.

Titterington, D. M. (1985). Common structure of smoothing techniques in statistics. *Int. Statist. Rev. 53*: 141–170.

Titterington, D. M. (1991). Choosing the regularization parameter in image reconstruction. In: Spatial Statistics and Imaging (A. Possolo, Ed.). IMS Lecture Notes #20, pp. 392–402.

Toutenburg, H. (1982). Prior Information in Linear Models. Wiley, New York.

van der Vaart, A. W. (1988). Statistical Estimation in Large Parameter Spaces. CWI Tract #44, Centre for Mathematics and Computer Science, Amsterdam.

Vapnik, V. (1982). Estimation of Dependences Based on Empirical Data. Springer, New York.

Vardi, Y., Shepp, L. A. and Kaufman, L. (1985). A statistical model for positron emission tomography (with comments). *J. Amer. Statist. Assoc. 80*: 8–37.

Vardi, Y. and Lee, D. (1993). From image deblurring to optimal investments: maximum likelihood solutions to positive linear inverse problems. *J. Roy. Statist. Soc. B, 55*: 569–612.

Wahba, G. (1977). Practical approximate solutions to linear operator equations when the data are noisy. *SIAM J. Numer. Anal. 14*: 651–667.

Wang, Y. (1995). Change-points via wavelets for indirect data. Tech. Report, Univ. Missouri-Columbia.

Wang, Y. (1997). Minimax estimation via wavelets for indirect long-memory data. *J. Stat. Pl. Inf. 64*: 45–55.

Zhang, C.-H. (1990). Fourier methods for estimating mixing densities and distributions. *Ann. Statist. 18*: 806–831.

17

Approaches for Semiparametric Bayesian Regression

ALAN E. GELFAND   University of Connecticut, Storrs, Connecticut

1. INTRODUCTION

Developing regression relationships is a primary inferential activity. We consider such relationships in the context of hierarchical models incorporating linear structure at each stage. Modern statistical work encourages less presumptive, i.e., nonparametric specifications for at least a portion of the modeling. That is, we seek to enrich the class of standard parametric hierarchical models by wandering nonparametrically near (in some sense) the standard class but retaining the linear structure. This enterprise falls within what is referred to as semiparametric modeling.

We focus here on nonparametric modeling of monotone functions associated with the model. Such monotone functions arise, for example, as the stochastic mechanism itself using the cumulative distribution function, as the link function in a generalized linear model, as the cumulative hazard function in survival analysis models, and as the calibration function in errors-in-variables models.

Nonparametric approaches for modeling such functions include discrete mixtures, Dirichlet processes, mixtures of Dirichlet processes, polya tree distributions, gamma processes, extended gamma processes and beta processes. We review these models focusing on key concepts, features and

This work was supported in part by NSF Grant DMS 96-25383.

points, downplaying mathematical detail. We then summarize the applications of these models in the literature to date.

The process of developing a functional form to explain a response variable through a set of explanatory variables is arguably the most common inferential activity in statistical work. In this regard, the simplicity, familiarity and ease of interpretation which accompanies linear relationships encourages the use of linear regression structures possibly on a transformed scale.

The limitations of linear forms are apparent, motivating fully nonparametric approaches to modeling functional form. Such so-called nonparametric curve fitting has an enormous literature by now but is not considered at all here. Rather, an alternative strategy is to add nonparametric bits to familiar parametric models possessing some sort of linear regression structure. That is, we enrich the class of standard parametric models for a given setting by wandering nonparametrically near (in some sense) the standard class. As a result, parts of the modeling are captured parametrically, e.g., at the least the linear structure while other aspects are captured nonparametrically. The entire enterprise falls within what is referred to as semiparametric regression modeling.

Here, we focus on work which takes a Bayesian viewpoint in modeling such semiparametric regression structure. The Bayesian approach to inference is attractive in incorporating prior information into the inference machinery through Bayes's theorem, resulting in a unifying, constructive inference methodology. Historically, nonparametric notions and Bayesian modeling have been viewed as incompatible. The difficulty is that, unlike the parametric case where the dimension of the parameter space is finite, nonparametric modeling requires an "infinite dimensional" parameter whence the Bayesian approach, in assuming all unknowns are random, requires an infinite dimensional stochastic specification. However, in the past twenty-or-so years there has been a considerable body of research which has formulated such specifications. Moreover, while initially such models languished as theoretically attractive but practically hopeless curiosities, recent advances in Bayesian computation enable the fitting of models incorporating these specifications and even extensions of these specifications. It is the objective of this chapter to review this work from a broad and conceptual perspective. The discussion will be current but no attempt is made to be exhaustive in scope or in the extant literature.

More specifically, we work exclusively with conditionally independent hierarchical models as in Kass and Steffey (1989). We consider adding nonparametric aspects to the stochastic mechanism itself using the cumulative distribution function, to the link function in generalized linear models, to the cumulative hazard function in survival analysis models and to the calibration function in errors-in-variables models. Note that all of these functions are monotone.

The various probabilistic specifications which yield Bayesian nonparametric models include discrete mixtures, Dirichlet processes, mixtures of Dirichlet processes, polya tree distributions, gamma processes, extended gamma processes and beta processes. While formally we could incorporate most of these within a so-called independent increment or Lévy process, it is more insightful to examine them individually. Hence, in the sequel, we shall devote a Section to discussion of each of these probability models. This discussion will focus on key concepts, features and points, downplaying rigorous mathematical detail. We then will conclude with a section reviewing their applicability to the modeling mentioned in the previous paragraph.

We conclude this section by noting that the basic object which we are attempting to model in all of the above is an unknown function, say $g(\cdot)$. The parametric Bayesian approach writes g as $g(\cdot\,;\theta)$, $\theta \in \Theta$ and then places a prior distribution over $\theta \in \Theta$. The nonparametric approach assumes only that $g \in \mathscr{G}$ where $\mathscr{G}$ is some class of functions. To be Bayesian requires a prior distribution over the elements of $\mathscr{G}$. In principle, $\mathscr{G}$ might be the class of all functions, all positive functions, all symmetric functions, all bounded functions, all continuous functions, all convex (concave) functions, possibly incorporating constraints, e.g., boundary conditions. As we noted above, the class that we require for all of the foregoing applications is the class of monotone functions. Again, from a Bayesian perspective, we model monotonicity a priori, hence it must emerge a posteriori. This contrasts with monotonic curve fitting procedures such as isotonic regression, Robertson, Wright and Dykstra (1988) or multidimensional scaling, Kruskal (1964).

2. MIXTURE MODELS

It is useful to notice that modeling a strictly monotone function g, is equivalent to modeling a cumulative distribution function. For instance, if the range of g is R^1 then $T(g(\cdot))$ where, e.g., $T(z) = k_1 e^{k_2 z}(1 + k_1 e^{k_2 z})^{-1}$ or $T(z) = 1 - \exp(-k_1 e^{k_2 z})$, with $k_1 > 0$, $k_2 > 0$, is a c.d.f. Here k_1 and k_2 are not model parameters but rather are constants chosen so that the resultant c.d.f. is well-behaved under the transformation, i.e., not increasing essentially to 1 before (to the left of) the practical domain of g, but also not staying essentially at 0 over this domain. Similarly, if the range of g is R^+ then, $T(g(\cdot))$ where $T(z) = k_1 z^{k_2}(1 + k_1 z^{k_2})^{-1}$ or $1 - \exp(-k_1 z^{k_2})$ is a c.d.f. Also, if the range of g is $[a, b]$ then $T(g(\cdot))$ where $T(z) = (z - a)(b - a)^{-1}$ is a c.d.f. Of course, if g is itself a c.d.f., then T is the identity function.

The mixture model approach models an unknown c.d.f. using a dense class of mixtures of standard distributions. For instance, Diaconis and Ylvisaker (1985) observe (though the result is clearly much older) that discrete mixtures of beta densities provide a dense class of models for densities on $[0, 1]$. (Similarly, discrete mixtures of gamma densities provide a dense class of models for densities on R^+.)

In particular, consider

$$T(g(\theta)) = \sum_{\ell=1}^{r} w_\ell IB(T(g_0(\theta)); c_\ell, d_\ell) \tag{1}$$

where r is the number of mixands, $w_\ell \geq 0$, $\sum w_\ell = 1$ are the mixing weights, $IB(u; c, d)$ denotes the incomplete beta function associated with the beta(c, d) density evaluated at u and g_0 is a centering function for g. Then Eq. (1) provides a generic member of the dense class. Inversion of Eq. (1), which is trivial, provides a generic g. Since g is determined by specification of $r, \mathbf{w}^{(r)} = (w_1, \ldots, w_r)$, $\mathbf{c}^{(r)} = (c_1, \ldots, c_r)$ and $\mathbf{d}^{(r)} = (d_1, \ldots, d_r)$, to introduce a distribution on g requires specification of a distribution of the form $f(r) \cdot f(\mathbf{w}^{(r)}, \mathbf{c}^{(r)}, \mathbf{d}^{(r)} | r)$.

If r is allowed to be random, then model dimension changes as r changes. Model fitting with random r could be implemented using reversible jump Markov chain Monte Carlo methods as in Green (1995). As a simpler alternative, Mallick and Gelfand (1994) suggest experimenting with various r, letting $\mathbf{w}^{(r)}$ be random given r but fixing $\mathbf{c}^{(r)}$ and $\mathbf{d}^{(r)}$ to provide a set of beta densities which "blanket" $[0, 1]$, e.g., $c_\ell = \ell$, $d_\ell = r + 1 - \ell$, $\ell = 1, 2, \ldots, r$. If $\mathbf{w}^{(r)}$ is a Dirichlet random variable with distribution $\text{Dir}(\alpha \mathbf{1}_{r \times 1})$ then Mallick and Gelfand show that $ET(g(\theta)) \approx T(g_0(\theta))$, i.e., g is roughly centered about g_0. Of course, implementing the discrete mixture modeling in this fashion is no longer "nonparametric". However, it is still nonparametric in motivation and does enrich the class of models for g near g_0. Also, this class of models is easy to work with in terms of inference. That is, since g is equivalent to $\mathbf{w}^{(r)}$, if we obtain samples from the posterior of $\mathbf{w}^{(r)}$ we can plug these into Eq. (1) to create samples from the posterior of $g(\theta)$ at any fixed θ. Attractively, the sampled curves are all continuous. The average of these curves provides an estimate of $g(\theta)$. At any fixed θ these curves provide realizations from the univariate posterior of g at θ. Again, the mean of these realizations provides an estimator of the posterior mean of g at θ.

3. DIRICHLET PROCESSES

Dirichlet processes have a substantial history in the literature by now. Anticipated in Freedman (1963), they were introduced by Ferguson (1973) with additional clarification provided by Blackwell and MacQueen (1973), Blackwell (1973), Antoniak (1974), Sethuraman and Tiwari (1982), and Sethuraman (1994).

A c.d.f. or equivalently, a probability measure G on Θ is said to follow a Dirichlet process, i.e., $G \sim DP(\alpha G_0)$ if, for any measurable finite partition of Θ, $B_1, B_2, \ldots, B_m$ the joint distribution of the random variables $(G(B_1)), G(B_2), \ldots, G(B_m))$ is $\text{Dir}(\alpha G_0(B_1), \ldots, \alpha G_0(B_m))$. Here G_0 is a specified probability measure and $\alpha > 0$ is a precision parameter. In fact, since $EG(B_\ell) = G_0(B_\ell)$ we express this as "$E(G) = G_0$." Also, since $\text{var}\, G(B_\ell)$ decreases in α, the larger α is the closer we expect G to be to G_0.

A criticism which has been leveled at Dirichlet process models is that they place mass on small class of distribution on Θ. In fact, the Dirichlet process assigns mass 1 to the subset of all discrete distributions on Θ, Ferguson (1973), Blackwell (1973). In particular, suppose we generate two independent sequences of iid random variables, $\{\theta_i\}$ and $\{z_i\}$ such that $\theta_i \sim G_0$ and $z_i \sim Be(1, \alpha)$. Let $w_i = z_i \prod_{j<i}(1 - z_j)$ and

$$G = \sum_{i=1}^{\infty} w_i \delta_{\theta_i}. \tag{2}$$

Here, δ_{θ_i} is a degenerate distribution at θ_i and we can easily show that $\sum w_i = 1$. Hence, G is a discrete probability measure. Then, Eq. (2) characterizes $DP(\alpha G_0)$ in the sense that $G \sim DP(\alpha G_0)$ if and only if almost every realization is of the form of Eq. (2). The realizations $\{w_i\}$ determine an infinite breaking of a stick (say of unit length). We first break off a piece of length w_1, then a piece of length w_2, etc. Freedman (1963) first considered distributions for $\{w_i\}$. An obvious practical approach to generating a G approximately from $DP(\alpha G_0)$ is to fix a large integer J and replace Eq. (2) with G_J, the resulting partial sum up to J. If we let J be random, Doss (1994) shows that we can generate G_J with sufficient accuracy to provide iid draws from G.

Another criticism of the Dirichlet process family of models for G is the negative correlation problem. By the properties of the Dirichlet distribution, for any disjoint B_1 and B_2 if $G \sim DP(\alpha G_0)$ then the correlation between $G(B_1)$ and $G(B_2)$ is negative. In particular, in certain applications we might wish that, say for adjacent intervals, the correlation is positive, e.g., if $P(a_1 < \theta < a_2)$ is large under G we would want $P(a_2 < \theta < a_3)$ to tend to be large under G.

Given a set of realizations $\theta^* = (\theta_1^* \ldots, \theta_n^*)$ from $G \sim DP(\alpha G_0)$, it is convenient that the posterior distribution of G given θ^* is also a Dirichlet process. More precisely, $G|\theta^* \sim DP(\alpha^* G_0^*)$ where $\alpha^* = \alpha + n$ and

$G_0^* = (\alpha + n)^{-1}(\alpha G_0 + \sum \delta_{\theta_i^*})$. Interpreted in a different way, for a new $\boldsymbol{\theta}_0$ from G, $P(\boldsymbol{\theta}_0 \in B|G) \equiv G(B)$ so that, integrating over G, its marginal distribution is G_0. However, given $\boldsymbol{\theta}^*$, $P(\boldsymbol{\theta}_0 \in B|G)|\boldsymbol{\theta}^*) = G(B)|\boldsymbol{\theta}^*$ so that, again integrating over G, its conditional distribution is G_0^*. That is, if $\boldsymbol{\theta}_1^*,\ldots,\boldsymbol{\theta}_n^*$ and $\boldsymbol{\theta}_0^*$ are all drawn independently from G, then $\boldsymbol{\theta}_0^*$ is conditionally independent of $\boldsymbol{\theta}^*$ given G. However, we can "marginalize" over G (whence the $\boldsymbol{\theta}_i^*$ are no longer independent since they shared a common random G) to obtain the conditional distribution of $\boldsymbol{\theta}_0^*$ given $\boldsymbol{\theta}^*$. This distribution is G_0^*. We return to this point in Sec. 4.

The prior and posterior distributions for G arising under a Dirichlet process can be given an illuminating interpretation in terms of a Polya urn scheme model, Blackwell and MacQueen (1973). Recall the set up for a Polya urn scheme. Initially we have an urn containing say M white and $N - M$ red balls. The random variable V is binary taking the value 0 if a red ball is drawn at random, 1 if a white ball is drawn. Then, a priori, $E(V) = M/N \equiv p$. Now if a white ball is drawn it is returned to the urn and an additional white ball is added, similarly if a red ball is drawn. Then

$E(V|\text{white ball drawn}) = (M + 1)/(N + 1)$

$\qquad = \dfrac{Np + 1}{N + 1}, E(V|\text{red ball drawn}) = M/N + 1 = Np/(N + 1).$

By analogy, suppose $G \sim DP(\alpha G_0)$ and we let $V = 1$ if the measurable event B occurs, $V = 0$ if not. Consider the random variable $E(V|G) = G(B)$. A priori, $E(V) = E(G(B)) = G_0(B)$. Now suppose θ_1^* is drawn from G. Then, with

$n = 1, G(B)|\theta_1^* \sim Be(\alpha^* G_0^*(B), (1 - \alpha^*)G_0^*(B))$

yielding

$E(V|\boldsymbol{\theta}_1^* \in B) = E(E(V|G)|\boldsymbol{\theta}_1^* \in B) = E(G(B)|\boldsymbol{\theta}_1^* \in B) = \dfrac{\alpha G_0(B) + 1}{\alpha + 1}$

while

$E(V|\boldsymbol{\theta}_1^* \notin B) = E(G(B)|\boldsymbol{\theta}_1^* \notin B) = \alpha G_0(B)/(\alpha + 1).$

We thus have a Polya urn scheme with $p = G_0(B)$ and $N = \alpha$.

4. MIXTURES OF DIRICHLET PROCESSES

There is a practical difficulty in working with Dirichlet processes. Simulation based model fitting is the customary tool for fitting Bayesian hierarchical models these days. Thus, we must be able to sample model unknowns.

When a parameter is finite dimensional we can do this either using iterative (Markov chain Monte Carlo) or noniterative (importance sampling) methods. With an infinite dimensional parameter such as G, this is impossible (as Eq. (2) reveals). That is, though G can be sampled to arbitrary accuracy (see, e.g., Sethuraman and Tiwari (1982) and the discussion below Eq. (2)) such finite approximation is problematic for simulation work.

For instance, consider the simple semiparametric setting where the observations y_i, $i = 1, \ldots, n$ are conditionally independent with distribution which depends upon a random G as well as covariates $\mathbf{x}_i$ incorporated linearly with unknown coefficient vector $\boldsymbol{\beta}$. (We use this illustrative model for the remainder of this Section.) That is, the likelihood takes the form

$$L(G, \boldsymbol{\beta}) = \prod_{i=1}^{n} f(y_i | G, \mathbf{x}_i^T \boldsymbol{\beta}). \tag{3}$$

As a prior suppose $G \sim DP(\alpha G_0)$ and $\boldsymbol{\beta} \sim f(\boldsymbol{\beta})$. To implement a Gibbs sampler, Gelfand and Smith (1990) for this model requires draws from $f(\boldsymbol{\beta} | G, \mathbf{y})$ and from $f(G | \boldsymbol{\beta}, \mathbf{y})$. If we can not sample G, we can not implement such a sampler.

A convenient remedy is offered by a mixture of Dirichlet processes (MDP) as proposed by Antoniak (1974). Dirichlet process mixing introduces a latent $\boldsymbol{\theta}_i$ for each y_i such that, e.g., in the above example

$$f(y_i | G, \mathbf{x}_i^T \boldsymbol{\beta}) = \int f(y_i | \boldsymbol{\theta}_i, \mathbf{x}_i^T \boldsymbol{\beta}) G(d\boldsymbol{\theta}_i). \tag{4}$$

In Eq. (4) $f(y_i | \boldsymbol{\theta}_i, \mathbf{x}_i^T \boldsymbol{\beta})$ is a standard parametric model while $G(d\boldsymbol{\theta}_i)$ can be usefully thought of as the conditional distribution of $\boldsymbol{\theta}_i$ given G. Integrating over $\boldsymbol{\theta}_i$ yields a marginal distribution for y_i which depends only on G and $\boldsymbol{\beta}$. In other words, the distribution for y_i arises by mixing with respect to G and since G comes from a Dirichlet process we refer to this formulation as Dirichlet process mixing.

With $\boldsymbol{\theta}_i$ iid from G, denoted as $f(\boldsymbol{\theta}_i | G)$ (rather than $G(d\boldsymbol{\theta}_i)$ in order to suggest the conditioning on G), plugging Eq. (4) into Eq. (3) yields the full Bayesian model

$$\prod_{i=1}^{n} f(y_i | \boldsymbol{\theta}_i, \mathbf{x}_i^T \boldsymbol{\beta}) \prod_{i=1}^{n} f(\boldsymbol{\theta}_i | G) f(G | G_0, \alpha) \cdot f(\boldsymbol{\beta}). \tag{5}$$

In Eq. (5) $f(G | G_0, \alpha)$ denotes $G \sim DP(\alpha G_0)$.

Recall from Section 3 that we can marginalize over G in Eq. (5) to obtain the joint distribution of $\boldsymbol{\theta}_1, \ldots, \boldsymbol{\theta}_n$ given G_0 and α. This distribution is built sequentially using appropriate G_0^*'s. That is, we draw θ_1 from G_0, we draw θ_2 given θ_1 from the distribution $(\alpha + 1)^{-1}(\alpha G_0 + \delta_{\theta_1}), \ldots$ up to

$$\theta_n | \theta_1, \ldots, \theta_{n-1} \sim (\alpha + (n-1))^{-1} \left(\alpha G_0 + \sum_{i=1}^{n-1} \delta_{\theta_i} \right).$$

As a result Eq. (5) can be replaced with

$$\prod_{i=1}^{n} f(y_i | \theta_i, \mathbf{x}_i^T \boldsymbol{\beta}) f(\theta_1, \ldots, \theta_n | G_0, \alpha) \cdot f(\boldsymbol{\beta}). \tag{6}$$

Using Dirichlet process mixing, we can run a Gibbs sampler without having to draw G, without having to introduce any approximations. In fact, the full conditional distribution for $\boldsymbol{\beta}$ is proportional to $\prod_{n=1}^{n} f(y_i | \theta_i, \mathbf{x}_i^T \boldsymbol{\beta}) f(\boldsymbol{\beta})$. The full conditional for each θ_i is also easily written down, Escobar and West (1995) i.e., for θ_i it is a distribution with a continuous piece and $n - 1$ atoms, proportional to $f(y_i | \theta_i, \mathbf{x}_i^T \boldsymbol{\beta}) \cdot f(\theta_i | \theta_j, j \neq i)$. Again, $f(\theta_i | \theta_j, j \neq i)$ has a distribution which is of the type G_0^*. Implementation of the Gibbs sampler requires making random draws from these full conditional distributions. When the likelihood, as a function of θ_i, is conjugate with G_0, as a function of θ_i, such sampling is routine as in Escobar and West. In the absence of conjugacy, MacEachern and Muller (1994) provide a clever augmentation strategy for such sampling which is illustrated in Mukhopadhyay and Gelfand (1997).

Note that in Eqns (5) and (6) primary interest is in $\boldsymbol{\beta}$. Introduction of G enriches the parametric formulation for $f(y | \theta, \mathbf{x}'\boldsymbol{\beta})$ in an nonparametric way. The latent θ_i's merely facilitate the model fitting but the marginal posterior distribution of $\boldsymbol{\beta}$ is what is desired. In Eq. (5) G_0 may be specified in terms of hyperparameters, i.e., $G_{0,\lambda}$. In addition, α may be assumed unknown. Hence, hyperpriors for λ and α would be added to the Bayesian specification.

Since G doesn't appear in Eq. (6), it seems as though we can not learn about G, e.g., about features of the posterior of G. In fact, this is not the case. Lo (1984) provides a careful mathematical development (including an explicit expression) for the posterior expectation of any functional of G. These expressions are intractable even for moderate n. Kuo (1986) presents a noniterative Monte Carlo integration approach using a clever importance sampling density. Gibbs sampling to obtain such expectations, as described in Escobar and West (1995), is perhaps more convenient and more flexible.

Gelfand and Mukhopadhyay (1995) show how a routine Monte Carlo integration can be obtained for the posterior expectation of linear functionals of $f(y | G, \mathbf{x}^T \boldsymbol{\beta})$ using the output of a Gibbs sampler run under Eq. (6). Denote such a functional by $H(f | G, \boldsymbol{\beta})$, suppressing $\mathbf{x}$. By the linearity,

$$E(H(f|G,\beta)|y) = \int H(f|G,\beta)f(G,\beta|y)$$

$$= \int\int\int H(f|\theta_0,\beta)\mathbf{f}(\theta_0|\theta)f(\theta|G)f(G|\beta,y)f(\beta|y)$$

$$= \int\int H(f|\theta_0,\beta)f(\theta_0|\theta)f(\theta|\beta,y)f(\beta|y)$$

$$= \int\int H(f|\theta_0,\beta)f(\theta_0|\theta)f(\theta,\beta|y), \tag{7}$$

where $\theta = (\theta_1,\ldots,\theta_n)$ and in the third equality we marginalize over G as usual. From Eq. (7) if $\theta'_\ell, \beta'_\ell, \ell = 1,\ldots,B$ are samples from $f(\theta,\beta|y)$ and $\theta'_{0,\ell}$ is a draw from $f(\theta_0|\theta' = (\theta'_1,\ldots,\theta'_n))$ (implemented, again using the appropriate G^*_0). A Monte Carlo integration for Eq. (7) is

$$B^{-1}\sum_{\ell=1}^{B}H(f|\theta'_{0,\ell},\beta'_\ell). \tag{8}$$

As a concluding comment we note that the Dirichlet process mixing set up is very convenient for prediction. In particular, consider the predictive distribution for new y_0 with covariates $\mathbf{x}_0, f(y_0|\mathbf{x}_0,y)$. Similar to the development of Eq. (7),

$$f(y_0|\mathbf{x}_0,y) = \int f(y_0|G,\mathbf{x}_0^T\beta)f(G,\beta|y)$$

$$= \int f(y_0|\theta_0,\mathbf{x}_0^T\beta)f(\theta_0|\theta)f(\theta,\beta|y). \tag{9}$$

Via Eq. (9), given samples from $f(\theta,\beta|y)$ we can readily sample the predictive distribution hence, examine any of its features. Specifically, if $\theta' = (\theta'_1,\ldots,\theta'_n)$, β' is drawn from $f(\theta,\beta|y)$, θ'_0 from $f(\theta_0|\theta')$ and y'_0 from $f(y_0|\theta'_0,\mathbf{x}_0^T\beta')$, then $y'_0 \sim f(y_0|\mathbf{x}_0,\mathbf{y})$.

5. POLYA TREES

Polya tree distributions provide an attractive generalization of Dirichlet processes. In particular, they can avoid the usual criticism of the Dirichlet process; they can be set up such that the random distribution G belongs to the set of continuous, or even absolutely continuous distributions, with probability 1. To date they have been viewed primarily as theoretical objects, Maulden, Sudderth and Williams, (1992); Lavine (1992, 1994) receiving little practical attention, but see Walker and Mallick (1996).

The basic idea is to develop a binary splitting of Θ, the support of θ. That is, first partition Θ into B_0 and B_1. Then, partition B_0 into B_{00} and B_{01}, partition B_1 into B_{10} and B_{11}, etc. In this way a tree-like structure arises from Θ and we denote the associated collection of sets $\{B_0, B_1, B_{00}, B_{01}, B_{10}, B_{11}, B_{000}, B_{001}, \ldots\}$ by Π. In order to complete the Polya tree distribution specification associated with Π it is necessary to assign a random distribution to the members of Π. This is most conveniently done in a sequential fashion, e.g.,

$$G(B_{1001}) = P(\theta \in B_{1001} | \theta \in B_{100}) \cdot P(\theta \in B_{100} | \theta \in B_{10})$$
$$\cdot P(\theta \in B_{10} | \theta \in B_1) \cdot P(\theta \in B_1).$$

Thus, if

$$P(\theta \in B_0) \equiv v_0 \sim Be(\alpha_0, \alpha_1),$$

if

$$P\theta \in B_{10} | \theta \in B_1) \equiv v_{10} \sim Be(\alpha_{10}, \alpha_{11}),$$

if

$$P(\theta \in B_{100} | \theta \in B_{10}) \equiv v_{100} \sim Be(\alpha_{100}, \alpha_{101}), \ldots,$$

we obtain

$$G(B_{1001}) = (1 - v_{1000}) \cdot v_{100} \cdot v_{10}(1 - v_0).$$

Denoting $\{\alpha_0, \alpha_1, \alpha_{00}, \alpha_{01}, \alpha_{10}, \alpha_{11}, \alpha_{000}, \alpha_{001}, \ldots\}$ by $\mathbf{A}$, the pair $(\Pi, \mathbf{A})$ defines a Polya tree distribution. A realization G from this distribution is, in fact, obtained from the set of v's since the latter yields probability assignments for all sets in Π.

In practice, we can only carry out this process to a finite number of branchings. That is, if we stop at level r, creating 2^r branches we obtain a "partially specified Polya tree", Lavine, (1992) which approximates a realization from $(\Pi, \mathbf{A})$. As a result, if θ is an observation from G having the Polya tree distribution $(\Pi, \mathbf{A})$, under the partially specified tree we can only ask to which of the $2^r B$'s does θ belong. Recall that similar approximation plagues the implementation of the Dirichlet process.

It is possible to center the Polya tree around a particular continuous G_0 on Θ by taking the partitions to align with percentiles of G_0. For instance, if $B_0 = (-\infty, G_0^{-1}(\frac{1}{2})]$ (hence $B_1 = (G_0^{-1}(\frac{1}{2}), \infty))$, if $B_{00} = (-\infty, G_0^{-1}(\frac{1}{4})], \ldots$ and $\alpha_0 = \alpha_1$, $\alpha_{00} = \alpha_{01}, \ldots$, then since $G(B_0) = v_0 \sim Be(\alpha_0, \alpha_0)$, $E(G(B_0)) = \frac{1}{2} = G_0(B_0)$. Also, e.g., $G(B_{00}) = v_0 v_{00}$ implies $E(G(B_{00})) = \frac{1}{4} = G_0(B_{00})$ and in fact, for any $B \in \Pi$, $E(G(B)) = G_0(B)$.

In this setting we can use a dyadic expansion to obtain a realization of θ from G centered at G_0. In particular, suppose the sequence of dependent random variables $\delta_1, \delta_2, \ldots$ such that $\delta_1 = 0$ w.p. $v_0, = 1$ w.p. $1 - v_0$. Then given say $\delta_1 = 0$, $\delta_2 = 0$ w.p. $v_{00}, = 1$ w.p. $1 - v_{00}$, etc. Let $\theta = G_0^{-1}(\sum_{j=1}^{\infty} \delta_j 2^{-j})$. Then θ is the desired realization. In application if we stop the tree at level r, then realizations are obtained with accuracy of measurement 2^{-r}, prior to inversion to the θ scale.

It is apparent that, by adjusting the α's, we need not confine ourselves to quantiles exclusively of the form $G_0^{-1}(j/2^r)$. It is also apparent that we obtain a Dirichlet process in the special case where $\alpha_{00} + \alpha_{01} = \alpha_0$, $\alpha_{10} + \alpha_{11} = \alpha_1$, $\alpha_{000} + \alpha_{001} = \alpha_{00}$, etc.

We see that specification of a Polya tree distribution requires more than that for a Dirichlet process. The former requires the set $\mathbf{A}$, the latter merely a precision and a G_0 from which all the α's in $\mathbf{A}$ can be computed. The Dirichlet process does provide insight into specification of $\mathbf{A}$. Under the Dirichlet process, as we go further down the branching, since the B's are nested, $G_0(B)$ decreases along any branch. Hence the precision associated with the v's decreases as we go further down the branching. As a prior, we obtain less informative specification as level increases, i.e., a posteriori the data exerts more influence on the conditional probabilities as level increases. If this seems unsatisfactory, we might instead choose the α's to increase as level decreases, e.g., $\alpha_{00} > \alpha_0$, $\alpha_{01} > \alpha_0$, etc. (see Walker and Mallick (1996), for further discussion).

Given n draws $\theta_1, \theta_2, \ldots, \theta_n$ from G having Polya tree distribution $(\Pi, \mathbf{A})$ the posterior distribution is again a polya tree distribution paralleling the discussion of Sec. 2. Specifically,

$$P(\theta \in B_0 | \theta_1, \ldots, \theta_n) \sim Be\left(\alpha_0 + \sum 1_{B_0}(\theta_i), \alpha_1 + \sum_{i=1}^{n} 1_{B_1}(\theta_i)\right)$$

so that $EG(B_0) = \alpha_0/(\alpha_0 + \alpha_1)$ is revised to

$$E(G(B_0) | \theta_1, \ldots, \theta_n) = \left(\alpha_0 + \sum_{i=1}^{n} 1_{B_0}(\theta_i)\right) \bigg/ (\alpha_0 + \alpha_1 + n).$$

Similar updating of the α's occurs for, e.g., $P(\theta \in B_{00} | \theta \in B_0, \theta_1, \ldots, \theta_n)$, $P(\theta \in B_{000} | \theta \in B_{00}, \theta_1, \ldots, \theta_n)$, etc., resulting in updated probabilities for each B in Π.

Applied in the context of a hierarchical generalized linear model, for illustration we could set, for $i = 1, 2, \ldots, n$, y_i given μ_i and ϕ_i to be distributed according to a customary one parameter exponential family having mean μ_i with dispersion parameter ϕ_i, McCullagh and Nelder (1989). (Often, as in the binomial and Poisson regression cases, ϕ_i is intrinsically specified.) Then with link function ℓ, we have $\ell(\mu_i) = \eta_i = \mathbf{x}_i^T \boldsymbol{\beta} + \theta_i$. Here we assume $\boldsymbol{\beta} \sim f(\boldsymbol{\beta})$ and that the θ_i are exchangeable according to G with G having a Polya tree distribution $(\Pi, \mathbf{A})$, equivalently $\eta_i - \mathbf{x}_i^T \boldsymbol{\beta} | G, \boldsymbol{\beta} \sim G$.

Assuming $\phi_i = \phi/w_i$ as usual, we can develop a Gibbs sampler to obtain draws from the posterior of $\boldsymbol{\beta}, \phi, \boldsymbol{\eta}$ (equivalently $\boldsymbol{\mu}$) and G. In particular, the only novelty here is that the full conditional distribution for G is obtained by updating the prior polya tree distribution as above with, given $\boldsymbol{\beta}, \phi$ and $\boldsymbol{\eta}$, the set of $\theta_1, \ldots, \theta_n$ such that $\theta_i = \eta_i - \mathbf{x}_i^T \boldsymbol{\beta}$, $i = 1, \ldots, n$.

6. GAMMA PROCESSES

We now return to the problem of modeling the monotonic function g, where the domain of g is assumed to be R^+. In fact, to suggest a time scale we write $g(t)$, $t \in R^+$, thinking of $g(t)$ as a stochastic process which is nondecreasing, nonnegative (so $g(0) = 0$) with independent increments. Such processes are referred to as Lévy processes dating to Lévy (1936).

In this Section we consider a particular Lévy process, the gamma process, as described in Kalbfleisch (1978). Consider an arbitrary finite partition of R^+ say $0 = a_0 < a_1 < a_2, \ldots, < a_k < a_{k+1} = \infty$. For a positive random variable T, let $q_\ell = \Pr(T \in [a_{\ell-1}, a_\ell) | T \geq a_{\ell-1})$ and let $r_\ell = -\log(1 - q_\ell)$. Then, since

$$1 - q_\ell = \Pr(T \geq a_\ell | T \geq a_{\ell-1}), \sum_{\ell=1}^{k} r_\ell = -\log \Pr(T \geq a_k) \equiv g(a_k).$$

Now, suppose we add randomness to the structure such that the r_i are independent. Then, clearly $g(0) = 0$, $g(t)$ increases in t and $g(t)$ arises as a sum of independent increments, i.e., g is a Lévy process.

Note that while the r's connect to g, the q's connect to the underlying distribution for T, i.e., if G is the c.d.f. associated with T, $q_\ell = (G(a_\ell) - G(a_{\ell-1}))/(1 - G(a_{\ell-1}))$. When the r_ℓ's are random, then g is but also G is as well. In fact, we have $G(t) = 1 - e^{-g(t)}$.

This observation helps to suggest a distributional model for the r_ℓ's. Given any c.d.f. G_0, let $g_0(t) = -\log(1 - G_0(t))$. Assume $r_\ell \sim \text{gamma}\,(c(g_0(a_\ell) - g_0(a_{\ell-1})), c)$, with the r_ℓ independent if they arise from disjoint sets in the partition. Then, $E(r_\ell) = g_0(a_\ell) - g_0(a_{\ell-1})$ and $var(r_\ell) = (g_0(a_\ell) - g_0(a_{\ell-1}))/c$. That is, c functions as a precision parameter. We refer to the stochastic process $g(t)$ as a gamma process and use the simplifying notation, $g(t) \sim \text{gamma}\,(cg_0(t), c)$. Since $E(g(t)) = g_0(t)$, we say that g is centered around g_0. Note the parallels of the above structure with that developed for the Dirichlet process.

Of course, modeling a random g is equivalent to modeling a random c.d.f., G. In fact, $g(t) = -\log(1 - G(t))$ is usually referred to as the cumulative hazard, and $g'(t)$ (if it exists) the hazard, yielding $r_\ell = g(a_\ell) - g(a_{\ell-1}) = \int_{a_{\ell-1}}^{a_\ell} g'(s) \, ds$, the hazard cumulated in the interval $[a_{\ell-1}, a_\ell)$, thus making the gamma process natural for use in the context of modeling survival data. Kalbfleisch (1978) notes that g, as well as the associated G, are almost surely discrete (as with the Dirichlet process). In practice, for example, in Clayton (1991), the gamma process is implemented by partitioning R^+ into finite intervals with r_ℓ interpreted as the constant hazard over the ℓth interval. Thus, when say $t \in [a_{\ell-1}, a_\ell)$, $g(t)$ takes the form $\sum_{j=1}^{\ell-1} r_j(a_j - a_{j-1}) + r_\ell(t - a_{\ell-1})$. Obviously, the hazard is a step function in the spirit of Breslow (1974) and, because the heights of the steps arise from independent increments, the hazard will tend to exhibit wild swings raising concerns about its clinical plausibility in applications. Note also that, because independence of the r_ℓ implies independence of the q_ℓ, we necessarily have $\Pr(T \in (a_{\ell-1}, a_\ell) | T \geq a_{\ell-1})$ independent of $\Pr(T \in (a_\ell, a_{\ell+1}) | T \geq a_\ell)$, an assumption which may not be appropriate in certain applications. By comparison, the Dirichlet process models the q_ℓ as independent beta random variables. In fact, q_ℓ corresponds to z_ℓ defined above in Eq. (2), i.e., $z_\ell = w_\ell / (1 - \sum_{j=1}^{\ell-1} w_j)$, whence if $q_\ell \sim \text{Be}(1, \alpha)$, $r_\ell \sim \text{Exp}(\alpha)$. When the random c.d.f. G is modeled through independent q_ℓ's, G is said to be neutral to the right, Doksum (1974).

It is worth noting that other authors have reacted to the problems associated with a hazard arising from independent step heights by suggesting a correlated prior process. That is, a natural way to "smooth" such a hazard is to model the log q_ℓ using a first-order autoregressive process. Gamerman (1991) and, more recently, Fahrmeier (1994) discuss parametric approaches while Arjas and Gasbarra (1994) adopt a nonparametric approach using a so-called martingale process.

In practice, the gamma process is used in conjunction with a Cox proportional hazards model, Cox (1972), writing the probability that the ith individual survives at least to time s as $\Pr(T_i \geq s) = \exp\{-g(s) \exp(\mathbf{x}_i^T \boldsymbol{\beta})\}$. Here g is referred to a the baseline cumulative hazard. Modeling g through the gamma process adds a nonparametric component to the hazard along with the parametric linear regression aspect associated with the covariate contribution. Kalbfleisch (1978) presents an elegant argument which shows that, with a flat prior on $\boldsymbol{\beta}$ and the gamma process specification for the r_ℓ, when c is near 0 (i.e., small precision for g about g_0) the marginal posterior density for $\boldsymbol{\beta}$ is essentially proportional to the customary partial likelihood, Cox (1975), used for making inference about $\boldsymbol{\beta}$, ignoring g.

Alternatively, fully Bayesian implementation of the gamma process prior for g along with a proportional hazards model, using Gibbs sampling is discussed in Clayton (1991). Now r_ℓ is replaced by $r_\ell(\mathbf{x}_i) = r_\ell \exp(\mathbf{x}_i^T \boldsymbol{\beta})$. A convenient approximation to the likelihood contribution for t_i considers $n_\ell^{(i)}$, the number of occurrences of the event for the ith individual in the interval $[a_{\ell-1}, a_\ell)$. Formally, the $n_\ell^{(i)}$ are independent with $n_\ell^{(i)} \sim P_0(r_\ell(\mathbf{x}_i))$ but since $n_\ell^{(i)}$ is, in fact, only 0 or 1, the likelihood associated with $t_i \in [a_{\ell-1}, a_\ell)$ is approximately $r_\ell(\mathbf{x}_i) \exp(\sum_j r_j(\mathbf{x}_i))$. The overall likelihood is thus conveniently conjugate with the independent gamma priors for the r_ℓ.

7. EXTENDED GAMMA PROCESSES

Suppose that $h(t)$ follows a gamma process, i.e., $h(t) \sim \text{gamma}(h_0(t), 1)$. (We can set the precision to 1 here without loss of generality, as we shall see.) That is, $h(b) - h(a) \sim \text{gamma}(h_0(b) - h_0(a), 1)$ for $a < b$ and h_0 is a strictly increasing centering function for h. We extend the gamma process by selecting $\beta(t)$ be a positive, continuous (for convenience) function. Define

$$g(t) = \int_0^t \frac{1}{\beta(s)} \, dh(s), \tag{10}$$

i.e., $g(t)$ arises by integrating $(\beta(s))^{-1}$ with respect to the sample paths of h. The stochastic process $g(t)$ defined by Eq. (9) is called an extended gamma process by Dykstra and Laud (1981). They observe that g is again an increasing independent increments process which they denote by $g(t) \sim \Gamma(h_0(\cdot), \beta(\cdot))$. In the case where $\beta(\cdot)$ is a constant β, $g(t) = h(t)/\beta \sim \text{gamma}(h_0(t), \beta)$ so that we simplify to a gamma process. However, by allowing the scale parameter to vary with time, the extended gamma process allows the modeler more freedom in capturing the uncertainty in the behavior of the hazard rate in each time interval.

One needs to specify $h_0(\cdot)$ and $\beta(\cdot)$ in order to determine an extended Gamma process. If one can offer an a priori guess for the hazard function as well as a function which guesses the amount of variation one might anticipate in the hazard at each t then Dykstra and Laud indicate how h_0 and β can be obtained from these two functions.

By contrast with the gamma process of Section 6, Dykstra and Laud envision g as modeling the hazard function itself. We are, therefore, restricting our modeling to the class of increasing baseline hazards (Damien, Laud and Smith (1995) indicate how this class may be extended to allow a possibly unknown change-point so that g increases before the change and then decreases thereafter.) Thus, the cumulative hazard is $\int_0^t g(s) \, ds$; the cumulative hazard is now absolutely continuous and no longer has independent increments.

In practice, we again seek to use the extended gamma process in conjunction with a Cox proportional hazards model, now writing $\Pr(T_i \geq t) = \exp\{-\int_0^t g(s) \, ds \cdot \exp(\mathbf{x}_i^T \boldsymbol{\beta})\}$. Using the extended gamma process as the prior on g, if we add an independent prior on $\boldsymbol{\beta}$, $f(\boldsymbol{\beta})$ we complete the Bayesian specification. However, the resultant posterior is computationally intractable. Laud, Damien and Smith (1996) suggest that a finite partition, $0 = a_0 < a_1 < a_2 < \dots a_k < a_{k+1} = \infty$ be chosen and define $\delta_\ell = g(a_\ell) - g(a_{\ell-1})$. Thus, δ_ℓ is the increase in the hazard in the interval $(a_{\ell-1}, a_\ell)$. The δ_ℓ's are independent but do not have explicit densities unless $\beta(\cdot)$ is constant. However, the Laplace transforms of the densities can be computed. Hence, if we approximate $\int_0^{a_k} g(s)ds$ by $\sum_{\ell=1}^k \delta_\ell (a_k - a_{\ell-1})$, we can simplify the likelihood for a (possibly right censored) sample to involve only a set of δ_ℓ's and $\boldsymbol{\beta}$. Laud, Damien and Smith then demonstrate how a Gibbs sampler may be developed to obtain draws from the joint posterior of the δ_ℓ's and $\boldsymbol{\beta}$.

8. BETA PROCESSES

Suppose T is a random variable with c.d.f. G, as before. If again g denotes the cumulative hazard, then a general definition for g would be through the differential (infinitesimal increment)

$$
\begin{aligned}
dg(s) &= P(T \in [s, s + ds] | T \geq s) \\
&= dG(s)/(1 - G(s^-)),
\end{aligned}
\tag{11}
$$

i.e., for the set $[a, b)$,

$$
g([a, b)) = \int_{[a,b)} dg(s) = \int_{[a,b)} \frac{1}{1 - G(s^-)} dG(s)
\tag{12}
$$

so that $g(t) \equiv g([0, t])$ is well defined and, in fact, is obviously nonnegative, right continuous and increasing. Also,

$$
G(b^-) - G(a^-) = \int_{[a,b)} dG(s) = \int_{[a,b)} (1 - G(s^-)) \, dg(s).
\tag{13}
$$

Clarification is provided if G is absolutely continuous. Then, from Eq. (11) $dg(s)/ds$ is the hazard associated with G whence, g is the cumulative hazard. As we have seen in previous Sections, we may not wish to insist that G be absolutely continuous; c.d.f.'s which are discrete with probability 1 may be of interest. Regardless, through Eqns (11) and (12) we can define a cumulative hazard even if G has no density function.

While G determines g, does g uniquely determine G? The answer is yes and Eq. (13) can be solved for g formally using the product integral, i.e.,

$$
G(t) = 1 - \prod_{[0,t)} (1 - dg(s)), t \geq 0
\tag{14}
$$

as in Gill and Johansen (1987). In particular, they show that if and only if g is continuous, then $\prod_{[a,b)}(1 - dg(s)) = \exp(-g[a,b))$ so that $g(t) = -\log(1 - G(t))$.

Insight into Eq. (14) can be gleaned by replacing the interval $[0, t)$ by the discrete set

$$\left\{0, \frac{t}{m}, \frac{2t}{m}, \ldots, \frac{(m-1)t}{m}, t\right\}.$$

It is straightforward to calculate that

$$\Pr\left(T > \frac{jt}{m}\right) = \prod_{\ell=0}^{j}\left(1 - g\left(\frac{\ell t}{m}\right)\right)$$

i.e.,

$$G\left(\frac{jt}{m}\right) = 1 - \prod_{\ell=0}^{j}\left(1 - g\left(\frac{\ell t}{m}\right)\right).$$

Passing to a suitable limit as $m \to \infty$ yields Eq. (14).

Since the cumulative hazard is the object of primary interest in modeling survival data and, since we may not wish to insist that the cumulative hazard is continuous, we should model $g(t)$ and not $-\log(1 - G(t))$. We recall that the latter is modeled, for instance, with both the Dirichlet processes and gamma processes.

Nonetheless, since g is nonnegative increasing, we might still attempt to create random realizations of g using a Lévy process. Hjort (1990) notes that we cannot employ an arbitrary Lévy process to model g. For instance, the gamma process yields g's which, through Eq. (14), do not produce proper c.d.f.'s with probability 1.

As a result, Hjort suggests using instead a beta process with independent increments. Such a process is defined, in an infinitesimal sense, through

$$dg(s) \sim \text{beta}(c(s)\, dg_0(s), \quad c(s)(1 - dg_0(s))). \tag{15}$$

In Eq. (15), $c(s)$ plays the role of a precision function which is allowed to vary with time, analogous to allowing the scale parameter to vary with time in the extended Gamma process. In Eq. (15), g_0 is a centering cumulative hazard, i.e., $E(dg(s)) = dg_0(s)$. While g need not be continuous, in practice we would use a standard differentiable cumulative hazard for g_0 (though Hjort, p. 1273, offers a technical benefit in allowing g_0 to have a finite number of discontinuities). Thus, we allow g to vary about g_0 through a nonparametric specification.

Hjort also notes that, while Eq. (15) may be used to motivate the formal development of an independent increments beta process, the resulting distribution which is assigned to $g([a, b))$ is only approximately a beta distribution. Nonetheless, the approximate beta distribution arising from Eq. (15) for g on intervals, is used in practical implementation of the beta process. For instance, Sinha (1996) employing the, by now familiar, finite partition of the time axis $0 = a_0 < a_1 \ldots < a_k < a_{k+1} = \infty$, assumes a priori that the $g([a_{\ell-1}, a_\ell)) = P(T \in [a_{\ell-1}, a_\ell) | T \geq a_{\ell-1})$ are independent beta random variables. If the observed event times are grouped according to the partition, the natural conjugacy leads to the posterior distribution for the $g([a_{\ell-1}, a_\ell))$ being a data updated beta distribution.

9. APPLICATIONS

A rather general parametric hierarchical model with linear structure takes the form

$$\mathbf{y}_i \sim f_1(\mathbf{y}_i | X_i \boldsymbol{\beta} + Z_i \boldsymbol{\alpha}_i, \boldsymbol{\phi}_i), \qquad i = 1, \ldots, n,$$

with the $\mathbf{y}_i$ independent and the $\boldsymbol{\alpha}_i$ independent such that

$$\boldsymbol{\alpha}_i \sim f_2(\boldsymbol{\alpha}_i | W_i \boldsymbol{\delta}_i, \boldsymbol{\Delta}_i).$$

Adding a prior specification for $\boldsymbol{\beta}, \boldsymbol{\delta}$, the $\boldsymbol{\phi}_i$ and $\boldsymbol{\Delta}_i$ completes the Bayesian modeling. (Usually the data $\mathbf{y} = (\mathbf{y}_1, \ldots, \mathbf{y}_n)$ will not identify all of these model parameters. Typically, we set $\boldsymbol{\phi}_i = \boldsymbol{\phi}$ and $\boldsymbol{\Delta}_i = \boldsymbol{\Delta}$.)

How can we add nonparametric aspects, using monotonic functions, to this model? A first possibility is to enrich f_1 or f_2. For instance, the individual level effects are usually modeled using normality. However, such an assumption might not allow tails heavy enough to accommodate outlying populations (i's). Also, for instance, in the case where α_i is just a simple univariate random effect, a form for f_2 permitting multimodal shape, reflecting possible partial exchangeability may be appropriate. As a result, several authors (see, e.g., West, Muller and Escobar (1994), and references therein) have replaced the second stage normal specification for f_2 with, say, $\boldsymbol{\alpha}_i - W_i \boldsymbol{\delta} \sim G$ with G from a Dirichlet process having a normal centering distribution. Recently, Walker and Mallick (1996) have proposed, in this setting, that G comes from a Polya tree distribution and have shown how to implement the requisite sampling to carry out the model fitting using the Gibbs sampler.

The first stage distribution f_1 may also be enriched. For example, in his thesis, Mallick (1994) shows how to do this through the discrete mixture-of-betas approach. Noting that the first stage is typically a generalized linear model (GLM) specification, i.e., $f_1(y|\theta, \phi)$ is a one parameter exponential family with canonical parameter θ, adding a dispersion parameter ϕ,

Mukhopadhyay and Gelfand (1997) remark upon its limitations. Such exponential family models are unimodal with dispersion controlled by the variance function (of the mean) relationship. As a result, they propose to Dirichlet process mix (DPM) the GLM yielding a class of DPMGLM's. For example, under a canonical link, writing the first stage linear structure as $\alpha + \mathbf{x}^T\boldsymbol{\beta}$ (suppressing i), we can obtain

$$f_1(y|\mathbf{x}^T\boldsymbol{\beta}, G) = \int f_1(y|\alpha + \mathbf{x}^T\boldsymbol{\beta})G(d\alpha). \tag{16}$$

The form of Eq. (16) retains the linear structure in the covariates, but utilizes a first stage nonparametric specification which is centered around the associated GLM. Mukhopadhyay and Gelfand note that general link functions can be handled and show how to fit such models using the Gibbs sampler via the augmented sampling approach of MacEachern and Muller (1994).

The link function in a GLM is a strictly monotone function, ℓ, which connects μ the mean of y to a transformed scale η on which a linear model is assumed, i.e., $\ell(\mu) = \eta$. Though usually specified (in fact, as the canonical link) there may be concern regarding the sensitivity of the analysis to the choice of ℓ. Parametric classes for ℓ have been discussed in the literature (see Czado (1992) for a discussion and review) but, in view of our foregoing discussion, the link is a natural candidate for nonparametric modeling. In the case of binary regression, the inverse link, $\ell^{-1}(\eta)$ is exactly a c.d.f. Newton, Czado and Chappel (1996) model ℓ^{-1} from a Dirichlet process prior. However, they note that in considering $\ell^{-1}(\mathbf{x}^T\boldsymbol{\beta})$, if ℓ^{-1} is allowed to be completely arbitrary, the intercept term is confounded with the location (median) of ℓ^{-1} and the overall scale of the regression coefficients is confounded with the scale of ℓ^{-1}. Since realizations from the Dirichlet process have random median and scale, Newton et al. propose modification of the Dirichlet process to yield realizations with specified median (in fact, 0) and specified scale, (say, 1). Basu and Mukhopadhyay (1994) suggest a mixture of Dirichlet processes approach for ℓ^{-1}, i.e., $\ell^{-1}(\eta) = \int \Phi(\eta/\sigma)G(d\sigma)$. That is, ℓ^{-1} arises as a scale mixture of normals with the scale $\sigma \sim G$ and G having a Dirichlet process prior. Mallick and Gelfand (1994) work with arbitrary GLM's, modeling ℓ^{-1} using the mixture-of-betas approach of Sec. 2. They suggest using the canonical link as the centering function. Also, they note that ℓ^{-1} is easier to work with than ℓ because simulation-based model fitting requires repeated evaluation of the likelihood. (This is true regardless of which nonparametric modeling approach is adopted.) Since the likelihood depends upon the $\mathbf{x}_i^T\boldsymbol{\beta}$ through ℓ^{-1} it is much too costly to repeatedly invert ℓ to do such evaluation.

In errors-in-variables models we customarily assume that at least some of the explanatory variables are observed with error or more generally that surrogate explanatory variables are observed rather than those of interest. (See Fuller (1987) for a general discussion.) In particular, we may write the covariate vector $\mathbf{x}$ as $\mathbf{x} = (\mathbf{x}_1, \mathbf{x}_2)$ where $\mathbf{x}_2$ is assumed to be observed without error but rather than observing $\mathbf{x}_1$ we observe $\mathbf{v}$.

In the typical case, we think of $\mathbf{x}_1$ as the actual value writing it as $\mathbf{x}_{1.act}$ and v as the observed value writing it as $\mathbf{x}_{1.obs}$. The regression model still seeks to explain the response y in terms of $\mathbf{x}_1$ and $\mathbf{x}_2$. However, the model now requires specification of the joint distribution of $\mathbf{x}_{1.act}$ and $\mathbf{x}_{1.obs}$. Since $\mathbf{x}_{1.obs}$ is known, in fact, we really only need specification regarding $f(\mathbf{x}_{1.act}|\mathbf{x}_{1.obs})$. Classical approaches require a separate validation or replication experiment to learn about the relationship between $\mathbf{x}_{1.act}$ and $\mathbf{x}_{1.obs}$. Usually an estimate of $E(x_{1.act}|x_{1.obs})$ is obtained and inserted into the original regression model in place of $\mathbf{x}_1$, Carroll (1992). In the Bayesian framework specification of $f(\mathbf{x}_{1.act}|\mathbf{x}_{1.obs})$ is required. This could be done fully parametrically, as in, e.g., Richardson and Gilks (1993). In the case that x_1 is univariate, $E(x_{1.act}|x_{1.obs})$ is called the calibration function. It would naturally be monotone, hence, a candidate for nonparametric modeling, using the identity as a centering function. Mallick and Gelfand (1996) employ the mixture-of-betas approach to model the calibration function. If the response model for y is a GLM, they also assume an unknown link, incorporating two unknown monotone functions in the model.

In the more general setting one would model the joint distribution of $\mathbf{x}_1$ and $\mathbf{w}$. This is usually taken to be a multivariate normal whence $E(\mathbf{x}_1|\mathbf{w})$ is available explicitly. Muller and Roeder (1997) propose nonparametrically enriching the class of models for this joint distribution through Dirichlet process mixing of the multivariate normal.

Survival models have attracted the greatest attention in terms of semiparametric modeling. Here the aim is to model the hazard function or, without assuming differentiability, the cumulative hazard. The two major approaches are through the proportional hazards model, Cox (1972) and the accelerated life model (also known as the accelerated failure time model) as in Cox and Oakes (1984), with the former predominant. The former yields a model for the cumulative hazard of the form $H(t)k(\mathbf{x}^T\boldsymbol{\beta})$ where both H and k are monotone. The latter models the cumulative hazard as $H(t \cdot k(\mathbf{x}^T\boldsymbol{\beta}))$ where again H and k are monotone. It is well known that these two modeling approaches agree if and only if $H(t) = t^\gamma$, see, e.g., Cox and Oakes (1984, p. 71). The presence of monotone functions implies that we can adopt our previous nonparametric approaches.

For the proportional hazards model $k(\eta)$ is customarily taken as e^η in which case only H needs to be modeled. Clayton (1991) proposes the use of a gamma process prior for H while Hjort (1990) and Sinha (1996) suggest the beta process prior. Recently Ibrahim, Chen and MacEachern (1996) assume that H is differentiable and model $h(t) = dH(t)/dt$ as a monotone function using the extended gamma process prior.

Often frailties (Clayton, 1991) are introduced so that the cumulative hazard associated with t_{ij}, the jth event time for the ith individual, is modeled as $H(t_{ij}) \exp(\alpha_i + \mathbf{x}_i^T \boldsymbol{\beta})$. The α_i are assumed exchangeable. Again Clayton uses the gamma process prior for H in this case with a parametric gamma distribution as the prior for the α_i's. Dos Santos, Davies and Francis (1995), in working with this same structure place a mixture of Dirichlet process prior on the α_i with a Weibull hazard for H. Recently, Walker and Mallick (1996) extend Clayton's work still using a gamma process prior for H but adopting a Polya tree distribution for the α_i.

Lastly, Gelfand and Mallick (1995) handle the case where both H and k are assumed unknown, incorporating two monotone functions using their mixture-of-betas approach.

Turning to the accelerated life model with $k(\eta) = e^\eta$, the cumulative hazard becomes $H(t \exp(\mathbf{x}^T \boldsymbol{\beta}))$. If, for t_i, $i = 1, \ldots, n$, $v_i = t_i \exp(\mathbf{x}_i^T \boldsymbol{\beta})$ then, $\log t_i = -\mathbf{x}_i^T \boldsymbol{\beta} + \log v_i$ (and $H(v_i) \sim \mathrm{Exp}(1)$). Hence, modeling H is equivalent to modeling the distribution for the v_i's. Christensen and Johnson (1988) and Johnson and Christensen (1989) assume the $v_i \sim G$ with G having a Dirichlet process prior distribution. They note the computational problems in working with Dirichlet processes and suggest finite approximations which appear to perform well. Kuo and Mallick (1997) avoid these problems using Dirichlet process mixing while, in his thesis, Mallick (1994) proposes the mixture-of-betas approach. Finally, Mallick and Walker (1997) assume $\log v_i$ is, a priori, from a Polya tree distribution but replace $\log t_i$ by an unknown monotone function $g(t_i)$ which is modeled using the mixture-of-betas approach.

REFERENCES

Antoniak, C. E. (1974). Mixtures of Dirichlet processes with applications to nonparametric problems. *Ann. Statist., 2*: 1152–1174.

Arjas, E. and Gasbarra, D. (1994). Nonparametric inference from right censored survival data using the Gibbs sampler. *Statistica Sinica, 4*: 505–524.

Basu, S. and Mukhopadhyay, S. (1994). Bayesian analysis of a random link function in binary response regression. Tech Rpt., Department of Statistics, University of Connecticut, Storrs.

Blackwell, D. (1973). Discreteness of Ferguson selections. *Ann. Statist., 1*: 356–358.

Blackwell, D. and MacQueen, J. B. (1973). Ferguson distributions via Polya urn schemes. *Ann. Statist., 1*: 353–355.

Breslow, N. (1974). Covariance analysis of censored survival data. *Biometrics, 30*: 89–99.

Carroll, R. J. (1992). Approaches to estimation with errors in predictors. Advances in GLM and Statistical Modeling, Springer-Verlag, New York, 40–44.

Christensen, R. and Johnson, W. (1988). Modeling accelerated failure time with a Dirichlet process. *Biometrika, 77*: 693–704.

Clayton, D. G. (1991). A Monte Carlo method for Bayesian inference in frailty models. *Biometrics, 47*: 467–485.

Cox, D .R. (1972). Regression models and life tables. *J. Roy. Statist. Soc., B, 34*: 187–220.

Cox, D. R. (1975). Partial likelihood. *Biometrika, 62*: 269–276.

Cox, D. R. and Oakes, D. (1984). The Analysis of Survival Data. Chapman and Hall, London.

Czado, C. (1992). On link selection in generalized linear models. In: Advances in GLIM and Statistical Modelling, Proceedings of the GLIM 92 Conference and the 7th International Workshop on Statistical Modelling. Lecture Notes in Statistics 78, Springer-Verlag, New York.

Damien, P., Laud, P. W. and Smith, A. F. M. (1995). Approximate random variate generation from infinitely divisible distributions with application to Bayesian inference. *J. Roy. Statist. Soc., B, 57*: 547–563.

Diaconis, P. and Ylvisaker, D. (1985). Quantifying prior opinion. In: Bayesian Statistics 2, eds. J. M. Bernardo et al., North-Holland, Amsterdam, 133–156.

Doksum, K. (1974). Tailfree and neutral probabilities and their posterior distributions. *Ann. Prob., 2*: 183–201.

dos Santos, D. M., Davies, R. B. and Francis, B. (1995). Nonparametric hazard versus nonparametric frailty distribution in modelling recurrence of breast cancer. *J. Stat. Plan. Inf., 47*: 111–127.

Doss, H. (1994). Bayesian nonparametric estimation for incomplete data via successive substitution sampling. *Ann. Statist., 22*: 1763–1786.

Dykstra, R. L. and Laud, P. (1981). A Bayesian nonparametric approach to reliability. *Amer. Statist., 9*: 356–367.

Escobar, M. and West, M. (1995). Bayesian density estimation and inference using mixtures. *J. Amer. Statist. Assoc., 90*: 577–588.

Fahrmeier, L. (1994). Dynamic modeling and penalized likelihood estimation for discrete time survival data. *Biometrika, 81*: 317–330.

Ferguson, T. S. (1973). A Bayesian analysis of some nonparametric problems. *Ann. Statist., 1*: 209–230.

Freedman, D. A. (1963). On the asymptotic behavior of Bayes estimates in the discrete case. *Ann. Math Statist., 34*: 1386–1403.

Fuller, W. A. (1987). Measurement Error Models. John Wiley & Sons, New York.

Gamerman, D. (1991). Dynamic Bayesian models for survival data. *Appl. Stats., 40*: 63–79.

Gelfand, A. E. and Smith, A. F. M. (1990). Sampling based approaches to calculating marginal densities. *J. Amer. Statist. Assoc., 85*: 398–409.

Gelfand, A. E. and Mallick, B. K. (1995). Bayesian analysis of semiparametric proportional hazards models. *Biometrics, 51*: 843–852.

Gelfand, A.E. and Mukhopadhyay, S. (1995). On nonparametric Bayesian inference for the distribution of a random sample. *Can. J. Statist., 23*: 411–420.

Gill, R. D. and Johansen, S. (1987). Produce integrals and counting processes. Research Rpt., CWI, Amsterdam.

Green, P. J. (1995). Reversible jump Markov chain Monte Carlo computation and Bayesian model determination. *Biometrika, 82*: 711–732.

Hjort, N. L. (1990). Nonparametric Bayes estimators based upon Beta processes in models for life history data. *Ann. Statist., 18*: 1259–1294.

Ibrahim, J. G., Chen, M-H. and MacEachern, S. (1996). Bayesian variable selection for proportional hazards models. Tech Rpt., Department of Biostatistics, Harvard School of Public Health, Cambridge, MA.

Johnson, W. and Christensen, R. (1989). Nonparametric Bayesian analysis of the accelerated failure time model. *Stat. and Prob. Letters, 8*: 179–184.

Kalbfleisch, J. D. (1978). Nonparametric Bayesian analysis of survival time data. *J. Roy. Statist. Soc., B, 40*: 214–221.

Kass, R. E. and Steffey, D. (1989). Approximate Bayesian inference in conditionally independent hierarchical models (parametric empirical Bayes). *J. Amer. Statist. Assoc., 84*: 717–726.

Kuo, L. (1986). Computation of mixtures of Dirichlet processes. *SIAM J. Sci. Statist. Comput., 7*: 60–71.

Kuo, L. and Mallick, B. K. (1997). Bayesian semiparametric inference for the accelerated failure time model. *Can. Jour. Stat. 25*: 471–486.

Kruskal, J. B. (1964). Multidimensional scaling. *Psychometrika, 29*: 1–42.

Laud, P. W., Damien, P. and Smith, A. F. M. (1996). Bayesian nonparametric and covariate analysis of failure time data. Tech. Rpt., Department of Mathematics, Imperial College, London.

Lavine, M. (1992). Some aspects of Polya tree distributions for statistical modeling. *Ann. Statist., 20*: 1222–1235.

Lavine, M. (1994). More aspects of Polya tree distributions for statistical modeling. *Ann. Statist., 22*: 1161–1176.

Lévy, P. (1936). Théorie de l'Addition des Variables Aléatoire. Gauthiers-Villars, Paris.

Lo, A. Y. (1984). On a class of Bayesian nonparametric estimates: I. Density estimates. *Ann. Statist., 12*: 351–357.

MacEachern, S. N. and Muller, P. (1994). Estimating mixture of Dirichlet process models. Tech Rpt., ISDS, Duke University, Durham.

Mallick, B. K. (1994). Bayesiain modeling incorporating unknown monotonic functions. Ph.D. Thesis, Department of Statistics, University of Connecticut, Storrs.

Mallick, B. K. and Gelfand, A. E. (1994). Generalized linear models with unknown link functions. *Biometrika, 81*: 237–245.

Mallick, B. K. and Gelfand, A. E. (1996). Semiparametric errors in variables models; a Bayesian approach, *J. Stat. Plan. Inf., 52*: 307–321.

Mallick, B. K. and Walker, S. G. (1997). Bayesian analysis of transformation models incorporating frailties. Tech. Rpt., Department of Mathematics, Imperial College, London.

Maulden, R. D., Suddeth, W. D. and Williams, S. C. (1992). Polya trees and random distributions. *Ann. Statist., 20*: 1203–1221.

McCullagh, P. and Nelder, J. A. (1989). Generalized Linear Models. Chapman and Hall, London.

Mukhopadhyay, S. and Gelfand, A.E. (1997). Dirichlet process mixed generalized linear models. *J. Amer. Statist. Assoc.*, *92*: 633–639.

Muller, P. and Roeder, K. (1997). A Bayesian semiparametric model for case-control studies with errors-in-variables. Biometrika *84*: 535–537.

Newton, M. A., Czado, C. and Chapell, R. (1996). Bayesian inference for semiparametric binary regression. *J. Amer. Statist. Assoc.*, *91*: 142–153.

Richardson, S. and Gilks, W. R. (1993). A Bayesian approach to measurement error problems in epidemiology using conditional independence models. *Amer. J. of Epidemiology*, *6*: 430–442.

Robertson, T., Wright, F. T. and Dykstra, R. L. (1988). Order-Restricted Statistical Inference. J. Wiley and Sons, New York.

Sethuraman, J. (1994). A constructive definition of Dirichlet priors. *Statistica Sinica, 4*: 639–650.

Sethuraman, J. and Tiwari, R. C. (1982). Convergence of Dirichlet measures and the interpretation of their parameter. In: Statistical Decision Theory and Related Topics III, Eds: Gupta, S. and Berger, J. O., Springer-Verlag, New York, 2, 305-315.

Sinha, D. (1996). Time-discrete Beta process model for interval censored survival data. Tech. Rpt., Department of Mathematics, University of New Hampshire, Durham.

Walker, S. G. and Mallick, B. K. (1996). Hierarchical generalized linear models and frailty models with Bayesian nonparametric mixing. Tech. Rpt., Department of Mathematics, Imperial College, London.

West, M., Muller, P. and Escobar, M. D. (1994). Hierarchical priors and mixture models, with applications in regression and density estimation. In *Aspects of uncertainty: A Tribute to D. V. Lindley*. Eds: A. F. M. Smith and P. Feeman, John Wiley and Sans, New York. 363–386.

18

Consistency Issues in Bayesian Nonparametrics

S. GHOSAL   Vrije Universiteit, Amsterdam, The Netherlands

J. K. GHOSH   Indian Statistical Institute, Calcutta, India

R. V. RAMAMOORTHI   Michigan State University, East Lansing, Michigan

1. INTRODUCTION

The basic Bayesian model consists of a parameter θ and a prior distribution Π for θ which reflects the investigator's belief regarding θ. This prior is updated by observing $X_1, X_2, \ldots, X_n$, which are modelled as iid P_θ given θ. The updating mechanism is Bayes theorem which results in changing Π to the posterior $\Pi(\cdot | X_1, X_2, \ldots, X_n)$.

One of the basic ingredients in the model is the prior distribution Π which should ideally correspond to the investigators's opinion about θ. However a complete elicitation of Π is still not feasible, at least for high or infinite dimensional problems. In practice, a prior is adopted which accords well with the investigator's belief and is also mathematically tractable. A pragmatic choice of the prior contains a subjective component as well as a technical one.

In this chapter we are mainly concerned with consistency of the posterior. Informally, the posterior is said to be consistent at a true value θ_0 if the following holds: Suppose $X_1, X_2, \ldots, X_n$ indeed arise from P_{θ_0}, then the posterior converges to the degenerate probability δ_{θ_0}. Alternatively with P_{θ_0} probability 1, the posterior probability of any neighborhood U of θ_0 converges to 1.

Why would a Bayesian be interested in consistency? Think of an experiment in which an experimenter generates observations from a known (to the experimenter) distribution. The observations are presented to a Bayesian. It would be embarrassing if, even with large data, the Bayesian fails to come close to finding the mechanism used by the experimenter. Consistency can be thought of as a validation of the Bayesian method. It can also be interpreted as requiring that the data, at least eventually, overrides the prior opinion. Alternatively two Bayesians, with two different priors, presented with the same data eventually agree. A result of this kind relating "merging of opinions" and posterior consistency is discussed in Diaconis and Freedman (1986a). In fact, Diaconis and Freedman (1986a) (henceforth abbreviated as DF) and the ensuing discussions contain a wealth of material pertaining to posterior consistency.

An early result in posterior consistency is due to Doob (1948), who showed that posterior consistency obtains on a set of prior measure 1. This result does not settle the question of consistency for a particular θ_0 of interest. In smooth finite dimensional problems, different methods show, for example Berk (1966), that consistency obtains at all parameter points. Freedman (1963) exhibits a prior and points of inconsistency for the infinite cell multinomial. He also showed that this phenomenon is quite general and the departure from consistency can be quite dramatic. This initiated a search for priors with good consistency properties in the nonparametric case. A major impetus to Bayesian nonparametrics came from the introduction of Dirichlet Process by Ferguson (1973). Recent years have seen a surge of new priors, especially in the context of densities and semi parametric inference. Walker et al. (1997) review and discuss in detail the interpretation of parameters appearing in the priors. Hjort (1996) contains both a review and some new constructions. However questions of consistency for such priors are difficult to settle and are only beginning to receive attention. A striking negative result is given in DF. A detailed systematic study of Bayesian nonparametrics with stress on asymptotics is available in a monograph under preparation by J. K. Ghosh and R. V. Ramamoorthi (1997).

This chapter presents a brief review of some of these developments. The focus will be on elucidation of different notions of convergence and on positive consistency results.

2. CONSISTENCY

Let **R** stand for real line and $\mathscr{B}$ stand for the Borel σ-algebra on **R**. The space of probability measures on **R** will be denoted by $\mathscr{M}$. The space $\mathscr{M}$ is equipped with the σ-algebra $\mathscr{B}(\mathscr{M})$—the smallest σ-algebra with respect to which the functions $\{P \mapsto P(B) : B \in \mathscr{B}\}$ are measurable. It is convenient to consider, as our parameter space, a subset $\mathscr{P}$ of $\mathscr{M}$. Unless otherwise stated, the σ-algebra associated with $\mathscr{P}$ will be the σ-algebra $\mathscr{B}(\mathscr{M})$, restricted to $\mathscr{P}$. A prior Π is a probability measure on $\mathscr{P}$. Denoting the random probability measure by $\mathbf{P}$, we will sometime write this as $\mathbf{P} \sim \Pi$ and will write "given P" for given $\mathbf{P} = P$.

Let $X_1, X_2, \ldots$ be a sequence of random variables which are given P, iid P. We will for notational simplicity use P for the distribution of each X_i, the joint distribution of $X_1, X_2, \ldots, X_n$ and the joint distribution of the entire sequence $X_1, X_2, \ldots$. It is convenient to think of $X_1, X_2, \ldots$ as being defined on $\Omega = \mathbf{R}^\infty$, with X_i as the ith co-ordinate map.

Let Π be a prior on $(\mathscr{M}, \mathscr{B}(\mathscr{M}))$ and given P, let $X_1, X_2, \ldots, X_n$ be iid P. These two together determine a joint distribution for $P, X_1, X_2, \ldots, X_n$. The resulting conditional distribution of P given $X_1, X_2, \ldots, X_n$ will be denoted by $\Pi(\cdot|X_1, X_2, \ldots, X_n)$. Of course, the conditional distribution is not unique but in most contexts there is a natural choice and it is this version that we will work with.

The sequence $\{\Pi(\cdot|X_1, X_2, \ldots, X_n), n \geq 1\}$ is a sequence of (random) probability measures on $\mathscr{M}$ and thus their convergence involves convergence of probabilities on the space of probabilities. Our interest is mainly in convergence to a degenerate measure δ_{P_0} and in this case consistency can be conveniently formulated by requiring that $\Pi(U|X_1, X_2, \ldots, X_n) \to 1$ for suitable neighborhoods of P_0. Since there are a variety of topologies available on $\mathscr{M}$, these lead to different notions of consistency. We next briefly discuss some of these.

2.1 Weak Consistency

A subset U of $\mathscr{M}$ is said to be a weak neighborhood of P_0 if it contains a set of the form $U = \{P : |\int f_i \, dP - \int f_i \, dP_0| < \varepsilon_i, i = 1, 2, \ldots, k\}$, where f_i, $i = 1, 2, \ldots, k$, are bounded continuous functions on $\mathbf{R}$.

DEFINITION 1. The sequence $\{\Pi(\cdot|X_1, X_2, \ldots, X_n), n \geq 1\}$ is said to be weakly consistent at P_0, if there exists a $\Omega_0 \subset \Omega$ with $P_0(\Omega_0) = 1$ such that for $\omega \in \Omega_0$, for every weak neighborhood U of P_0, $\Pi(U|X_1, X_2, \ldots, X_n) \to 1$ as $n \to \infty$.

When $\mathscr{M}$ is given the weak topology, $\mathscr{M}$ itself becomes a complete separable metric space and this in turn induces a weak topology on the space of probability measures on $\mathscr{M}$. Weak consistency corresponds to the convergence of the posterior in this topology. Since $\mathscr{M}$ is a complete separable metric space, it is meaningful to talk of the (topological) support of a prior Π—the smallest closed set with Π measure 1. It is not hard to see that $\mathscr{B}(\mathscr{M})$ is the Borel σ-algebra when $\mathscr{M}$ is viewed as a metric space.

2.2 K-Consistency

If P_1, P_2 are probability measures on $\mathbf{R}$, define the Kolomogorov distance

$$d_K(P_1, P_2) = \sup_{t \in \mathbf{R}} |P_1(-\infty, t) - P_2(-\infty, t)|.$$

A K-neighborhood of P_0 is a set of the form $\{P : d_K(P_0, P) < \varepsilon\}$.

DEFINITION 2. The sequence $\{\Pi(\cdot|X_1, X_2, \ldots, X_n), n \geq 1\}$ is said to be K-consistent at P_0 if there exists a $\Omega_0 \subset \Omega$ with $P_0(\Omega_0) = 1$ such that for $\omega \in \Omega_0$, for every K-neighborhood U of P_0, $\Pi(U|X_1, X_2, \ldots, X_n) \to 1$ as $n \to \infty$.

This consistency is of course motivated by the Glivenko-Cantelli theorem. Under the metric d_K, $\mathscr{M}$ is neither separable nor complete.

2.3 Consistency in Uniform Neighborhoods

Another popular metric on $\mathscr{M}$ is the total variation metric

$$d_T(P_1, P_2) = \sup_{B \in \mathscr{B}} |P_1(B) - P_2(B)|.$$

A total variation neighborhood of P_0 is a set of the form

$$\{P : d_T(P_0, P) < \varepsilon\}.$$

DEFINITION 3. The sequence $\{\Pi(\cdot|X_1, X_2, \ldots, X_n), n \geq 1\}$ is said to be consistent at uniform neighborhoods of P_0 if there exists a $\Omega_0 \subset \Omega$ with $P_0(\Omega_0) = 1$, such that for $\omega \in \Omega_0$, for every total variation neighborhood U of P_0,

$$\Pi(U|X_1, X_2, \ldots, X_n) \to 1 \qquad \text{as } n \to \infty.$$

The space $\mathscr{M}$ under the total variation metric is non-separable and the associated measure theoretic problems renders it uninteresting when the parameter space is all of $\mathscr{M}$. However, if the prior is supported by a set of densities, then the total variation metric coincides with the L_1-metric. This space is separable and in this case consistency over uniform neighborhoods is preferred over weak or K-consistency.

Let Π be a prior on $\mathcal{M}$. The expected value under Π, $E_\Pi(\mathbf{P})$ is the probability measure defined by $\hat{P}(B) = \int P(B)\Pi(dP)$. We will refer to the expectation of $\mathbf{P}$ under $\Pi(\cdot|X_1, X_2, \ldots, X_n)$ as the Bayes estimate or as the predictive distribution. It is not hard to see that if the posterior is consistent at P_0 in any of the above senses, then the Bayes estimate converges to P_0 in the corresponding topology.

3. GENERAL CONSISTENCY THEOREMS

We begin with a consistency result for the multinomial.

THEOREM 1. Let Π be a prior on $\mathcal{M}(\chi)$, where χ is a finite set. Then the posterior is consistent at all P_0 in the support of Π.

This result appears as Theorem 1 in Freedman (1963). Here is a brief outline of the argument involved. Let $\chi = \{1, 2, \ldots, k\}$, so that $\mathcal{M}(\chi) = \{P = (P(1), P(2), \ldots, P(k)) : P(i) \geq 0, \sum P(i) = 1\}$. Let P_0 be in the support of Π.

For any $\varepsilon > 0$, let

$$K_\varepsilon = \left\{ P : \sum_{j=1}^{k} P_0(j) \log \frac{P_0(j)}{P(j)} < \varepsilon \right\}.$$

Finite dimensionality and compactness of $\mathcal{M}(\chi)$ ensures two things:

(i) Every neighborhood V of P_0 contains a set of the form K_ε and more crucially, every K_ε contains a neighborhood of P_0. Thus if P_0 is in the support of Π, then $\Pi(K_\varepsilon) > 0$ for all $\varepsilon > 0$.

(ii) The convergence of

$$\frac{1}{n} \sum_{i=1}^{n} \log \frac{P_0(X_i)}{P(X_i)} \quad \text{to} \quad \sum_{j=1}^{k} P_0(j) \log \frac{P_0(j)}{P(j)}$$

is uniform in P in the sense, for any $\varepsilon > 0$,

$$\sup_{P \in K_\varepsilon} \left| \frac{1}{n} \sum_{i=1}^{n} \log \frac{P_0(X_i)}{P(X_i)} - \sum_{j=1}^{k} P_0(j) \log \frac{P_0(j)}{P(j)} \right| \to 0 \qquad \text{a.s. } P_0.$$

$$\inf_{P \in K_\varepsilon^c} \frac{1}{n} \sum_{i=1}^{n} \log \frac{P_0(X_i)}{P(X_i)} > \frac{\varepsilon}{2} \to 0 \qquad \text{a.s. } P_0.$$

For any neighborhood V of P_0, let $\varepsilon_1 > 2\varepsilon_2$ and let $K_{\varepsilon_1}, K_{2\varepsilon_2}$ be contained in V. Then the posterior probability of V^c satisfies

$$\Pi(V^c | X_1, X_2, \ldots, X_n) = \frac{\displaystyle\int_{V^c} \exp\left[\sum_{i=1}^{n} \log \frac{P(X_i)}{P_0(X_i)}\right] \Pi(dP)}{\displaystyle\int_{\mathscr{M}(\chi)} \exp\left[\sum_{i=1}^{n} \log \frac{P(X_i)}{P_0(X_i)}\right] \Pi(dP)}$$

$$\leq \frac{\displaystyle\int_{K_{\varepsilon_1}^c} \exp\left[\sum_{i=1}^{n} \log \frac{P(X_i)}{P_0(X_i)}\right] \Pi(dP)}{\displaystyle\int_{K_{\varepsilon_2}} \exp\left[\sum_{i=1}^{n} \log \frac{P(X_i)}{P_0(X_i)}\right] \Pi(dP)}$$

$$= \frac{I_{1n}(X_1, X_2, \ldots, X_n)}{I_{2n}(X_1, X_2, \ldots, X_n)} \quad \text{(say)}$$

The uniform convergence mentioned in (ii) yields

$$\limsup_{n \to \infty} e^{n\varepsilon_2} I_{1n} = 0 \quad \text{a.s. } P_0.$$

Since $\Pi(K_{\varepsilon_2}) > 0$, an application of Fatou's lemma gives

$$\liminf_{n \to \infty} e^{n\varepsilon_2} I_{2n} = \infty \quad \text{a.s. } P_0.$$

These two together show that $\Pi(V^c | X_1, X_2, \ldots, X_n) \to 0$ a.s. P_0.

When χ is infinite, even weak consistency can fail to occur in the support of Π. Freedman (1963) provided a dramatic example of this when $\chi = \{1, 2, \ldots\}$.

Theorem 1 provides a passage to consistency for tail free priors. A prior Π on $\mathscr{M}$ is tail free with respect to a sequence of partitions $T = \{\mathbf{T}_n, n \geq 1\}$ if, $T = \{\mathbf{T}_n, n \geq 1\} = \{\{B_\varepsilon : \varepsilon \in \{0, 1\}^n\}, n \geq 1\}$ is a nested sequence of partitions as described in Sec. 7, and if

$$\{P(B_0)\}, \{P(B_{00} | B_0), P(B_{10} | B_1)\}, \ldots, \{P(B_{\varepsilon 0} | B_\varepsilon) : \varepsilon \in \{0, 1\}^k\}, \ldots$$

are all independent. As we will see later, Dirichlet process and Polya tree process are tail free.

THEOREM 2. If Π is tail free with respect to $T = \{\mathbf{T}_n, n \geq 1\}$, then (a suitable version of) the posterior is weakly consistent.

The idea behind Theorem 2 is simple. Every weak neighborhood is determined by $\{B_\varepsilon : \varepsilon \in \{0, 1\}^k\}$ for some k. Tail free property ensures that the posterior distribution of $\{P(B_\varepsilon) : \varepsilon \in \{0, 1\}^k\}$ given $X_1, X_2, \ldots, X_n$ is same as the posterior given the 2^k-cell multinomial with cells $\{B_\varepsilon : \varepsilon \in \{0, 1\}^k\}$. This puts us in the framework work of Theorem 1.

When the prior is not tail free, a useful tool to establish weak consistency is a theorem of Schwartz (1965). In this theorem too, like Theorem 1, the Kullback–Leibler numbers play an important role. A similar theorem for the infinite cell multinomial already appears in Freedman (1963).

Let $L(\mu)$ be the set of all densities with respect to a σ-finite measure μ. The Kullback–Leibler divergence of f from f_0, where both are in $L(\mu)$ is defined as $K(f_0, f) = \int f_0 \log(f_0/f)\, d\mu$. The Kullback–Leibler neighborhood $\{f : K(f_0, f) < \varepsilon\}$ will be denoted by $K_\varepsilon(f_0)$.

Let U be a set and $f_0 \in U$. Say that there exists a uniformly consistent sequence of tests for testing $H_0 : f = f_0$ vs. $H_1 : f \in U^c$, if there exists a sequence of tests $\phi_n(X_1, X_2, \ldots, X_n)$ such that as $n \to \infty$,

$$E_{f_0} \phi_n(X_1, X_2, \ldots, X_n) \to 0$$

and

$$\inf_{f \in U^c} E_f \phi_n(X_1, X_2, \ldots, X_n) \to 1.$$

THEOREM 3. (Schwartz). Let Π be a prior on $L(\mu)$. If $f_0 \in L(\mu)$ and $U \subset L(\mu)$ satisfy

1. $\Pi(K_\varepsilon(f_0)) > 0$ for all $\varepsilon > 0$;
2. there exists a uniformly consistent sequence of tests for testing $H_0 : f = f_0$ vs. $H_1 : f \in U^c$;

then $\Pi(U|X_1, X_2, \ldots, X_n) \to 1$ a.s. P_0.

The Schwartz theorem retains the flavor of Theorem 1. In Theorem 1, $\Pi(K_\varepsilon(f_0)) > 0$ was derived as a consequence of f_0 being in the support of Π. Condition (2) essentially plays the role of the compactness of $\mathcal{M}(\chi)$.

It is not hard to see that if U is a weak neighborhood of f_0, then there exists a uniformly consistent sequence of tests for testing $H_0 : f = f_0$ vs. $H_1 : f \in U^c$. Since the weak neighborhoods have a countable base, Schwartz's theorem immediately yields the following corollary.

COROLLARY 1. If $\Pi(K_\varepsilon(f_0)) > 0$ for all $\varepsilon > 0$, then the posterior is weakly consistent at f_0.

We consider two examples below. The first of these shows that the condition of Schwartz is not necessary for consistency. The second example shows that if the Schwartz condition does not hold then consistency cannot be expected in general even for finite dimensional examples. In the second example, P_0 is in the weak support of the prior.

Let $X_1, X_2, \ldots, X_n$ be iid $U(0, \theta)$ where $\theta \in \Theta = (0, 1]$. In this example the Kullback–Leibler divergence of every $U(0, \theta)$ from $U(0, 1)$ is ∞. Thus the Schwartz condition fails but if Π is a prior with support all of $[0, 1]$, then it is easy to see that the posterior is consistent at $f_0 = U(0, 1)$.

If the above example is modified by setting $\Theta = (0, 1] \cup (2, 3)$, then it can be shown, as in Ghosh and Ramamoorthi (1997), that there is a prior with $f_0 = U(0, 1)$ in its weak support such that the posterior fails to be consistent at $f_0 = U(0, 1)$. In this example too, Schwartz's condition fails to hold.

If U is a strong neighborhood, then Le Cam (1973) and Barron (1989) show that, in general, there will not exist a uniformly consistent sequence of tests for testing $H_0 : f = f_0$ vs. $H_1 : f \in U^c$. The role of uniformly consistent sequence of tests is greatly clarified by the following theorem of Barron which is discussed in Barron et al. (1996) (henceforth abbreviated as BSW).

THEOREM 4.   Let Π be a prior on $L(\mu)$, $f_0 \in L(\mu)$ and U is a neighborhood of f_0. Assume $\Pi(K_\varepsilon(f_0)) > 0$ for all $\varepsilon > 0$. Then the following are equivalent:

1.   There exists a β_0 such that

$$P_{f_0}\{\Pi(U^c | X_1, X_2, \ldots, X_n) > e^{-n\beta_0} \quad \text{infinitely often}\} = 0.$$

2.   There exist subsets V_n, W_n of $L(\mu)$, positive numbers $c_1, c_2, \beta_1, \beta_2$ and a sequence of tests $\{\phi_n(X_1, X_2, \ldots, X_n)\}$ such that
     (a) $U^c = V_n \cup W_n$
     (b) $\Pi(W_n) \leq c_1 e^{-n\beta_1}$
     (c) $P_{f_0}\{\phi_n(X_1, X_2, \ldots, X_n) > 0 \quad \text{infinitely often}\} = 0$
           and

$$\inf_{f \in V_n} E_f \phi_n \geq 1 - c_2 e^{-n\beta_2}.$$

The last theorem can be used to develop sufficient conditions for posterior consistency on uniform neighborhoods. BSW provide such a condition using bracketing metric entropy. Motivated by the result of BSW and Theorem 4, we can, Ghosal et al. (1997b) prove the following:

Let $\mathscr{F} \subset L(\mu)$. For $\delta > 0$, the L_1-metric entropy of $\mathscr{F}$, denoted by $J(\delta, \mathscr{F})$, is $\log a(\delta)$, where $a(\delta)$ is the minimum over all k such that there exist $f_1, f_2, \ldots, f_k$ in $L(\mu)$ with $\mathscr{F} \subset \cup_{i=1}^{k}\{f : \|f - f_i\|_1 < \delta\}$.

THEOREM 5.   Let Π be a prior on $L(\mu)$ with $\Pi(\mathscr{F}) = 1$. Suppose $f_0 \in L(\mu)$ and $\Pi(K_\varepsilon(f_0)) > 0$ for all $\varepsilon > 0$. If for each $\varepsilon > 0$ there is a

$\delta < \varepsilon/4$, $c_1, \beta_1 > 0$, $\beta < \varepsilon^2/8$ and also $\mathscr{F}_n \subset \mathscr{F}$ such that, for all n large,

1. $\Pi(\mathscr{F}_n^c) < c_1 e^{-n\beta_1}$,
2. $J(\delta, \mathscr{F}_n) < n\beta$,

then the posterior is consistent at f_0 on uniform neighborhoods.

4. DIRICHLET PROCESSES

Dirichlet processes were introduced in the statistical context by Ferguson (1973), where he developed their basic properties and used it to provide a Bayesian interpretation of many popular nonparametric procedures. A recent review of Dirichlet process is the review article by Ferguson et al. (1996). Another good source is the text by Schervish (1995). We provide a brief summary of some of the properties for later use in this chapter.

Let α be a finite measure on $\mathbf{R}$. A probability measure D_α on $\mathscr{M}$ is said to be a Dirichlet process with parameter α if for every partition $B_1, B_2, \ldots, B_k$ of $\mathbf{R}$ by Borel sets, the vector $(P(B_1), P(B_2), \ldots, P(B_k))$ has a k-variate Dirichlet distribution with parameter $(\alpha(B_1), \alpha(B_2), \ldots, \alpha(B_k))$.

1. $E_{D_\alpha}(P(A)) = \alpha(A)/\alpha(\mathbf{R})$.
   It is convenient to write $\bar{\alpha}$ for the probability $\alpha(\cdot)/\alpha(\mathbf{R})$. Thus a natural choice for $\bar{\alpha}$ is the prior belief of the distribution of X.
2. If $P \sim D_\alpha$ and given P, $X_1, X_2, \ldots, X_n$ are iid P, then the posterior given $X_1, X_2, \ldots, X_n$ is $D_{\alpha + \Sigma \delta_{X_i}}$.
3. Properties (1) and (2) immediately show that the Bayes estimate of P given $X_1, X_2, \ldots, X_n$ is

   $$\frac{\alpha(\mathbf{R})}{\alpha(\mathbf{R}) + n} \bar{\alpha} + \frac{n}{\alpha(\mathbf{R}) + n} P_n,$$

   where P_n is the empirical distribution arising from $X_1, X_2, \ldots, X_n$. Since as $\alpha(\mathbf{R}) \to 0$ the Bayes estimate goes to the empirical distribution, $\alpha(\mathbf{R})$ can be thought of as a "prior sample size" or as a measure of belief in the prior. Sethuraman and Tiwari (1982) have pointed out the need for some care in this interpretation.
4. D_α gives mass 1 to the set of discrete distributions. Thus D_α is concentrated on a small set. On the other hand the topological support of D_α is $\{P : \text{support}(P) \subset \text{support}(\alpha)\}$. If α has (topological) support all of $\mathbf{R}$, then D_α will have as its support all of $\mathscr{M}$.
5. If α_1 and α_2 are two distinct nonatomic measures, then $D_{\alpha_1} \perp D_{\alpha_2}$. A slight extension shows that with a D_α prior, the prior and posterior are singular with respect to each other.

6. Let $P \sim D_\alpha$ and given P, $X_1, X_2, \ldots, X_n$ be iid P. The marginal distribution of $X_1, X_2, \ldots, X_n$ can be interpreted as a Polya urn scheme, see Blackwell and MacQueen (1973) and Mauldin et al. (1992). The distribution of X_1 is $\bar{\alpha}$, the distribution of X_2 given $X_1 = x_1$ is $(\alpha + \delta_{x_1})/(\alpha(\mathbf{R}) + 1)$ and the distribution of X_{n+1} given $X_1 = x_1$, $X_2 = x_2, \ldots, X_n = x_n$ is $(\alpha + \sum_1^n \delta_{x_i})/(\alpha(\mathbf{R}) + n)$. One consequence of this expression is that even when α is nonatomic, the probability of getting coincidences among $X_1, X_2, \ldots, X_n$ is positive. Suppose α has density g with respect to Lebesgue measure. Then the joint density of $X_1, X_2, \ldots, X_n$ is given by

$$\frac{1}{[\alpha(\mathbf{R})]^{(n)}} \prod_{j:I_j=1} g(x_j) \tag{1}$$

where $I_j = 1$ if $x_j \notin \{x_1, x_2, \ldots, x_{j-1}\}$ and 0 otherwise and $[\alpha(\mathbf{R})]^{(n)}$ is the ascending factorial $\alpha(\mathbf{R})(\alpha(\mathbf{R}) + 1)(\alpha(\mathbf{R}) + 2) \cdots (\alpha(\mathbf{R}) + n - 1)$. The above expression is to be viewed as a density with respect to $\bar{\lambda} = \sum \lambda_{B_1 B_2 \ldots B_k}$, where $B_1, B_2, \ldots, B_k$ is a partition of $\{1, 2, \ldots, n\}$ and $\lambda_{B_1 B_2 \ldots B_k}$ is the k-dimensional Lebesgue measure on $\{(x_1, x_2, \ldots, x_n) : x_i = x_j$ if $i, j \in$ the same B_l and $x_i \neq x_j$ otherwise$\}$.

7. The Dirichlet process is tail free with respect to every partition and hence the posterior is weakly consistent. In fact, it can be shown that it is K-consistent.

5. MIXTURES OF DIRICHLET PROCESS

Mixtures of Dirichlet processes considered by Antoniak (1974) provide greater flexibility. In this model, a hyper parameter θ is first chosen according to a prior μ and given θ, $P \sim D_{\alpha(\theta)}$ and given (P, θ), $X_1, X_2, \ldots, X_n$ are iid P. For example, one may be ready to say that the expected value of P is normal but not be able to specify (μ, σ). In such cases it may be appropriate to take $\theta = (\mu, \sigma)$ and $\alpha_\theta = \alpha_\theta(\mathbf{R})N(\mu, \sigma)$.

Since $\Pi(P|\theta, X_1, X_2, \ldots, X_n)$ is $D_{\alpha(\theta) + \sum \delta_{x_i}}$ and

$$\Pi(P|X_1, X_2, \ldots, X_n) = \int \Pi(P|\theta, X_1, X_2, \ldots, X_n)\, \Pi(d\theta|X_1, X_2, \ldots, X_n),$$

the posterior distribution of P given $X_1, X_2, \ldots, X_n$ is again a mixture of Dirichlet with μ changing to $\mu^* = \Pi(d\theta|X_1, X_2, \ldots, X_n)$ and $\alpha(\theta)$ changing to $\alpha(\theta) + \sum \delta_{X_i}$. We need an expression for $\Pi(d\theta|X_1, X_2, \ldots, X_n)$ to evaluate the posterior.

If μ has a density $h(\theta)$ and if $\alpha(\theta)$ has a density $g_\theta(x)$, then using Eq. (1), the joint density of $(\theta, X_1, X_2, \ldots, X_n)$ is

$$\frac{h(\theta)}{[\alpha(\theta)(\mathbf{R})]^{(n)}} \prod_{j:I_j=1} g_\theta(x_j)$$

and the conditional density of θ given $X_1, X_2, \ldots, X_n$ becomes

$$c(X_1, X_2, \ldots, X_n) h(\theta) \prod_{j:I_j=1} g_\theta(x_j).$$

If the true distribution P_0 has a density then with P_0 probability 1, the I_j's are all equal to 1 and the conditional density of θ given $X_1, X_2, \ldots, X_n$ becomes

$$c(X_1, X_2, \ldots, X_n) h(\theta) \prod_{j=1}^{n} g_\theta(x_j). \tag{2}$$

Mixtures are no longer tail-free and in general one has inconsistency. Ferguson et al. (1996) has a nice example illustrating this phenomenon. Consistency issues in the case of mixtures of Dirichlet is studied in detail in Freedman and Diaconis (1983), where the following result is available.

THEOREM 6.   If $\alpha(\theta)(\mathbf{R}) \leq M$ for all θ, then the mixture of $D_{\alpha(\theta)}$ yields a weakly consistent posterior.

The mixture model, apart from being interesting in its own right, also arises in Bayesian models for semiparametric problems. Location model discussed in the next Section is one example. Another interesting type of problems is the binary regression model studied by Newton (1994).

6. LOCATION MODEL

Consider the location parameter problem where μ is a prior on the location parameter θ and P is independently chosen according to D_α; given (θ, P), $X_1, X_2, \ldots, X_n$ are iid P_θ, where $P_\theta(A) = P(A - \theta)$. This setup falls in the mixture model with $\alpha_\theta(A) = \alpha(A - \theta)$, except that our interest is in the posterior distribution of θ given $X_1, X_2, \ldots, X_n$, rather than in the posterior distribution of P. In other words, θ is the parameter of interest and not just an index in the mixture distribution.

If μ has a density $h(\theta)$ and if α has a density $g(x)$, when $X_1, X_2, \ldots, X_n$ are all distinct and Eq. (1) yields

$$\Pi(\theta | X_1, X_2, \ldots, X_n) = \frac{h(\theta) g(x_1 - \theta) g(x_2 - \theta) \cdots, g(x_n - \theta)}{\int h(t) g(x_1 - t) g(x_2 - t) \cdots, g(x_n - t) dt}.$$

Barron, in his discussion of DF, notes from this expression that even though θ is the location parameter of an unknown P, the posterior acts as though the X_i's come from a fixed known density g. So, one would not, in general, expect consistency of the posterior.

Since the location model is not identifiable without some assumptions like symmetry, it is appropriate to consider a prior on $\mathcal{M}^s$—the space of distributions on $\mathbf{R}$, symmetric around 0. DF use a symmetrized Dirichlet process on $\mathcal{M}^s$. Symmetrized Dirichlet processes were first studied by Dalal (1979).

The expression for the posterior distribution for θ is somewhat complicated (Lemma 3.1 of Diaconis and Freedman (1986b)), but DF show that asymptotic analysis of the previous case essentially holds also for symmetric P. To be more precise, in the symmetric location model, if α is the Cauchy distribution then, when the true parameter is $(0, P_0)$ where P_0 has a trimodal density with compact support, for approximately half the samples the posterior concentrates near a mode δ and another half concentrates near $-\delta$. The number δ can be made as large as one wants by choosing P_0 suitably. This is an example of how dramatic deviations from consistency can be. A similar phenomenon was observed by Doss (1985) under a different setup.

This is a surprising result. Addition of a single parameter seems to destroy the attractive features of the Dirichlet. That inconsistency should occur for a P_0 with simple structure also comes as a surprise.

In Ghosal et al (1998), we consider the location model and prove the following:

THEOREM 7.   Let λ be a prior on $\mathcal{M}^s$—the set of all symmetric densities and μ be a prior on $\mathbf{R}$, f_0 be a symmetric density and θ_0 be a real number. Assume that λ, μ satisfy the following:

1. μ gives positive mass to every open neighbourhood of θ_0;
2. for every $\varepsilon > 0$ and all sufficiently small $|t|$, $\lambda(K_\varepsilon(f_{0,t}^*)) > 0$, where $f_{0,t}(x) = f_0(x - t)$ and $f_{0,t}^*(x) = (f_{0,t}(x) + f_{0,t}(-x))/2$.

If f_0 satisfies either of the following conditions, then the posterior is weakly consistent at (θ_0, f_0) (and as a consequence, the posterior of θ is consistent at θ_0 in the usual topology):

1. $\displaystyle \lim_{t \to 0} \int f_0 \log \frac{f_0}{f_{0,t}} \, dx = 0$;

2. f_0 is continuous and has compact support.

For Polya tree priors with expectation measure α, a sufficient condition for $\lambda(K_\varepsilon(f_{0,t}^*)) > 0$ will be $\int f_{0,t} \log(f_{0,t}/\alpha) < \infty$, or equivalently

$\int f_0 \log(f_0/\alpha_t) < \infty$ for all sufficiently small $|t|$, a little stronger than simply $\int f_0 \log(f_0/\alpha) < \infty$

Thus, if the assumptions of the above theorem hold, then consistency does occur in the DF type of distributions. Since for any density f, $K(f, P) = \infty$ whenever P is discrete and since D_α gives mass 1 to discrete densities, clearly the assumptions do not hold for the Dirichlet. In the next section we will show that Polya tree priors can be chosen to satisfy the assumptions. In Sec. 8.2, we will show that if the Dirichlet process is smoothed by convoluting with a (normal) kernel, then also the assumptions are satisfied.

Of course, the theorem does not preclude inconsistency but it does appear that in this case points of inconsistency, if they exist, will be quite complicated.

7. POLYA TREE PRIORS

Polya tree priors are generalization of Dirichlet process. These were discussed by Ferguson (1974). For a recent explication of their role in nonparametrics we refer to Lavine (1992, 1994). Other related references are Mauldin et al. (1992) and Schervish (1995).

Let E_j be the set of all sequences of 0's and 1's of length j and let $E^* = \bigcup_j E_j$.

Let $T = \{T_n, n \geq 1\}$ be a sequence of nested partitions of $\mathbf{R}$ into intervals such that $\bigcup_j T_j$ generates the Borel σ-algebra. A bit more explicitly, let $T_j = \{B_\varepsilon : \varepsilon \in E_j\}$. At the $(j+1)$th stage, each B_ε is partitioned into $B_{\varepsilon 0}$ and $B_{\varepsilon 1}$. We want each B_ε to be an interval and the σ-algebra generated by $\bigcup_{\varepsilon \in E} \cdot B_\varepsilon$ be the Borel σ-algebra on $\mathbf{R}$.

A prior Π on $\mathcal{M}$ is said to be a Polya tree prior with respect to the partition T and with parameters $\{\alpha_\varepsilon : \varepsilon \in E^*\}$, written as $P \sim PT(T, \alpha)$, if under Π

1. $\{P(B_{\varepsilon 0}|B_\varepsilon) : \varepsilon \in E^*\}$ are a set of independent random variables;
2. for each $\underline{\varepsilon} \in E^*$, $P(B_{\underline{\varepsilon}0}|B_{\underline{\varepsilon}}) \sim \text{Beta}(\alpha_{\varepsilon 0}, \alpha_{\varepsilon 1})$.

Here are some properties of Polya tree priors:

1. The expected value of P under $PT(T, \alpha)$ is the probability measure $\hat{P}$ defined by, if $\varepsilon_k = (\varepsilon_1, \varepsilon_2, \ldots, \varepsilon_k)$, $\varepsilon_k 0 = (\varepsilon_k, 0)$ and $\varepsilon_k 1 = (\varepsilon_k, 1)$,

$$\hat{P}(B_{\varepsilon_k}) = \prod_{i=1}^{k} \frac{\alpha_{\varepsilon_i}}{\alpha_{\varepsilon_{i-1}0} + \alpha_{\varepsilon_{i-1}1}}.$$

2. If $P \sim PT(\mathbf{T}, \boldsymbol{\alpha})$ and given P if $X_1, X_2, \ldots, X_n$ are iid P, then the posterior distribution is again a Polya tree with parameters $\boldsymbol{\alpha} = \{\alpha'_\varepsilon : \varepsilon \in E^*\}$, where $\alpha'_\varepsilon = \alpha_\varepsilon + \sum_{i=1}^n I_{B_\varepsilon}(X_i)$.

The above two properties together easily yield an expression for the Bayes estimate. Being tail free, weak consistency is immediate for Polya trees.

Unlike the Dirichlet, by choosing the parameters carefully, a Polya tree can be made to sit on densities. Results like the following appear in Mauldin et al. (1992) and Schervish (1995).

THEOREM 8. Suppose that λ is a continuous probability measure such that $\lambda(B_{\varepsilon_k}) = 2^{-k}$ for all k and further $\alpha_{\varepsilon_k} = a_k$. If $\sum a_k^{-1} < \infty$, then the resulting polya tree gives mass 1 to the set of all distributions absolutely continuous with respect to λ.

In the discussion of the location model and strong consistency we required priors on densities which give positive mass to Kullback–Leibler neighborhoods. The following theorem proved in Ghosal et al. (1998) shows that this can be achieved for Polya tree priors. A weaker result also appears in Lavine (1994).

THEOREM 9 Suppose that λ is a continuous probability measure such that $\lambda(B_{\varepsilon_k}) = 2^{-k}$ for all k and further $\alpha_{\varepsilon_k} = a_k$. If $\sum a_k^{-1/2} < \infty$ then the then for any density f_0 (with respect to λ) with $\int f_0 \log f_0 < \infty$, we have, for all $\varepsilon > 0$, $PT(\boldsymbol{T}, \boldsymbol{\alpha})(K_\varepsilon(f_0)) > 0$.

Thus Polya tree priors (for suitable α_ε) provides an example of priors mentioned in Theorem 7. As for consistency in uniform neighborhoods, BSW construct an appropriate sieve and show that if $\alpha_{\varepsilon_k} = 8^k$, then the posterior is strongly consistent. The result of BSW suggests using such a prior for the location model. We expect that if the tails of the prior μ for the location parameter decay rapidly then this approach would yield consistency (of the pair (θ, f)) in uniform neighborhoods for a wide class of "true" distributions.

8. DIRICHLET MIXTURES

While Polya tree priors can be made to sit on densities, it is not possible to ensure the densities in the support to have enough smoothness properties. Priors on smooth families of densities can be constructed via Dirichlet mixtures, a method suggested by Lo (1994).

Let Θ be a parameter set, typically $\mathbf{R}$ or $\mathbf{R}^2$. Let $K(x,\theta)$ be a kernel, i.e., for each θ, $K(x,\theta)$ is a probability density in x and jointly measurable in (x,θ). For any probability P on Θ, let $K(x,P) = \int K(x,\theta)P(d\theta)$. Lo's method consists of choosing a mixture $K(x,P)$ at random according to a Dirichlet prior. Note that when $K(x,\theta)$ is a location family $K(x,P)$ is just the convolution $K * P$. Typically, such location families will also have a scale parameter σ, which plays a role similar to that of window length in kernel density estimation. We look at the structure of the posterior before looking at specific kernels.

Formally, we have a hierarchical Bayes model which consists of $P \sim D_\alpha$ and given P, $X_1, X_2, \ldots, X_n$ are iid $K(\cdot, P)$; our interest is in the posterior distribution P given $X_1, X_2, \ldots, X_n$. This is the approach taken in West (1992) and West et al. (1994).

It is convenient to view the observations $X_1, X_2, \ldots, X_n$ as arising from P and $(\theta_1, X_1), (\theta_2, X_2), \ldots, (\theta_n, X_n)$, where $P \sim D_\alpha$ and given P, the pairs $(\theta_1, X_1), (\theta_2, X_2), \ldots, (\theta_n, X_n)$ are independent with θ_i having the distribution P and given θ_i, X_i is an observation from $K(\cdot, \theta_i)$. The latent variables θ_i, while unobserved, serve as a useful tool in describing and simulating the posterior. Indeed, West and others have set up efficient computing algorithms based on Gibbs sampling.

Letting $\Pi(P|X_1, X_2, \ldots, X_n)$ stand for the posterior distribution of P given $X_1, X_2, \ldots, X_n$ and by $H(\boldsymbol{\theta}|X_1, X_2, \ldots, X_n)$, the distribution of $(\theta_1, \theta_2, \ldots, \theta_n)$ given $X_1, X_2, \ldots, X_n$,

$$\Pi(P|X_1, X_2, \ldots, X_n) = \int \Pi(P|(\theta_1, X_1), (\theta_2, X_2), \ldots, (\theta_n, X_n))$$
$$\times H(d\boldsymbol{\theta}|X_1, X_2, \ldots, X_n).$$

Since P and the X_i's are conditionally independent given the θ_i's,

$$\Pi(P|(\theta_1, X_1), (\theta_2, X_2), \ldots, (\theta_n, X_n)) = D_{\alpha + \Sigma \delta_{\theta_i}},$$

and hence

$$\Pi(P|X_1, X_2, \ldots, X_n) = \int D_{\alpha + \Sigma \delta_{\theta_i}}(P)H(d\boldsymbol{\theta}|X_1, X_2, \ldots, X_n).$$

Let $\alpha(\mathbf{R}) = M$ and $\bar{\alpha} = \alpha/M$. Denote by $\hat{G}_n(\cdot, \boldsymbol{\theta})$ the empirical distribution based on $(\theta_1, \theta_2, \ldots, \theta_n)$. The Bayes estimate of $P(\cdot)$ then becomes

$$\frac{M}{M+n}\bar{\alpha}(\cdot) + \frac{n}{M+n}\int \hat{G}_n(\cdot, \boldsymbol{\theta})H(d\boldsymbol{\theta}|X_1, X_2, \ldots, X_n)$$

and the Bayes estimate of the density at x, $E(K(x,\theta)|X_1, X_2, \ldots, X_n)$, becomes

$$\frac{M}{M+n}K_0(x) + \frac{n}{M+n}\frac{1}{n}\sum_{i=1}^{n}\int K(x,\theta_i)H(d\boldsymbol{\theta}|X_1, X_2, \ldots, X_n),$$

where $K_0(x)$ is the prior expectation $\int K(x, \theta) \, \bar{\alpha}(d\theta)$. The Bayes estimate is thus composed of a part attributable to the prior and a part due to observations. Ferguson (1983) remarks that the second term in the above convex combination, namely $n^{-1} \sum_{i=1}^{n} \int K(x, \theta_i) H(d\theta | X_1, X_2, \ldots, X_n)$ can be viewed as a partially Bayesian estimate with the influence of the prior guess removed. The evaluation of the above quantities depend on $H(\theta | X_1, X_2, \ldots, X_n)$. The joint prior for $(\theta_1, \theta_2, \ldots, \theta_n)$ is, from Eq. (1),

$$\frac{\alpha(d\theta_1)}{\alpha(\mathbf{R})} \times \frac{(\alpha(d\theta_2) + \delta_{\theta_1})}{\alpha(\mathbf{R}) + 1} \times \cdots \times \frac{(\alpha(d\theta_n) + \sum_{i=1}^{n-1} \delta_{\theta_i})}{\alpha(\mathbf{R}) + n}.$$

Further, the likelihood given $(\theta_1, \theta_2, \ldots, \theta_n)$ is $\prod_{i=1}^{n} K(X_i, \theta_i)$. Hence H can be written down using standard Bayes formula. Some algebra yields the following expression for the Bayes estimate of the density at x:

$$\frac{M}{M+n} \int K(x, \theta) \bar{\alpha}(d\theta) + \frac{n}{M+n} \sum_p W(p)$$

$$\times \sum_{i=1}^{N(p)} \frac{e_i}{n} \frac{\int K(x, \theta) \prod_{l \in C_i} K(X_l, \theta) \alpha(d\theta)}{\int \prod_{l \in C_i} K(X_l, \theta) \alpha(d\theta)},$$

where $p = \{C_1, C_2, \ldots, C_n\}$ is a partition of $\{1, 2, \ldots, n\}$, $N(p)$ being the cardinality of p, e_i is the number of elements in C_i,

$$W(p) = \frac{\phi(p)}{\sum_{p'} \phi(p')}, \quad \text{and} \quad \phi(p) = \prod_{i=1}^{N(p)} \left\{ (e_i - 1)! \int \prod_{l \in C_i} K(X_l, \theta) \alpha(d\theta) \right\}.$$

Consistency of the Bayes estimates in these models is studied in Ghorai and Rubin (1982).

A different kind of application of Dirichlet mixtures is made by Brunner and Lo (1989), who estimate a decreasing density on the positive half-line by using a Dirichlet mixture of uniform densities. By a well-known theorem of Khintchine and Shepp, any decreasing density on the positive half-line is a given by a mixture $\int \theta^{-1} I\{0 \leq x \leq \theta\} P(d\theta)$ and conversely. Brunner and Lo (1989) induce a prior on the space of these densities by putting a Dirichlet prior on P. A symmetric (about zero) strongly unimodal density, which is precisely a symmetrization of a decreasing density on the positive half-line, is often a reasonable model for error distribution. Brunner and Lo (1989) also consider the estimation of a symmetric (about an unknown location) strongly unimodal density on the real line, by using the above Dirichlet mixture of uniform and some prior on the location of the symmetry. Brunner (1992) used a similar idea when the densities are not necessarily symmetric.

Brunner (1995) applied the idea of Brunner and Lo (1989) to the linear regression problem.

8.1   Random Histograms

The simplest kernel is $K(x, \theta) = h^{-1}$ if both x and θ are between $(ih, (i + 1)h)$. This corresponds to choosing a histogram with bins $(ih, (i + 1)h)$. A useful method suggested by Gasparini (1992) is to start with a prior μ with density $m(h)$ for h and given h, pick a histogram with bin width h along the following lines: Given h, let $\alpha_h = M_h \bar{\alpha}_h$ be a finite measure on integers and given h, let $P \sim D_{\alpha_h}$. Define the random histogram by

$$ f_{h,P} = \sum_{i=-\infty}^{\infty} \frac{P(\{i\})}{h} I_{[ih,(i+1)h]}. $$

If $n_i(h)$ is the number of $X_1, X_2, \ldots, X_n$ in the bin $[ih, (i + 1)h]$, it is not hard to see that the posterior is of the same form as the prior with α_h updated to $\alpha_h + \sum \delta_{n_i(h)}$ and $m(h)$ changing to

$$ m^*(h) = \frac{m(h) \prod_{i=1}^{\infty} (\alpha_h(\{i\}))^{(n_i(h)-1)}}{M_h + n}. $$

The predictive density with no observation is given by $\hat{f}(x) = \int f_h(x) m(h)\, dh$, where $f_h(x) = h^{-1} \sum_{i=-\infty}^{\infty} \bar{\alpha}_h(\{i\}) I_{[ih,(i+1)h]}(x)$.

In view of the conjugate property, the predictive density given observations $X_1, X_2, \ldots, X_n$ can be easily written down. For any f_0, let $f_{0,h}$ be the approximation f_0 by histograms of bin width h, i.e., if x is in $(ih, (i + 1)h]$ then $f_{0,h}(x) = (1/h) \int_{ih}^{(i+1)h} f(y)\, dy$. The calculations in Gasparini (1992) can be used to show:

THEOREM 10   Suppose the prior satisfies

1. α_h is a probability measure for all h;
2. For each h, there exists a constant K_h such that

$$ \frac{\alpha_h(\{j - 1\})}{\alpha_h(\{j\})} < K_h \qquad \text{for} \qquad j = \cdots, -1, 0, 1, 2, \cdots. $$

Then the posterior is weakly consistent at any f_0 satisfying $\int x^2 f_0(x)\, dx < \infty$ and $\lim_{h \to 0} \int f_0(x) \log(f_{0,h}/f_0)\, dx = 0$.

Under additional conditions, Gasparini (1992) shows that the Bayes estimate is strongly consistent. We expect that the techniques of BSW and the

Sec. 3 would enable in identifying densities for which the posterior is consistent on uniform neighborhoods.

Choice of m which is positive in a neighborhood of 0 will allow for a wide variability in the choice of histograms and will ensure that the prior has all of L_1 as support. If f_0 is one's prior belief about the density, then an appropriate choice of $\bar{\alpha}_h$ would be

$$\bar{\alpha}_h(\{i\}) = \int_{ih}^{(i+1)h} f_0(x)\, dx.$$

Of course, this choice will lead to a prior expected density which may not equal f_0 but can be viewed as an approximation to f_0. A different Bayesian histogram is proposed by Hartigan (1996). Another method is to consider for each k, a mixture of k many beta random variables. A suitable choice of the parameters of the beta leads to choosing the distribution function at random through Bernstein polynomials. This method was proposed by Diaconis and has been investigated by Petrone (1997).

8.2   Dirichlet Mixtures of Normal Densities

A natural choice for the kernel is the normal density $\phi_\sigma(x - \theta) = (1/\sqrt{2\pi}\sigma)\exp[-(x - \theta)^2/2\sigma^2]$. The σ in the kernel is analogous to the window length in density estimation. It may be chosen empirically or in a fully Bayesian way by selecting a prior on σ. We start looking at the case when σ is fixed. If P is chosen according to D_α then, as before, this set up yields the pair $(\theta_1, X_1), (\theta_2, X_2), \ldots, (\theta_n, X_n)$ of latent variables $\theta_1, \theta_2, \ldots, \theta_n$ which are iid P, and $X_i \sim N(\theta_i, \sigma^2)$.

The posterior calculations can be carried out as before. A convenient choice of $\bar{\alpha}$ is the conjugate prior $N(\mu, \tau^2)$. If $\alpha = M\bar{\alpha}$, Ferguson (1983) argues that as $M \to \infty$, the Bayes estimate with the influence of the prior removed, converges to $1/n \sum_{i=1}^n f(x|X_i)$, where $f(x|X_i)$ is the Bayes estimate of $K(x, \theta)$ based on a single observation X_i and when θ has the prior $\bar{\alpha}$. In particular when

$$\bar{\alpha} = N(\mu, \tau^2), f(x|X_i) = N\left(x \,\middle|\, \frac{\mu\sigma^2 + \tau^2 X_i}{\sigma^2 + \tau^2}, \frac{\sigma^2 + 2\tau^2}{\sigma^2 + \tau^2}\right).$$

Ferguson (1983) notes that "this yields a variable kernel estimate with constant window size, but centered at a point between X_i and μ as is typical of shrinkage estimates".

The sample $X_1, X_2, \ldots, X_n$ may also be viewed as conditionally independent (given $\theta_1, \theta_2, \ldots, \theta_n$) $N(\theta_i, \sigma^2)$ variables, where the means $\theta_1, \theta_2, \ldots, \theta_n$ are drawn from an uncertain P which is itself distributed as D_α. Given

$\theta_1, \theta_2, \ldots, \theta_n$, the next value θ_{n+1} is a new value with probability $M/(M+n)$ and is one of the previous ones with probability $n/(M+n)$. Thus if M is small, typically the n observations arise from few, much fewer than n, normal populations. This view is adopted by West, Escobar and others (West (1992), West et al. (1994)), where they effectively demonstrate the use of the Dirichlet mixture model in many applications.

In general, we expect consistency with the mixture only when the bandwidth is allowed to take arbitrarily small values. Suppose that the bandwidth is also given a prior having 0 in its support. Below we present some results of Ghosal et al. (1997b) regarding consistency of these mixtures.

Since the prior is on densities, consistency on uniform neighborhoods is the appropriate notion. The tool we use is Theorem 5 which involves two parts—the positive prior mass for Kullback–Leibler neighborhoods and sieves with suitable metric entropy. The first two results are concerned about the first issue, among which the Theorem 11 is about consistency at compactly supported densities and has a neat form. For this result, actually we neither need that the mixture is Dirichlet nor the kernel is normal. The only facts used are that the compactly supported density is in the weak support of the distribution of the mixtures and the kernel is positive and continuous.

THEOREM 11. Let f_0 be a density having compact support contained in the support of α. Suppose that $\lim_{\sigma \to 0} \int f_0 \log(f_0/f_{0,\sigma}) = 0$, where $f_{0,\sigma} = \phi_\sigma * f_0$. Then $\Pi(K_\varepsilon(f_0)) > 0$ for all $\varepsilon > 0$.

The analogue of Theorem 12 when α has support all of $\mathbf{R}$ is somewhat involved and as in Shyamalkumar (1996), requires that the tail behavior of f_0 and α be related. Loosely speaking, a result of Doss and Sellke (1982) shows that there are functions l_1, u_1 and l_2, u_2 such that with probability 1 under D_α, the upper tail $P(x, \infty)$ is greater than $l_1(x)$ and $P(x + k\log x, \infty)$ is less than $u_1(x)$. The functions l_2, u_2 deal similarly with the lower tails. For any $\sigma > 0$, set

$$L_\sigma(x) = \begin{cases} \phi_\sigma(k \log x)(l_1(x) - u_1(x)), & \text{if } x > 0, \\ \phi_\sigma(k \log x)(l_2(x) - u_2(x)), & \text{if } x < 0. \end{cases}$$

THEOREM 12. If

1. $\displaystyle\lim_{\sigma \to 0} \int f_0 \log(f_0/f_{0,\sigma}) = 0$;

2. $\displaystyle\lim_{a \to \infty} \int_{-\infty}^{\infty} f_0(x) \log \left(\frac{f_{0,\sigma}(x)}{\int_{-a}^{a} \phi_\sigma(x - \theta) f_0(\theta) d\theta} \right) dx = 0$,

3. for all σ,

$$\lim_{M \to \infty} \int_{|x|>M} f_0(x) \log\left(\frac{f_0(x)}{L_\sigma(x)}\right) = 0,$$

then $\Pi(K_\varepsilon(f_0)) > 0$ for all $\varepsilon > 0$.

For example, when α is double exponential, we may choose any $k > 2$ and the requirements of the theorem are satisfied if f_0 has finite moment generating function in an open interval containing $[-1, 1]$. If α is normal, the condition needed is the integrability of $x(\log x)e^{x^2/2}$ with respect to f_0.

For strong consistency, we estimate the L_1-metric entropies of sets of the form $\{\phi_\sigma * P : P[-a, a] = 1\}$ and $\{\phi_\sigma * P : P[-a, a] \geq 1 - \delta\}$. Then using Theorem 5, we can establish the following two consistency theorems.

THEOREM 13. Let α have compact support. Suppose that the prior μ for σ has bounded support and satisfies $\mu\{\sigma < t\} \leq c_1 \exp[-c_2/t]$ for some c_1, c_2. Then $\Pi(K_\varepsilon(f_0)) > 0$ for all $\varepsilon > 0$ implies that the posterior is strongly consistent at f_0.

THEOREM 14. Suppose the prior μ for σ has bounded support. For any $\delta > 0$, let a_n be such that for some β_1, for all large n,

$$D_\alpha\{P : P[-a_n, a_n] < 1 - \delta\} < e^{-n\beta_1}.$$

For an $\eta > 0$, suppose that there is a sequence $\sigma_n \downarrow 0$ be such that $a_n/\sigma_n < n\eta$ and $\mu\{\sigma < \sigma_n\} \leq e^{-n\beta_0}$ for some $\beta_0 > 0$. Then $\Pi(K_\varepsilon(f_0)) > 0$ for all $\varepsilon > 0$ implies that the posterior is strongly consistent at f_0.

If for example, α is chosen as a normal distribution and σ^2 is given an inverse gamma prior, then a_n in Theorem 14 is of the order $\sqrt{n}$ and $\sigma_n = C/\sqrt{n}$ for a suitable (large) C (depending on δ) satisfies the conditions of Theorem 14.

It seems more natural to consider the location and scale of the base measure of Dirichlet prior for the mixing distribution as unknown hyperparameter with some specified prior. Under some conditions, consistency continues to hold for such priors.

Another way of handling σ is to treat the pair (θ, σ) as the a hyper parameter and consider a Dirichlet prior for the distribution of (θ, σ). This would yield Dirichlet mixtures of normal over both the location and scale parameters. Ferguson (1983) carries out an analysis when the $\bar{\alpha}$ is the normal-gamma conjugate prior.

Interestingly, Theorems 11 and 12 have an important implication in the location problem discussed in Sec. 6. If we smooth the (symmetrized) Dirichlet process used by DF by a normal kernel with variance h^2 and h is chosen from a prior distribution on $(0, \infty)$ having 0 in its support, then the posterior distribution of the location parameter is consistent if the true error density is symmetric, satisfies conditions of Theorem 7 and $K(f_0, f_0 * \phi_h) \to 0$ as $h \to 0$, where $\phi_h(\cdot)$ stands for the normal density with mean 0 and variance h^2. This follows from Theorem 7 by observing the following facts:

(1) Since the normal kernel is symmetric, smoothing a symmetrized Dirichlet process is same as symmetrizing a smoothed Dirichlet process;

(2) If Π is a prior and f_0 is a symmetric density with $\Pi(K_\varepsilon(f_0)) > 0$, then the symmetrization $\tilde{\Pi}$ of Π also satisfies $\tilde{\Pi}(K_\varepsilon(f_0)) > 0$ (see Lemma 4.1 of Ghosal et al. (1998)). The posterior density for the location parameter is a smooth function in this case, as opposed to the posterior based on a Polya tree prior. On the other hand, computation is much more involved in this case.

9. GAUSSIAN PROCESS PRIORS

These priors introduced by Lenk (1988) are generalizations of a construction of Leonard (1978). The idea is to start with a Gaussian process $\{Z(x, \omega), x \in I\}$ on an interval I and define a random density on I through

$$f(x, \omega) = \frac{\exp[Z(x, \omega)]}{\int_I \exp[Z(t, \omega)] \, dt}.$$

The Gaussian process has as parameters the mean $\mu(x)$ and the covariance kernel $\sigma(x, y)$. Lenk introduces an additional parameter ξ to obtain a conjugate family.

Let $\mu(x)$ be a continuous mean function and $\sigma(x, y)$ be continuous and positive definite and let $\{Z(x, \omega) : x \in I\}$ be a Gaussian process with mean $\mu(x)$ and covariance kernel $\sigma(x, y)$. It is convenient to introduce the intermediate process $W(x, \omega) = \exp[Z(x, \omega)]$. Denote the distribution of W by $LN(\mu, \sigma, 0)$ (LN stands for lognormal). For each ξ define a process or equivalently a probability measure $LN(\mu, \sigma, \xi)$ on $(\mathbf{R}^+)^I$ by

$$\frac{dLN(\mu, \sigma, \xi)}{dLN(\mu, \sigma, 0)}(\omega) = \frac{[\int_I W(x, \omega) \, dx]^\xi}{C(\mu, \sigma, \xi)}.$$

The function on $(\mathbf{R}^+)^I$ defined by $f(x, \omega) = W(x)/\int W(t)\,dt$ gives a random density and the distribution of this density under $LN(\mu, \sigma, \xi)$ is denoted by $LNS(\mu, \sigma, \xi)$.

Suppose $f \sim LNS(\mu, \sigma, \xi)$ and given f, $X_1, X_2, \ldots, X_n$ are iid f. Then the posterior is $LNS(\mu^*, \sigma, \xi^*)$, where

$$\mu^*(x) = \mu(x) + \sum_{i=1}^{n} \sigma(x_i, x), \qquad \xi^* = \xi - n.$$

This expression, though not entirely convincingly, allows ξ to be thought of as a "pseudo sample size". Similarly if $\sigma(x, y)$ is of the form $\rho(|x - y|)$, then ρ might be considered as a strength of belief about the prior.

In Ghosh and Ramamoorthi (1997), consistency is shown when the Gaussian process is a standard Brownian motion. Consistency issues in general case is under investigation.

10. CENSORED DATA

Dirichlet process and Polya trees provide an elegant framework for the Bayesian analysis of right censored data. The model under consideration consists of $X_1, X_2, \ldots, X_n$ and $Y_1, Y_2, \ldots, Y_n$ positive iid random variables with distributions F and G respectively. The X's correspond to life times and the Y's to censoring times; the observations are (Z_i, δ_i), where $Z_i = (X_i \wedge Y_i)$, $\delta_i = I_{[X_i < Y_i]}, i = 1, 2, \ldots, n$. The goal is to make inference on F, the distribution of the life time.

Susarla and van Ryzin (1976) investigate the case when $F \sim D_\alpha$ and later Blum and Susarla (1977) show that the posterior distribution of F can be obtained as a mixture of Dirichlet process. The mixture representation is cumbersome and an alternative representation can be obtained as a Polya tree with the partition depending on $(Z_1, \delta_1), (Z_2, \delta_2), \ldots, (Z_n, \delta_n)$. We describe this representation below.

Let $(Z_1, \delta_1), (Z_2, \delta_2), \ldots, (Z_n, \delta_n)$ be the observations and the distinct values of the censored observations, arranged in increasing order, be denoted by $Z^* = (Z_{(1)}, Z_{(2)}, \ldots, Z_{(k)})$. Construct a sequence $T(Z^*)$ of nested partitions as follows:

$$B_0 = (0, Z_{(1)}], \qquad B_1 = (Z_{(1)}, \infty),$$

$$B_{10} = (Z_{(1)}, Z_{(2)}], \qquad B_{11} = (Z_{(2)}, \infty),$$

and in general if $\mathbf{1}_m$ is a string of m 1's, then for $m \leq k - 1$,

$$B_{\mathbf{1}_m 0} = (Z_{(m)}, Z_{(m+1)}]$$

and

$$B_{1_m1} = (Z_{(m+1)}, \infty).$$

The other B_ε are partitioned arbitrarily into two intervals so that the partition $T(Z^*)$ satisfies the conditions in Sec. 7.

THEOREM 15. Let $F \sim D_\alpha$. Then the posterior is Polya tree with respect to the partition $T(Z^*)$ and with parameters

$$\alpha^*_{\varepsilon_1,\varepsilon_2,...,\varepsilon_k} = \alpha(B_{\varepsilon_1,\varepsilon_2,...,\varepsilon_k}) + \sum_{i=1:\delta_i=1}^{n} I_{B_{\varepsilon_k}}(Z_i) + C_{\varepsilon_k},$$

where C_{ε_k} is the number of pairs $(Z_i, 0)$'s such that the interval (Z_i, ∞) is contained in B_{ε_k}.

Note that if $B_{\varepsilon_k} = (Z_k, \infty)$, then C_{ε_k} is just the number of individuals on test at time Z_k.

The Polya tree representation of the posterior gives consistency.

THEOREM 16. If $F \sim D_\alpha$, then the posterior distribution of F is weakly consistent.

This and other related consistency results appear in Rajagopalan (1997).

Important classes of priors such as the simple homogeneous process (Ferguson and Phadia (1979)), and the extended gamma process (Dykstra and Laud (1981)) have been used for analyzing right censored data. These priors are neutral to the right in the sense of Doksum (1974). Another interesting class of priors for right censored data is the class of beta process introduced by Hjort (1990). These have been further generalized and studied by Muliere and Walker (Muliere and Walker (1997), Walker and Muliere (1997)). We expect the posterior consistency to hold in these cases.

11. NON-INFORMATIVE PRIORS

Except for the choice $\alpha(\mathbf{R}) \to 0$ for a Dirichlet process, there is very little development towards a notion (or notions) of non-informative priors for infinite dimensional models. This section summarizes Ghosal et al. (1997a), where some tentative proposals were made.

Let $\mathscr{F}$ be a family of densities equipped with, say, the Hellinger metric $d_H(f, g)$ defined by $d_H^2(f, g) = \int (\sqrt{f} - \sqrt{g})^2 dx$. Assume further that $\mathscr{F}$ is compact. For each $\varepsilon > 0$, let $\mathscr{F}_\varepsilon$ be an ε-sieve, i.e., a maximal set with the property $\mathscr{F}_\varepsilon \subset \mathscr{F}$ and $d_H(f, g) > \varepsilon$ for every f, g in $\mathscr{F}_\varepsilon$. Since $\mathscr{F}$ is compact, $\mathscr{F}_\varepsilon$ is a finite set. Denote by $D(\varepsilon, \mathscr{F})$ the cardinality of $\mathscr{F}_\varepsilon$.

Since $\mathscr{F}_\varepsilon$ is, in a sense, an ε-approximation of $\mathscr{F}$, a natural approach would be to take the uniform distribution on $\mathscr{F}_\varepsilon$, say Π_ε, as an approximation to whatever might be considered as the "uniform" distribution on $\mathscr{F}$. In practice, ε should depend on the sample size to reflect the disposition to entertain more complex models when a large sample is available. This approach, while attractive, would, in view of the dependence of the prior on sample size, run into problems of incoherence in the sense of Heath and Sudderth (1978).

Another approach would be to take any limit point Π^* of $\{\Pi_\varepsilon : \varepsilon > 0\}$ as a non informative prior. If Π^* is unique, it is precisely the uniform distribution defined and studied by Dembski (1990). In Ghosal et al. (1997a), it is shown that in finite dimensional regular models this approach leads to Jeffreys' prior.

In keeping with the spirit of the mixture models studied earlier, yet another alternative is to view ε as a hyper parameter and consider a hierarchical prior λ for ε. In Ghosal et al. (1997a), the following consistency result is proved for such priors.

THEOREM 17. Let $\mathscr{F}$ be a family of densities where $\mathscr{F}$, metrized by the Hellinger distance, is compact. Let ε_n be a positive sequence satisfying the condition $\sum_{n=1}^{\infty} n^{1/2}\varepsilon_n < \infty$. Let $\mathscr{F}_n$ be an ε_n-sieve in $\mathscr{F}$, μ_n be the uniform distribution on $\mathscr{F}_n$ and μ be the probability on $\mathscr{F}$ defined by $\mu = \sum_{n=1}^{\infty} \lambda_n \mu_n$, where λ_n's are positive numbers adding upto unity. If for any $\beta > 0$

$$\lim_{n \to \infty} e^{\beta n} \frac{\lambda_n}{D(\varepsilon_n, \mathscr{F}_n)} = \infty, \tag{3}$$

then the posterior distribution based on the prior μ and iid observations $X_1, X_2, \ldots, X_n$ is consistent at every $f_0 \in \mathscr{F}$.

An example where this theorem is applicable is the following class of densities considered in density estimation, see, e.g., Wong and Shen (1995):

$$\mathscr{F} = \left\{ f = g^2 : g \in C^r[0,1], \int g^2(x)dx = 1, \|g^{(j)}\|_{\sup} \le L_j, j = 1, \ldots, r, \right.$$

$$\left. |g^{(r)}(x_1) - g^{(r)}(x_2)| \le L_{r+1}|x_1 - x_2|^m \right\},$$

where r is a positive integer, $0 \le m \le 1$ and L_j's are fixed constants. By Theorem XV of Kolmogorov and Tihomirov (1961), $D(\varepsilon, \mathscr{P}) \le \exp[c\varepsilon^{-1/(r+m)}]$. Hence the hierarchical prior constructed in Theorem 17 leads to consistent posterior. With a little modification of the construction of the prior, in this case it is possible to achieve a rate of convergence of the order $n^{-(r+m)/(2(r+m)+1)}$ of the posterior distribution, which is optimal.

Diaconis and Freedman (1993) consider a binary regression problem and show that a hierarchical prior based on uniform distributions (on certain finite dimensional sets) leads to a consistent posterior under a condition on the rate of decay of the hierarchical weights. This prior is somewhat similar in spirit to what we considered above.

While we have demonstrated the feasibility of obtaining consistency for a variety of popular priors, consistency by itself is not adequate. Given a consistency result, one would want results on rates of convergence or simulations to get an idea of how large an n is required to get convergence to δ_{P_0} to a desired level of accuracy. More precisely, given P_0, ε, η and U, one would want an n_0 such that for $n > n_0$, $\Pi(U|X_1, X_2, \ldots, X_n) \geq 1 - \varepsilon$ with P_0-probability greater than $1 - \eta$. Such results or simulations would be relatively easy to get for tail free priors and weak neighborhoods U. Similar results in the context of Theorem 3 or 5 will require more work. We will return to these topics elsewhere.

REFERENCES

Antoniak, C. (1974). Mixtures of Dirichlet processes with application to Bayesian non-parametric problems. *Ann. Statist.* 2: 1152–1174.

Barron, A. R. (1989). Uniformly powerful goodness of fit tests. *Ann. Statist.* 17: 107–124.

Barron, A. R., Schervish, M. and Wasserman, L. (1996). The consistency of posterior distributions in non parametric problems. Technical report, Carnegie Mellon University. *Ann. Statist.* To appear.

Berk, R. (1966). Limiting behavior of the posterior distribution when the model is incorrect. *Ann. Math. Statist.* 37: 51–58.

Blackwell, D. and MacQueen, J. B. (1973). Ferguson distributions via Polya urn schemes. *Ann. Statist.* 1: 353–355.

Blum, J. and Susarla, V. (1977). On the posterior distribution of a Dirichlet process given randomly right censored observations. *Stoch. Processes Appl.* 5: 207–211.

Brunner, L. J. (1992). Bayesian nonparameteric methods for data from a unimodal density. *Statist. Probab. Lett.* 14: 195–199.

Brunner, L. J. (1995). Bayesian linear regression with error terms that have symmetric unimodal densities. *J. Nonparameteric Statist.* *4*: 335–348.

Brunner, L. J. and Lo, A. Y. (1989). Bayes methods for a symmetric unimodal density and its mode. *Ann. Statist.* *17*: 1550–1566.

Dalal, S. R. (1979). Dirichlet invariant processes and application to nonparametric estimation of symmetric distributions. *Stoch. Process Appl.* *9*: 99–107.

Dembski, W. A. (1990). Uniform probability. *J. Theoret. Probab.* *3*: 611–626.

Diaconis, P. and Freedman, D. (1986a). On the consistency of Bayes estimates (with discussion). *Ann. Statist.* *14*: 1–67.

Diaconis, P. and Freedman, D. (1986b). On inconsistent Bayes estimates of location. *Ann. Statist.* *14*: 68–87.

Diaconis, P. and Freedman, D. (1993). Nonparametric binary regression: a Bayesian approach. *Ann. Statist.* *21*: 2108–2137.

Doksum, K. A. (1974). Tailfree and neutral random probabilities and their posterior distributions. *Ann. Probab.* *2*: 183–201.

Doob, J. L. (1948). Application of the theory of martingales. *Coll. Int. du CNRS, Paris,* 22–28.

Doss, H. (1985). Bayesian nonparametric estimation of the median. II. Asymptotic properties of the estimates. *Ann. Statist.* *13*: 1445–1464.

Doss, H. and Sellke, T. (1982). The tails of probabilities chosen from a Dirichlet prior. *Ann. Statist.* *10*: 1302–1305.

Dykstra, R. L. and Laud, P. W. (1981). A Bayesian nonparameteric approach to reliability. *Ann. Statist.* *9*: 356–367.

Ferguson, T. S. (1973). A Bayesian analysis of some nonparametric problems. *Ann. Statist.* *1* 209–230.

Ferguson, T. S. (1974). Prior distribution on the spaces of probability measures. *Ann. Statist.* *2*: 615–629.

Ferguson, T. S. (1983). Bayesian density estimation by mixtures of Normal distributions. In *Recent advances in Statistics* (Rizvi M., Rustagi, J. and Siegmund, D., Eds.) 287–302.

Ferguson, T. S. and Phadia, E. G. (1979). Bayesian nonparameteric estimation based on censored data. *Ann. Statist.* *7*: 163–186.

Ferguson, T. S., Phadia, E. G. and Tiwari, R. (1996). Bayesian nonparametric inference. In *Current issues in Statistical inference. Essays in honor of D. Basu* (Ghosh, M. and Pathak, P. K., Eds.) 127–150.

Freedman, D. (1963). On the asymptotic distribution of Bayes estimates in the discrete case I. *Ann. Math. Statist. 34*: 1386–1403.

Freedman, D. and Diaconis, P. (1983). On inconsistent Bayes estimates in the discrete case. *Ann. Statist. 11*: 1109–1118.

Gasparini, M. (1992). *Bayes Nonparametrics for biased sampling and density estimation.* Ph. D. thesis, University of Michigan.

Ghorai, J. K. and Rubin, H. (1982). Bayes risk consistency of nonparametric Bayes density estimates. *Austral. J. Statist. 24*: 51–66.

Ghosal, S., Ghosh, J. K. and Ramamoorthi, R. V. (1997a). Noniformative priors via sieves and consistency. In *Advances in Statistical Decision Theory and Applications* (S. Panchapakesan and N. Balakrishnan, Eds.) Birkhauser, Boston, 1997, 119–132.

Ghosal, S., Ghosh, J. K. and Ramamoorthi, R. V. (1997b). Posterior consistency of Dirichlet mixtures in density estimation. Technical report # WS-490, Vrije Universiteit, Amsterdam. *Ann. Statist.* To appear.

Ghosal, S., Ghosh, J. K. and Ramamoorthi, R. V. (1998). Consistent semi-parametric estimation about a location parameter. *J. Statist. Plan. Inf.* To appear.

Ghosh, J. K. and Ramamoorthi R. V. (1997). *Lecture notes on Bayesian asymptotics.* Under preparation.

Hartigan, J. A. (1996). Bayesian histograms. In *Bayesian statistics 5* (Bernardo J. et al. Eds.) 211–222.

Heath, D. and Sudderth, W. (1978). On finitely additive priors, coherence and extended admissibility. *Ann. Statist. 6*: 333–345.

Hjort, N. L. (1990). Nonparametric Bayes estimators based on beta processes in models for life history data. *Ann. Statist. 18*: 1259–1294.

Hjort, N. L. (1996). Bayesian approaches to non- and semiparametric density estimation. In *Bayesian statistics 5* (Bernardo J. et al., Eds.) 223–253.

Kolmogorov, A. N. and Tihomirov. V. M. (1961). ε-entropy and ε-capacity of sets in function spaces. *Amer. Math. Soc. Transl. Ser. 2*: 17 277–364. (Translated from Russian: *Uspekhi Mat. Nauk 14*: 3–86, (1959).)

Lavine, M. (1992). Some aspects of Polya tree distributions for statistical modeling. *Ann. Statist. 20*: 1222–1235.

Lavine, M. (1994). More aspects of Polya tree distributions for statistical modeling. *Ann. Statist. 22*: 1161–1176.

Le Cam, L. (1973). Convergence of estimates under dimensionality restrictions. *Ann. Statist. 1*: 38–53.

Lenk, P. J. (1988). The logistic normal distribution for Bayesian, nonparametric, predictive densities. *J. Amer. Statist. Assoc. 83*: 509–516.

Leonard, T. (1978). Density estimation, stochastic processes, and prior information. *J. Roy. Statist. Soc., Ser. B 40*: 113–146.

Lo, A. Y. (1994). On a class of Bayesian nonparametric estimates I: Density estimates. *Ann. Statist. 12*: 351–357.

Mauldin, R. D., Sudderth, W. D. and Williams, S. C. (1992). Polya trees and random distributions. *Ann. Statist. 20*: 1203–1221.

Muliere, P. and Walker, S. G. (1997). A Bayesian nonparametric approach to survival analysis using Polya trees. *Scand. J. Statist. 24*: 231–240.

Newton, M. A. (1994). A diffuse prior limit in semiparametric binary regression. Technical Report # 936, Department of Statistics, University of Wisconsin, Madison.

Rajagopalan, K. Srikanth (1997). *Posterior consistency in some Bayesian nonparametric problems.* Ph. D. Thesis, Michigan State University.

Schwartz, L. (1965). On Bayes procedures. *Z. Wahrsch. Verw. Gebiete 4*: 10–26.

Sethuraman, J. and Tiwari, R. (1982). Convergence of Dirichlet measures and interpretation of their parameters. In *Statistical Decision Theory and Related Topics. III 2* (Gupta, S. S. and Berger, J. O., Eds.), Academic Press, New York, 305–315.

Schervish, M. J. (1995). *Theory of Statistics.* Springer Series in Statistics. Springer-Verlag, New York.

Shyamalkumar, N. D. (1996). *Contributions to Bayesian Nonparametrics and Bayesian Robustness.* Unpublished Ph. D. Thesis, Purdue University.

Petrone, S. (1997). Bayesian density estimation using Bernstein polynomials. Preprint.

Susarla V. and van Ryzin, J. (1976). Nonparametric Bayesian estimation of survival curves from incomplete observations. *J. Amer. Statist. Assoc.* *71*: 897–902

Walker, S. G., Damien, P., Laud, P. W. and Smith, A. F. M. (1997). Bayesian nonparametric inference for random distributions and related functions. Preprint.

Walker, S. G. and Muliere, P. (1997). Beta-Stacy processes and a generalization of the Polya-urn scheme. *Ann. Statist. 25*: 1762–1780.

West, M. (1992). Modeling with Mixtures. In *Bayesian Statistics 4*: (J. M. Bernardo et al., Eds.) 503–524.

West, M., Muller, P. and Escobar, M. D. (1994). Hierarchical priors and mixture models, with applications in regression and density estimation. In *Aspects of uncertainty: A Tribute to D. V. Lindley.* Eds: A. F. M. Smith and P. Feeman, John Wiley and Sans, New York. 363–386.

Wong, W. H. and Shen, X. (1995). Probability inequalities for likelihood ratios and convergence rates of sieve MLEs. *Ann. Statist. 23*: 339–362.

19

Breakdown Theory for Estimators Based on Bootstrap and Other Resampling Schemes

GUTTI JOGESH BABU Pennsylvania State University, University Park, Pennsylvania

1. INTRODUCTION

The computer intensive resampling method called bootstrap has become very popular in the mid-eighties and early nineties. Its success lies in the ease of estimation of the sampling distribution, standard error and confidence intervals, with little or no assumptions about the distribution of the underlying population. Among other desirable properties, it derives its strength from the second-order accuracy in estimating the sampling distributions of a wide class of commonly used statistics. Although the bootstrap is generally applied as a nonparametric tool, there are very few studies on robustness of the method.

In this chapter, the bootstrap method is examined using the notion of breakdown point in robustness, in the context of estimation of the variance of an estimator and of confidence intervals. Since the bootstrap utilizes all the data points, in general, the bootstrap estimator of variance of a statistic is not robust even for robust statistics. An alternative resampling procedure known as the half-sample method is explored, to get around this problem.

Another related area is the study of breakdown points for bootstrap quantiles. The quantiles are essential for deriving the confidence intervals.

Research supported in part by NSA grant MDA904-97-1-0023 and NSF grant DMS-9626189.

Effects of Winsorization before resampling on the breakdown points are discussed.

The bootstrap method has become very popular in statistical inference ever since the appearance of the paper by Efron (1979). Efron's work generated an unprecedented enthusiasm among mathematical statisticians as well as data analysts. In a series of papers, Babu and Singh provided theoretical support to Efron's bootstrap. See Babu (1984, 1986), Babu and Singh (1983, 1984a, 1984b), Singh (1981), and Singh and Babu (1990). See also Babu and Bose (1988), Bose and Babu (1991), and Linder and Babu (1994). The work on the asymptotic theory of the bootstrap in the early eighties resulted in establishing the superiority of the bootstrap approximation for a wide class of statistics. Babu and Bai (1996) developed a unified theory for various resampling procedures including the Bayesian bootstrap. The texts by Efron (1982), Efron and Tibshirani (1993), Hall (1992), Shao and Tu (1995), and the review articles by Babu (1989) and Babu and Rao (1993) are some of the selected references to the literature on bootstrap methodology. In spite of active contributions from numerous researchers on several different aspects of the bootstrap methodology, very little is known about robustness, Huber (1981) of bootstrap estimators.

Standard error or sampling variance and confidence bounds are two widely used measures to assess the accuracy in statistical inference. Jackknife and Bootstrap are two of the most popular methods employed in practice to estimate the variance of an estimator, especially when a closed form expression is not available. Henceforth the term estimation of variance will be used to mean estimation of the variance of the sampling distribution of an estimator. Jackknife and bootstrap methods of estimation of variance are briefly described in the next two sections (Sec. 2 and Sec. 3). Does the bootstrap method lead to robust estimation of variance and/or confidence intervals, at least for robust estimators? This question is examined in detail in this chapter.

One measure of robustness is the celebrated concept of breakdown point (see Sec. 4). Stromberg (1997) discusses the breakdown points of jackknife and bootstrap covariance estimators. It is shown in Sec. 5 that the bootstrap estimator of variance of a robust statistic is not necessarily robust. The present article focuses on non-smooth but robust statistics such as the sample median, where jackknife leads to an inconsistent estimation of variance, while the bootstrap estimator is consistent. (The sample median is considered a non-smooth statistic as it cannot be obtained as a smooth function of the sample mean of multivariate random variables.) An alternative resampling procedure called the half-sample method is explored in Sec. 6, and shown to lead to robust estimation of variance. Babu (1992) investigated the half-sample method and related sub-sample methods in detail. The half-sample method has some advantages over the bootstrap, especially in the case of non-homogeneous populations. In particular, Babu (1992) has established that the half-sample method is robust in estimating the parameters of a linear regression model when the errors are heterogeneous. The half-sample method does not share some of the second-order properties enjoyed by the bootstrap method (see Babu (1992), p. 708). Advantages of the half-sample method in robust estimation of variance are detailed in Sec. 6.

The bootstrap has also become a popular method for estimating confidence intervals, see Babu and Bose (1988), and Hall (1992). Quantiles are essential in deriving confidence intervals. The recent results of Singh (1996) on breakdown points for bootstrap quantiles of robust statistics are discussed in Sec. 7. His suggestions to improve the breakdown points are also briefly discussed.

2. JACKKNIFE ESTIMATION OF VARIANCE

Let $\hat{\theta}_n$ be an estimator of θ based on n iid random vectors $X_1, \ldots, X_n$, i.e., $\hat{\theta}_n = f_n(X_1, \ldots, X_n)$, for some function f_n. Let

$$\hat{\theta}_{n,-i} = f_{n-1}(X_1, \ldots, X_{i-1}, X_{i+1}, \ldots, X_n)$$

be the corresponding recomputed statistic based on all but the i-th observation. The jackknife estimator of the variance of $\hat{\theta}_n$ is then given by

$$\mathrm{Var}_J(\hat{\theta}_n) = \frac{n-1}{n} \sum_{i=1}^{n} (\hat{\theta}_{n,-i} - \hat{\theta}_n)^2.$$

For most statistics, jackknife estimation is consistent, i.e.,

$$\mathrm{Var}_J(\hat{\theta}_n)/\mathrm{Var}(\hat{\theta}_n) \to 1,$$

as $n \to \infty$. Hence it is very attractive in practice; more so, due to its computational and conceptual simplicity. However, consistency does not always hold; for example the jackknife method fails for non-smooth statistics, such as the sample median. If $\hat{\theta}_n$ denotes the sample median, then in general,

$$\mathrm{Var}_J(\hat{\theta}_n)/\mathrm{Var}(\hat{\theta}_n) \to \left(\frac{1}{2}\chi_2^2\right)^2$$

in distribution, where χ_2^2 denotes a chi-square random variable with 2 degrees of freedom (see Efron (1982), Sec. 3.4). So in this case, the jackknife method does not lead to a consistent estimator of the variance.

3. BOOTSTRAP ESTIMATION OF VARIANCE

To describe the bootstrap method, let $X_1, \ldots, X_n$ denote independent random vectors with a common distribution function F. Let $\hat{\theta}_n$ be an estimator of θ based on $X_1, \ldots, X_n$, i.e., $\hat{\theta}_n = f_n(X_1, \ldots, X_n)$, for some function f_n. Suppose $X_1^*, \ldots, X_n^*$ is a sample from the empirical distribution function F_n of $X_1, \ldots, X_n$. That is, $X_1^*, \ldots, X_n^*$ are sampled with replacement from $X_1, \ldots, X_n$. Let $\theta_n^* = f_n(X_1^*, \ldots, X_n^*)$. The bootstrap variance Var*, is simply the conditional variance of θ_n^* under the empirical measure, given the data $X_1, \ldots, X_n$. In general it can be shown that

$$\text{Var}^*(\theta)_n^* / \text{Var}(\hat{\theta}_n) \to 1,$$

as $n \to \infty$. In principle, $\text{Var}^*(\theta_n^*)$ is completely known, as it is a known function of the observations $X_1, \ldots, X_n$.

For many simple statistics, it is possible to evaluate the bootstrap variance analytically, by finding a closed form. For example, if $\hat{\theta}_n$ denotes the sample mean $\bar{X}_n = n^{-1} \sum_{i=1}^{n} X_i$, then its bootstrap variance is $n^{-2} \sum_{i=1}^{n} (X_i - \bar{X}_n)^2$. In the general case, when there is no such simple closed formula, the bootstrap variance is approximated by the Monte Carlo method as follows.

Using a computer, first generate B bootstrap samples of size n each, by simple random sampling with replacement:

$$X_1^{*(1)}, \quad \ldots, \quad X_n^{*(1)}$$

$$X_1^{*(2)}, \quad \ldots, \quad X_n^{*(2)}$$

$$\ldots \quad \ldots \quad \ldots$$

$$X_1^{*(B)}, \quad \ldots, \quad X_1^{*(B)}.$$

Then compute the corresponding statistics $\theta_i^*; i = 1, \ldots, B$. If B is of the order of $(n \log n)^2$, then in general, the sample variance of $\theta_1^*, \ldots, \theta_B^*$,

$$\text{Var}_B(\theta_n^*) = \frac{1}{B} \sum_{i=1}^{B} (\theta_i^* - \bar{\theta}_B^*)^2,$$

provides a good approximation to the bootstrap variance, where

$$\bar{\theta}_B^* = \frac{1}{B} \sum_{i=1}^{B} \theta_i^*.$$

While the jackknife method is not suitable for estimation of variance of non-smooth statistics such as the sample median or sample quantiles, the bootstrap method provides a conceptually simple way of estimation. This raises the question whether the bootstrap estimator of the variance of a robust statistic is robust. In other words, is $\text{Var}^*(\theta_n^*)$ robust for robust statistics $\hat{\theta}_n$?

4. BREAKDOWN POINT

To answer the above question, the concept of breakdown in robustness, attributable to Hampel (1971, 1974), is introduced here. The breakdown point is perhaps the most widely used measure of robustness in modern statistical literature. In order to describe the concept of breakdown point, let $X_1, \ldots, X_n$ have a common distribution, and let $\hat{\theta}_n = f_n(X_1, \ldots, X_n)$ be a statistic. Let m be the least number of X_i destabilizing $\hat{\theta}_n$. That is, m is the minimum number of data points that need to be replaced by worst possible outliers to move the statistic beyond any bound. Then the breakdown point is defined as m/n. In the case of the sample mean,

$$\hat{\theta}_n = \frac{1}{n} \sum_{i=1}^{n} X_i = \frac{1}{n} \sum_{i=1}^{n} X_{(i)},$$

the breakdown point is $1/n$, where $X_{(1)} < X_{(2)} < \cdots < X_{(n)}$ is the ordering of the data $X_1, \ldots, X_n$. The result holds because replacement of the single data point $X_{(n)}$ by a very large real number destabilizes the estimate $\hat{\theta}_n$ of the mean. Singh (1993) provides a detailed discussion of paradoxes in robustness, in particular of breakdown points. On the other hand, for the sample median $m_n = X_{(r)}$, where

$$r = \frac{n}{2} \qquad \text{if } n \text{ is an even integer}$$

$$= \frac{(n+1)}{2} \qquad \text{if } n \text{ is an odd integer},$$

the breakdown point is $r/n = \frac{1}{2}$ or $\frac{1}{2} + 1/2n$, depending on whether n is an even or an odd integer.

For example if $n = 5$, increasing the largest two observations without any bound will not affect m_5, but changing three points will. So the breakdown point here is $\frac{3}{5}$. In case $n = 4$, decreasing the smallest observation without any bound will not affect m_4. But if two observations are moved, then m_4 will change, so the breakdown point is $\frac{1}{2}$. Therefore the breakdown point of the sample median is approximately $\frac{1}{2}$, when the sample size n is large.

5. BOOTSTRAP VARIANCE OF MEDIAN

Although the sample median is a robust estimator of location, as indicated by the asymptotically maximum possible value for its breakdown point, its bootstrap variance estimator is not robust. To see this let m_n^* denote the sample median of a sample, $X_1^*, \ldots, X_n^*$, from the empirical distribution F_n of $X_1, \ldots, X_n$. It is easy to see that

$$m_n^* \leq X_{(i)} \quad \text{if and only if} \quad Z_{n,i}^* = \#\{1 \leq j \leq n : X_j^* \leq X_{(i)}\} \geq (n/2),$$

where $X_{(i)}$ is the ith order statistic. If P^* denotes the probability measure induced by the bootstrap sampling scheme, then the distribution of $Z_{n,i}^*$ under P^* is binomial with n and $p = i/n$. Simple integration by parts leads to

$$p_{i,n} = P^*\left(m_n^* = X_{(i)}\right) = n\binom{n-1}{r-1} \int_{(i-1)/n}^{i/n} u^{r-1}(1-u)^{n-r}\, du,$$

where r is as defined in Sec. 4, and

$$\mathrm{Var}^*(m_n^*) = \frac{1}{2} \sum_{i=1}^{n} \sum_{j=1}^{n} (X_{(i)} - X_{(j)})^2 p_{i,n} p_{j,n}$$

$$\geq (X_{(n)} - X_{(n-1)})^2 p_{n,n} p_{n-1,n} \to \infty,$$

as $X_{(n)}$ is increased without any bound. Consequently the breakdown point of the bootstrap variance is $1/n$. One reason for this is that, with positive probability, however small, the bootstrap median m_n^* can take the value $X_{(n)}$. The breakdown point does not pay any attention to the probability of occurrence of outliers, however small it may be.

Ghosh et al. (1984) have shown that if the tail of the population distribution is not heavy, i.e., $\mathrm{E}|X_1|^\varepsilon < \infty$, for some $\varepsilon > 0$, then

$$\mathrm{Var}^*(m_n^*)/\mathrm{Var}(m_n) \to 1,$$

for almost all sample sequences. Babu (1986) has improved upon this result by establishing the above result under the weaker assumption of

$$\mathrm{E}(\log(1 + |X_1|)) < \infty.$$

He showed that the condition cannot be relaxed and that the bootstrap variance explodes to infinity as $n \to \infty$, for almost all samples, if

$$\mathrm{E}\left(\frac{\log(1 + |X_1|)}{\log\log(3 + |X_1|)}\right) = \infty.$$

In summary, the bootstrap estimator of variance of the sample median, $\mathrm{Var}^*(m_n^*)$ is consistent but not robust. However, recall that the breakdown point for the sample median is $\frac{1}{2}$. Thus the outliers hold unusually strong influence on the bootstrap variance, even when they have no effect on the statistic itself. Similar conclusions can be drawn for the sample quantiles.

Thus in general, the bootstrap estimator of variance of a robust estimator is not robust.

Now it is natural to ask if there is a resampling procedure that gives robust estimates of variance of robust estimators. The answer is yes, and a method called the half-sample method is discussed in the next Section.

6. HALF-SAMPLE METHOD

Estimation procedures based on random collection of subsamples of a sample have been in use for a long time. For example, Mahalanobis (1946) uses a method under the name of interpenetrating samples. Efron (1979) discusses Hartigan's (1969, 1975) work on subsample methods and compares it with the then newly developed, bootstrap method. He notes the first-order asymptotic equivalence of the two methods in several cases.

The half-sample method consists of sampling half the number of data points without replacement from the original sample, $X_1, \ldots, X_n$ first, and then basing estimation on this sub-collection. For example, if the basic sample drawn is $X_1, \ldots, X_n$, $n = 2r$, then a sample of size r is drawn without replacement from $X_1, \ldots, X_n$. Suppose the resample consists of the data

$$X_{n_1}, \ldots, X_{n_r}, \quad \text{where} \quad 1 \leq n_1 < \cdots < n_r \leq n.$$

Then the half-sample estimator of the median $m_{n.H}$, is the median of the points $X_{n_1}, \ldots, X_{n_r}$. Note that if all the original data points, $X_1, \ldots, X_n$ are distinct, then unlike bootstrap samples, the resampled points $X_{n_1}, \ldots, X_{n_r}$ are also distinct.

To understand the basic difference between the half-sample method and the bootstrap method, let $n = 15$ and let the original data be arranged in an increasing order

$$X_{(1)} < X_{(2)} < \cdots < X_{(14)} < X_{(15)}.$$

The size of the half-sample is $[n/2] = [15/2] = 7$, where $[x]$ denotes the greatest integer not exceeding x. Suppose a particular drawing of the half-sample consists of the seven points

$$X_{(2)}, X_{(3)}, X_{(7)}, X_{(10)}, X_{(11)}, X_{(13)}, X_{(14)}.$$

For this realization, $m_{15,H} = X_{(10)}$. It is trivial to note that, irrespective of the values of the original sample, and for any realization of the half-sample, $m_{15.H}$ can never be equal to $X_{(1)}, X_{(2)},$ or $X_{(3)}$. So the half-sample median avoids the least three values of the original data. This is in sharp contrast to the bootstrap method, where the bootstrap median can take any of the values of the original sample. Consequently, the least three values do not contribute to the estimation of the variance of $m_{15.H}$. But $X_{(4)}$ does contribute to the estimation of the variance, since for some sub-samples of size 7, $m_{15.H}$ can take the value $X_{(4)}$. So moving four values to the extreme would substantially alter the value of $m_{15.H}$. Hence the breakdown point in this case is $\frac{4}{15}$. See Babu (1992) for details on the half-sample method of estimation of the variance of quantiles.

To describe the half-sample distribution P_H of the half-sample median $m_{n.H}$ given the data $X_1, \ldots, X_n$, let

$$
h = \frac{1}{2}\left(\left[\frac{n}{2}\right] + 1\right) \quad \text{if}\left[\frac{n}{2}\right] \text{ is odd}
$$

$$
= \frac{1}{2}\left[\frac{n}{2}\right] \qquad \text{if}\left[\frac{n}{2}\right] \text{ is even.}
$$

It is shown in Babu (1992) that the sampling distribution of the half-sample median is given by

$$
p_{i.n.H} = P_H\left(m_{n.H} = X_{(i)}|X_1, \ldots, X_n\right)
$$

$$
= \frac{\binom{i-1}{h-1}\binom{n-i}{[n/2]-h}}{\binom{n}{[n/2]}},
$$

if $h \le i \le n + h - [n/2]$ and $p_{i.n.H} = 0$, otherwise. This leads to the half-sample estimator of variance

$$
\text{Var}_H(m_{n.H}) = \frac{1}{2}\sum_{i,j=h}^{n+h-[n/2]} (X_{(i)} - X_{(j)})^2 p_{i.n.H} p_{j.n.H}.
$$

As in the case of bootstrap estimation, the variance estimator depends only on the sample data points. However it is not as easy to compute as the bootstrap estimator of variance of the sample median, as $p_{i.n.H}$ are difficult to compute when n is large.

Note that the order statistics $X_{(i)}$ for $i < h$ do not influence the variance estimator. On the other hand the hth order statistic $X_{(h)}$ has strong influence on the estimation of variance. Even though for large n, the probability $p_{h.n.H}$ is very small, the variance term $\text{Var}_H(m_{n.H})$ can be made to grow without any bounds by moving the hth order statistic, i.e., by moving h observations in one direction. Hence the breakdown point here is h/n, which approaches $\frac{1}{4}$ as $n \to \infty$. This marked improvement in robustness of the half-sample method over the bootstrap is achieved by sacrificing computational simplicity. In addition, have any other desirable properties been sacrificed? Babu

(1992) shows that the half-sample variance estimator is as efficient as the bootstrap estimator. In fact, it is shown that under very general conditions, the relative efficiency,

$$\frac{E(\text{Var}^*(m_n^*) - \text{Var}(m_n))}{E(\text{Var}_H(m_{n.H}) - \text{Var}(m_n))} \to 1,$$

as $n \to \infty$. This holds even when X_i are not necessarily identically distributed. Recall that m_n^* is the bootstrap median and m_n is the sample median.

One of the main reasons for robustness of the half-sample method is that each observation appears in the resampling scheme at most once. On the other hand, in the case of bootstrap method, a data point can appear more than once in the resample.

The same analysis leads to the low breakdown point $1/n$ of the bootstrap estimator of the variance of a sample quantile (see p. 712 of Babu 1992). On the other hand, the breakdown point of the half-sample estimator of the variance of a p-th sample quantile, $0 < p < 1$, is given by k/n, where

$$k = p\left[\frac{n}{2}\right] \qquad \text{if it is an integer}$$

$$= \left[p\left[\frac{n}{2}\right]\right] + 1 \qquad \text{otherwise.}$$

See Theorem 5.1 of Babu (1992). Consequently the asymptotic breakdown point is $p/2$. Similar situation occurs for robust statistics such as the trimmed mean, and certain M and L estimators.

7. BREAKDOWN POINTS FOR BOOTSTRAP QUANTILES

The bootstrap method is, to a large extent used in the estimation of variance and of confidence intervals. Estimation of bootstrap quantiles of a statistic T_n are essential for computing bootstrap confidence intervals, see Babu and Bose (1988), and Hall (1992). In this Section, the breakdown of bootstrap quantiles, as opposed to the sample quantiles are considered. Singh (1996) establishes the relation between the breakdown point of a statistic T_n and the tth bootstrap quantile. He also suggests improvements for certain L and M estimators via Winsorization. To discuss this further, consider 10 percent trimmed mean T_{20} based on a sample of size $n = 20$, see Singh (1996). Even if the largest of the observations $X_{(20)}$ is quite large, the value of the statistic T_{20} is unaffected, as $X_{(20)}$ does not enter into the computation of T_{20}. If the bootstrap method is used to estimate the confidence interval, the extreme value $X_{(20)}$ could appear in the resampled set one or more times. It is quite conceivable that the bootstrap sample consists of 20 copies of the largest value. Even though the probability of this event is extremely low, the outlier has a very strong influence on the bootstrap trimmed mean T^*_{20}. Due to 5 percent trimming on each side, T^*_{20} is free of $X_{(20)}$ if it appears at most once in the bootstrap sample. Suppose Z denotes the number of times $X_{(20)}$ appears in a bootstrap sample. If $Z > 1$, for a particular bootstrap sample, then $X_{(20)}$ will affect the value of T^*_{20}.

Note that the different bootstrap realizations lead to different values of T^*_{20} and to different value of Z. To analyze the bootstrap quantiles, it is necessary to understand how often T^*_{20} is influenced by the extreme value in the sample. Clearly, under the measure induced by the bootstrap sampling scheme, the random variable Z has a binomial distribution $\mathrm{Bin}(20, 0.05)$. Hence $X_{(20)}$ influences $100(1 - p)$ percent of all the bootstrap samples, where $p \approx 0.736$ denotes the probability that $Z \leq 1$. For such samples T^*_{20} can be made arbitrarily large by letting $X_{(20)}$ increase without any bound. Consequently, the bootstrap quantile Q^*_t of T^*_{20} will tend to ∞ for all $p < t < 1$. So for such t, a single data point has a vast influence; leading to the upper breakdown point $= 1/20$. That is, at least 5% of the data have to approach ∞ in order to carry the statistic to ∞. If $t \leq p$, then Q^*_t is not affected by any outliers other than those influencing T_{20}. So the upper breakdown point of Q^*_t is at least as large as that of T_{20} for $t \leq p$.

Singh (1996) builds upon these arguments and establishes the following result. Let T_n be a robust statistic, based on a sample of size n, with upper breakdown point b/n, i.e., b is the smallest number of observations that need to grow without any bound in order to force T_n to approach ∞. Let T^*_n denote the bootstrap estimator of T_n, and let

$$Q^*_t = \min\{x : P^*(T^*_n \leq x) \geq t\}$$

denote the tth quantile of the bootstrap distribution. Then the upper breakdown point b_t for Q^*_t is m/n, where

$$m = \min\{j : 1 \leq j \leq n, P(\mathrm{Bin}(n, j/n) \geq b) > 1 - t\}.$$

Singh (1996) also establishes that

$$nb_t - b + \frac{z_t \sqrt{b(n-b)}}{\sqrt{n}}$$

is asymptotically bounded, where z_t is the tth quantile of the standard normal distribution.

Singh (1996) suggests Winsorization prior to bootstrapping to improve the breakdown point of the bootstrap quantiles. To Winsorize a β-fraction of the data from each end, replace all the data points not exceeding the ℓth order statistic $X_{(\ell)}$ by the $(\ell + 1)$-th order statistic and all the data larger than

$(n - \ell + 1)$-th order statistic by the $(n - \ell)$-th order statistic, where $\ell = [n\beta]$, i.e.,

$$X_i^0 = \begin{cases} X_{(\ell+1)} & \text{if} \quad X_i \leq X(\ell) \\ X_{(n-\ell)} & \text{if} \quad X_i \geq X(n - \ell + 1) \, . \\ X_i & \text{otherwise} \end{cases}$$

Resampling from the Winsorized data produces bootstrap samples that are free of outliers. He establishes that the upper breakdown point b_t^* for Q_t^* is given by

$$b_t^* = \max\left(b_t, \frac{([n\beta] + 1)}{n}\right)$$

for $(1-2\alpha)100$ percent-trimmed means, where $0 < \beta < \alpha < \frac{1}{2}$. For additional details on breakdown theory for bootstrap quantiles and the effects of Winsorization, see Singh (1996).

8. CONCLUDING REMARKS

The half-sample estimator of the variance of a sample quantile is shown to have a very high breakdown point. This property is not shared by the bootstrap estimator. Extensions of these results to other robust statistics are under investigation and will be reported elsewhere.

Singh's (1996) general formula for the computation of the breakdown point, for a bootstrap quantile of a statistic, is briefly discussed. For a class of L and M estimators, Winsorization before bootstrapping seems to improve the breakdown value. A detailed account of the current investigations by Singh and his colleagues will be reported elsewhere.

Acknowledgments

The author would like to thank the referees for their valuable remarks which led to a considerable improvement of the chapter.

9. REFERENCES

Babu, G. J., Bootstrapping statistics with linear combinations of chi-squares as weak limit. *Sankhyā, Series A 46*: 85–93, 1984.

Babu, G. J., A note on bootstrapping the variance of the sample quantile. *Ann. Inst. Statist. Math., A 38*: 439–443, 1986.

Babu, G. J., Applications of Edgeworth expansions to bootstrap—a review. In: Dodge Y, ed. Statistical Data Analysis and Inference. Amsterdam: Elsevier Science Publishers B. V., 1989, pp. 223–237.

Babu, G. J., Subsample and half-sample methods. *Ann. Inst. Statist. Math.* *44*: 703–720, 1992.

Babu, G. J. and Z. D. Bai. Mixtures of global and local Edgeworth expansions and their applications. *J. Mult. Analysis 59*: 282–307, 1996.

Babu, G. J. and A. Bose. Bootstrap confidence intervals. *Statistics & Probab. Letters 7*: 151–160, 1988.

Babu, G. J. and C. R. Rao. Bootstrap Methodology. In: Rao C. R., ed. Handbook of Statistics 9, "Computational Statistics." Amsterdam: Elsevier Science Publishers B. V., 1993, pp. 627–659.

Babu, G. J. and K. Singh. Inference on means using the bootstrap. *Ann. Statist. 11*: 999–1003, 1983.

Babu, G. J. and K. Singh. Asymptotic representations related to jackknifing and bootstrapping L-statistics. *Sankhyā, Series A 46*: 195–206, 1984a.

Babu, G. J. and K. Singh. On one term Edgeworth correction by Efron's bootstrap. *Sankhyā, Series A 46*: 219–232, 1984b.

Bose, A. and G. J. Babu. Accuracy of the bootstrap approximation. *Probab. Theory and Related Fields 90*: 301–316, 1991.

Efron, B. Bootstrap methods: Another look at the jackknife. *Ann. of Statis. 7*: 1–26, 1979.

Efron, B. The jackknife, the bootstrap, and other resampling plans. CBMS-NSF Regional Conference Series in Applied Mathematics, Volume 38, Philadelphia: SIAM, 1982.

Efron, B. and R. J. Tibshirani. An Introduction to the bootstrap. New York: Chapman & Hall, 1993.

Ghosh, M. W. Parr, K. Singh and G. J. Babu. A note on bootstrapping the sample median. *Ann. of Statist. 12*: 1130–1135, 1984.

Hall, P. The bootstrap and Edgeworth expansion. New York: Springer-Verlag, 1992.

Hampel, F. R. A general qualitative definition of robustness. *Ann. of Math. Stat. 42*: 1887–1896, 1971.

Hampel, F. R. The influence curve and its role in robust estimation. *J Amer. Statist. Assoc. 69*: 383–393, 1974.

Hartigan, J. A. Using subsample values as typical values, *J. Amer. Stat. Assoc. 64*: 1303–1317, 1969.

Hartigan, J. A. Necessary and sufficient conditions for asymptotic joint normality of a statistic and its subsample values. *Ann. of Statist. 3*: 573–580, 1975.

Huber, P. Robust Statistics. New York: John Wiley, 1981.

Linder, E. and G. J. Babu. Bootstrapping the linear functional relationship with known error variance ratio. *Scand. J. Statist. 21*: 21–39, 1994.

Mahalanobis, P. C. Report on the Bihar crop survey: rabi season 1943–1944. *Sankhyā, Series A 7*: 269-280, 1946.

Shao, J. and D. Tu. Jackknife and bootstrap. New York: Springer-Verlag, 1995.

Singh, K. On the asymptotic accuracy of Efron's bootstrap. *Ann. of Statist. 9*: 1187–1195, 1981.

Singh, K. Paradoxes in Robustness. In: Ghosh, J. K., Mitra, S. K., Parthasarathy, K. R. and Prakasarao, B. L., ed. A Raghu Raj Bahadur Festschrift. New Delhi: Wiley Eastern Ltd. 1993, pp. 531–537.

Singh, K. Breakdown Theory for bootstrap quantiles. Technical Report #96-015, Department of Statistics, Rutgers University, New Brunswick, NJ, 1996.

Singh, K. and G. J. Babu. On Asymptotic optimality of the bootstrap. *Scand. J. Statist. 17*: 1–9, 1990.

Stromberg, A. J. Robust covariance estimates based on resampling. *J. Stat. Plan. and Inf. 57*: 321-334, 1997.

20

On Second-Order Properties of the Stationary Bootstrap Method for Studentized Statistics

S. N. LAHIRI   Iowa State University, Ames, Iowa

1.   INTRODUCTION

Politis and Romano (1994) proposed a resampling method, called the stationary bootstrap. In this paper, we show that under appropriate conditions, the stationary bootstrap method provides second-order correct approximations to the distributions of a broad class of studentized and normalized statistics, and thus outperforms the normal approximation.

Bootstrap methods for time series data typically requires resampling blocks of observations instead of resampling a single observation at a time. Starting with the advent of the moving block bootstrap of Künsch (1989) and Liu and Singh (1992), several methods of blocking have been proposed in the literature. In a recent work, Politis and Romano (1994) proposed a block bootstrap method, which they called the "stationary bootstrap" (SB). Unlike the previously known block bootstrap methods, the SB has the desirable property that for data coming from a stationary time series, the bootstrap observations generated by the SB method also forms a stationary sequence. The main artifact of the method is to resample blocks of *random* lengths (suitably), in contrast to resampling blocks of a fixed length. The additional randomness introduced in the resampling mechanism smooths out nonstationarity effects of fixed length blocks and yields a stationary sequence of bootstrap observations. Politis and Romano (1994) showed that the SB provides valid approximations to the unknown sampling distributions of many commonly used estimators. In this chapter, we refine their result and show that in spite of the additional randomness of the block length variables, the SB enjoys the second-order correctness property as does the bootstrap method of Efron (1979) under independence, cf. Singh (1981), and as do certain other block bootstrap methods based on non-random block lengths under weak dependence.

We investigate second order properties of the SB method for normalized and studentized versions of estimators under the 'smooth function model' of Bhattacharya and Ghosh (1978), cf. Hall (1992). A brief description of the model is given in Sec. 2 below. Concisely, the smooth function model deals with estimators that can be represented as smooth functions of certain sample means. As noted in Sec. 2, this class is rich enough to contain many commonly used estimators, such as the sample variance, the sample auto-correlation, Yule–Walker estimators, M-estimators, etc. The main results of the paper show that under some conditions, the bootstrap approximation provided by the SB for *normalized* as well as *studentized* versions of such estimators is better than the normal approximation obtained from the Central Limit Theorem for dependent variables. As a result, tests and confidence intervals constructed using the SB are more accurate than those obtained using the limiting normal distribution.

An important tool required for establishing the second-order optimality of the SB is the development of valid two-term Edgeworth expansions for the bootstrap versions of the normalized and the studentized statistics. Under the SB resampling scheme, the block lengths are random and have a *nondegenerate* conditional distribution given the data. As a result, bootstrap versions of the statistics of interest are defined in terms of a *random number* of conditionally independent blocks. This makes the derivation of the relevant Edgeworth expansions somewhat nontrivial. We develop an iterated conditioning argument to obtain valid expansions for the bootstrapped statistics.

Block bootstrap methods for dependent observations have been formulated by Hall (1985), Carlstein (1986), Künsch (1989), Politis and Romano (1992, 1994), Liu and Singh (1992), among others. As pointed out earlier, all except Politis and Romano (1994) consider nonrandom block lengths only. Properties of block bootstrap methods with nonrandom block lengths have been investigated by Davison and Hall (1993), Hall, Horowitz and Jing (1995), Götze and Künsch (1996), Lahiri (1993b, 1995, 1996a), among others. Second-order optimality of the moving block bootstrap (MBB) of Künsch (1989) and Liu and Singh (1992) has been established by Lahiri (1991) for normalized estimators and by Götze and Künsch (1996) and Lahiri (1996a) for studentized estimators. Though the article Lahiri (1991) was entitled 'second-order optimality of stationary bootstrap', it actually dealt with the MBB. The phrase 'stationary bootstrap' was used to mean 'bootstrap for stationary' dependent observations, just as the ordinary bootstrap method, cf. Efron (1979) for iid data has been often termed as the "iid bootstrap". Since Lahiri (1991) predates Politis and Romano (1994), this possibly would not cause too much of a confusion for the reader.

The rest of the chapter is organized as follows. In Sec 2, we describe the stationary bootstrap method and introduce the smooth function model. In Sec 3, we state the assumptions and main results of the paper. A number of auxiliary lemmas are presented in Sec 4, some of which may be of independent interest. Proofs of the main results are given in Sec 5.

2. THE STATIONARY BOOTSTRAP

We begin with a description of the SB method. Let $X_1, \ldots, X_n$ be a sample of size n from a sequence $X_1, X_2, \ldots$ of stationary random vectors taking values in $\mathbb{R}^d$, and let $T_n = t_n(X_1, \ldots, X_n; \theta)$ be a random variable based on the observations $X_1, \ldots, X_n$ and a certain population parameter θ. Suppose, we are interested in estimating (some characteristics of) the sampling distribution G_n of T_n. For example, we may have $T_n = \sqrt{n}(\hat{\theta}_n - \theta)$ for an estimator $\hat{\theta}_n$ of a parameter of interest θ and we might be interested in estimating the scaled mean squared error of the estimator $\hat{\theta}_n$, given by

$$nMSE(\hat{\theta}_n) = nE(\hat{\theta}_n - \theta)^2 = \int x^2 \, dG_n(x).$$

The SB method provides estimators of population quantities like, $nMSE(\hat{\theta}_n)$ (which is a functional of G_n) and more generally, of the entire sampling distribution G_n without any parametric model assumptions.

To apply the SB method, first we define a new time series $\{X_{ni} : i \geq 1\}$ by periodic extension of the observed data set $X_1, \ldots, X_n$ as follows. Let $i \geq 1$ be an integer. Then there exist integers $k \geq 0$ and $1 \leq j \leq n$ such that $i = kn + j$. We define the variable X_{ni} by setting $X_{ni} = X_j$. Thus, $X_{ni} = X_i$ for $1 \leq i \leq n$, $X_{ni} = X_{n-i}$ for $n + 1 \leq i \leq 2n$, and so on. The sequence $\{X_{ni} : i \geq 1\}$ may be thought of as obtained by wrapping the data $X_1, \ldots, X_n$ around a circle, and relabelling them as $X_{n1}, X_{n2}, \ldots$ as the circle is traversed, starting from X_1.

Next for a positive integer l, define the blocks $\mathcal{B}(i, l), i \geq 1$ as $\mathcal{B}(i, l) = \{X_{ni}, \ldots, X_{n.i+l-1}\}$. Bootstrap observations under the SB method are obtained by selecting a random number of blocks from the collection $\{\mathcal{B}(i, l), i = 1, \ldots, n, l \geq 1\}$. To do this, we generate random variables $I_1, \ldots, I_n$ and $L_1, L_2, \ldots, L_n$ such that conditional on the observations $X_1, \ldots, X_n$,

(i) $I_1, \ldots, I_n$ are independent and identically distributed (iid) with

$$P_*(I_1 = i) = \frac{1}{n}, i = 1, \ldots, n;$$

(ii) $L_1, \ldots, L_n$ are iid random variables having the geometric distribution with parameter $p \in (0, 1)$, i.e.,

$$P_*(L_1 = l) = p(1 - p)^{l-1}, l = 1, 2, \ldots ;$$

and

(iii) the collections $\{I_1, \ldots, I_n\}$ and $\{L_1, \ldots, L_n\}$ are independent.

Here, and in the following, we write P_* and E_* to denote, respectively, the conditional probability and the conditional expectation, given $X_1, \ldots, X_n$. Also, for notational simplicity, we supress the dependence of the variables $I_1, \ldots, I_n, L_1, \ldots, L_n$, and of the parameter p on n.

Thus, it follows from the description above that the expected block length E_*L_1 under the SB method is p^{-1}. We shall assume that p goes to zero at a certain rate with n. This ensures that the expected block size under the SB tends to infinity with the sample size, just as do the block sizes of other block bootstrap methods. However, unlike block bootstrap methods based on nonrandom block lengths, since the block length variables $L_1, \ldots, L_n$ under the SB method are random, we need to resample a *random* number of blocks to generate a set of n bootstrap samples under the SB method. Let $K = \inf\{k \geq 1 : L_1 + \cdots + L_K \geq n\}$. Then, the SB method selects the K blocks $\mathscr{B}(I_1, L_1), \ldots, \mathscr{B}(I_K, L_K)$. Note that there are altogether $N_1 \equiv L_1 + \cdots + L_K$ elements in the resampled blocks $\mathscr{B}(I_1, L_1), \ldots, \mathscr{B}(I_K, L_K)$. Arranging these elements in a series, we get the bootstrap observations $X_1^*, \ldots, X_{N_1}^*$. To define the SB version of the random variable $T_n = t_n(X_1, \ldots, X_n; \theta)$, we may use the first n or all N_1 of the X_i^*'s. Thus, the SB versions of T_n are given by

$$T_n^* \equiv t_{N_1}(X_1^*, \ldots, X_{N_1}^*; \hat{\theta}_n)$$

and

$$\tilde{T}_n^* \equiv t_n(X_1^*, \ldots, X_n^*; \hat{\theta}_n),$$

where $\hat{\theta}_n$ is an estimator of θ based on $X_1, \ldots, X_n$. Since the difference $N_1 - n$ is negligible compared to n, T_n^* and $\tilde{T}_n^*$ tend to have very similar large sample properties. For technical reasons, we would use T_n^* as the SB version of T_n in the rest of the paper. The SB estimator of the unknown sampling distribution G_n of T_n is then defined as the conditional distribution $\hat{G}_n$, say of T_n^*, given $X_1, \ldots, X_n$. Thus, for a functional $\Lambda(\cdot)$ of the sampling distribution G_n, the SB estimator of $\Lambda(G_n)$ is given by $\Lambda(\hat{G}_n)$. For example, the SB estimators of $E(T_n)$ and $\mathrm{var}(T_n)$ are respectively given by $E_*T_n^* = \int x d\hat{G}_n(x)$ and $E_*(T_n^* - E_*T_n^*)^2 = \int (x - \int y d\hat{G}_n(y))^2 d\hat{G}_n(x)$.

In this paper, we shall consider random variables $T_n = t_n(X_1, \ldots, X_n; \theta)$ that are the normalized or studentized versions of estimators specified by the *smooth function model*. Let $\mu = EX_1$ denote the expected value of X_1 and let $\bar{X}_n = n^{-1} \sum_{i=1}^{n} X_i$ denote the sample mean of $X_1, \ldots, X_n$. The smooth function model prescribes that the parameter of interest θ and its estimator $\hat{\theta}_n$ have representations of the form

$$\theta \equiv H(\mu)$$

and

$$\hat{\theta}_n \equiv H(\bar{X}_n),$$

where H be a smooth real valued function on $\mathbb{R}^d$. Though it looks like a very restrictive model, considering suitable transformations of the original observations, many important population parameters and their commonly used estimators may be expressed in this form. We illustrate the scope of the smooth function model with the following examples.

EXAMPLE 1   Suppose, $\theta \equiv \mathrm{cov}(Y_1, Y_{1+k})$, the kth order lag-covariance of a univariate, stationary sequence $\{Y_n\}$. A natural estimator of θ based on $Y_1, \ldots, Y_{n+k}$ is given by

$$\hat{\theta}_n = n^{-1} \sum_{i=1}^{n} Y_i Y_{i+k} - (\bar{Y}_n)^2,$$

where $\bar{Y} = n^{-1} \sum_{i=1}^{n} Y_i$. To express θ and $\hat{\theta}_n$ in the form specified by the smooth function model, we define $X_i = (Y_i, Y_i Y_{i+k})'$, $i \geq 1$ and

$$H(x_1, x_2) = x_2 - x_1^2, \qquad (x_1, x_2) \in \mathbb{R}^2.$$

Then, it follows that

$$\theta = H(EX_1) \qquad \text{and} \qquad \hat{\theta}_n = H(\bar{X}_n).$$

EXAMPLE 2   Let $\{Y_n\}$ be a sequence of stationary random variables and let parameter of interest θ be a functional of the joint distribution of $\{Y_n\}$ satisfying

$$E\psi(Y_1, \ldots, Y_{k+1}; \theta) = 0,$$

for some function $\psi : \mathbb{R}^{k+1} \to \mathbb{R}$, $k \geq 1$. Then, a generalized M-estimator (cf. Bustos (1982)) of θ based on $Y_1, \ldots, Y_{n+k}$ is defined as a solution to the equation

$$n^{-1} \sum_{i=1}^{n} \psi(Y_i, \ldots, Y_{i+k}; t) = 0.$$

It can be shown that under some regularity conditions on the function ψ and the sequence $\{Y_n\}$, there exists a function $H : \mathbb{R}^k \to \mathbb{R}$ such that with $X_i \equiv (Y_i, \ldots, Y_{i+k-1})'$, $i \geq 1$,

$$\theta = H(EX_1) \qquad \text{and} \qquad \hat{\theta}_n = H(\bar{X}_n) \qquad \text{on a set } A_n,$$

where $P(A_n^c)$ goes to zero rapidly with n. The results obtained in this chapter are applicable to this class of estimators as well.

Next we define the normalized and the studentized versions of the estimator $\hat{\theta}_n \equiv H(\bar{X}_n)$ given by the smooth function model. Note that under some standard conditions,

$$\sqrt{n}(\hat{\theta}_n - \theta) \longrightarrow^d N(0, \tau_\infty^2) \qquad \text{as} \qquad n \to \infty,$$

where $\tau_\infty^2 = \sum_{i=-\infty}^{\infty} \operatorname{cov}(a'X_1, a'X_{1+i})$, $a = (D_1 H(\mu), \ldots, D_d H(\mu))'$ and where $D_j H$ denotes the partial derivative of H with respect to the jth coordinate, $1 \leq j \leq d$. Therefore, in the ideal case where one knows the asymptotic variance τ_∞^2, an asymptotically pivotal quantity for the parameter θ is given by

$$T_n^\dagger = \sqrt{n}(\hat{\theta}_n - \theta)/\tau_\infty.$$

We refer to $T_n^\dagger$ as the "normalized" version of the estimator $\hat{\theta}_n$. When τ_∞^2 is known, one can use $T_n^\dagger$ to construct large sample tests and confidence intervals for the parameter θ.

In most applications, however, it is rare that the quantity τ_∞^2 is known. Hence, one needs to estimate it using the data. To that end, let $Y_i = \hat{a}_n' X_i$, $i \geq 1$, where $\hat{a}_n = (D_1 H(\bar{X}_n), \ldots, D_d H(\bar{X}_n))'$. Define $\hat{\tau}_n^2 \equiv n^{-1} \sum_{i=1}^{n} (Y_i - \bar{Y}_n)^2 + 2 \sum_{j=1}^{l_1} \sum_{i=1}^{n} (Y_i - \bar{Y}_n)(Y_{i+j} - \bar{Y}_n)$, where $l_1 \equiv l_{1n}$ is an integer between 1 and n. We use $\hat{\tau}_n^2$ as an estimator of τ_∞^2. If $l_1 \to \infty$ with n at a suitable rate, then $\hat{\tau}_n^2 \longrightarrow_p \tau_\infty^2$, and

$$T_n \equiv \sqrt{n}(\hat{\theta}_n - \theta)/\hat{\tau}_n$$

is asymptotically pivotal with the $N(0, 1)$ as its limit distribution. We refer to T_n as the "studentized" version of the statistic $\hat{\theta}_n$. T_n serves as the common choice of an asymptotically pivotal quantity for setting confidence intervals and tests for the parameter θ in the general case when the asymptotic variance τ_∞^2 remains unknown.

Next we specialize the description of the SB method and define bootstrap versions of $T_n^\dagger$ and T_n. Let $\bar{X}_n^* = (X_1^* + \cdots + X_{N_1}^*)/N_1$ denote the sample average of the bootstrap values, and let $U_{1i}^* = \sum_{j=I_i}^{I_i + L_i - 1} X_j^* / \sqrt{L_i}$ denote the block sum of the ith resampled block $\mathcal{B}(I_i, L_i)$, multiplied by the square-root of the block length. Then, the bootstrap version of $T_n^\dagger$ is given by

$$T_n^{\dagger*} = \sqrt{N_1}(H(\bar{X}_n^*) - H(\bar{X}_n))/\hat{\tau}_n.$$

Similarly, we define the bootstrap version of the studentized statistic T_n as

$$T_n^* = \sqrt{N_1}(H(\bar{X}_n^*) - H(\bar{X}_n))/\tau_n^*$$

where $\tau_n^{*2} = K^{-1}\sum_{i=1}^{K}(a_n^{*\prime}U_{1i}^* - a_n^{*\prime}\bar{U}_{1K})^2$ is the sample variance of the bootstrap variables $a_n^{*\prime}U_{11}^*, \ldots, a_n^{*\prime}U_{1K}^*$, each defined in terms of the elements of an individual resampled block, and where a_n^* is the $d \times 1$ vector of first order partial derivatives of $H(x)$ at $x = \bar{X}_n^*$.

Note that we have used the centering $H(\bar{X}_n)$ in the definitions of the bootstrapped statistics $T_n^{\dagger*}$ and T_n^*, which is apparently different from the definitions of the bootstrap versions of $T_n^\dagger$ and T_n under other block bootstrap methods. It is known (cf. Lahiri (1992)) that for the existing block bootstrap methods based on *nonrandom* block lengths, centering $H(\bar{X}_n^*)$ at $H(\bar{X}_n)$ in the definition of a bootstrapped statistic does not generally yield second-order accurate approximations. Because of the nonstationarity of the bootstrap observations under these resampling methods, the conditional expectation of the bootstrap sample mean is usually different from $\bar{X}_n$, and therefore, the natural choice of centering $H(\bar{X}_n)$ does not work. Consequently, one needs to modify the definitions of bootstrapped statistics suitably to attain higher order accuracy. An advantage of the SB method is that no such additional tuning of the bootstrapped statistics $T_n^{\dagger*}$ and T_n^* is required. The conditional expectation of $\bar{X}_n^*$ under the SB is $\bar{X}_n$ for all n, and the natural choice of the centering can be used for applying the SB. Indeed, in the next Section we show that with the above definitions, the conditional distributions of $T_n^{\dagger*}$ and T_n^* (given the data) provide second-order accurate approximations to the unknown sampling distributions of the asymptotically pivotal quantities $T_n^\dagger$ and T_n.

3. MAIN RESULTS

For describing the results, we need to introduce some notation. Let $\mathbb{Z}_+ = \{0, 1, \ldots, \}$ denote the set of all nonnegative integers. For $\alpha = (\alpha_1, \ldots, \alpha_d)' \in (\mathbb{Z}_+)^d, x = (x_1, \ldots, x_d)' \in \mathbb{R}^d$, let $|\alpha| = \alpha_1 + \ldots + \alpha_d$, $\alpha! = \prod_{i=1}^{d}\alpha_i!$, $x^\alpha = \prod_{i=1}^{d}(x_i)^{\alpha_i}$, $\|x\| = (x_1^2 + \cdots + x_d^2)^{1/2}$, and let D^α denote the αth order partial differential operator $\partial^{\alpha_1 + \cdots + \alpha_d}/\partial x_1^{\alpha_1} \ldots \partial x_d^{\alpha_d}$. For a matrix A, let A' denote the transpose of A and for a square matrix A, let $\|A\|$ denote the spectral (or Euclidean) norm of A defined by $\|A\| = \sup\{\|Ax\| : \|x\| = 1\}$. Write $c_\alpha = D^\alpha H(\mu)/\alpha!$, $\hat{c}_\alpha = D^\alpha H(\bar{X}_n)/\alpha!, \alpha \in (\mathbb{Z}_+)^d$.

Write Σ_∞ for the limiting covariance matrix of the normalized sample mean $V_{1n} \equiv n^{1/2}(\bar{X}_n - \mu)$. Let Z_∞ denote a random vector having the $N(0, \Sigma_\infty)$ distribution on $\mathbb{R}^d$. For any collection of random variables $\{W_i : i \in I\}$ on a given probability space, let $\sigma\langle W_i : i \in I\rangle$ denote the σ-field generated by $\{W_i : i \in I\}$. Let $p_{2n}(x)$ be the quadratic polynomial in d-variables $x \in \mathbb{R}^d$, defined by

$$p_{2n}(x) = \sum_{|\alpha|=2} c_\alpha x^\alpha / \tau_n - \sum_{|\alpha|=1} \sum_{|\beta|=1} \{(\alpha+\beta)! c_{\alpha+\beta}\} \cdot \{s_{11}(\beta)/\tau_n^3\}(a'x)x^\alpha$$

where $s_{11}(\beta) \equiv \lim_{n\to\infty} \mathrm{cov}(a'V_{1n}, V_{1n}^\beta) = \sum_{j=-\infty}^\infty \mathrm{cov}(a'X_1, X_{1+j}^\beta)$. Also, let $\hat{p}_{2n}(x)$ be the quadratic polynomial obtained from $p_{2n}(x)$ by replacing a, c_α, $s_{11}(\beta)$, and τ_n^2 by $\hat{a}_n$, $\hat{c}_\alpha$, $\hat{s}_{11}(\beta) \equiv E_*(\hat{a}_n' U_{11}^*)(U_{11}^*)^3$, and $\hat{\tau}_{1n}^2 \equiv E_*(a'U_{11}^*)^2$. For notational convenience, we shall also write $p_{2n}(x) = \sum_{|\alpha|=2} d_\alpha x^\alpha$ and $\hat{p}_{2n}(x) = \sum_{|\alpha|=2} \hat{d}_\alpha x^\alpha$.

For proving second-order correctness of the SB, we need to obtain Edgeworth expansions for the bootstrapped statistics $T_n^\dagger{}^*$ and T_n^*. In a significant paper, Götze and Hipp (1983) obtained Edgeworth expansions for sums of weakly dependent random vectors under suitable regularity conditions. Here we derive an expansion for the bootstrap versions of $T_n^\dagger$ and T_n under similar conditions. Suppose that $X_1, X_2, \ldots$ are defined on a common probability space $(\Omega, \mathcal{F}, P)$ and a sequence $\mathcal{D}_0, \mathcal{D}_{\pm 1}, \mathcal{D}_{\pm 2}, \ldots$ of sub-σ-fields of $\mathcal{F}$ is given. Also, for any $-\infty \leq r \leq s \leq \infty$, let $\mathcal{D}_r^s$ denote the σ-field generated by $\{\mathcal{D}_i : r \leq i \leq s\}$. We will use the following assumptions to prove the results of this chapter.

ASSUMPTIONS

(A.1) $H : \mathbb{R}^d \to \mathbb{R}$ is four times continuously differentiable in a neighborhood of μ, and $a \equiv (D_1 H(\mu), \ldots, D_d H(\mu))' \neq 0$.

(A.2) There exists $\delta > 0$ such that for all $n, k = 1, 2, \ldots$ with $k > \delta^{-1}$, there exists a $\mathcal{D}_{n-k}^{n+k}$-measurable random vector $\tilde{X}_{n,k}$ such that

$$E \parallel X_n - \tilde{X}_{n,k} \parallel \leq \delta^{-1} \exp(-\delta k).$$

(A.3) There exists $\delta > 0$ such that for all $n, k = 1, 2, \ldots,$ $A \in \mathcal{D}_{-\infty}^n, B \in \mathcal{D}_{n+k}^\infty$,

$$|P(A \cap B) - P(A)P(B)| \leq \delta^{-1} \exp(-\delta k).$$

(A.4) There exists $\delta > 0$ such that for all $n, k, r = 1, 2, \ldots$ and $A \in \mathcal{D}_{k-r}^{k+r}$,

$$E|P(A|\mathcal{D}_j : j \neq n) - P(A|\mathcal{D}_j : 0 < |n-j| \leq k+r)| \leq \delta^{-1} \exp(-\delta k).$$

(A.5) There exists $\delta > 0$ such that for all $n, k = 1, 2, \ldots, \delta^{-1} < k < n$, and for all $t \in \mathbb{R}^d$ with $\| t \| > \delta$,

$$E|E(\exp(it'(X_{n-k} + \cdots + X_{n+k}))|\mathcal{D}_j : j \neq n)| \leq e^{-\delta}.$$

(A.6) $E \| X_1 \|^{16+\delta} < \infty$ for some $\delta > 0$.

Formulation of assumptions (A.2)–(A.5) in terms of the σ-fields $\mathcal{D}_j$'s, instead of the random vectors X_j's themselves provides more flexibility in verification of these conditions. Choosing $\mathcal{D}_j = \sigma\langle X_j \rangle \equiv$ the σ-field generated by X_j, renders verification of (A.2) and (A.4) rather trivial but makes the verification of the conditional Cramer condition (A.5) quite difficult. In a specific problem, a different choice of $\mathcal{D}_j$ (from $\sigma\langle X_j \rangle$) is often more suitable for verifying (A.2)–(A.5). For example, if $X_j = \sum_{k=0}^{\infty} c_k Y_{j-k}$ for some sequence of iid random vectors $\{Y_n\}$ and constants $\{c_n\}$, then a natural choice of $\mathcal{D}_j$ is $\sigma\langle Y_j \rangle$ rather than $\sigma\langle X_j \rangle$. We refer the interested reader to Bose (1988) and Götze and Hipp (1983, 1994), where conditions (A.2)–(A.5) are verified for a number of important dependent models. We also point out that the results of this paper can be proved under other possible combinations of moment and mixing conditions. Indeed, analogs of Theorems 3.1–3.3 below remain valid under a *polynomial* rate of decay of the strong mixing coefficient in assumption (A.3) and under a suitable moment condition (cf. Lahiri, (1996b)). However, we do not pursue such extensions here in order to keep the proofs simple.

The first result of this paper establishes a valid Edgeworth expansion for the bootstrapped studentized statistic T_n^* under assumptions (A.1)–(A.6). Define the Edgeworth expansion $\hat{\Psi}_n$ of T_n^* by its Fourier transform as

$$\int \exp(itx) \, d\hat{\Psi}_n(x)$$

$$= \exp(-t^2/2)[1 + n^{-1/2}(it)$$

$$\times \left\{ \sum_{|\alpha|=2} \hat{d}_\alpha E_* S(I_1, L_1)^\alpha \cdot p - (2\hat{\tau}_n^3)^{-1} E_*((\hat{a}_n' S(I_1, L_1))(a' U_{1i}^*)^2) \right\}$$

$$+ (3!)^{-1}(it)^3 \{ n^{-1/2} p E_*(\hat{a}_n' S(I_1, L_1)/\hat{\tau}_n)^3 + 3p^2(n^{1/2}\hat{\tau}_n^2)^{-1}$$

$$\times \sum_{|\alpha|=1} \sum_{|\beta|=1} \sum_{|\gamma|=1} \sum_{|\iota|=1} \hat{d}_{\alpha+\beta} \hat{c}_\gamma \hat{c}_\iota (\mathrm{cov}_*(S(I_1, L_1)^\alpha, S(I_1, L_1)^\gamma)$$

$$\times \mathrm{cov}_*(S(I_1, L_1)^\beta, S(I_1, L_1)^\iota)$$

$$+ \mathrm{cov}_*(S(I_1, L_1)^\alpha, S(I_1, L_1)^\iota) \cdot \mathrm{cov}_*(S(I_1, L_1)^\beta, S(I_1, L_1)^\gamma)) \}]$$

for $t \in \mathbb{R}$.

Then, we have the following result.

THEOREM 3.1   Assume that assumptions (A.1)–(A.6) hold and that $n^{\varepsilon} < p^{-1} < \varepsilon^{-1} n^{1/3}$ for some $\varepsilon > 0$. Then,

$$\sup_{x \in \mathbb{R}} |P_*(T_n^* \leq x) - \hat{\Psi}_n(x)| = o_p(n^{-1/2}) \quad \text{as} \quad n \to \infty.$$

Theorem 3.1 shows that the expansion for the bootstrapped statistic remains valid as long as the expected block length p^{-1} grows to infinity at least as fast as a positive power of the sample size n. This can be relaxed to a lower bound of the form "$p^{-1} > (\log n)^{1+\varepsilon}$ for some $\varepsilon > 0$" under stronger moment conditions on X_1. In the same vein, the upper bound on "p^{-1}" can be relaxed to "$p^{-1} = O(n^{(1-\varepsilon)/2})$ for some $\varepsilon \in (0,1)$", provided certain higher order absolute moments of X_1 are finite (depending on the value of ε). Politis and Romano (1994) show that the optimal expected block length (minimizing the mean squared error (MSE)) for estimating the *variance* of the estimtor $\hat{\theta}_n$ satisfies $p^{-1} = C_0 n^{1/3}(1 + o(1))$ as $n \to \infty$ where $C_0 > 0$ is a constant. Thus, the expansion for the bootstrapped statistic T_n^* remains valid for such optimal values of the blocking parameter p solely under the present set of conditions. The MSE optimal values of block lengths for estimating *distribution functions* are typically of a smaller order (cf. Hall, Horowitz and Jing (1995)) and hence, the conditions presented in the chapter cover such values of p as well.

Under assumptions (A.1)–(A.6), the unbootstrapped statistic T_n also admits a similar Edgeworth expansion with a remainder term of the order $o(n^{-1/2})$. Comparing the expansions for T_n and T_n^*, we obtain the following result.

THEOREM 3.2   Assume that the conditions of Theorem 3.1 hold and that $n^{\varepsilon} < l_1 < \varepsilon^{-1} n^{1/3}$ for some $\varepsilon > 0$. Then,

$$\sup_{x \in \mathbb{R}} |P_*(T_n^* \leq x) - P(T_n \leq x)| = o_p(n^{-1/2}) \quad \text{as} \quad n \to \infty.$$

Thus, it follows that under assumptions (A.1)–(A.6), the SB approximates the sampling distribution of the studentized statistic T_n with an accuracy $o_p(n^{-1/2})$. On the other hand, the results of Götze and Künsch (1996) and Lahiri (1996a) imply that under assumptions (A.1)–(A.6),

$$\sup_{x \in \mathbb{R}} |P(T_n \leq x) - \Phi(x)| = O(n^{-1/2}),$$

and that the order of magnitude of the remainder term $O(n^{-1/2})$ in general, cannot be improved. Therefore, it follows that the SB provides a more accurate approximation than the usual large sample normal approximation for T_n under the conditions of the paper. Indeed, it follows from the proof of Theorem 3.2 that not only does the SB captures the limiting distribution of T_n, but it also approximates the second-order terms (i.e., terms of order $n^{-1/2}$) in the Edgeworth expansion of T_n consistently and thus, outperforms the normal approximation. A simple consequence of this result is that inference procedures based on the SB approximation $P_*(T_n^* \leq x)$ would be more accurate than those based on the limiting normal distribution.

A similar second-order optimality of the SB holds for the normalized statistic $T_n^\dagger$. Though $T_n^\dagger$ has a limited statistical application, the second-order optimality of the SB for $T_n^\dagger$ can be established under a reduced moment condition.

THEOREM 3.3 Assume that assumptions (A.1)–(A.5) hold and that $E\|X_1\|^{8+\delta} < \infty$ for some $\delta > 0$. If $n^\varepsilon < p^{-1} < \varepsilon^{-1} n^{1/3}$ for some $\varepsilon > 0$, then

$$\sup_{x \in \mathbb{R}} |P_*(T_n^{\dagger*} \leq x) - P(T_n^\dagger \leq x)| = o_p(n^{-1/2}) \; .$$

4. LEMMAS

For any function $g : \mathbb{R}^k \to \mathbb{R}$, $k \geq 1$, write $\|g\|_\infty = \sup\{|g(x)| : x \in \mathbb{R}^k\}$. Let 1_A and/or $1(A)$ denote the indicator function of a set A. Let $S(i,l) = \sum_{j=i}^{i+l-1} X_{nj}$, (where $\{X_{nj} : j \geq 1\}$ denotes the circular extension of the sequence $\{X_1, \ldots, X_n\}$), and let $T(i,l) = \sum_{j=i}^{i+l-1} X_j$, $i \geq 1, l \geq 1$. Define

$$S_n = X_1 + \cdots + X_n, \qquad n \geq 1,$$

$$U_j(l) = T(j,l)/l, \quad U_{1j}(l) = U_j l^{1/2}, \qquad j \geq 1,$$

$$U_j^*(l) = S(I_j, l)/l, \quad U_{1j}^*(l) = U_j^*(l) l^{1/2}, \quad U_{1j}^* = U_{1j}^*(L_j) \quad j = 1, \ldots, b.$$

Without loss of generality, assume that the underlying probability space $(\Omega, \mathscr{F}, P)$ is rich enough to support the resampling variables $(I_1, L_1), \ldots,$ (I_n, L_n) for all n. Write P_{**} (and E_{**}) to denote the conditional probability (and expectation, respectively) given $X_1, X_2, \ldots$ and $\{L_1, \ldots, L_n\}_{n \geq 1}$. Next, for a nonnegative definite matrix A of order $k \geq 1$, let Φ_A (and ϕ_A) denote the $N(0, A)$ distribution (and its Lebesgue density, respectively) on $\mathbb{R}^k$. Write I_k for the identity matrix of order $k \geq 1$. For any random variable W, we denote its νth cumulant by $\chi_\nu(W)$, $\nu \in \mathbb{Z}_+^d$. Let $C, C(\cdot)$ denote generic constants, depending on their arguments (if any) but not on n. For notational simplicity, any such dependence on population quantities like, $\|\Sigma_\infty^{-1}\|$ or $E\|X_1\|^s$, that remain fixed throughout the proof of a result will be suppressed. Unless otherwise specified, limits in order symbols are taken letting $n \to \infty$. Also, in the proofs below, unless there is any chance of confusion, we set $\mu = 0$.

LEMMA 4.1 Let $W_1, \ldots, W_n$ be independent random variables with $EW_i = 0$ and $E|W_i|^s < \infty$, $1 \le i \le n$ for some $s > 2$. Then, there exist positive constants C_1 and C_2, depending only on s, such that

$$P(|W_1 + \cdots + W_n| \ge C_1 \sigma_n (\log n)^{1/2})$$

$$\le C_2 \sigma_n^{-s} (\log n)^{-s/2} \sum_{i=1}^{n} E|W_i|^s + n^{-(s-1)/2}$$

where $\sigma_n^2 = \sum_{i=1}^{n} EW_i^2$.

Proof. Follows, for example, from Corollary 4, Sec. 4 of Fuk and Nagaev (1971). We omit the details.

LEMMA 4.2 Let $W_1, \ldots, W_n$ be independent $\mathbb{R}^k$-valued random vectors with $EW_j = 0$, $n^{-1} \sum_{j=1}^{n} EW_j W_j' = I_k$ and $\zeta_s \equiv n^{-1} \sum_{j=1}^{n} E\|W_j\|^s < \infty$ for $s = 4$. Then, for any $\varepsilon > 0$ and any Borel set B in $\mathbb{R}^k$,

$$\left| P\left(n^{-1/2} \sum_{i=1}^{n} W_i \in B \right) - \Psi_n^{\ddagger}(B) \right|$$

$$\le C(d)(1 + \zeta_4)[n^{-1} + \{\alpha_n \varepsilon^{-2d} + n^{d+4} \varepsilon^{-8d} \exp(-\varepsilon^{-1})\} + \Phi_{I_k}((\partial B)^{2\varepsilon})]$$

whenever $n \ge C(k) \cdot \zeta_3^2$. Here,

$$\alpha_n = \sum_{1 \le j_1, \ldots, j_{d-4} \le n} \sup\left\{ \prod_{j \ne j_1, \ldots, j_{d+4}} |E \exp(it'W_j)| : (16\zeta_3)^{-1} \le \|t\| \le \varepsilon^{-4} \right\},$$

$\Psi_n^{\ddagger}(\cdot)$ is a signed measure on $\mathbb{R}^q$ defined by its Fourier transform as

$$\int_{\mathbb{R}^q} \exp(it'x) \, d\Psi_n^{\ddagger}(x) = \left(1 + \sum_{|\nu|=3} \left(n^{-1} \sum_{j=1}^{n} \chi_\nu(W_j) \right) (it)^\nu / \nu! \right) \cdot \exp(-\|t\|^2/2),$$

$t \in \mathbb{R}^q$, and ∂B denotes the boundary of the set B.

Proof. Lemma 4.2 can be proved by retracing the steps in the proofs of Theorem 20.1 in Bhattacharya and Ranga Rao (1986). Details are omitted.

For stating the next lemma, define

$$\hat{w}_n(t, s; l) = E_* \exp(i[t' U_{11}^*(l) + s\{(a' U_{11}^*(l))^2 - E_*(a' U_{11}^*(l))^2\}]),$$

and

$$w_\infty(s,t) = E\exp(it'Z_\infty + s\{(a'Z_\infty)^2 - E(a'Z_\infty)^2\}),$$

$t \in \mathbb{R}^d,\, s \in \mathbb{R},\, l \geq 1.$

LEMMA 4.3 Let $\Delta_n(\lambda) \equiv \sup\{|\hat{w}_n(t,s;l) - w_\infty(t,s)| : \|t\| + |s| < p^{-\lambda}, (2p)^{-1} < l < 2p^{-1}\}, \lambda > 0$. Assume that the conditions of Theorem 3.3 hold. Then, for any $\lambda > 0$, there exists a sequence $\delta_n \downarrow 0$ such that

$$P(\Delta_n(\lambda) > \delta_n) = o(1).$$

Proof. Let

$$\hat{w}_{1n}(t,s;l) = N^{-1}\sum_{j=1}^{N}\exp(i[t'U_{1j}(l) + s\{(a'U_{1j}(l))^2$$

$$- N^{-1}\sum_{r=1}^{N}(a'U_{1r}(l))^2\}]),\, t \in \mathbb{R}^d,\, s \in \mathbb{R},\, 1 \leq l < n,$$

where

$$N = n - l + 1.$$

Then, $\hat{w}_{1n}(t,s;l)$ is the characteristic function of the corresponding variables under the moving block bootstrap method. It is easy to check that

$$\sup\{|\hat{w}_n(t,s;l) - \hat{w}_{1n}(t,s;l)| : t \in \mathbb{R}^d, s \in \mathbb{R}, (2p)^{-1} < l < 2p^{-1}\}$$

$$\leq C(np)^{-1}.$$

Also, approximating the variables X_i, $1 \leq i \leq n$ by the variables $\tilde{X}_{i,m}$ of condition (A.2) with $m = (\log n)^2$ and using Lemma 5.1 of Lahiri (1992) (with $m = l + 2(\log n)^2$), for any $s \in \mathbb{R}$, $t \in \mathbb{R}^d$, and any integer r, we get,

$$P(|\hat{w}_{1n}(t,s;l) - w_\infty(t,s)| > C(\log l)^{-1})$$

$$\leq C(r)(\log l)^{2r}n^{-2r}[n^r l^{2r} + n^{2r}\exp(-C(\delta)(\log n)^2)]$$

$$\leq C(r,\varepsilon)n^{-r}p^{2r}(\log n)^{2r}$$

uniformly in $(2p)^{-1} \leq l \leq 2p^{-1}$.

Now, using the arguments in the proof of Lemma 5.3 of Lahiri (1992), one can show that

$$P(\Delta_n(\lambda) > C(E\|Z_\infty\|^2) \cdot |\log p|^{-1})$$

$$\leq \sum_{1 \leq 2/p \leq 4} P(\sup\{|\hat{w}_{1n}(t,s;l) - w_\infty(t,s)| : \|t\| + |s| \leq p^{-\lambda}\}$$

$$> C(E\|Z_\infty\|^2) \cdot |\log p|^{-1})$$

$$\leq 2p^{-1} \cdot (|\log p| \cdot p^\lambda)^d \cdot C(r,\varepsilon)n^{-r}p^{2r}|\log n|^{2r}$$

$$= o(1) \,,$$

provided r is chosen suitably large.

LEMMA 4.4   Under the conditions of Theorem 3.3, as $n \to \infty$,

(i) $\max_{|\nu|=2} |pE_*(S(I_1, L_1))^\nu - EZ_\infty^\nu| \longrightarrow_p 0$;

(ii) $\max_{|\nu|=3} |E_*S(I_1, L_1)^\nu p - n^{-1}E(S_n)^\nu| \longrightarrow_p 0$;

(iii) $\max_{\substack{|\alpha|=2 \\ |\nu|=1}} |E_*S(I_1, L_1)^\nu (U_{11}^*)^\alpha - n^{-1}E(S_n)^{\nu+\alpha}| \longrightarrow_p 0$;

(iv) $E_*\|S(I_1, L_1)\|^4 p^2 - E\|Z_\infty\|^4 \longrightarrow_p 0$;

(v) $E_*\|U_{11}^*\|^8 - E\|Z_\infty\|^8 \longrightarrow_p 0$, provided assumption (A.6) holds.

Proof.   Since L_1 has the Geometric distribution with parameter p, it follows that for any sequence $t_n \to \infty$,

$$P_*(L_1 > t_n p^{-1}) = \sum_{i > t_n/p} p(1-p)^{i-1}$$

$$\leq (1-p)^{t_n/p-1}$$

$$= \exp([p^{-1}t_n - 1]\log(1-p))$$

$$= O(\exp(-t_n)). \tag{4.1}$$

Also, for any sequence $s_n \to 0$,

$$P^*(L_1 < p^{-1}s_n) = \sum_{i < p^{-1}s_n} p(1-p)^{i-1}$$

$$\leq 1 - (1-p)^{s_n/p}$$

$$= 1 - \exp(s_n p^{-1}\log(1-p))$$

$$= 1 - \exp(s_n p^{-1}[-p + O(p^2)])$$

$$= 1 - \exp(-s_n(1 + o(1)))$$

$$= O(s_n) \quad \text{as} \quad n \to \infty. \tag{4.2}$$

Next using properties of the circular extension, one can show that on the set $\{L_1 \le p^{-1}(\log n)^2\}$, for any $\nu \in (\mathbb{Z}_+)^d$,

$$\left| E_{**}S(I_1, L_1)^\nu - n^{-1}\sum_{i=1}^n T(i, L_1)^\nu \right|$$

$$\le C(d, \nu)n^{-1}\sum_{i \le (\log n)^2/p} \{\|X_1 + \cdots + X_i\|^{|\nu|}$$

$$+ \|X_{n-L_1-i+1} + \cdots + X_{n-L_1+1}\|^{|\nu|} + \|X_{n+1} + \cdots + X_{n+i}\|^{|\nu|}\}$$

$$\equiv n^{-1}\hat{Q}_n(L_1, \nu), \qquad \text{say.} \tag{4.3}$$

Hence, for any $\varepsilon > 0$, any $\nu \in (\mathbb{Z}_+)^d$ with $|\nu| \le 4$,

$$P(|p^{|\nu|/2}E_*S(I_1, L_1)^\nu - E(Z_\infty^\nu)| > \varepsilon)$$

$$\le P(|E_*\{E_{**}S(I_1, L_1)^\nu p^{|\nu|/2} - E(Z_\infty)^\nu\}| > \varepsilon,$$

$$\qquad \|X_i\| < n^{1/r} \quad \text{for} \quad i = 1, \ldots, n + p^{-1}(\log n)^2)$$

$$+ 2n\,P(\|X_1\| > n^{1/r})$$

$$\le P(|E_*\{E_{**}S(I_1, L_1)^\nu p^{|\nu|/2} - E(Z_\infty^\nu)\}1(L_1 \le p^{-1}(\log n)^2)| > \varepsilon,$$

$$\qquad \max\{\|X_i\| : 1 \le i \le n + p^{-1}(\log n)^2\} \le n^{1/r})$$

$$+ P(n^{C(r)}P_*(L_1 > p^{-1}(\log n)^2) > \varepsilon)$$

$$+ 2n\,P(\|X_1\| > n^{1/r})$$

$$\le P\left(\left|\sum_{1 \le l \le (\log n)^2/p} \left(n^{-1}\sum_{i=1}^n T(i, l)^\nu p^{|\nu|/2} - EZ_\infty^\nu\right) \cdot p(1-p)^{l-1}\right| > \varepsilon\right)$$

$$+ P\left(n^{-1}\sum_{1 \le l \le (\log n)^2/p} \hat{Q}_n(l, \nu)p^{(|\nu|+2)/2}(1-p)^{l-1} > \varepsilon\right)$$

$$+ 2n\,P(\|X_1\| > n^{1/r})$$

$$\le P\left(\left|\sum_{1 \le l \le (\log n)^2/p} \left(n^{-1}\sum_{i=1}^n (S(i, l)^\nu - ES_l^\nu)p^{|\nu|/2}\right) \cdot p(1-p)^{l-1}\right| > \varepsilon/2\right)$$

$$+ \varepsilon^{-1}n^{-1}\sum_{1 \le l \le (\log n)^2/p} p^{(|\nu|+2)/2}(1-p)^{l-1}E\hat{Q}_n(l, \nu)$$

$$+ 2n\,P(\|X_1\| > n^{1/r}) \tag{4.4}$$

for all n such that $n^{C(r)} \exp(-(\log n)^2) \leq \varepsilon$, and

$$\left| \sum_{1 \leq l \leq (\log n)^2/p} \{p^{|\nu|/2} E(S_l)^\nu - E(Z_\infty)^\nu\} \cdot p(1-p)^{l-1} \right| \leq \varepsilon/2 .$$

Note that by the mixing and moment conditions,

$$|E(S_l)^\nu| \leq C \cdot l^{|\nu|/2} \qquad \text{for} \quad l \geq 1$$

and

$$\max\{|l^{-|\nu|/2} ES_l^\nu - EZ_\infty^\nu| : p^{-1}(\log n)^{-2} \leq l \leq p^{-1}(\log n)^2\}$$
$$= O(p^{-1/2}(\log n)^{-1}),$$

for all $|\nu| \leq 8$, by Lemma 3.3 and Theorem 2.3 of Lahiri (1993a). Hence, using Eq. (4.4) and the above inequalities, one gets, for all $n \geq n_o(\varepsilon, \nu)$,

$$P(|p^{|\nu|/2} E_* S(I_1, L_1)^\nu - E(Z_\infty^\nu)| > \varepsilon)$$
$$\leq C(\varepsilon) \sum_{1 \leq l \leq (\log n)^2/p} p^{|\nu|} \cdot E\left(n^{-1} \sum_{i=1}^n S(i,l)^\nu - ES_l^\nu\right)^2 \cdot p(1-p)^{l-1}$$
$$+ C(\varepsilon) \left(n^{-1} \sum_{1 \leq l \leq (\log n)^2/p} p^{(|\nu|+2)/2} (1-p)^{l-1}\right) \left(\sum_{1 \leq i \leq (\log n)^2/p} i^{|\nu|/2}\right)$$
$$+ 2n P(\|X_1\| > n^{1/r})$$
$$\leq C(\varepsilon, \zeta_{8+\delta})[n^{-1}(\log n)^2/p + n^{-1} p^{|\nu|/2} \{(\log n)^2/p\}^{(|\nu|+2)/2}$$
$$+ 2n \cdot o(n^{-(8+\delta)/r})]$$
$$= o(1) \quad \text{as} \quad n \to \infty, \tag{4.5}$$

if we take $r = 8$, say.

Next note that for $|\nu| = 3$,

$$\max\{|l^{-1} E(S_l)^\nu - n^{-1} E(S_n)^\nu| : p^{-1}(\log n)^{-2} \leq l \leq p^{-1}(\log n)^2\}$$
$$= o(1) \quad as \quad n \to \infty.$$

Hence using the arguments leading to Eq. (4.5), for $|\nu| = 3$, we have

$$P(|E_* S(I_1, L_1)^\nu p - n^{-1} E(S_n)^\nu| > \varepsilon)$$
$$\leq C(\varepsilon) \sum_{1 \leq l \leq (\log n)^2/p} \left\{ p^2 E\left(n^{-1} \sum_{i=1}^n T(i,l)^\nu - E(S_l)^\nu\right)^2 \right.$$
$$\left. + n^{-1} p \left(\sum_{1 \leq i \leq (\log n)^2/p} i^{3/2}\right)\right\} p(1-p)^{l-1} + 2n P(\|X_1\| > n^{1/8})$$
$$\leq C(\varepsilon, \zeta_{8+\delta})[n^{-1}(\log n)^2/p^2 + n^{-1} p^{-3/2}(\log n)^5] + o(1)$$
$$= o(1) \quad \text{as} \quad n \to \infty.$$

This proves parts (i), (ii) and (iv). Proofs of parts (iii) and (v) are similar and hence, are omitted.

LEMMA 4.5 There exists a numerical constant $C_1 > 0$ such that for all $n \geq C_1^{-1}$,

$$P(K \leq (np) - (np)^{1/2} \log n) \leq \exp(-C_1(\log n)^2)$$

and

$$P(K \geq (np) + (np)^{1/2} \log n) \leq \exp(-C_1(\log n)^2).$$

Proof. We only prove the first inequality. The other part can be proved by using similar arguments. Let r be the largest integer not exceeding $np - (np)^{1/2} \log n$. Then, by the definition of K and the formula for the moment generating function of a geometric random variable,

$$P(K \leq r) = P(L_1 + \cdots + L_r \geq n)$$
$$= P(\exp(t(L_1 + \cdots + L_r)) > \exp(tn))$$
$$\leq e^{-tn}(pe^t/[1 - qe^t])^r$$

for all $0 < t < -\log q$, where $q \equiv 1 - p$.

Let $f(t) = \log\{e^{-tn}(pe^t(1 - qe^t)^{-1})^r\}, 0 < t < -\log q$. Then, it is easy to see that $f(t)$ attains its minimum at $t_o \equiv \log[(n - r)/n] - \log q \in (0, -\log q)$. Next, using the fact that $r(np)^{-1} = 1 - o(1)$ as $n \to \infty$, and using Taylor's expansion, after some algebra, one can show that

$$P(K \leq r) \leq \exp(f(t_o))$$

$$= \exp\left(-\frac{3}{2}np \cdot \eta^2 + 0(n(p\eta)^2 + np\eta^3)\right),$$

where $\eta \equiv (np)^{-1/2} \log n$. This completes the proof of Lemma 4.5.

5. PROOFS

We continue to use the notation and the conventions adopted in earlier sections. In particular, we set $\mu = 0$ unless otherwise stated.

Proof of Theorem 3.1. To prove the theorem, it is enough to show that there exist a real number $n_o \geq 1$ and a sequence of sets $\{A_n\}$ such that

(i) $A_n \in \sigma\langle X_1, \ldots, X_n \rangle$ for all $n \geq 1$,

(ii) $\lim_{n \to \infty} (A_n^c) = 0$, and

(iii)   for all $n \geq n_o$,

$$\sup_{x \in \mathbb{R}} |P_*(T_n^* \leq x) - \hat{\Psi}_n(x)| 1_{A_n} \leq C(np)^{-1}. \tag{5.1}$$

Assuming that we have made an appropriate choice of the sets $\{A_n\}$ satisfying (i) and (ii), we need to obtain a bound on the difference $\|P_*(T_n^* \leq x) - \hat{\Psi}_n(x)\|_\infty$ on the set A_n for $n \geq n_o$. To do this, we will first derive an Edgeworth expansion for T_n^* upto the desired degree of accuracy, by further conditioning with respect to the block length variables $L_1, \ldots, L_n$, and then integrate the resulting expression (with respect to the joint distribution of $L_1, \ldots, L_n$ given $X_1, \ldots, X_n$) on A_n to arrive at inequality (5.1). Recall that P_{**} (and E_{**}) denotes the conditional probability (and expectation, respectively) given $X_1, X_2, \ldots$ and $\{L_1, \ldots, L_n\}_{n \geq 1}$. Hence, it follows that for any $\mathscr{F}$-measurable random variable Y,

$$E_* Y = E_*(E_{**} Y) . \tag{5.2}$$

In particular, for any set $A_n^* \in \sigma\langle L_1, \ldots, L_n; X_1, \ldots, X_n \rangle$,

$$P_*(T_n^* \leq x) = E_*\{P_{**}(T_n^* \leq x) \cdot 1_{A_n^*}\} + E_*\{P_{**}(T_n^* \leq x) 1_{A_n^{*c}}\}$$

so that

$$\sup_{x \in \mathbb{R}} |P_*(T_n^* \leq x) - E_*\{P_{**}(T_n^* \leq x) 1_{A_n^*}\}| 1_{A_n} \leq P_*(A_n^{*c}) \cdot 1_{A_n}. \tag{5.3}$$

Next note that given $X_1, X_2, \ldots$, and $\{L_1, \ldots, L_n\}_{n \geq 1}$, $S(I_1, L_1), \ldots, S(I_n, L_n)$ are independent (but not necessarily identically distributed) random variables with

$$P_{**}(S(I_j, L_j) = S(i, L_j)) = n^{-1}, \qquad 1 \leq i \leq n,$$

for all $1 \leq j \leq n$. Hence, using properties of the circularly extended sequence $\{X_{ni} : i \geq 1\}$ one can show (cf. Politis and Romano (1994)), that

$$E_{**} S(I_j, L_j) = L_j \cdot \bar{X}_n$$

for all $1 \leq j \leq n$. Consequently, it follows that

$$E_{**} \bar{X}_n^* = \bar{X}_n . \tag{5.4}$$

To derive an Edgeworth expansion for T_n^* under P_{**}, first we obtain a suitable stochastic expansion for T_n^* and then use the transformation technique of Bhattacharya and Ghosh (1978) and the Edgeworth expansion theory for sums of independent random vectors (cf. Bhattacharya and Ranga Rao (1986)) to the resulting stochastic expansion. Without loss of generality, let $\delta > 0$ be such that $(D_1 H(x), \ldots, D_d H(x))' \neq 0$ for all $\|x - \mu\| \leq 2\delta$. Hence, using Taylor's expansion and some simple algebra, one can show that

$$\tau_n^{*2} = \tilde{\tau}_n^2 + K^{-1} \sum_{i=1}^{K} \{(\hat{a}_n' U_{1i}^*)^2 - E_{**}(\hat{a}_n' U_{1i}^*)^2\}$$

$$+ 2 \sum_{|\alpha|=1} \sum_{|\beta|=1} [D^{\alpha+\beta} H(\bar{X}_n)](\bar{X}_n^* - \bar{X}_n)^\alpha \left(K^{-1} \sum_{i=1}^{K} E_{**} S_{1i}^*(\beta) \right) + Q_{1n}^*$$

$$(5.5)$$

where $S_{1i}^*(\beta) = (\hat{a}_n' U_{1i}^*)(U_{1i}^*)^\beta$, $\tilde{\tau}_n^2 = K^{-1} \sum_{i=1}^{K} E_{**}(\hat{a}_n' U_{1i}^*)^2$, and with $h_{jn}^* \equiv \sup\{|D^\alpha H(\bar{X}_n + x)| : \|x\| \leq \|\bar{X}_n^* - \bar{X}_n\|, |\alpha| \leq j\}$, $j \geq 1$, the remainder term Q_{1n}^* admits the bound

$$|Q_{1n}^*| \leq C \sum_{|\beta|=1} \left| K^{-1} \sum_{i=1}^{K} (S_{1i}^*(\beta) - E_{**} S_{1i}^*(\beta)) \right| \cdot \|\bar{X}_n^* - \bar{X}_n\|$$

$$+ C \cdot h_{3n}^* \sum_{|\beta|=1} \left(K^{-1} \sum_{i=1}^{K} |S_{1i}^*(\beta)| \right) \cdot \|\bar{X}_n^* - \bar{X}_n\|^2$$

$$+ C \cdot h_{3n}^{*2} \cdot \left(K^{-1} \sum_{i=1}^{K} \|U_{1i}^*\|^2 \right) \{\|\bar{X}_n^* - \bar{X}_n\|^2 + \|\bar{X}_n^* - \bar{X}_n\|^4\}$$

$$+ C \cdot h_{3n}^{*2} (\bar{U}_{1K}^*)^2,$$

$$(5.6)$$

provided $\{\|\bar{X}_n - \mu\| \leq \delta\}$, where $\bar{U}_{1K}^* = K^{-1} \sum_{i=1}^{K} U_{1i}^*$. Let

$$A_{1n}^* = \left\{ \sum_{i=1}^{K} E_{**} \|S(I_i, L_i)\|^4 < 2E\|Z_\times\|^4 n/p, \right.$$

$$\left\| K^{-1} \sum_{i=1}^{K} (E_{**} S(I_i, L_i) S(I_i, L_i)' - L_i \Sigma_\times) \right\| < (p\|\Sigma_\times^{-1}\|)^{-1}/4,$$

$$\left. \tilde{\tau}_n^2 > \tau_\times^2/2, |K - np| \leq (np)^{1/2} \log n \right\}.$$

Then, on A_{1n}^*,

$$\sum_{i=1}^{K} E_{**}\|S(I_i, L_i)\|^2 = \text{trace}\left\{\sum_{i=1}^{K} E_{**}S(I_i, L_i)S(I_i, L_i)'\right\}$$

$$\leq C(d)\left\{\left\|\sum_{i=1}^{K} L_i\Sigma_\infty\right\| + K(2p\|\Sigma_\infty^{-1}\|)^{-1}\right\}$$

$$\leq C(d) \cdot \|\Sigma_\infty\| \cdot \left(\sum_{i=1}^{K} L_i\right)$$

$$\leq C(d, \|\Sigma_\infty\|)n^{-1}\left(\sum_{i=1}^{K} L_i\right)^2.$$

Next, recall the fact that for square matrices B_1 and B_2, if B_2 is nonsingular and $\|B_1 - B_2\| \cdot \|B_2^{-1}\| < 1/2$, then B_1^{-1} exists and $\|B_1^{-1}\| < 2\|B_2^{-1}\|$. Since on A_{1n}^*,

$$\left\{\left\|\left(\sum_{i=1}^{K} L_i\Sigma_\infty\right)^{-1}\right\|\right\} \cdot \{K(p\|\Sigma_\infty^{-1}\|)^{-1}/4\} = \left(\sum_{i=1}^{K} L_i\right)^{-1} K/4p$$

$$\leq K/(4np)$$

$$< 1/2$$

for n large, it follows that $(\sum_{i=1}^{K} E_{**}S(I_i, L_i)S(I_i, L_i)')^{-1}$ exists and its norm is bounded above by $2\|\Sigma_\infty^{-1}\|/n$. Therefore, by Lemma 4.1, on the set A_{1n}^*,

$$P_{**}(\|\bar{X}_n^* - \bar{X}_n\| > C(d, \|\Sigma_\infty\|)n^{-1/2}(\log n)^{1/2})$$

$$= P_{**}\left(\left\|\sum_{i=1}^{K}(S(I_i, L_i) - E_{**}S(I_i, L_i))\right\|\right.$$

$$\left. > C(d, \|\Sigma_\infty\|)\left(n^{-1}\left(\sum_{i=1}^{K} L_i\right)^2 \log n\right)^{1/2}\right)$$

$$\leq P_{**}\left(\left\|\sum_{i=1}^{K}(S(I_i, L_i) - E_{**}S(I_i, L_i))\right\|\right.$$

$$\left. > C(d)\left(\sum_{i=1}^{K} E_{**}\|S(I_i, L_i)\|^2 \log n\right)^{1/2}\right)$$

$$\leq C(d, \|\Sigma_\infty^{-1}\|)\left\{\sum_{i=1}^{K}(E_{**}\|S(I_i, L_i)\|^4)n^{-2}(\log n)^{-2} + n^{-3/2}\right\}$$

$$\leq C(d, \|\Sigma_\infty^{-1}\|, E\|Z_\infty\|^4)(np)^{-1}(\log n)^{-2}. \tag{5.7}$$

By similar arguments, one can show that on the set

$$A_{2n}^* \equiv A_{1n}^* \cap \left\{ \sum_{i=1}^{K} \|E_{**}U_{1i}^*\|^8 \le 2nE\|Z_\times\|^8/p \right\},$$

$$P_{**}\left(\left| \sum_{i=1}^{K}(\|U_{1i}^*\|^2 - E_{**}\|U_{1i}^*\|^2) \right| > C(E\|Z_\times\|^4) \cdot (np\log n)^{1/2} \right)$$

$$\le C(d)\left[\sum_{i=1}^{K} E_{**}\|U_{1i}^*\|^8 (np\log n)^{-2} + (np)^{-3/2} \right]$$

$$\le C(d, E\|Z_\times\|^8)(np)^{-1}(\log n)^{-2}; \tag{5.8}$$

$$P_{**}\left(\sum_{|j|=1} \left| \sum_{i=1}^{K}(S_{1i}^*(\beta) - E_{**}S_{1i}(\beta)) \right| > C(d, E\|Z_\times\|^4) \cdot (np\log n)^{1/2} \right)$$

$$\le C(d, E\|Z_\times\|^8)(np)^{-1}(\log n)^{-2}; \tag{5.9}$$

and

$$P_{**}\left(\left\| \sum_{i=1}^{K}(U_{1i}^* - L_i^{1/2}\bar{X}_n) \right\| > C(d, \|\Sigma_\times\|)(np\log n)^{1/2} \right)$$

$$\le C(d, \|Z_\times\|^4)(np)^{-1}(\log n)^{-2}. \tag{5.10}$$

Therefore, by Eqns (5.5)–(5.10), on A_{2n}^*,

$$P_{**}(|Q_{1n}^*| > C(d, \|\Sigma_\times\|, E\|Z_\times\|^4)(np)^{-1}\log n)$$

$$\le C(d, E\|Z_\times\|^8)(np)^{-1}(\log n)^{-2}, \tag{5.11}$$

provided $\|\bar{X}_n - \mu\| \le C(d, \|\Sigma_\times\|)(n^{-1}\log n)^{1/2}$.

Next we further simplify the second term by replacing $\hat{a}_n$ with a. Note that if $X_1, \ldots, X_n$ are such that $\|\bar{X}_n - \mu\| < C(d, \|\Sigma_\times\|)n^{-1/2}(\log n)^{1/2}$, then on the set A_{2n}^*

$$P_{**}\left(\left| K^{-1}\sum_{i=1}^{K}[\{(\hat{a}_n' U_{1i}^*)^2 - E_{**}(\hat{a}_n' U_{1i}^*)^2\} - \{(a' U_{1i}^*)^2 - E_{**}(a' U_{1i}^*)^2\}] \right| \right.$$

$$\left. > C(d, E\|Z_\times\|^4, \|\Sigma_\times\|)n^{-1}p^{-1/2}\log n \right)$$

$$= P_{**}\left(\left|\sum_{i=1}^{K}\left[\sum_{|\alpha|=1}\sum_{|\beta|=1}(\hat{a}_n - a)^{\alpha}(\hat{a}_n + a)^{\beta}\{(U_{1i}^*)^{\alpha+\beta} - E_{**}(U_{1i}^*)^{\alpha+\beta}\}\right]\right|\right.$$

$$\left. > C(d, E\|Z_{\infty}\|^4, \|\Sigma_{\infty}\|)p^{1/2}\log n\right)$$

$$\leq \sum_{|\alpha|=1}\sum_{|\beta|=1}P_{**}\left(\left|\sum_{i=1}^{K}(U_{1i}^*)^{\alpha+\beta} - E_{**}(U_{1i}^*)^{\alpha+\beta}\right|\cdot\right.$$

$$\left. > C(d, E\|Z_{\infty}\|^4)(np)^{1/2}(\log n)^{1/2}\right)$$

$$\leq C(d, E\|Z_{\infty}\|^8)\cdot(np)^{-1}(\log n)^{-2}. \tag{5.12}$$

Thus, from Eqns (5.5), (5.11) and (5.12), we have

$$\tau_n^{*2} = \tilde{\tau}_n^2 + K^{-1}\sum_{i=1}^{K}\{(a'U_{1i}^*)^2 - E_{**}(a'U_{1i}^*)^2\}$$

$$+ 2\sum_{|\alpha|=1}\sum_{|\beta|=1}D^{\alpha+\beta}H(\bar{X}_n)\cdot(\bar{X}_n^* - \bar{X}_n)^{\alpha}\cdot\left(K^{-1}\sum_{i=1}^{K}E_{**}S_{1i}^*(\beta)\right) + Q_{2n}^*$$

$$\equiv \tilde{\tau}_n^2 + \tau_{1n}^* + Q_{2n}^*, \qquad \text{say} \tag{5.13}$$

where, on the set $A_{2n}^* \cap \{\|\bar{X} - \mu\| < C(d, \|\Sigma_{\infty}\|)(n^{-1}\log n)^{1/2}\}$,

$$P_{**}(|Q_{2n}^*| > C(d, \|\Sigma_{\infty}\|, E\|Z_{\infty}\|^4)(np)^{-1}\log n)$$

$$\leq C(d, E\|Z_{\infty}\|^8)(np)^{-1}(\log n)^{-2}. \tag{5.14}$$

Let $A_{1n} = \{\|\bar{X}_n - \mu\| < C(d, \|\Sigma_{\infty}\|)(n^{-1}\log n)^{1/2}\}$. Then, it is easy to check that

$$\tau_n^{*-1} = \tilde{\tau}_n^{-1} - (2\tilde{\tau}_n^3)^{-1}\tau_{1n}^* + Q_{3n}^*$$

where, $|Q_{3n}^*| \leq C(\tau_{\infty})\{|\tau_n^{*2} - \tilde{\tau}_n^2|^2 + |Q_{2n}^*|\}$ on the set $A_{2n}^* \cap A_{1n}$. Hence, using the above expansion for τ_n^{*-1} and expanding $H(\bar{X}_n^*)$ into a Taylor's series around $\bar{X}_n$, we have

$$T_n^* = \sqrt{N_1}(H(\bar{X}_n^*) - H(\bar{X}_n))/\tau_n^*$$

$$= \hat{a}_n'V_{1n}^*/\tilde{\tau}_n - (2\tilde{\tau}_n^3)^{-1}(\hat{a}_n'V_{1n}^*)\tau_{1n}^* + n^{-1/2}\sum_{|\alpha|=2}\tilde{c}_{\alpha}(V_{1n}^*)^{\alpha}/\tilde{\tau}_n + Q_{4n}^*$$

$$\equiv T_{1n}^* + Q_{4n}^*, \qquad \text{say} \tag{5.15}$$

where $V_{1n}^* = \sqrt{N_1}(\bar{X}_n^* - \bar{X}_n)$, and by Eqns (4.3), (4.4) and Lemma 4.1,

$$P_{**}(|Q_{4n}^*| > C(d, \|\Sigma_{\times}\|, E\|Z_{\times}\|^4)(np)^{-1}(\log n)^{3/2})$$

$$\leq P_{**}(\|V_{1n}^*\| \cdot |Q_{3n}^*| > C(d, \|\Sigma_{\times}\|, E\|Z_{\times}\|^4)(np)^{-1}(\log n)^{3/2})$$

$$+ P_{**}(\|V_{1n}^*\|^2 \|\tau_n^{*2} - \tilde{\tau}_n^2\| > C(d, \|\Sigma_{\times}\|, E\|Z_{\times}\|^4) \cdot n^{1/2}(np)^{-1}(\log n)^{3/2})$$

$$+ P_{**}(\|V_{1n}^*\|^3 > C(d, \|\Sigma_{\times}\|)p^{-1}(\log n)^{3/2})$$

$$+ P_{**}(|\tau_n^{*2} - \tilde{\tau}_n^2|^2 > C(d, E\|Z_{\times}\|^4)(np)^{-1}(\log n))$$

$$\leq C(d, \|\Sigma_{\times}^{-1}\|, E\|Z_{\times}\|^8)(np)^{-1}(\log n)^{-2}, \qquad (5.16)$$

provided $(L_1, \ldots, L_n)$ and $(X_1, \ldots, X_n)$ lie in $A_{2n}^* \cap A_{1n}$. Thus, T_{1n}^* gives the desired stochastic expansion for T_n^*.

Note that T_{1n}^* is a quadratic polynomial in $(d+1)$ variables V_n^*, where $V_n^* = (V_{1n}^{*\prime}, V_{2n}^*)$ with $V_{2n}^* \equiv K^{-1/2} \sum_{i=1}^K \{(a'U_{1i}^*)^2 - E_{**}(a'U_{1i}^*)^2\}$. Therefore, an Edgeworth expansion for the distribution function of T_{1n}^* can be derived from an expansion of $P_{**}(V_n^* \in B)$ for suitable Borel sets B in $\mathbb{R}^{d+1}$. To that end, let $\Sigma_V^* = \text{cov}_{**}(V_n^*)$ and $V_i^{**} = ((K/N_1)^{1/2}(S(I_i, L_i) - E_{**}S(I_i, L_i))', ((a'U_{1i}^*)^2 - E_{**}(a'U_{1i}^*)^2))', 1 \leq i \leq K$. Also, let $V_{\times} = (Z_{\times}', (a'Z_{\times})^2 - \tau_{\times}^2))'$ and $\Sigma_V = \text{cov}(V_{\times})$. Write $A_{3n}^* = A_{2n}^* \cap \{\|\Sigma_V^* - \Sigma_V\| < \|\Sigma_V^{-1}\|/4\}$. Then, conditional on $\{L_1, \ldots, L_n\}$ and $\{X_1, \ldots, X_n\}$ in $A_{3n}^* \cap A_{1n}$, Σ_V^* is nonsingular, and by Lemma 4.2,

$$\left| P_{**}(\Sigma_V^{*-1/2} V_n^* \in B) - \Psi_n^{\dagger}(B) \right|$$

$$\leq C(d, E\|Z_{\times}\|^8)[K^{-1} + \alpha_n^* K^{2d} + \Phi_{I_{d+1}}((\partial B)^{2\varepsilon})] \qquad (5.17)$$

for any Borel set $B \subset \mathbb{R}^{d+1}$ and for all $n \geq C(d, E\|Z_{\times}\|^8)$, where $\varepsilon = K^{-1}$, $\Psi_n^{\dagger}$ is as in Lemma 4.2 with $W_j = \Sigma_V^{*-1/2} V_{j*}, j = 1, \ldots, K$, and

$$\alpha_n^* = \sum_{1 \leq j_1, \ldots, j_{d+4} \leq K} \sup \left\{ \prod_{j \neq j_1, \ldots, j_{d+4}} |E_{**} \exp(it'\Sigma_V^{*-1/2} V_j^{**})| : C(d, E\|Z_{\times}\|^8) \right.$$

$$\left. < \|t\| < K^4 \right\}.$$

Let $A_{4n}^* = A_{3n}^* \cap \{\sum_{i=1}^K 1(|L_i - p^{-1}| < p^{-1/2}) \geq (\log n)^2\}$. Thus, on A_{4n}^*, at least $(\log n)^2$ of the blocks have lengths larger than $p/2$, for large n. Therefore, by Lemma 4.3 (with a $\lambda > 0$ satisfying $p^{-\lambda} > (2np)^4$) and by Eq. (5.17), for any Borel set B in $\mathbb{R}^{d+1}$, we get

$$|P_{**}(\Sigma_V^{*-1/2} V_n^* \in B) - \Psi_n^{\dagger}(B)|$$

$$\leq C(d, E\|Z_{\times}\|^8)[\sup\{(|w_{\times}(t,s)| + \delta_n) : C(d, E\|Z_{\times}\|^8)$$

$$< \|(t',s)'\| < p^{-\lambda}\}^{(\log n)^2}$$

$$+ (np)^{-1} + \Phi_{I_{d+1}}((\partial B)^{2\varepsilon})]$$

$$\leq C(d, E\|Z_{\times}\|^8)[(np)^{-1} + \Phi_{I_{d+1}}((\partial B)^{2\varepsilon})].$$

Next let $\tilde{p}_{2n}(v) \equiv \sum_{|\alpha|=2} \tilde{d}_\alpha v^\alpha$, $v \in \mathbb{R}^d$ be the polynomial obtained from $\hat{p}_{2n}(\cdot)$ by replacing $E_* S_{11}^*(\beta)$ and $\hat{\tau}_{1n}^2$ by $K^{-1} \sum_{i=1}^{K} E_{**} S_{1i}^*(\beta)$ and $\tilde{\tau}_n^2$, respectively. Then, using the transformation technique of Bhattacharya and Ghosh (1978), it follows that on the set $A_{3n}^* \cap A_{1n}$,

$$\sup_x |P_{**}(T_{1n}^* \leq x) - \tilde{\Psi}_n(x)| \leq C(d, E\|Z_{\times}\|^8, \mathscr{D}_H) \cdot (np)^{-1} \tag{5.18}$$

for all n such that $C(d, \|\Sigma_{\times}\|)n^{-1/2}(\log n)^{1/2} < \delta$, where $\mathscr{D}_H \equiv \sup\{|D^\alpha H(x)| : \|x - \mu\| < \delta, |\alpha| \leq 3\}$ and $\tilde{\Psi}_n(x)$ is defined by its Fourier transform as

$$\int e^{itx} d\tilde{\Psi}_n(x) = \exp(-t^2/2)\Bigg[1 + N_1^{-1/2}(it)\Bigg\{E_{**}\tilde{p}_{2n}(V_{1n}^*) - (2\tilde{\tau}_n^3)^{-1}$$

$$\times E_{**}\Bigg(\hat{a}_n' V_{1n}^* \cdot \Bigg(K^{-1}\sum_{i=1}^{K}(a'U_{1i}^*)^2 - E_{**}(a'U_{1i}^*)^2\Bigg)\Bigg)\Bigg\}$$

$$+ (it)^3(3!)^{-1}\Bigg\{E_{**}(\hat{a}_n' V_{1n}^*/\tilde{\tau}_n)^3 + 3(N_1^{5/2}\tilde{\tau}_n^2)^{-1}$$

$$\times \sum_{|\alpha|=1}\sum_{|\beta|=1}\sum_{|\gamma|=1}\sum_{|\iota|=1}\sum_{i=1}^{K}\sum_{j=1}^{K}\hat{d}_{\alpha+\beta}\tilde{c}_\gamma\tilde{c}_\iota\Bigg(\text{cov}_{**}(S(I_i,L_i)^\alpha, S(I_i,L_i)^\gamma)$$

$$\times \text{cov}_{**}(S(I_j,L_j)^\beta, S(I_j,L_j)^\iota)$$

$$+ \text{cov}_{**}(S(I_i,L_i)^\alpha, S(I_i,L_i)^\iota) \cdot \text{cov}_{**}(S(I_j,L_j)^\beta, S(I_j,L_j)^\gamma)\Bigg)\Bigg\}\Bigg]$$

for $t \in \mathbb{R}$. Thus, $\tilde{\Psi}_n$ gives the Edgeworth expansion for the conditional distribution of T_{1n}^* given $\{X_1, \ldots, X_n; L_1, \ldots, L_n\}$. This also coincides with the Edgeworth expansion of the bootstrapped statistic T_n^* upto the second-order, i.e., upto the terms of order $O_p(n^{-1/2})$. To see this, note that

$$\sup_x |P_{**}(T_n^* \leq x) - P_{**}(T_{1n}^* \leq x)|$$

$$\leq P_{**}(|Q_{4n}^*| > C(d, \|\Sigma_\infty\|, E\|Z_\infty\|^4)(np)^{-1}(\log n)^{3/2})$$

$$+ 2 \sup_x |P_{**}(T_{1n}^* \leq x) - \tilde{\Psi}_n(x)|$$

$$+ \sup\{|\tilde{\Psi}_n(x+y) - \tilde{\Psi}_n(x)| : \|y\|$$

$$< C(d, \|\Sigma_\infty\|, E\|Z_\infty\|^4)(np)^{-1}(\log n)^{3/2}, \; x \in \mathbb{R}\}. \tag{5.19}$$

Hence, there exists an integer $n_o \geq 1$ such that for all $n \geq n_o$,

$$\sup_x |P_{**}(T_n^* \leq x) - \tilde{\Psi}_n(x)| 1_{A_{4n}^* \cap A_{1n}} \leq \tilde{Q}_{5n} 1_{A_{4n}^* \cap A_{1n}},$$

where, on the set $A_{4n}^* \cap A_{1n}$, by Eqns (5.16), (5.18) and (5.19),

$$|\tilde{Q}_{5n}| \leq C(d, \Sigma_\infty, E\|Z_\infty\|^8, \mathscr{D}_H)(np)^{-1}(\log n)^{3/2}.$$

Let $A_{5n}^* = A_{4n}^* \cap \{|E_{**}(V_{1n}^*)^\nu - N_L^{-|\nu|/2} KE_*(S(I_i, L_i))^\nu| \leq C(d, E\|Z_\infty\|^{2|\nu|}) (np)^{-1/2}(\log n)^{3/2} : |\nu| = 2, 3\} \cap \{|\sum_{i=1}^k L_i^3| \leq 2Kp^{-3}\}$. Then, from Eq. (5.18), it follows that

$$\sup_x |P_*(T_n^* \leq x) - \hat{\Psi}_n(x)|$$

$$= \sup_x |E_*\{P_{**}(T_n^* \leq x)\} - \hat{\Psi}_n(x)|$$

$$\leq \sup_x E_* |P_{**}(T_n^* \leq x) - \tilde{\Psi}_n(x)| 1_{A_{5n}^*}$$

$$+ \sup_x E_* |\tilde{\Psi}_n(x) - \hat{\Psi}_n(x)| 1_{A_{5n}^*} + \left(\sup_x |\hat{\Psi}_n(x)| + 1\right) P_*(A_{5n}^{*c})$$

$$\leq C(d, \Sigma_\infty, E\|Z_\infty\|^8, \mathscr{D}_H)\{(np)^{-1}(\log n)^{3/2} + P_*(A_{5n}^{*c})\}.$$

To complete the proof of the theorem, it is now enough to show that

$$P(A_{1n}^c) = o(1) \quad \text{as} \quad n \to \infty \tag{5.20}$$

and that on the set A_{1n},

$$P_*(A_{4n}^{*c}) \leq C(d, E\|Z_\infty\|^8, \mathscr{D}_H)(np)^{-1}(\log n)^{3/2} . \tag{5.21}$$

Let $p_{1n} \equiv P_*(|L_1 - p^{-1}| \leq p^{-1/2})$. Then, by Eqns (4.3) and (4.4), $p_{1n} = 1 - O(p^{1/2})$ as $n \to \infty$. Hence, there exists a $n_o \geq 1$ such that for all

$n \geq n_o$, by Wald's identity,

$$P_* \left(\sum_{i=1}^{K} 1(|L_i - p^{-1}| < p^{-1/2}) \leq (\log n)^2, K \geq np/2 \right)$$

$$\leq P_* \left(\left| \sum_{i=1}^{K} \{1(|L_i - p^{-1}| < p^{-1/2}) - p_{1n}\} \right| > np/4 \right)$$

$$\leq 16(np)^{-2} E_* \left(\sum_{i=1}^{K} \{1(|L_i - p^{-1}| < p^{-1/2}) - p_{1n}\} \right)^2$$

$$= (np)^{-2} (E_* K)(p_{1n}(1 - p_{1n}))$$

$$\leq C(np)^{-2}(np)p^{-1/2}, \tag{5.22}$$

and by using properties of the Geometric distribution,

$$P_* \left(\sum_{i=1}^{K} L_i^3 \leq CKp^{-3}, np/2 \leq K \leq 2np \right)$$

$$\leq P_* \left(\left| \sum_{i=1}^{K} (L_i^3 - EL_i^3) \right| > Cnp^{-2} \right)$$

$$\leq C(np^{-2})^{-2}(np)E_* L_1^6$$

$$\leq C(np)^{-1}. \tag{5.23}$$

Now using Eqns (5.22), (5.23) and Lemmas 4.1, 4.4 and 4.5, one obtains Eqns (5.20) and (5.21). This completes the proof of Theorem 3.1.

Proof of Theorem 3.2. From the results of Götze and Künsch (1996) and Lahiri (1996a), it follows that under assumptions (A.1)–(A.6),

$$\sup_{x \in \mathbb{R}} |P(T_n \leq x) - \Psi_n(x)| = o(n^{-1/2})$$

where the Edgeworth expansion Ψ_n of T_n is defined by its Fourier transform as

$$\int \exp(itx) d\Psi_n(x) = \exp(-t^2/2) \left[1 + n^{-1/2}(it) \left\{ \sum_{|\alpha|=2} d_\alpha E(S_n/\sqrt{n})^\alpha \right. \right.$$

$$\left. \left. - (2n^{3/2}\tau_n^3)^{-1} E \left[(a'S_n) \left(\sum_{i=1}^{n} (a'X_i)^2 \right) \right] \right. \right]$$

$$+ 2 \sum_{j=1}^{l_1} \sum_{i=1}^{n-j} (a' X_i)(a' X_{i+j}) \Big) \Big] \Big\}$$

$$+ (3!)^{-1} (it)^3 \Big\{ nE(a' S_n / \tau_n)^3$$

$$+ 3(\tau_n^2 n^{5/2})^{-1} \sum_{|\alpha|=1} \sum_{|\beta|=1} d_{\alpha + \beta}$$

$$\cdot [2\text{cov}(a' S_n, S_n^\alpha) \cos(a' S_n, S_n^\beta)] \Big\} \Big]$$

for $t \in \mathbb{R}$. Now comparing the co-efficients of Ψ_n and Ψ_n^*, and using Lemmas 4.1, 4.4 and 4.5, one can easily complete the proof of Theorem 3.2.

Proof of Theorem 3.3. Using arguments similar to the proof of Theorem 3.1, one can obtain a sufficiently close stochastic expansion for the boot-strapped statistic $T_n^{\dagger *}$, which is given by a second degree polynomial in the d-dimensional random variable V_{1n}^*. Next using Lemma 5.2, the transformation technique of Bhattacharya and Ghosh (1978), and the conditioning arguments in the proof of Theorem 3.1, one can derive an Edgeworth expansion for $T_n^{\dagger *}$. Also, in view of Lemma 4.2, a similar expansion holds for the normalized statistic $T_n^{\dagger}$ under assumptions (A.1)–(A.5) if, in addition, $E\|X_1\|^{4+\delta} < \infty$ for some $\delta > 0$. Hence, Theorem 3.3 follows by comparing the coefficients of the resulting Edgeworth expansions and applying Lemmas 4.1, 4.4, and 4.5 as in the proof of Theorem 3.2. We omit the details.

Acknowledgment

The author would like to thank a referee for a careful reading of an earlier draft of the paper and for some constructive suggestions that improved the presentation of the paper.

REFERENCES

Bhattacharya, R. N. and Ghosh, J. K. (1978). On the validity of the formal Edgeworth expansion. *Ann. Statist.* 7: 434–451.

Bhattacharya, R. N. and Ranga Rao, R. (1986). *Normal Approximations and Asymptotic Expansions*. R. E. Krieger Publishing Company, Malabar, Florida.

Bose, A. (1988). Edgeworth correction by bootstrap in autoregressions. *Ann. Statist.* 16: 1709–1722.

Bustos, O. (1982). General M-estimates for Contaminated p-th Order Autoregressive Processes: Consistency and Asymptotic Normality. *Z. Wahrsch. verw. Gebiete 59*: 491–504.

Carlstein, E. (1986). The Use of Subseries Methods for Estimating the Variance of a General Statistic from a Stationary Time Series. *Ann. Statist., 14*: 1171–1179.

Davison, A. C. and Hall, P. (1993). On Studentizing and Blocking Methods for Implementing the Bootstrap with Dependent Data. *Aust. J. Statist., 35*: 215–224.

Efron, B. (1979). Bootstrap Methods: Another Look at the Jackknife. *Ann. Statist., 7*: 1–26.

Fuk, D. K. and Nagaev, S. V. (1971). Probability inequalities for sums of independent random variables. *Theory Probabl. Appl. 16*: 643–660.

Götze, F. and Hipp, C. (1983). Asymptotic Expansions for Sums of Weakly Dependent Random Vectors. *Zieb Wahr. verw. Gebiete 64*: 211–239.

Götze, F. and Hipp, C. (1994). Asymptotic Distribution of Statistics in Time Series. *Ann. Statist. 22*: 2062–2088.

Götze, F. and Künsch, H. R. (1996). Blockwise Bootstrap for Dependent Observations: Higher Order Approximations for Studentized Statistics. *Ann. Statist. 24*: 1914–1933.

Hall, P. (1985). Resampling a Coverage Pattern. *Stoch. Proc. Appl., 20*: 231–246.

Hall, P. (1992). *The Bootstrap and Edgeworth Expansion.* Springer-Verlag. New York.

Hall, P., Horowitz, J. L. and Jing, B-Y. (1995). On Blocking Rules for the Bootstrap with Dependent Data. *Biometrika 82*: 561–574.

Künsch, H. R. (1989). The Jackknife and the Bootstrap for General Stationary Observations. *Ann. Statist., 17*: 1217–1261.

Lahiri, S. N. (1991). Second-Order Optimality of Stationary Bootstrap. *Statist. Probab. Lett., 11*: 335–341.

Lahiri, S. N. (1992). Edgeworth correction by "Moving Block" Bootstrap for Stationary and Nonstationary data. In *Exploring the Limits of Bootstrap* (Eds. R. Lepage and L. Billard). Wiley, New York, 183–214.

Lahiri, S. N. (1993a). Refinements in the Asymptotic Expansions for Sums of Weakly Dependent Random Vectors. *Ann. Probab., 21*: 791–799.

Lahiri, S. N. (1993b). On the moving Block Bootstrap under long range dependence. *Statist. Probab. Lett. 18*: 405–413.

Lahiri, S. N. (1995). On the Asymptotic Behaviour of the Moving Block Bootstrap for normalized sums of heavy-tail random variables. *Ann. Statist. 23*: 1331–1349.

Lahiri, S. N. (1996a). On Edgeworth Expansion and Moving Block Bootstrap for Studentized *M*-Estimators in Multiple Linear Regression Models. *J. Mult. Analysis. 56*: 42–59.

Lahiri, S. N. (1996b). Asymptotic expansions for sums of random vectors under polynomial mixing rates. *Sankhya, Ser. A. 58*: 206–224.

Liu, R. Y. and Singh, K. (1992). Moving Blocks Jackknife and Bootstrap Capture Weak Dependence. In *Exploring the Limits of Bootstrap* (Eds. R. Lepage and L. Billard). Wiley, New York, 225–248.

Politis, D. and Romano, J. P. (1992). A Circular Block Resampling Procedure for Stationary Data. In *Exploring the Limits of Bootstrap* (Eds. R. Lepage and L. Billard), Wiley, New York, 263–270.

Politis, D. and Romano, J. P. (1994). The Stationary Bootstrap. *J. Amer. Statist. Assoc. 89*: 1303–1313.

Singh, K. (1981). On the asymptotic accuracy of Efron's bootstrap. *Ann. Statist. 6*: 1187–1195.

.

21

Convergence to Equilibrium of Random Dynamical Systems Generated by IID Monotone Maps, with Applications to Economics

RABI BHATTACHARYA*   Indiana University, Bloomington, Indiana

MUKUL MAJUMDAR   Cornell University, Ithaca, New York

1. INTRODUCTION

Consider a random dynamical system (S, Γ, P) where S is the state space (for example, some subset of a finite dimensional Euclidean space), Γ an appropriate family of maps on S into itself and P is a probability measure on (some σ–field of) Γ.

The evolution of the system can be described as follows: initially, the system is in some state x, an element α_1 of Γ is chosen randomly according to the probability measure P and the system moves to the state $X_1 = \alpha_1(x)$ in period one. Again, independently of α_1, an element α_2 of Γ is chosen according to the probabilty measure P and the state of the system in period two is obtained as $X_2 = \alpha_2(\alpha_1(x))$. In general, starting from some x in S, one has

$$X_{n+1}(x) = \alpha_{n+1}(X_n(x))$$

* Research supported by NSF Grant DMS-9504557.

where the maps (α_n) are independent and identically distributed according to the measure P. The initial point x can also be chosen (independently of (α_n)) as a random variable X_0. The sequence X_n of states obtained in this manner is a Markov process and has been of particular interest in dynamic economics. It may be noted that every Markov process (with an arbitrarily given transition probability) may be constructed in this manner, provided S is a Borel subset of a complete separable metric space (Kiefer (1986) p. 8, or Bhattacharya and Waymire (1990) p. 228).

For describing "convergence to a long run steady state", perhaps the most widely used results identify conditions under which there is some time invariant probability measure π such that, no matter what the initial X_0 is, X_n converges in distribution to π. That is, the n–step transition probability $p^{(n)}(x, dy)$, starting from an initial state x, converges weakly to a probability measure π, irrespective of x. The class of processes for which we study such convergence corresponds to monotone nondecreasing α_n on an interval, or on appropriate subsets $S \subset \mathbb{R}^k$. Excepting for Sec. 7, where examples of multidimensional S are considered, the present chapter emphasizes the case of an interval.

For compact intervals $S = [a, b]$ and α_n iid nondecreasing and continuous, Dubins and Freedman (1966) derived necessary and sufficient conditions for convergence to a unique steady state, or invariant probability, in terms of a 'splitting condition' (see Sec. 2). Yahav (1975) obtained an extension to the case of nondecreasing α_n, without the requirement of continuity. Generalizations to noncompact intervals and to multidimension, without the requirement of continuity, were obtained in Bhattacharya and Lee (1988). The latter article also contains a functional central limit theorem for such processes. The present chapter provides an expository account of this theory (Secs. 2, 3, 7) and presents a number of significant applications, especially to economics (Secs. 4–6). Apart from these applications, there are some new results here.

In Sec. 4 we provide applications of the theory to random iterations of two quadratic maps $F_{\theta_i}(x) := \theta_i x(1 - x)(i = 1, 2)$ on $S = (0, 1)$. The quadratic maps arise as models in environmental biology and natural sciences. They are also among the most extensively studied families of dynamical systems, Devaney (1989). A choice between two such maps may arise from randomness in the environment.

The applications to economics in Sec. 5 on economic growth arise from two growth–mechanisms, one "favorable" and the other "unfavorable," chosen at random at each period of time by the random environment.

In Sec. 6 quadratic maps $F_\theta(x) := \theta x(1 - x)(1 \le \theta \le 4)$ are shown to arise in an economy as an optimal choice of a transition function $h(x)$ of the 'stock' in the next period, given a stock x in the current period. Initially, the economy has a capital stock x_0 from which an output $f(x_0)$ arises, a part c_1 of which is consumed and the rest, $x_1 := f(x_0) - c_1$, is to be used as capital in period one to produce an output $f(x_1)$. A part c_2 of $f(x_1)$ is consumed and the rest, $x_2 := f(x_1) - c_2$ used as capital for the next period, and so on. The consumption c_{t+1} in period $t + 1$ yields a one–period return, or felicity, $w(x_t, c_{t+1})$. The object of the planner is to choose a sequence of stocks x_t, or equivalently, a sequence of consumptions $c_t (t = 1, 2, \ldots)$, that maximizes the long–term discounted return $\sum_{t=0}^{\infty} \delta^t w(x_t, c_{t+1})$. The optimal choice, as embodied in the transition function h, depends on f and w. Certain economically meaningful classes of f and w lead to monotone nondecreasing h, while others yield quadratic $h(x) = \theta x(1 - x)$, Benhabib and Nishimura (1985); Majumdar and Nermuth (1982); Majumdar and Mitra (1994a, b). Under uncertainty then, the theory of Secs. 2, 3, and 4 become applicable.

2. CONVERGENCE TO EQUILIBRIUM OF MARKOV PROCESSES GENERATED BY IID MONOTONE MAPS

Let S be a nonempty subset of an Euclidean space and Γ a set of maps on S to S. Let $\mathscr{S}$ be the Borel σ-field of S. Endow Γ with a σ-field Σ such that the map $(\gamma, x) \to \gamma(x)$ on $(\Gamma \times S, \Sigma \otimes \mathscr{S})$ into $(S, \mathscr{S})$ is measurable. Let P be a probability measure on (Γ, Σ). Take $(\alpha_n)_{n=1}^{\infty}$ to be a sequence of independent, identically distributed random functions from Γ with common distribution P. For a given random variable X_0 (with values in S), independent of the sequence $(\alpha_n)_{n=1}^{\infty}$, define

$$X_1 = \alpha_1(X_0) \equiv \alpha_1 X_0.$$
$$X_n = \alpha_n(X_{n-1}) \equiv \alpha_n \alpha_{n-1} \cdots \alpha_1 X_0. \tag{2.1}$$

Then X_n is a Markov process with transition probability $p(x, dy)$ given by

$$p(x, A) = P(\gamma \in \Gamma : \gamma(x) \in A), \quad x \in S, \quad A \in \mathscr{S} \tag{2.2}$$

We shall write $X_n(x)$ for X_n in case $X_0 = x$. But, to simplify notation we shall write $X_n = \alpha_n \cdots \alpha_1 X_0$ for general X_0.

Let $\mathbf{B}(S)$ be the linear space of all bounded real valued measurable functions on S. The transition operator T on $\mathbf{B}(S)$ is defined by

$$(Tf)(x) = \int_S f(y) p(x, dy) \qquad f \in \mathbf{B}(S). \tag{2.3}$$

Its adjoint T^* is defined on the space $M(S)$ of all finite signed measures on $(S, \mathscr{S})$ by

$$(T^*\mu)(A) = \int_S p(x, A)\mu(dx) = \int_{\Gamma} \mu(\gamma^{-1}A) P(d\gamma) \qquad \mu \in M(S). \tag{2.4}$$

Let $\mathscr{P}(\mathscr{S})$ denote the set of all probability measures on $(S, \mathscr{S})$. An element π of $\mathscr{P}(S)$ is invariant for $p(x, dy)$ (or for the Markov process X_n) if it is a fixed point of T^*, i.e., $T^*\pi = \pi$. We write $p^{(n)}(x, dy)$ for the n–step transition probability, with $p^{(1)} \equiv p$. Then $p^{(n)}(x, dy)$ is the distribution of $\alpha_n \ldots \alpha_1 x$.

Define T^{*n} as the nth iterate of T^*:

$$T^{*n}\mu = T^{*(n-1)}(T^*\mu) \ (n \geq 2), \quad T^{*1} = T^*, \quad T^{*0} := \text{Identity} \qquad (2.5)$$

Note that

$$(T^{*n}\mu)(A) = \int_S p^{(n)}(x, A)\mu(dx) \qquad (A \in \mathscr{S}), \qquad (2.6)$$

so that $T^{*n}\mu$ is the distribution of X_n when X_0 has distribution μ. To express T^{*n} in terms of the common distribution P of the iid maps, let Γ^n denote the usual Cartesian product $\Gamma \times \Gamma \times \cdots \times \Gamma$ (n terms), and let P^n be the product probability $P \times P \times \cdots \times P$ on $(\Gamma^n, \mathscr{S}^{\otimes n})$ where $\mathscr{S}^{\otimes n}$ is the product σ–field on Γ^n. Thus P^n is the (joint) distribution of $\alpha = (\alpha_1, \alpha_2, \ldots, \alpha_n)$. For $\gamma = (\gamma_1, \gamma_2, \ldots, \gamma_n) \in \Gamma^n$ let γ denote the composition

$$\gamma := \gamma_n \gamma_{n-1} \cdots \gamma_1 \qquad \text{for} \qquad \gamma = (\gamma_1, \gamma_2, \ldots, \gamma_n) \in \Gamma^n. \qquad (2.7)$$

Then, since $T^{*n}\mu$ is the distribution of $X_n = \alpha_n \ldots \alpha_1 X_0$, one has $(T^{*n}\mu)(A) = \text{Prob}(X_n \in A) = \text{Prob}(X_0 \in \alpha^{-1}A)$, where $\alpha = \alpha_n\alpha_{n-1} \ldots \alpha_1$. Therefore, by the independence of α and X_0,

$$(T^{*n}\mu)(A) = \int_{\Gamma^n} \mu(\gamma^{-1}A)P^n(d\gamma) \qquad (A \in \mathscr{S}, \mu \in \mathscr{P}(S)). \qquad (2.8)$$

The main result of this Section is Theorem 2.1 below, which is an extension of a result of Dubins and Freedman (1966), Yahav (1975), Bhattacharya and Lee (1988). Further generalizations are given in the last section of this article. To state Theorem 2.1, we take $S \subset \mathbb{R}^1$ to be an interval, finite or infinite and closed, semi–closed or open. Let $\alpha_n (n \geq 1)$ be an iid sequence of monotone nondecreasing random maps on S into S. We will assume the following splitting condition

(H)   There exist $x_0 \in S, \delta_i > 0 (i = 1, 2)$ and a positive integer N such that

   (1) $\text{Prob}(\alpha_N\alpha_{N-1} \cdots \alpha_1 x \leq x_0 \ \forall x \in S) \geq \delta_1$,

and

   (2) $\text{Prob}(\alpha_N\alpha_{N-1} \cdots \alpha_1 x \geq x_0 \ \forall x \in S) \geq \delta_2$.

Note that conditions (1) and (2) in **(H)** may be expressed, respectively, as

$$P^N(\{\gamma \in \Gamma^N : \gamma^{-1}(-\infty, x] = S\}) \geq \delta_1 \qquad \forall x \geq x_0, \qquad (2.9)$$

and

$$P^N(\{\gamma \in \Gamma^N : \gamma^{-1}(-\infty, x] = \varnothing\}) \geq \delta_2 \qquad \forall x < x_0. \qquad (2.10)$$

Here $\gamma = \gamma_N \gamma_{N-1} \cdots \gamma_1$.

Denote by $d(\mu, \nu)$ the Kolmogorov distance on $\mathscr{P}(S)$. That is, if F_μ denotes the distribution function (d.f.) of $\mu \in \mathscr{P}(S)$, then

$$d(\mu, \nu) := \sup_{x \in S} |F_\mu(x) - F_\nu(x)| \equiv \sup_{x \in \mathbb{R}^1} |F_\mu(x) - F_\nu(x)|,$$

$$(\mu, \nu \in \mathscr{P}((S))). \qquad (2.11)$$

REMARK 2.1   It should be noted that convergence in the distance d on $\mathscr{P}(S)$ implies weak convergence in $\mathscr{P}(S)$.

THEOREM 2.1   Assume that a splitting condition (H) holds.

(a)   Then the distribution $T^{*n}\mu$ of $X_n := \alpha_n \cdots \alpha_1 X_0$ converges to a probability measure π on S exponentially fast in the Kolmogorov distance d, whatever be the initial state x. Indeed,

$$d(T^{*n}\mu, \pi) \leq (1 - \delta)^{[n/N]} \qquad \forall \mu \in \mathscr{P}(S), \qquad (2.12)$$

where $\delta := \min\{\delta_1, \delta_2\}$ and $[y]$ denotes the integer part of y.

(b) π in (a) is the unique invariant probability of the Markov process X_n.

Proof. Suppose (H) holds, i.e., (2.9), (2.10) hold. First assume α_n are continuous. Let $x \geq x_0$, and denote by Γ_x the set of all $\gamma \in \Gamma^N$ appearing within the curly brackets in (2.9). Then, for all $\gamma \in \Gamma_x$, $\gamma^{-1}(-\infty, x] = S$, so that for all μ and ν in $\mathscr{P}(S)$ one has $\mu \circ \gamma^{-1}(-\infty, x] - \nu \circ \gamma^{-1}(-\infty, x] = 1 - 1 = 0$. Therefore,

$$|(T^{*N}\mu)(-\infty, x] - (T^{*N}\nu)(-\infty, x]|$$

$$= \left| \int_{\Gamma^x} [(\mu \circ \gamma^{-1})(-\infty, x] - (\nu \circ \gamma^{-1})(-\infty, x]] P^N(d\gamma) \right|$$

$$= \left| \int_{\Gamma_x} + \int_{\Gamma^N \backslash \Gamma_x} \right| = \left| \int_{\Gamma^N \backslash \Gamma_x} [(\mu \circ \gamma^{-1})(-\infty, x] - (\nu \circ \gamma^{-1})(-\infty, x]] P^N(d\gamma) \right|$$

$$\leq (1 - \delta_1) d(\mu, \nu) \qquad (\mu, \nu \in \mathscr{P}(S)). \qquad (2.13)$$

Note that, for the last inequality, we have used the facts (i) $P^N(\Gamma^N \backslash \Gamma_x) \leq 1 - \delta_1$, and (ii) $|\mu(\gamma^{-1}(-\infty, x]) - \nu(\gamma^{-1}(-\infty, x])| = |\mu(-\infty, y] - \nu(-\infty, y]| \leq d(\mu, \nu)$, where y is some real number. Similarly, if $x < x_0$ then, letting Γ_x denote the set in curly brackets in (2.10), we arrive at the same relations as in (2.13), excepting that the extreme right side is now $(1 - \delta_2) d(\mu, \nu)$. For

this note that $(\mu \circ \gamma^{-1})(-\infty, x] = \mu(\varnothing) = 0 = (\nu \circ \gamma^{-1})(-\infty, x] \ \forall \gamma \in \Gamma_x$.
Combining these two inequalities we get

$$d(T^{*N}\mu, T^{*N}\nu) \leq (1 - \delta)d(\mu, \nu) \qquad (\mu, \nu \in \mathscr{P}(S)). \tag{2.14}$$

Also, note that

$$d(T^*\mu, T^*\nu) = \sup_{x \in S}\left|\int_\Gamma [\mu(\gamma^{-1}(-\infty, x]) - \nu(\gamma^{-1}(-\infty, x])]P(d\gamma)\right|$$

$$\leq \int_\Gamma d(\mu, \nu)P(d\gamma) = d(\mu, \nu). \tag{2.15}$$

That is, T^{*N} is a strict contraction and T^* is a contraction. As a consequence, $\forall n > N$, one has

$$d(T^{*n}\mu, T^{*n}\nu) = d(T^{*N}(T^{*(n-N)}\mu), T^{*N}(T^{*(n-N)}\nu))$$

$$\leq (1 - \delta)d(T^{*(n-N)}\mu, T^{*(n-N)}\nu) \leq \cdots$$

$$\leq (1 - \delta)^{[n/N]}d(T^{*(n-[n/N]N)}\mu, T^{*(n-[n/N]N)}\nu)$$

$$\leq (1 - \delta)^{[n/N]}d(\mu, \nu). \tag{2.16}$$

Now suppose that we are able to show that $\mathscr{P}(S)$ is a complete metric space under d. Since T^{*N} is a strict contraction, it will then follow, from the contraction mapping principle (Friedman (1982) p. 119), that T^{*N} has a unique fixed point π in $\mathscr{P}(S)$. In this case $T^{*N}T^*\pi = T^*T^{*N}\pi = T^*\pi$, so that $T^*\pi$ is also a fixed point of T^{*N}, and by the uniqueness of the fixed point of T^{*N} one has $T^*\pi = \pi$. Thus π is a fixed point of T^*. Since every fixed point of T^* is also a fixed point of T^{*N}, π is the unique fixed point of T^*. Now take $\nu = \pi$ in (2.16) to get the desired relation in (2.12).

Assuming the completeness of $\mathscr{P}(S)$ under d, let us now consider the general nondecreasing case, without the requirement of continuity of α_n. The only place where we made use of continuity in the above proof is in the assumption that $\gamma^{-1}(-\infty, x]$ is of the form $(-\infty, y]$ for some y. If γ is not continous, but monotone nondecreasing, then $\gamma^{-1}(-\infty, x]$ may be of the form $(-\infty, y]$ or $(-\infty, y)$. However, $|\mu(-\infty, y) - \nu(-\infty, y)| \leq d(\mu, \nu)$ in view of the right continuity of d.f.'s. Therefore, the theorem holds if we are able to prove the completeness of $\mathscr{P}(S)$, which is shown below.

Let μ_n be a Cauchy sequence in $(\mathscr{P}(S), d)$, and let G_n be the d.f. of μ_n. Then $\sup_{x \in \mathbb{R}^1} |G_n(x) - G_m(x)| = \sup_{x \in S}|G_n(x) - G_m(x)| \to 0$ as $n, m \to \infty$. By the completeness of $\mathbb{R}^1$, there exists a function $G(x)$ such that $\sup_{x \in \mathbb{R}^1} |G_n(x) - G(x)| \to 0$. It is simple to check that G is the d.f. of a probability measure ν, say, on $\mathbb{R}^1$. We need to check that $\nu(S) = 1$. For this, given $\varepsilon > 0$ find n_ε such that $\sup_x |G_n(x) - G(x)| < \varepsilon/2 \ \forall n \geq n_\varepsilon$. Now find $y_\varepsilon \in S$ such that $G_{n_\varepsilon}(y_\varepsilon) > 1 - \varepsilon/2$. It follows that $G(y_\varepsilon) > 1 - \varepsilon$. In the same manner, find $x_\varepsilon \in S$ such that $G(x_\varepsilon) < \varepsilon$. Then $\nu(S) \geq G(y_\varepsilon) - G(x_\varepsilon) > 1 - 2\varepsilon \quad \forall \varepsilon > 0$, so that $\nu(S) = 1$. $\qquad\square$

REMARK 2.2. Suppose that α_n are strictly increasing a.s. Then if the initial distribution μ is nonatomic (i.e., $\mu(\{x\}) = 0 \quad \forall x$ or, equivalently, the d.f. of μ is continuous), then $\mu \circ \gamma^{-1}$ is nonatomic $\forall \gamma \in \Gamma$ (outside a set of zero P-probability). It follows that if X_0 has a continuous d.f., then so has X_1 and, in turn, X_2 has a continous d.f, and so on. Since, by Theorem 2.1, this sequence of continuous d.f.'s (of $X_n(n \geq 1)$) converges uniformly to the d.f. of π, the latter is continuous. Thus π is nonatomic if α_n are strictly increasing a.s.

REMARK 2.3. Suppose that α_n is a monotonically nonincreasing iid sequence on an interval S. Then $\beta_1 := \alpha_2\alpha_1$, $\beta_2 := \alpha_4\alpha_3, \ldots,$ $\beta_n := \alpha_{2n}\alpha_{2n-1}, \ldots$ are iid monotone nondecreasing. If then the splitting condition (H) holds for $\beta_n(n \geq 1)$, instead of $\alpha_n(n \geq 1)$, then the Markov process $Y_n := X_{2n}(n \geq 0)$ has a unique invariant probability π, and the convergence to π of the distribution $T^{*2n}\mu$ of X_{2n} is exponentially fast in the d–metric. Since it is still true in this case, as may be easily checked, that $d(T^*\mu, T^*\nu) \leq d(\mu, \nu)$ one has $d(T^{*2n+1}\mu, T^{*2n+1}\nu) = d(T^*(T^{*2n}\mu),$ $T^*(T^{*2n}\nu)) \leq d(T^{*2n}\mu, T^{*2n}\nu)$. It follows that the distribution $T^{*n}\mu$ of X_n converges to π exponentially fast and, in particular, π is the unique invariant distribution of X_n. An example of this kind appears in Sec. 4.

The next result shows that barring a trivial case, on compact intervals $S = [a, b](a < b)$ the splitting condition (H) is necessary for the convergence in distribution of $p^{(n)}(x, dy)$ to an invariant probability π for all $x \in [a, b]$. This result is an extension to possibly discontinuous α_n of a result of Dubins and Freedman (1966), Bhattacharya and Lee (1988).

THEOREM 2.2 Let $\alpha_n(n \geq 1)$ be iid nondecreasing on $S = [a, b]$. Suppose, for every $x \in [a, b]$, $p^{(n)}(x, dy)$ converges weakly to a probability π. If π is not degenerate then the splitting condition (H) holds.

Proof. Recall that $p^{(n)}(x, dy)$ is the distribution of $X_n(x) = \alpha_n \ldots \alpha_1 x$ and, therefore, of $Y_n(x) = \alpha_1 \ldots \alpha_n x$. Note that $Y_{n+1}(a) = \alpha_1\alpha_2 \ldots \alpha_n\alpha_{n+1}a \geq \alpha_1\alpha_2 \ldots \alpha_n a = Y_n(a)$ (since $\alpha_{n+1}a \geq a$). Let $\underline{Y}$ be the limit of the nondecreasing sequence $Y_n(a)(n \geq 1)$. Then $p^{(n)}(a, dy)$ converges weakly to the distribution $\underline{\pi}$, say, of $\underline{Y}$. Similarly, $Y_{n+1}(b) \leq Y_n(b) \quad \forall n$, and the nonincreasing sequence $Y_n(b)$ converges to a limit $\overline{Y}$, so that $p^{(n)}(b, dy)$ converges weakly to the distribution $\overline{\pi}$ of $\overline{Y}$. By the hypothesis of the theorem, $\underline{\pi} = \overline{\pi} = \pi$. Since $\underline{Y} \leq \overline{Y}$, it follows that $\underline{Y} = \overline{Y}$ (a.s.) $= Y$, say. Assume π is not degenerate. Then there exist $c, d \in (a, b), c < d$ such that $\text{Prob}(Y < c) > 0$ and

$\text{Prob}(Y > d) > 0$. Since $Y_n(a) \uparrow Y$ and $Y_n(b) \downarrow Y$, there exists a positive integer N such that (i) $\delta_1 := \text{Prob}(Y_N(b) < c) > 0$, and (ii) $\delta_2 := \text{Prob}(Y_N(a) > d) > 0$. But then $\text{Prob}(X_N(x) < c \ \forall x \in [a,b]) = \text{Prob}(Y_N(x) < c \ \forall x \in [a,b]) = \text{Prob}(Y_N(b) < c) = \delta_1$, and $\text{Prob}(X_N(x) > d \ \forall x \in [a,b]) = \text{Prob}(Y_N(x) > d \ \forall x \in [a,b]) = \text{Prob}(Y_N(a) > d) = \delta_2$. Choose any x_0 in (c,d). The splitting condition (**H**) now holds. □

REMARK 2.4. If $p^{(n)}(x, dy)$ converges weakly to a degenerate π, $\pi(\{z_0\}) = 1$, say, $\forall x$, then $\underline{Y} = \overline{Y} = z_0$ a.s. If π is invariant, then

$$1 = \pi(\{z_0\}) = \int p(x, \{z_0\})\pi(dx) = p(z_0, \{z_0\})\pi(\{z_0\})$$

$$= p(z_0, \{z_0\}) = \text{Prob}(\alpha_1 z_0 = z_0).$$

Hence z_0 is a fixed point of all $\gamma \in \Gamma$, outside a set of zero P–probability.

3. CENTRAL LIMIT THEOREM

Here we state a central limit theorem for monotone dynamical systems, Bhattacharya and Lee (1988).

THEOREM 3.1. Let the hypothesis of Theorem 2.1 hold. Then for every function f on S which may be expressed as the difference between two nondecreasing functions and which belongs to $L^2(S, \pi)$, the sequence of stochastic processes defined by

$$Y_n(t) := n^{-1/2} \sum_{j=0}^{k} \left(f(X_j) - \int f d\pi \right) \qquad \text{for } t = \frac{k}{n} \qquad (k = 0, 1, \ldots) \quad (3.1)$$

and $Y_n(\cdot)$ linearly interpolated in $[k/n, (k+1)/n]$, converges in distribution to a Brownian motion on $[0, \infty)$ with zero drift and variance parameter $\sigma^2 = \int g^2 d\pi - \int (Tg)^2 d\pi$, where T is the transition operator in Eq. (2.3), and g is an element of $L^2(S, \pi)$ satisfying $(T - I)g = f - \int f d\pi$.

If S is a bounded interval, then Theorem 3.1 applies to the function $f(x) = x$, and leads to weak convergence of such functionals as $n^{-1/2} \max\{X_0 + \cdots + X_j : 0 \le j \le n\}$ to the distribution of the maximum of the limiting Brownian motion on $[0, 1]$.

More generally, Theorem 3.1 applies to all bounded functions of bounded variation on S, since such a function may be expressed as the difference of two bounded nondecreasing functions.

Note that Theorem 3.1 holds under an arbitrary initial distribution.

The complete proof of Theorem 3.1 is a little involved and may be found in Bhattacharya and Lee (1988). One of the main steps is to show that for every nondecreasing f in $L^2(S, \pi)$ there is a g in $L^2(S, \pi)$ such that $(T - I)g = f - E_\pi f(E_\pi f := \int f(y)\pi(dy))$, i.e., $f - E_\pi f$ is in the range of $T - I$. Here I is the identity operator. A natural choice is

$$g = -\sum_{n=0}^{\infty} T^n(f - E_\pi f) \qquad (T^0 := I). \tag{3.2}$$

To show that Eq. (3.2) makes sense, one needs to show that the right side converges in $L^2(S, \pi)$. This is shown in Bhattacharya and Lee (1988). To explain why it is important to find such a g, write

$$\sum_{j=0}^{n} [f(X_j) - E_\pi f] = \sum_{j=0}^{n} [(Tg)(X_j) - g(X_j)]$$

$$= \sum_{j=0}^{n} [(Tg)(X_j) - g(X_{j+1})] + g(X_{n+1}) - g(X_0)$$

$$\simeq \sum_{j=0}^{n} [(Tg)(X_j) - g(X_{j+1})],$$

where, in the last step, $g(X_{n+1}) - g(X_0)$ is neglected, since on dividing by $\sqrt{n}$ it goes to zero in probability. By the Markov property, the last expression in Eq. (3.3) is a sum of martingale differences, i.e., $E[(Tg)(X_j) - g(X_{j+1})|\sigma\{X_0, \dots, X_j\}] = 0$. Now apply the martingale central limit theorem as given e.g., in Billingsley (1968), Theorem 23.1, or Bhattacharya and Waymire (1990), pp. 508–513. A central limit theorem for f such that $f - E_\pi f$ is in the range of $T - I$ was obtained by Gordin and Lifsic (1978) in the case of a discrete parameter Markov process. In the continuous parameter case a corresponding central limit theorem was proved in Bhattacharya (1982), where $T - I$ is replaced by the infinitesimal generator of the process.

4. RANDOM ITERATION OF TWO QUADRATIC MAPS

The quadratic family of maps $F_\theta(x) = \theta x(1 - x)$ on the unit interval $[0, 1]$ $(0 \le \theta \le 4)$ has in recent years occupied a position of distinction in the modern theory of dynamical systems and chaos, Devaney (1989). Originally investigated as models of environmental biology, they now serve as prototypes of general one–dimensional dynamical systems. Most recently, this entire family has been rigorously generated from a dynamic control problem in economics (see Sec. 6).

Our goal in this section is to analyze a class of problems in which, at each point of time n, one of two parameter values θ_1, θ_2 is chosen at random with probabilities β, $1 - \beta$ respectively $(0 < \beta < 1)$.

Note that the only fixed point of the quadratic map F_θ, for $0 < \theta \le 1$, is 0, and the nth iterate $F_\theta^n(x) \to 0$ for all x, as $n \to \infty$. For $1 < \theta \le 4$, there is a second fixed point of F_θ, namely, $q_\theta = 1 - 1/\theta$. It may be shown that, for $1 < \theta \le 3$, this fixed point is attracting, while 0 is repelling (Devaney (1989) Sec. 1.5). Thus $F_\theta^n(x) \to q_\theta \; \forall x \in (0,1)$, provided $1 < \theta \le 3$. It is also not difficult to show that a period-two orbit $\{p_\theta, r_\theta\}$ appears, $p_\theta < q_\theta < r_\theta$, for $\theta > 3$. The points p_θ, r_θ are attracting fixed points for F_θ^2, while 0 and q_θ are repelling, if $3 < \theta < 1 + \sqrt{6}$.

For some purposes below one should keep in mind a few elementary facts:

(1) F_θ is increasing on $[0, \frac{1}{2}]$ and decreasing on $(\frac{1}{2}, 1]$;
(2) $F_\theta(x) > x$ for $x \in (0, q_\theta)$ and $F_\theta(x) < x$ for $x \in (q_\theta, 1)$;
(3) for $1 < \theta \le 2$, $q_\theta \le \frac{1}{2}$; and,
(4) as a consequence of (1)–(3), if $1 < \theta_1 < \theta_2 \le 2$, then $q_{\theta_1} \le F_{\theta_i}(x) \le q_{\theta_2}$ for $x \in [q_{\theta_1}, q_{\theta_2}]$ $(i = 1, 2)$. Finally,
(5) $F_\theta(x) \le \theta/4 \quad \forall x \in [0, 1]$.

We now consider random iterations with $\alpha_n = F_{\theta_1}$ or F_{θ_2} with probabilities β and $1 - \beta$, respectively, where $1 < \theta_1 < \theta_2$ $(0 < \beta < 1)$. Three cases are considered here.

Case 1. $1 < \theta_1 < \theta_2 \le 2$. From facts (1), (3) above it follows that $F_{\theta_i}(i = 1, 2)$ is increasing on $[q_{\theta_1}, q_{\theta_2}]$. From (4) it follows that F_{θ_i} leaves $[q_{\theta_1}, q_{\theta_2}]$ invariant. Thus one may take $S = [q_{\theta_1}, q_{\theta_2}]$ as the state space of the Markov process $X_n := \alpha_n \ldots \alpha_1 X_0$. Since $F_{\theta_1}^n(q_{\theta_2}) \to q_{\theta_1}$ and $F_{\theta_2}^n(q_{\theta_2}) \to q_{\theta_2}$, as $n \to \infty$, for $0 < \varepsilon \le (q_{\theta_1} + q_{\theta_2})/4$ there exists a positive integer N such that $F_{\theta_1}^N(S) \subset [q_{\theta_1}, q_{\theta_1} + \varepsilon], F_{\theta_2}^N(S) \subset [q_{\theta_2} - \varepsilon, q_{\theta_2}]$. Thus the splitting condition **(H)** holds with x_0 any point in $[q_{\theta_1} + \varepsilon, q_{\theta_2} - \varepsilon]$ and $\delta_1 = \beta^N$, $\delta_2 = (1 - \beta)^N$. Then Theorem 2.1 applies. Note also that if the initial state x lies in $(0, 1) \backslash [q_{\theta_1}, q_{\theta_2}]$, there exists a finite integer $m(\omega)$ such that $X_n(x) \in S = [q_{\theta_1}, q_{\theta_2}] \quad \forall n \ge m(\omega)$, for a.e. ω. Thus $p^{(n)}(x, dy)$ converges weakly to the unique invariant probability π on S, for all $x \in (0, 1)$.

Case 2. $2 < \theta_1 < \theta_2 \le 3, \theta_1 \in [8/\theta_2(4 - \theta_2), \theta_2]$. In this case, F_{θ_i} is decreasing on $[\frac{1}{2}, \theta_2/4] = S$, $F_{\theta_i}(\frac{1}{2}) \le \theta_2/4$ since $2 < \theta_1 < \theta_2 \le 3$. The additional restriction on θ_1 ensures that $F_{\theta_2}(\theta_2/4) \ge F_{\theta_1}(\theta_2/4) \ge \frac{1}{2}$. Thus S is an invariant interval for F_{θ_i}, and F_{θ_i} is monotone decreasing on S $(i = 1, 2)$. Then $F_{\theta_i} F_{\theta_j}$ is increasing $(i, j = 1, 2)$, so that $X_{2n}(n = 0, 1, 2, \ldots)$ is a Markov process obtained as actions of random iterated nondecreasing maps. Since q_{θ_i} is an attractive fixed point $(i = 1, 2)$, arguing as in Case 1 one verifies the splitting condition (**H**). Since the Markov process $X_{2n+1}(n = 0, 1, 2, \ldots)$ has the same transition probability as $X_{2n}(n \geq 0)$, it follows that $p^{(n)}(x, dy)$ converges weakly to a unique invariant probability π, $\forall x \in S$. Since q_{θ_i} is attracting, and $F_{\theta_i}^n(x) \in [\frac{1}{2}, \theta_2/4]$ $\forall n \geq m(\omega)$, say, this weak convergence holds for all $x \in (0, 1)$.

Case 3. $2 < \theta_1 \leq 3 < \theta_2 \leq 1 + \sqrt{5}$, $\theta_1 \in [8/\theta_2(4 - \theta_2), \theta_2]$. Just as in Case 2, F_{θ_i} is decreasing on $[\frac{1}{2}, \theta_2/4]$, and leaves $[\frac{1}{2}, \theta_2/4] := S$ invariant ($i = 1, 2$). To ensure that this is not vacuous, one should check that $F_{\theta_2}(\theta_2/4) \geq \frac{1}{2}$. This holds if and only if $2 \leq \theta_2 \leq 1 + \sqrt{5}$. We have considered part of this range in Case 2. The present case considers the remaining range of θ_2's and the associated set of θ_1 values. Two situations may arise. Let $p_{\theta_2} < r_{\theta_2}$ be the period–two points of F_{θ_2}. Suppose $q_{\theta_1} < p_{\theta_2} (< q_{\theta_2} < r_{\theta_2})$. Then $F_{\theta_1}^n(S)$ lies within an arbitrarily small neighborhood of q_{θ_1} if n is chosen sufficiently large. On the other hand, since the period–two orbit is attracting, with $F_{\theta_2}^{2n}(\frac{1}{2}) \uparrow p_{\theta_2}$ and $F_{\theta_2}^{2n}(\theta_2/4) \downarrow r_{\theta_2}$, it follows that $F_{\theta_2}^{2n}(S)$ is bounded away from q_{θ_1} for sufficiently large n. Thus the splitting condition (**H**) holds, and Theorem 2.1 applies.

Now suppose $p_{\theta_2} < q_{\theta_1}(< q_{\theta_2} < r_{\theta_2})$. Let $\varepsilon > 0$ be such that $p_{\theta_2} < q_{\theta_1} - \varepsilon < q_{\theta_1} + \varepsilon < q_{\theta_2}$. There exists n_0 such that $F_{\theta_1}^n(S) \subset (q_{\theta_1} - \varepsilon, q_{\theta_1} + \varepsilon)$, for $n \geq n_0$. Then $F_{\theta_2}F_{\theta_1}^{n+1}(S) \subset (q_{\theta_2}, \theta_2/4)$, while $F_{\theta_2}^2 F_{\theta_1}^n(S) \subset (\frac{1}{2}, q_{\theta_2})$ $\forall n \geq n_0$. Thus (**H**) holds with $N = n_0 + 2$, $x_0 = q_{\theta_2}$, and Theorem 2.1 applies.

REMARK 4.1.   The invariant probabilities π in Cases (1)–(3) are non-atomic. Some of them have supports on Cantor sets, and some have full supports, Bhattacharya and Rao (1993).

5.   APPLICATIONS TO MODELS OF ECONOMIC GROWTH AND SURVIVAL

A large class of economic models dealing with problems of intertemporal allocation of resources under uncertainty led to the study of random dynamical systems. See Sargent (1987), Majumdar, Mitra and Nyarko (1989) for reviews and references. As an application of our results we shall provide a sketch of an example somewhat heuristically.

Case 1.   Let Γ be a set consisting of two functions, denoted by F and G, from $\mathbb{R}_+$ into $\mathbb{R}_+$. We impose the following restrictions:

(Ai) $F(0) = G(0) = 0$;

(Aii) F and G are continuous and increasing functions such that $G(x) > F(x)$ for all $x > 0$;

(Aiii)   There are points $b > a > 0$ such that $F(a) = a, G(b) = b$;

$F(x) > x$ on $(0, a)$ and $F(x) < x$ on (a, ∞)

$G(x) > x$ on $(0, b)$ and $G(x) < x$ on (b, ∞)

We shall comment briefly on the role of some of the assumptions later. It will be helpful to keep Fig. 1 in mind in the subsequent discussion. Consider now a random dynamical system

$X_{n+1} = \alpha_n(X_n)$

where (α_n) is an iid sequence of elements from Γ chosen according to probabilities $p, 1 - p (0 < p < 1)$ for the elements F and G respectively.

In a descriptive macroeconomic growth model, X_n is interpreted as per capita output in period n, and the transition α_n occurs according to the growth–mechanism described by Solow (1956), with G and F representing "favorable" and "unfavorable" production environments. In the context of dynamic optimization under uncertainty discussed in the next Section, the transition α_n may be interpreted as optimal transition determined by solving some suitable dynamic optimization problem. For example, the Ramsey (1928) problem of optimal savings when the technology is subject to random shocks is discussed in detail in Majumdar, Mitra and Nyarko (1989). Note that for all initial x in $(0, a)$, $X_n(x)$ will eventually be in the interval

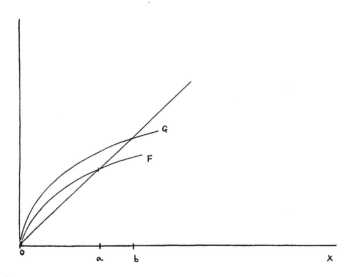

Figure 1.

$[a, b]$ with probability one. Indeed, the process does not "return" to any point in $(0, a)$ if it "visits" it. Similarly, the process leaves (b, ∞) with probability one when it starts from initial x in (b, ∞). Hence, for studying the long run behavior it is enough to restrict attention to the interval $[a, b]$; it is easy to check that if the process is in this interval in any period, it stays there for all subsequent periods. Observe that given any sufficiently small $\varepsilon > 0$, there is a finite positive integer N_1 such that $X_{N_1}(a) > b - \varepsilon$ with probability no less than $(1 - p)^{N_1}$, since $X_{N_1}(a) > b - \varepsilon$ if G occurs for N_1 consecutive periods. By monotonicity of G, it follows that for all initial x in $[a, b]$, $X_{N_1}(x) > b - \varepsilon$ with probability no less than $(1 - p)^{N_1}$. Similarly, given $\varepsilon > 0$, there is a finite positive integer N_2 such that for all initial x in $[a, b]$, $X_{N_2}(x) < a + \varepsilon$ with probability no less than p^{N_2}. We can thus verify the splitting condition (**H**) in Sec. 2. Taking S to be the interval $[a, b]$, we can apply the global stability result Theorem 2.1 and the central limit theorem, Theorem 3.1.

Case 2. The role of assumption (Aiii) is made clearer if we consider a random dynamical system that has been studied in the context of economic survival, Majumdar and Radner (1992) and Roy (1995 and 1996). Maintain the assumptions (Ai) and (Aii), but, instead of (Aiii), assume:

(Aiv)   (i)   There are two points $0 < b' < a$ such that $b' = F(b'), a = F(a)$; moreover, $F(x) < x$ on $(0, b')$, $F(x) > x$ on (b', a) and $F(x) < x$ for all $x > a$;

        (ii)   There are two points $0 < a' < b$ such that $a' = G(a')$ and $b = G(b)$; moreover $G(x) < x$ on $(0, a')$, $G(x) > x$ on (a', b) and $G(x) < x$ for all $x > b$;

The graphs of F and G enable us to analyze the qualitative properties of the process

$$X_{(n+1)}(x) = \alpha_n(X_n(x)).$$

See also Fig. 2.

Again, we proceed somewhat informally. For all initial x in $(0, a')$, $X_n(x)$ converges to 0 with probability one. This is the case in which "survival" is impossible. For all initial $x \geq b'$, we can apply the analysis in Case 1 to characterize the behavior of $X_n(x)$. Of interest is the case in which the initial x is in the interval (a', b'). Here, with positive probability, $X_n(x)$ will fall below a'. However, there is also a positive probability that $X_n(x)$ will go above b'. Hence, with a positive probability $X_n(x)$ converges to 0 for all x in (a', b'). Also, with a positive probability $X_n(x)$ enters the interval $[a, b]$ and converges to an invariant distribution.

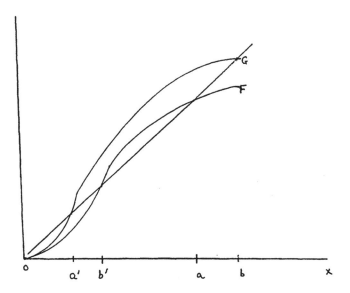

Figure 2.

6. APPLICATIONS TO DYNAMIC OPTIMIZATION: MONOTONICITY AND QUADRATIC FAMILY

Consider a discrete-time aggregative model of development planning in which an economy E is specified by a gross output function $f : \mathbb{R}_+ \to \mathbb{R}_+$, a felicity (return) function $w : \mathbb{R}_+^2 \to \mathbb{R}_+$ and a discount factor $\delta \in (0,1)$. The following assumptions on f are used.

(F1) $f(0) = 0, f$ is continuous on $\mathbb{R}_+$.
(F2) f is nondecreasing and concave on $\mathbb{R}_+$.
(F3) There is $K > 0$ such that $f(x) > x$ when $0 < x < K$ and $f(x) < x$ when $x > K$.

To describe some of the remaining assumptions as well as our results, it is convenient to define a set $\Omega = \{(x,z) \in \mathbb{R}_+^2 : z \le f(x)\}$. The following assumptions on w are used.

(W1) $w(x,c)$ is continuous on $\mathbb{R}_+^2$.
(W2) $w(x,c)$ is nondecreasing in x given c, and nondecreasing in c given x on $\mathbb{R}_+^2$.

Furthermore, if $x > 0$, $w(x,c)$ is strictly increasing in c on Ω.

A program from $x > 0$ is a sequence $(x_t)_0^\infty$ satisfying

$$x_0 = x, \ 0 \leq x_{t+1} \leq f(x_t) \quad \text{for } t \geq 0. \tag{6.1}$$

The consumption sequence $(c_{t+1})_0^\infty$ is given by

$$c_{t+1} = f(x_t) - x_{t+1} \quad \text{for } t \geq 0. \tag{6.2}$$

It is easy to verify that for every program $(x_t)_0^\infty$ from $x \geq 0$ we have

$$x_t, c_{t+1} \leq K(x) := \max\{K, x\} \quad \text{for all } t \geq 0. \tag{6.3}$$

In particular, if $x \in [0, K]$, then $x_t, c_{t+1} \leq K$ for all $t \geq 0$.
A program $(\hat{x}_t)_0^\infty$ from $x \geq 0$ is optimal if

$$\sum_{t=0}^\infty \delta^t w(\hat{x}_t, \hat{c}_{t+1}) \geq \sum_{t=0}^\infty \delta^t w(x_t, c_{t+1}) \tag{6.4}$$

for every program $(x_t)_0^\infty$ from x.

The problem of development planning captured by the model can be informally described as follows: a particular economy starts with a capital stock x_0 which gives rise to the output $f(x_0)$ in the first period; the planner decides on the part of this output to be used as capital in the first period (denoted by x_1), and the rest $[f(x_0) - x_1]$ is used up as consumption in the first period. The choice of x_1 generates the output $f(x_1)$ in the second period, and the decision–making process is repeated. As a result of consuming c_{t+1} in period $t + 1$, the economy derives a return $w(x_t, c_{t+1})$. The aim of the planner is to select a program, i.e., a sequence of capital stocks that maximizes a discounted sum of one period returns among all programs from the same initial stock of capital.

Note that the "standard" Ramsey model of development planning in which the one period return depends solely on consumption is a special case of this model. The general formulation is motivated by a number of remarks, see, e.g., Koopmans (1967) Arrow and Kurz (1970) on the special importance of capital stock and infrastructure to initiate an industrial development process.

6.1 Monotonicity: Simple Dynamics

It is often convenient to look at the ("reduced–form utility") function $U := \Omega \to \mathbb{R}$ defined by

$$U(x, z) \equiv w(x, f(x) - z). \tag{6.5}$$

A compactness argument can be used to show that given any $x \geq 0$ there is some optimal program $(\hat{x}_t)_{t=0}^\infty$ from x. Furthermore, the optimal program is unique. We can define a value function $V : \mathbb{R}_+ \to \mathbb{R}$ by

$$V(x) = \sum_{t=1}^{\infty} \delta^t w(\hat{x}_t, \hat{c}_{t+1}) \tag{6.6}$$

and the optimal transition function

$$h(x) = \hat{x}_1 \tag{6.7}$$

where $\{\hat{x}_t\}$ is the optimal program from $x \geq 0$. The properties of V and h are summarized in the following.

THEOREM 6.1

(i) The value function V is the unique continuous real–valued function on $[0, K]$ satisfying the functional equation of dynamic programming

$$V(x) = \max_{(x,z)\in\Omega} [u(x, z) + \delta V(z)]. \tag{6.8}$$

Further, V is concave and nondecreasing on $\mathbb{R}_+$.

(ii) The transition function h satisfies the following property: for each $x \in \mathbb{R}$, $h(x)$ is the unique solution to the constrained maximization problem

maximize $u(x, z) + \delta V(z)$ subject to $(x, z) \in \Omega$

Furthermore, h is continuous on $\mathbb{R}_+$.

The problem of identifying conditions under which the function h is monotone non-decreasing has been much discussed, Benhabib and Nishimura (1985), Majumdar and Nermuth (1982), Dechert and Nishimura (1983), Majumdar and Mitra (1994a), and Mitra and Ray (1984). Instead of presenting the most general technical conditions (which involve the notion of "supermodular" functions studied by Topkis (1968), see also Ross (1983), we present a set of sufficient conditions in the "differentiable" case. Let $\mathbb{R}_{++}$ denote the set of all positive numbers. Also, let w_{12}, w_{22} be the usual second-order partial derivatives of w w.r.t. the first and second arguments.

THEOREM 6.2   Suppose f is continuously differentiable on $\mathbb{R}_{++}$ and w is twice continuously differentiable on the interior of Ω (denoted by Ω^0) and

$$-w_{22}(x, c)f'(x) \geq w_{12}(x, c) \qquad \text{for all } (x, c) \in \Omega^0. \tag{6.9}$$

Then h is nondecreasing on $\mathbb{R}_+$.

Observe, finally, that if w depends only on c (as in the "standard" Ramsey type model), $w_{12} = 0$, so that Eq. (6.9) is satisfied if $w_{22} < 0$.

6.2 Complex Dynamics

We consider a class of economies indexed by a parameter μ (where $\mu \in I = [1, 4]$). Each economy in this family has the same gross output function (satisfying (F1)–(F3)) and the same discount factor $\delta \in (0, 1)$. The economies in this family differ in the specification of their felicity or one–period return functions: $w : \mathbb{R}_+^2 \times I \to \mathbb{R}_+$ (w depending on the parameter μ). For a fixed $\mu \in [1, 4]$, the one–period return function $w(., ., \mu)$ can be shown to satisfy (W1)–(W3).

The numerical specifications are as follows:

$$f(x) = \begin{cases} (16/3)x - 8x^2 + (16/3)x^4 & \text{for } x \in [0, 0.5) \\ 1 & \text{for } x \geq 0.5, \end{cases}$$

$$\delta = 0.0025.$$

The function w is specified in a more involved manner. To ease the writing denote $L \equiv 98$, $a \equiv 425$, $X \equiv [0, 1]$; recall the family

$$h(x, \mu) = \mu x(1 - x) \qquad \text{for } x \in X, \mu \in I$$

and define $u : X^2 \times I \to \mathbb{R}$ by

$$u(x, z, \mu) \equiv ax - 0.5\,Lx^2 + zh(x, \mu) - 0.5\,z^2$$
$$- \delta\{az - 0.5\,Lz^2 + 0.5\,[h(z, \mu)]^2\}$$

Define $D \subset X^2$ by

$$D = \{(c, x) : c \leq f(x)\}$$

and a function $w : D \times I \to \mathbb{R}_+$ by

$$w(c, x, \mu) = u[x, g(x) - c, \mu] \quad \text{for } (c, x) \in D, \mu \in I.$$

The definition of $w(., ., \mu)$ can be extended to the domain Ω as follows: for $(c, x) \in \Omega$ with $x > 1$ (so that $f(x) = 1, c \leq 1$) define

$$w(c, x, \mu) = w(c, 1, \mu).$$

Finally, define $w(., ., \mu)$ on $\mathbb{R}_+^2$ as follows: for $(c, x) \in \mathbb{R}_+^2$ with $c > f(x)$, let

$$w(c, x, \mu) = w(f(x), x, \mu).$$

It can be shown, Majumdar and Mitra (1994b) that the optimal transition function for this family is

$$h(x, \mu) = \mu x(1 - x) \quad \text{for } x \in X, \mu \in I.$$

REMARK 6.1. That the quadratic family can be obtained as optimal transition functions in a class of discounted dynamic programming problems can be viewed as a result on the "inverse optimal problem" studied in dynamic optimization theory, Sorger (1992), Boldrin and Montrucchio (1986), which can be informally described as follows. Suppose that the state space S is a given compact set of reals, and a function $h : S \to S$ is also given. Identify a function $U : S \times S \to R$, and a discount factor $\delta \in (0, 1)$ such that the function h is the optimal transition function of the following discounted dynamic programming problem:

$$\text{``maximize} \sum_{t=0}^{\infty} \delta^t U(k_t, k_{t+1}) \quad \text{subject to} \quad (k_t, k_{t+1}) \in S \times S.\text{''}$$

Sorger has used the function

$$U(x, y) = -\left[\frac{|y - h(x)|^2}{2} \right] - \left(\frac{a}{2} \right)|x|^2 + \left(\frac{a\delta}{2} \right)|y|^2 + bx - b\delta y$$

where $a > 0$, and b are appropriate real numbers. Restrictions on h, a, b, and δ are needed when U is required to satisfy particular differentiability or concavity properties.

6.3   Uncertainty

Several routes have been taken to study the role of uncertainty. One possibility is to use the techniques of discounted (stochastic) dynamic programming, reviewed in detail in Majumdar, Mitra and Nyarko (1989). Here the effects of random shocks to the technology are studied formally by considering a class of gross output functions from which a particular one is chosen in each period according to some probability distribution. In the standard model the return function w depends only on consumption, and the objective of the planner is to maximize the sum of discounted one period expected returns. The optimal transition is described in terms of a random dynamical system $x_{t+1} = h(x_t)$ where h comes from an appropriate class of functions. Two different cases have emerged. First, in the "classical case" where all gross output functions are nondecreasing and concave (and satisfy some other conditions), and the return function w is strictly concave in consumption, all the possible transition functions can be shown to be continuous and monotone. However, in the second "non–classical" case where the gross output functions are "S–shaped," h can be shown to be monotonic, but continuity may not be obtained. Our results in Sections 2 and 3 apply directly in such frameworks.

   Introduction of uncertainty in a chaotic optimization model has not been done in any systematic manner, and remains a promising direction of future research in mathematical economics. One can think of a variation in the deterministic model in which the planner is not sure about the preferences, and would like to see the effects of random shifts from period to period. Or, one may think of implementing the planned allocation, and would like to track an essential randomness in the implementation procedure. In these cases, when "randomness" is captured in terms of a choice from the quadratic family, the mathematical results in Sec. 5 are of particular interest.

7. GENERALIZATIONS

In this Section we obtain some generalizations of Theorem 2.1. For a relatively simple extension consider a metric space (S, ρ). Let $\mathscr{S}$ be its Borel sigmafield, and $\alpha_n (n \geq 1)$ an iid sequence of random maps on S with common distribution P on (Γ, Σ). We continue to use the same notation as in Sec. 2. Let $\mathscr{A} \subset \mathscr{S}$, and define

$$d(\mu, \nu) := \sup_{A \in \mathscr{A}} |\mu(A) - \nu(A)| \qquad (\mu, \nu \in \mathscr{P}(S)). \tag{7.1}$$

Consider the following hypothesis $(\mathbf{H_1})$:

(1) $(\mathscr{P}(S), d)$ is a complete metric space.
(2) $\forall \gamma \in \Gamma^n$ and $\forall A \in \mathscr{A}$,

$$|\mu(\gamma^{-1}A) - \nu(\gamma^{-1}A)| \leq d(\mu, \nu) \qquad (\mu, \nu \in \mathscr{P}(S)). \tag{7.2}$$

(3) There exist $\delta > 0$ and a positive integer N such that, $\forall A \in \mathscr{A}$,

$$\text{Prob}(\alpha_N \cdots \alpha_1 x \in A \quad \forall x \in S) \geq \delta > 0$$

or $\tag{7.3}$

$$\text{Prob}(\alpha_N \cdots \alpha_1 x \in A^c \quad \forall x \in S) \geq \delta > 0.$$

PROPOSITION 7.1. Under the hypothesis $(\mathbf{H_1})$ there exists a unique invariant probability π for the Markov process $X_n := \alpha_n \cdots \alpha_1 X_0$, and one has

$$d(T^{*n}\mu, \pi) \leq (1 - \delta)^{[n/N]} \tag{7.4}$$

for every (initial) distribution μ of X_0.

The proof of Proposition 7.1 is analgous to the proof of Theorem 2.1 and will appear in Bhattacharya and Majumdar (1998).

Theorem 2.1 is a special case of Proposition 7.1, if one takes S to be an interval, $\mathscr{A} = \{(-\infty, y] \cap S : y \in \mathbb{R}^1\}$, α_n nondecreasing. If $A = (-\infty, y] \cap S$ with $y \geq x_0$, then the first relation in (7.3) is satisfied, in view of hypothesis $(\mathbf{H})$ of Section 2 (or, relation in (2.9)). If $A = (-\infty, y] \cap S$ with

$y < x_0$, then (2.10) implies the second relation in (7.3). Conditions (1) and (2) of $(\mathbf{H}_1)$ have been checked in Sec. 2.

A simple application of Proposition 7.1 of a different nature is the following. Suppose, for an arbitrary S and iid α_n, there is $x_0 \in S$ such that the constant function $\gamma_0(x) := x_0$ $(\forall x \in S)$ has positive probability $P(\{\gamma_0\}) = \delta > 0$. Then $(\mathbf{H}_1)$ holds with $\mathscr{A} = \mathscr{S}$ (the class of all Borel sets) and $N = 1$, since either $x_0 \in A$, or $x_0 \in A^c$, for every $A \in \mathscr{S}$.

For generalizations of Theorem 2.1 to multidimension, we will use a different route, which is useful in proving the CLT outlined in Sec. 3. In the rest of this Section we present a brief exposition of the method given in Bhattacharya and Lee (1988), along with some amendments and extensions, for an appropriate generalization of Theorem 2.1 to $\mathbb{R}^k (k > 1)$.

Let S be a Borel subset of $\mathbb{R}^k$, α_n an iid sequence of measurable non-decreasing random maps on S into itself. Here $\gamma = (\gamma^{(1)}, \gamma^{(2)}, \ldots, \gamma^{(k)})$ is a nondecreasing map on S into S if $(x^{(1)}, x^{(2)}, \ldots, x^{(k)}) \rightarrow \gamma^{(i)}(x^{(1)}, x^{(2)}, \ldots, x^{(k)})$ is nondecreasing in each coordinate $x^{(1)}, \ldots x^{(k)}$ $(i = 1, 2, \ldots, k)$. In other words, we consider the partial order on $\mathbb{R}^k$: if $x = (x^{(1)}, x^{(2)}, \ldots x^{(k)})$ and $y = (y^{(1)}, y^{(2)}, \ldots, y^{(k)})$, then $x \leq y$ if and only if $x^{(i)} \leq y^{(i)}$ $\forall i = 1, 2, \ldots, k$. The significant distinction and the main source of difficulty here is that, unlike the one–dimensional case, this is only a partial order. It is somewhat surprising then that one still has an appropriate and broad generalization of Theorem 2.1. For this let $\mathscr{A}$ be the class of all sets A of the form

$$A = \{y \in S : \phi(y) \leq x\}$$

$$(x \in \mathbb{R}^k, \phi : S \rightarrow S \text{ continuous and nondecreasing}). \qquad (7.5)$$

A generalization on $\mathscr{P}(S)$ of the Kolmogorov distance in Eq. (2.11) is

$$d(\mu, \nu) := \sup_{A \in \mathscr{A}} |\mu(A) - \nu(A)|, \qquad (\mu, \nu \in \mathscr{P}(S)). \qquad (7.6)$$

Note that d is only a pseudo–metric, a priori.

Consider the following splitting condition:

$(\mathbf{H}_2)$ There exist $\delta_i > 0 (i = 1, 2), x_0 \in S$, and a positive integer N such that:
(1) $P^N(\{\gamma \in \Gamma^N : \gamma x \leq x_0 \quad \forall x \in S\}) = \delta_1$, and
(2) $P^N(\{\gamma \in \Gamma^N : \gamma x \geq x_0 \quad \forall x \in S\}) = \delta_2$.
     Here for $\gamma = (\gamma_1, \gamma_2, \ldots, \gamma_N) \in \Gamma^N, \gamma = \gamma_N \gamma_{N-1} \ldots \gamma_1$.

We will show that, under $(\mathbf{H}_2)$, Theorem 2.1 extends to a large class of Borel sets $S \subset \mathbb{R}^k$. The problems that one faces here are: (a) T^{*N} may not be a strict contraction on $\mathscr{P}(S)$ for the metric d, (b) $(\mathscr{P}(S), d)$ may not be a complete metric space, and (c) convergence in d may not imply weak convergence. Problems (b) and (c) are related, and are resolved by the following.

LEMMA 7.2 Suppose S is a closed subset of $\mathbb{R}^k$. Then $(\mathscr{P}(S), d)$ is a complete metric space, and convergence in d implies weak convergence.

Proof. Let $P_n (n \geq 1)$ be a Cauchy sequence in $(\mathscr{P}(S), d)$. Let $\hat{P}_n$ be the extension of P_n to $\mathbb{R}^k$, i.e., $\hat{P}_n(B) := P_n(B \cap S), B \in \mathscr{B}(\mathbb{R}^k)$. By taking ϕ in Eq. (7.5) to be the identity map, it is seen that the distribution functions (d.f.'s) F_n of $\hat{P}_n$ are Cauchy in the uniform distance for functions on $\mathbb{R}^k$. The limit F on $\mathbb{R}^k$ is the distribution function of a probability measure $\hat{P}$ on $\mathbb{R}^k$, and $\hat{P}_n$ converges weakly to $\hat{P}$ on $\mathbb{R}^k$. On the other hand, $P_n(A)$ converges to some $P_\infty(A)$ uniformly $\forall A \in \mathscr{A}$. It follows that $P_\infty(A) = \hat{P}(A)$ for all $A = (-\infty, x] \cap S$ $(\forall x \in \mathbb{R}^k)$. To show that $\hat{P}(S) = 1$ recall that since S is closed and $P_n(S) = 1$ $\forall n$, by Alexandroff's theorem (Billingsley, (1968) pp. 11, 12) one has

$$\hat{P}(S) \geq \varliminf_{n \to \infty} \hat{P}_n(S) = \varliminf_{n \to \infty} P_n(S) = 1.$$

Let P be the restriction of $\hat{P}$ to S. Then P_n converges weakly to P on S. To see this let O be an arbitrary open subset of S. Then $O = U \cap S$, where U is an open subset of $\mathbb{R}^k$, and

$$\underline{\lim} P_n(O) = \underline{\lim} \hat{P}_n(U) \geq \hat{P}(U) = P(O),$$

showing, again by Alexandroff's theorem, that P_n converges weakly to P. Now $P_n \circ \phi^{-1}$ converges weakly to $P \circ \phi^{-1}$ for every continuous nondecreasing ϕ on S. This implies $(P_n \circ \phi^{-1})^\wedge$ converges weakly to $(P \circ \phi^{-1})^\wedge$. Therefore, the d.f. of $(P_n \circ \phi^{-1})^\wedge$ converges to that of $(P \circ \phi^{-1})^\wedge$ at all points of continuity of the latter. But this set of points is dense and, since P_n is Cauchy in $(\mathscr{P}(S), d)$, the convergence is uniform. Therefore, the d.f. of $(P_n \circ \phi^{-1})^\wedge$ converges to that of $(P \circ \phi^{-1})^\wedge$ uniformly on $\mathbb{R}^k$. Thus $P(A) = P_\infty(A)$ $\forall A \in \mathscr{A}$, and $\sup_{A \in \mathscr{A}} |P_n(A) - P(A)| \to 0$. $\qquad\square$

REMARK 7.1. The above Lemma rectifies an error in Bhattacharya and Lee (1988) (see Correction (1997), ibid.), where the completeness of $(\mathscr{P}(S), d)$ was asserted to hold for all topologically complete subsets S of $\mathbb{R}^k$. The following counter–example is due to Prof. B. V. Rao (1992). Let $C \subset [0, 1]$ be the classical Cantor "middle–third" set, and $S = \mathbb{R} \setminus C$. Then S is topologically complete, since every open set is topologically complete (Kelley (1955), p. 207). Let μ be the cantor distribution with support C, and μ_n is μ shifted by $1/3^n$. That is, if X is a random variable with distribution μ, μ_n is the distribution of $X + 1/3^n$. It is simple to see that $\mu_n(\mathbb{R} \setminus C) = 1$, and the distribution function F_n of μ_n converges to the distribution function F of μ uniformlay on $\mathbb{R}$ (since, considered as probability measures on $\mathbb{R}$, μ_n converges weakly to μ and F is continuous). But $\mu \notin \mathscr{P}(S)$. Thus, although μ_n is a Cauchy sequence in the d–metric it has no limit in $\mathscr{P}(S)$.

REMARK 7.2.   Of course, there are sets S other than closed subsets of $\mathbb{R}^k$ for which $(\mathscr{P}(S), d)$ is complete. For example, if S is an open rectangle $(a_1, b_1) \times \cdots \times (a_k, b_k)$ then there exists a strictly increasing function ϑ_0 on S onto $\mathbb{R}^k$, where ϑ_0 is a homeomorphism. Then S may be considered a relabeling of $\mathbb{R}^k$, preserving the partial order and continuous nondecreasing maps. Since $(\mathscr{P}(\mathbb{R}^k), d)$ is complete, so is $(\mathscr{P}(S), d)$ in this case. Similarly, for the rectangle $S = (a_1, b_1] \times (a_2, b_2] \times \cdots \times (a_k, b_k]$ completeness of $(\mathscr{P}(S), d)$ holds, since there is a strictly increasing homeomorphism between it and the closed set $(-\infty, b_1] \times (-\infty, b_2] \times \cdots \times (-\infty, b_k]$.

Problem (a) mentioned above is taken care of by defining a new metric on $\mathscr{P}(S)$. Let $\mathscr{G}_a$ denote the set of all real–valued nondecreasing measurable maps f on S such that $0 \leq f \leq a$ $(a > 0)$. Consider the metric

$$d_a(\mu, \nu) := \sup_{f \in \mathscr{G}_a} \left| \int f \, d\mu - \int f \, d\nu \right| \qquad (\mu, \nu \in \mathscr{P}(S)). \tag{7.7}$$

Clearly,

$$d_a(\mu, \nu) = a d_1(\mu, \nu) \tag{7.8}$$

Also, the topology under d_1 is stronger than that under d,

$$d_1(\mu, \nu) \geq d(\mu, \nu) \tag{7.9}$$

To see this note that the indicator 1_{A^c} of the complement of every set $A \in \mathscr{A}$ belongs to $\mathscr{G}_a$.

We are now ready to prove our main theorem of this Section.

THEOREM 7.3   Let S be a Borel subset of $\mathbb{R}^k$, and $\alpha_n (n \geq 1)$ an iid sequence of nondecreasing random maps on S. If $(\mathbf{H_2})$ holds, then

$$d_1(T^{*n}\mu, T^{*n}\nu) \leq (1 - \delta)^{[n/N]} d_1(\mu, \nu) \quad (\mu, \nu \in \mathscr{P}(S)), \tag{7.10}$$

where $\delta := \min\{\delta_1, \delta_2\}$. If, in addition, S is such that $(\mathscr{P}(S), d)$ is complete, then there exists a unique invariant probability π on S for the Markov process $X_n := \alpha_n \alpha_{n-1} \cdots \alpha_1 X_0$, and

$$d_1(T^{*n}\mu, \pi) \leq (1 - \delta)^{[n/N]}, \qquad (\mu \in \mathscr{P}(S)). \tag{7.11}$$

Proof.   Let $\Gamma_1 := \{\gamma \in \Gamma^N : \gamma(S) \subset (-\infty, x_0] \cap S\}$, $\Gamma_2 := \{\gamma \in \Gamma^N : \gamma(S) \subset [x_0, \infty) \cap S\}$ where $\gamma = \gamma_N \ldots \gamma_1$ for $\gamma = (\gamma_1, \gamma_2, \ldots, \gamma_N) \in \Gamma^N$. Let $f \in \mathscr{G}_1$.

Write

$$\int_S f \, d(T^{*N}\mu) - \int_S f \, d(T^{*N}\nu)$$

$$= \int_S \left\{ \int_{\Gamma^N} f(\gamma x) P^N(d\gamma) \right\} \mu(dx) - \int_S \left\{ \int_{\Gamma^N} f(\gamma x) P^N(d\gamma) \right\} \nu(dx)$$

$$= \sum_{i=1}^{4} \left\{ \int_S h_i(x)\mu(dx) - \int_S h_i(x)\nu(dx) \right\}, \tag{7.12}$$

where

$$h_1(x) := \int_{\Gamma_1 \backslash (\Gamma_1 \cap \Gamma_2)} f(\gamma x) P^N(d\gamma), \quad h_2(x) := \int_{\Gamma_2 \backslash (\Gamma_1 \cap \Gamma_2)} f(\gamma x) P^N(d\gamma),$$

$$h_3(x) := \int_{\Gamma^N \backslash (\Gamma_1 \cup \Gamma_2)} f(\gamma x) P^N(d\gamma), \quad h_4(x) := \int_{\Gamma_1 \cap \Gamma_2} f(\gamma x) P^N(d\gamma). \tag{7.13}$$

Note that, on $\Gamma_1 \cap \Gamma_2$, $\gamma(x) = x_0$, so that the integrals of h_4 in Eq. (7.12) cancel each other. Also,

$$\int_S h_2(x)\mu(dx) - \int_S h_2(x)\nu(dx) = \int_S h_2'(x)\nu(dx) - \int_S h_2'(x)\mu(dx),$$

$$h_2'(x) := \int_{\Gamma_2 \backslash (\Gamma_1 \cap \Gamma_2)} (1 - f(\gamma x)) P^N(d\gamma). \tag{7.14}$$

Now $h_1(x)$ and $h_3(x)$ are nondecreasing, as is $a_2 - h_2'(x)$, where

$$0 \le h_1(x) \le f(x_0)(P^N(\Gamma_1) - P^N(\Gamma_1 \cap \Gamma_2)) = a_1,$$

$$0 \le h_2'(x) \le (1 - f(x_0))(P^N(\Gamma_2) - P^N(\Gamma_1 \cap \Gamma_2)) = a_2,$$

$$0 \le h_3(x) \le 1 - P^N(\Gamma_1) - P^N(\Gamma_2) + P^N(\Gamma_1 \cap \Gamma_2) = a_3.$$

Hence $h_1 \in \mathscr{G}_{a_1}$, $a_2 - h_2' \in \mathscr{G}_{a_2}$, $h_3 \in \mathscr{G}_{a_3}$, and, since

$$\left| \int h_i \, d\mu - \int h_i \, d\nu \right| \le d_{a_i}(\mu, \nu) = a_i d_1(\mu, \nu)(i = 1, 3),$$

and

$$\left| \int (a_2 - h_2') \, d\mu - \int (a_2 - h_2') \, d\nu \right| \le d_{a_2}(\mu, \nu) = a_2 d_1(\mu, \nu),$$

it follows that

$$\left| \int_S f\, d(T^{*N}\mu) - \int_S f\, d(T^{*N}\nu) \right| = \left| \int_S h_1\, d\mu - \int_S h_1\, d\nu \right.$$

$$\left. + \int_S (a_2 - h_2')d\mu - \int_S (a_2 - h_2')d\nu + \int_S h_3 d\mu - \int_S h_3 d\nu \right|$$

$$\leq (a_1 + a_2 + a_3)d_1(\mu, \nu), \tag{7.15}$$

where

$$a_1 + a_2 = f(x_0)(P^N(\Gamma_1) - P^N(\Gamma_1 \cap \Gamma_2))$$

$$+ (1 - f(x_0))(P^N(\Gamma_2) - P^N(\Gamma_1 \cap \Gamma_2))$$

$$\leq \max\{P^N(\Gamma_1), P^N(\Gamma_2)\} - P^N(\Gamma_1 \cap \Gamma_2),$$

and $a_3 = 1 - P^N(\Gamma_1) - P^N(\Gamma_2) + P^N(\Gamma_1 \cap \Gamma_2)$. It follows that $a_1 + a_2 + a_3 \leq 1 - \delta_2$ if $\delta_1 = P^N(\Gamma_1) \geq P^N(\Gamma_2) = \delta_2$, and $a_1 + a_2 + a_3 \leq 1 - \delta_1$ if $\delta_1 \leq \delta_2$. Thus $a_1 + a_2 + a_3 \leq 1 - \delta$. Hence we have derived

$$d_1(T^{*N}\mu, T^{*N}\nu) \leq (1 - \delta)d_1(\mu, \nu), \qquad (\mu, \nu \in \mathscr{P}(S)). \tag{7.16}$$

Also, $\forall f \in \mathscr{G}_1$,

$$\left| \int_S f d(T^*\mu) - \int_S f d(T^*\nu) \right| = \left| \int_S g(x)\mu(dx) - \int_S g(x)\nu(dx) \right|,$$

where $g(x) := \int_\Gamma f(\gamma x)P(d\gamma)$, so that $g \in \mathscr{G}_1$. Thus

$$d_1(T^*\mu, T^*\nu) \leq d_1(\mu, \nu), \qquad (\mu, \nu \in \mathscr{P}(S)). \tag{7.17}$$

Combining (7.16) and (7.17) one arrives at (7.10). In particular, this implies

$$d(T^{*n}\mu, T^{*n}\nu) \leq (1 - \delta)^{[n/N]}, \qquad (\mu, \nu \in \mathscr{P}(S)). \tag{7.18}$$

Fix $x \in S$, and a positive integer r arbitrarily, and let $\mu = \delta_x$, $\nu = T^{*r}\delta_x$ in (7.18), where δ_x is the point mass at x. Then $T^{*n}\mu = p^{(n)}(x, dy)$, $T^{*n}\nu = p^{(n+r)}(x, dy)$, and (7.18) yields

$$d(p^{(n)}(x, dy), p^{(n+r)}(x, dy)) \leq (1 - \delta)^{[n/N]} \to 0 \quad \text{as} \quad n \to \infty. \tag{7.19}$$

That is, $p^{(n)}(x, dy)$ $(n \geq 1)$ is a Cauchy sequence in $(\mathscr{P}(S), d)$. By the assumption of completeness, there exists $\pi \in \mathscr{P}(S)$ such that $d(p^{(n)}(x, dy), \pi(dy)) \to 0$ as $n \to \infty$. Letting $r \to \infty$ in (7.19) one gets

$$d(p^{(n)}(x, dy), \pi(dy)) \leq (1 - \delta)^{[n/N]}. \tag{7.20}$$

Since this is true $\forall x$, one may average the left side with respect to $\mu(dx)$ to get (7.11).

It remains to show that π is the unique invariant probability for $p(x, dy)$. This is proved in the Lemma below. $\qquad\square$

LEMMA 7.4 Let $(\mathbf{H_2})$ hold and let $(\mathscr{P}(S), d)$ be complete. Then π in (7.11) is the unique invariant probability for $p(x, dy)$, and $p^{(n)}(x, dy)$ converges weakly to π as $n \to \infty$, $\forall x \in S$.

Proof. The fact that $p^{(n)}(x, dy)$ converges weakly to π $\forall x$ follows from (the proof of) Lemma 7.2. Suppose first that there exist $a = (a_1, a_2, \ldots, a_k)$ and $b = (b_1, b_2, \ldots, b_k)$ in S such that $a \le x \le b$ $\forall x \in S$. Let $Y_n(x) := \alpha_1 \alpha_2 \ldots \alpha_n(x)$. Then, as in the proof of Theorem 2.2, $Y_n(a) \uparrow \underline{Y}$ as $n \uparrow \infty$, and $Y_n(b) \downarrow \overline{Y}$ as $n \uparrow \infty$. Therefore, $p^{(n)}(a, dy)$ converges weakly to the distribution of $\underline{Y}$, which must be π. Similarly, $p^{(n)}(b, dy)$ converges weakly to the distribution of $\overline{Y}$, also π. Since $\underline{Y} \le \overline{Y}$, it now follows that $\underline{Y} = \overline{Y}$. Let α be independent of $\{\alpha_n : n \ge 1\}$, and have the same distribution P. Then, $\alpha Y_n(z)$ has the distribution $p^{(n+1)}(z, dy)$. Hence, $\forall x \in \mathbb{R}^k$,

$$p^{(n+1)}(b, S \cap [a, x]) = \mathrm{Prob}(\alpha Y_n(b) \le x) \le \mathrm{Prob}(\alpha \overline{Y} \le x)$$

$$= \int p(z, S \cap [a, x])\pi(dz) = \mathrm{Prob}(\alpha \underline{Y} \le x)$$

$$\le \mathrm{Prob}(\alpha Y_n(a) \le x) = p^{(n+1)}(a, S \cap [a, x]). \tag{7.21}$$

In view of (7.11), the two extreme sides of (7.21) both converge to $\pi(S \cap [a, x])$ uniformly $\forall x$. Therefore,

$$\int p(z, S \cap [a, x])\pi(dz) = \pi(S \cap [a, x]) \qquad \forall x \in \mathbb{R}^k. \tag{7.22}$$

This implies that π is invariant. For the two sides of Eq. (7.22) are the d.f.'s of $\alpha_1 X_0$ and X_0, respectively, where X_0 has distribution π, and invariance of π just means $\alpha_1 X_0$ and X_0 have the same distribution. If π' is another invariant probability, then $\forall n$

$$\int p^{(n)}(z, S \cap [a, x])\pi'(dz) = \pi'(S \cap [a, x]) \qquad \forall x. \tag{7.23}$$

But the integrand on the left converges to $\pi(S \cap [a, x])$ as $n \to \infty$, by (7.11). Therefore, the left side converges to $\pi(S \cap [a, x])$, showing that $\pi' = \pi$.

Finally, suppose there do not exist a and/or $b \in S$ which bound S from below and/or above. In this case one can find an increasing homeomorphism of S onto a bounded set S_1. It is then enough to consider S_1.

Let $a_i := \inf\{x_i : x = (x_1, \ldots, x_k) \in S_1\}$, $1 \leq i \leq k$, and $b_i := \sup\{y_i : y = (y_1, \ldots, y_k) \in S_1\}$. Let $a = (a_1, a_2, \ldots, a_k)$, $b = (b_1, b_2, \ldots, b_k)$. For simplicity, assume $N = 1$. Let $\Gamma_1 := \{\gamma \in \Gamma : \gamma x \leq x_0 \ \forall x \in S_1\}$, $\Gamma_2 := \{\gamma \in \Gamma : \gamma x \geq x_0 \ \forall x \in S_1\}$. Extend $\gamma \in \Gamma_1 \backslash (\Gamma_1 \cap \Gamma_2)$ to $S_1 \cup \{a, b\}$ by setting $\gamma(a) = a$, $\gamma(b) = x_0$. Similarly, extend $\gamma \in \Gamma_2 \backslash (\Gamma_1 \cap \Gamma_2)$ by setting $\gamma(a) = x_0$, $\gamma(b) = b$. Extend $\gamma \in \Gamma_1 \cap \Gamma_2$ by setting $\gamma(a) = x_0 = \gamma(b)$. For $\gamma \notin \Gamma_1 \cup \Gamma_2$, let $\gamma(a) = a$, $\gamma(b) = b$. On the space $S_1 \cup \{a, b\}$ the splitting condition ($\mathbf{H_2}$) holds, and the proof given above shows that π is the unique invariant probability on $S_1 \cup \{a, b\}$. Since $\pi(S_1) = 1$, the proof is now complete. $\qquad \square$

REMARK 7.3. Lemma 7.4 could have been avoided if one could show that $(\mathscr{P}(S), d_1)$ is complete. For in that case the contraction mapping principle yields a fixed point of T^*. Identifying the class of sets S for which this completeness holds is a nontrivial task. Recently Chakraborty and Rao (1998) have analyzed this problem for dimensions $k = 1, 2$, with ideas which are relevant for higher dimensions as well.

REMARK 7.4. The estimates in (7.10), (7.11) play a crucial role in the proof of the functional CLT in Section 3 and its generalization to multi-dimension. See Bhattacharya and Lee (1988) for details.

We close this section by pointing out that the estimate (7.10) in Theorem 7.3 holds, under the splitting condition ($\mathbf{H_2}$), for every Borel set $S \subset \mathbb{R}^k$. By extending the probability measures to the closure $\overline{S}$ of S one may conclude, in view of Lemma 7.2, that $T^{*n}\mu$ converges to a probability measure π on $\overline{S}$ weakly and in the d-metric, for all $\mu \in \mathscr{P}(S)$. To be precise we have the following result.

PROPOSITION 7.5. Let S be a Borel subset of $\mathbb{R}^k$, and $\alpha_n (n \geq 1)$ an iid nondecreasing sequence of random maps on S such that the splitting condition ($\mathbf{H_2}$) holds. Then there exists a unique probability π on $\overline{S}$ such that $p^{(n)}(x, dy)$, viewed as a probability measure on $\overline{S}$, converges weakly to π $\forall x \in S$. Indeed, $\forall \mu \in \mathscr{P}(S)$, the extension of $T^{*n}\mu$ to $\overline{S}$ converges exponentially fast to π in the d-metric on $\mathscr{P}(\overline{S})$, i.e.,

$$\sup\nolimits_{A \in \overline{\mathscr{A}}} |(T^{*n}\mu)(A) - \pi(A)| \leq (1 - \delta)^{[n/N]} \qquad (\mu \in \mathscr{P}(S))$$

where $\overline{\mathscr{A}}$ is the class of all sets of the form Eq. (7.5) with S replaced by $\overline{S}$.

Acknowledgment

The authors wish to thank the referees for their thorough reading of the manuscript and for their kind suggestions.

REFERENCES

Arrow, K. J. and Kurz, M. Public Investment, The Rate of Return and Optimal Fiscal Policy. Baltimore, MD: Johns Hopkins University Press, 1970.

Benhabib, J. and Nishimura, K. Competitive equilibrium cycles. *J. Econ. Theory 35*: 284–306, 1985.

Bhattacharya, R. N. On the functional central limit theorem and the law of iterated logarithm for Markov processes. *Z. Wahr. Verw. Geb. 60*: 185–201, 1982.

Bhattacharya, R. N. and Lee, O. Asymptotics of a class of Markov processes which are not in general irreducible. *Ann. Probab. 16*: 1333–1347, 1988. (Correction, *Ann. Probab. 25*: 1541–1543, 1997).

Bhattacharya, R. N. and Majumdar, M., On a theorem of Dubins and Freedman (to appear), 1998.

Bhattacharya, R. N. and Rao, B. V. Random iterations of two quadratic maps. Stochastic Processes: A Festschrift in Honour of Gopinath Kallianpur. Cambanis S., Ghosh J. K., Karandikar R. L. and Sen P. K., ed. New York: Springer, 13–21, 1993.

Bhattacharya, R. N. and Waymire, E. C. Stochastic Processes with Applications. New York: Wiley, 1990.

Billingsley, P. Convergence of Probability Measures. New York: Wiley, 1968.

Boldrin, M., and L. Montrucchio. On the indeterminacy of capital accumulation paths. *J. Econ. Theory 40*: 26–39, 1986.

Chakraborty, S. and Rao, B. V. Completeness of the Bhattacharya metric in the space of probabilities. *Statist. Probab. Lett. 36*: 321–326, 1998.

Dechert, W. D. and Nishimura, K. A complete characterization of optimal growth path in an aggregated model with a non–concave production function. *J. Econ. Theory 31*: 332–54, 1983.

Devaney, R. L. An Introduction to Chaotic Dynamical Systems. Second edition. New York: Addison–Wesley, 1989.

Dubins, L. E. and Freedman, D. A. Invariant probabilities for certain Markov processes. *Ann. Math. Stat. 37*: 837–848, 1966.

Friedman, A. Foundations of Modern Analysis. New York: Dover, 1982.

Gordin, M. I. and Lifsic, B. A. The central limit theorem for stationary ergodic Markov processes. *Dokl. Akad. Nauk. SSSR. 19*: 392–393, 1978.

Kelley, J. L. General Topology. New York: Van Nostrand, 1995.

Kifer, Y. Ergodic Theory of Random Transformations. Boston: Birkhauser, 1986.

Koopmans, T. C. Objectives, constraints and outcomes in optimal growth models. *Econometrica 35*: 1–15, 1967.

Majumdar, M. and Mitra, T. Periodic and chaotic programs of optimal intertemporal allocation in an aggregative model with wealth effect. *Economic Theory 4*: 649–676, 1994a.

Majumdar, M. and Mitra, T. Robust ergodic chaos in discounted dynamic optimization models. *Economic Theory 4*: 677–688, 1994b.

Majumdar, M, Mitra, T. and Nyarko, Y. Dynamic optimization under uncertainty: non–convex feasible set. Joan Robinson and Modern Economic Theory. Feiwel, G. R. ed. London: MacMillan, 545–590, 1989.

Majumdar, M. and Nermuth, M. Dynamic optimizations in non–convex models with irreversible investments: monotonicity and turnpike results. *Zeitschrift für Nationalökonomie 42*: 339–62, 1982.

Majumdar, M. and Radner, R. Survival under Production Uncertainty. Equilibrium and Dynamics. Majumdar, M., ed. London: MacMillan, 179–200, 1992.

Mitra, T. and Ray, D. Dynamic optimization on a non–convex feasible set: some general results for non–smooth technologies. *Zeitschrift für Nationalökonomie 44*: 151–75, 1984.

Ramsey, F. A mathematical theory of saving. *Economic Journal 38*: 543–59, 1928.

Rao, B. V. Personal communication, 1992.

Ross, S. M. Introduction to Stochastic Dynamic Programming. New York: Academic Press, 1983.

Roy, S. Theory of dynamic portfolio choice for survival under uncertainty. *Mathematical Social Sciences 38*: 171–194, 1995.

Roy, S. Economic models of survival under uncertainty. Proceedings of the First World Congress of Nonlinear Analysis. Lakshmikantam, V., ed. New York: Walter de Gruyter, 1996.

Sargent, T. J. Dynamic Macroeconomic Theory. Harvard University Press, 1987.

Solow, R. M. A contribution to the theory of economic growth. *Quarterly Journal of Economics 70*: 65–94, 1956.

Sorger, G. Minimum Impatience Theorems for Recursive Economic Models. Heidelberg: Springer Verlag, 1992.

Topkis, D. Minimizing a submodular function on a lattice. *Operations Research 26*: 305–21, 1968.

Yahav, J. A. On a fixed point theorem and its stochastic equivalent. *J. Appl. Probab. 12*: 605–611, 1975.

22

Chi-Squared Tests of Goodness-of-Fit For Dependent Observations

KAMAL C. CHANDA Texas Tech University, Lubbock, Texas

1. INTRODUCTION

Tests of goodness-of-fit of a sequence of observations to a given distribution or a family of distributions usually requires that the sequence of observations constitutes a set of independent and identically distributed (iid) random variables. The situation is considerably different if the data set is a realization from a stationary stochastic process (SSP). For such SSP's, effect of dependence on these goodness-of-fit tests can be substantial (see for example Bartlett (1951), Chanda (1981), Gastwirth and Rubin (1975), Gleser and Moore (1983), Moore (1982) and Withers (1975)).

Gasser (1975) conducted a finite sample simulation study to explore the effect of dependence on the validity of a test for normality based on Pearson's chi-square statistics. For primarily Gaussian autoregressive schemes the Pearson test using the asymptotic distribution for iid appeared to reject normality too frequently.

Chanda (1981) has investigated the asymptotic properties of Pearson's chi-squared statistics for strong mixing processes with particular application to linear processes. He has also discussed the effect of estimation of parameters on the asymptotic distribution of these chi-squared statistics. Moore (1982) has discussed in detail the asymptotic properties of these statistics for Gaussian processes with parameters which are unknown and have to be estimated from the given sample. He has also considered the situation where data-dependent cells have been used to construct the test statistic. He has shown, in particular, that for Gaussian SSP's with positive autocorrelations the chi-squared statistics is stochastically larger in the asymptotic sense than that for the iid case.

The purpose of the present chapter is to explore theoretically the effects of dependence on the Pearson chi-squared test when the SSP belongs to a general class which include linear (both Gaussian and non Gaussian), bilinear (distinctly nonGaussian) and Volterra processes. We shall mainly restrict ourselves to SSP's for which the univariate distributions are absolutely continuous. There is a vast area of SSP's for which the univariate distributions are discrete and the corresponding formulation of the Pearson's goodness-of-fit tests requires entirely different treatment (see for example Bartlett (1951), Billingsley (1961) and Goodman (1958)).

2. CHI-SQUARED GOODNESS-OF-FIT TESTS

Let $\{X_t, t \in \mathbb{Z}\}$ be a SSP and let the data $\{X_1, \ldots, X_n\}$ from this SSP be distributed into k cells $B_i = (a_{i-1}, a_i)(1 \le i \le r)$ with boundaries $-\infty = a_0 < a_1 < \cdots < a_{r-1} < a_r = \infty$. Write $p_i(\theta) = F(a_i, \theta) - F(a_{i-1}, \theta)$ where $F(x, \theta)$ denotes the distribution function (d.f.) of X_1. $F(x, \theta)$ is supposed to be known except for a parameter θ which may be unknown. Let N_i represent the number of observations (in the data set) which fall into B_i. Set $\mathbf{V}_n = \mathbf{V}_n(\theta)$ to be the r-vector for which the ith component is $(N_i - np_i(\theta))/(np_i(\theta))^{1/2}$. If θ is not known we estimate θ by $\hat{\theta} = \hat{\theta}(X_1, \ldots, X_n)$ and we then use the r-vector $\hat{\mathbf{V}}_n = \mathbf{V}_n(\hat{\theta})$ of which the ith component is $(N_i - n\hat{p}_i)/(n\hat{p}_i)^{1/2}$ where $\hat{p}_i = p_i(\hat{\theta})$. We shall consider two groups of null hypotheses H_0, viz., (i) when F is fully specified and the parameter θ is known and (ii) when F is fully specified but the parameter θ is to be estimated from chi-squared statistics

$$\chi^2 = \sum_{i=1}^{r} (N_i - np_i)^2 / np_i \tag{2.1}$$

and

$$\hat{\chi}^2 = \sum_{i=1}^{r} (N_i - n\hat{p}_i)^2 / n\hat{p}_i \tag{2.2}$$

where we write $p_i = p_i(\theta_0)$, θ_0 being the given value θ and $\hat{p}_i = p_i(\hat{\theta})$.

In this Section we shall be primarily concerned with exploring the large sample properties of χ^2 as defined in Eq. (2.1). We shall restrict ourselves to the class of SSP's $\{X_t\}$ which can be expressed as

$$X_t = \varepsilon_t + \sum_{r=1}^{\infty} W_{rt} \tag{2.3}$$

where the infinite sum on the right side of Eq. (2.3) converges in probability and $\{W_{rt}, t \in \mathbb{Z}\}$ is strictly stationary for every $r \geq 1$. We assume that W_{rt} can be written as a function $g_r(\varepsilon_{t-1}, \ldots, \varepsilon_{t-r-q})$ where q is a fixed known integer ≥ 0, $\{\varepsilon_t, t \in \mathbb{Z}\}$ is a sequence of iid random variables and $E(|W_{rt}|^\gamma) < Mc_r^\gamma$ for some $\gamma \in (0,1]$ and for some $\{c_r, r \geq 1\}$ for which $\sum_{r=1}^{\infty} rc_r^\lambda < \infty$, with $\lambda = \gamma/(1+\gamma)$. M is used here and elsewhere as a generic symbol to denote a finite positive constant independent of n and other parameters whose nature will be clear from the contexts in which they appear.

The interesting aspect of Eq. (2.3) is that the entire class of ARMA processes and most of the bilinear and Volterra processes belong to the type in Eq. (2.3) and the conditions stated above are not hard to check in any of these situations. We consider below some specific cases.

EXAMPLE 1   Let $\{X_t\}$ be a linear process defined by

$$X_t = \sum_{r=0}^{\infty} g_r \varepsilon_{t-r} \tag{2.4}$$

where $\{g_r\}$ is such that for some $\theta \geq 1, \sum_{r=v}^{\infty} H_r^{1/(2+\gamma)} \leq Mv^{-\theta}$ for $1 \leq v < \infty$, $(H_r^\gamma = \sum_{s=r}^{\infty} |g_s|^\gamma)$ and $E|\varepsilon_1|^\gamma < \infty$ for some $\gamma > 0$ and if $\gamma \geq 1$ then $E(\varepsilon_1) = 0$. Note that $\sum_{r=v}^{\infty} H_r^{1/(2+\gamma)} \leq \sum_{r=v}^{\infty} r|g_r|^\lambda$ where $\lambda = \gamma/(2+\gamma)$ so that if $\sum_{r=v}^{\infty} r|g_r|^\lambda < Mv^{-\theta}(\theta > 0)$ then the condition above holds. Note that $|g_r| < M\rho^r$ for some $\rho \in (0,1)$ if $\{X_t\}$ is an ARMA process and the relation in Eq. (2.3) is satisfied.

EXAMPLE 2   Let $\{X_t\}$ be a bilinear process defined by

$$X_t + \sum_{j=1}^{p} \phi_j X_{t-j} = \varepsilon_t + \sum_{j=1}^{q} \theta_j \varepsilon_{t-j} + \sum_{j=1}^{p} \beta_j X_{t-j} \varepsilon_{t-1}, \tag{2.5}$$

where $\phi_1, \ldots, \phi_p, \theta_1, \ldots, \theta_q$ and $\beta_1, \ldots, \beta_p$ are unknown parameters and $p \geq 0, q \geq 0$ are known. It is well-known (see, for example, Bhaskara Rao et al. (1983) and Chanda (1991)) that X_t can be rewritten a.s. as

$$X_t = \varepsilon_t + \sum_{r=1}^{\infty} W_{rt}$$

where $\quad W_{rt} = \sum_{j=1}^{q} \theta_j \varepsilon_{t-j} + \mathbf{\eta}^T (\mathbf{A} + \mathbf{B}\varepsilon_{t-1})\Theta\varepsilon_{t-1}\quad$ if $\quad r = 1\quad$ and
$= \mathbf{\eta}^T \Pi_{j=1}^{r}(\mathbf{A} + \mathbf{B}\varepsilon_{t-j})\Theta\varepsilon_{t-r}$ if $r \geq 2$,

$$\mathbf{\eta} = [1, 0, \ldots, 0]^T, \mathbf{A}^{p \times p} = \begin{bmatrix} -\phi_1 & -\phi_2 & \cdots & -\phi_{p-1} & -\phi_p \\ 1 & 0 & \cdots & 0 & 0 \\ \vdots & & & \vdots & \vdots \\ 0 & 0 & \cdots & 1 & 0 \end{bmatrix}$$

$$\mathbf{B}^{p \times p} = \begin{bmatrix} \beta_1 & \beta_2 & \cdots & \beta_p \\ 0 & 0 & \cdots & 0 \\ \vdots & \vdots & & \vdots \\ 0 & 0 & \cdots & 0 \end{bmatrix}, \quad \Theta^{p \times (q+1)} = \begin{bmatrix} 1 & \theta_1 & \cdots & \theta_q \\ 0 & 0 & \cdots & 0 \\ \vdots & & & \vdots \\ 0 & 0 & \cdots & 0 \end{bmatrix},$$

and $\varepsilon_t = [\varepsilon_t, \ldots, \varepsilon_{t-q}]^T$.

It is easy to see that W_{rt} involves $\varepsilon_{t-1}, \ldots, \varepsilon_{t-r-q}, r \geq 1$. If we now assume that $E(\varepsilon_1) = 0$, $0 < E(\varepsilon_1^2) = \sigma_\varepsilon^2 < \infty$ and that the spectral radius ρ_0 of the matrix $\mathbf{A} \otimes \mathbf{A} + \sigma_\varepsilon^2 \mathbf{B} \otimes \mathbf{B}$ is less than unity, then for every $\gamma, 0 < \gamma \leq 1$, $E|W_{rt}|^\gamma \leq M\rho_0^{(r-1)\gamma/2}$ which immediately establishes the validity of Eq. (2.3) with $c_r = \rho_0^{(r-1)/2}$.

EXAMPLE 3   Let $\{X_t\}$ be a Volterra process defined by

$$X_t = \varepsilon_t + \sum_{r=1}^{\infty} g_r \varepsilon_{t-r} + \sum_{r_1, r_2 = 1}^{\infty} g_{r_1, r_2} \varepsilon_{t-r_1} . \varepsilon_{t-r_2}$$

$$+ \cdots + \sum_{r_1, \ldots, r_q = 1}^{\infty} g_{r_1, r_2, \ldots, r_q} \varepsilon_{t-r_1} \varepsilon_{t-r_2} \cdots \varepsilon_{t-r_q} \tag{2.6}$$

where $\{\varepsilon_t\}$ is an iid sequence of r.v.'s with $E|\varepsilon_1|^\gamma < \infty$. Then we can write

$$c_r = \sum_{j=1}^{q} \sum_{Srj} |g_{u_1, \ldots, u_j}|^\gamma (r \geq 1),$$

where

$$Srj = \{(r_1, \ldots, r_j) : 1 \leq r_1, \ldots, r_j < \infty, \max(r_1, \ldots, r_j) = r\}.$$

If, now, we let $\mathbf{Y}_t$ denote a k-vector for which the ith component is $Y_{it} = (I_i(X_t) - p_i)/p_i^{1/2}$ where $I_i(x)$ is the indicator function of $B_i (1 \leq i \leq k)$ then we can write

$$\chi^2 = \mathbf{V}_n^T \mathbf{V}_n,$$

$$\mathbf{V}_n = n^{-1/2} \sum_{t=1}^{n} \mathbf{Y}_t. \tag{2.7}$$

Note that $E(\mathbf{V}_n) = 0$ and $E(\mathbf{V}_n \mathbf{V}_n^T) = \Lambda_n$ where $\Lambda_n = [\lambda_{ijn}]$,

$$\lambda_{ijn} = \sum_{v=-n+1}^{n-1} (1 - |v|/n)\lambda_{ij}(v)$$

$$= \lambda_{ij}(0) + \sum_{v=1}^{n-1} (1 - v/n)(\lambda_{ij}(v) + \lambda_{ji}(v)),$$

$$\lambda_{ij}(v) = (p_i p_j)^{-1/2}(p_{ij}(v) - p_i p_j)$$

and

$$p_{ij}(v) = P(X_t \in B_i, \quad X_{t+v} \in B_j) \tag{2.8}$$

under $F(x, \theta_0)$, $(1 \leq i, j \leq k)$. If $\{X_t\}$ is a sequence of iid random variables then, of course, $\lambda_{ij}(v) = 0$ for $v > 0$, $1 \leq i, j \leq k$, and $\Lambda_n = \mathbf{I} - \delta\delta^T$ where δ is a k-vector for which the ith element is $p_i^{1/2}(1 \leq i \leq k)$. It can be shown quite easily that rank $(\Lambda_n) = k - 1$. Even for the SSP $\{X_t\}$, $\delta^T \Lambda_n = \mathbf{0}$ because $\sum_{i=1}^{k} p_{ij}(v) = p_j$ so that rank $(\Lambda_n) \leq k - 1$.

We now assume that the following conditions hold.

C1.   For all i $(1 \leq i \leq k)$, $p_i > 0$ and $\sum_{i=1}^{k} p_i = 1$.
C2.   If φ_ε denotes the characteristic function (ch.f.) of ε_1 then φ_ε is L_1-integrable.

Let α be an arbitrary k-vector of real numbers and set

$$T_{t\alpha} = \alpha^T \mathbf{Y}_t \tag{2.9}$$

Then $\alpha^T \mathbf{V}_n = n^{-1/2} \sum_{t=1}^{n} T_{t\alpha}$ and we can prove the following Theorem.

THEOREM 2.1   Let the relation in Eq. (2.3) and conditions C1 and C2 hold. Then under $F(x, \theta_0)$,

$$\mathscr{L}(\mathbf{V}_n) \to \mathscr{N}(\mathbf{0}, \Lambda) \tag{2.10}$$

as $n \to \infty$ where $\Lambda = [\lambda_{ij}]$ and λ_{ij} is as defined in Eq. (2.8).

Before we establish the result of Theorem 2.1 we need to prove a few lemmas.

LEMMA 2.2   Let U, V, W be arbitrary random variables and let J be an arbitrary interval and K be an arbitrary finite interval. Assume that for any real numbers $a, b(a \le b)$, $P(a < V \le b) \le M(b - a)$ where M is a finite positive constant independent of a and b. Then for every $\eta > 0$

$$P(U \in J, V + W \in K) - P(U \in J, V \in K)| \le M(\eta + P(|W| > \eta)).$$

(2.11)

Proof.   Note that if $K = (c, d)$ then

$$P(U \in J, c + \eta < V < d - \eta) - P(|W| > \eta) \le P(U \in J, V + W \in K)$$
$$\le P(U \in I, c - \eta < V < d + \eta) + P(|W| > \eta).$$ (2.12)

Again since

$$P(U \in J, c - \eta < V < d + \eta) - P(U \in J, c < V < d)$$
$$\le P(c - \eta < V \le c) + P(d \le V < d + \eta) \le M\eta,$$

and

$$|P(U \in J, c < V < d) - P(U \in J, c + \eta < V < d - \eta)|$$
$$\le P(c < V \le c + \eta) + P(d - \eta \le V < d) \le M\eta,$$

the result in Eq. (2.11) follows from Eq. (2.12) and the last two inequalities above.

COROLLARY 2.3   Let the conditions of Lemma 2.2 hold. Then for every $\eta > 0$.

$$|P(V + W \in K) - P(V \in K)| \le M(\eta + P(|W| > \eta))$$ (2.13)

Proof.   The result follows from Lemma 2.3 if we set $J = (-\infty, \infty)$.

LEMMA 2.4   Let U, V, W be random variables such that U, V are independent, and let J be an arbitrary interval and K an arbitrary finite interval. Assume that for any real numbers a, b $(a \le b)$, $P(a < V \le b) \le M(b - a)$ with M as in Lemma 2.2. Then for every $\eta > 0$

$$|P(U \in J, V + W \in K) - P(U \in J)P(V + W \in K)| \le M(\eta + P(|W > \eta))$$

(2.14)

Proof.   It is easy to see that if $K = (c, d)$ then

$$P(U \in J, c + \eta < V < d - \eta) - P(|W| > \eta)$$
$$\le P(U \in J, V + W \in K, |W| \le \eta)$$
$$\le P(U \in J, c - \eta \le V \le d + \eta),$$
$$0 \le P(U \in J, V + W \in K, |W| > \eta) \le P(|W| > \eta).$$ (2.15)

Now note that U, V are independent random variables and that $|P(c \pm \eta \leq V \leq d \mp \eta)| - P(c < V + W < d)| \leq P(c - \eta < V \leq d + \eta) - P(c < V \leq d) + |P(V + W \in K) - P(V \in K)| \leq M(\eta + P(|W| > \eta)$ by the condition of Lemma 2.4 and the result of Lemma 2.2. Therefore, if we apply the inequality above to Eq. (2.12) we immediately get Eq. (2.14).

LEMMA 2.5   Let relations in Eq. (2.3) and conditions C1 and C2 hold. Then, as $n \rightarrow \infty$.

$$\Lambda_n \rightarrow \Lambda \tag{2.16}$$

where the (i, j)th element λ_{ij} of Λ is a power series which converges absolutely.

Proof.   Note that subject to convergence

$$\lambda_{ij} = \lambda_{ij}(0) + \sum_{v=1}^{\infty} (\lambda_{ij}(v) + \lambda_{ji}(v)). \tag{2.17}$$

We shall now show that $\sum_{v=1}^{\infty} |\lambda_{ij}(v)| < \infty$ $(1 \leq i, j \leq k)$, which will imply that $|\lambda_{ij}| \leq M$. Now observe that for every $v \geq q + 1$, we have, by relation in Eq. (2.3) and condition C2,

$$|\lambda_{ij}(v)| \leq M|P(X_t \in B_i, X_{t+v} \in B_j) - P(X_t \in B_i)P(X_{t+v} \in B_j)|$$
$$\leq M|P(U \in B_i, V + W \in B_j) - P(U \in B_i)P(V + W \in B_j)| \tag{38}$$

where M is a finite positive constant independent of t, v, i and j, and $U := X_t$, $V =: X_{t+v,v-q-1} := \varepsilon_{t+v} + \sum_{r=1}^{v-q-1} W_{r,t+v}$, $W := X^*_{t+v,v-q-1} := X_{t+v} - X_{t+v,v-q-1}$. Since V involves the random variables $\varepsilon_{t+v}, \varepsilon_{t+v-1}....\varepsilon_{t+1}$ (by conditions on W_{rt}) and U is a function of random variables $\varepsilon_t, \varepsilon_{t-1}, \ldots, U$ and V are independent. Also, since ε_{t+v} and $V - \varepsilon_{t+v}$ are independent random variables, and, therefore, $\varphi(u) = \varphi_\varepsilon(u)\varphi^*(u)$ where φ, φ^* denote respectively the ch.f.'s of V and $V - \varepsilon_{t+v}$, we have that $|\varphi(u)| \leq |\varphi_\varepsilon(u)|$. Hence by condition C2, the conditions of Lemma 2.4 hold and we conclude that

$$|\lambda_{ij}(v)| \leq M(\eta + Q_v(\eta)) \tag{2.18}$$

where $Q_v(\eta) = P(|X^*_{t+v,v-q}| > \eta)$. Now note that

$$Q_v(\eta) \leq \eta^{-\gamma} \sum_{r=v-q+1}^{\infty} E|W_{r,t+v}|^\gamma \leq M\eta^{-\gamma} \sum_{r=v-q+1}^{\infty} c_r^\gamma.$$

If now we choose $\eta = (\sum_{r=v-q+1}^{\infty} c_r^{\gamma})^{1/(1+\gamma)}$, then from Eq. (2.18) it follows that for $v \geq q + 1$

$$|\lambda_{ij}(v)| \leq M \left(\sum_{r=v-q+1}^{\infty} c_r^{\gamma} \right)^{1/(1+\gamma)} \leq M \sum_{r=v-q+1}^{\infty} c_r^{\lambda}$$

and

$$\sum_{v=q+1}^{\infty} |\lambda_{ij}(v)| \leq M \sum_{r=1}^{\infty} r c_r^{\lambda} < \infty \tag{2.19}$$

The proof of Lemma 2.5 is now complete.

For any arbitrary positive integer m let

$$U_{nm\alpha} := n^{-1/2} \sum_{t=1}^{n} T_{tm\alpha}$$

$$U^{*}_{nm\alpha} := n^{-1/2} \sum_{t=1}^{n} T^{*}_{tm\alpha}$$

where $T_{tm\alpha} := \alpha^T Y_{tm}, T^{*}_{tm\alpha} := \alpha^T Y^{*}_{tm}, Y^{*}_{tm}, = Y_t - Y_{tm}, Y_{tm}$ is a k-vector of which the ith element is $(I_i(X_{tm}) - p_{im})/p_i^{1/2}$, $X_{tm} = \sum_{r=0}^{m} W_{rt}$ and $p_{im} = P(X_{tm} \in B_i)$. We also set $U_{n\alpha} = \alpha^T V_n$ so that $U^{*}_{nm\alpha} = U_{n\alpha} - U_{nm\alpha}$. We now prove the following Lemma.

LEMMA 2.6   Let the conditions of Lemma 2.5 hold. Then, for every finite $m \geq 1$,

$$\mathscr{L}(U_{nm\alpha}) \to \mathscr{N}(0, \sigma_{m\alpha}^2) \tag{2.20}$$

as $n \to \infty$, where $\sigma_{m\alpha}^2$ is a finite positive constant.

Proof.   First note that $\{T_{tm}, \ t \in \mathbb{Z}\}$ is an $(m+q)$-dependent and uniformly bounded SSP. Also $E(U_{nm}) = 0$ and $V(U_{nm}) = \alpha^T \Lambda_{nm} \alpha$ where

$$\Lambda_{nm} = [\lambda_{ijnm}],$$

$$\lambda_{ijnm} = \lambda_{ijm}(0) + \sum_{v=1}^{m+q} (1 - v/n)(\lambda_{ijm}(v) + \lambda_{jim}(v))$$

and

$$\lambda_{ijm}(v) = (p_i p_j)^{-1/2}(P(X_{tm} \in B_i, X_{t+v.m} \in B_j) - p_{im} p_{jm}).$$

Therefore, $nV(U_{nm})/n^{1/3} \to \infty$ as $n \to \infty$, and hence by Theorem 7.3.1 in Chung (1974) Eq. (2.19) holds, with $\sigma_{m\alpha}^2 = \alpha^T \Lambda_m \alpha$ where $\Lambda_m = [\lambda_{ijm}]$ and $\lambda_{ijm} = \lambda_{ijm}(0) + \sum_{v=1}^{m+q}(\lambda_{ijm}(v) + \lambda_{jim}(v))$.

LEMMA 2.7   Let the conditions of Lemma 2.5 hold. Then

$$V(U_{nm\alpha}^*) \le M_{m\alpha}, \tag{2.21}$$

where $M_{m\alpha}$ is a finite positive constant which depends on m and α but not on n, and $M_{m\alpha} \to 0$ as $m \to \infty$.

Proof.   We can show by routine analysis that $V(U_{nm\alpha}^*) = \alpha^T \Lambda_{nm}^* \alpha$ where $\Lambda_{nm}^* = [\lambda_{ijnm}^*]$,

$$\lambda_{ijnm}^* = \lambda_{ijm}^*(0) + \sum_{v=-n+1}^{n-1} (1 - v/n)(\lambda_{ijm}^*(v) + \lambda_{jim}^*(v))$$

and $\lambda_{ijm}^*(v) = (p_i p_j)^{-1/2}[(p_{ij}(v) - p_i p_j) - P(X_{im} \in B_i, X_{t+v} \in B_j) + p_{im} p_j]$. Now by using argument similar to that in Lemma 2.6 we conclude that for $v > m + q, |\lambda_{ijm}^*(v)| \le M \sum_{r=v-q}^{\infty} c_r^\lambda$ which in turn implies that

$$|\lambda_{ijnm}^*| \le \sum_{v=m+q+1}^{\infty} |\lambda_{ijm}^*(v)| \le M \sum_{r=m+1}^{\infty} r c_r^\lambda \tag{2.22}$$

which is independent of n and $\to 0$ as $m \to \infty$ by condition on $\{W_{rt}\}$. Therefore, Eq. (2.21) holds.

LEMMA 2.8   Let the conditions of Lemma 2.5 hold. Then

$$\lim_{m \to \infty} \sigma_{m\alpha}^2 = \sigma_\alpha^2.$$

Proof.   Note that

$$\left| \sum_{v=1}^{m+q} \lambda_{ijm}(v) - \sum_{v=1}^{\infty} \lambda_{ij}(v) \right| \le \sum_{v=1}^{m+q} |\lambda_{ijm}(v) - \lambda_{ij}(v)| + \sum_{v=m+q+1}^{\infty} \lambda_{ij}(v).$$

Also using argument similar to that in Lemma 2.6 we can show that for any $v \ge 1$ $|\lambda_{ijm}(v) - \lambda_{ij}(v)| \le M \sum_{r=m+1}^{\infty} c_r^\lambda$. If, now, we apply this result and the inequality in (2.19) we conclude immediately that the result of Lemma 2.8 holds with $\sigma_\alpha^2 = \alpha^T \Lambda \alpha$ where $\Lambda = [\lambda_{ij}]$ and λ_{ij} is as defined in Eq. (2.14).

Proof of Theorem 2.1.   Note that $\alpha^T V_n = U_{nm} + U_{nm}^*$. Therefore, by Lemmas 2.6–2.8 and Corollary 7.7.1 in Anderson (1971) we obtain the relation $\mathcal{L}(\alpha^T V_n) \to \mathcal{N}(0, \sigma_\alpha^2)$ as $n \to \infty$. Since α is arbitrary, the result in (2.10) of Theorem 2.1 follows immediately.

THEOREM 2.9   Let the conditions of Theorem 2.1 hold. Then

$$\mathcal{L}(\chi^2) \to \mathcal{L}\left(\sum_{j=1}^{k-1} \tau_j \chi_j^2\right), \tag{2.23}$$

where $\chi_j^2(1 \le j \le k-1)$ are iid chi-square variables with one degree of freedom, and $\tau_1, \ldots, \tau_{k-1}, 0$ are the eigenvalues of Λ.

Proof.   Note that for any $v \ge 1$,

$$\sum_{i=1}^{k} p_i^{1/2} \lambda_{ij}(v) = p_j^{-1/2} \sum_{i=1}^{k} (p_{ij}(v) - p_i p_j) = 0.$$

Also $\sum_{i=1}^{k} p_i^{1/2} \lambda_{ij}(0) = p_j^{-1/2} \sum_i (p_{ij}(0) - p_i p_j) = 0$ because $p_{ij}(0) = \delta_{ij} p_j$ where δ_{ij} is the Kronecker's delta. Similarly, $\sum_{i=1}^{k} p_i^{1/2} \lambda_{ji}(v) = 0$ for $v \ge 0$. Therefore $\pi^T \Lambda = 0$. In other words rank $(\Lambda) \le k-1$. The result in Eq. (2.23) of Theorem 2.9 follows easily.

Observe that although $\lambda_{ij}(0)(1 \le i, j \le k)$ are known because p_i's are given, the quantities $\lambda_{ij}(v)(v \ne 0, 1 \le i, j \le k)$ are not usually known. If these are given as additional conditions under the hypothesis H_0 then we replace χ^2 by $\chi^{*2} = V_n^T \Lambda^- V_n$ where Λ^- is the (unique) g-inverse of Λ. It can then be shown that

$$\mathcal{L}(\chi^{*2}) \to \mathcal{L}(\chi^2(d)) \tag{2.24}$$

where $\chi^2(d)$ is distributed as a chi-square variable with d degrees of freedom with $d = \operatorname{rank}(\Lambda) \le k-1$.

In general it is difficult to compute λ_{ij} even if analytic expressions for $p_{ij}(v)$ are known. For example, if $\{X_t\}$ is a linear process (Example 1 in this Section) which is Gaussian with zero mean, unit variance and autocorrelation function ρ_v then it is well-known that $p_{ij}(v) = \sum_{u=0}^{\infty} \rho_v^u c_{iu} c_{ju}/u!$ where c_{iu} are easy to calculate. Even if $\{\rho_v\}$ is completely known, λ_{ij} is hard to compute. It is, therefore, desirable to investigate if there exist estimates $\hat{\lambda}_{ij}$ of λ_{ij} which can be used to modify χ^2 so that the new statistic has a simple asymptotic null distribution. A possible construction of such an estimate $\hat{\lambda}_{ij}$ is given below.

Let for $v \ge 0$,

$$\hat{\lambda}_{ij}(v) = (n-v)^{-1} \sum_{t=1}^{n-v} Y_{it} Y_{j,t+v} \tag{2.25}$$

and set $\hat{\lambda}_{ij}(-v) = \hat{\lambda}_{ji}(v)$.

Let $\{\ell_n\}$ be a sequence of positive integers such that $\ell_n \to \infty$ but $\ell_n/n \to 0$ as $n \to \infty$, and define

$$\hat{\lambda}_{ij} = \sum_{v=-\ell_n}^{\ell_n} k(v/\ell_n)\hat{\lambda}_{ij}(v) \tag{2.26}$$

where $k(x) = k(-x), k(0) = 1, |k(x)| \le M$ for all $|x| \le 1$, and $(1 - k(x))/|x| \to 1$ as $x \to 0$. Assume that $\sum_{r=1}^{\infty} r^2 C_r^{\lambda} < \infty$ where $C_r = \sum_{s=r}^{\infty} c_s$. Then

$$\hat{\lambda}_{ij} \to \lambda_{ij} \text{ in probability}, \tag{2.27}$$

as $n \to \infty (1 \le i, j \le k)$.

We conjecture that the proof of Eq. 2.27 can be constructed by using arguments similar to those in Theorems 9.3.3 and 9.3.4 in Anderson (1971).

3. EFFECTS OF ESTIMATION OF PARAMETERS ON Λ

We now investigate different methods of estimating θ and the effects of such estimation on Λ. (We still assume that $F(x,\theta)$ is completely known except for the parameter θ which is unknown).

We assume that the following conditions hold.

C3.   Under $F(x,\theta_0)$, the estimator $\hat{\theta}$ satisfies the relation $\hat{\theta} - \theta_0 = O_p(n^{-1/2})$.

C4.   For all i $(1 \le i \le k)$, $p_i(\theta)$ is continuously differentiable at $\theta = \theta_0$.

Let θ be an m-vector $(m < k)$ of which the ith component is θ_i and let $\mathbf{B}$ be the $k \times m$ matrix $[p_i^{-1/2}\partial p_i/\partial \theta_j]$ where $\partial p_i/\partial \theta_j = \partial p_i(\theta)/\partial \theta_j$ evaluated at $\theta = \theta_0$.

LEMMA 3.1   Let the conditions of Theorem 2.1 and conditions C3 and C4 hold. Then under $F(x,\theta_0)$,

$$\hat{\mathbf{V}}_n = \mathbf{V}_n - \mathbf{B}n^{1/2}(\hat{\theta} - \theta_0) + o_p(1). \tag{3.1}$$

Proof. The result follows easily from the fact that $(N_i - n\hat{p}_i)/(n\hat{p}_i)^{1/2} = (N_i - np_i)/(np_i)^{1/2} - n^{1/2}(\hat{p}_i - p_i)p_i^{-1/2} + o_p(1) = (N_i - np_i)/(np_i)^{1/2} - n^{1/2}\mathbf{b}_i(\hat{\theta} - \theta_0) + o_p(1)$ where $\mathbf{b}_i(1 \le i \le k)$ is the ith row vector of $\mathbf{B}$.

As we shall see later on, most of the estimators $\hat{\theta}$ used for the SSP $\{X_t, t \in \mathbb{Z}\}$ satisfying the relation in Eq. (2.3) above have, asymptotically, the form specified by the following condition.

C5.   Under $F(x, \boldsymbol{\theta}_0)$,

$$n^{1/2}(\hat{\boldsymbol{\theta}} - \boldsymbol{\theta}_0) = n^{-1/2} \sum_{t=p+1}^{n} h(X_t, \ldots, X_{t-p}) + o_p(1), \tag{3.2}$$

for some p which is an integer ≥ 0 and $\mathbf{h}_t := \mathbf{h}(X_t, \ldots, X_{t-p})$ is an m-vector with zero mean and a finite covariance matrix.

If conditions C1–C5 hold then it is easy to show that under $F(x, \boldsymbol{\theta}_0)$

$$\hat{\mathbf{V}}_n = n^{-1/2} \sum_{t=1}^{n-p} (\mathbf{Y}_t - \mathbf{Bh}_t) + o_p(1). \tag{3.3}$$

Now set

$$\bar{T}_{t\alpha} = \boldsymbol{\alpha}^T(\mathbf{Y}_t - \mathbf{Bh}_t) \tag{3.4}$$

where $\boldsymbol{\alpha}$ is defined in Eq. (2.6). Then $\boldsymbol{\alpha}^T \hat{\mathbf{V}}_n = n^{-1/2} \sum_{t=1}^{n-p} \bar{T}_{t\alpha} + o_p(1)$. We can, now, prove

THEOREM 3.2   Let the conditions of Theorem 2.1 hold. Also let conditions C3–C5 hold. Then

$$\mathcal{L}(\hat{\mathbf{V}}_n) \to \mathcal{N}(\mathbf{0}, \boldsymbol{\Lambda}), \tag{3.5}$$

as $n \to \infty$ where $\bar{\boldsymbol{\Lambda}} = [\bar{\lambda}_{ij}] = \bar{\boldsymbol{\Lambda}}(0) + \sum_{v=1}^{\infty}(\bar{\boldsymbol{\Lambda}}(v) + \bar{\boldsymbol{\Lambda}}(-v))$, $\bar{\boldsymbol{\Lambda}}(v) = \boldsymbol{\Lambda}(v) - E(\mathbf{Y}_t \mathbf{h}_{t+v}^T)\mathbf{B}^T - \mathbf{B}E(\mathbf{h}_t \mathbf{Y}_{t+v}^T) + \mathbf{B}E(\mathbf{h}_t \mathbf{h}_{t+v}^T)\mathbf{B}^T$ and $\boldsymbol{\Lambda}(v) = [\lambda_{ij}(v)]$.

Proof.   For any arbitrary positive integer m we define

$$\begin{aligned}\bar{U}_{nm\alpha} &:= n^{-1/2} \sum \bar{T}_{tm\alpha}, \\ \bar{U}^*_{nm\alpha} &:= n^{-1/2} \sum \bar{T}^*_{tm\alpha},\end{aligned} \tag{3.6}$$

where $\bar{T}_{tm\alpha} := \boldsymbol{\alpha}^T(\mathbf{Y}_{tm} - \mathbf{Bh}_{tm})$, $\bar{T}^*_{tm\alpha} := \boldsymbol{\alpha}^T(\mathbf{Y}^*_{tm} - \mathbf{Bh}^*_{tm})\mathbf{h}_{tm} := \mathbf{h}(X_{tm}, \ldots, X_{t-p.m})$, $\mathbf{h}^*_{tm} := \mathbf{h}_t - \mathbf{h}_{tm}$. We also set $\bar{U}_{n\alpha} := \boldsymbol{\alpha}^T \hat{\mathbf{V}}_n$ so that $\bar{U}^*_{nm\alpha} = \bar{U}_{n\alpha} - \bar{U}_{nm\alpha}$. Now use arguments similar to those that lead to the proofs of Lemmas 2.5–2.8 (we need extended versions of Lemmas 2.2–2.4) and the proof of Eq. (3.4) follows.

4.   AN EXAMPLE

In this Section we consider a special case in order to demonstrate how the details of Sects. 2 and 3 can be worked out. Let

$$X_t = \varepsilon_t + b\varepsilon_{t-1}X_{t-2} \tag{4.1}$$

where $\{\varepsilon_t; t \in \mathbb{Z}\}$ is a sequence of iid Gaussian random variables, with

$E(\varepsilon_1) = 0$, $0 < \sigma_\varepsilon^2 = V(\varepsilon_1) < \infty$ and $0 < \pi^2 < 1$ where $\pi = b\sigma_\varepsilon$. It is then easy to show that $X_t = \varepsilon_t + \sum_{r=1}^{\infty} W_{rt}$ where $W_{rt} = b^r \varepsilon_{t-1} \varepsilon_{t-3} \cdots \varepsilon_{t-2r+1} \varepsilon_{t-2r}$ $(r \geq 1)$ and for every $\gamma \in (0, 1]$, $E|W_{rt}|^\gamma \leq M b^{r\gamma} \nu_\gamma^r \leq M|\pi|^{r\gamma}$ where $\nu_\gamma = E|\varepsilon_1|^\gamma$, so that we can take $c_r = |\pi|^r$, and observe that $\sum_{r=1}^{\infty} r c_r^\lambda < \infty$. Let $\boldsymbol{\theta} = (b, \sigma_\varepsilon)$ and if $F(x, \boldsymbol{\theta})$ denote the d.f. of X_1 then one can easily show that condition C2 holds and the density function $f(x, \theta)$ of X_1 is given as a solution of the equation

$$f(x, \theta) = \int \phi(x/q(y))(q(y))^{-1} f(y, \theta) \, dy \qquad (4.2)$$

where $\phi(x) = (2\pi)^{-1/2} \int_{-\infty}^{x} \exp(-y^2/2) \, dy$ and $q(y) = \sigma_\varepsilon \sqrt{1 + b^2 y^2}$. Then following Grahn (1995) we can estimate $\beta_{11} := \sigma_\varepsilon^2$, $\beta_{12} := b^2 \sigma_\varepsilon^2 := \pi^2$ and $\beta_2 := b\sigma_\varepsilon^2$ by minimizing $\sum_{t=3}^{n} (X_t^2 - \beta_{11} - \beta_{12} X_{t-2}^2)^2$ and $\sum_{t=3}^{n} (X_t X_{t-1} - \beta_2 X_{t-2})^2$. If we set $\hat{\boldsymbol{\theta}} = (\hat{b}, \hat{\sigma}_\varepsilon^2)$, $\boldsymbol{\theta}_0 = (b_0, \sigma_\varepsilon^{0^2})$ then $\hat{b} = \hat{\beta}_2/\hat{\beta}_{11}$, and $\hat{\sigma}_\varepsilon^2 = \hat{\beta}_{11}$, where $\hat{\beta}_{11}, \hat{\beta}_2$ are the least squares estimators defined above. Some routine but tedious calculations will then show that

$$n^{1/2}(\hat{\boldsymbol{\theta}} - \boldsymbol{\theta}_0) = n^{-1/2} \sum_{t=3}^{n} \mathbf{h}(X_t, X_{t-1}, X_{t-2}) + o_p(1),$$

where $\mathbf{h}$ is a 2×1 vector with the first element

$$= [(1 - \pi_0^2)(X_t X_{t-1} - \beta_2^0 X_{t-2}) X_{t-2}/\sigma_\varepsilon^{0^2} - b_0(X_t^2 - \beta_{11}^0 - \beta_{12}^0 X_{t-2}^2)$$
$$\times (\mu_4 - \mu_2 X_{t-2}^2)(1 - \pi_0^2)^2(1 - 3\pi_0^4)/2\sigma_\varepsilon^4]/\beta_{11}^0,$$

and the second element $= (1 - \pi_0^2)(X_t X_{t-1} - \beta_2^0 X_{t-2}) X_{t-2}/\sigma_\varepsilon^{0^2}$ where $\mu_s = E(X_1^s)$. It is easy to express μ_s in terms of b^0 and $\sigma_\varepsilon^{0^2}$. In fact $\mu_2 = \sigma_\varepsilon^{0^2}/(1 - \pi_0^2)$ and $\mu_4 = 3\sigma_\varepsilon^{0^4}(1 + \pi_0^2)/(1 - \pi_0^2)$, $\pi_0 = b_0 \sigma_\varepsilon^0$.

REFERENCES

Anderson, T. W. The Statistical Analysis of Time Series. New York: John Wiley & Sons, Inc., 1971.

Bartlett, M. S. The frequency goodness-of-fit test for probability chains. *Proc. Camb. Phil. Soc.* 47: 86–95, 1951.

Bhaskara Rao, M., Subba Rao T., and Walker, A. M. On the existence of some bilinear time series models. *J. Time. Ser. Anal.* 4: 95–110, 1983.

Billingsley, Patrick. Statistical Inference for Markov Processes. Chicago and London: The University of Chicago Press, 1961.

Chanda, Kamal C. Chi-square goodness-of-fit tests based on dependent observations. In: C. Taillie et al. (eds). Statistical Distributions in Scientific Work, 5. Dordrecht, Holland: D. Reidel Publishing Company, 1981, pp 35–49.

Chanda, Kamal C. Stationarity and central limit theorem associated with bilinear time series models. *J. Time Ser. Anal. 12*: 301–313, 1991.

Chung, Kai Lai. A Course on Probability Theory. (2nd edition). New York: Academic Press, 1974.

Gasser, T. Goodness-of-fit tests for correlated data. *Biometrika 62*: 563–570, 1975.

Gastwirth, J. L. and Rubin, H. The asymptotic distribution theory of the empiric c.d.f. for mixing processes. *Ann. Statist. 3*: 809–824, 1975.

Gleser, Leon J. and Moore, David S. The effect of dependence on chi-squared and empiric distribution tests of fit. *Ann. Statist. 4*: 1100–1133, 1983.

Goodman, L. A. Asymptotic distributions of "psi-squared" goodness of fit criteria for mth order Markov chains. *Ann. Math. Statist. 29*: 1123–1133, 1958.

Grahn, T. A conditional least squares approach to bilinear time series estimation. *J. Time Ser. Anal. 16*: 509–529, 1995.

Moore, David S. The effect of dependence on chi-squared tests of fit. *Ann. Statist. 10*: 1163–1171, 1982.

Withers, C. S. Convergence of empirical processes of mixing rv's on [0, 1]. *Ann. Statist. 3*: 1101–1108, 1975.

23

Positive and Negative Dependence with Some Statistical Applications

GEORGE G. ROUSSAS University of California, Davis, California

1. INTRODUCTION

The purpose of this chapter is to provide a selective, as opposed to an exhaustive, review of some modes of positive and negative dependence, and also to describe certain statistical applications.

The probabilistic theory of positive and negative dependence has been developed to a remarkable degree over the last few years, but statistical inference is lagging behind. Positive and negative dependence, in their different versions, have found wide applicability in reliability, statistical mechanics, probability/stochastic processes, and statistics. Many more applications are to be anticipated in a host of other areas and, in particular, those areas where spatial statistics play an important role. Such areas are, for example, analysis of agricultural field experiments, geostatistical analysis, image analysis, oceanographic applications, signal processing in radar and sonars, and stereology. It is expected that results on nonparametric functional estimation will significantly enhance the applicability of

This work was supported in part by a research grant from the University of California, Davis.

positive and negative dependence models. An introductory snapshot for such applications can be found in the report Panel on Spatial Statistics and Image Processing (1991).

There are several versions of positive and negative dependence which arose in connection with applications in systems reliability, statistical mechanics, percolation models in probability/stochastic processes, and statistics. It appears that positive and negative association are the modes of positive and negative dependence which have found significant applicability in the areas just cited. The papers in the review section of this work provide a fair picture of the state of the art.

This paper is organized in six Sections as follows. Following the Introduction, a Section is devoted to a review of some past work. It is not claimed that this review is exhaustive; rather, it reflects the interests of this author in this particular line of research. In Sec. 3, the required definitions are given, and also relations or lack thereof between some of the concepts defined are discussed. The Central Limit Theorem for positively or negatively associated random variables is presented in the following Section. Actually, the proof of the theorem itself is not given, but the underlying basic ideas and facts are outlined to a considerable extent. Moment inequalities are important in their own right and indispensable in estimation. Such moment inequalities are discussed in Sec. 5, and this discussion is followed by the presentation of several estimation problems. The last Section of the chapter contains some recent results on efficient estimation of a smooth distribution function, quantile estimation, Berry–Essen bounds, weak convergence of certain stochastic processes, and the Kaplan–Meier estimate. A basic assumption throughout is that the underlying random variables are either positively or negatively associated. This assumption may be relaxed to positive or negative quadrant dependence when the interest lies only in the efficient estimation of a distribution function by a smooth kernel-type estimate.

Finally, it should be mentioned that all limits are taken as $n \to \infty$ whether this is explicitly mentioned or not.

2. A BRIEF REVIEW OF PAST WORK

The stochastic models referred to below consist primarily of discrete time-parameter stochastic processes, exhibiting some kind of multivariate positive or negative dependence, including the special but most important cases of (positive or negative) association. In some instances, the index is a multi-dimensional parameter, in which cases we are dealing with random fields.

Positive and negative association models and their variations have found significant applicability in many diverse areas, including systems reliability, statistical mechanics, probability/stochastic processes, and statistics. Undoubtedly, further applications will be found in the future. Future applications are anticipated to be in atmospheric, geological and oceanographic fields, as well as certain biological and ecological problems, in particular, in connection with models where the stipulation of negative dependence seems to be natural.

It appears that the first paper on positive dependence was that of Harris (1960), where the author established a lower bound for critical probabilities in the context of a certain percolation model. The conditions expressing this bound (see Lemma 4.1 and its Corollary) is a case of positive association. The next paper, dealing in a systematic manner with some modes of positive and negative dependence in a bivariate framework, was Lehmann's (1966) paper. For a pair of random variables, the author introduced several kinds of positive and negative dependence, established various properties, and applied his results to some statistical situations, such as the slippage problem, regression, and hypotheses testing problems. However, positive or negative dependence, formulated in a bivariate context, is not sufficient for obtaining significant results and enlarging the scope of applications. This apparent weakness was rectified to a large extent by a truly seminal paper, that of Esary, Proschan and Walkup (1967). These authors introduced the notion of association—to be termed hereafter positive association—derived an abundance of basic properties, discussed equivalent characterizations and special cases of particular interest, and presented several applications in probability and statistics.

Several years later, Fortuin, Kasteleyn and Ginibre (1971) arrived at the concept of positive association in an entirely different context; namely, in studying Ising spin systems, and certain percolation and random-cluster models in statistical mechanics. Their basic results (see Propositions 1 and $1'$ on pp. 91 and 95) are expressions of positive association, and they are now referred to in the literature as FKG-inequalities.

Since 1971, there has been a significant output in the relevant literature. The papers Lebowitz (1972) and Simon (1973) are wholly motivated by statistical mechanics problems. The first author studies systems in statistical mechanics whose Hamiltonians satisfy the FKG-inequalities. An important class of such systems is the class of Ising spin systems with ferromagnetic pair interactions and Hamiltonians of a certain form. The second author utilized FKG-inequalities and FKG-type inequalities in discussing Gibbs states. Preston (1974) emphasized the fundamental role that the FKG-inequalities play in the study of correlation functions for the Ising and other lattice models in statistical mechanics, and proved generalized versions of these inequalities. A systematic study of FKG-type inequalities, in a general mathematical setting and mostly from probabilistic viewpoint, was undertaken by Kemperman (1977).

The implications of positive dependence in systems reliability were investigated thoroughly by Barlow and Proschan (1975) in their classical book on reliability. For the employment of positive association in deriving bounds on a system reliability (see Secs. 2 and 3, pp. 29–39, and Sec. 4, pp. 142–152 in the reference just cited). Since that time, there have been quite a few papers focusing on positive dependence, in general, and positive association in particular. Among these works, Newman's (1980) paper occupies a special place. The motivation of this paper stemmed from statistical mechanics, but the results obtained are probabilistic in nature and very rigorous. Indeed, these results have been heralded as one of the noteworthy advances in probability in recent years. More specifically, Newman (1980) considered a strictly stationary random field of positively associated random variables with finite variance and susceptibility, and established weak convergence of partial sums, suitably normalized (see Theorem 2 on p. 122). The proof of this result was facilitated by Theorem 1 on p. 121, which is of independent interest and usefulness.

Various aspects of positive dependence were addressed in the papers by Block and Ting (1981), and Newman and Wright (1981, 1982). The latter authors established an invariance principle for second-order positively associated random variables and also some classical submartingale inequalities for sequences of random variables with positively associated increments. Rüschendorf (1981) established equivalence between two kinds of positive dependence, while Shaked (1982) explained how reliability bounds for coherent structures may be derived under a positive dependence condition, which is weaker than positive association. Other contributions are those by Pitt (1982), Joag-Dev, Perlman and Pitt (1983), and Wood (1983), who proved a Berry–Essen theorem under positive association. Cox and Grimmett (1984) established a central limit theorem for random fields of positively associated random variables subject to certain conditions on their moments and covariances, without requiring stationarity. Their result is further appropriately modified to fit into a certain percolation model. Burton and Waymire (1985) obtained an extension of Newman's (1980) basic result to certain stationary positively associated random measures. Burton, Dabrowski and Dehling (1986) defined the notion of weak association and established a weak convergence result under such a condition. In a series of papers, Birkel (1988a,b, 1989) established moment inequalities for partial sums of positively associated random variables, elaborated on the convergence rate in the central limit theorem for such random variables, and also established a law of large numbers. Additional contributions are those by Glaz (1990), who proved Bonferroni-type and product-type inequalities in the presence of dependence; Newman (1990); Subramanyan (1990), who studied positive quadrant dependence in 3-dimensions and compared its features with those of the 2-dimensional case; and Wright (1990), who obtained certain results pertaining to stochastic processes with positively associated increments.

Yu (1993) gave a comprehensive account of weak convergence for empirical and quantile processes under positive association, and Cai (1995) undertook a thorough investigation of a host of problems under the same mode of dependence, replacing the empirical distribution function by a a smooth version of it. Jogdeo (1978) has given conditions under which an m-dimensional positively associated random vector W and on n-dimensional random vector X form an $(n + m)$-dimensional positively associated random vector (X, W). Langberg, Proschan and Quinzi (1978) provide necessary and sufficient conditions under which a given set of n random variables (r.v.'s) $T_1, \ldots, T_n$, representing life lengths, can be essentially replaced by a set of independent r.v.'s. The dependence of $T_1, \ldots, T_n$ is not specified, and it may be positive association. The paper by Karlin and Rinott (1980a) is concerned primarily with the concept of multivariate total positivity of order two, which implies positive association (see Theorem 4.2 on p. 484), and which, according to the authors, is easier to verify than positive association. They also derive a wealth of results and discuss some applications in statistics. Lindqvist (1988) has extended the definition of positive association to probability measures defined on a partially ordered Polish space, and has given a number of equivalent characterizations of this property. Also, he has derived an abundance of relevant results, and has shown how the classical basic results in Esary, Proschan and Walkup (1967) carry over to this more general case.

Bagai and Prakasa Rao (1995) study positively associated r.v.'s and derive a Strong Law of Large Numbers, and a version of Birkel's (1988a) moment inequality. Then they apply these results to obtain consistency of a kernel estimate of a probability density function (p.d.f.) and of the associated failure rate. More explicitly, let f_n be certain real-valued functions defined on $\Re$ such that $\mathscr{E}f_n(X_1) = 0$ and $\mathrm{Var}[f_n(X_1)] < \infty$, and set $S_{n,n} = \sum_{j=1}^{n} f_n(X_j)$. Then, under suitable conditions, $S_{n,n}/n \to 0$ a.s.

With respect to negative dependence, there are the papers by Ebrahimi and Ghosh (1981), where they argue (see Sec. 5) that negative dependence, holding in many common multivariate distributions, has an impact on certain reliability problems. Block, Savits and Shaked (1982) undertook a systematic study of negative dependence by introducing certain suitably motivated kinds of such dependences and studying their interrelationships and their relation to other concepts of negative dependence. They also showed that many multivariate distributions satisfy one of their basic modes of negative dependence. Joag-Dev and Proschan (1983) introduced the notion of negative association, compared it with other proposed concepts of negative dependence, and justified their claim that negative association possesses certain advantages over competing notions of negative dependence. They emphasized that negative association is not simply dual to that of positive association, but it differs in important respects. The authors also derived as a byproduct of their main results that many well-known multivariate distributions (notably, multinomial, convolution of unlike multinomials, multivariate hypergeometric, Dirichlet, Dirichlet compound multinomial) are negatively associated. Finally, the authors discussed a situation (contingency tables), where both positive and negative association exist. The papers of Joag-Dev (1983), Newman (1984), Glaz (1990), and Hollander, Proschan and Sconing (1990) are concerned with both positive and negative dependence, and Samuel-Cahn (1991) established a so-called prophet inequality under a certain negative dependence condition. Karlin and Rinott (1980b) introduced the concepts of multivariate reverse rule of order two and strong multivariate reverse rule of order two, derived closure results, established several inequalities of statistical interest, and concluded that the same distributions cited above satisfy the property of being strongly multivariate reverse of order two. Finally, an early paper, Brindley and Thompson (1972), dealing with negative dependence among other things, points out such a kind of dependence as a suitable model for certain biological and ecological situations concerning the coexistence of various competing species.

Applications of positive and negative dependence in reliability have already been mentioned, and certain representative references have been cited. Also, a relatively extensive review of the usage of positive association in statistical mechanics has been made, in particular, as it pertains to Ising spin systems with ferromagnetic pair interactions and Hamiltonians of a specific type, as well as other random-cluster models, Gibbs states, and boson field fluctuations in Yukawa quantum field theory models. These subjects are fully covered by the references given. Finally, various applications of positive and negative dependence, and in particular, association, in probability/stochastic processes have been exemplified by a number of references briefly discussed. Asymptotic normality in a random field framework and under either positive or negative association is discussed in Roussas (1994). Furthermore, a Hoeffding-type inequality for negatively associated random variables is presented in Roussas (1996).

Applications of positive or negative dependence in statistics may be classified broadly into two categories. In one category are papers which have focused primarily on the following aspects: in establishing that many of the common multivariate distributions fall into such classes (Karlin and Rinott

(1980a,b), Ebrahimi and Ghosh (1981), Block, Savits and Shaked (1982), Pitt (1982), Joag-Dev and Proschan (1983), and Joag-Dev, Perlman and Pitt (1983)); in concluding independence of positively associated random variables which are uncorrelated, or in attaining asymptotic independence for positively associated random variables, see, for example, Newman (1980, 1984), Joag-Dev (1983); in studying the behavior of information and censorship under kinds of dependence, which include positive association, Hollander, Proschan and Sconing (1990); in establishing that the conditional distribution of certain order statistics possesses some kind of negative dependence, Joag-Dev (1990). The second class consists of papers concerned primarily with inference issues. The first paper in this class seems to be that by Lehmann (1966). The papers by Bagai and Prakasa Rao (1991, 1995) concern themselves with estimation of the survival function, and kernel-type density and failure rate estimation under positive association. Kernel estimates under positive association is also the subject of the paper Roussas (1991). A variation of this problem in a random field framework is discussed in Roussas (1993). In Cohen and Sackrowitz (1992), the authors introduce a new mode of positive dependence and compare it with other existing kinds of positive dependence. They puruse their work in Cohen and Sackrowitz (1994). We will provide additional comments on these papers in a subsequent Section.

In Roussas (1995), the asymptotic normality of a smooth estimate of a distribution function is established for random fields and under either positive or negative association. Finally, Cai and Roussas (1998) study the behavior of the above mentioned smooth estimate of a distribution function under quadrant dependence in a different direction, namely, from asymptotic efficiency/deficiency viewpoint. Some further results on the estimation of quantiles and the Kaplan–Meier estimate are presented in Sec. 6. The paper, Granger (1993), refers to time series with applications in economics and finance. In this paper, it is argued among other things, that if a series X_t is a Gaussian I(1) series which started long enough ago, then the random variables $X_t, \ldots, X_{t-k}$ are positively associated. Observations as this one may be exploited profitably in a time series framework. The monograph Szekli (1995) deals with stochastic ordering and dependence, and devotes one section to each, positive association and negative association (Secs. 3.1 and 3.5, respectively). Finally, in the ninth volume of his monumental work, Yoshihara (1997), the author devotes two long chapters (Chap 4 and 5) to association of r.v.'s and measures (both positive and negative), and applications. In the wealth of information provided, attention is drawn, in particular, to Secs. 4.1.1, 4.1.2, 4.1.3, 4.2.1, 4.2.2, 4.5.1, 4.5.2 and 5.3.1, where most commonly used material is discussed.

3. DEFINITIONS AND SOME INTERRELATIONSHIPS OF DEPENDENCES

As has been mentioned in Sec. 2, there have been many versions of positive and negative dependence proposed in the literature. Some of these dependences are stronger than others, whereas others are not comparable, in the sense that one of them does not imply another. Below, we define some of the most popular modes of positive and negative dependence found in the literature. In these definitions, all r.v.'s to be considered are real-valued and are defined on an underlying probability space $(\Omega, \mathscr{A}, P)$. Although the definitions are given in terms of r.v.'s, they also hold for the respective distributions.

DEFINITION 3.1 Two r.v.'s X_1 and X_2 are said to be positively quadrant dependent (PQD), if

$$P(X_1 > x_1, X_2 > x_2) - P(X_1 > x_1)P(X_2 > x_2) \geq 0 \quad \text{for all } x_1, x_2 \in \Re;$$

(3.1)

and negatively quadrant dependent (NQD), if the "$\geq$" in Eq. (3.1) is reversed.

DEFINITION 3.2 For any index set I, the r.v.'s $\{X_i; i \in I\}$ are said to be pairwise positively quadrant dependent (PPQD), if

$$P(X_i > x_i, X_j > x_j) - P(X_i > x_i)P(X_j > x_j) \geq 0 \quad \text{for all } i, j \in I,$$

(3.2)

with $i \neq j$ and all $x_i, x_j \in \Re$; and they are said to be pairwise negatively quadrant dependent (PNQD), if the "$\geq$" in Eq. (3.2) is reversed.

DEFINITION 3.3. The r.v.'s $X_1, \ldots, X_n$ are said to be positively upper orthant dependent (PUOD), if, for every $c_i \in \Re$, $i = 1, \ldots, n$,

$$P(X_i \geq c_i, \quad i = 1, \ldots, n) \geq \prod_{i=1}^{n} P(X_i \geq c_i);$$

(3.3)

and they are said to be negatively upper orthant dependent (NUOD), if the "$\geq$" in Eq. (3.3) is reversed. The r.v.'s are said to be positively lower orthant dependent (PLOD), if

$$P(X_i \leq c_i, \quad i = 1, \ldots, n) \geq \prod_{i=1}^{n} P(X_i \leq c_i);$$

(3.4)

and they are said to be negatively lower orthant dependent (NLOD), if the "$\geq$" in Eq. (3.4) is reversed. The r.v.'s are said to be positively (negatively) orthant dependent (POD, NOD), if they are both PUOD(NUOD) and PLOD, (NLOD).

DEFINITION 3.4   For any finite index set I, the r.v.'s $\{X_i; i \in I\}$ are said to be linearly positively quadrant dependent (LPQD), if for any nonempty and disjoint subsets A and B of I and for any positive λ_i's, the r.v.'s $\sum_{i \in A} \lambda_i X_i$ and $\sum_{j \in B} \lambda_j X_j$ are PQD; and they are said to be linearly negatively quadrant dependent (LNQD), if the r.v.'s $\sum_{i \in A} \lambda_i X_i$ and $\sum_{j \in B} \lambda_j X_j$ are NQD.

DEFINITION 3.5   For a finite index set I, the r.v.'s $\{X_i; i \in I\}$ are said to be positively associated (PA), if for any real-valued coordinatewise increasing functions G and H defined on $\Re^I$,

$$\text{Cov}\left[G(X_i, i \in I), \quad H(X_j, j \in I)\right] \geq 0, \tag{3.5}$$

provided $\mathscr{E}G^2(X_i, i \in I) < \infty$, $\mathscr{E}H^2(X_j, j \in I) < \infty$. These r.v.'s are said to be negatively associated (NA), if for any nonempty and disjoint subsets A and B of I, and any coordinatewise increasing functions G and H with $G : \Re^A \to \Re$ and $H : \Re^B \to \Re$ with $\mathscr{E}G^2(X_i, i \in A) < \infty$, $\mathscr{E}H^2(X_j, j \in B) < \infty$,

$$\text{Cov}\left[G(X_i, i \in A), \quad H(X_j, j \in B)\right] \leq 0. \tag{3.6}$$

If I is not finite, the r.v.'s $\{X_i; i \in I\}$ are said to be PA or NA, if any finite subcollection is a set of PA or NA r.v.'s, respectively.

For $i = 1, \ldots, n$, let $\mathcal{X}_i$ be a totally ordered set and let $\mathcal{X} = \mathcal{X}_1 \times \cdots \times \mathcal{X}_n$ be a lattice under the operations $\wedge$ and $\vee$, where, for $\mathbf{x} = (x_1, \ldots, x_n)$ and $\mathbf{y} = (y_1, \ldots, y_n)$, $\mathbf{x}, \mathbf{y} \in \mathcal{X}$, $\mathbf{x} \vee \mathbf{y} = (\max(x_i, y_i), i = 1, \ldots, n)$, $\mathbf{x} \wedge \mathbf{y} = (\min(x_i, y_i), i = 1, \ldots, n)$. It is also assumed that the random vector $\mathbf{X}$ referred to below has a p.d.f. f. Then:

DEFINITION 3.6   The random vector $\mathbf{X}$ is said to be multivariate totally positive of order 2 (MTP$_2$), if its p.d.f. is MTP$_2$; namely,

$$f(\mathbf{x} \vee \mathbf{y}) f(\mathbf{x} \wedge \mathbf{y}) \geq f(\mathbf{x}) f(\mathbf{y}) \text{ for every } \mathbf{x}, \mathbf{y} \in \mathcal{X}; \tag{3.7}$$

The random vector $\mathbf{X}$ is said to be multivariate reverse rule of order 2 (MRR$_2$), if the "$\geq$" in Eq. (3.7) is reversed.

REMARK 3.1

(i)   Relation (3.1) is equivalent to

$$P(X_1 \leq x_1, X_2 \leq x_2) - P(X_1 \leq x_1)P(X_2 \leq x_2) \geq 0.$$

(ii) From the definitions, if follows that either PA (NA) or LPQD (LNQD) r.v.'s are PPQD (PNQD). The converse is not true. Also LPQD (LNQD) does not imply PA (NA) (see Examples 3.1 and 3.2 below).

(iii) It is seen that MTP_2 (MRR_2) implies PA(NA).

(iv) It should be pointed out that, by the continuity of the probability, the ">" in the definitions can be consistently replaced by "$\geq$", when applicable.

The concept of PQD (NQD) was introduced by Lehmann (1966); the concept of LPQD (LNQD) was introduced and discussed by Newman (1984), whereas the concepts of PA (or just association in the original terminology) and NA are due to Esary, Proschan and Walkup (1967) and Joag-Dev and Proschan (1983), respectively. Many other modes of negative dependence have been defined and studied by Ebrahimi and Ghosh (1981), Block, Savits and Shaked (1982) and Joag-Dev (1990).

The following two examples are taken from Joag-Dev (1983) (see pp. 1038–1039), and they are meant to demonstrate the fact that pairwise quadrant dependence does not imply either linear quadrant dependence or association, and also that linear quadrant dependence does not imply association.

EXAMPLE 3.1   The r.v.'s X_1, X_2, X_3 take on the values $0, 1, 2$ each with the following joint probabilities:

Values	Probabilities	Values	Probabilities	Values	Probabilities
(0,0,0)	3/14	(0,0,1)	0	(0,0,2)	0
(0,1,0)	0	(0,1,1)	1/14	(0,1,2)	0
(0,2,0)	1/14	(0,2,1)	2/14	(0,2,2)	0
(1,0,0)	2/14	(1,0,1)	1/14	(1,0,2)	0
(1,1,0)	1/14	(1,1,1)	0	(1,1,2)	0
(1,2,0)	0	(1,2,1)	3/14	(1,2,2)	0
(2,0,0)	0	(2,0,1)	0	(2,0,2)	0
(2,1,0)	0	(2,1,1)	0	(2,1,2)	0
(2,2,0)	0	(2,2,1)	0	(2,2,2)	0

It is checked that the r.v.'s X_1, X_2, X_3 are PPQD. However, they are not LPQD. Indeed,

$$P(X_1 \geq 1, X_2 + X_3 \geq 2) = \frac{3}{14}, \quad P(X_1 \geq 1) = \frac{1}{2}, \quad P(X_2 + X_3 \geq 2) = \frac{1}{2},$$

so that $P(X_1 \geq 1, X_2 + X_3 \geq 2) - P(X_1 \geq 1)P(X_2 + X_3 \geq 2) = -\frac{1}{28}$.

The same example shows that X_1, X_2, X_3 are not PA, because, by taking $G(x_1, x_2, x_3) = I(x_1 \geq 1)$, $H(x_1, x_2, x_3) = I(x_2 + x_3 \geq 2)$, we have that G and H are coordinatewise increasing, but $\text{Cov}[G(X_1, X_2, X_3), H(X_1, X_2, X_3)] = -\frac{1}{28}$ as above.

EXAMPLE 3.2 The r.v.'s X_1, X_2, X_3 take on the values $0, 1, 2$ each with the following joint probabilities:

Values	Probabilities	Values	Probabilities	Values	Probabilities
(0,0,0)	84/360	(0,0,1)	16/240	(0,0,2)	0
(0,1,0)	18/360	(0,1,1)	12/240	(0,1,2)	0
(0,2,0)	9/360	(0,2,1)	12/240	(0,2,2)	0
(1,0,0)	18/360	(1,0,1)	12/240	(1,0,2)	0
(1,1,0)	18/360	(1,1,1)	0	(1,1,2)	0
(1,2,0)	15/360	(1,2,1)	8/240	(1,2,2)	0
(2,0,0)	9/360	(2,0,1)	12/240	(2,0,2)	0
(2,1,0)	15/360	(2,1,1)	8/240	(2,1,2)	0
(2,2,0)	30/360	(2,2,1)	16/240	(2,2,2)	0

It is argued that the r.v.'s X_1, X_2, X_3 are LPQD. However, they are not PA. Indeed, let $G(x_1, x_2, x_3) = I(x_3 \geq 1)$, $H(x_1, x_2, x_3) = I(x_1 \geq 1, x_2 \geq 1)$, so that G and H are coordinatewise increasing, but

$$\text{Cov}[G(X_1, X_2, X_3), H(X_1, X_2, X_3)] = P(X_1 \geq 1, X_2 \geq 1, X_3 \geq 1)$$
$$- P(X_3 \geq 1)P(X_1 \geq 1, X_2 \geq 1) = \frac{2}{15} - \frac{7}{50} = -\frac{1}{150}.$$

Applications of PA r.v.'s have been mentioned extensively in Sec. 2. This Section is concluded with some examples and applications of NA r.v.'s. This material can be found in Joag-Dev and Proschan (1983), and theorems and other results cited below refer to this paper. To start with, let us recall that the p.d.f. f of a real-valued r.v. X is said to be a Polya frequency function of order 2 (PF_2), if, for all $x_1 \leq x_2$ and $y_1 \leq y_2$,

$$\begin{vmatrix} f(x_1 - y_1) & f(x_1 - y_2) \\ f(x_2 - y_1) & f(x_2 - y_2) \end{vmatrix} \geq 0.$$

The p.d.f. is PF_2, if $\log f$ is concave. Many of the common p.d.f.'s are PF_2. In particular, so is the p.d.f. of $N(\mu, \sigma^2)$ for all $\sigma > 0$, see Efron (1965).

Fact 3.1. (Theorem 2.8 of Joag-Dev and Proschan, 1983). If $X_1, \ldots, X_k$ are independent r.v.'s with PF_2 p.d.f.'s, then the joint conditional distribution of $X_1, \ldots, X_k$, given $\sum_{i=1}^{k} X_i$, is NA (a.s.).

DEFINITION 3.7. The distribution of the vector $\mathbf{X} = (X_1, \ldots, X_k)$ is a permutation distribution, if $\mathbf{X}$ takes as values all $k!$ permutations of $\mathbf{x} = (x_1, \ldots, x_k)$ with probability $1/k!$, where $x_1, \ldots, x_k (k > 1)$ are real numbers.

Fact 3.2 (Theorem 2.11 of Joag-Dev and Proschan, 1983). A permutation distribution is NA.

Application 3.1. (Sections 3.2(a) and 3.1(c) in Joag-Dev and Proschan, 1983). For $n \leq N$, let $X_1, \ldots, X_n$ be a random sample without replacement from a finite population consisting of the values $x_1, \ldots, x_N$. Then $X_1, \ldots, X_n$ are NA, being a subset of $X_1, \ldots, X_N$, which has a permutation distribution and hence is NA. In particular, the multivariate hypergeometric distribution is NA.

Application 3.2. (Sections 3.2(b), (c) in Joag-Dev and Proschan, 1983). Let $X_1, \ldots, X_k$ be a random sample from a population, and let R_i be the rank of X_i, $i = 1, \ldots, k$. Then the vector $(R_1, \ldots, R_n)$ has a permutation distribution and hence is NA.

As a concrete application of this last result to a selection procedure, suppose that m judges independently rank n individuals, and that several "good" candidates are to be selected. The criterion of goodness requires that the rank R_i is at least as large as a constant $c(m, n)$. Then

$$P[R_i \geq c(m, n), \, i = 1, \ldots, k] \leq \prod_{i=1}^{k} P[R_i \geq c(m, n)] \tag{3.8}$$

Calculating the exact probability on the left-hand side in Eq. (3.8) may be an exceedingly complicated operation. Inequality (3.8) provides an upper bound for the required probability, as the calculation of the probabilities on the right-hand side is usually much easier to carry out.

Fact 3.3. (Section 3.1(a) in Joag-Dev and Proschan, 1983). If the r.v.'s $X_1, \ldots, X_k$ are jointly multinomially distributed, then they are NA.

Fact 3.4. (Section 3.4 in Joag-Dev and Proschan, 1983). If the r.v.'s $X_1, \ldots, X_k$ are jointly normally distributed and are negatively correlated, then they are NA.

Fact 3.5. (Section 3.5 A_1 in Joag-Dev and Proschan, 1983). Consider a $k \times r$ contingency table with fixed marginals. Then the conditional (marginal) distribution of each column (row) vector, given that the column (row) total is fixed, is NA.

In the paper Cohen and Sackrowitz (1992) cited in Sec. 2, the authors built on the concept of MTP_2 (see Definition 3.6) and recall the following two definitions before they proceed with the introduction of a weaker concept. Namely, the n-dimensional random vector $\mathbf{Y}$ is said to be stochastically greater than or equal to the random vector $\mathbf{X}$, denoted by $\mathbf{X} \leq^P \mathbf{Y}$, if $\mathscr{E}h(\mathbf{X}) \leq \mathscr{E}h(\mathbf{Y})$ for all nondecreasing real-valued functions h defined on $\mathfrak{R}^n$ for which the expectations exist. The random vector $\mathbf{X}$ is said to be conditionally increasing in sequence (CIS), if for $j = 1, \ldots, n-1$, $[X_{j+1}|X_1 = x_1, \ldots, X_j = x_j] \leq^P [X_{j+1}|X_1 = x_1^*, \ldots, X_j = x_j^*]$ for $x_i \leq x_i^*$, $i = 1, \ldots, j$. Finally, the r.v.'s $X_1, \ldots, X_n$ are said to be weak conditionally in sequence (WCIS), if for $j = 1, \ldots, n-1$, $[X_{j+1}, \ldots, X_n|X_1 = x_1, \ldots, X_j = x_j] \leq^P [X_{j+1}, \ldots, X_n|X_1 = x_1, \ldots, X_j = x_j^*]$ for $x_j \leq x_j^*$. The authors point out that MTP_2 implies CIS, which, in turn, implies WCIS; finally, WCIS implies PA, and all these implications are strict. This is demonstrated by concrete examples presented, and one example referred to in Tong (1980). The fact that WCIS implies PA is to be interpreted as follows: If there is a permutation $\{j_1, \ldots, j_n\}$ of $\{1, 2, \ldots, n\}$ for which $X_{j_1}, \ldots, X_{j_n}$ are WCIS, then $X_1, \ldots, X_n$ are PA. The side benefit that PA is implied by any one of MTP_2 or CIS or WCIS is that any one of them may serve as a means of verifying PA. The concept of PA, as defined, is rather intractable analytically, whereas any one of the concepts MTP_2 or CIS or WCIS is much more tractable analytically. The authors point out that, with regard to statistical applications, PA is frequently a tool which can often be used in establishing unbiasedness of classes of test. They pursue work along these lines in Cohen and Sackrowitz (1994).

4. THE CENTRAL LIMIT THEOREM UNDER POSITIVE OR NEGATIVE ASSOCIATION

As already mentioned in Sec. 2, the Central Limit Theorem (CLT) for PA r.v.'s was obtained by Newman (1980) in a random field framework. Variations of it are discussed in Cox and Grimmett (1984) and Roussas (1994). Rates of convergence in the CIT for a random field of PA r.v.'s were obtained by Bulinskii (1995). An outline of this theorem will be presented below, both for the PA and the NA case. The notation to be used is that employed in Roussas (1994). In the process, it will be pointed out how the proof of the method of proving the CLT under association differs from the familiar process of establishing such a result. Also, some auxiliary results, interesting on their own right, will be stated.

For a positive integer d, let $\mathscr{Z}^d$ be the lattice of all points in $\Re^d$ having integer coordinates, and for each $\mathbf{n} = (n_1, \ldots, n_d)$ in $\mathscr{Z}^d$, let $X_{\mathbf{n}}$ be a real-valued r.v. Thus, what we are dealing with here is a random field $\{X_{\mathbf{n}}, \mathbf{n} \in \mathscr{Z}^d\}$ whose elements take values in $\Re$. The basic assumption on this random field is that it consists of either positively or negatively associated r.v.'s which have finite second moment and are covariance invariant; that is, for any $\mathbf{u}, \mathbf{v}$ in $\mathscr{Z}^d$, $\mathrm{Cov}(X_{\mathbf{u}}, X_{\mathbf{v}}) = C(\mathbf{u} - \mathbf{v})$ for some $C : \mathscr{Z}^d \to \Re$. In particular, $C(\mathbf{0}) = \sigma^2(X_{\mathbf{n}})$ will be denoted by σ^2, and, by centering the $X_{\mathbf{n}}'s$ at their expectations, it will be assumed that $\mathscr{E}X_{\mathbf{n}} = 0$, $\mathbf{n} \in \mathscr{Z}^d$.

For $\mathbf{u}$ and $\mathbf{v}$ in $\Re^d$, $\mathbf{u} \leq \mathbf{v}$, means, of course, that $u_i \leq v_i$, $i = 1, \ldots, d$, and these inequalities are all strict for $\mathbf{u} < \mathbf{v}$. For any two such points, define the box $B_{\mathbf{u}}^{\mathbf{v}}$ in $\mathscr{Z}^d$ as follows:

$$B_{\mathbf{u}}^{\mathbf{v}} = \{\mathbf{n} \in \mathscr{Z}^d; \mathbf{u} < \mathbf{n} \leq \mathbf{v}\} = \{\mathbf{n} \in \mathscr{Z}^d; u_i < n_i \leq v_i, i = 1, \ldots, d\}. \qquad (4.1)$$

For $\mathbf{n} = (n_1, \ldots, n_d)$, $\|\mathbf{n}\|$ denotes the product $n_1 \ldots n_d$, and the symbol $|B_{\mathbf{u}}^{\mathbf{v}}|$ denotes the cardinality of the box $B_{\mathbf{u}}^{\mathbf{v}}$, so that $|B_{\mathbf{u}}^{\mathbf{v}}| = \|\mathbf{v} - u\|$. The norm, $|\mathbf{n}|$, of $\mathbf{n}$ is taken to be $|\mathbf{n}| = \max\{|n_i|; i = 1, \ldots, d\}$.

The random field formulation of the discussion here makes our results especially useful to spatial stochastic processes. Such processes arise in analysis of atmospheric applications, environmental science, geostatistical analysis, image analysis, oceanographic applications, and signal processing in radar and sonars. Such applications are discussed in a recent publication, Panel on Spatial Statistics and Image Processing (1991). Also, for an expert account of some statistical aspects of spatial processes, the reader is referred to Ripley (1988) and Cressie (1991).

In order to formulate the main result of this paper, let $N = 1, 2, \ldots$, let $\mathbf{k}(N)$ be in $\mathscr{Z}^d$ with $0 < k_i(N) \to \infty$, as $N \to \infty$, for $i = 1, \ldots, d$, let $X_{\mathbf{n}}(N)$ be observations taken over the box $B_{\mathbf{0}}^{\mathbf{k}(N)}$, and define $\hat{S}_{\mathbf{0}}^{\mathbf{k}(N)}$ by

$$\hat{S}_0^{\mathbf{k}(N)} = \|\mathbf{k}(N)\|^{-1/2} \sum_{\mathbf{n} \in B_0^{\mathbf{k}(N)}} X_{\mathbf{n}}(N). \tag{4.2}$$

Since $\mathbf{k}(N) > 0$, in all that follows, we shall consider only boxes $B_{\mathbf{u}}^{\mathbf{v}}$ with $0 < \mathbf{u} < \mathbf{v}$; so all $\mathbf{n}$ will belong to $\mathcal{Z}_+^d$ consisting of $\mathbf{n}'s$ in $\mathcal{Z}^d$ with $\mathbf{n} > 0$. Next, for $\mathbf{u}$ and $\mathbf{v}$ in $\mathcal{Z}_+^d$, define by $\mathbf{uv} = (u_1 v_1, \ldots, u_d v_d)$, and in all that follows, omit explicit expression of dependence on N of relevant quantities for notational simplicity. Thus, in the sequel, we shall often write $\mathbf{k}$ instead of $\mathbf{k}(N)$. For $\mathbf{r} \in \mathcal{Z}_+^d$, consider the box defined by Eq. (4.1) with $\mathbf{u} = \mathbf{k}(\mathbf{r} - 1)$ and $\mathbf{v} = \mathbf{kr}$; that is,

$$\begin{aligned} B_{\mathbf{k}(\mathbf{r}-1)}^{\mathbf{kr}} &= \{\mathbf{n} \in \mathcal{Z}_+^d; \mathbf{k}(\mathbf{r} - 1) < \mathbf{n} \le \mathbf{kr}\} \\ &= \{\mathbf{n} \in \mathcal{Z}^d; k_i(r_i - 1) < n_i \le k_i r_i, \, i = 1, \ldots, d\}, \end{aligned} \tag{4.3}$$

so that $B_{\mathbf{k}0}^{\mathbf{k}1} = B_0^{\mathbf{k}}$. Also, let

$$\hat{S}_{\mathbf{k}(\mathbf{r}-1)}^{\mathbf{kr}} = \|\mathbf{k}\|^{-1/2} \sum X_{\mathbf{n}} \quad \text{and} \quad S_{\mathbf{k}(\mathbf{r}-1)}^{\mathbf{kr}} = Y_{\mathbf{r}}^{\mathbf{k}} = \|\mathbf{k}\|^{-1/2} \hat{S}_{\mathbf{k}(\mathbf{r}-1)}^{\mathbf{kr}} \tag{4.4}$$

where the summation in $\mathbf{n}$ is over $B_{\mathbf{k}(\mathbf{r}-1)}^{\mathbf{kr}}$.

The assumptions to be used throughout this Section are the following:

Assumptions. For $N = 1, 2, \ldots$, let $\{X_{\mathbf{n}}(N)\}$, $\mathbf{n} \in B_0^{\mathbf{k}(N)}$, with $k_i(N) \to \infty$, $i = 1, \ldots, d$, as $N \to \infty$, be r.v.'s which satisfy the following requirements:

(A1) They have finite second moment, are centered at their expectations and are covariance invariant, so that $\text{Cov}(X_{\mathbf{u}}(N), X_{\mathbf{v}}(N)) = C(\mathbf{u} - \mathbf{v})$ for some $C: \mathcal{Z}^d \to \Re$ independent of N; set $C(\mathbf{0}) = \sigma^2$.

(A2) For all $\mathbf{n} \in \mathcal{Z}_+^d$ and all $N = 1, 2, \ldots, \mathcal{E}|X_{\mathbf{n}}(N)|^{2+\delta} \le C$ for some $\delta > 0$.

(A3) Either they are PA and $\sum_{\mathbf{n} \in \mathcal{Z}^d} C(\mathbf{n}) < \infty$, or they are NA and $\sum_{\mathbf{n} \in \mathcal{Z}^d} C(\mathbf{n}) > 0$.

The main result of this Section establishes joint asymptotic normality for the set of r.v.'s $\{\hat{S}_{\mathbf{k}(\mathbf{r}-1)}^{\mathbf{kr}}, \mathbf{r} \in \Lambda\}$, where Λ is any finite nonempty subset of $\mathcal{Z}_+^d$. The limit is taken as the number N of observed r.v.'s tends to ∞, the mean vector of the limiting normal distribution is zero and its covariance matrix is the diagonal matrix with diagonal elements equal to A. In particular, by taking $\Lambda = \{\mathbf{r}\}$, one obtains the asymptotic distribution of $\hat{S}_{\mathbf{k}(\mathbf{r}-1)}^{\mathbf{kr}}$. More precisely, one has

THEOREM 4.1. Let Λ be a nonempty finite subset of $\mathcal{Z}_+^d$, let $N \to \infty$, and suppose Assumptions (A1)–(A3) are fulfilled. Then, with $A = \sum_{\mathbf{n} \in \mathcal{Z}^d} C(\mathbf{n})$,

$$\mathscr{E} \exp\left(i \sum_{\mathbf{r} \in \Lambda} t_{\mathbf{r}} \hat{S}^{\mathbf{k}(N)\mathbf{r}}_{\mathbf{k}(N)(\mathbf{r}-1)} \right) \to \exp\left(-\frac{1}{2} A \sum_{\mathbf{r} \in \Lambda} t_{\mathbf{r}}^2 \right), \quad t_{\mathbf{r}} \in \Re.$$

In particular, for any $\mathbf{r} \in \mathscr{Z}^d_+$,

$$\hat{S}^{\mathbf{k}(N)\mathbf{r}}_{\mathbf{k}(N)(\mathbf{r}-1)} \xrightarrow{d} N(0, A) \quad \text{and} \quad \hat{S}^{\mathbf{k}(N)}_{\mathbf{0}} \xrightarrow{d} N(0, A).$$

REMARK 4.1.

(i)   Under PA, the assumption $A < \infty$ is stipulated. Under NA, it is
      implied by the finiteness of σ^2, but it may be equal to 0. Indeed, it is
      explained below (see number 7) that $A \geq 0$. Since $C(\mathbf{n}) \leq 0$ for $\mathbf{n} \neq \mathbf{0}$,
      it follows that $0 \leq A \leq \sigma^2$. Furthermore, $\sum_{\mathbf{n} \in \mathscr{Z}^d} |C(\mathbf{n})| = 2\sigma^2 - A$, as
      is easily seen.

(ii) Under PA, the theorem is a version of Newman's celebrated CLT (The-
      orem 2 in Newman (1980)) without the assumption of translation invar-
      iance and without the restriction that the boxes in Eq. (4.3) be cubic
      boxes. The quantity A is referred to as susceptibility in Newman (1980).

(iii) The theorem with $\Lambda = \{\mathbf{1}\}$ and under PA is a version of Theorem 1.2
      in Cox and Grimmett (1984), where they do not assume covariance
      invariance, but at the expense of boundedness away from 0 of
      the variances of $X_{\mathbf{n}}(N)$. They also assume boundedness (from
      above) of the absolute third moments of $X_{\mathbf{n}}(N)$, which, clearly, can
      be replaced by boundedness of the moments $\mathscr{E}|X_{\mathbf{n}}(N)|^{2+\delta}$ for some
      $0 < \delta(\leq 1)$.

(iv) The theorem for the NA case was established for the first time in
      Roussas (1994).

The details of the proof of Theorem 4.1 may be found in Roussas (1994).
Below, we give a brief outline of the main points of the proof, and the
auxiliary results which lead to it.

1. First, it is to be pointed out that monotone functions of associated r.v.'s
    are likewise associated; here the term associated r.v.'s is used collectively
    for both PA and NA r.v.'s.

2. In the theorem, for convenience, omit N. Then from the result stated in
    the theorem, it follows that the variance and covariances of $\hat{S}^{\mathbf{kr}}_{\mathbf{k}(\mathbf{r}-1)}$ must
    behave appropriately. More precisely, one must have:

$$\text{Var}(\hat{S}^{\mathbf{kr}}_{\mathbf{k}(\mathbf{r}-1)}) \to A, \ \text{Cov}(\hat{S}^{\mathbf{pr}}_{\mathbf{p}(\mathbf{r}-1)}, \hat{S}^{\mathbf{kr}}_{\mathbf{k}(\mathbf{r}-1)}) \to A$$

and $$\text{Cov}(\hat{S}^{\mathbf{kr}}_{\mathbf{k}(\mathbf{r}-1)}, \hat{S}^{\mathbf{ks}}_{\mathbf{k}(\mathbf{s}-1)}) \to 0.$$

This is, indeed, the case, as it can be seen by suitably exploiting the
property stated in numbers 1.

Next, for any two r.v.'s X and Y, set

$$H_{X,Y}(x,y) = P(X \leq x, Y \leq y) - P(X \leq x)P(Y \leq y)$$
$$= P(X > x, Y > y) - P(X > x)P(Y > y), \quad x, y \in \Re.$$

Then, in Lehmann (1966), the following equality (attributed to Hoeffding) is established:

3. If the r.v.'s X and Y have finite second moments, then

$$\text{Cov}(X, Y) = \int_{-\infty}^{\infty} \int_{-\infty}^{\infty} H_{X,Y}(x,y)\, dx\, dy.$$

Now, if f and g are complex-valued functions defined on $\Re$ with bounded derivatives f' and g', then the equality in number 3 generalizes as follows:

4. $$\text{Cov}(f(X), g(Y)) = \int_{-\infty}^{\infty} \int_{-\infty}^{\infty} f'(x)g'(y)H_{X,Y}(x,y)\, dx\, dy.$$

This relation was stated and used in Newman (1980) (see relation (2.2)), and also in Newman (1984) (see relation (4.10)), where boundedness of f' and g' is replaced by absolute continuity of f and g and finiteness of the second moments of $f(X)$ and $g(Y)$. Actually, this is a special case of a more general result obtained in Yu (1993) (see Theorem 2.3).

From number 4, the following inequality follows easily.

5. Let f and g be as above and suppose that f' and g' are bounded in the sup-norm. Then, if the r.v.'s X and Y are quadrant dependent (in particular, associated), it holds:

$$|\text{Cov}(f(X), g(Y))| \leq \|f'\|_\infty \|g'\|_\infty |\text{Cov}(X, Y)|.$$

By way of this inequality for covariances, one can establish an inequality for characteristic functions. More precisely, the following inequality holds.

6. If $Y_1, \ldots, Y_m$ are associated r.v.'s with finite second moments, then, for any $r_1, \ldots, r_m$ in $\Re$:

$$\left| \mathscr{E} \exp\left(i \sum_{k=1}^{m} r_k Y_k \right) - \prod_{k=1}^{m} \mathscr{E} \exp(i r_k Y_k) \right|$$

$$\leq \frac{1}{2} \sum_{\substack{1 \leq k, \ell \leq m \\ k \neq \ell}} |r_k r_\ell| |\text{Cov}(Y_k, Y_\ell)|$$

$$= \sum_{1 \leq k < \ell \leq m} |r_k r_\ell| |\text{Cov}(Y_k, Y_\ell)|.$$

For the PA case, this inequality was proved by Newman (see Theorem 1 in Newman (1980)), using induction and a rather ingenious technique. The proof for the NA case follows closely that for the PA case, and the details are presented in Roussas (1994) (see Proposition 3.1).

7. That $A \geq 0$ follows by the fact that A is the limit of a sequence of variances (see Proposition 2.1 in Roussas (1994)).

The proof of Theorem 4.1 is carried out by establishing the Feller–Lindeberg condition (see Lòeve (1963), p. 280). In doing so, the results stated above, and, in particular, result number 6, are instrumental.

5. MOMENT INEQUALITIES AND SOME ESTIMATION PROBLEMS

In this Section, we discuss some cases of (nonparametric) estimation of distribution functions (d.f.'s), p.d.f.'s, their derivatives, as well as the associated hazard rates. Optimal asymptotic properties of proposed estimates hinge on moment and probability inequalities, some of which will be presented here. To this end, Birkel (1988a) derives moment inequalities with rates for partial sums of PA r.v.'s. More specifically, if $\{X_j; j \in \mathcal{N}\}$, $\mathcal{N} = \{1, 2, \ldots\}$, is a sequence of PA r.v.'s with $\mathscr{E}X_j = 0$ and $\mathscr{E}|X_j|^{r+\delta} < \infty$ for some $r > 2$ and $\delta > 0$, and if $S_k = \sum_{j=1}^{k} X_j$, then conditions are given under which the following type of an inequality holds: $\sup\{\mathscr{E}|S_{n+m} - S_m|^r; m \in \mathcal{N} \cup \{0\}\} = \mathcal{O}(n^{r/2})$. A generalized version of this inequality, required for estimation purposes, was obtained by Bagai and Prakasa Rao (1995). More explicitly, let f_n be certain real-valued functions defined on $\mathfrak{R}$, let the PA r.v.'s depend on an index $\alpha \in J$, some index set, and suppose that $\mathscr{E}f_n(X_j(\alpha)) = 0$ for all $n \geq 1$, all $j \geq 1$ and all $\alpha \in J$; also, suppose that $\sup_{n \geq 1} \sup_{\alpha \in J} \sup_{j \geq 1} \mathscr{E}|f_n(X_j(\alpha))|^{r+\delta} < \infty$ for some $r > 2$ and some $\delta > 0$. Then, under some further appropriate conditions,

$$\sup_{m \geq 1} \sup_{\alpha \in J} \sup_{k \geq 0} \mathscr{E}|S_{n+k,m}(\alpha) - S_{k,m}(\alpha)|^r \leq Bn^{r/2},$$

for some constant $B > 0$.

Bagai and Prakasa Rao (1991) give conditions under which the empirical survival d.f., based on a segment of n r.v.'s from a stationary sequence of PA r.v.'s, enjoys several desirable properties. If $\bar{F}$ is the survival function of X_1, that is, $\bar{F}(x) = P(X_1 > x)$, and $\bar{F}_n$ is the empirical survival function based on $X_1, \ldots, X_n$, that is, $\bar{F}_n(x) = n^{-1} \sum_{j=1}^{n} I(X_j > x)$, then, under suitable conditions, the following results hold: For some $r > 1$, there exists a constant

$C > 0$ such that, for every $\varepsilon > 0$ and every $n \geq 1$

$$\sup\{P[|\bar{F}_n(x) - \bar{F}(x)| > \varepsilon]; \quad x \in \Re\} \leq C\varepsilon^{-2r}n^{-r}.$$

In particular, then, for every x, $\bar{F}_n(x) \to \bar{F}(x)$ a.s. Furthermore, under appropriate conditions, $\sup\{P[|\bar{F}_n(x) - \bar{F}(x)| > \varepsilon]; x \in J\} \to 0$ a.s. over any compact subset J of $\Re$, and, for every x for which $0 < F(x) < 1$, the sequence of r.v.'s $\{\sqrt{n}[\bar{F}_n(x) - \bar{F}(x)]/\sigma(x)\}$, $n \geq 1$, converges in distribution to the standard normal r.v., where $\sigma^2(x) = \bar{F}(x)[1 - \bar{F}(x)] + 2\sum_{j=2}^{\infty}[P(X_1 > x, X_j > x) - \bar{F}^2(x)]$. Now, let $f_n(x)$ be the usual kernel estimate of the p.d.f. $f(x)$ of $X_1, \ldots, X_n$, coming from a stationary sequence of PA r.v.'s, and suppose that the support of f is a closed interval $I = [a, b]$ in $\Re$. Then, under suitable conditions, $f_n(x) \to f(x)$ a.s. on I, and also $\sup[|f_n(x) - f(x)|; x \in I] \to 0$ a.s. Similar results also hold for the usual estimate $r_n(x)$ of the hazard rate $r(x)$; namely, $r_n(x) \to r(x)$ a.s. on I, and $\sup[|r_n(x) - r(x)|; x \in J] \to 0$ a.s., where J is any compact subset of $\{x \in \Re; \bar{F}(x) > 0\}$. Similar results have been obtained in Roussas (1991) under a slightly different set of conditions. In addition to obtaining estimates for the d.f., the p.d.f. and the hazard rate, estimates of the derivatives of the p.d.f. are also given and their strong uniform convergence is established with rates. A typical result for the (kernel) estimate $f_n^{(r)}(x)$ of the rth derivative of $f(x)$ is as follows:

$$\sup\left[h_n^{-\nu}|f_n^{(r)}(x) - f^{(r)}|; \quad x \in I\right] \to 0 \quad \text{a.s.},$$

where I is a closed interval supporting f, and $0 < \nu < 1$ and satisfies some further conditions.

Results similar to those obtained in Roussas (1991) have been carried over to a random field framework in Roussas (1993), under both PA and NA. A brief account of these results is as follows by means of the notation used in Roussas (1994). Let $F_{\|\mathbf{k}(N)\|}$ be the empirical d.f. based on $X_\mathbf{n}$, $\mathbf{n} \in B_0^{\mathbf{k}(N)}$, defined by

$$F_{\|\mathbf{k}(N)\|}(x) = \|\mathbf{k}(N)\|^{-1} \sum_{\mathbf{n} \in B_0^{\mathbf{k}(N)}} I(X_\mathbf{n} \leq x), \qquad x \in \Re. \tag{5.1}$$

For a bandwidth $0 < h_{\|\mathbf{k}(N)\|}$ and a known p.d.f. (kernel) K, defined on $\Re$, let $f_{\|\mathbf{k}(N)\|}$ be the usual kernel estimate of f; namely,

$$f_{\|\mathbf{k}(N)\|}(x) = \|\mathbf{k}(N)\|^{-1} h_{\|\mathbf{k}(N)\|}^{-1} \sum_{\mathbf{n} \in B_0^{\mathbf{k}(N)}} K\left(\frac{x - X_\mathbf{n}}{h_{\|\mathbf{k}(N)\|}}\right)$$

$$= h_{\|\mathbf{k}(N)\|}^{-1} \int_{-\infty}^{\infty} K\left(\frac{x - t}{h_{\|\mathbf{k}(N)\|}}\right) dF_{\|\mathbf{k}(N)\|}(t). \tag{5.2}$$

For the sake of convenience, in the expressions defined by Eqns (5.1) and (5.2) and others to be introduced below, we shall often write $\|\mathbf{k}\|$ rather than $\|\mathbf{k}(N)\|$. Also, let $f_{\|\mathbf{k}\|}^{(r)}(x)$ be the kernel estimate of the rth derivative of $f(x)$ given by

$$
f_{\|\mathbf{k}\|}^{(r)}(x) = \|\mathbf{k}\|^{-1} h_{\|\mathbf{k}\|}^{-(r+1)} \sum_{\mathbf{n} \in B_0^{\mathbf{k}}} K^{(r)} \left(\frac{x - X_{\mathbf{n}}}{h_{\|\mathbf{k}\|}} \right)
$$

$$
= h_{\|\mathbf{k}\|}^{-(r+1)} \int_{-\infty}^{\infty} K^{(r)} \left(\frac{x - t}{h_{\|\mathbf{k}\|}} \right) dF_{\|\mathbf{k}\|}(t), \quad x \in \Re. \tag{5.3}
$$

Then, under NA and suitable regularity conditions, it is shown that:

$$
\sup_{x \in I} \psi_{\|\mathbf{k}\|} \left| F_{\|\mathbf{k}\|}(x) - F(x) \right| \to 0 \quad \text{a.s.,} \tag{5.4}
$$

where $\psi_{\|\mathbf{k}\|}$ is a norming factor subject to a certain summability condition. Also, for a specific choice of $\psi_{\|\mathbf{k}\|}$,

$$
\sup_{x \in I} \|\mathbf{k}\|^{\nu \rho} \left| f_{\|\mathbf{k}\|}(x) - f(x) \right| \to 0 \quad \text{a.s.,} \tag{5.5}
$$

for $0 < \nu < 1$ and $\rho > 0$ subject to certain conditions; I is the support of f, which is assumed to be a closed interval. For the estimate of the rth derivative, a convergence such as that in relation (5.5) is also true; namely,

$$
\sup_{x \in I} \|\mathbf{k}\|^{\nu \rho} \left| f_{\|\mathbf{k}\|}^{(r)}(x) - f^{(r)}(x) \right| \to 0 \quad \text{a.s.} \tag{5.6}
$$

Finally, the estimate $r_{\|\mathbf{k}\|}(x)$ of the hazard rate $r(x)$ satisfies the convergence

$$
\sup_{x \in J_c} \|\mathbf{k}\|^{\nu \rho} \left| r_{\|\mathbf{k}\|}(x) - r(x) \right| \to 0 \quad \text{a.s.} \tag{5.7}
$$

Under PA, the convergences in relation (5.4), (5.5), (5.6) and (5.7) still hold true. Here, however, the general norming factor $\psi_{\|\mathbf{k}\|}$, and, in particular, $\|\mathbf{k}\|$, ν and ρ are subject to different restrictions.

In Roussas (1995) and in the framework of a random field, a d.f. F is estimated by a smooth d.f. $\hat{F}_{\|\mathbf{k}(N)\|}$ given by

$$
\hat{F}_{\|\mathbf{k}(N)\|} = \frac{1}{\|\mathbf{k}(N)\|} \sum_{\mathbf{n} \in B_0^{\mathbf{k}(N)}} G \left(\frac{x - X_{\mathbf{n}}}{h_{\|\mathbf{k}(N)\|}} \right), \qquad 0 < h_{\|\mathbf{k}(N)\|} \to 0, \text{ as } N \to \infty,
$$

where G is a known d.f. Then, under either PA or NA and some further regularity conditions, the asymptotic normality of $\hat{F}_{\|\mathbf{k}(N)\|}(x)$ is established. More precisely, it is shown that:

$$
\|\mathbf{k}(N)\|^{1/2} \left[\hat{F}_{\|\mathbf{k}(N)\|}(x) - \mathscr{E} \hat{F}_{\|\mathbf{k}(N)\|}(x) \right] \xrightarrow{d} N(0, A(x)), \qquad x \in \Re, \text{ as } N \to \infty,
$$

$$
\tag{5.8}
$$

where $A(x)$ is defined by

$$0 < A(x) = \sum_{\mathbf{n} \in \mathcal{Y}^d} [P(X_0 < x, X_\mathbf{n} < x) - F^2(x)], \qquad x \in \Re.$$

The proof of the convergence in relation (5.8) consists, essentially, in establishing the validity of the Feller–Lindeberg criterion. The details, however, involve a substantial amount of work.

6. SOME RECENT RESULTS

The standard way of estimating a d.f. is that of using the empirical d.f. An alternative to it is to employ as an estimate a smooth kernel-type d.f. This is, in particular, appealing when the underlying d.f. is known to have a p.d.f. These two estimates share most of their asymptotic properties, and the smooth estimate may be preferable, under certain conditions, on efficiency/deficiency considerations. To be more specific, let F_n be the empirical d.f. based on $X_1, \ldots, X_n$, and let $\hat{F}_n$ be the smooth d.f. referred to above. Let $\mathrm{MSE}(F_n(x))$ be the mean squared error of $F_n(x)$ and likewise for $\hat{F}_n(x)$, and define $i(n)$ as follows:

$$i(n) = \min\{k \in \{1, 2, \ldots\}; \mathrm{MSE}(F_k(x)) \leq \mathrm{MSE}(\hat{F}_n(x))\}.$$

Then, typically, $i(n)/n$ tends to 1, as $n \to \infty$, so that one cannot choose one estimate over the other in terms of sample size and by means of mean squared error. Under certain conditions, however, it is seen that $(i(n) - n)/(nh_n)$ tends to a positive constant; here h_n is the bandwidth of the kernel employed. It follows that the deficiency $i(n) - n$ of $F_n(x)$ with respect to $\hat{F}_n(x)$ is substantial, and, actually, tends to ∞, as $n \to \infty$. In terms then of deficiency, the smooth estimate $\hat{F}_n(x)$ is preferable to the empirical d.f. $F_n(x)$. Results of this nature are discussed in Cai and Roussas (1998), under the basic assumption of quadrant dependence. They carry over to the present framework original results established by Reiss (1981) and Falk (1983) for independent identically distributed (iid) r.v.'s, and supplemented, in a certain sense, by Mammitzsch (1984). Some convergence properties of smooth estimates of a d.f. are discussed in early papers by Nadaraya (1964), Watson and Leadbetter (1964), Roussas (1969), Winter (1973,1979), and Yamato (1973). The papers by Azzalini (1981), Lejeune and Sarda (1992) and Sarda (1993) are also relevant. A decisive tool under the present kind of dependence is an Essen-type inequality for two dimensional d.f.'s obtained by Sadikova (1996).

In Cai and Roussas (1997), the authors address the issue of estimating the quantiles of a d.f. The basic assumption of dependence made here is that of

PA or NA, and the proposed estimates are constructed by means of the kernel-type estimate $\hat{F}_n$ discussed above. To be more precise, for $0 < p < 1$, let $\xi_p = F^{-1}(p)$, and define the estimate $\hat{\xi}_{pn}$ by:

$$\hat{\xi}_{pn} = \hat{F}_n^{-1}(p) = \inf\{x \in \Re; \hat{F}(x) \geq p\}.$$

Then, under suitable regularity conditions, it is shown that $\hat{\xi}_{pn}$ is strongly consistent and asymptotically normal. It is also shown that a certain stochastic process, generated by $\hat{\xi}_{pn}$, is weakly convergent. Some related papers, mostly in the iid framework, are those by Nadaraya (1964), Yamato (1973), Parzen (1979), Falk (1985), Falk and Reiss (1989), Ralescu (1992), Ralescu and Sun (1993), and Xiang (1995).

There is a vast literature on Berry-Esseen bounds for iid r.v.'s when the empirical d.f. is involved. Under PA, Wood (1983) obtained such a bound with maximum order of magnitude $O(n^{-1/5})$ when the r.v.'s are indexed by integers, and Shao (1986) with order $O(n^{-1/4})$ in a random field framework. Finally, Birkel (1986b) obtained a rate $O(n^{-1/2} \log n)$, under certain conditions, and showed that this rate cannot be improved. In Cai and Roussas (1995a), the empirical d.f. is replaced by the smooth kernel-type d.f. estimate $\hat{F}_n(x)$ referred to above, and, under the basic assumption of PA or NA and additional suitable conditions, Berry-Esseen bounds with rates are obtained for $\sqrt{n}[\hat{F}_n(x) - \mathcal{E}\hat{F}_n(x)]$. The rates obtained depend on the assumptions made and hinge heavily on the quantity $u(r) = \sum_{j=r}^{\infty} |\text{Cov}(X_1, X_{j+1}|^{1/3}$. The reference Hall (1982) was quite helpful in some derivations.

Berry–Esseen bounds for $\alpha_n(x) = \sqrt{n}[\hat{F}_n(x) - \mathcal{E}\hat{F}_n(x)]$ yield asymptotic normality for $\alpha_n(x)$ for each $x \in \Re$. On the other hand, looking at the stochastic process $\{\alpha_n(x); x \in \Re\}$, one may be interested in investigating it from weak convergence viewpoint. This process was considered in Cai and Roussas (1995b) and its weak convergence was established, under the basic assumption of PA or NA. As a corollary, the weak convergence of the stochastic process $\{\beta_n(x); x \in \Re\}$ also follows, where $\beta_n(x) = \sqrt{n}[\hat{F}_n(x) - F(x)]$. In the derivations involved, certain general results obtained in Yu (1993) proved quite helpful. Most recently, Louhichi (1998) also established a weak convergence result of the stochastic process generated by the empirical d.f. based on PA r.v.'s. The author, commenting on PA and mixing, shows, by means of examples, that the classes of PA processes and mixing processes are distinct but not disjoint. When the underlying r.v.'s are iid and $\hat{F}_n$ is replaced by the empirical d.f. F_n, a wealth of results, both of probabilistic as well as of statistical interest, may be found in Shorack and Wellner (1986).

In Bozorgnia, Patterson and Taylor (1993), the authors derive an abundance of results in the framework of negative dependence, in general, and

NA, in particular. These results include WLLN and SLLN. The authors employ the term negatively dependent (ND) to mean that the r.v.'s involved are NOD (see Definition 3.3). For bounded (by 1, say) ND r.v.'s $X_1, \ldots, X_n$, and for $t_i \geq 0$, $1 \leq i \leq n$ (or $t_i \leq 0$, $1 \leq i \leq n$), the authors establish the following side result; namely,

$$\frac{e^{2t_{(n)}}}{n-1} \leq \max_{i,j} \text{Cov}(e^{t_i X_i}, e^{t_j X_j}) \leq 0.$$

This double inequality shows that, for large n, the correlation becomes very small between some r.v.'s. The question then may arise as to whether an infinite sequence of r.v.'s can have an infinite number of ND r.v.'s. The answer to this question is in the affirmative and is provided by the following example, see Example 3.1 in Bozorgnia, Patterson and Taylor (1993). Let Y_i, $i \geq 1$, be iid r.v.'s, such that $P(Y_i = 1) = P(Y_i = -1) = 1/2$, $i \geq 1$. Define $X_{2i-1} = Y_i$ and $Y_{2i} = -Y_i$. Then the r.v.'s X_i, $i \geq 1$, are ND.

In the monograph Szekli (1995) cited in Sec. 2 the author devotes one section to each PA and NA, Secs. 3.1 and 3.5, respectively, in the framework of stochastic ordering and dependence. The author compares PA to other modes of positive dependence, and emphasizes, in particular, the fact that MTP_2 implies PA. Thus, MTP_2, which is analytically more tractable than PA, may be employed as a tool of establishing PA. Regarding NA, the author states that it has the distinct advantage over other kinds of negative dependence in that nondecreasing function over disjoint sets preserve NA. The author observes that NA arises when r.v.'s are subjected to conditioning; also, NA arises naturally via permutation distributions.

In a recent paper, Sarkar and Chang (1997) show that, in testing the overall hypothesis $H_0 = \bigcap_{i=1}^{n} H_i$ of n hypotheses $H_1, \ldots, H_n$, suitable PA test statistics allow control of the probability of type I error.

Another recent result concerning PA and NA r.v.'s was obtained by Hu and Hu (1998). As suggested by the authors, this result may be of use to geneticists engaged in animal breeding. For a brief description, let **a** and **b** be two n-dimensional vectors, and declare the vector **a** to be majorized by the vector **b**, denoted by $\mathbf{b} \rangle^m \mathbf{a}$, if $\sum_{i=1}^{n} a_i = \sum_{i=1}^{n} b_i$ and $\sum_{i=1}^{k} a_{(i)} \geq \sum_{i=1}^{k} b_{(i)}$, $k = 1, \ldots, n$; $a_{(i)}$ and $b_{(i)}$ are the ordered a_i and b_i, $i = 1, \ldots, n$. Now, let $X_1, \ldots, X_n$ be NA r.v.'s, and let $X_1^*, \ldots, X_n^*$ be independent r.v.'s such that X_i^* and X_i have the same distributions, $i = 1, \ldots, n$. Let $F_{(i)}$ and $F_{(i)}^*$ be the d.f.'s of the ordered statistics $X_{(i)}$ and $X_{(i)}^*$, respectively, $i = 1, \ldots, n$, and let h be any monotone function for which the expectations below exist. Then it is shown that, for every $t \in \Re$,

$$\left(F_{(1)}(t), \ldots, F_{(n)}(t)\right) \rangle^m (F_{(1)}^*(t), \ldots, F_{(n)}^*(t)),$$

and

$$(\mathscr{E}h(X_{(1)}),\ldots,\mathscr{E}h(X_{(n)})))^{'''}(\mathscr{E}h(X_{(1)}^*),\ldots,\mathscr{E}h(X_{(n)}^*)).$$

The majorization is reversed, if the r.v.'s $X_1,\ldots,X_n$ are PA.

Thus, if $X_1,\ldots,X_n$ stand for the phenotypes of animals, the selection differential $\sum_{i=n-k+1}^{n} X_{(i)}$, used by geneticists in selecting the best k of n animals, is more significant when based on NA r.v.'s rather than independent samples (and less so, when based on PA r.v.'s); it is assumed that the X_i's have common expectation.

This Section is concluded with a brief reference to the Kaplan–Meier estimate. When the underlying observations are iid but incomplete, in the sense that they are subject to censoring, then the problem of estimating their d.f. F has received extensive attention during the last 40 or so years. The basic result in this respect and standard point of reference is the Kaplan-Meier (1958) paper. From among the numerous papers and monographs which ensued, we mention only a handful; they are those by Breslow and Crowley (1974), Gill (1980, 1981, 1983), Stute and Wang (1993), and Stute (1994). Voelkel and Crowley (1984) considered the problem in a semi-Markov process framework in conjunction with cancer research clinical trials, and Ying and Wei (1994) explored consistency and asymptotic normality of the Kaplan-Meier estimate under φ-mixing. The last two references are the only ones that we are aware of, where dependence is present.

Ying and Wei (1994) have analyzed a real data set from a tumorigenesis in a litter-matched experiment. All experimental rats are sacrificed at the end of 104 weeks. Let X_1 be the observation from a drug-treated rat, and let X_2 and X_3 be the responses from the litter-matched controls. The main interest here lies in estimating the common marginal d.f. F of the tumor appearance time (in weeks) for the controls. It is argued that, if the failure times are highly positively correlated in each stratum of highly stratified survival times, one would expect that the variance estimate provided by the authors would tend to be much larger than the standard estimate with independent observations. This did not materialize in the specified data set due, perhaps, to weak litter effect, see Ying and Wei (1994, p. 22). The persuasive argument, advanced by Ying and Wei (1994), that one would expect larger variance estimates under highly positive correlation than under independence, is quite appropriate for the case of PA discussed here.

When the mode of dependence is either PA or NA, then the Kaplan–Meier estimate may be shown to be uniformly strongly consistent; certain rates may also be provided. These results may be found in Cai and Roussas (1996). Furthermore, the stochastic process, generated by the Kaplan-Meier estimate, is shown to be weakly convergent. Also, a strongly consistent estimate is obtained for the asymptotic variance of the Kaplan–Meier estimate.

Finally, most recent results include a Bernstein–Hoeffding probability inequality for PA r.v.'s, Ioannides and Roussas (1998), and a CLT for the kernel estimate of the marginal p.d.f. of stationary and PA r.v.s., Roussas (1998). This last result is distinct from previously obtained CLT's or weak convergence results in that here the kernel, used in the estimation process, is not a monotone function.

REFERENCES

Azzalini, A. A note on the estimation of a distribution function and quantiles by a kernel method. *Biometrika 68*: 326–328, 1981.

Bagai, I. and Prakasa Rao, B. L. S. Estimation of the survival function for stationary associated processes. *Statist. Probab. Lett. 12*: 385–391, 1991.

Bagai, I. and Prakasa Rao, B. L. S. Kernel-type density and failure rate estimation for associated sequences. *Ann. Inst. Statist. Math. 47*: 253–266, 1995.

Barlow, R. E. and Proschan, F. Statistical Theory of Reliability and Life Testing: Probability Models. New York: Holt, Rinehart and Winston, 1975.

Birkel, T. Moment bounds for associated sequences. *Ann. Probab. 16*: 1184–1193, 1988a.

Birkel, T. On the convergence rate in the central limit theorem for associated processes. *Ann. Probab. 16*: 1685–1698, 1988b.

Birkel, T. A note on the strong law of large numbers for positively dependent random variables. *Statist. Probab. Lett. 7*: 17–20, 1989.

Block, H. W. and Ting, M. L. Some concepts of multivariate dependence. *Commun. Statist. Theory Meth. A10(8)*: 749–762, 1981.

Block, H. W., Savits, T. H. and Shaked, M. Some concepts of negative dependence. *Ann. Probab. 10*: 765–772, 1982.

Bozorgnia A., Patterson, R. F. and Taylor, R. L. Limit theorems for negatively dependent random variables. Unpublished manuscript, 1993.

Breslow, N. and Crowley, J. A large sample study of the life table and product limit estimators under random censorship. *Ann. Statist. 2*: 437–453, 1974.

Brindley Jr, E. C. and Thompson Jr., W. A. Dependence and aging aspects of multivariate survival. *J. Amer. Statist. Assoc. 67*: 822–830, 1972.

Bulinskii, A. V. Inequalities for the moments of sums of associated multi-indexed variables. *Theory Probab. Appl. 38*: 342–349, 1993.

Bulinskii, A. V. Rate of convergence in the central limit theorem for fields of associated random variables. *Theory Probab. Appl. 40*: 136–144, 1995.

Burton, R. M. and Waymire, E. Scaling limits for associated random measures. *Ann. Probab. 13*: 1267–1278, 1985.

Burton, R. M., Dabrowski. A. R. and Dehling, H. An invariance principle for weakly associated random vectors. *Stochastic Process Appl. 23*: 301–306, 1986.

Cai, Z. W. Statistical inference under dependence. PhD. dissertation, Division of Statistics, University of California, Davis, CA, 1995.

Cai, Z. W. and Roussas, G. G. Berry–Esseen bounds for smooth estimator of a distribution function under association. Technical Report No. 306, University of California, Davis, CA, 1995a. Also, J. Nonparametric Statistist. (to appear).

Cai, Z. W. and Roussas, G. G. Weak convergence for a smooth estimator of a distribution function under association. Technical Report No. 307, University of California, Davis, CA, 1995b. Also, Stochastic Anal. Appl. (to appear).

Cai, Z. W. and Roussas, G. G. Kaplan–Meier estimator under association. Technical Report No. 310, University of California, Davis, CA, 1996. Also, J. Multivariate Anal. (to appear).

Cai, Z. W. and Roussas, G. G. Smooth estimate of quantiles under association. *Statist. Probab. Lett. 36*: 275–287, 1997.

Cai, Z. W. and Roussas, G. G.. Efficient estimation of a distribution function under quadrant dependence. *Scand. J. Statist. 25*: 211–224, 1998.

Cohen, A. and Sackrowitz, H. B. Some remarks on a notion of positive dependence, association, and unbiased testing. In: Shaked, M. Tong, Y. L., eds. Stochastic Inequalities. IMS Lecture-Notes Monograph Series *22*, 1992.

Cohen, A. and Sackrowitz, H. B. Association, and unbiased tests in statistics. In: Shaked, M. Shanthikumar, J. G., eds. Stochastic Orders and Their Applications. Boston: Academic Press, 1994.

Cox, J. T. and Grimmett, G. Central limit theorem for associated random variables and the percolation model. *Ann. Probab. 12*: 514–528, 1984.

Cressie, N. Statistics for Spatial Data. New York: Wiley, 1991.

Ebrahimi, N. and Ghosh, M. Multivariate negative dependence. *Commun. Statist. Theory Meth. A10(4)*: 307–337, 1981.

Efron, B. Increasing properties of Polya frequency functions. *Ann. Math. Statist. 36*: 272–279, 1965.

Esary, J. D. Proschan, F. and Walkup, D. W. Association of random variables, with applications. *Ann. Math. Statist. 38*: 1466–1474, 1967.

Falk, M. Relative efficiency and deficiency of kernel type estimators of smooth distribution functions. *Statist. Neerlandica 37*: 73–83, 1983.

Falk, M. Asymptotic normality of the kernel quantile estimator. *Ann. Statist. 13*: 428–433, 1985.

Falk, M. and Reiss, R. D. Weak convergence of smoothed and non-smoothed bootstrap quantile estimates. *Ann. Probab. 17*: 362–371, 1989.

Fortuin, C. M., Kasteleyn, P. W. and Ginibre, J. Correlation inequalities on some partially ordered sets. *Commun. Math. Phys. 22*: 89–103, 1971.

Gill, R. D. Censoring and Stochastic Integrals. Mathematical Centre Tracts No. 124, Mathematisch Centrum, Amsterdam, 1980.

Gill, R. D. Testing with replacement and the product limit estimator. *Ann. Statist. 9*: 853–860, 1981.

Gill, R. D. Large sample behavior of the product limit estimator on the whole line. *Ann. Statist. 11*: 49–56, 1983.

Glaz, J. A comparison of Bonferroni-type and product-type inequalities in presence of dependence. In: Block, H. W., Sampson, A. R., Savits, T. H., eds. Topics in Statistical Dependence. IMS Lecture Notes–Monograph Series *16*: 223–235, 1990.

Granger, C. W. J. Positively related processes and cointegration. In: Subba, T., ed. Developments in Time Series Analysis. New York: Chapman & Hall, 1993.

Hall, P. Rates of Convergence in the Central Limit Theorm. London: Pitman, 1982.

Harris, T. E. A lower bound for the critical probability in a certain percolation process. *Proc. Camb. Phil. Soc. 56*: 13–20, 1960.

Hollander, M., Proschan, F. and Sconing, J. Information, censoring and dependence. In: Block, H. W., Sampson, A. R., Savits, T. H, eds. Topics in Statistical Dependence. IMS Lecture Notes-Monograph Series *16*: 257–268, 1990.

Hu, T. and Hu, J. Comparison of order statistics between dependent and independent random variables. *Statist. Prob. Lett. 37*: 1–6, 1998.

Ioannides, D. A., and Roussas, G. G. Exponential inequality for associated random variables. Technical Report No. 333, University of California, Davis, CA, 1998. Also, Statist. Probab. Lett. (to appear).

Joag-Dev, K. Independence via uncorrelatedness under certain dependence structures. *Ann. Probab. 11*: 1037–1041, 1983.

Joag-Dev, K. Conditional negative dependence in stochastic ordering and interchangeable random variables. In: Block, H. W, Sampson, A. R, Savits, T. H, eds. Topics in Statistical Dependence. IMS Lecture Notes-Monograph Series *16*: 295–298, 1990.

Joag-Dev, K. and Proschan, F. Negative association of random variables, with applications. *Ann. Statist. 11*: 286–295, 1983.

Joag-Dev, K., Perlman, M. D. and Pitt, L. D. Association of normal random variables and Slepian's inequality. *Ann. Probab. 11*: 451–455, 1983.

Jogdeo, K. On a probability bound of Marshall and Olkin. *Ann. Statist. 6*: 232–234, 1978.

Kaplan, E. L. and Meier, P. Nonparametric estimation from incomplete observations. *J. Amer. Statist. Assoc. 53*: 457–481, 1958.

Karlin, S. and Rinott, Y. Classes of ordering of measures and related correlation inequalities: I. Multivariate totally positive distributions. *J. Multivariate Anal. 10*: 467–498, 1980a.

Karlin, S. and Rinott, Y. Classes of ordering of measures and related correlation inequalities: II. Multivariate reverse rule distributions. *J. Multivariate Anal. 10*: 499–516, 1980b.

Kemperman, J. H. B. On the FKG-inequality for measures on a partially ordered space. *Nederl. Akad. Wetensch. Indag. Math. Proce. Ser A80(4)*: 313–331, 1977.

Langberg, N., Proschan, F., and Quinzi, A. J. Converting dependent models into independent ones, preserving essential features. *Ann. Probab. 6*: 174–181, 1978.

Lebowitz, J. L. Bounds on the correlations and analyticity properties of ferromagnetic Ising spin systems. *Commun. Math. Phys. 28*: 313–321, 1972.

Lehmann, E. L. Some concepts of dependence. *Ann. Math. Statist. 37*: 1137–1153, 1966.

Lejeune, M. and Sarda, P. Smooth estimators of distribution and density functions. *Comput. Statist. Data Anal. 14*: 457–471, 1992.

Lindqvist, B. H. Association of probability measures on partially ordered spaces. *J. Multivariate Anal. 26*: 111–132, 1988.

Loève, M. Probability Theory, 3rd. ed. New Jersey: Van Nostrand, 1963.

Louhichi, S. Weak convergence for empirical processes of associated sequences. Technical Report No. 98.36, Université de Paris-Sud Mathematiques, 1998.

Mammitzsch, V. On the asymptotically optimal solution within a certain class of kernel type estimators. *Statist. Decisions 2*: 247–255, 1984.

Nadaraya, E. A. Some new estimates for distribution function. *Theory Probab. Appl. 9*: 497-500, 1964.

Newman, C. M. Normal fluctuations and the FKG inequalities. *Commun. Math. Phys. 74*: 119–128, 1980.

Newman, C. M. Asymptotic independence and limit theorems for positively and negatively dependent random variables. In: Tong, Y. L., ed. Inequalities in Statistics and Probability. IMS Lecture Notes-Monograph Series *5*: 127–140, 1984.

Newman, C. M. Ising models and dependent percolation. In: Block, H. W, Sampson, A. R, Savits, T. H, eds. Topics in Statistical Dependence. IMS Lecture Notes-Monograph Series *16*: 395–401, 1990.

Newman,C. M. and Wright, A. L. An invariance principle for certain dependent sequences. *Ann. Probab. 9*: 671–675, 1981.

Newman, C. M. and Wright, A. L. Associated random variables and martingale inequalities. *Z. Wahrsch. Verw. Gebiete 59*: 361–371, 1982.

Panel on Spatial Statistics and Image Processing. In: Spatial Statistics and Digital Image Analysis. Washington, DC: National Academy Press, 1991.

Parzen, E. Nonparametric statistical data modeling. *J. Amer. Statist. Assoc.* *74*: 105–121, 1979.

Pitt, L. D. Positively correlated normal variables are associated. *Ann. Probab. 10*: 496–499, 1982.

Preston, C. J. A generalization of the FKG inequalities. *Commun. Math. Phys. 36*: 233–241, 1974.

Ralescu, S. S. A remainder estimate for the normal approximation of perturbed sample quantiles. *Statist. Probab. Lett. 14*: 293–298, 1992.

Ralescu, S. S. and Sun, S. Necessary and sufficient conditions for the asymptotic normality of perturbed sample quantiles. *J. Statist. Plann. Inference 35*: 55–64, 1993.

Reiss, R. D. Nonparametric estimation of smooth distribution functions. *Scand. J. Statist. 8*: 116–119, 1981.

Ripley, B. D. Statistical Inference for Spatial Processes. Cambridge: Cambridge University Press, 1988.

Roussas, G. G. Nonparametric estimation of the transition distribution function of a Markov process. *Ann. Math. Statist. 40*: 1386–1400, 1969.

Roussas, G. G. Kernel estimates under association: Strong uniform consistency. *Statist. Probab. Lett. 12*: 393–403, 1991.

Roussas, G. G. Curve estimation in random fields of associated processes. *J. Nonparametric Statist. 2*: 215–224, 1993.

Roussas, G. G. Asymptotic normality of random fields of positively or negatively associated processes. *J. Multivariate Anal. 50*: 152–173, 1994.

Roussas, G. G. Asymptotic normality of a smooth estimate of a random field distriubtion function under association. *Statist. Probab. Lett. 24*: 77–90, 1995.

Roussas, G. G. Exponential probability inequalities with some applications. In: Ferguson, T. S., Shapley, L. S., MacQueen, J. B., eds. Statistics, Probability and Game Theory. IMS Lecture Notes-Monograph Series 30: 303–319, 1996.

Roussas, G. G. Asymptotic normality of the kernel estimate of a probability density function under association. Technical Report No. 334, University of California, Davis, CA, 1998.

Rüschendorf, L. Weak association of random variables. *J. Multivariate Anal. 11*: 448–451, 1981.

Sadikova, S. M. Two dimensional analogues of an inequality of Esseen with applications to the central limit theorem. *Theory Probab. Appl. 11*: 325–335, 1966.

Samuel-Cahn, E. Prophet inequalities for bounded negatively dependent random variables. *Statist. Probab. Lett. 12*: 213–216, 1991.

Sarda, P. Smoothing parameter selection for smooth distribution functions. *J. Statist. Plann. Inference. 35*: 65–75, 1993.

Sarkar, S. K. and Chang, C. K. The Simes method for multiple hypothesis testing with positively dependent test statistics. *J. Amer. Statist. Assoc. 92*: 1601–1608, 1997.

Shaked, M. A general theory of some positive dependence notions. *J. Multivariate Anal. 12*: 199–218, 1982.

Shao, Q. M. A Berry–Esseen inequality and an invariance principle for associated random fields. *Chinese J. Appl. Probab. Statist. 5*: 1–8, 1986.

Shorack, G. R. and Wellner, J. A. Empirical Processes with Applications to Statistics. New York: Wiley, 1986.

Simon, B. Correlation inequalities and the mass gap in $P(\varphi)_2$: I. Domination by the two point function. *Commun. Math. Phys. 31*: 127–136, 1973.

Stute, W. and Wang, J. L. A strong law under random censorship. *Ann. Statist. 21*: 1591–1607, 1993.

Stute, W. Convergence of Kaplan–Meier estimator in weighted sup-norms. *Statist. Probab. Lett. 20*: 219–223, 1994.

Subramanyam, K. Some comments on positive quadrant dependence in three dimensions. In: Block, H. W., Sampson, A. R., Savits, T. H., eds. Topics in Statistical Dependence. IMS Lecture Notes-Monograph Series *16*: 443-449, 1990.

Szekli, R. Stochastic ordering and Dependence in Applied Probability. New York: Springer-Verlag, 1995.

Tong, Y. L. Probability Inequalities in Multivariate Distributions. New York: Academic Press, 1980.

Voelkel, J. and Crowley, J. Nonparametric inference for a class of semi-Markov processes with censored observations. *Ann. Statist. 12*: 142–160, 1984.

Watson, G. S. and Leadbetter, M. R. Hazard analysis II. *Sankhyā Ser A26*: 101–116, 1964.

Winter, B. B. Strong uniform consistency of integrals of density estimators. *Canad. J. Statist. 1*: 247–253, 1973.

Winter, B. B. Convergence rate of perturbed empirical distribution functions. *J. Appl. Probab. 16*: 163–173, 1979.

Wood, T. A Berry–Esseen theorem for associated random variables. *Ann. Probab. 4*: 1042–1047, 1983.

Wright, L. A. A strong limit theorem for process with associated increments. In: Block, H. W., Sampson, A. R., Savits, T. H., eds. Topics in Statistical Dependence. IMS Lecture Notes-Monograph Series *16*: 481–487, 1990.

Xiang, X. Deficiency of the sample quantile estimator with respect to kernel quantile estimators for censored data. *Ann. Statist. 23*: 836–854, 1995.

Yamato, H. Uniform convergence of an estimator of a distribution function. *Bull. Math. Statist. 15*: 69–78, 1973.

Ying, Z. and Wei, L. J. The Kaplan–Meier estimate for dependent failure time observations. *J. Multivariate Anal. 50*: 17–29, 1994.

Yoshihara, K. I. Weakly Dependent Stochastic Sequences and Their Applications IX: Poisson Approximations and Associated Processes. Tokyo: Sanseido Co., Ltd., 1997.

Yu. H. A Glivenko–Cantelli lemma and weak convergence for empirical processes of associated sequences. *Probab. Theory Relat. Fields 95*: 357–370, 1993.

24

Second-Order Information Loss Due to Nuisance Parameters: A Simple Measure

BRUCE G. LINDSAY   Pennsylvania State University, University Park, Pennsylvania

RICHARD WATERMAN   University of Pennsylvania, Philadelphia, Pennsylvania

1. INTRODUCTION

This chapter is about evaluating the loss of efficiency in maximum likelihood estimation due to the estimation, and subsequent "plugging–in" of nuisance parameters. These effects are necessarily observed at the second-order because the information in the efficient score and plug-in score are equivalent to first-order. We provide a two part partition of the information loss by means of an orthogonal decomposition of the nuisance parameter derivative of the efficient score. One part of the decomposition can be interpreted as the information loss due to the need to bias correct the m.l.e. The second part is due to the increased variability that a plug-in score has compared to the efficient score. We provide several worked examples. Our approximate formula for second-order information loss provides correct second-order information calculations for the two classic Neyman–Scott problems. We show the relationship of our computational formulas with two types of conditional structures, as well as with statistical curvature in the normal curved exponential family. A final example shows the relevance of this calculation in adaptive estimation.

It is now nearly 50 years since Neyman and Scott (1948) established that maximum likelihood estimation could have severe failings when estimating a parameter of interest in the presence of many nuisance parameters. In particular, they demonstrated that with infinitely many nuisance parameters, the maximum likelihood estimator could fail to be consistent, and even if consistent, could fail to be the most efficient estimator. We return to this problem to show that with a little mathematical prodding, one can derive simple and meaningful ways of assessing the statistical impact of nuisance parameters.

One can decompose the mean squared error of a maximum likelihood estimator into squared bias and variance. If we do so in a setting in which there is a parameter of interest ψ and a nuisance parameter λ, then the variance and bias for maximum likelihood estimator $\hat{\psi}$ of ψ can be decomposed yet further. Some of the variability and bias would occur even if we knew λ exactly, but there are additional effects due to the fact that λ is unknown, and must be estimated. We will call these latter "nuisance parameter" effects; they are the focus of this chapter.

In particular, we wish to show that there is a simple concept of second-order information loss in the presence of nuisance parameters. It extends the first-order information loss due to λ that can be calculated from the Fisher's information matrix. It is straightforward to calculate and so can be used as a means of assessing the magnitude of the statistical difficulties faced when the nuisance parameter is poorly estimated. Formally, it is created by the evaluation of the information in a set of bias-corrected maximum likelihood equations. It can be interpreted as the asymptotic correlation between the optimal "efficient" score that underlies the first-order theory and the actual score equations used for estimation.

We note that our approach focuses on the "invariant" part of the information lost due to nuisance parameters. That is, it is invariant under "interest-preserving" transformations of the parameter space, those that preserve the identity of the parameter of interest. In this sense it reflects information because it does not incorporate those aspects of second-order error that are due to the particular parameterization used.

Among other features, we will show that it provides a measure of the "conditionality of the problem". Its relevance is most clear when there are many nuisance parameters in a problem, with only a single parameter of interest, so that the nuisance parameters could severely affect first-order inference about the parameter of interest. Thus we will adopt a Neyman and Scott (1948) framework for motivation for some of our focal points. We will also use two of their famous examples to exemplify how our information calculations give the "right" answer.

In our Neyman–Scott type framework, the data will consist of a two way array of independent variables X_{ij} having densities $f(x_{ij}; \psi, \lambda_i)$, with rows indexed $i = 1, \ldots, p$, and varying numbers of observations per row, $j = 1, \ldots, n_i$. Thus each row is a "cluster" or "stratum", with a cluster-specific parameter λ_i. In this context, the first-order information about ψ in the presence of λ, namely J_1, for the ith stratum has the standard form

$$J_1 = n_i I_{11.2} = n_i(I_{11} - I_{12}I_{22}^{-1}I_{21})$$

where I_{ij} is the relevant block of the partitioned Fisher's information matrix. Our objective is demonstrate that there is a meaningful definition of the second-order information about ψ in any particular row that has the form:

$$J_2 = J_1 - C_{\text{Bias}} - C_{\text{Var}} + O(1/n)$$

The terms C_{Bias} and C_{Var} are both $O(1)$ in n_i, and correspond to two distinct costs: C_{Bias} can be interpreted as the cost of bias-correcting the maximum likelihood estimator while C_{Var} arises from the variability of the nuisance parameter estimate. The formulas will be given in Sec. 2. We will write $J_2^* = J_1 - C_{\text{Bias}} - C_{\text{Var}}$ for the case when we drop the $O(1/n)$ term.

We note that this information analysis is for the second-order information loss of the bias-corrected maximum likelihood estimator. It should be noted that if one models the nuisance parameters λ_i as coming from a distribution, following the proposal of Kiefer and Wolfowitz (1956), then it is possible to reduce the variability cost of the nuisance parameters. See Lindsay (1985) for an extended treatment of how such an empirical Bayes treatment can improve efficiency over the "fixed effects" maximum likelihood estimator we consider here.

Our first task, in Sec. 2, is to create the asymptotic framework that motivates our work. In Sec. 3, we consider the bias that arises in maximum likelihood estimation, and develop the natural way to eliminate the bias to the second-order, together with the corresponding cost involved with this correction. Secs. 4, 5, and 6 consider the implications of the calculations in a variety of models, including those with and without various kinds of conditional structures.

2. SETTING UP THE PROBLEM

We start by reviewing some relevant first-order asymptotic theory, which will be used to motivate our considerations in the second-order theory. We will adopt the convention here of deriving our results in the context of what we will call the "scalar" case, in which both ψ and λ are scalar parameters. This will enable us to avoid complicated arrays, while not significantly changing the nature of the problem. We will write the key formulas in the vector case, however. Many similar calculations are provided in the multi-dimensional case in Waterman and Lindsay (1996a, b, 1997) which provide much of the foundation of this analysis. A detailed treatment of the calculations as well as background material can be found in Waterman (1993).

2.1 The First-Order Theory

Our initial focus is on the asymptotics in a single row, in which $n_i \to \infty$. In the ensuing developments, when we are doing calculations within a single row, we will drop the strata subscript, i, and write n for n_i and so on. Whenever i does reappear as a subscript, it will refer to the corresponding variable as calculated from row i.

Within a fixed row, define the score function with respect to the parameter of interest, ψ, as $U_0 = \partial_\psi \{\log f(x; \psi, \lambda)\}$ and the score function with respect to λ as $V_1 = \partial_\lambda \{\log f(x; \psi, \lambda)\}$ and denote the operation of projection onto the closed linear space spanned by $\{V_1\}$ by Π_1. That is, $\Pi_1 S$, for any random variable S, equals $\hat\rho V_1$, where $\hat\rho$ minimizes $E[S - \rho V_1]^2$. The efficient score $U_1 = U_1(\psi, \lambda)$ is obtained by removing from U_0 its projection onto the space spanned by V_1 and therefore has the representation

$$U_1 = U_0 - \Pi_1 U_0.$$

This can be explicitly calculated as

$$U_1 = U_0 - \frac{E(U_0 V_1)}{E(V_1 V_1)} V_1.$$

Note that $\mathrm{Var}(U_1) = J_1$, so that the efficient score reflects the first-order information about the parameter ψ in the presence of λ.

This is reflected in the first-order asymptotic theory as follows: let $\hat\lambda(\psi)$ be the maximum likelihood estimator for λ when ψ is fixed—it can be obtained by solving $V_1(\psi, \lambda) = 0$ in λ. The plug-in score is defined to be $\hat U_1(\psi) = U_1(\psi, \hat\lambda(\psi))$, and the maximum likelihood estimator for ψ can be obtained by equating this score to zero. If we were to estimate ψ and λ from a single stratum, with n going to infinity, then we could analyze the maximum likelihood estimator of ψ through its plug-in score equation $\hat U_1(\psi) = 0$. In the first-order asymptotic theory for the stratum, we have that the centered maximum likelihood estimator $(\hat\psi - \psi)$ is asymptotically equivalent to $J_1^{-1} U_1$. We note that it follows that the maximum likelihood estimator is equivalent to the solution to the efficient score equation $U_1(\psi, \lambda) = 0$, where we have altered the plug-in score equation by replacing the maximum likelihood estimator $\hat\lambda(\psi)$ with the true value of λ.

We will later show that $U_1(\psi, \lambda)$ and $\hat U_1(\psi)$ are first-order equivalent, in that

$$n^{-1/2}[U_1(\psi, \lambda) - \hat U_1(\psi)] \to_{\mathrm{prob}} 0$$

However, the equivalence of inference from the plug-in score and the efficient score fails at the second-order.

In Sec. 3.2 we look more closely at this convergence and find the key terms that reveal second-order effects. In particular, the plug-in score differs from the efficient score in that it is not necessarily mean zero, and it has a variance not necessarily equal to J_1. It is this distinction that causes the second-order departure of maximum likelihood estimation from its ideal properties.

In the multistrata model, the same basic observations apply. One can find the maximum likelihood estimator of ψ by solving the plug-in equations:

$$\sum U_{1i}(\psi, \hat{\lambda}_i(\psi)) = 0. \tag{1}$$

If we naively applied first-order theory within each row the estimator would be asymptotically equivalent to the solution to the efficient score equations:

$$\sum U_{1i}(\psi, \lambda_i) = 0. \tag{2}$$

The difference between the plug-in score and the efficient score now gets accumulated over the strata, magnifying its effects.

2.2 The Formulas

The preceding remarks suggest that a key to the problem of second-order information loss is in the dependence of the efficient score on the nuisance parameter. If it depends only weakly on this parameter, then it is intuitively clear that the plug-in score should not deviate much from the efficient score, and so the cost should be low.

This notion can be made more precise by defining the random vector $W = U'_1$, where $'$ refers to differentiation with respect to the nuisance parameter. It can be verified that W has mean zero—it follows because U_1 is orthogonal to V_1. We will call W the cost score. This variable clearly determines the nature and the extent of the local dependence of the efficient score on λ, and in particular $W = 0$ if the efficient score does not depend on the nuisance parameter.

If the parameters ψ and λ are both vectors, the cost score is the $\dim(\psi) \times \dim(\lambda)$ matrix $W = \nabla_\lambda U_1^T$.

We can decompose W by projection into two uncorrelated terms $W_b = \Pi_1 W$ and $W_v = W - W_b$. In the scalar case, we let $\kappa = E[WV_1]/E[V_1^2]$, so that

$$W_v = W - \kappa V_1 \quad \text{and} \quad W_b = \kappa V_1.$$

We define the bias cost to be:

$$C_{\text{Bias}} = \frac{1}{2} E(W_b I_{22}^{-1} W_b^T),$$

and note that it is largest when the cost score is highly correlated with the nuisance score V_1. In the scalar case, we can reexpress this as $C_{\text{bias}} = \kappa^2/2$. This will be the cost of bias-correcting the maximum likelihood estimator.

On the other hand, the variance cost will be

$$C_{\text{Var}} = E(W_v I_{22}^{-1} W_v^T),$$

which is large if W has a large component orthogonal to the nuisance score space. This is the cost in estimation due to the additional variability in the plug-in score over of the efficient score.

We note the curiosity that the overall information cost $C_{\text{bias}} + C_{\text{var}}$ is off by the factor of $1/2$ in C_{bias} from simply being the normalized variance of the cost score W. Our intuition is that some of the bias cost of the m.l.e. gets recovered by using bias correction.

The simplicity of these formulas makes them a handy tool for a preliminary analysis of the secondary costs in models with many parameters, including semiparametric models. In addition to measuring the overall cost, the magnitude of the separate components separately indicate potential problems with bias and inflated variance. We do note, however, that these simple formula arise from disregarding terms of smaller order, and so can give negative informations when calculated with small values of n. We provide an extensive set of examples later in the chapter; we first provide some further theoretical justification for their relevance.

2.3   Information in Estimating Functions

The starting point of our strategy is that the maximum likelihood equation for ψ, as given in Eq. (1), has the form $H(\psi) = 0$ for a particular function H. This links our subject to that of estimating functions, the theory of which we now briefly review.

We will start by considering information concepts in the simplest case that is when there are no nuisance parameters. For an estimating function, say $H(\psi)$ that is mean zero, we define the information for vector ψ as

$$\text{Info}(H) = E(\nabla_\psi H) E(HH^T)^{-1} E(\nabla_\psi H)^T. \tag{3}$$

If the function H is sum of independent terms, then the usual theory of M-estimation indicates that the asymptotic variance of the estimator satisfying $H(\psi) = 0$ is equal to the inverse of the information, which gives it statistical content as being a measure of information.

Following the arguments of Godambe (1960), differentiation of the mean zero identity $E_\psi[H(\psi)] = 0$ with respect to ψ leads to the information identity $E(\nabla_\psi H + H U_0) = 0$. If we then replace $E(\nabla_\psi H)$ by $-E(H U_0)$ in Eq. (3), we can recognize that in the scalar case, the information equals the Fisher information $E(U_0^2)$ times the squared correlation of H and U_0. It immediately follows that the information in an estimating equation H is maximized when it is a linear function of U_0.

On the other hand, if there exists a nuisance parameter λ, and we are evaluating a zero-unbiased function $H(\psi)$ for its information, we can differentiate a mean zero identity $E_{\psi,\lambda}[H(\psi)]$ with respect to λ and obtain $E[\nabla_\lambda H + H V_1] = 0$. Since H is free of λ, the last equality shows that H is orthogonal to V_1, which in turn implies that we can measure information by correlation with the efficient score:

$$\text{Info}(H) = E(U_1 H^T) E(H H^T)^{-1} E(H U_1^T).$$

This is the formula we will later use to calculate information.

2.4 The Strategy

We return to the estimating equation for the maximum likelihood estimator:

$$\sum U_{1i}(\psi, \hat\lambda_i(\psi)) = 0.$$

If the plug-in scores $\hat U_{1i}$ were mean zero, then one could apply the foregoing theory of estimating functions (or M-estimation) to derive the asymptotic distribution of the solution as $p \to \infty$. However, there is too much bias in these equations to use them directly.

The first part of the strategy is to replace the plug-in score $\hat U_1(\psi)$ with a bias-corrected version $\hat U_2(\psi) = U_2(\psi, \hat\lambda_\psi)$. When this is done, solving

$$\sum \hat U_{2i}(\psi) = 0$$

gives an estimating function that produces a bias-corrected maximum likelihood estimator, and the bias in the scores is sufficiently reduced that one can proceed with an asymptotic information calculation, as described above. Thus, we desire to calculate $\text{Info}(\hat U_2)$.

Since the presence of the maximum likelihood estimator $\hat\lambda_\psi$ in $\hat U_2$ makes calculating $\text{Info}(\hat U_2)$ potentially tedious, and because we wish simple formulas for the second-order in n, we create a surrogate estimating function $U_2^\dagger$ that is information equivalent, in the sense that $\text{Info}(\hat U_2) =$

Info($U_2^\dagger$) + $O(1/n)$. Finally, we simplify yet further by showing that
Info($U_2^\dagger$) = J_2^* + $O(1/n)$.

REMARK   An alternative to this strategy would be to analyze the asymptotic mean square of the maximum likelihood estimator in some appropriate fashion that separates out the invariant aspects. We have chosen our present course because we believe it is more straightforward. However, we could here follow the decomposition of the mean squared error of the bias-corrected maximum likelihood estimator such as developed by Efron (1975) and later advanced by Amari (1982).

If we did so in the asymptotic setting where there are a finite number of strata and the stratum sample sizes becoming infinite at equal rates, the resulting mean squared error for ψ would have components due to parameterization effects and statistical curvature. With a slight modification, the inverse of the information we calculate here can be shown equal (to the right order in n) to the "invariant" components of mean squared error—those that do not depend on the choice of the ψ parameterization. The distinction is that C_{var} must be augmented with a correction term that is $O(1/p)$, where p is the number of strata, and so washes out in an infinite strata calculation. However, it could be significant for a small number of strata, especially as to the case of a single stratum. Of our two components, C_{bias} corresponds to parameterization effects that are unavoidable because the parameter of interest must be preserved and C_{var}^* corresponds to the effects of statistical curvature.

The second-order optimality of the bias corrected m.l.e. then establishes that the information quantities we calculate here, as adjusted above, provide second-order lower bounds for information loss.

3.   CORRECTING THE m.l.e. FOR BIAS

We start with a familiar example from the literature to illustrate the key issues in bias correction.

3.1   The Many Means Example

We suppose that the ith stratum consists of n_i independent observations from the normal density with parameters (μ_i, σ^2). When viewed across the strata, we have many means μ_i and a single variance σ^2. The σ^2 parameter will be considered the parameter of interest, and so corresponds to ψ.

Before we turn to our bias arguments, we illustrate some of our basic calculations on this example. It is straight forward to calculate the scores in the ith stratum as

$$\frac{\partial \log f(x_i; \sigma^2, \mu_i)}{\partial \sigma^2} = U_0 = -\frac{n}{2\sigma^2} + \frac{1}{2\sigma^4} \sum_{j=1}^{n} (x_{ij} - \mu_i)^2$$

$$\frac{\partial \log f(x_i; \sigma^2, \mu_i)}{\partial \mu_i} = V_1 = \frac{1}{\sigma^2} \sum_{j=1}^{n} (x_{ij} - \mu_i).$$

We can therefore calculate the information matrix to be:

$$I = \begin{bmatrix} n/2\sigma^4 & 0 \\ 0 & n/\sigma^2 \end{bmatrix}.$$

Direct calculation then shows that $U_1 = U_0$, and so

$$W = U_1' = -\frac{1}{\sigma^4} \sum_{j=1}^{n} (x_{ij} - \mu_i).$$

We can then calculate $\kappa = E[V_1 W]/E[V_1^2] = -1/\sigma^2$. Because W is perfectly correlated with V_1, we must have $W_v = 0$ and $C_{var} = 0$. We also have $C_{bias} = \kappa^2/2 = 1/2\sigma^4$. It follows that

$$J_2^* = \frac{n}{2\sigma^4} - \frac{1}{2\sigma^4} = \frac{n-1}{2\sigma^4}.$$

Thus our approximate information calculation correctly recovers the degree of freedom lost by estimating the mean.

We return to the bias problem. If the expectation of U_0 is taken when μ_i is replaced by it maximum likelihood estimator, $\bar{x}_i$, then $E\{U_0(\sigma^2, \bar{x}_i)\} = -1/(2\sigma^2)$ indicating that the plug-in score is biased. The score bias accumulates over the strata to such an extent that the maximum likelihood estimator for σ^2 is inconsistent when the stratum sample sizes are held fixed while $p \to \infty$. In the case where all the n_i are equal to n, $\hat{\sigma}^2 \to \{(n-1)/n\}\sigma^2$. Thus the asymptotic bias is substantial for small n.

Waterman and Lindsay (1997) proposed a rectangular array asymptotics to get a closer understanding of the relative roles of the magnitudes of n_i and p. To illustrate their approach, if one used an asymptotics in which there are p rows, each with γp observations, then as $p \to \infty$, the maximum likelihood estimator is still consistent, but it has the limiting property that $\sqrt{N}(\hat{\sigma}^2 - \sigma^2)$, where N is the total sample size $\gamma p \cdot p$, is normal with mean $-\gamma^{-1}\sigma^2$ and variance $2\sigma^4$. That is, bias shows up in the mean of the limiting distribution. However, the asymptotic variance is exactly as it would be for a regular problem.

This example raises the question as to whether there is a systematic way to eliminate the bias at this level of asymptotics. We next review the methodology developed by Waterman and Lindsay (1996a) that achieves this bias elimination within the context of estimating functions.

3.2 Bias Removal and the U_2 Score

The efficient score U_1 generally has magnitude $O_p(n^{1/2})$. Although U_1 is mean zero, the substitution of the nuisance parameter estimate generally induces a bias of order $O(1)$ in the plug-in score. This bias is such that it dominates in the calculation of mean squared error unless its magnitude is reduced to $O(n^{-1/2})$. Fortunately there is a natural extension of the preceding developments that leads to an estimating equation with the appropriate degree of bias removal.

We define the second-order Bhattacharyya scores, for vector λ case, to be:

$$V_{2(ij)} = \frac{(\partial^2/\partial\lambda_i\partial\lambda_j)f(x;\psi,\lambda)}{f(x;\psi,\lambda)}.$$

Higher order scores, such as $V_{3(ijk)}$, can be similarly defined, but are not needed here. Returning to the scalar λ case, we define the projection operator Π_2 to be the action of projection onto the space spanned by V_1 and V_2. The second-order projected score is defined to be $U_2 = U_0 - \Pi_2 U_0$.

It was shown by Small and McLeish (1989) that the second-order score eliminates bias in the sense that $U_2(\psi,\hat{\lambda}_c) = \hat{U}_2$ has bias of order $O(n^{-1/2})$. We offer a summary of the necessary calculations to establish the bias results, as well as identify other second-order effects.

First, since both nuisance score V_1 and the variance component of the cost score, W_v are iid sums, and they are orthogonal, one can determine the joint limiting distribution to be:

$$nE(V_1^2)^{-1/2}\binom{W_v}{V_1} = \binom{Z_{1n}}{Z_{2n}} \to_{\text{dist}} \binom{Z_1}{Z_2},$$

where Z_1 and Z_2 are independent mean zero normal variables with respective variances C_{var} and 1.

With the help of these variables, we can describe more precisely how the plug-in score $\hat{U}_1$ departs from the efficient score U_1 by capturing their difference to the next order of magnitude. It can be shown that

$$\hat{U}_1 - U_1 - Z_{1n}Z_{2n} + \frac{\kappa}{2} Z_{2n}^2 \to_{\text{prob}} 0.$$

The proof is a straightforward asymptotic expansion argument. Here the term κ is as defined earlier, equal to $E[WV_1]/E[V_1^2]$, with $C_{\text{bias}} = \kappa^2/2$. We can rewrite this convergence in the form:

$$\hat{U}_1 - U_1 \to_{\text{dist}} Z_1 Z_2 + \frac{\kappa}{2} Z_2^2$$

Thus we see that if κ, and hence C_{bias}, is 0, then the asymptotic mean of $\hat{U}_1 - U_1$ is zero, but otherwise there is a bias term of magnitude $\kappa/2$. Notice that this second-order convergence implies in an elementary fashion the first-order equivalence

$$n^{-1/2}[\hat{U}_1 - U_1] \to_{\text{prob}} 0.$$

The score function $\hat{U}_2$ very precisely removes the bias term in the sense that

$$\hat{U}_2 - U_1 \to Z_1 Z_2 + \frac{\kappa}{2} Z_2^2 - \frac{\kappa}{2},$$

and so gives a score function that is closer on average to the true efficient score U_1. However, if we want to represent $\hat{U}_2$ with a simpler score type variable, we could do so with less variation by using U_2, because:

$$\hat{U}_2 - U_2 \to Z_1 Z_2.$$

If there is no variance cost, so that $C_{\text{var}} = 0$, we must have Z_1 degenerate at zero, and so in this case the $\hat{U}_2$ score is asymptotically equivalent to U_2 at the second-order, and we could use $\text{Info}(U_2)$ for second-order information, as we establish in the next Section.

However, in general C_{var} is not zero, in which case U_2 is not sufficiently close to $\hat{U}_2$ to accurately reflect information to the second-order, so we first find a better surrogate.

3.3 The Surrogate Estimating Function

The above asymptotic statements lead us directly to a method for constructing a simple surrogate for the estimating function $\hat{U}_2$, as we have

$$\hat{U}_2 - \left\{ U_1 + Z_{1n} Z_{2n} + \frac{\kappa}{2} [Z_{2n}^2 - 1] \right\} \to_{\text{prob}} 0.$$

Thus we identify the term in brackets, written in terms of scores,

$$U_2^\dagger = U_1 + \frac{V_1 W_v}{n I_{22}} + \frac{\kappa}{2} \left[\frac{V_1^2}{n I_{22}} - 1 \right].$$

as a simpler surrogate for $\hat{U}_2$.

The next proposition shows that if an estimating equation, H, is close to the efficient score U_1, then the information in H has a simple second-order form: (The subscript n is introduced to emphasize the fact that the functions under consideration depend on n and that the limits are taken as n approaches infinity.)

PROPOSITION 3.1   Assume that the estimating function H_n can be written as

$$H_n = U_{1n} + \Delta_n,$$

and that the following two conditions are satisfied: that

$$E[\Delta_n^2] \to C$$

and that

$$\frac{E[U_{1n}\Delta_n]}{E[U_{1n}^2]} \to 0.$$

Then $\mathrm{Info}(U_{1n}) - \mathrm{Info}(H_n) \to C$.

Proof.   By the definition of information and from the construction of H_n we have that

$$\mathrm{Info}(H_n) = \frac{E^2(U_{1n}H_n)}{E(H_n^2)} = \frac{E^2\{U_{1n}(U_{1n} + \Delta_n)\}}{E\{(U_{1n} + \Delta_n)^2\}},$$

which can be rewritten as

$$\frac{[E\{(U_{1n} + \Delta_n)^2 - \Delta_n(U_{1n} + \Delta_n)\}]^2}{E\{(U_{1n} + \Delta_n)^2\}}.$$

Expanding the square in the numerator and canceling by the denominator gives

$$E\{(U_{1n} + \Delta_n)^2\} - 2E\{\Delta_n(U_{1n} + \Delta_n)\} + \frac{E^2(\Delta_n U_{1n} + \Delta_n^2)}{E\{(U_{1n} + \Delta_n)^2\}}.$$

The third term is $o(1)$ since the numerator is at most $O(1)$ whilst the denominator is $O(n)$. On simplifying the first two terms we obtain

$$\mathrm{Info}(H_n) = E(U_{1n}^2) - E(\Delta_n^2) + o(1),$$

which is the desired result. $\square$

We can apply this proposition as follows. In our case, we can write

$$\Delta_n = \hat{U}_2 - U_1 = Z_{1n}Z_{2n} + \frac{\kappa}{2}[Z_{2n}^2 - 1] + o_p(1)$$

and so under some further regularity, we have

$$E(\Delta_n^2) \rightarrow E\left[Z_1 Z_2 + \frac{\kappa}{2}\left[Z_2^2 - 1\right]\right]^2 = C_{\text{bias}} + C_{\text{var}}.$$

This leads us to the desired conclusion: $\text{Info}(\hat{U}_2) = J_2^* + o(1)$.

4. TYPE I CONDITIONAL PROBLEMS

We next consider the interpretation of our calculations in an important class of examples for which there is an alternative method for bias reduction.

In a subset of nuisance parameter problems there exists a complete and sufficient statistic, say S, for the nuisance parameter λ_i when ψ is fixed, where S does not depend on ψ. We call such a problem a Type I conditional problem to distinguish it from the case when there exists a complete and sufficient statistic $S(\psi)$ that does depend on ψ, which we call Type II. The many means problem has Type I conditional structure, with the sufficient statistic for the mean μ_i being $S_i = \bar{x}_i$, which does not depend on $\psi = \sigma^2$.

In all Type I models one can create a conditional score function $U_c(\psi)$ by differentiating the logarithm of the conditional density of the data given S. The conditional score U_c is free of the nuisance parameters and is mean zero and so provides an unbiased estimating function, the solution to which is called the conditional maximum likelihood estimator. The conditional score can be written as

$$U_c = U_0 - E[U_0|S].$$

The conditional maximum likelihood was developed extensively by Andersen (1970). In the many means problem, the corresponding conditional maximum likelihood estimator is exactly the bias corrected maximum likelihood estimator $n/(n-1)\hat{\sigma}^2$ when the stratum sample sizes are equal to n.

In such problems it seems reasonable to presume that the conditional score is optimal, and so our J_2 formula should agree with the conditional information to the right order of magnitude. (For more on the optimality of conditional methods, see Lindsay (1983), in which it is established that even if the nuisance parameters are assumed to come from a distribution, it is not possible to construct more efficient estimators.)

Waterman and Lindsay (1996a, b) extensively considered the relationship of the corrected score U_2 and the conditional score U_c. We briefly summarize some of their key findings.

The first surprise is that in every Type I model with exponential family structure, there is in fact no variance cost associated with using $\hat{U}_2$; that is, $W = U_1'$ is a linear function of V_1 so that $U_1' - \kappa V_1 = 0$. (This is a special case of a more general result that says that the tth derivative of U_t is a linear function of V_t^*, where V_t^* is an orthogonalized version of the tth Bhattacharyya basis function.)

It follows that $\hat{U}_2$ is asymptotically equivalent to U_2. The information in U_2 was calculated by Waterman and Lindsay (1996a) and shown to satisfy:

$$\text{Info}(U_2) = J_1 - C_{\text{bias}} + O(1/n) = J_2^* + O(1/n).$$

They also establish the relationship between the U_2 score and the conditional score, showing that $E(U_c - U_2)^2$ is $O(1/n)$, so that $\text{Info}(U_c) = \text{Info}(U_2) + O(1/n)$. Thus J_2^* captures the information in U_c to second-order.

This establishes our basic information formula as being appropriate for capturing the second-order information in Type I conditional problems. As a simple illustration, note that in the many means example, the information in the conditional score function equals $J_1 - C_{\text{Bias}} = (n-1)/2\sigma^4$ exactly, capturing the loss of a degree of freedom from estimating the nuisance parameter. We conclude this Section with some further illustrations of the calculations.

4.1 Paired Poisson

In some special cases the U_1 score function is equal to U_c. An example of this is the paired Poisson problem, where we observe two independent Poissons with means $\exp(\psi + \lambda)$ and $\exp(\lambda)$ respectively. Here, $X + Y$ is the complete sufficient statistic for λ and $V_1 = x + y - E(X + Y)$. The conditional distribution of X given $X + Y$ is well known to be Binomial with parameters $N = X + Y$ and $p = \exp(\psi)/1 + \exp(\psi)$. Using this fact it is easy to show that

$$E(U_0 \mid X + Y) = (X + Y)\frac{\exp(\psi)}{1 + \exp(\psi)} - \exp(\psi + \lambda).$$

This is a linear function of V_1, so that $U_1 = U_c$ and therefore $J_1 = J_2^* = \text{Info}(U_2)$, so there is no second-order cost to estimation of the nuisance parameter.

4.2 Paired Bernoulli

Moving now to the building block for many commonly used statistical models (e.g., logistic regression, Rasch models), we consider a paired Bernoulli experiment. We observe two independent Bernoulli variables with odds of success $\exp(\psi + \lambda)$ and $\exp(\lambda)$ respectively, so that the parameter of interest is the log-odds ratio ψ. This is again a Type 1 model with $X + Y$ being the complete and sufficient statistic for λ. One interesting aspect of this model is that $U_2 = U_c$.

It can be shown, preferably by symbolic computation, that

$$\kappa = \frac{2\, e^{\lambda + \upsilon}\left(1 + e^{\lambda}\right)\left(e^{\upsilon} - 1\right)\left(1 + e^{\lambda + \upsilon}\right)}{\left(1 + e^{\upsilon} + 4\, e^{\lambda + \upsilon} + e^{2\,(\lambda + \upsilon)} + e^{2\,\lambda + \upsilon}\right)^2}.$$

Clearly κ is equal to zero when $\psi = 0$, and in this situation U_1 also equals $U_2 = U_c$. One can indeed verify our earlier claim that $U_1' = \kappa V_1$, so $W_v = 0$, although the details of this verification are rather complicated. So, as predicted, the only cost is $C_{\text{bias}} = \frac{1}{2}\kappa^2$.

To get a feeling for the speed in which the second-order asymptotics kicks in we can let the Bernoulli grow into a Binomial. The following calculations are for paired Binomial experiments with sample sizes of 1 through 5, 10, and 20. All calculations are exact, and the parameters are set as $\psi = 0.5$ and $\lambda = -0.25$ (see Table 1).

We note that our second-order approximation only captures half the information loss for $n = 1$, but improves rapidly and quite generally captures the overall magnitude.

In this particular example it could be argued that the second-order information loss is negligible and that from a practical perspective of no consequence. However, the information loss is a function of the curvature κ, which for Bernoulli problems is an increasing function of the asymmetry of the probabilities. If we had considered parameter values for $\psi = 2$ and $\lambda = -1$, then Table 1 would have had a first line as shown in Table 2 so that the second-order loss in information would have been of the order of 50%.

Table 1. Information Calculations for a Paired Binomial Experiment. Parameter Values Are $\psi = 0.5$ and $\lambda = -0.25$.

Trials	J_1	J_2	$J_1 - J_2$	$C_{\text{bias}} = \frac{1}{2}\kappa^2$
1	0.123067	0.119319	0.00374818	0.00193296
2	0.246134	0.243584	0.00255057	0.00193296
3	0.369201	0.366896	0.00230507	0.00193296
4	0.492268	0.490069	0.00219923	0.00193296
5	0.615335	0.613195	0.00214026	0.00193296
10	1.23067	1.22864	0.00203133	0.00193296
20	2.46134	2.45936	0.00198093	0.00193296

Table 2. Information Calculations for a Paired Bernoulli Experiment. Parameter Values Are $\psi = 2.0$ and $\lambda = -1.0$.

Trials	J_1	J_2	$J_1 - J_2$	$\frac{1}{2}\kappa^2$
1	0.098306	0.0637076	0.0345984	0.026694

5. TYPE II CONDITIONAL MODELS

We turn now to conditional models that display a different kind of structure because the sufficient statistic $S(\psi)$ for the nuisance parameter depends on the parameter of interest. There is an interesting subclass of these models which are in the exponential family and characterized by densities of the form

$$f(x, \psi, \lambda) = \exp\{\lambda S(\psi) - \lambda\tau(\psi, \lambda)\}.$$

Such a structure arises when the parameter of interest ψ is the ratio of natural parameters. In this case, the conditional score function depends on the nuisance parameter, but only through a weighting factor ($'$ denotes differentiation with respect to ψ):

$$U_c = \lambda[S'(\psi) - E_c\{S'(\psi) \mid S(\psi)\}].$$

Notice the important point that with this type of conditioning, the conditional score depends on λ, and we will find that in this class, unlike the Type I, we will have $C_{\text{var}} > 0$.

One can show that in this class of models, U_1 is equal to U_c, and that W, its derivative with respect to lambda is therefore free of lambda. Furthermore, W is orthogonal to V_1 so that $C_{\text{Bias}} = 0$ and the information loss is therefore entirely in C_{var}. We consider some important examples with this kind of Type II structure.

5.1 The Many Variances Problem

We return to a second important Neyman–Scott example. Suppose now that each stratum consists of n_i independent normal variables with common mean μ and variance σ_i^2 that depends on the stratum. We now consider the mean parameter to be of interest, and the nuisance parameters are the many variances σ_i^2.

We have earlier laid out in the many means problem the necessary first-order calculations—one only needs to reverse the role of the two parameters. The U_0 score function is therefore

$$U_0 = \frac{1}{\sigma_i^2} \sum_{j=1}^{n_i} (x_{ij} - \mu).$$

It is straightforward to show that because U_0 is linear in μ and the nuisance parameter enters only as a weight that $U_0 = U_1 = U_2$. Furthermore, one can show $U_0 = U_c$. We can further calculate that

$$W = -\frac{1}{\sigma_i^4} \sum_{j=1}^{n_i} (x_{ij} - \mu).$$

Since W is orthogonal to V_1, the score for σ^2, it follows that $C_{\text{bias}} = 0$, and we can calculate $C_{\text{var}} = E[W^2]/I_{22} = 2/\sigma^2$. Therefore the corrected information about μ is

$$J_2^* = \frac{n}{\sigma^2} - \frac{2}{\sigma^2} = \frac{n-2}{\sigma^2}.$$

We can compare this with the estimating function analysis of the actual maximum likelihood estimator. In this case, the plug-in score is unbiased, $E\{U_0(\mu, \hat{\sigma}_{i\,|i}^2)\} = 0$, so that one can proceed to show that the maximum likelihood estimator of μ is consistent. One can therefore apply estimating function methodology to find its information, which for a single stratum is $\text{Info}(\hat{U}_0) = J_2^*$, so our second-order formula is exact in this case.

5.2 Paired Exponentials

For another example from this class of models, consider two exponential random variables with means $E(X) = 1/\psi\lambda$ and $E(Y) = 1/\lambda$. A complete sufficient statistic for the nuisance parameter λ is given by $\psi X + Y$, which depends on ψ; our characterization of a Type II problem. The score functions U_0 and V_1 are

$$U_0 = \frac{1}{\psi} - \lambda X,$$

$$V_1 = \frac{2}{\lambda} - (\psi X + Y).$$

The U_1 score function is simply

$$\lambda \left(\frac{y - \psi x}{2\psi} \right),$$

so that W is simply the term within the brackets. Once again $W = W_v$ and a straight forward calculation shows that

$$E(W_v^2) = \frac{1}{2\psi^2\lambda^2},$$

so that C_{Var} is equal to $1/4\ \psi^2$. The first-order information $J_1 = n/2\psi^2$, so that

$$J_2^* = \frac{n - \frac{1}{2}}{\psi^2},$$

reflecting the loss in some sense of "half a degree of freedom."

In this model there is a marginal distribution that we can exploit to obtain an exact calculation of the information in the maximum likelihood estimator, which is based on $\hat{U}_1$. The key fact is that the marginal likelihood of $X_./Y_.$, where · denotes summation, is free of λ. The information in $\hat{U}_1$ can be calculated as $-E\{(\partial/\partial\psi)\hat{U}_1\}$ because this is the score function in the marginal likelihood and therefore information unbiased. One can show that

$$\frac{\partial}{\partial\psi}\ \hat{U}_1(\psi) = -\frac{2n}{\psi^2}\ \frac{\psi X_.}{(\psi X_. + Y_.)^2},$$

and noting that the random variable $\psi X_./(\psi X_. + Y_.)^2$ has a beta distribution with parameters (n, n) allows the information to be readily calculated as

$$E\left(-\frac{\partial}{\partial\psi}\ \hat{U}_1(\psi)\right) = \frac{n - (n/2n + 1)}{2\psi^2}.$$

Thus we see that once again J_2^* has captured the information terms to the right order, although it is not exact in this instance.

6. MODELS WITHOUT CONDITIONAL STRUCTURE

We next turn to models that lack conditional structure, but by their simple nature can offer us a new set of insights into the loss from nuisance parameters. We first consider models for pairs of independent normal variables with common known variance. The nuisance parameter effects arise because parameters ψ and λ are nonlinear functions of the normal means.

6.1 Circular Normal

We illustrate with a simple example. The stratum consists of a pair of random variables, X and Y, with distributions $N(\rho\cos(\phi), 1)$ and $N(\rho\sin(\phi), 1)$.

We alternatively consider the parameter of interest to be radius ρ and angle ϕ. (The model corresponds to a bivariate normal distribution with the unknown mean being parameterized in polar coordinates.)

The two score functions are $U_\rho = r\cos(\hat{\phi} - \phi) - \rho$ and $U_o = \rho r \sin(\hat{\phi} - \phi)$, where r and $\hat{\phi}$ are the observed values of the radius and angle respectively. The Fisher's information matrix is

$$\begin{pmatrix} I_{\rho\rho} & I_{\rho\phi} \\ I_{\phi\rho} & I_{\phi\phi} \end{pmatrix} = \begin{pmatrix} 1 & 0 \\ 0 & \rho^2 \end{pmatrix}.$$

The parameters are orthogonal, (the off diagonal elements of the Fisher's information matrix are zero) which makes the calculations simpler, in particular $U_1 = U_0$ for both versions of the parameter of interest.

First consider the case where ρ is the parameter of interest, ψ. The Fisher's information for n observations is $J_1 = n \cdot 1$, and $W = -r\sin(\hat{\phi} - \phi)$, which is perfectly correlated with V_1, so $C_{\text{Bias}} = 1/(2\rho^2)$ and $C_{\text{Var}} = 0$. In particular, we have

$$J_2^* = n - \frac{1}{2\rho^2}.$$

Thus we will have bias problems with maximum likelihood in this setting, and they will be aggravated as ρ becomes small, but there is no further variance cost in estimating the parameter ρ.

On the other hand, if ϕ is the parameter of interest, we can calculate $J_1 = \rho^2$ and $W = r\sin(\hat{\phi} - \phi)$. The latter is orthogonal to V_1 so $C_{\text{Bias}} = 0$ and $C_{\text{Var}} = 1$. This gives

$$J_2^* = n\rho^2 - 1.$$

and indicates that a multistrata m.l.e. will not be bothered by bias, but will have severe variability problems if the nuisance parameters ρ_i are sufficiently small.

An alternative way of comparing the information calculations is to consider the relative information, defined here as J_2^*/J_1. For the two versions of the circular normal problem the relative information is $1 - 1/(2n\rho^2)$ and $1 - 1/(n\rho^2)$ respectively, which clearly shows the role of the curvature.

6.2 Adaptive Estimation

A curious feature of the location problem is the phenomenon of adaptive estimation. It has been shown by Bickel (1982) that if we have the semi-parametric model $f(x - \theta)$ where f is an unknown density that is symmetric about zero, then it is possible to estimate θ as accurately with f unknown as with f known. That is, it is possible to achieve the asymptotic variance formula $E[f'(x)/f(x)]^2$, here and hereafter setting $\theta = 0$ to simplify formulas.

We can here think of this as a nuisance parameter problem, with the parameter of interest being θ, and the nuisance parameter being the infinite dimensional space of functions $f(\cdot)$. The concept of efficient score has been adapted to such a semiparametric setting, and in this case the efficient score is $U_1 = -f'(x)/f(x)$, which also equals U_0 (the prime here refers to x differentiation). Thus, based on first-order information calculations, there is no cost to not knowing the function-valued parameter f.

However, it is clear that U_1 depends very strongly on the nuisance function, so we would expect some second-order costs. One simple way to consider the higher cost involved from the estimation of the density f is to consider a simple nuisance parameter model of the form

$$h(x, \theta, \pi) = (1 - \pi)f(x - \theta) + \pi g(x - \theta),$$

where f and g are two symmetric densities, and the mixing parameter π is the nuisance parameter. The efficient score is here

$$U_0 = U_1 = \frac{-h'}{h}$$

The nuisance parameter score is

$$V_1 = \frac{g - f}{h}.$$

The cost score is

$$W = -\partial_\pi \frac{h'}{h} = \frac{h \cdot (g' - f') - h'(g - f)}{h^2} = \frac{fg' - f'g}{h^2}.$$

In this case, $C_{\text{Bias}} = 0$, but $C_{\text{Var}} = E_h[(U_f - U_g)gf/h^2]^2/E[V_1^2]$, where $U_f = f'/f$ and $U_g = g'/g$.

As an illustration of these calculations, if f is standard normal and g is standard Cauchy, which are rather dramatically different, then at $\pi = \frac{1}{2}$, we can calculate numerically that $J_1 = n(0.67)$, whereas $C_{\text{var}} = 0.23/0.27 = 0.85$, corresponding to only the loss of about $0.85/0.67 = 1.27$ observations at the second-order. One might compare this with our earlier calculation in the many variances problem that showed a second-order cost of two observations as the price for not knowing the normal variance.

7. DISCUSSION

We have shown that by decomposing the variance of the cost score W, one can obtain simple and meaningful measures of the second-order costs of nuisance parameters. This leaves open the question of whether they could fruitfully be used in a more direct manner, such as a route to improving confidence interval statements. We rather suspect that is it so, but have our doubts that in the modern age, such second-order improvements would compete well against more computer-intensive methods, such as a parametric bootstrap, which would incorporate all the aspects of finite-sample properties, not just the ones we have focussed on.

We rather think that our analyses are more fruitfully used for understanding several more theoretical questions.

One, which we raised in our last example, relates to whether these methods could be fruitfully used to gain further understanding of the costs of joint estimation in semiparametric models, where the dimension of the nuisance parameter could be sufficiently large as to raise very large second-order costs.

Another question relates to modeling a large set of nuisance parameters as coming from a random effects distribution as opposed to a fixed effects Neyman–Scott type structure: when is this beneficial? This is a case where the benefits in a single stratum do not come at the first-order, but perhaps a second-order analysis incorporating this aspect could provide some insights.

In our analysis we have used the natural (from an analytic viewpoint) estimator for λ. By definition we are not interested in λ, so would be happy if we could use a simpler estimate for it; two immediate candidates are method of moments estimates and one step approximations to the maximum likelihood estimator. If we are in a random effects environment, we could used posterior expectations of the nuisance parameters. We are interested in the efficiency consequences of such methods.

Another area of interest is the subject of approximate conditionality. We know that in Type I models, those in which conditional distributions exist that are free of the nuisance parameters, then $C_{\mathrm{Var}} = 0$. We are interested to explore the extent to which the statement $C_{\mathrm{var}} = 0$, or equivalently $W \perp V_1$, is a substitute for conditionality.

We finally note that we have presented but one of many approaches to the Neyman–Scott problem. An alternative perspective is that offered by Bickel et al. (1993) who consider semi-parametric statistical models, in which the strata specific parameters are modeled as realizations from an unspecified distribution.

REFERENCES

Amari, S. Differential geometry of curved exponential families—curvatures and information loss. *Ann. Statist., 10*: 357–385, 1982.

Andersen, E. B. Asymptotic properties of conditional maximum likelihood estimators. J. R. Statist. Soc. B, 32:283–301, 1970.

Bickel, P. J. On adaptive estimation. *Ann. Statist., 10*: 647–671, 1982.

Bickel, P. J., Klaassen, C. A. J., Ritov, Y., and Wellner, J. A. *Efficient and adaptive estimation for semiparametric models*. The Johns Hopkins University Press, Baltimore. 1993.

Efron, B., Defining the curvature of a statistical problem (with applications to second-order efficiency). *Ann. Statist., 3*: 1189–1117, 1975.

Godambe, V. P. An optimum property of regular maximum likelihood estimation. *Ann. Math. Statist., 31*: 1208–1211, 1960.

Kiefer, J., and Wolfowitz, J. Consistency of the maximum likelihood estimator in the presence of infinitely many nuisance parameters. *Ann. Math. Statist., 27*: 887–906, 1956.

Lindsay, B. G. Efficiency of the conditional score in a mixture setting. *Ann. Statist., 11*: 486–497, 1983.

Lindsay, B. G. Using empirical partially Bayes inference for increased efficiency. *Ann. Statist., 13*: 914–931, 1985.

Neyman, J., and Scott, E. L. Consistent estimates based on partially consistent observations. *Econometrica, 16*: 1–32, 1948.

Small, C. G., and McLeish, D. L. Projection as a method for increasing sensitivity and eliminating nuisance parameters. *Biometrika, 76*: 693–703, 1989.

Waterman, R. P. Projective Score Methods. PhD. dissertation, Pennsylvania State University, University Park, PA, 1993.

Waterman, R. P., and Lindsay, B. G. The accuracy of projected score methods in approximating conditional scores. *Biometrika, 83*: 1,14, 1996a.

Waterman, R. P., and Lindsay, B. G. A simple and accurate method for approximate conditional inference in generalized linear models. *J. Roy. Statist. Soc. B, 58*: 177,188, 1996b.

Waterman, R. P., and Lindsay, B. G. Projected score methods for nuisance parameters: Asymptotics and Neyman–Scott problems. Unpublished manuscript, 1997.

Appendix: THE PUBLICATIONS OF MADAN L. PURI

Books and Monographs

1970

1. Nonparametric Techniques in Statistical Inference. Cambridge University Press, England, 623 pages. (Editor).

1971

2. Nonparametric Methods in Multivariate Analysis. John Wiley and Sons, New York, 440 pages. Co-author: P. K. Sen. Revised Reprint by Krieger Publication Company, Florida (1993).

1975

3. Statistical Inference and Related Topics. Academic Press, New York, 352 pages. (Editor).
4. Stochastic Processes and Related Topics. Academic Press, New York, 315 pages. (Editor).

1982

5. Nonparametric Statistical Inference, Vol I. North Holland Publishing Company, Amsterdam, 463 pages. Co-editors: B. V. Gnedenko (Moscow State University, Moscow, U.S.S.R.) and Istvan Vincze (Hungarian Academy of Sciences, Budapest, Hungary).
6. Nonparametric Statistical Inference, Vol. II. North Holland Publishing Company, Amsterdam, 446 pages. Co-editors: B. V. Gnedenko (Moscow State University, Moscow, U.S.S.R.) and Istvan Vincze (Hungarian Academy of Sciences, Budapest, Hungary).

1985

7. Nonparametric Methods in General Linear Models. John Wiley and Sons, New York, 399 pages. Co-author: P. K. Sen.

1987

8. New Perspectives in Theoretical and Applied Statistics. John Wiley and Sons, New York, 544 pages. Co-editors: J. P. Vilaplana (University of Bilbao, Spain) and Wolfgang Wertz (Technische Universität, Vienna, Austria).
9. Mathematical Statistics and Probability Theory: Volume A, Theoretical Aspects. D. Reidel Publication Company, Holland, 326 pages. Co-editors: Pal Révész (Hungarian Academy of Sciences, Budapest, Hungary) and Wolfgang Wertz (Technische Universität, Vienna, Austria).

1993

10. Statistical Sciences and Data Analysis: Proceedings of the Third Pacific Area Statistical Conference. VSP International Science Publishers, The Netherlands, 570 pages. Co-editors: Kameo Matusita and Takesi Hayakawa.

1994

11. Recent Advances in Statistics and Probability: Proceedings of the 4th International Meeting of Statistics in the Basque Country, Spain. VSP International Science Publishers, Utrecht, The Netherlands, 458 pages. Co-editor: J. P. Vilaplana.

Research Papers

1964

12. *Asymptotic efficiency of a class of c–sample tests.* The Annals of Mathematical Statistics, 35 (1964), 102–121.

1965

13. *On the estimation of contrasts in linear models.* The Annals of Mathematical Statistics, 36, 198–202. Co-author: S. Bhuchongkul.
14. *Some distribution–free k–sample rank tests of homogeneity against ordered alternatives.* Communications on Pure and Applied Mathematics, 18, 51–63.
15. *On the combination of independent two–sample tests of a general class.* Review of the International Statistical Institute, 33, 229–241.
16. *On some tests of homogeneity of variances.* Annals of the Institute of Statistical Mathematics, 17, 323–330.
17. *On a class of testing procedures in linear models.* The Journal of Indian Statistical Association, 3, 1–8.

1966

18. *On a class of multivariate multi–sample rank–order tests.* Sankhyā Series A, 28, 353–376. Co-author: P. K. Sen.

1967

19. *Multi–sample analogues of some one–sample tests.* The Annals of Mathematical Statistics, 38, 523–549. Co-author: K. L. Mehra.
20. *On the theory of rank–order tests for location in the multivariate one–sample problem.* The Annals of Mathematical Statistics, 38, 1216–1228. Co-author: P. K. Sen.
21. *On robust estimation in incomplete block designs.* The Annals of Mathematical Statistics, 38, 1587–1591. Co-author: P. K. Sen.
22. *On some optimum nonparametric procedures in two–way layouts.* Journal of the American Statistical Association, 62, 1214–1229. Co-author: P. K. Sen.
23. *Combining independent one–sample tests of significance.* Annals of the Institute of Statistical Mathematics, 19, 285–300.
24. *On some rank tests for homogeneity of scale parameters.* Trabajos de Estadistica, 28, 353–376. Co-author: P. S. Puri.

1968

25. *On Chernoff–Savage tests for ordered alternatives in randomized blocks.* The Annals of Mathematical Statistics, 39, 967–972. Co-author: P. K. Sen.
26. *Nonparametric confidence regions for some multivariate location problems.* Journal of the American Statistical Association, 62, 1373–1378. Co-author: P. K. Sen.
27. *Selection procedures based on ranks: Scale parameter case.* Sankhyā, Series A, 30, 291–302. Co-author: P. S. Puri.
28. *On a class of rank–order estimates of contrasts in MANOVA.* Sankhyā, Series A, 30, 31–36. Co-author: P. K. Sen.
29. *On a class of multivariate multisample rank order tests II. Tests for homogeneity of dispersion matrices.* Sankhyā, Series A, 30, 1–22. Co-author: P. K. Sen.
30. *Multi–sample scale problem: Unknown location parameter.* Annals of the Institute of Statistical Mathematics, 22, 99–106.

1969

31. *Multiple decision procedures based on ranks for certain problems in analysis of variance.* The Annals of Mathematical Statistics, 40, 619–632. Co-author: P. S. Puri.

32. *A class of rank order tests for a general linear hypothesis.* The Annals of Mathematical Statistics, 40, 1325–1343. Co-author: P. K. Sen.
33. *Analysis of covariance based on general rank scores.* The Annals of Mathematical Statistics, 40, 610–618. Co-author: P. K. Sen.
34. *Rank order tests for multivariate paired comparisons.* The Annals of Mathematical Statistics, 40, 2101–2117. Co-author: H. Shane.
35. *On a class of rank–order tests for the identity of two multiple regression surfaces.* Zeitschrift für Wahrscheinlichkeitstheorie und verw. Geb., 12, 1–8. Co-author: P. K. Sen.
36. *On the asymptotic normality of one–sample rank order test statistics.* Teoria Veroyatnostey i ee Primeneniya, (also in Theory of Probability and its Applications, Translation by SIAM) 14, 167–173. Co-author: P. K. Sen.
37. *The Van Elteren W Test and non–null hypotheses.* Review of the International Statistical Institute, 37, 166–175.
38. *On the asymptotic theory of rank order tests for experiments involving paired comparisons.* Annals of the Institute of Statistical Mathematics, 20, 163–173. Co-author: P. K. Sen.
39. *On robust nonparametric estimation in some multivariate linear models.* Proceedings of the Second International Symposium on Multivariate Analysis, Dayton, Ohio, 33–52, Academic Press. Co-author: P. K. Sen.
40. *Structure of distribution–free multivariate paired comparison tests.* Proceedings of the 37th Session of the International Statistical Institute, 320–322.

1970

41. *Asymptotic theory of likelihood ratio and rank order tests in some multivariate linear models.* The Annals of Mathematical Statistics, 41, 87–100. Co-author: P. K. Sen.
42. *On the estimation of location parameters in the multivariate one–sample and two–sample problems.* Metrika, 16, 58–73. Co-author: P. K. Sen.
43. *On a class of rank–order tests for independence in multivariate distributions.* Sankhyā, Series A, 32, 271–298. Co-author: P. K. Sen.
44. *Statistical inference based on incomplete blocks designs.* Nonparametric Techniques in Statistical Inference, Cambridge University Press, 131–153. Co-author: H. Shane.

1972

45. *On the robustness of rank order tests and estimates in the generalized multivariate one–sample location problem.* Zeitschrift fur Wahrscheinlichkeitstheorie und Verw. Geb., 22, 226–241-author: P. K. Sen.

46. *On some selection procedures in two-way layouts.* Zeitschrift fur Wahrscheinlichkeitstheorie und Verw. Geb., 22, 242–250. Co-author: P. K. Sen.
47. *Some aspects of nonparametric inference.* Review of the International Statistical Institute, 40, 298–327.
48. *Rank tests for some linear hypotheses in paired comparisons designs.* Sankhyā, Series A, 34, 257–264.

1973

49. *On parallel sum and difference of matrices.* Journal of Mathematical Analysis and Applications, 44, 92–97. Co-author: S. K. Mitra.
50. *A note on ADF tests for subhypotheses in multiple linear regression.* The Annals of Statistics, 1, 533–536. Co-author: P. K. Sen.

1974

51. *Joint asymptotic multinormality for a class of rank order statistics in multivariate paired comparisons.* Journal of Multivariate Analysis, 4, 88–105. Co-author: C. T. Russell.
52. *The asymptotic behavior of the minimax risk for multiple decision problems.* Sankhyā, Series A, 36, 1–12. Co-author: O. Krafft.
53. *Asymptotic distribution of rank statistics for experiments involving incomplete block designs.* Annals of the Institute of Statistical Mathematics, 26, 421–446. Co-author: N. L. Wykoff.
54. *Asymptotic results for rank statistics in some multivariate problems.* Bulletin of the International Statistical Institute, 35(2), 285–293.

1975

55. *Order of normal approximation for rank test statistics distribution.* Annals of Probability, 3, 526–533. Co-author: J. Jureckova.
56. *Distribution–free approaches to general linear models.* Proceedings of the International Symposium in Statistical Design and Linear Models, Fort Collins, Colorado; North Holland Publishing Company, 459–474. Co-author: P. K. Sen.
57. *Miscellaneous problems of rank order theory.* Proceedings of the International Conference on Asymptotic Methods of Statistics and Their Applications in Physical and Social Sciences, Charles University, Prague, Czechoslovakia, 273–311.

1976

58. *Augmenting Shapiro–Wilk Test for normality.* Contributions to Applied Statistics, Birkhauser Verlag, 129–139. Co-author: C. R. Rao.
59. *On linear combination of order statistics.* Essays in Probability and Statistics, 433–449. Co-author: F. Eicker.

60. *Remainder term estimate in the distribution of rank statistics under alternatives.* Proceedings of the Statistics Days Conference, 136–142. Co-author: H. Bergström.

1977

61. *Convergence and remainder terms in linear rank statistics.* Annals of Statistics, 5, 671–680. Co-author: H. Bergström.
62. *Asymptotically distribution–free aligned rank order tests for composite hypotheses for general linear models.* Zeitschrift fur Wahrscheinlich-keitstheorie und verw. Geb., 39 (1977), 175–186. Co-author: P. K. Sen.
63. *Problems of association for bivariate circular data and new test of independence.* Proceedings of the Fourth International Symposium on Multivariate Analysis, North Holland Publishing Company, 513–522. Co-author: J. S. Rao.

1978

64. *A spherical correlation coefficient robust against scale.* Biometrika, 65, 391–395. Co-author: K. V. Mardia.

1979

65. *Local maxima of the sample functions of the N–parameter Bessel process.* Stochastic Processes and Their Applications, 9, 137–145. Co-author: L. T. Tran.
66. *Analysis of central place theory.* Bulletin of the International Statistical Institute, 47(2), 93–110. Co-authors: R. Edwards and K. V. Mardia.
67. *Realization of ℓ_p by spaces of random variables.* Applicable Analysis, 8, 337–347. Co-author: S. S. Sheu.
68. *Shorted operators and generalized inverses of matrices.* Linear Algebra and its Applications, 25, 45–56. Co-author: S. K. Mitra.
69. *Rank order estimates in the case of grouped data.* Sankhyā, Series B, 41, 239–259. Co-author: A. R. Padmanabhan.
70. *Asymptotic behavior of stochastic linear rank statistics.* Bulletin of the International Statistical Institute, 47(4), 292–297. Co-author: N. S. Rajaram.
71. *A simple test for goodness-of-fit based on spacing with some efficiency comparisons.* Contributions to Statistics – Jaroslav Hajek Memorial Volume, Academia Prague, Czechoslovak Academy of Sciences, 197–209. Co-authors: J. S. Rao and Y. Yoon.
72. *Sample size, parameter rates and contiguity – the i.i.d. case.* Communication in Statistics – Theory and Methods, A8 (1), 71–83. Co-authors: M. G. Akritas and G. G. Roussas.

73. *Matrices G satisfying simultaneous equations $A * MAG = A * M$ and $G * NGA = G * N$.* Journal of Indian Statistical Association, 17, 103–108. Co-authors: C. G. Khatri and S. K. Mitra.

1980

74. *Empirical distribution functions and functions of order statistics for mixing random variables.* Journal of Multivariate Analysis, 10, 405-425. Co-author: L. T. Tran.

75. *Asymptotic normality and convergence rates of linear rank statistics under alternatives.* Banach Center Publications, 6, 267–277. Co-author: N. S. Rajaram.

1981

76. *Invariance principles for rank statistics for testing independence.* Contributions to Probability Theory. Academic Press, New York, 267–282. Co-author: L. T. Tran.

77. *A note on predicting the results of chess championship.* Behavioral Science, 26, 85–87. Co-author: Robert Bartoszynski.

78. *Complex Planar Splines.* Journal of Approximation Theory, 31, 383–402. Co-author: Gerhard Opfer.

79. *Some remarks on strategy in playing tennis.* Behavioral Science, 26, 379–387. Co-author: Robert Bartoszynski.

80. *Différentielle d'une fonction floue.* C.R. Acad. Sc. Paris, 293, Série I, 237–239. Co-author: Dan A. Ralescu.

81. *A nonparametric test of ordered alternatives in the case of skewed data, with a biomedical application.* Statistics and Related Topics, North Holland Publishing Company, 279–283. Co-authors: A. R. Padmanabhan and A. K. Md. Ehsanes Saleh.

1982

82. *Shorted matrices — an extended concept and some applications.* Linear Algebra and its Applications, 42, 57–79. Co-author: S. K. Mitra.

83. *On estimating intersubject variability of choice probabilities under observability constraints.* Journal of Mathematical Psychology, 25, 175–184. Co-author: Robert Bartoszynski.

84. *Integration on fuzzy sets.* Advances in Applied Mathematics, 3, 430–434. Co-author: Dan Ralescu.

85. *A possibility measure is not a fuzzy measure.* Fuzzy Sets and Systems, 7, 311–313. Co-author: Dan A. Ralescu.

86. *On the degeneration of the variance in the asymptotic normality of signed rank statistics.* Statistics and Probability: Essays in Honor of C. R. Rao. North Holland Publishing Company, 591–608. Co-author: Stefan Ralescu.

87. *Stochastic integrals and rank statistics.* Nonparametric Statistical Inference, Volume II, North Holland Publishing Company, 729–747. Co-author: N. S. Rajaram.

88. *The asymptotic distribution theory of one sample signed rank statistic.* Statistical Decision Theory and Related Topics III, Volume 2, Academic Press, New York, 213–232. Co-author: Stefan S. Ralescu.

1983

89. *Theory of nonparametric statistics for rounded–off data with applications.* Mathematische Operations-Forschung und Statistik, Series Statistics, 14, 301–349. Co-author: A. R. Padmanabhan.

90. *Strong law of large numbers for Banach space valued random sets.* Annals of Probability, 11, 222–224. Co-author: Dan A. Ralescu.

91. *On the order of magnitude of cumulants of von Mises functionals and related statistics.* Annals of Probability, 11, 346–354. Co-author: R. N. Bhattacharya.

92. *Strong law of large numbers with respect to a set–valued probability measure.* Annals of Probability, 11, 1051–1054. Co-author: Dan A. Ralescu.

93. *Differentials of fuzzy functions.* Journal of Mathematical Analysis and Applications, 91, 552–558. Co-author: Dan A. Ralescu.

94. *Complex Chebyshev polynomials and generalizations with an application to the optimal choice of interpolating knots in complex planar splines.* Journal of Approximation Theory, 37, 89–101. Co-author: Gerhard Opfer.

95. *The fundamental bordered matrix of linear estimation and the Duffin–Morley general linear electromechanical systems.* Applicable Analysis, 14, 241–258. Co-author: S. K. Mitra.

96. *Improving upon the best invariant estimator in multivariate location problems.* The Australian Journal of Statistics, 25, 453–462. Co-author: Dan A. Ralescu.

1984

97. *On Berry–Esséen rates, a law of the iterated logarithm and an invariance principle for the proportion of the sample below the sample mean.* Journal of Multivariate Analysis, 14, 231–247. Co-author: Stefan S. Ralescu.

98. *Centering of signed rank statistics with continuous score generating function.* Teoria Veroyatnostei i ee Primeneniya (also in Theory of Probability and Its Applications, Translation by SIAM), 29, 580–584. Co-author: Stefan S. Ralescu.

99. *Law of large numbers and central limit theorem for fuzzy random variables.* Cybernetics and Systems Research, 2, 525–529. Co-authors: E. P. Klement and D. A. Ralescu.

100. *The Hausdorff α–dimensional measures of the level sets and the graph of the N–parameter Wiener process.* Metrika, 31, 275–283. Co-author: L. T. Tran.

101. *Convergence of generalized inverses with applications to asymptotic hypothesis testing.* Sankhyā, Series A, 46, 277–286. Co-authors: Carl T. Russell and Thomas Mathew.

102. *Rank order tests for the parallelism of several regression surfaces.* Journal of Statistical Planning and Inference, 10, 43–57. Co-author: Ching-Yuan Chiang.

103. *Edgeworth expansions for signed linear rank statistics with regression constants.* Journal of Statistical Planning and Inference, 10, 137–149. Co-author: Munsup Seoh.

104. *Edgeworth expansions for signed linear rank statistics under near location alternatives.* Journal of Statistical Planning and Inference, 10, 289–309. Co-author: Munsup Seoh.

105. *Asymptotic normality of the lengths of a class of nonparametric confidence intervals for regression parameters.* Canadian Journal of Statistics, 12, 217–228. Co-author: T. J. Wu.

106. *Linear minimax estimators for estimating a function of the parameter.* The Australian Journal of Statistics, 26, 277–283. Co-authors: Dan A. Ralescu and Stefan S. Ralescu.

107. *Rank procedures for testing subhypotheses in linear regression.* Annals of the Institute of Statistical Mathematics, 36, 35–50. Co-author: Ching-Yuan Chiang.

108. *Locally most powerful rank tests for the two–sample problem with censored samples.* Journal of Organizational Behavior and Statistics, 1, 189–196. Co-author: A. K. MD. Ehsanes Saleh.

1985

109. *Linear serial rank tests for randomness against ARMA alternatives.* Annals of Statistics, 13, 1156–1181. Co-authors: Marc Hallin and Jean-François Ingenbleek.

110. *Cramér–type large deviations for generalized rank statistics.* Annals of Probability, 13, 115–125. Co-authors: Stefan S. Ralescu and Munsup Seoh.

111. *The concept of normality for fuzzy random variables.* Annals of Probability, 13, 1373–1379. Co-author: Dan A. Ralescu.

112. *Limit theorems for random compact sets in Banach space.* Mathematical Proceedings of the Cambridge Philosophical Society, 97, 151–158. Co-author: Dan A. Ralescu.

113. *On the rate of convergence in the central limit theorem for signed rank statistics.* Advances in Applied Mathematics, 6, 23–51. Co-author: Stefan S. Ralescu.

114. *Asymptotic distributions of likelihood ratio criteria for testing latent roots and latent vectors of a covariance matrix under an elliptic population.* Biometrika, 72, 331–338. Co-author: T. Hayakawa.

115. *Gaussian approximations of signed linear rank statistics process.* Journal of Statistical Planning and Inference, 11, 277–312. Co-author: T. J. Wu.

116. *A sharpening of the remainder term in the higher dimensional central limit theorem for multilinear rank statistics.* Journal of Multivariate Analysis, 17, 148–167. Co-authors: Manfred Denker and Uwe Rösler.

117. *Asymptotic expansions of the distributions of some test statistics.* Annals of the Institute of Statistical Mathematics, 37, 95–108. Co-author: Takesi Hayakawa.

118. *On the rate of convergence to asymptotic normality for generalized linear rank statistics.* Annals of the Institute of Statistical Mathematics, 37, 51–69. Co-author: Munsup Seoh.

119. *Berry–Esséen theorems for signed linear rank statistics with regression constants.* Limit Theorems in Probability Statistics, (P. Révész, Editor). North Holland Publishing Company, 875–905. Co-author: Munsup Seoh.

120. *Central limit theorem for linear rank statistics process.* Statistical Theory and Data Analysis, (K. Matusita, Editor). North Holland Publishing Company, 103–145. Co-author: Mark A. Carlson.

121. *On the Cramér–Fréchet–Rao inequality for translation parameter in the case of finite support.* Mathematische Operationsforschung und Statistik, Series Statistics, 16, 495–506. Co-author: I. Vincze.

122. *Tests de rangs linéaires pour une hypothèse de bruit blanc.* C.R. Acad. Sc. Paris, 301, Série I, 49-52. Co-authors: Marc Hallin and Jean- François Ingenbleek.

123. *Solutions fortes des équations différentielles stochastiques pour des processus à indice multidimensionnel.* C.R. Acad. Sc. Paris, 301, Série I, 845–848. Co-author: Markus Dozzi.

124. *Tests de rangs quadratiques pour une hypothèse de bruit blanc.* C.R. Acad. Sc. Paris, 301, Série I, 935–938. Co-authors: Marc Hallin and Jean-François Ingenbleek.

1986

125. *Maximum likelihood estimation for stationary point processes.* Proceedings of the National Academy of Sciences, U.S.A., 83, 541–545. Co-author: Pham D. Tuan.

126. *Fuzzy random variables.* Journal of Mathematical Analysis and Applications, 114, 409–422. Co-author: Dan A. Ralescu.
127. *The order of normal approximation for signed linear rank statistics.* Teoria Veroyatnostei i ee Primeneniya (also in Theory of Probability and Its Applications, Translation by SIAM), 31, 156–163. Co-author: Tiee-Jian Wu.
128. *Gaussian random sets in Banach space.* Teoria Veroyatnostei i ee Primeneniya (also in Theory of Probability and Its Applications, Translation by SIAM), 31, 598–601. Co-authors: Dan A. Ralescu and Stefan S. Ralescu.
129. *Strong solutions of stochastic differential equations for multiparameter processes.* Stochastics, 17, 19–41. Co-author: Markus Dozzi.
130. *Central limit theorem for perturbed empirical distribution functions evaluated at a random point.* Journal of Multivariate Analysis, 19, 273–279. Co-author: Stefan S. Ralescu.
131. *Limit theorems for fuzzy random variables.* Proceedings of the Royal Society of London, 407, 171–182. Co-authors: E. P. Klement and D. A. Ralescu.
132. *Bhattacharyya bound of variances of unbiased estimators in nonregular cases.* Annals of the Institute of Statistical Mathematics, 39, 35–44. Co-authors: M. Akahira and K. Takeuchi.
133. *Almost sure linearity for signed rank statistics in the non–i.i.d. case.* Acta Mathematica Hungarica, 48, 273–284. Co-author: Stefan S. Ralescu.
134. *Les tests de rangs dans l'analyse des séries chronologiques.* Cahiers du C.E.R.O., 28, 41–55. Co-authors: Marc Hallin and Jean-François Ingenbleek.
135. *Tests de rangs localement optimaux pour une hypothèse de bruit blanc multivarié.* C.R. Academy of Sciences, Paris, 303, Série I, 901–904. Co-authors: Marc Hallin and Jean-François Ingenbleek.

1987

136. *Linear and quadratic serial rank tests for randomness against serial dependence.* Journal of Time Series Analysis, 8, 409–424. Co-authors: Marc Hallin and Jean-François Ingenbleek.
137. *Asymptotic normality of a class of nonparametric statistics.* Econometric Theory, 3, 313–347. Co-author: Munsup Seoh.
138. *Linear rank statistics, bounded operators and weak convergence.* New Perspectives in Theoretical and Applied Statistics, John Wiley and Sons, New York, 171–206. Co-authors: Manfred Denker and Gerhard Keller.

139. *Locally asymptotically optimal tests for randomness.* New Perspectives in Theoretical and Applied Statistics, John Wiley and Sons, New York, 87–99. Co-author: Marc Hallin.

140. *Berry–Esséen theorems for signed rank statistics under near location alternatives.* Studia Scientiarum Mathematicarum Hungarica, 20, 197–211. Co-author: Munsup Seoh.

141. *Tests of subhypotheses in linear regression based on rank order estimates.* Studia Scientiarum Mathematicarum Hungarica, 20, 237–247. Co-author: Ching-Yuan Chiang.

142. *Limit theorems for random central order statistics.* Adaptive Statistical Procedures and Related Topics, IMS Lecture Notes and Monograph Series, 8, 447–475. Co-author: Stefan Ralescu.

143. *On admissible estimation in exponential families with imprecise information.* In Statistical Decision Theory and Related Topics IV, (Editors: Shanti S. Gupta and James O. Berger), Springer Verlag, 403–408. Co-author: Stefan Ralescu.

144. *Convergence faible de la statistique serielle linéaire de rang en condition de dépendance avec applications aux séries chronologiques et processus de Markov.* C.R. Academy of Sciences, Paris, 304, Série I, 583–586. Co-author: Michel Harel.

145. *Convergence faible du processus empirique multidimensionnel corrigé en condition de mélange.* C.R. Academy of Sciences, Paris, 305, Série I, 93–95. Co-author: Michel Harel.

146. *Weak convergence of weighted multivariate empirical processes under mixing conditions.* Proceedings of the Sixth Pannonian Symposium on Mathematical Statistics, Volume A: Mathematical Statistics and Probability (Editors: M. L. Puri, P. Révész and W. Wertz), Reidel Publishing Company, Holland, 121–141. Co-author: Michel Harel.

1988

147. *Locally asymptotically rank–based procedures for testing autoregressive moving average dependence.* Proceedings of the National Academy of Sciences, U.S.A., 85, 2031–2035. Co-author: Marc Hallin.

148. *Optimal rank–based procedures for time series analysis: testing an ARMA model against other ARMA models.* Annals of Statistics, 16, 402–432. Co-author: Marc Hallin.

149. *On time–reversibility and the uniqueness of moving average representations for non–Gaussian stationary time series.* Biometrika, 75, 170–171. Co-authors: Marc Hallin and Claude Lefevre.

150. *Asymptotic behavior of multi–response permutation procedures.* Advances in Applied Mathematics, 9, 200–210. Co-author: Manfred Denker.

151. *Adaptive nonparametric procedures and applications.* Journal of the Royal Statistical Society, Series C (Applied Statistics), 37, 205–218. Co-authors: N.J. Hill and A.R. Padmanabhan.

152. *Asymptotic properties of linear functions of order statistics.* Journal of Statistical Planning and Inference, 18, 203–223. Co-author: Cheun-Der Lea.

153. *An invariance principle for processes indexed by two parameters and some statistical applications.* Probability and Mathematical Statistics, 9, 25–76. Co-author: Manfred Denker.

154. *Asymptotic properties of perturbed empirical distribution functions evaluated at a random point.* Journal of Statistical Planning and Inference, 19, 201–215. Co-author: Cheun-Der Lea.

155. *L'espace $\tilde{D}_k$ et la convergence faible du processus empirique indexé par rectangles en condition de mélange.* C.R. Academy of Sciences, Paris, 306, Série I, 207–210. Co-author: Michel Harel.

156. *On locally asymptotically maximin tests for ARMA processes.* In Statistical Theory and Data Analysis II (Editor: K. Matusita), 495–499. North Holland Publishing Company. Co-author: Marc Hallin.

157. *Comportement limite de la U–statistique, de la V–statistique et d'une statistique de rang à un échantillon pour des processus absolument réguliers non stationnaires.* C.R. Academy of Sciences, Paris, 306, Série I, 625-628. Co-author: Michel Harel.

158. *Tests de rangs signés localement optimaux pour une hypothèse de dépendance ARMA.* C.R. Academy of Sciences, Paris, 307, Série I, 355–358. Co-author: Marc Hallin.

159. *Convergence faible de la statistique sérielle linéaire de rang avec des fonctions de scores et des constantes de régression non bornées en condition de mélange.* C.R. Academy of Sciences, Paris, 307, Série I, 617–620. Co-author: Michel Harel.

1989

160. *Asymptotic expansions for sums of nonidentically distributed Bernoulli random variables.* Journal of Multivariate Analysis, 28, 282–303. Co-authors: Paul Deheuvels and Stefan Ralescu.

161. *Asymptotically most powerful rank tests for multivariate randomness against serial dependence.* Journal of Multivariate Analysis, 30, 34–71. Co-authors: Marc Hallin and Jean François Ingenbleek.

162. *Central limit theorems under alternatives for a broad class of nonparametric statistics.* Journal of Statistical Planning and Inference, 22, 271–294. Co-author: Munsup Seoh.

163. *Limiting behavior of U–statistics, V–statistics and one sample rank order statistics for nonstationary absolutely regular processes.* Journal of Multivariate Analysis, 30, 181–204. Co-author: Michel Harel.

164. *On the rate of convergence in normal approximation and large deviation probabilities for a class of statistics.* Teoria Veroyatnostei i ee Primeneniya (also in Theory of Probability and its Applications, Translation by SIAM), 33, 735–750. Co-author: Munsup Seoh.

165. *A new semigroup technique in Poisson approximation.* Semigroup Forum, 38, 189–201. Co-authors: Paul Deheuvels and Dietmar Pfeifer.

166. *Weak convergence of the U–statistic and weak invariance of the one sample rank order statistic for Markov processes and ARMA models.* Journal of Multivariate Analysis, 31, 258–265. Co-author: Michel Harel.

167. *Convergence faible de la U–statistique generalisée pour des processes nonstationnaires absolument réguliers.* C.R. Academy of Sciences, Paris, 309, Série I, 135–138. Co-author: Michel Harel.

1990

168. *Weak convergence of serial rank statistics under dependence with applications in time series and Markov processes.* Annals of Probability, 18, 1361–1387. Co-author: Michel Harel.

169. *Weak invariance of generalized U–statistics for non–stationary absolutely regular processes.* Stochastic Processes and Their Applications, 34, 341–360. Co-author: Michel Harel.

170. *The space $\tilde{D}_k$ and weak convergence for the rectangle–indexed processes under mixing.* Advances in Applied Mathematics, 11, 443–474. Co-author: Michel Harel.

171. *Asymptotic normality of L–statistics based on decomposible time series.* Journal of Multivariate Analysis, 35, 260–275. Co-authors: K. C. Chanda and F. H. Ruymgaart.

172. *A strong invariance principle concerning J–upper order statistics for stationary m–dependent sequences.* Journal of Statistical Planning and Inference, 25, 43–51. Co-author: George Haiman.

173. *Weak convergence of the serial linear rank statistics with unbounded scores and regression constants under mixing conditions.* Journal of Statistical Planning and Inference, 25, 163–186. Co-author: Michel Harel.

174. *Information.* In: Acting Under Uncertainty: Multidisciplinary Conceptions (Editor: George von Furstenberg). Kluwer Academic Publishers, Boston/Dordrecht/Lancaster, 451–469. Co-author: Howard L. Resnikoff.

175. *Measure of information and contiguity.* Statistics and Probability Letters, 9, 223–228. Co-author: Istvan Vincze.

176. *Information and Mathematical Statistics.* Proceedings of the Fourth Prague Symposium on Asymptotic Statistics, Charles Univer-

sity, Prague (Editors: P. Mandl and M. Hušková) August 29, 1988–September 2, 1988, 447–456. Co-author: Istvan Vincze.

1991

177. *A local algorithm for constructing nonnegative cubic splines.* Journal of Approximation Theory, 64, 1–16. Co-authors: Bernd Fischer and Gerhard Opfer.

178. *Time series analysis via rank order theory: signed–rank tests for ARMA models.* Journal of Multivariate Analysis, 39, 1–29. Co-author: Marc Hallin.

179. *Convergence theorem for fuzzy martingales.* Journal of Mathematical Analysis and Applications, 160, 107–122. Co-author: Dan Ralescu.

180. *Limiting behavior of one sample rank–order statistics with unbounded scores for nonstationary absolutely regular processes.* Journal of Statistical Planning and Inference, 27, 1–23. Co-author: Michel Harel.

181. *Weak invariance of the multidimensional rank statistic for nonstationary absolutely regular processes.* Journal of Multivariate Analysis, 36, 204–221. Co-author: Michel Harel.

182. *On Hilbert space valued U–statistics.* Teoria Veroyatnostei i ee Primeneniya (also in Theory of Probability and Its Applications, Translation by SIAM), 36, 604–605. Co-author: V. V. Sazonov.

183. *Théoréme central limite de l'estimateur de densité et de l'erreur quadratique intégrée pour des variables aléatoires nonstationnaires.* C.R. Academy of Sciences, Paris, 312, Série 1, No. 1, 145–148. Co-author: Michel Harel.

184. *Invariance faible de la statistique de rang multidimensionnelle. Applications.* C.R. Academy of Sciences, Paris, 312, Série 1, No. 4, 349–352. Co-author: Michel Harel.

1992

185. *Rank tests for time series analysis.* In: New Directions in Time Series Analysis (the IMA Volume in Mathematics and Its Applications) 45, 111–153. (Editors: David Brillinger, Emanuel Parzen, Murray Rosenblatt et al.). Springer–Verlag, New York/Heidelberg/Berlin. Co-author: Marc Hallin.

186. *Rank-based estimation of ratio of scale parameters and applications.* Journal of Statistical Planning and Inference, 31, 23–49. Co-author: A. R. Padmanabhan.

187. *The Neyman–Pearson probability ratio and information.* In: Data Analysis and Statistical Inference, Festschrift in Honor of Friedhelm Eicher (Editors: Siegfried Schach and Götz Trenkler), 53–64. Verlag Josef Eul, Bergisch Gladbach, Köin. Co-author: Istvan Vincze.

188. *Asymptotic normality of two–sample linear rank statistics under U–statistic structure.* Journal of Statistical Planning and Inference, 32, 89–110. Co-author: Manfred Denker.

189. *Some asymptotic results for a broad class of nonparametric statistics.* Journal of Statistical Planning and Inference, 32, 165–196. Co-author: Marc Hallin.

1993

190. *A strong invariance principle concerning the Jth upper order statistics for stationary Gaussian sequences.* Annals of Probability, 21, 86–135. Co-author: George Haiman.

191. *The time spent by the Wiener process in a narrow tube before leaving a wide tube.* Proceedings of the American Mathematical Society, 117, 529–536. Co-author: Antonia Földes.

192. *Weak convergence of the simple linear rank statistic under mixing conditions in the nonstationary case.* Teoria Veroyatnostei i ee Primeneniya (also in Theory of Probability and Its Applications, Translation by SIAM) 38, 579–599. Co-author: Michel Harel.

193. *Nonparametric prediction for random fields.* Stochastic Processes and Their Applications, 48, 139–156. Co-author: Frits Ruymgaart.

194. *Asymptotic behavior of L–statistics for a large class of time series.* Annals of the Institute of Statistical Mathematics, 45, 687–701. Co-author: Frits H. Ruymgaart.

195. *On the central limit theorem in Hilbert space with application to U-statistics.* In: Statistical Sciences and Data Analysis: Proceedings of the Third Pacific Area Statistical Conference. VSP International Science Publishers, Holland. (Editors: Kameo Matusita, Madan L. Puri and Takesi Hayakawa), 407–413. Co-author: V. V. Sazonov.

196. *Asymptotics of the perturbed sample quantile for a sequence of m-dependent stationary random process.* In: Statistical Sciences and Data Analysis: Proceedings of the Third Pacific Area Statistical Conference. VSP International Science Publishers, Holland. (Editors: Kameo Matusita, Madan L. Puri and Takesi Hayakawa), 415–426. Co-author: Shan Sun.

1994

197. *Weak convergence of weighted empirical U–statistics processes for dependent random variables.* Journal of Multivariate Analysis, 48, 297–314. Co-authors: Colm Art O'Cinneide and Michel Harel.

198. *Aligned rank tests for linear models with autocorrelated error terms.* Journal of Multivariate Analysis, 50, 175-237. Co-author: Marc Hallin.

199. *Law of the iterated logarithm for perturbed empirical distribution functions evaluated at a random point for nonstationary random variables.* Journal of Theoretical Probability, 7, 831–855. Co-author: Michel Harel.

200. *Berry–Esséen rate in asymptotic normality for perturbed sample quantiles.* Metrika, 41, 83–98. Co-author: Munsup Seoh.

201. *On some limit laws for aligned perturbed empirical distribution functions.* Statistics and Probability Letters, 21, 317–321. Co-author: Manfred Denker.

202. *Change curves in the presence of dependent noise.* In: Change-Point Problems, a volume in the IMS Lecture Notes and Monograph Series, Volume 23, 242–254. Co-author: Frits H. Ruymgaart.

1995

203. *Nonparametric methods for stratified two-sample designs with application to multiclinic trials.* Journal of the American Statistical Association, 90, 1004–1014. Co-authors: Edgar Brunner and Shan Sun.

204. *Extremes of Markov sequences.* Journal of Statistical Planning and Inference, 45, 185–201. Co-authors: George Haiman and Maxime Kiki.

205. *A multivariate Wald–Wolfowitz rank test against serial dependence.* The Canadian Journal of Statistics, 23, 55–65. Co-author: Marc Hallin.

206. *Higher order asymptotic theory for normalizing transformations.* Annals of the Institute of Statistical Mathematics, 17, 581–600. Co-author: Masanobu Taniguchi.

207. *Asymptotic expansions in statistics: a review of methods and applications.* In: Advances in Econometrics and Quantitative Economics: Essays in honor of C. R. Rao. Blackwell Publishers, Oxford, United Kingdom. (Editors: G. S. Maddala, Peter C. B. Phillips and T. N. Srinavasan), 88–122. Co-author: R. N. Bhattacharya.

208. *Comportement asymptotique des estimateurs de la fonction de regression pour des variables aleatoires dependantes et nonstationnaires.* C.R. Academy of Sciences, Paris, 320, Série I, 75–78. Co-author: Michel Harel.

1996

209. *Nonparametric approach for non-Gaussian vector stationary processes.* Journal of Multivariate Analysis, 56, 259–283. Co-authors: Masanobu Taniguchi and Masao Kondo.

210. *Valid Edgeworth expansions of M–estimators in regression models with weakly dependent residuals.* Econometric Theory, 12, 331–346. Co-author: Masanobu Taniguchi.

211. *Conditional U–statistics for dependent random variables.* Journal of Multivariate Analysis, 57, 84–100. Co-author: Michel Harel.

212. *Asymptotic normality of nearest neighbor regression function estimates based on nonstationary dependent observations.* Special Issue on Multivariate Statistical Analysis in Honor of Professor Minoru Siotani, American Journal of Mathematical and Management Sciences, 15, 255–290. Co-author: Michel Harel.

213. *Weak convergence of sequences of first passage processes and applications.* Stochastic Processes and Their Applications, 62, 327–345. Co-author: Stefan Ralescu.

214. *Characterization of weak convergence for perturbed empirical and quantile processes under φ–mixing.* Journal of Statistical Planning and Inference, 53, 285–295. Co-authors: H. J. A. Degenhardt, Shan Sun and Martien van Zuijlen.

215. *Nonparametric density estimators based on nonstationary absolutely regular random sequences.* Journal of Applied Mathematics and Stochastic Analysis, 9, 233–254. Co-author: Michel Harel.

1997

216. *Nonparametric methods in design and analysis of experiments.* In: Handbook of Statistics, volume 13, Design of Experiments. (Editors: S. Ghosh and C. R. Rao) North Holland, Amsterdam, Netherlands, 631–703. Co-author: Edgar Brunner.

217. *Normal approximation of U–statistics in Hilbert spaces.* Teoria Veroyatnostei i ee Primeneniya (also in Theory of Probability and its Applications, Translation by SIAM), 41, 481–504. Co-authors: Yu. V. Borovskich and V. V. Sazonov.

1998

218. *Autoregressive quantiles and related rank scores processes for generalized random coefficient autoregressive processes.* Journal of Statistical Planning and Inference, 68, 271–294. Co-author: Michel Harel.

219. *Records and 2-block designs of 1-dependent stationary sequences under local dependence.* Annales de l'institut Henri Poincaré, 34, 481–503. Co-authors: George Haiman, Nelly Mayeur and Valery Nevzorov.

Index

829

Milton Keynes UK
Ingram Content Group UK Ltd.
UKHW031536071024
449327UK00023B/1839